Praise for the Book

T0186457

"This volume contains 19 chapters covering the use of a multitude of new technological tools available to study elasmobranch fishes. Co-authored by more than 60 active shark researchers, *Shark Research* summarizes the state of the science in shark study. This volume belongs in the library of anyone with a serious interest in elasmobranch research."

Dr. John A. (Jack) Musick
Professor Emeritus, Virginia Institute of Marine Science

"The future of shark research is here. Advanced sampling technologies and analytical techniques are already changing the landscape of many fields of shark research. From UAVs to AVEDs, AUVs to ROVs, BRUVs to MBESs, CT scans to MRIs, NIRS to photogrammetry, genomics to eDNA, to name only a few of these platforms and techniques, this book offers a compendium of state-of-the-art technologies and methods that are quickly becoming commonplace and will continue to evolve and revolutionize how we study these animals. Through its 19 chapters, the book describes how decreasing costs of electronics and increased miniaturization, quality, power, and types of sensory platforms are leading to accumulation of larger datasets, which in tandem with increasing collaborative initiatives, computing power, and advances in computer science and modeling techniques will result in a new understanding of crucial aspects of elasmobranch ecology and behavior. This book should be of interest to students, academics, and professionals working on this and other groups of marine animals to keep abreast of the latest applications of advanced sampling technologies and analytical techniques that are being used to study elasmobranchs."

Dr. Enric Cortés
Southeast Fisheries Science Center, National Marine Fisheries Service, National Oceanic and Atmospheric Administration

"If you are keen to see how new technologies and applications are shaping modern shark research, then this is a must-have book. I will be recommending this book to anyone interested in marine science. Its impressive coverage of topics from environmental DNA to social science applications provides the reader with a more holistic view of shark research."

Dr. Will White
Senior Curator, CSIRO Australian National Fish Collection

"The editors are to be congratulated for publishing a synoptic book that highlights the use of rapidly developing, novel, technological methods to study the ecology of sharks and rays. These editors are excellent chondrichthyan biologists who have helped pioneer some of these new techniques. Its

19 chapters are organized into four basic sections: (1) trophic ecology, (2) habitat utilization and behavioral ecology, (3) life history studies and population ecology, and (4) citizen and social science. With this advanced tool kit now available under one cover, it will enable advanced studies that were heretofore impossible but nonetheless important. This book captures the revolutionary growth in technology of such processes as tracking and monitoring movements and activities, imaging structures and organisms, sensing ecophysiological parameters, and using biochemical, genetic, and other methods to help researchers more rapidly advance knowledge of sharks and rays. This will break barriers that have heretofore hindered scientific progress toward understanding the ecology and conservation of chondrichthyan fishes, and other organisms as well."

Dr. Gregor M. Cailliet
*Professor Emeritus, Moss Landing Marine Laboratories;
Program Director Emeritus, Pacific Shark Research Laboratory*

"Innovative technologies are rapidly advancing the field of shark research. This must-have book features leading innovators in the field who have contributed informative chapters summarizing the current state of research in a variety of fields. Whether you're a student aspiring to study sharks, a professional, or just keenly interested in the current state of shark research, this is the book for you."

Dr. David A. Ebert
*Director, Pacific Shark Research Center,
Moss Landing Marine Laboratories*

"Besides bringing exciting and important new research findings, tools, and techniques to the table, Dr. Jeff Carrier's most recent contribution, *Shark Research: Emerging Technologies and Applications for the Field and Laboratory*, also provides a keen roadmap for the future of shark science."

Dr. Toby S. Daly-Engel
Assistant Professor, Department of Ocean Engineering and Marine Sciences, Florida Institute of Technology

"This timely volume provides an outstanding overview of how technological advances enable researchers to address formerly intractable questions. As we look back at this volume in a decade or two, the featured technology that is currently *avant-garde* will be *de rigueur*, but these early adopters will be recognized for applying this technology to the development of entirely new methods of inquiry."

Dr. Stephen M Kajiura
Professor of Biological Sciences, Florida Atlantic University

Shark Research

Emerging Technologies and Applications for the Field and Laboratory

CRC
MARINE BIOLOGY
SERIES

The late Peter L. Lutz, Founding Editor
David H. Evans and Stephen Bortone, Series Editors

Biology of Marine Birds
E.A. Schreiber and Joanna Burger

Biology of the Spotted Seatrout
Stephen A. Bortone

The Biology of Sea Turtles, Volume II
Peter L. Lutz, John A. Musick, and Jeanette Wyneken

Early Stages of Atlantic Fishes: An Identification Guide for the Western Central North Atlantic
William J. Richards

Biology of the Southern Ocean, Second Edition
George A. Knox

Biology of the Three-Spined Stickleback
Sara Östlund-Nilsson, Ian Mayer, and Felicity Anne Huntingford

Biology and Management of the World Tarpon and Bonefish Fisheries
Jerald S. Ault

Methods in Reproductive Aquaculture: Marine and Freshwater Species
Elsa Cabrita, Vanesa Robles, and Paz Herráez

Sharks and Their Relatives II: Biodiversity, Adaptive Physiology, and Conservation
Jeffrey C. Carrier, John A. Musick, and Michael R. Heithaus

Artificial Reefs in Fisheries Management
Stephen A. Bortone, Frederico Pereira Brandini, Gianna Fabi, and Shinya Otake

Biology of Sharks and Their Relatives, Second Edition
Jeffrey C. Carrier, John A. Musick, and Michael R. Heithaus

The Biology of Sea Turtles, Volume III
Jeanette Wyneken, Kenneth J. Lohmann, and John A. Musick

The Physiology of Fishes, Fourth Edition
David H. Evans, James B. Claiborne, and Suzanne Currie

Interrelationships Between Coral Reefs and Fisheries
Stephen A. Bortone

Impacts of Oil Spill Disasters on Marine Habitats and Fisheries in North America
J. Brian Alford, Mark S. Peterson, and Christopher C. Green

Hagfish Biology
Susan L. Edwards and Gregory G. Goss

Marine Mammal Physiology: Requisites for Ocean Living
Michael A. Castellini and Jo-Ann Mellish

Shark Research: Emerging Technologies and Applications for the Field and Laboratory
Jeffrey C. Carrier, Michael R. Heithaus, and Colin A. Simpfendorfer

For more information about this series, please visit: https://www.crcpress.com/CRC-Marine-Biology-Series/book-series/CRCMARINEBIO?page=&order=pubdate&size=12&view=list&status=published,forthcoming

Shark Research

Emerging Technologies and Applications for the Field and Laboratory

Edited by

Jeffrey C. Carrier
Michael R. Heithaus
Colin A. Simpfendorfer

CRC Press
Taylor & Francis Group
Boca Raton London New York

CRC Press is an imprint of the
Taylor & Francis Group, an **informa** business

CRC Press
Taylor & Francis Group
6000 Broken Sound Parkway NW, Suite 300
Boca Raton, FL 33487-2742

First issued in paperback 2020

ISBN-13: 978-1-138-03292-7 (hbk)
ISBN-13: 978-0-367-89343-9 (pbk)

Library of Congress Cataloging-in-Publication Data

Names: Carrier, Jeffrey C., editor.
Title: Shark research : emerging technologies and applications for the field
and laboratory / editors, Jeffrey C. Carrier, Michael R. Heithaus, and
Colin A. Simpfendorfer.
Description: Boca Raton : Taylor & Francis, 2019. | Series: Marine biology |
Includes bibliographical references.
Identifiers: LCCN 2018007870 | ISBN 9781138032927 (hardback : alk. paper)
Subjects: LCSH: Sharks--Research.
Classification: LCC QL638.9 .S45397 2019 | DDC 597.3072--dc23
LC record available at http://lccn.loc.gov/2018007870

Visit the Taylor & Francis Web site at
http://www.taylorandfrancis.com

and the CRC Press Web site at
http://www.crcpress.com

Contents

The remarkable pace of advancements in technology, particularly in the last two decades, has contributed to the development of a toolbox that greatly enhances the range of investigations into the biology and life history of elasmobranchs. In our preface to *The Biology of Sharks and Their Relatives* in 2004, Jack Musick, Mike Heithaus, and I hinted at the potential impact of these advances when we noted that

> … virtually every area of research associated with these animals has been strongly impacted by the revolutionary growth in technology, and the questions we can now ask are very different than those reported in [Perry] Gilbert's work not so long ago. A careful reading of the chapters we have presented in this work will show conclusions based on emergent technologies that have revealed some long-hidden secrets of these animals. Modern immunological and genetic techniques, satellite telemetry and archival tagging, modern phylogenetic analysis, GIS, and bomb dating are just a few of the techniques and procedures that have become a part of our investigative lexicon.

Even then we did not anticipate the magnitude of expansion that was to occur in the 15 years since that volume was produced and the improvements that would occur to existing methods. Now, biologists, field biologists and laboratory biologists alike, are faced with a bewildering array of techniques and instruments with which to investigate almost every aspect of elasmobranch biology. From traditional studies of comparative morphology to satellite tracking and the almost limitless uses of DNA for examining species relatedness and assessing variability within and between populations, the questions that can be asked and the data that can be obtained for analysis are providing new insights and understanding of this ancient line of aquatic vertebrates.

The dilemma facing investigators, with such an extensive array of tools and techniques, is which investigative approach is most appropriate for a particular line of inquiry. Knowing how to use technology also assumes that the right choices are made with regard to selecting instruments and methodologies that will provide answers that are relevant to a particular line of inquiry. This applies not only to the technology or approach that is applied but also to the analytical methods with which the collected data are analyzed. Increases in computer power and statistical methods have progressed at rates similar to those of the technology applied to study these animals. One example of this is network analysis (Chapter 18), which until recently had not been applied to sharks or rays but is now a fundamental tool in the analysis of data across a number of data collection techniques.

When we began this project, our goal was straightforward: We intended to feature chapters presenting the various techniques and applications we identified as being among the most useful approaches to broadening the ways we could better investigate and understand the biology and life history attributes of elasmobranch species. Although some technologies, such as acoustic tracking, have been present for many years, miniaturization, data storage, and battery technology, as well as advanced approaches to analysis of increasingly large volumes of data, have helped to improve upon these tried and tested techniques and long-accepted approaches. Outlining these changes was as important as introducing newer, more novel investigative approaches. We were also fully aware that no single volume could hope to present every possible technique, instrument, or technological advancement. Perhaps later volumes will expand on our initial attempts.

Each chapter is designed to identify the types of studies that are appropriate for the use of the various technologies presented within each chapter, the kinds of results that can be expected from their use, and what information the studies reveal that advances our understanding of elasmobranch biology. Most certainly these techniques are equally applicable to studies of other marine groups, as well.

Of equal importance, we also believed that each chapter should include a discussion of where such techniques are inappropriate, not likely to succeed, or are otherwise probably not applicable to the study of elasmobranch biology. Choosing an inappropriate study methodology simply leads to wasted time and dashed expectations. We hoped that our treatments would prevent investigators from making such mistakes or having unrealistic expectations. In that sense, the chapters serve as a rudimentary "how to," at least with respect to making more informed choices about a particular approach to address questions of biological interest. We expected that such information would prove useful to students just beginning their formal studies of elasmobranch biology while also serving as a guide for more seasoned scientists seeking to apply new techniques to ongoing studies. Our hope is that we have succeeded in serving both groups.

Our authors are a diverse group, all of whom have strong records of scholarship and all of whom have served as pioneers and leaders in applying these technologies to their own investigations. They thus provide a knowledge base from practical experience that we expect to serve as a valuable resource for our readers. We hope the information and "advice" we have assembled will accomplish that goal.

Jeffrey C. Carrier, PhD, is professor emeritus of biology at Albion College, Michigan, where he was a faculty member from 1979 to 2010. He earned a bachelor of science degree in biology in 1970 from the University of Miami and completed a doctorate in biology from the University of Miami in 1974. While at Albion College, Dr. Carrier received multiple awards for teaching and scholarship and held endowed professorships in biology. His primary research interests center on various aspects of the physiology and ecology of nurse sharks in the Florida Keys. His most recent work investigated the reproductive biology and mating behaviors of this species in a long-term study from an isolated region of the Florida Keys. Dr. Carrier's projects with acoustic telemetry, animal-borne video, ultrasound and endoscopy, and baited remote underwater video systems drive his interest in applications of technology to the study of the biology of sharks and their relatives. Dr. Carrier has been a long-time member of the American Elasmobranch Society, the American Society of Ichthyologists and Herpetologists, Sigma Xi, the Society for Animal Behavior, and the Council on Undergraduate Research. He served multiple terms as president of the American Elasmobranch Society and received several distinguished service awards from the society. He holds an appointment as an adjunct research scientist with Mote Marine Laboratory's Center for Shark Research. In addition to his publications in the scientific literature, he has written and edited five previously published books on sharks and their biology.

Michael R. Heithaus, PhD, is a professor in the department of biological sciences and dean of the College of Arts, Sciences and Education at Florida International University (FIU) in Miami, Florida, where he has been a faculty member since 2003. He received his bachelor of arts degree in biology from Oberlin College, Ohio, in 1995 and his doctorate from Simon Fraser University, Burnaby British Columbia, in 2001. He was a postdoctoral scientist and staff scientist at the Center for Shark Research and also served as a research fellow at the National Geographic Society's Remote Imaging Department. At FIU, Dr. Heithaus served as the director of the Marine Sciences Program before becoming the director of the School of Environment, Arts, and Society. Dr. Heithaus is a behavioral and community ecologist. His main research interests are in understanding the ecological roles and importance of large predators, especially their potential to impact community structure through nonconsumptive effects. His work also explores the factors influencing behavioral decisions, especially of large marine taxa, including marine mammals, sharks and rays, and sea turtles, and the importance of individual variation in behavior in shaping ecological interactions. Dr. Heithaus is the co-lead of the Global FinPrint project, a worldwide survey of elasmobranchs on coral reefs. His lab is engaged in marine conservation and research projects around the world, including ongoing long-term projects in Shark Bay, Australia, and the coastal Everglades of southwest Florida.

Colin A. Simpfendorfer, PhD, is a professor in the College of Science and Engineering at James Cook University, Queensland, Australia, and currently serves as the associate dean for Research. He has also worked at the Center for Shark Research at Mote Marine Laboratory, Sarasota, Florida, and the Shark Fisheries Section of the Western Australian Department of Fisheries, Perth, Australia. He received his bachelor of science degree in marine biology and zoology in 1986 and doctorate in fisheries science in 1993, both from James Cook University. He has spent his career studying the life history, ecology, status, and conservation of sharks and rays with the principle aim of providing scientific information for improving their management. He regularly provides scientific advice to governments, nongovernmental organizations, and industry. He has been at the forefront of applying new technology and approaches to sharks and rays, including early work on the analysis of acoustic telemetry data, using eDNA as a means of surveying for critically endangered sawfish, and he is a principle investigator for the Global FinPrint project, surveying sharks and rays on coral reefs globally. Dr. Simpfendorfer is an author of over 200 peer-reviewed scientific papers on sharks and rays and has trained more than 30 master of science and doctoral students, some of whom have authored or co-authored chapters in this book. He is currently the co-chair of the IUCN Shark Specialist Group, which works to improve the conservation status of this important group of ocean predators by assessing their status, developing conservation plans, and delivering quality scientific information to decision makers. He also serves on Australia's national Threatened Species Scientific Committee.

Contributors

Allen H. Andrews
Pacific Islands Fisheries Science Center
National Marine Fisheries Service
Honolulu, Hawaii

Judith Bakker
School of Environment and Life Sciences
University of Salford
Salford, United Kingdom

Mike Cappo
Australian Institute of Marine Science
Townsville, Queensland, Australia

Diego Cardeñosa
School of Marine and Atmospheric Science
Stony Brook University
Stony Brook, New York
and
Fundación Colombia Azul
Bogotá DC, Colombia

Jeffrey C. Carrier
Department of Biology
Albion College
Albion, Michigan

Demian D. Chapman
Department of Biological Sciences
Florida International University
North Miami, Florida

Andrew Chin
College of Science and Engineering, and
Centre for Sustainable Tropical Fisheries and Aquaculture
James Cook University
Townsville, Queensland, Australia

Christopher M. Clark
Lab for Autonomous and Intelligent Robotics
Harvey Mudd College
Claremont, California

Madalyn K. Cooper
College of Science and Engineering, and
Centre for Sustainable Tropical Fisheries and Aquaculture
James Cook University
Townsville, Queensland, Australia

Amy Diedrich
College of Science and Engineering, and
Centre for Sustainable Tropical Fisheries and Aquaculture
James Cook University
Townsville, Queensland, Australia

Christine Dudgeon
Molecular Fisheries Laboratory
School of Biomedical Sciences
The University of Queensland
St Lucia, Queensland, Australia

Nicholas K. Dulvy
Earth to Ocean Research Group
Simon Fraser University
Burnaby, British Columbia, Canada

Kevin Feldheim
Pritzker Laboratory for Molecular Systematics and
   Evolution
Field Museum of Natural History
Chicago, Illinois

Luciana C. Ferreira
Australian Institute of Marine Science
Indian Ocean Marine Research Centre
University of Western Australia
Crawley, Western Australia, Australia

Pierre Feutry
CSIRO Oceans and Atmosphere
Hobart, Tasmania, Australia

William J. Foley
Animal Ecology and Conservation
University of Hamburg
Hamburg, Germany
and
and Research School of Biology
The Australian National University
Canberra, Australian Capital Territory, Australia

Karin Gerhardt
College of Science and Engineering, and
Centre for Sustainable Tropical Fisheries and Aquaculture
James Cook University
Townsville, Queensland, Australia

Adrian C. Gleiss
Centre for Aquatic and Ecosystems Research
Harry Butler Institute
Murdoch University
Murdoch, Western Australia, Australia

Jordan Goetze
School of Molecular and Life Sciences
Curtin University
Western Australia, Australia
and
Marine Program
Wildlife Conservation Society
Bronx, New York

Tristan Guttridge
Bimini Biological Field Station Foundation
South Bimini, Bahamas

Euan S. Harvey
School of Molecular and Life Sciences
Curtin University
Perth, Western Australia, Australia

Michael R. Heithaus
Marine Sciences Program
Florida International University
North Miami, Florida

Michelle R. Heupel
Australian Institute of Marine Science
Townsville, Queensland, Australia

Jason Holmberg
Wild Me
Portland, Oregon

Roger Huerlimann
Centre for Tropical Water and Aquatic Ecosystem Research
    (TropWATER), and
College of Science and Engineering
James Cook University
Townsville, Queensland, Australia

David M.P. Jacoby
Institute of Zoology
Zoological Society of London
London, United Kingdom

Vanessa Jaiteh
Centre for Fish and Fisheries Research and Asia Research
    Centre
Murdoch University
Perth, Western Australia, Australia
and
Coral Reef Research Foundation
Koror, Palau

Steven T. Kessel
Daniel P. Haerther Center for Conservation and Research
John G. Shedd Aquarium
Chicago, Illinois

Jeremy J. Kiszka
Marine Sciences Program
Florida International University
North Miami, Florida

Alison A. Kock
South African National Parks
Pretoria, South Africa
and
South African Institute of Aquatic Biodiversity
Grahamstown, South Africa

Agnes Le Port
Centre for Tropical Water and Aquatic Ecosystem Research
    (TropWATER), and
College of Science and Engineering
James Cook University
Townsville, Queensland, Australia

Karissa O. Lear
Centre for Aquatic and Ecosystems Research
Harry Butler Institute
Murdoch University
Murdoch, Western Australia, Australia

Elodie Lédée
Fish Ecology and Conservation Physiology Lab
Carleton University
Ottawa, Ontario, Canada

Christopher G. Lowe
Shark Lab
California State University
Long Beach, California

Gregory E. Maes
Laboratory for Cytogenetics and Genome Research
Centre for Human Genetics
University of Leuven
Leuven, Belgium

Kate L. Mansfield
Marine Turtle Research Group
University of Central Florida
Orlando, Florida

Stefano Mariani
School of Environment and Life Sciences
University of Salford
Salford, United Kingdom

Andrea D. Marshall
Marine Megafauna Foundation
Truckee, California
and
Wild Me
Portland, Oregon

Jordan K. Matley
Center for Marine and Environmental Studies
University of the Virgin Islands
Charlotte Amalie, St. Thomas, U.S. Virgin Islands

Mark G. Meekan
Australian Institute of Marine Science
Indian Ocean Marine Research Centre
University of Western Australia
Crawley, Western Australia, Australia

Carl G. Meyer
Hawaii Institute of Marine Biology
University of Hawaii at Manoa
Kaneohe, Hawaii

Lauren Meyer
College of Science and Engineering
Flinders University
Bedford Park, South Australia, Australia

Johann Mourier
PSL Research University
EPHE–UPVD–CNRS, CRIOBE USR 3278
Perpignan, France

Samantha E.M. Munroe
Australian Rivers Institute
Griffith University,
Nathan, Queensland, Australia

Lisa J. Natanson
Apex Predators Program
National Marine Fisheries Service
Narragansett, Rhode Island

Jennifer R. Ovenden
Molecular Fisheries Laboratory
School of Biomedical Sciences
The University of Queensland
St Lucia, Queensland, Australia

Yannis P. Papastamatiou
Marine Sciences Program
Florida International University
North Miami, Florida

Michelle S. Passerotti
Department of Biological Sciences
University of South Carolina
Columbia, South Carolina

Nicholas Payne
School of Natural Sciences
Trinity College Dublin
Dublin, Ireland

Gretta Pecl
Institute for Marine and Antarctic Studies, and
Centre for Marine Socioecology
University of Tasmania
Hobart, Tasmania, Australia

Simon J. Pierce
Marine Megafauna Foundation
Truckee, California
and
Wild Me
Portland, Oregon

Cassandra L. Rigby
Centre for Sustainable Tropical Fisheries and Aquaculture,
and
College of Science and Engineering
James Cook University
Townsville, Queensland, Australia

Julia Santana-Garcon
Department of Global Change Research
Institut Mediterrani d'Estudis Avançats
Spanish Research Council
Mallorca, Spain

Benjamin J. Saunders
School of Molecular and Life Sciences
Curtin University
Perth, Western Australia, Australia

Colin A. Simpfendorfer
Centre for Sustainable Tropical Fisheries and Aquaculture,
and
College of Science and Engineering
James Cook University
Townsville, Queensland, Australia

Adam P. Summers
Department of Biology, SAFS
Friday Harbor Laboratories
University of Washington
Seattle, Washington

Michele Thums
Australian Institute of Marine Science
Indian Ocean Marine Research Centre
University of Western Australia
Crawley, Western Australia, Australia

Rowan Trebilco
Antarctic Climate and Ecosystems Cooperative Research
    Centre
University of Tasmania
Hobart, Tasmania, Australia

Yuuki Y. Watanabe
National Institute of Polar Research
Tokyo, Japan

Connor F. White
Shark Lab
California State University
Long Beach, California

Nicholas M. Whitney
Anderson Cabot Center for Ocean Life
New England Aquarium
Boston, Massachusetts

Sabine P. Wintner
KwaZulu-Natal Sharks Board
Umhlanga Rocks
South Africa
and
Biomedical Resource Unit
University of KwaZulu-Natal
Durban, South Africa

Kara E. Yopak
Department of Biology and Marine Biology
UNCW Center for Marine Science
University of North Carolina Wilmington
Wilmington, North Carolina

CHAPTER 1

Dietary Biomarkers in Shark Foraging and Movement Ecology

Samantha E.M. Munroe
Australian Rivers Institute, Griffith University, Nathan, Queensland, Australia

Lauren Meyer
College of Science and Engineering, Flinders University, Bedford Park, South Australia, Australia

Michael R. Heithaus
Marine Sciences Program, Florida International University, North Miami, Florida

CONTENTS

1.1 INTRODUCTION

Knowledge of the trophic and spatial ecology of elasmobranchs is widely recognized as a critical component of successful management and conservation programs (Mourier et al., 2016; Oh et al., 2017; White et al., 2017). Dietary biomarkers have emerged as an effective and affordable tool in the field of foraging ecology (Graham et al., 2010; Newsome et al., 2010; Peterson and Fry, 1987). Although shark research has been relatively slow to incorporate biomarker approaches, recent years have seen a substantial increase in the number of elasmobranch biomarker studies (Hussey et al., 2012). Examples can be found across the analytical spectrum, including dietary reconstructions (e.g., Stewart et al., 2017), resource partitioning (e.g., Kinney et al., 2011), and movement (e.g., Munroe et al., 2015). Biomarkers are now considered a standard and valuable technique in shark ecology that can provide much-needed answers to significant environmental questions.

Despite the range of dietary biomarkers from which to choose (Pethybridge et al., 2018), all analyses are based on the same universal premise: biochemically speaking, you are what you eat (Fry, 2006; Parrish, 2013). The chemical composition of abiotic and biotic resources varies at the bottom of the food chain via different biogeochemical processes (Boutton, 1991; Dalsgaard et al., 2003; Ostrom et al., 1997). Consumers assimilate and reflect the chemical composition of their diet in relatively predictable and measurable ways (Beckmann et al., 2013a; Ramos and González-Solís, 2012). As a result, the chemical composition of shark tissues can be used to estimate their trophic role and dietary preferences (e.g., Barría et al., 2017; Daly et al., 2013; McMeans et al., 2012). Biomarker values vary over space and time; thus, they can also be used to measure shark movement between biochemically distinct and distant food webs (e.g., Carlisle et al., 2015; Shipley, 2017).

Biomarker analysis is a particularly appealing approach in shark ecology because other methods used to study the diet and movement of species can be impractical when applied to highly mobile marine predators. For example, stomach content analysis can provide a precise assessment of shark diet (Bethea et al., 2004; Cortés, 1999), but shark studies are regularly plagued by a high number of empty stomachs, techniques are labor intensive and highly invasive, results underestimate easily digestible prey, and successful captures provide only a snapshot of what a shark has eaten most recently (Baker et al., 2014; Simpfendorfer et al., 2001). Electronic tracking techniques have revolutionized the field of animal ecology and can provide explicit information on shark movement patterns (Heupel et al., 2006; Hussey et al., 2015), but these techniques can be prohibitively costly and labor and time intensive, and they do not provide data on shark resource consumption (Hammerschlag et al., 2011; Murphy and Jenkins, 2010). By comparison, biomarker analysis provides a long-term, integrated interpretation of shark diet; therefore, a great deal of information can be procured from only a few tissue samples. Basic analysis only requires 1 to 2 grams of tissue, so sampling procedures are typically nonlethal and minimally invasive. Analysis is also relatively inexpensive, making it an affordable option in most situations. As a result, biomarker analysis is an excellent alternative or supplement to techniques such as stomach content analysis and electronic tracking when working with mobile, rare, or endangered species or in remote and isolated areas.

Unfortunately, misuse of this deceivingly simple technique has led to problems with sample processing and data interpretation across all fields. Although biomarker dynamics have been relatively well studied in some taxa, the most basic assumptions about biomarker integration in sharks have yet to be tested (Hussey et al., 2012; Shipley et al., 2017). Shark physiology and behavior can also hinder more precise trophic analyses, such as population connectivity and dietary source determination (Kim and Koch, 2012; Kim et al., 2012). Therefore, despite its enormous potential, shark ecologists must carefully tailor biomarker research according to taxa-specific limitations.

In this chapter, we introduce the two most commonly used dietary biomarkers in shark ecology, stable isotope analysis (SIA) and fatty acid analysis (FAA). Of these two, SIA currently dominates shark literature. This is in large part because SIA techniques and methods have had more time to evolve and are currently more refined and better developed. For this reason, the majority of this chapter will review SIA applications. However, FAA is set to become an indispensable approach in shark research because it can provide far more detailed dietary analyses. It is important to consider the benefits and limitations of both options before embarking on any foraging study. In fact, these two techniques are increasingly being combined to investigate a wider range of ecological questions (e.g., Couturier et al., 2013). We begin with a discussion of the basic principles that govern stable isotope and fatty acid distribution and integration in marine food webs. We also discuss the most commonly used tracers and tissues in shark biomarker research and address critical aspects of sample collection, preparation, and laboratory analysis. Finally, we review the most relevant biomarker applications in shark ecology, current limitations in shark analysis, and directions for future research.

1.2 PRINCIPLES OF BIOMARKER ECOLOGY

1.2.1 Stable Isotope Analysis

Stable isotope foraging ecology uses isotope variation in the environment to define the dietary and movement patterns of species. Stable isotopes differ from radioactive isotopes (e.g., ^{14}C) because they have no half-life and do not decay into other elements. As a result, stable isotopes persist in the environment and can be tracked as they transition between different abiotic and biotic molecules and throughout the food web (Fry, 2006). Most stable isotopes have two variants: the abundant light isotope and the less common heavier isotope. Stable heavy isotopes make up only a small percentage of the total concentration in nature ($\sim <1\%$) but are still found in high enough quantities that they can be accurately measured (De Groot, 2004; McKinney et al., 1950). Stable isotope concentrations are reported in delta (δ) notation as the ratio of heavy to light isotopes (i.e., rare to abundant) and are measured as deviations from known and internationally accepted standards. Because heavy isotopes are quite rare, isotope ratios are small, and results are expressed as *parts per thousand* (‰) or *per mill* using the following calculation: $\delta^b X = [(R_{sample}/R_{standard}) - 1] \times 1000$, where X is the element, b is the mass of the heavy isotope, and R_{sample} and $R_{standard}$ are the ratios of heavy to light isotopes (e.g., $^{13}C/^{12}C$,

^{15}N/^{14}N, ^{34}S/^{32}S) in the sample and standard, respectively. Decreases in the δ value signify decreases in the heavy isotope content of a sample. The two primary dietary isotope tracers in foraging ecology are the heavy isotopes of carbon (δ^{13}C) and nitrogen (δ^{15}N).

All isotopes of a given element have nearly identical chemical characteristics and are functionally equivalent in most reactions (Peterson and Fry, 1987); however, small mass-dependent differences in the physical and chemical behavior of isotopes cause isotopic separation during biogeochemical processes (Hoefs and Hoefs, 1997; White, 2013). This phenomenon, where isotopes are unequally distributed between different substances or phases, is known as *isotope fractionation*. Isotope fractionation is primarily driven by equilibrium and kinetic effects (Criss, 1999). Equilibrium fractionation is the separation of isotopes between substances or phases that have reached chemical equilibrium. Heavier isotopes have lower mobility and diffusion velocity and thus tend to concentrate in the substance or phase with the greatest bond strengths. For example, when water vapor condenses into liquid in a closed system, the isotopically heavy water molecules (^{2}H and ^{18}O) tend to concentrate in the liquid phase, whereas the lighter water molecules (^{1}H and ^{16}O) concentrate in the vapor phase (Gat, 2000). Kinetic fractionation separates isotopes during incomplete, relatively fast, unidirectional biogeochemical reactions. Molecules with heavier isotopes form stronger chemical bonds than molecules with lighter isotopes. As a result, it requires less energy to disassociate light isotope molecules and they react more quickly during biogeochemical processes. This leads to isotopic separation between the product and the reactant. Lighter isotopes are used to produce the biologically mediated product, and the heavy isotopes remain in the original source or substrate; for example, plants preferentially use isotopically light CO_2 (^{12}C) during photosynthesis because it requires less energy to disassociate ^{12}C bonds than ^{13}C bonds in biochemical processes (Evans et al., 1986; Smith, 1972). Changes in heavy isotope abundance as a result of isotope fractionation is generally expressed as a fractionation factor (α) and is calculated as follows: $\alpha = \delta^b X_{product}/\delta^b X_{substrate}$. Isotope fractionation is also commonly quantified as an enrichment or discrimination factor: $\Delta \delta X = \delta^b X_{product} - \delta^b X_{substrate}$.

Isotopic fractionation is the major process that creates variation in stable isotope ratios among different abiotic compartments, primary producers, and consumers. For example, δ^{13}C fractionation factors in primary producers vary depending on their photosynthetic fixation pathways (C_3, C_4, CAM) and local environmental conditions (Farquhar et al., 1989; France, 1995; Madhavan et al., 1991). As a result, δ^{13}C values vary substantially at the base of the food chain and can be used to identify and distinguish different primary producers such as seagrass, mangroves, and plankton (Fry et al., 1977; Peterson and Fry, 1987). Another significant type of fractionation is *trophic fractionation* or *diet–tissue fractionation*.

Animals preferentially eliminate lighter isotopes in waste products, and as a result consumer tissues have higher stable isotope values than their diet (DeNiro and Epstein, 1978, 1981). *Trophic discrimination factors* or *diet–tissue-discrimination factors* (DTDFs) are equal to the difference between the isotope value of the consumer and its diet: $\Delta \delta X_{DTDF} = \delta^b X_{consumer} - \delta^b X_{prey}$. As a result of these diverse but consistent fractionation processes, unique isotope ratios are characteristic of specific primary producers, prey species, food webs, and foraging locations.

Elasmobranchs, like all consumers, assimilate and modify the isotopic ratios of their diet (Hussey et al., 2011; Logan and Lutcavage, 2010). More specifically, the isotope values of shark tissues (e.g., muscle, blood, skin) are the weighted average of the isotope values of their prey. Therefore, shark isotope values can be compared to the isotope values of primary producers, food webs, or prey to estimate sources of primary production, shark foraging locations, and the types of prey they consume. However, due to such factors as trophic fractionation, shark isotope values will never be identical to prey or habitat values, and these changes must be taken into account when comparing the isotope values of sharks and their dietary resources.

1.2.2 Fatty Acid Analysis

Fatty acids (FAs) are a diverse group of macromolecules that can be used to differentiate consumer dietary niches and specific prey items (Parrish, 2013). They are the primary component of oils and fats (i.e., lipids) in tissues and are grouped into several major lipid classes, including triacylglycerols, wax esters, phospholipids, and sterols. FAs are critical to biochemical functions, serving as metabolic fuel sources, structural components in cell membranes, and gene regulators (Sargent et al., 2002). FAs are classified and named based on their chemical structure and are often referred to by these "chemical codes" within the literature. They fit within three functional groups: (1) saturated fatty acids (SFAs), which lack double bonds within the carbon backbone; (2) monounsaturated fatty acids (MUFAs), which contain one double bond; and (3) polyunsaturated fatty acids (PUFAs), which contain multiple double bonds. The nomenclature of individual FAs outlines the chemical structure, beginning with the number of carbons in the chain (usually 14 to 24), followed by the number and location of the double bonds. As an example, 22:6ω3 refers to a fatty acid with a backbone containing 22 carbons with 6 double bonds, the first of which is 3 carbons from the end of the chain. The chemical complexity of these molecules can make the nomenclature and literature initially intimidating, but using fatty acid analysis (FAA) in foraging ecology does not necessarily require an advanced understanding of organic biochemical processes.

Fatty acids can be employed as biochemical markers for two reasons: First, like stable isotopes, they chemically reflect the base of the food chain (Dalsgaard et al., 2003);

however, differentiating FAs is not based on fractionation but instead on fatty acid production specific to distinct primary and secondary producers (Ackman et al., 1968). For example, diatoms produce 20:5ω3, macroalgae and seagrass produce 18:3ω3, and dinoflagellates and zooplankton produce 22:6ω3 (Falk-Petersen et al., 2000; Kelly and Scheibling, 2011). Second, because vertebrates are unable to create or interconvert most of the FAs that are vital to maintaining cellular functions, they must take these directly from their food sources with minimal modification (Dalsgaard et al., 2003; Iverson, 2009). Therefore, a consumer's diet can be determined based on the similarity between predator and prey FA profiles. Because there are many more FAs than dietary isotopes (>60), FAA can identify distinct food resources and habitats in much finer detail.

Individual FAs and the subsequent full profiles (the relative proportions of all fatty acids in a sample) are also tissue and taxa dependent. Certain FA groups are preferentially integrated into distinct tissues according to their biochemical functions; for example, chondrichthyan livers have higher MUFA concentrations than muscle, which contain more PUFAs (Beckmann et al., 2014; McMeans et al., 2012; Pethybridge et al., 2011). This is true for both the predator of interest and for potential prey items, thus entangling physiology and diet at each trophic level. Despite this additional layer of potential confusion, FA routing means that FAA has the capacity to distinguish prey contributions to consumer diet by analyzing distinct tissue types. Seal blubber, for example, has particularly high levels of 22:5ω3, which was used to identify ringed seals as a key prey item of Greenland sharks, *Somniosus microcephalus* (McMeans et al., 2012).

Similar to stable isotopes, which undergo predictable fractionation processes, some FAs can also undergo bioconversion (Beckmann et al., 2013a; Iverson, 2009; Sargent et al., 2002). Thus, predator fatty acids will never precisely mirror those of their prey (Beckmann et al., 2013b; McMeans et al., 2012). Rates of bioconversion between trophic links have gone largely unexplored in sharks. Currently, these bioconversion processes remain a limitation rather than a tool and make both qualitative assessment and quantitative modeling efforts challenging.

1.3   COMMON BIOMARKERS IN SHARK ECOLOGY

1.3.1   Stable Isotopes

The most commonly used isotope tracers in shark foraging ecology are $\delta^{13}C$ and $\delta^{15}N$. These tracers are usually analyzed together because they provide two complementary pieces of dietary information. $\delta^{15}N$ undergoes substantial trophic fractionation and increases at each consumer level (Deniro and Epstein, 1981; Hussey et al., 2010); therefore, it is an extremely useful indicator of shark trophic position

(TP) and dietary selectivity (Dicken et al., 2017; Ferreira et al., 2017). $\delta^{15}N$ can also vary between abiotic sources and foraging locations at the base of the food web (Wada and Hattori, 1976); for example, wastewater treatment inputs can be significant sources of elevated nitrogen and $\delta^{15}N$ in coastal habitats and species (Costanzo et al., 2001; Savage, 2005; Schlacher et al., 2005). Consequently, $\delta^{15}N$ can also be used to determine species foraging locations (Hadwen and Arthington, 2007), although the interactive effects of trophic enrichment and $\delta^{15}N$ baseline variability on shark isotope ratios can confound qualitative and quantitative TP determination (see Section 1.5.1, Trophic Position).

$\delta^{13}C$ values vary substantially at the base of the food chain between primary producers and different habitats with distinct inorganic carbon sources (Bouillon et al., 2011; Peterson and Fry, 1987); however, $\delta^{13}C$ trophic fractionation is considerably less than that for $\delta^{15}N$ (DeNiro and Epstein, 1978; Hussey et al., 2010). As a result, $\delta^{13}C$ ratios are relativity constant up the food chain and can be used as a more direct indicator of the sources of primary production for a food web (e.g., Heithaus et al., 2013) and the dietary and habitat selection of consumers (e.g., Rosas-Luis et al., 2017). In shark foraging studies, $\delta^{13}C$ values are most often used to distinguish benthic and pelagic, freshwater and marine, and inshore and offshore sources to animal diet (McMahon et al., 2013). For example, Burgess et al. (2016) measured $\delta^{13}C$ in giant manta ray (*Mobula birostris*) muscle tissue, surface zooplankton, and mesopelagic species and found that the majority of *M. birostris* diet was subsidized by mesopelagic sources. However, $\delta^{13}C$ patterns in marine systems are complex and variable, and environmental trends in isotope distribution should be independently verified in each study if possible.

Another useful dietary isotopic tracer is the heavy stable isotope of sulfur, $\delta^{34}S$, although it is rarely used to study sharks (Munroe, unpublished data). Limited work has shown that, similar to $\delta^{13}C$, $\delta^{34}S$ can distinguish the dietary contributions of different producers and food webs (Connolly et al., 2004; Fry et al., 1982); however, $\delta^{34}S$ has a highly complex marine cycle, and these trends are often inconsistent among studies (Connolly et al., 2004). Therefore, although $\delta^{34}S$ may help to distinguish different habitats and prey, $\delta^{34}S$ values themselves can be difficult to interpret in high-level consumers.

Other potential stable isotope tracers include oxygen ($\delta^{18}O$) and strontium ($\delta^{87}Sr$); however, these tracers are not necessarily dietary tracers and to date have only been applied to shark fossils in paleoecological studies (Fischer et al., 2013). $\delta^{18}O$ is typically measured in the hard parts of fish and are used as a proxy with which to measure local environmental conditions. This is because $\delta^{18}O$ fractionation during biomineralization is temperature dependent. As a result, $\delta^{18}O$ in otoliths can be used to estimate population connectivity and movement between habitats defined by different climates or temperature ranges (Currey et al., 2014; Darnaude et al., 2014; Fraile et al., 2015). The use of $\delta^{18}O$ with sharks and rays is limited, though, because the hydroxyapatite that makes up their skeleton lacks oxygen; thus, shark skeletons do not contain the same record of information that is stored in the aragonite that composes teleost otoliths. $\delta^{87}Sr$ values are also measured in fish hard parts and change in accordance with a wide range of factors, including salinity, temperature, and lithology (Capo et al., 1998). In aquatic ecology, it is most commonly used to estimate the natal origins and movement of fish in complex river systems (Brennan, 2015; Humston et al., 2017; Kennedy, 2000).

1.3.2 Fatty Acids

Given the large number of potential FA tracers (>60), it not possible to describe each one in detail here, so this section will focus on the primary groups of FAs most commonly used to define consumer foraging patterns. It is important to note, however, that the validity of individual FAs as tracers is largely unexplored for many higher order taxa, sharks in particular. With only a few controlled feeding studies (Beckmann et al., 2013a,b), much of our understanding stems from retrospective analyses, whereby the individual FAs driving differences between sharks are then compared to FAs from other taxa. Furthermore, most FAA studies use the entire FA profile to identify distinct trophic groups (Dunstan et al., 1988; Pethybridge et al., 2011). Only recently have studies begun to examine individual FAs in greater detail (Every et al., 2016).

Certain FAs, particularly some PUFAS, are known as *essential fatty acids* (EFAs) and can only be obtained from dietary sources (Iverson, 2009). They are transferred up the food chain with minimal modification and can be traced to known trophic sources; for example, EFA 22:6ω3 (docosehexaenoic acid, or DHA) was used to distinguish captive Port Jackson sharks (*Heterodontus portusjacksoni*) that were fed different diets (Beckmann et al., 2013a). Because the biosynthesis of certain monounsaturated and saturated fatty acids is also limited, they retain their use as bioindicators. For example, the shorter chain MUFAs 16:1ω7 and 18:1ω9 are indicative of phytoplankton-, mangrove-, and macroalgae-based food webs, whereas the long-chain MUFAs 20:1ω9, 20:1ω11, and 22:1ω9 stem from zooplankton (Falk-Petersen et al., 2000; Kelly and Scheibling, 2011) and are thus valuable indicators of mesopelagic feeding (Pethybridge et al., 2014). SFAs (16:0 and 18:0, in particular) often dominate the muscle and liver FA profiles of many elasmobranchs. They have been used to distinguish demersal sharks feeding on mesopelagic and benthopelagic fish, squid, and crustaceans (Pethybridge et al., 2011).

Furthermore, certain FAs are preferentially retained in different types of tissues and thus can highlight the consumption of prey with starkly contrasting physiology (e.g., marine mammal blubber vs. teleost muscle vs. cephalopod mantle). For example, 18:1ω9 and 20:1ω9 are especially high in blubber (Ackman et al., 1968; Waugh et al., 2014) and can indicate marine mammals as a prey type (Pethybridge

et al., 2014; Schaufler et al., 2005). As promising as these indicators are, especially for elucidating the diet of complex predators feeding on a diverse range of organisms, they should be treated with caution as they remain entangled in trophic pathways, prey physiology, and potential bioconversion within the consumer.

1.4 SAMPLE ANALYSIS

1.4.1 Tissues

Tissue selection is a crucial component of biomarker study design. In elasmobranch isotopic studies, the most frequently collected tissues are skin, muscle, whole blood, plasma, and to a lesser extent liver and calcified cartilage (i.e., hard parts). More recently, one study also sampled epidermal mucus (Burgess et al., 2017). The main issue to consider when selecting a tissue for SIA is that different tissues assimilate isotope ratios from food at different rates (Tieszen et al., 1983). Therefore, different tissues will represent a shark's diet over different periods of time. The amount of time that is required for full isotopic assimilation, or for the isotope ratios of an animal's previous diet to be replaced by the ratios of its current diet, is referred to as *isotopic turnover*. Tissues that are structural components and/or have low protein turnover, such as muscle, have slower isotopic turnover rates than more active tissues, such as plasma and liver (Vander Zanden et al., 2015b). As a result, muscle and plasma isotope values are often compared to observe changes in shark diet or habitat use over time (Kinney et al., 2011; Munroe et al., 2015).

Isotopic turnover in sharks is relatively slow for all tissues. Logan and Lutcavage (2010) performed a diet-switching experiment on captive sandbar sharks (*Carcharhinus plumbeus*) and found that complete isotope turnover for whole blood required 200 to 300 days, while muscle turnover required 300 to 500 days. Kim et al. (2012) found similar results, in that isotopic turnover in captive leopard sharks (*Triakis semifasciata*) was 300 days for plasma and >700 days in muscle. In general, isotopic turnover in adult shark muscle tissue takes ≥1 year. For this reason, shark muscle values reflect the long-term foraging patterns of individuals and should only be used to evaluate equally long-term dietary and movement trends. Plasma and liver can be used to monitor more recent dietary trends, potentially at a seasonal scale. Turnover times are considerably shorter in fast-growing, more metabolically active juveniles (Malpica-Cruz et al., 2012), and juvenile dietary patterns can be investigated over notably shorter timeframes (Matich et al., 2015). However, some neonate and young of year sharks will also reflect the isotope values of their mothers (Belicka et al., 2012; Olin et al., 2011). Potential maternal isotopic influences will need to be taken into account when examining any ontogenetic changes in juvenile shark isotope values.

Hard parts, more than any other tissue, provide a tangible link to the past. In contrast to soft tissues, hard parts (e.g., vertebra, teeth, spines) become metabolically (i.e., biochemically) inert after formation (Clement, 1992; Dean et al., 2015). Therefore, hard parts can be used to infer shark dietary or habitat use patterns at the time when that part was originally formed (McMillan et al., 2017). Vertebrae are particularly useful because sharks deposit annual growth-related bands which can be used to age each individual (Cailliet et al., 2006). Successive bands become metabolically inert after they are deposited; therefore, vertebrae can provide chronological isotopic records of shark foraging patterns over the lifetime of the individual (Tillett et al., 2011). Eye lenses are also being used as potential recorders of stable isotope histories (Wallace et al., 2014). Similar to vertebrae, lens fiber cells are deposited in successive concentric circles that become metabolically inert after formation (Dahm et al., 2007; Nicol and Somiya, 1989). This technique has only recently been validated for a few shark species, but early work has shown isotopic variation in shark lens layers can identify ontogenetic variation in foraging locations over regional scales (Quaeck, 2017).

The most commonly collected tissues for FAA are subdermal tissue, muscle, liver, and blood plasma. Unlike SIA, the primary issue regarding tissue selection is the difference in tissue physiology, which dictates the underlying FA profiles (Pethybridge et al., 2010). Specific groups of fatty acids (i.e., MUFAs vs. PUFAs) are preferentially sequestered into different tissues, which in turn serve distinct functional roles; for example, shark muscle tissue has higher relative concentrations of PUFAs (including EFAs) compared to liver, which has a greater proportion of high-energy MUFAs (Davidson et al., 2014; Pethybridge et al., 2014). How tissue-specific physiology confounds shark FAA remains largely unknown. Regardless, a range of tissue types have been successfully used to assess shark diet. Despite some debate about which tissues most accurately reflect shark prey (Beckmann et al., 2013a; McMeans et al., 2012; Pethybridge et al., 2011), the selection of appropriate tissue should be based on availability, and, where possible, multiple tissue types should be used to confirm hypotheses (Davidson et al., 2011; Pethybridge et al., 2011, 2014; Schaufler et al., 2005).

Similar to isotopes, FA turnover also varies between shark tissues. Beckmann et al. (2013b) found that the FA profiles of *Heterodontus portusjacksoni* blood, serum, and liver took 12 weeks to reach equilibrium with a new diet, whereas muscle took 18 weeks. However, tissue-based differences in FA profiles are far more significant than those influenced by turnover rates alone, and as such tissue FA profiles cannot be directly compared to examine dietary changes over time. Moreover, FA turnover rates are all relatively fast compared to stable isotopes and are better suited to assess fine-scale dietary shifts over much shorter time scales (Wai et al., 2011).

1.4.2 Sample Collection and Preservation

Biomarker samples need to be carefully and quickly preserved to prevent degradation (Meyer et al., 2017). The two most common methods for long-term sample preservation and storage, regardless of biomarker, are freezing ($-20°C$) or submersion in ethanol. Freezing is preferred because it has a minimal effect on biomarker values (but see Wolf et al., 2016) and will maintain ratios and FA profiles for prolonged periods of time (Meyer et al., 2017). Unfortunately, keeping samples frozen is not always a viable option. Storing samples in ethanol is a practical alternative (Hobson et al., 1997), but lipids are soluble in ethanol and submerging the sample will, in essence, begin the lipid extraction process (see Section 1.4.4, Lipid Extraction). Samples may also exchange carbon with ethanol (Edwards et al., 2002). As a result, ethanol can alter stable isotope ratios (Sweeting et al., 2006). The change is usually small (Hobson et al., 1997) but may be as high as 1.5‰ (Kaehler and Pakhomov, 2001; Sarakinos et al., 2002). The effect of ethanol also varies among taxa, tissues, and individuals (Arrington and Winemiller, 2002; Kim and Koch, 2012); therefore, if samples are stored in ethanol small differences in isotope ratios between samples (<1%) should be considered insignificant and ignored (Kim and Koch, 2012; Sweeting et al., 2006). Tissues preserved in ethanol are still viable for FAA; however, ethanol can complicate the lipid extraction process. If the ethanol is discarded and replaced (as is common with samples retained for genetic studies), the extracted lipids will be lost and the sample will no longer be viable. Ethanol can also diminish growth-banding patterns in vertebrae, which can increase the likelihood of errors in age estimation (Wintner et al., 2002).

Tissue samples should be dried and analyzed as soon as is reasonable to ensure the longevity of the samples and reliability of the analyses. Soft-tissue isotope samples can be dried via freeze drying or oven drying. Both techniques are suitable for bulk (i.e., whole tissue) isotope analysis, but oven drying will destroy amino and fatty acids, making it impossible to use oven-dried samples for compound-specific (see Section 1.5.5, Compound-Specific Stable Isotope Analysis) or fatty acid studies. Thus, FAA can be performed on frozen or freeze-dried tissues only. Finally, vertebrae and other hard parts do not necessarily have to be oven or freeze dried prior to analysis. In fact, air drying hard parts in a fume hood is often ideal because oven drying can crack samples and confound aging analysis (Kim and Koch, 2012; Smith et al., 2013b).

1.4.3 Urea Extraction

Sharks and rays store large amounts of urea to help maintain osmotic equilibrium with their environment (Goldstein and Forster, 1971; Olson, 1999; Shuttleworth, 1988). Nitrogen waste products such as urea typically have low $\delta^{15}N$. As a result, urea can artificially depress $\delta^{15}N$ values, making shark $\delta^{15}N$ incomparable to other taxa (Fisk et al., 2002;

Hussey et al., 2010; Kim and Koch, 2012). Urea concentrations also vary among species, tissues, and individuals and are affected by environmental conditions (Ballantyne, 1997; Pillans et al., 2005). Therefore, urea must always be removed from shark tissues prior to isotope analysis (Carlisle et al., 2016; Churchill et al., 2015). Urea can be removed from muscle tissue by rinsing samples with deionized water (Li et al., 2016; Marcus et al., 2017), although this procedure is not recommended for blood components (Kim and Koch, 2012).

1.4.4 Lipid Extraction

Lipids are isotopically light, or low in ^{13}C, compared to carbohydrates and proteins (DeNiro and Epstein, 1977). Lipid content also varies substantially among different tissues, species, and individuals (Davidson et al., 2011; Meyer et al., 2017; Pethybridge et al., 2014); therefore, lipids may have to be removed from tissue samples prior to isotope analysis in order to standardize $\delta^{13}C$ comparisons (Post et al., 2007). However, shark tissues, with the exception of liver, usually have low lipid content, so extraction may not be required. In general, it is always preferable to perform as few chemical treatments on any sample as possible. Bulk C:N ratios can be used as an indicator of lipid content and help to determine if lipid extraction is necessary (Post et al., 2007). Lipid extraction is commonly performed using a modified Bligh and Dyer (1959) method and is offered at an additional cost in most SI laboratories. Conveniently, Bligh and Dyer is also the preferred method of lipid extraction for FAA, so it is wise to retain lipid extracts from isotope samples for future FA studies (Marcus et al., 2017). Lipid extraction will also affect bulk $\delta^{15}N$ values. If lipid extraction is performed, samples should be divided into two portions and $\delta^{15}N$ ratios should be determined separately with untreated, urea-extracted samples (Kim and Koch, 2012; Marcus et al., 2017). Another potential option is to apply a mathematical normalization to shark $\delta^{13}C$ that adjusts values for lipid content in non-treated samples (Post et al., 2007). Mathematical normalizations can help to streamline tissue preparation procedures and maintain $\delta^{15}N$ values.

1.4.5 Special Considerations for Calcified Structures

Hard parts contain mineralized inorganic carbon as well as the carbon found within the protein structures. Because inorganic carbon $\delta^{13}C$ can distort protein $\delta^{13}C$ values, mineralized inorganic carbon may have to be removed prior to analysis. Traditionally, HCl is used to dissolve and evaporate inorganic $\delta^{13}C$; however, the addition of HCl can change $\delta^{15}N$ values and will also rapidly dissolve collagen along with the inorganic matrix (Kim and Koch, 2012). Kim and Koch (2012) adapted an alternative technique originally described by Tuross et al. (1988) that uses EDTA (an organic chelating agent that binds to Ca^{2+} ions) to treat powdered calcified samples. This approach helps to maximize the amount of collagen available for analysis while also removing inorganic material. It is important to note, however, that some shark hard parts are minimally calcified and treatment may be unnecessary. A subset of samples should be tested with and without treatment prior to a full analysis.

1.5 ECOLOGICAL APPLICATIONS

1.5.1 Trophic Position

Elasmobranch biomarker analysis is most often used to determine shark trophic position. As previously discussed, $\delta^{15}N$ increases substantially with each trophic level and is an excellent tracer to estimate the TP of sharks and rays. Although some FAs may reflect trophic position similar to $\delta^{15}N$, these are untested and are not currently recommended to determine shark TP. $\delta^{15}N$ values can be qualitatively compared among tissues, individuals, age classes, or species, where groups with higher $\delta^{15}N$ values are deemed to be at a higher trophic level. Malpica-Cruz et al. (2013) found that the $\delta^{15}N$ values of shortfin mako shark (*Isurus oxyrinchus*) and white shark (*Carcharodon carcharias*) soft tissues varied substantially as a function of size. Small individuals experienced a rapid increase in $\delta^{15}N$ of 2 to 3‰ at approximately 85 cm and 150 cm total length for *I. oxyrinchus* and *C. carcharias*, respectively. This change strongly suggested an increase in TP with size for each species. Estrada et al. (2006) performed incremental $\delta^{15}N$ analysis on *C. carcharias* vertebrae and produced individual chronological records of trophic information. Data showed significant enrichment in $\delta^{15}N$ with increasing distance from the vertebrae center, indicating a positive correlation between body size and trophic position (Figure 1.1). Thus, with little additional information, $\delta^{15}N$ can help to decipher the relative trophic patterns of sharks at the individual and population level; however, qualitative comparisons assume that differences in $\delta^{15}N$ between groups are almost entirely due to trophic fractionation and changes in TP, and that there is little or no variation in $\delta^{15}N$ baseline values between each group of interest. This is an unlikely assumption in most circumstances as sharks are highly mobile and forage between food webs with distinct $\delta^{15}N$ baseline values. Moreover, $\delta^{15}N$ baseline values will change over time. Shark movement patterns and local $\delta^{15}N$ fluctuations must be considered when comparing $\delta^{15}N$ between target groups.

$\delta^{15}N$ can also be used to calculate more precise estimates of shark TP. Among the various TP equations, the simplest one is as follows:

$$TP_{elasmobranch} = \frac{\left(\delta^{15}N_{elasmobranch} - \delta^{15}N_{baseline\,organism}\right)}{\left(\Delta^{15}N\right)} + TP_{baseline\,organism}$$

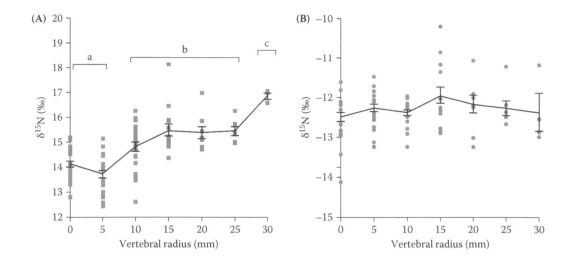

Figure 1.1 Stable isotope fractionation during growth in the white shark ($n = 27$). (A) $\delta^{15}N$ fractionation (‰) vs. vertebral radius (mm). Significantly different groups are indicated by lowercase letters and corresponding spanner lines above the symbols. (B) $\delta^{13}C$ fractionation vs. vertebral radius. Squares (A) and circles (B) indicate individual values; the solid line connects mean values ± SE. (Estrada, J.A. et al., *Ecology*, 87, 829–834, 2006.)

where $\delta^{15}N_{baseline\ organism}$ and $TP_{baseline\ organism}$ are the $\delta^{15}N$ value and expected TP for a baseline organism in the ecosystem (e.g., prey species, primary producer), $\delta^{15}N_{elasmobranch}$ is the $\delta^{15}N$ value of the shark, and $\Delta^{15}N$ is the DTDF (Post, 2002). Note that most TP calculations assume the DTDF remains constant between each trophic level. Unlike qualitative approaches, TP calculations standardize trophic comparisons between elasmobranchs because they can account for changes in DTDFs and $\delta^{15}N$ baseline values over different temporal and spatial scales. Speed et al. (2012) calculated the TP of four shark species found at Ningaloo Reef, Western Australia, to compare their trophic role. Results indicated that $\delta^{15}N$ values were important for distinguishing species and size ranges; however, TP calculations showed that the maximum difference was only 0.6 trophic steps (TP = 4.3–3.7), which suggested considerable trophic overlap between species.

Accurate TP estimates, however, hinge on a sound knowledge of local ecosystems and isotope dynamics. It is important to select baseline organisms and shark tissues that are appropriate for the spatial and temporal scale of the study; for example, due to long turnover rates, shark muscle values will likely reflect the dietary contributions of prey from outside the study area or time period. TP estimates are also heavily dependent on DTDFs. The most commonly used $\Delta^{15}N$ value is 3.4‰, which was originally put forward by Post (2002) and was derived by averaging the $\Delta^{15}N$ values from past studies. However, these studies were dominated by birds, mammals, and teleosts, and Post (2002) also clearly acknowledged there was substantial variation across taxa. Unfortunately, there are no validated DTDFs for elasmobranchs, although a few studies have tried to calculate more specific values. Hussey et al. (2010) estimated the $\Delta^{15}N$ of muscle and liver in three sand tiger sharks (*Carcharias*

taurus) and a lemon shark (*Negaprion brevirostris*) in public aquaria. They found that the mean $\Delta^{15}N$ value in muscle was 2.29‰, but the liver $\Delta^{15}N$ value was considerably lower, with a mean of 1.5‰. Kim et al. (2012) determined that the $\Delta^{15}N$ values for captive *Triakis semifasciata* ($n = 6$) were 2.2‰ and 3.7‰ in plasma and muscle, respectively. However, Malpica-Cruz et al. (2012) found significantly smaller $\Delta^{15}N$ values in neonate and young of the year *T. semifasciata* ($n = 16$), with values ranging from 1.08‰ to 1.76‰. These studies indicate that elasmobranch DTDFs are highly variable among species, age classes, and tissues and that no single DTDF is suitable for all elasmobranchs (Olin et al., 2013).

$\Delta^{15}N$ values (as well as $\Delta^{13}C$) will also vary with diet quality and type (Caut et al., 2009; Rosenblatt and Heithaus, 2012), as well as with trophic level, where the most appropriate species-specific $\Delta^{15}N$ values generally decrease with increasing trophic level (Olin et al., 2013). This violates a key assumption of most standard TP equations and can be problematic if baseline organisms are at a much lower trophic level than the predator. To avoid misleading results, calculate TP using a range of probable $\Delta^{15}N$ values. The best options for baseline organisms are secondary consumers that integrate local $\delta^{15}N$ values over a period of at least several months. Mixing-model TP equations, which account for the variable contributions of different prey to shark diet, are also available and will likely be more appropriate in most situations (Post, 2002). Compound-specific stable isotope analysis (see Section 1.5.5) can provide more precise TP estimates than bulk calculations. Ultimately, although $\delta^{15}N$ can be a highly useful way to examine the diet and trophic level of shark species, TP calculations and comparisons should be interpreted with caution and factors such as $\Delta^{15}N$ must be carefully considered prior to data analysis.

1.5.2 Niche Breadth and Specialization

Biomarkers are increasingly used as a proxy to estimate the trophic niche of aquatic species. The variety of distinct and contradictory definitions for the niche concept cannot be properly explored here (Poisot et al., 2011), but within the context of shark dietary analysis *niche breadth* is defined as a measurement of all the resources used by a species, relative to the resources available within the environment (Colwell and Futuyma, 1971; Munroe et al., 2014). Hutchinson (1957) proposed that niche breadth could be quantitatively measured using an *n*-dimensional hypervolume, where the dimensions are environmental variables (e.g., prey that are available), and the multidimensional hypervolume is the environmental space occupied by the species (e.g., prey that are consumed). The hypervolume can be plotted on a Cartesian coordinate system where the axes are environmental variables. Biomarkers that are measured in primary producers or prey items reflect the range of available dietary resources, while the values of consumers reflect the specific resources a species actually uses. As a result, biomarkers can be used as a proxy to measure the trophic niche of sharks and rays; however, the biochemical niche of a species is not equivalent to its ecological niche, which includes a range of variables and factors that biomarkers cannot reliable detect. Thus, biomarker and ecological niches are two separate entities and are not interchangeable.

Stable isotopes are commonly used to investigate the trophic niche of sharks, where the niche is the area in isotopic space occupied by the elasmobranch population (i.e., hypervolume) and isotope values are used as coordinates (i.e., environmental variables) (Newsome et al., 2007). These estimates are best referred to as *isotopic niches* to avoid confusion with niches based on other diet data. Groups (e.g., species, age classes, sexes) with smaller isotopic niches are presumed to consume a less diverse array of resources and are therefore more specialized than populations with larger isotopic niches (but see caveats described below). The relative position and size of isotopic niches are then used to estimate shark foraging patterns, resource partitioning, and trophic specialization (Tilley et al., 2013; Vaudo and Heithaus, 2011). Heithaus et al. (2013) used $\delta^{13}C$ and $\delta^{15}N$ to estimate the isotopic niche overlap of dolphins, large sharks, and small elasmobranchs in Shark Bay, Australia. Each group occupied a unique area in isotopic space, suggesting limited resource use overlap and low competition for prey (Figure 1.2). The relative position of each population in isotopic space also helped delineate resource use patterns. Elasmobranchs had the highest $\delta^{13}C$ values, which suggested that seagrass-based ($\delta^{13}C$-enriched) food webs were more important to elasmobranchs than resident dolphins.

This approach relies on two key assumptions (Bearhop et al., 2004): First, prey species must be isotopically distinct. If prey isotope values overlap or are poorly resolved, measuring niche breadth is a pointless exercise. The amount of

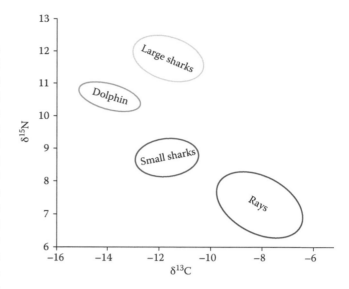

Figure 1.2 Standard ellipse areas corrected for sample size (SEAc) of large predators based on guild-level analyses. (Based on Heithaus, M.R. et al., *Mar. Ecol. Prog. Ser.*, 481, 225–237, 2013.)

variation that might occur between populations is directly tied to the variation in isotopic values of potential prey resources; therefore, the ecosystem context of isotope values must be considered. Second, the tissue that is analyzed must reflect a time period over which population niche width (i.e., intraspecific variation) is expressed. In a population of individual generalists, individual variability in isotopic values will often only manifest over shorter timeframes. Shark tissues, particularly muscle, have long turnover times. Individual shark muscle ratios will eventually converge via the averaging of isotope values from prey consumed over long periods of time. The muscle values of highly migratory sharks will also reflect contributions from distant food webs. As a result, niche breadth models of shark muscle tissue can be misleading. To avoid these problems, it is important to sample and compare multiple tissues with different turnover rates. Fast-turnover tissues such as plasma should be used whenever possible. Finally, it is important to consider that isotopes are not measuring diets but overall trophic interactions. High isotopic specialization suggests that there is specialization in the types of food webs (e.g., seagrass vs. phytoplankton) that individuals are foraging in, and not specific prey types being consumed. To obtain prey-specific detail, mixing models, more specific tracers, or stomach content analyses are required.

Isotopic niche breadth can be measured using a variety of different approaches (Jackson et al., 2011; Layman et al., 2007, 2012). The total area (TA) metric uses a convex hull to encompass all of the individuals within a population (Layman et al., 2007), so TA is a useful measure of total isotopic spread in a population. A Bayesian standard elliptical area (SEA$_b$) is fit around the densest isotope values within a population and explicitly accounts for the uncertainty associated with isotope ratios (Jackson et al., 2011). As a result, SEA_b is less sensitive to sample size and is a better measure of core isotopic niche breadth size. Espinoza et al. (2015) used $\delta^{13}C$ and $\delta^{15}N$ of muscle tissue to examine the feeding ecology of four common demersal elasmobranchs off the Pacific Coast of Costa Rica. The authors compared the isotopic niches of juveniles and adults of each species using convex hulls and SEA_b (Figure 1.3). The authors found that, for most species, juveniles and adults had distinct isotopic niches, with adults consuming prey at higher trophic levels. TA size, however, clearly increased with sample size, whereas SEA_b resulted in more reasonable comparisons between age classes with unequal sample sizes. Empirical evaluations of both TA and SEA_b methods

have shown that a minimum of 30 samples are needed to ensure that SEA_b size and position are indicative of the true isotopic niche of the population (Syväranta et al., 2013). TA estimates require a much greater sample size to reach a similar level of precision. Rarefaction analysis can be used to determine if sample sizes adequately measure TA (Vaudo and Heithaus, 2011).

Layman et al. (2007) proposed several other metrics that help elucidate aspects of trophic diversity within populations using stable isotopes; for example, the mean distance of individual stable isotopic values to the centroid (CD) of the isotopic niche provides an estimate of the average trophic diversity within a population. The density and evenness of individual packing within an isotopic niche can be calculated using nearest neighbor distances (NDDs) and

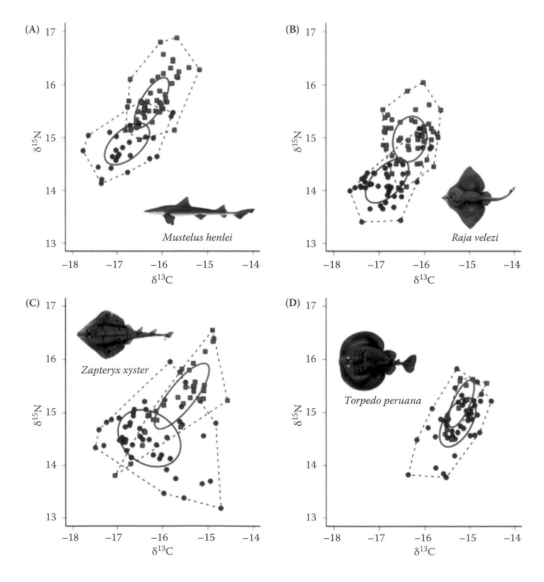

Figure 1.3 Isotopic niche overlap plot of immature (black filled circles) and mature (red filled squares) individuals of (A) *Mustelus henlei*, (B) *Raja velezi*, (C) *Zapteryx xyster*, and (D) *Torpedo peruana*. Convex hulls are indicated by black dashed lines. Standard Bayesian ellipses are indicated by solid colored lines. (Espinoza, M. et al., *J. Exp. Mar. Biol. Ecol.*, 470, 12–25, 2015.)

the standard deviation of NDD (SDNDD), respectively. These metrics were used to show that batoids foraging on a common sandflat in Australia had similar degrees of trophic diversity and evenness within populations (Vaudo and Heithaus, 2011). Tiger sharks (*Galeocerdo cuvier*) in the same ecosystem had higher CDs than other shark species, suggesting greater trophic diversity (Heithaus et al., 2013). Batoid species with low sample sizes showed higher values across metrics, suggesting that adequate sample sizes are important to avoid drawing inaccurate conclusions.

The same principles and approaches apply to assessing niche breadth using FA profiles. Broadly speaking, greater intraspecific variation in FA profiles implies populations having larger biomarker niches. Pethybridge et al. (2010) found that dogfish (*Squalus acanthias*) had the greatest relative variation in both muscle and liver FA profiles compared with 17 other deep-sea chondrichthyans, indicating that the *S. acanthias* population consumed the widest range of trophic resources. In contrast, the gulper shark (*Centrophorus zeehaani*) had the lowest level of variation in liver FA profiles, indicating that it had a limited habitat distribution and a smaller trophic niche. More recently, Every et al. (2017) used ellipse fitting in multidimensional space to calculate fatty acid niche areas for sympatric sharks from northern Australia. Despite the novelty of applying FA datasets to methods traditionally reserved for isotopic data, this is an emerging area of work that should be explored further given the overlapping foundations of both biomarker groups. In some cases, FA may actually be the preferable approach. FA turnover is much faster than SI turnover, regardless of tissue; therefore, there is a smaller chance of FA profile convergence, and profiles are more likely to represent recent and local feeding patterns.

Biomarkers, especially isotopes, can also be used to measure individual specialization. Most approaches use either hard parts that incorporate the stable isotope values of an individual at a particular point in time (e.g., vertebrae) (Christiansen et al., 2015) or by measuring the isotopic values of multiple tissues that incorporate isotopic values over different time frames (e.g., Matich et al., 2011). Repeated sampling of the same tissue from an individual captured multiple times is another potential option, although recapture rates of most elasmobranchs are too low to obtain sufficient sample sizes. In simplest terms, if isotopic values remain constant over time, this would imply that the individual is a specialist; if values change over time, this would imply that the individual is a generalist (Bearhop et al., 2004). Statistical methods for measuring individual specialization rely on measuring the proportion of variance in isotopic variables that is attributable to variation within an individual (WIC) and variation between individuals (BIC) (Bolnick et al., 2003; Matich et al., 2011). If individuals have generalized diets, the WIC component of variation should be higher, and the specialization ratio (BIC:WIC) should increase as the degree of individual

specialization increases. Matich et al. (2011) measured $\delta^{13}C$ values of plasma (shortest isotopic turnover rate), whole blood, muscle, and fin clips (longest turnover rate) from juvenile bull sharks (*Carcharhinus leucas*) in Florida and tiger sharks (*Galeocerdo cuvier*) in Australia. General linear models (GLMs) were used to calculate specialization indices. Although at the population level both populations appear to be trophic generalists, specialization ratios revealed higher degrees of individual specialization within the *C. leucas* population than the *G. cuvier* population. Individual *C. leucas* appeared to adopt one of several foraging tactics, including specializing in foraging from marine or estuarine/freshwater food webs, exhibiting a stable mix for foraging across habitats, or adopting more temporally variable diets (Matich et al., 2011). Similar to population-level assessments, in order to employ these analyses the prey species must be isotopically distinct, and the tissues must reflect a time period over which inter- and intraspecific variation is expressed.

1.5.3 Diet Reconstruction and Mixing Models

Diet reconstruction is arguably the most popular application for biomarkers today. Despite the wide range of techniques and models available, diet reconstruction itself is a relatively simple concept. Elasmobranch biomarker values are the weighted average of the biomarker values of their prey. As a result, biomarkers can be used to determine the relative contributions of different food sources to elasmobranch diet. Isotope diet reconstructions require the isotope values of the consumer, values for each potential prey item or resource (known as isotopic sources or end members), and DTDFs to account for changes in isotope values between the consumer and its diet (Parnell et al., 2013). Isotopic sources must be sufficiently distinct, or well resolved, to distinguish their separate contributions to shark diet (Phillips et al., 2014). Traditionally, dietary contributions to shark populations have been qualitatively assessed and have considered only a few broad resource groups. Fisk et al. (2002) used $\delta^{13}C$ to examine the dietary patterns of *Somniosus microcephalus*. Pelagic habitats typically have lower $\delta^{13}C$ values than benthic food webs, and the authors found that shark tissues had low $\delta^{13}C$, indicating that *S. microcephalus* consumed a greater proportion of pelagic than benthic prey. More recently, Matich and Heithaus (2014) used $\delta^{13}C$ values of whole blood and plasma from juvenile bull sharks (*Carcharhinus leucas*) collected in a Florida estuary to observe seasonal changes in prey consumption. Because plasma has a faster turnover rate than whole blood, plasma was used to detect comparatively recent dietary patterns. Matich and Heithaus (2014) found that plasma $\delta^{13}C$ was higher than whole blood, which suggested that juvenile sharks exhibited seasonal dietary shifts between marine and freshwater prey. This particular study emphasizes the value of multi-tissue analysis, which allows researchers to observe diet-switching behavior over time.

More often, however, researchers want to create detailed reconstructions that specify the dietary contributions of particular prey from of a potentially large array of options. Isotope mixing models are a highly effective approach that can provide more precise dietary reconstructions. Mixing models determine the proportional contributions of sources (e.g., prey) to a mixture (Moore and Semmens, 2008; Parnell et al., 2013; Phillips, 2012). In this case, the "mixture" is the isotope value of the elasmobranch. Mixing models calculate solutions based on a mass-balance approach, where the contributions of potential prey must sum to 100% (Parnell et al., 2013; Phillips, 2012). The simplest linear mixing models can only uniquely distinguish the contributions of a maximum of $n + 1$ sources, where n is the number of isotopes included in the study (Phillips et al., 2014). When the number of sources is greater than the number of isotopes, the model is underdetermined and will return an infinite number of solutions (Boecklen et al., 2011). Shark isotope studies usually only include $\delta^{13}C$ and $\delta^{15}N$ (Hussey et al., 2012), so there will almost always be more significant dietary sources than isotopes in a given environment. These underdetermined models are common when working with sharks because top predators often consume a wide range of resources. Some mixing models (e.g., IsoSource, SIAR) overcome this limitation by reporting a range of likely contribution solutions and thus do not require complete isotopic resolution (Parnell et al., 2013; Phillips, 2012). Bayesian mixing models (SIAR) can be particularly useful when working with shark populations because they account for the intrinsic uncertainty linked with multiple sources, fractionation, and isotope ratios (Moore and Semmens, 2008). Navarro et al. (2014) used a combination of stomach content and stable isotope analysis to study the trophic ecology of the deep-sea kitefin shark (*Dalatias licha*). The authors used $\delta^{13}C$ and $\delta^{15}N$ in a Bayesian isotope mixing model to estimate the contributions of small sharks, teleosts, crustaceans, and cephalopods to the *D. licha* diet. The model indicated that small elasmobranchs constituted the majority of the *D. licha* diet (Figure 1.4), and these results were supported by stomach content analysis. This study demonstrates mixing models can be an accurate and valuable tool in shark foraging ecology.

Model accuracy, however, depends on a number of key factors, and Bayesian or IsoSource models should not be viewed as an easy or infallible way to address complex food web questions. Similar to niche breadth models, diet reconstructions using shark muscle ratios will not capture dietary trends over small spatial and temporal scales. Mixing models also assume that every source in the model contributes to the consumer's diet and that all significant dietary sources have been included in the model (Parnell et al., 2013; Phillips, 2012). Even when using models designed to cope with too many sources, underdetermined model outputs are less precise, the range of possible solutions increases, and models will increasingly indicate the predator consumes an equal proportion of each prey category, although in reality this may not be the case (Boecklen et al., 2011; Brett, 2014). If

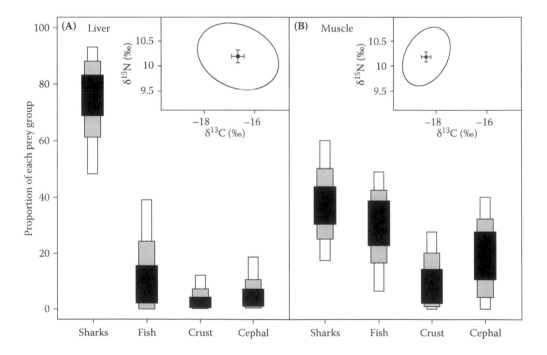

Figure 1.4 Results of the SIAR Bayesian isotope mixing model (95, 75, and 50% credibility intervals) showing estimated prey contributions (SHARKS, small sharks; FISH, fin-fishes; CRUST, crustaceans; CEPHAL, cephalopods) of the diet of *Dalatias licha* in the western Mediterranean sea based on liver (a) and muscle (b) isotopic values. Mean and SE of $\delta^{15}N$ and $\delta^{13}C$ and the standard ellipse areas of liver (upper-right corner of panel a) and muscle (upper-right corner of panel b) are also shown. (Navarro, J. et al., *Mar. Biol.*, 161, 1697–1707, 2014.)

important sources are not included, some models will simply fail to compute a result altogether (Smith et al., 2013a). It is critical to understand that underdetermined models will always provide less reliable results, and models become less reliable with each additional source (Brett, 2014; Phillips and Gregg, 2003; Phillips et al., 2014). It is important to keep potential dietary sources to a realistic minimum and only include sources if there is reasonable evidence to suggest that the prey sources are indeed components of the shark diet. DTDFs also have a large effect on model outputs and should be carefully considered prior to analysis (Bond and Diamond, 2011) (see Section 1.5.1, Trophic Position, for expanded discussion).

Fatty acid analysis can also be applied to shark diet reconstructions, whereby trophic links are presumed based on the similarity of predator and prey profiles. As with stable isotopes, fatty acid profiles must be distinct in order to distinguish different dietary sources (McMeans et al., 2012; Pethybridge et al., 2011). However, because there are far more fatty acids than stable isotopes, it is possible to distinguish a greater variety of more specific sources. Unlike isotope approaches that predominantly use only two isotopes, fatty acid methods usually include ~20 biomarkers, which necessitates the use of multivariate statistical analyses. FAA dietary assessments rely on ordination plots, including multidimensional scaling, principal coordinate analysis, and constrained ordination plots, where the prey groups that are most closely clustered with the predator groups are considered the primary prey species for that consumer (Rohner et al., 2013; Semeniuk et al., 2007). For example, a study using fatty acid profiles of *Somniosus microcephalus* from the same location as those studied in Fisk et al. (2002) confirmed the reliance on pelagic resources, but also indicated predation on halibut and ringed seals in particular (McMeans et al., 2012). As with SIA, fatty acid dietary assessments can be used to observe ontogenetic (e.g., Wai et al., 2011), spatial (e.g., Belicka et al., 2012; McMeans et al., 2013), and temporal (e.g., Every et al., 2017) changes in diet. Mixing models can also theoretically be applied to shark fatty acid datasets. One such technique is quantitative fatty acid signature analysis (QFASA). This technique was originally developed for pinnipeds (Iverson et al., 2004) and has since been applied to fish and other taxa (Magnone et al., 2015). These models are based on the same principles as those developed with SIA and can provide quantitative estimates of the proportions of prey species in the diets of individual predators. However, QFASA, similar to SI mixing models and DTDFs, must account for lipid metabolism and deposition in the target predator. This information is not currently available for elasmobranchs. Calibration coefficients can be determined using captive feeding trials (Iverson et al., 2004), but as with DTDFs they will likely vary between species.

Fatty acid analysis and SIA can also be combined to create more accurate and precise reconstructions. McMeans et al. (2013) combined FAA and SIA to identify the location-specific

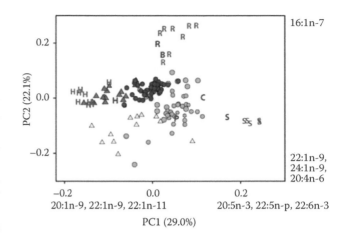

Figure 1.5 Principal component analyses of Greenland shark (*Somniosus microcephalus*) fatty acid profiles from Kongsfjorden, Svalbard, Norway (dark-green circles, muscle; light-green circles, plasma) and Cumberland Sound, Canada (dark-blue triangles, muscle; light-blue triangles, plasma), as well as Greenland halibut (*Reinhardtius hippoglossoides*) (H), ringed seals (*Pusa hispida*) (R), bearded seals (*Erignathus barbatus*) (B), plaice (*Hippoglossoides platessoides*) (P), starry skate (*Amblyraja radiate*) (S), and Atlantic cod (*Gadus morhua*) (C) from Cumberland Sound, Canada (gray) and Kongsfjorden, Svalbard, Norway (red). The amount of variance explained and fatty acids that loaded significantly (i.e., >0.60) on each PC axis are shown. (McMeans, B.C. et al., *Mar. Biol.*, 160(5), 1223–1238, 2013.)

foraging ecology of *Somniosus microcephalus* in Norway and Canada, respectively. Here, the use of ordination plots of the fatty acid profiles showcased specific differences in prey consumption, with the Canadian *S. microcephalus* feeding more on halibut (rich in 20:1ω9) and the Norwegian sharks feeding on ringed seals (high in 16:1ω7) and benthic skates (high in 22:1ω9 and 20:5ω3) (Figure 1.5). SIA results quantified trophic position and feeding location; thus, the authors were able to build a clearer picture of the distinct ecological role *S. microcephalus* plays within these two food webs than would have been possible with either technique alone.

Ultimately, biomarker dietary reconstructions can provide incredible insights into the foraging patterns of sharks and rays that are far more difficult to achieve using traditional methods. Nonetheless, underdetermined isotope mixing models are a prevalent problem in shark ecology, and research questions should be tailored with this limitation in mind (e.g., Burgess et al., 2016). If underdetermined models cannot be avoided, then additional tracers or data, such as $\delta^{34}S$ or stomach contents, should also be included (Hussey et al., 2012; Layman et al., 2012). In cases where potential sources are poorly resolved and have similar isotope values, it may be necessary to combine sources into broad resource categories (Phillips et al., 2005). Model simplification may make outputs less precise, but it will make the results more ecologically and mathematically robust. Looking forward,

FAA can provide more precise dietary reconstructions, and it is important to continue to develop this technique for future shark research. Currently, however, poorly understood tissue-specific fatty acid profiles and bioconversion processes can interfere with this approach. Preliminary studies are recommended to help ensure that prey or resource groups are sufficiently resolved and ecologically appropriate for each model.

1.5.4 Movement and Migration

The biochemical values of food webs vary over time and space in accordance with changes in biogeochemical and oceanographic processes. These differences are transferred up the food chain and are ultimately reflected in the biochemical values of the elasmobranchs that forage within these distinct habitats. In simplest terms, a shark's biomarker signature being consistent with the local food web implies that the individual is resident to the foraging location in which it was sampled. In contrast, when a shark's biomarker signature is inconsistent with the local food web, this would imply that the individual recently acquired food from a different, biochemically distinct location. Therefore, biomarker values can act as intrinsic tags to study shark and ray movement at local, regional, and even global scales (Hobson, 1999, 2008).

Binary or relatively specific assessments of population or foraging connectivity can be achieved using isotope mixing models. Carlisle et al. (2012) used a combination of electronic tags and Bayesian mixing models to study the migratory patterns of *Carcharodon carcharias* in the northeastern Pacific (NEP). Initially, the authors used electronic tagging data to identify focal regions inhabited by *C. carcharias*. These regions included coastal areas off the western United States and offshore pelagic habitats. They then isotopically characterized these different focal regions by collecting SI values for all known *C. carcharias* prey in the NEP from the literature.

Finally, the authors developed a modified Bayesian mixing model (constrained by prior information on shark movement and diet) where the isotope values for each focal region were used as the isotopic sources, and *C. carcharias* muscle ratios were included as the mixture. Results showed relatively equal dietary contributions from coastal and offshore regions, indicating that *C. carcharias* foraged in both areas. These results strongly suggest significant *C. carcharias* migratory connectivity between these two distinct regions. In this case, muscle values were a useful metric because their ratios were the weighted average of prey consumption across distant areas visited over long periods of time. McCauley et al. (2012) also used a Bayesian mixing model to establish the spatial foraging patterns of the blacktip reef shark (*Carcharhinus melanopterus*) and gray reef shark (*Carcharhinus amblyrhynchos*) at the Palmyra Atoll. The authors established isotopic end members for three distinct habitats in and near the atoll, specifically the lagoon, the forereef, and the surrounding pelagic habitat. Isotopic end members were determined using the mean isotope values of resident predatory fish in each habitat. Results indicated that both species consumed prey in forereef and pelagic habitats. Thus, these species likely help to energetically link distinct resources in different habitats.

Isoscapes (shortened form of "isotope landscapes") are another, more geographically explicit way to estimate animal foraging patterns across diverse and distant environments. Isoscapes are spatially continuous predictions of isotope ratios projected over a geographic coordinate system (West et al., 2010). These models are incredibly useful for visualizing isotope variance in aquatic environments because it is simply not plausible to directly sample isotope values for every possible coordinate in the ocean. Isoscapes can fill in the spatial gaps in our regional and global maps and can be constructed using biogeochemical (process-based) models (Figure 1.6) or via spatial interpolation of measured values (measure-based).

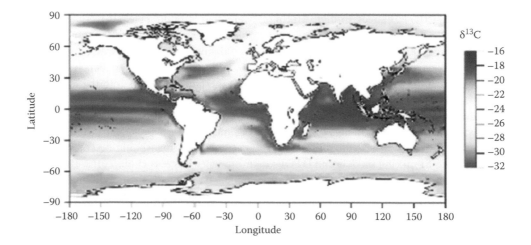

Figure 1.6 Process-based modeled, annually averaged surface water distribution of the carbon isotope composition of phytoplankton ($\delta^{13}C_{PLK}$, ‰). Annual average $\delta^{13}C_{PLK}$ values are calculated using a monthly climatology for the period from 2001 to 2010. (Magozzi, S. et al., *Ecosphere*, 8, e01763, 2017.)

Process-based approaches use environmental variables (e.g., temperature, depth) and known biochemical processes (e.g., fractionation) to predict isotope ratios over space and time (Magozzi et al., 2017; Somes et al., 2010; West et al., 2008). Measure-based isoscapes predict the isotope values of undefined locations using inverse distance weighting (i.e., kriging), where the isotope value of the undefined coordinate is the weighted average of nearby measured coordinates (Jaeger et al., 2010; MacKenzie et al., 2014). Measure-based isoscapes are constructed using reference organisms or abiotic compartments that represent local baseline isotope values.

When an isoscape has been established, it can be used to investigate the spatial foraging patterns of marine species. In a recent global-scale study, Bird et al. (2018) assembled records of the $\delta^{13}C$ values of muscle from approximately 5000 sharks. Samples were originally collected across large latitudinal gradients in oceanic, shelf, and deep water foraging habitats. Bird and his coworkers compared the $\delta^{13}C$ values of shark muscle to a global $\delta^{13}C$ phytoplankton ($\delta^{13}C_{PLK}$) process-based isoscape model (Magozzi et al., 2017). The latitudinal $\delta^{13}C$ gradients of coastal sharks mimicked those predicted by the isoscape (Figure 1.7). This indicated that, in general, coastal sharks do not undertake large latitudinal migrations and obtain the majority of their carbon from the local food webs where they were sampled. By comparison, the latitudinal $\delta^{13}C$ gradients of oceanic sharks were much shallower

than those predicted from local $\delta^{13}C_{PLK}$. These inconsistent latitudinal trends indicate that oceanic sharks did not obtain their prey from local food webs. Instead, results suggest oceanic sharks acquire large proportions of their carbon from a narrower latitudinal range between 30 and 50 degrees.

The movement and migratory patterns of sharks can also be investigated using probabilistic assignment to origin models. Assignment models use the isotope ratios of tissue samples to predict the geographic area in which the tissue was formed (Hobson et al., 2010; Wunder, 2010). If a sample's isotopic signature "matches" a specific location, then the sample is assigned to that location. Thus, these models can generate geographically specific assignments that indicate the likely migratory origin for a sample (i.e., individual). Assignment models typically use the stable isotope ratios of hard parts because they reflect the isotope values of the foraging location where they were originally formed; otherwise, the isotope values of different tissues with different turnover rates are measured. There are several different assignment model approaches (Wunder, 2012). Nominal assignment frameworks, such as classification trees, assign individuals to one of several possible locations that have been predetermined by the researcher as likely areas of migratory origin. The isotopic values of predetermined locations can be defined by reference organisms or abiotic compartments; however, this model requires significant prior knowledge of a species' likely migration patterns. Continuous assignment frameworks use isoscapes to assign individuals to a general area (i.e., probability distribution) and do not require a predetermined list of potential locations.

Assignment models, however, are rarely applied in open marine environments, and we are aware of just three studies where marine animals have been assigned to likely geographic origins using continuous or isoscape-based frameworks (Torniainen et al., 2017; Trueman et al., 2017; Vander Zanden et al., 2015a). This is in large part due to do the difficulties associated with developing marine isoscapes and the limited number of reference samples that have been collected across the global ocean (McMahon et al., 2013). Nonetheless, the limited work that is available indicates that, in the future, marine animal assignments will be a powerful and accurate technique in migratory studies. Trueman et al. (2017), for example, used a jellyfish-based isoscape of the North Sea to assign scallops of known origin and herring with well-known population level distributions. Results were extremely promising; 75% of sampled scallops were accurately assigned with a mean linear error on the order of 10^2 km. When the assignment model was applied to herring, it produced ecologically realistic assignments that were validated by available fisheries survey data.

Unfortunately, any potential shark isotope assignment models will be uniquely problematic. Assignment models rely on the assumptions that the original isoscape or reference organism is an appropriate metric with which to estimate animal migration and that ratios are relatively static

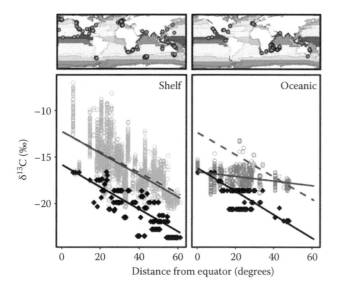

Figure 1.7 Linear regressions between latitude and carbon isotopic composition measured in shark muscle ($\delta^{13}C_S$, solid gray), modeled in phytoplankton from capture locations ($\delta^{13}C_{PLK}$, solid black), and predicted for shark muscle assuming local residency and a pelagic phytoplankton-supported food web ($\delta^{13}C_E$, dashed red). Individual $\delta^{13}C_S$ data points are displayed using open circles; predicted $\delta^{13}C_{PLK}$ values from shark capture locations are black diamonds. Maps indicate individual shark sample distributions overlaid with the $\delta^{13}C_{PLK}$ isoscape. (Bird, C. et al., *Nat. Ecol. Evol.*, 2, 299–305, 2018.)

through time (Graham et al., 2010; Wunder, 2010). Sharks, however, consume a wide range of dietary resources and travel large distances over short periods of time. Individual specialization within populations is common, and oceanic isotope ratios are highly dynamic. For these reasons, these significant assumptions will likely be difficult to justify. Moreover, shark tissue turnover rates are exceptionally long, particularly in adults, so the isotopic values of some tissues may simply not provide the level of precision required for accurate assignments (Hobson et al., 2010). Continuous frameworks also assume that the transfer or integration of isotope ratios from the isoscape to the consumer is relatively predicable and consistent (Wunder, 2010). Similar to diet reconstruction models, isoscape values must be rescaled prior to assignment to account for trophic fractionation between the consumer and its diet (Hobson et al., 2010). This can be done by applying a fixed offset or DTDF, but selecting the correct offset for shark species, as has been discussed in other sections of this chapter, can be highly problematic. Isoscape calibration using known values from each unique study is the best option (Wunder, 2010), but this may not always be possible.

To date, most migratory biomarker research has relied on SIA, but recent work has shown that fatty acids can also provide insight into shark movement and migration; for example, high levels of 20:4ω6 highlighted diel vertical movements in whale sharks (*Rhincodon typus*) (Marcus et al., 2016). There is also growing potential to use FA metabolism to reflect energy expenditure and, in turn, large-scale migrations (Osako et al., 2006), but this has yet to be validated in elasmobranchs. One of the most promising applications in fatty acid movement analysis is the *FATscape*. Similar to isoscapes, FATscapes are geographic contour maps of distinct FA bioregions that can be used to trace the spatial and temporal foraging strategies of individuals as they move across distinct food webs (Pethybridge et al., 2015). Although this approach was only recently validated with albacore tuna (*Thunnus alalunga*) (Pethybridge et al., 2015), this work showcases the novel use of predictive mixing models with fatty acids. These FA applications will certainly prove useful in future elasmobranch movement studies as they have the potential to provide a greater level of precision than stable isotopes alone.

To improve the quality of any shark movement or migration study, models should be constrained by prior knowledge of shark movement patterns. Catch, tagging, or genetic data can help narrow the geographic range of assignment or mixing models and help to identify the most likely migratory pathways. It is also important to recognize that biomarker values can vary substantially at small and local scales. Local scale variation can sometimes dwarf isotopic variation at regional or latitudinal scales. This could confound movement analysis, and localized dietary specialization may be misinterpreted as long-distance migration; additional tracking data may be necessary to support or supplement biomarker results and allow for more robust conclusions. Finally, no spatial foraging model, no matter how well validated or informed, will provide pinpoint accuracy. If exact coordinates are required, some form of electronic tracking is the only option, but electronic tracking cannot tell us if a shark is actually consuming resources in the area it inhabits. Biomarkers provide this important information.

1.5.5 Compound-Specific Stable Isotope Analysis

Thus far, we have focused our discussion on the ecological applications of bulk isotope analysis and how fractionation, particularly trophic fractionation, can differentiate dietary sources and delineate food web structure. However, analytical techniques are now available that determine the isotopic ratios of individual compounds, a process referred to as compound-specific stable isotope analysis (CSIA) (Krummen et al., 2004; Sessions, 2006). CSIA can reduce much of the uncertainty associated with bulk SIA applications because it yields a wider selection of isotopic values and can provide more detailed ecological information (McMahon et al., 2010, 2016). Moreover, the biochemical processes that affect individual compounds are often far fewer than those that affect the bulk tissue as a whole. Therefore, CSIA allows researchers to unscramble the confounding effects of trophic enrichment and source variation over space and time. The compound-specific ratios of amino acids (AAs) are of particular interest in shark foraging ecology because they provide useful information on the TP and foraging patterns of high-level predators.

For the purposes of TP calculation, AAs are divided into two categories: trophic AAs and source AAs. The $\delta^{15}N$ ratios of trophic AAs fractionate significantly with each consumer level, but source AAs undergo minimal fractionation and retain the $\delta^{15}N$ ratio of the original baseline dietary source (Chikaraishi et al., 2009; McClelland and Montoya, 2002). This is because each type of AA has unique governing metabolic processes. Unlike trophic AAs, source AA nitrogens are not interchangeable with the metabolic pool, so source $\delta^{15}N$ ratios will be similar to the original dietary AAs (O'Connell, 2017). As a result, trophic AAs are an excellent proxy with which to measure trophic position, and source AAs can be used to identify the $\delta^{15}N$ sources at the base of the food web. Bulk isotope analysis normally provides a $\delta^{15}N$ ratio that is an average of these two AAs, but by measuring each AA separately it is possible to calculate the TP of an individual without knowing the external isotope values of baseline organisms (Lorrain et al., 2009; McMahon and McCarthy, 2016). CSIA trophic position can be calculated as follows:

$$TP_{CSIA} = 1 + (\delta^{15}N_{TAA} - \delta^{15}N_{SAA}) - \beta/\Delta^{15}N_{(TAA-SAA)}$$

where TP_{CSIA} is the trophic position of the sample; $\delta^{15}N_{TAA}$ and $\delta^{15}N_{SAA}$ are the $\delta^{15}N$ of the trophic and source AA, respectively; β is the isotopic difference between the trophic and source AA in the primary producer; and $\Delta^{15}N_{(TAA-SAA)}$ is the trophic discrimination factor for the trophic AA (Chikaraishi et al., 2009). In general, trophic AAs are represented by glutamic acid (Glu) and source AAs are represented by phenylalanine (Phe).

Trophic and source AAs can also be used to differentiate the effects of trophic enrichment and enriched baseline $\delta^{15}N$ sources on shark tissue ratios. Dale et al. (2011) used a combination of stomach content, bulk, and CSIA to study the foraging habits of the brown stingray (*Dasyatis lata*) in a coastal nursery. The nursery receives large volumes of $\delta^{15}N$-enriched wastewater from local human settlements, but as *D. lata* grow they move farther away from shore and away from these facilities. Bulk SIA indicated that $\delta^{15}N$ declined with size. Taken at face value, this would indicate that *D. lata* TP strangely decreased with ontogeny. Site-specific CSIA told a very different story. Juvenile *D. lata* within the bay had enriched source AAs as a result of enriched $\delta^{15}N$ wastewater entering the area. Larger *D. lata* in offshore areas assimilated $\delta^{15}N$ from relatively depleted oceanic food webs and therefore had much lower source AA ratios. CSIA TP indicated that, in fact, *D. lata* TP increased with size, and average bulk isotope trends were the product of anthropogenic exposure to wastewater effluent.

Compound-specific stable isotope analysis of $\delta^{13}C$ is also a highly valuable tool in diet reconstruction. In this case, AAs are divided into two categories: essential and nonessential. Because essential AAs can only be synthesized by primary producers and bacteria (Borman et al., 1946; Reeds, 2000), consumers must obtain essential AAs from their diet. As a result, essential AAs are directly integrated into animal tissues with minimal fractionation (Howland et al., 2003; McMahon et al., 2010). In addition, individual essential AAs have unique and highly variable $\delta^{13}C$ ratios across different species; therefore, not unlike FAA, species-specific $\delta^{13}C$ ratios of essential AAs can be used to profile or fingerprint specific primary producers, food webs, and foraging locations (Larsen et al., 2009; McMahon et al., 2016). Where bulk analysis can differentiate broad resource categories (Abrantes and Barnett, 2011; Borrell et al., 2011; MacNeill et al., 2005), CSIA $\delta^{13}C$ has the potential to distinguish primary producers among a wide range of species. Thus, essential AAs can be used to more precisely identify the producers and food webs that support consumer populations (Arthur et al., 2014; Larsen et al., 2013). It is important to note that FA CSIA of $\delta^{13}C$ has also been used to identify consumer carbon sources and resource partitioning (Oxtoby et al., 2016), but FA CSIA methods are in their infancy compared to AA CSIA and have not yet been successfully applied to elasmobranchs.

Similar to bulk $\delta^{15}N$ calculations, CSIA TP calculations can be confounded by poorly resolved DTDFs. The standard β value for $\delta^{15}N_{Glu}-\delta^{15}N_{phe}$ TP calculations is 7.6‰, but recent research has shown that β is affected both by diet quality and mode of nitrogen excretion (Germain et al., 2013; McMahon et al., 2015), so β is highly variable among taxa and between trophic levels (McMahon and McCarthy, 2016). Moreover, few labs are currently equipped to perform these analyses, which is also more labor intensive, time consuming, and expensive than bulk analysis. For these reasons, although CSIA can be incredibly useful in shark ecological research, a pragmatic preliminary strategy may be to first determine bulk isotope ratios and preserve some samples for compound-specific analysis if needed.

1.6 SUMMARY AND CONCLUSIONS

Sharks are notoriously elusive and difficult to capture in high numbers, and without the help of expensive technological aids it can be extremely difficult to directly observe and monitor their behavior in the wild. Biomarker analysis provides an affordable way to study the movement and diet of sharks over variable timeframes and across vast oceanic expanses. A range of analytical platforms have made biomarker analysis intuitive and accessible (Jackson et al., 2011), and at the same time our understanding of stable isotope and fatty acid environmental dynamics is becoming increasingly refined. Although it was not explicitly explored in this chapter, biomarker analysis is also a useful complementary technique that can be used to support tracking or visual data (Matich and Heithaus, 2014; Papastamatiou et al., 2010). Thus, biomarker analysis is primed to become an important staple of shark ecological research. Stable isotope analysis will likely continue to be the preferred technique in shark biological studies because specific applications are currently more well developed, and in general there is a better understanding of stable isotope assimilation throughout the food web. However, the detailed results that fatty acid analysis can provide will no doubt lead to significant breakthroughs in the field of foraging ecology, and it will likely be a focal component of future shark ecological research.

Unfortunately, shark-specific biomarker dynamics are still poorly understood relative to other taxa. Elasmobranchs are also not ideal subjects for most biomarker applications. Long isotopic turnover times and wide foraging ranges currently exclude sharks from more specific interpretations. The good news is these limitations provide us with boundless opportunities for creative solutions and discoveries as we seek to fill the knowledge gaps in shark biochemical processes. Techniques such as CSIA will also help us to overcome traditional roadblocks and may pave the way toward a more accurate and meaningful understanding of shark foraging ecology.

ACKNOWLEDGMENTS

We thank Clive Trueman, Heidi Pethybridge, Bronwyn Gillanders, and Jonathan Smart for their helpful advice and contributions.

REFERENCES

Abrantes KG, Barnett A (2011) Intrapopulation variations in diet and habitat use in a marine apex predator, the broadnose sevengill shark *Notorynchus cepedianus*. *Mar Ecol Prog Ser* 442:133–148.

Ackman RG, Tocher C, McLachlan J (1968) Marine phytoplankten fatty acids. *J Fish Board Can* 25:1603–1620.

Arrington DA, Winemiller KO (2002) Preservation effects on stable isotope analysis of fish muscle. *Trans Am Fish Soc* 131:337–342.

Arthur KE, Kelez S, Larsen T, Choy CA, Popp BN (2014) Tracing the biosynthetic source of essential amino acids in marine turtles using $\delta^{13}C$ fingerprints. *Ecology* 95:1285–1293.

Baker R, Buckland A, Sheaves M (2014) Fish gut content analysis: robust measures of diet composition. *Fish Fisher* 15:170–177.

Ballantyne JS (1997) Jaws: the inside story. The metabolism of elasmobranch fishes. *Comp Biochem Physiol B Biochem Mol Biol* 118:703–742.

Barría C, Navarro J, Coll M (2017) Trophic habits of an abundant shark in the northwestern Mediterranean Sea using an isotopic non-lethal approach. *Estuar Coast Mar Sci*, in press.

Bearhop S, Adams CE, Waldron S, Fuller RA, Macleod H (2004) Determining trophic niche width: a novel approach using stable isotope analysis. *J Anim Ecol* 73:1007–1012.

Beckmann CL, Mitchell JG, Seuront L, Stone DA, Huveneers C (2013a) Experimental evaluation of fatty acid profiles as a technique to determine dietary composition in benthic elasmobranchs. *Physiol Biochem Zoo* 86:266–278.

Beckmann CL, Mitchell JG, Stone DA, Huveneers C (2013b) A controlled feeding experiment investigating the effects of a dietary switch on muscle and liver fatty acid profiles in Port Jackson sharks *Heterodontus portusjacksoni*. *J Exp Mar Biol Ecol* 448:10–18.

Beckmann CL, Mitchell JG, Stone DA, Huveneers C (2014) Inter-tissue differences in fatty acid incorporation as a result of dietary oil manipulation in Port Jackson sharks (*Heterodontus portusjacksoni*). *Lipids* 49:577–590.

Belicka LL, Matich P, Jaffé R, Heithaus MR (2012) Fatty acids and stable isotopes as indicators of early-life feeding and potential maternal resource dependency in the bull shark *Carcharhinus leucas*. *Mar Ecol Prog Ser* 455:245–256.

Bethea DM, Buckel JA, Carlson JK (2004) Foraging ecology of the early life stages of four sympatric shark species. *Mar Ecol Prog Ser* 268:245–264.

Bird CS, Veríssimo A, Magozzi S, Abrantes KG, Aguilar A, Al-Reasi H, Barnett A, et al. (2018) A global perspective on the trophic geography of sharks. *Nat Ecol Evol* 2:299–305.

Bligh EG, Dyer WJ (1959) A rapid method for total lipid extraction and purification. *Can J Biochem Physiol* 37:911–917.

Boecklen WJ, Yarnes CT, Cook BA, James AC (2011) On the use of stable isotopes in trophic ecology. *Annu Rev Ecol Evol Syst* 42:411–440.

Bolnick DI, Svanbäck R, Fordyce JA, Yang LH, Davis JM, Hulsey CD, Forister ML (2003) The ecology of individuals: incidence and implications of individual specialization. *Am Nat* 161:1–28.

Bond AL, Diamond AW (2011) Recent Bayesian stable-isotope mixing models are highly sensitive to variation in discrimination factors. *Ecol Appl* 21:1017–1023.

Borman A, et al. (1946) The role of arginine in growth, with some observations on the effects of argininic acid. *J Biol Chem* 166:585–594.

Borrell A, Cardona L, Kumarran RP, Aguilar A (2011) Trophic ecology of elasmobranchs caught off Gujarat, India, as inferred from stable isotopes. *ICES J Mar Sci* 68:547–554.

Bouillon S, Connolly R, Gillikin D (2011) Use of stable isotopes to understand food webs and ecosystem functioning in estuaries. In: Wolanski E, McLusky D (eds) *Treatise on Estuarine and Coastal Science*, vol 7. Academic Press, Waltham, MA, pp 143–173.

Boutton T (1991) Stable carbon isotope ratios of natural materials: II. Atmospheric, terrestrial, marine, and freshwater environments. In: Coleman DC, Fry B (eds) *Carbon Isotope Techniques*. Academic Press, San Diego, CA, pp 173–185.

Brennan SR (2015) Strontium isotopes delineate fine-scale natal origins and migration histories of Pacific salmon. *Am Assoc Advance Sci* 1:e1400124–e1400124.

Brett MT (2014) Resource polygon geometry predicts Bayesian stable isotope mixing model bias. *Mar Ecol Prog Ser* 514:1–12.

Burgess KB, Couturier LIE, Marshall AD, Richardson AJ, Weeks SJ, Bennett MB (2016) *Manta birostris*, predator of the deep? Insight into the diet of the giant manta ray through stable isotope analysis. *R Soc Open Sci* 3(11):160717.

Burgess KB, Guerrero M, Richardson AJ, Bennett MB, Marshall AD (2017) Use of epidermal mucus in elasmobranch stable isotope studies: a pilot study using the giant manta ray (*Manta birostris*). *Mar Freshwater Res* 69(2):336–342.

Cailliet GM, Smith WD, Mollet HF, Goldman KJ (2006) Age and growth studies of chondrichthyan fishes: the need for consistency in terminology, verification, validation, and growth function fitting. In: Carlson J, Goldman K (eds) *Special Issue: Age and Growth of Chondrichthyan Fishes: New Methods, Techniques and Analysis*. Springer, Netherlands, pp 211–228.

Capo RC, Stewart BW, Chadwick OA (1998) Strontium isotopes as tracers of ecosystem processes: theory and methods. *Geoderma* 82:197–225.

Carlisle AB, Kim SL, Semmens BX, Madigan DJ, Jorgensen SJ, Perle CR, Anderson SD, et al. (2012) Using stable isotope analysis to understand the migration and trophic ecology of northeastern Pacific white sharks (*Carcharodon carcharias*). *PLoS ONE* 7:e30492.

Carlisle AB, Goldman KJ, Litvin SY, Madigan DJ, Bigman JS, Swithenbank AM, Kline Jr TC, Block BA (2015) Stable isotope analysis of vertebrae reveals ontogenetic changes in habitat in an endothermic pelagic shark. *Proc R Soc B Biol Sci* 282(1799):20141446.

Carlisle AB, Litvin SY, Madigan DJ, Lyons K, Bigman JS, Ibarra M, Bizzarro JJ (2016) Interactive effects of urea and lipid content confound stable isotope analysis in elasmobranch fishes. *Can J Fish Aquat Sci* 74:419–428.

Caut S, Angulo E, Courchamp F (2009) Variation in discrimination factors (δ^{15}N and δ^{13}C): the effect of diet isotopic values and applications for diet reconstruction. *J Appl Ecol* 46:443–453.

Chikaraishi Y, et al. (2009) Determination of aquatic food-web structure based on compound-specific nitrogen isotopic composition of amino acids. *Limnol Oceanogr Methods* 7:740–750.

Christiansen H, Fisk A, Hussey N (2015) Incorporating stable isotopes into a multidisciplinary framework to improve data inference and their conservation and management application. *Afr J Mar Sci* 37:189–197.

Churchill DA, Heithaus MR, Dean Grubbs R (2015) Effects of lipid and urea extraction on δ^{15}N values of deep-sea sharks and hagfish: can mathematical correction factors be generated? *Deep-Sea Res Part II Top Stud Oceanogr* 115:103–108.

Clement J (1992) Re-examination of the fine structure of endoskeletal mineralization in Chondrichthyans: implications for growth, ageing and calcium homeostasis. *Mar Freshwater Res* 43:157–181.

Colwell RK, Futuyma DJ (1971) On the measurement of niche breadth and overlap. *Ecology* 52:567–576.

Connolly RM, Guest MA, Melville AJ, Oakes JM (2004) Sulfur stable isotopes separate producers in marine food-web analysis. *Oecologia* 138:161–167.

Cortés E (1999) Standardized diet compositions and trophic levels of sharks. *ICES J Mar Sci* 56:707–717.

Costanzo SD, O'Donohue MJ, Dennison WC, Loneragan NR, Thomas M (2001) A new approach for detecting and mapping sewage impacts. *Marine Pollut Bull* 42:149–156.

Couturier LI, Rohner CA, Richardson AJ, Marshall AD, Jaine FR, Bennett MB, Townsend KA, et al. (2013) Stable isotope and signature fatty acid analyses suggest reef manta rays feed on demersal zooplankton. *PLoS ONE* 8:e77152.

Criss RE (1999) *Principles of Stable Isotope Distribution*. Oxford University Press, New York.

Currey LM, Heupel MR, Simpfendorfer CA, Williams AJ (2014) Inferring movement patterns of a coral reef fish using oxygen and carbon isotopes in otolith carbonate. *J Exp Mar Biol Ecol* 456:18–25.

Dahm R, Schonthaler HB, Soehn AS, Van Marle J, Vrensen GF (2007) Development and adult morphology of the eye lens in the zebrafish. *Exp Eye Res* 85:74–89.

Dale JJ, Wallsgrove NJ, Popp BN, Holland KN (2011) Nursery habitat use and foraging ecology of the brown stingray *Dasyatis lata* determined from stomach contents, bulk and amino acid stable isotopes. *Mar Ecol Prog Ser* 433:221–236.

Dalsgaard J, John MS, Kattner G, Müller-Navarra D, Hagen W (2003) Fatty acid trophic markers in the pelagic marine environment. *Adv Mar Biol* 46:225–340.

Daly R, Froneman PW, Smale MJ (2013) Comparative feeding ecology of bull sharks (*Carcharhinus leucas*) in the coastal waters of the southwest Indian ocean inferred from stable isotope analysis. *PLoS ONE* 8:e78229.

Darnaude AM, Sturrock A, Trueman CN, Mouillot D; EIMF, Campana SE, Hunter E (2014) Listening in on the past: what can otolith δ^{18}O values really tell us about the environmental history of fishes? *PLoS ONE* 9:e108539.

Davidson B, Sidell J, Rhodes J, Cliff G (2011) A comparison of the heart and muscle total lipid and fatty acid profiles of nine large shark species from the east coast of South Africa. *Fish Physiol Biochem* 37:105–112.

Davidson BC, Nel W, Rais A, Namdarizandi V, Vizarra S, Cliff G (2014) Comparison of total lipids and fatty acids from liver, heart and abdominal muscle of scalloped (*Sphyrna lewini*) and smooth (*Sphyrna zygaena*) hammerhead sharks. *SpringerPlus* 3:521.

De Groot PA (2004) *Handbook of Stable Isotope Analytical Techniques*, vol 1. Elsevier, Amsterdam.

Dean MN, Ekstrom L, Monsonego-Ornan E, Ballantyne J, Witten PE, Riley C, Habraken W, Omelon S (2015) Mineral homeostasis and regulation of mineralization processes in the skeletons of sharks, rays and relatives (Elasmobranchii). *Semin Cell Dev Biol* 46:51–67.

DeNiro MJ, Epstein S (1977) Mechanism of carbon isotope fractionation associated with lipid synthesis. *Science* 197:261–263.

DeNiro MJ, Epstein S (1978) Influence of diet on the distribution of carbon isotopes in animals. *Geochim Cosmochim Acta* 42:495–506).

DeNiro MJ, Epstein S (1981) Influence of diet on the distribution of nitrogen isotopes in animals. *Geochim Cosmochim Acta* 45:341–351.

Dicken ML, Hussey NE, Christiansen HM, Smale MJ, Nkabi N, Cliff G, Wintner SP (2017) Diet and trophic ecology of the tiger shark (*Galeocerdo cuvier*) from South African waters. *PLoS ONE* 12:e0177897.

Dunstan GA, Sinclair AJ, O'Dea K, Naughton JM (1988) The lipid content and fatty acid composition of various marine species from southern Australian coastal waters. *Comp Biochem Physiol B Biochem Mol Biol* 91:165–169.

Edwards MS, Turner TF, Sharp ZD (2002) Short-and long-term effects of fixation and preservation on stable isotope values (δ^{13}C, δ^{15}N, δ^{34}S) of fluid-preserved museum specimens. *Copeia* 2002:1106–1112.

Espinoza M, Munroe SEM, Clarke TM, Fisk AT, Wehrtmann IS (2015) Feeding ecology of common demersal elasmobranch species in the Pacific coast of Costa Rica inferred from stable isotope and stomach content analyses. *J Exp Mar Biol Ecol* 470:12–25.

Estrada JA, Rice AN, Natanson LJ, Skomal GB (2006) Use of isotopic analysis of vertebrae in reconstructing ontogenetic feeding ecology in white sharks. *Ecology* 87:829–834.

Evans J, Sharkey T, Berry J, Farquhar G (1986) Carbon isotope discrimination measured concurrently with gas exchange to investigate CO_2 diffusion in leaves of higher plants. *Funct Plant Biol* 13:281–292.

Every SL, Pethybridge HR, Crook DA, Kyne PM, Fulton CJ (2016) Comparison of fin and muscle tissues for analysis of signature fatty acids in tropical euryhaline sharks. *J Exp Mar Biol Ecol* 479:46–53.

Every SL, Pethybridge HR, Fulton CJ, Kyne PM, Crook DA (2017) Niche metrics suggest euryhaline and coastal elasmobranchs provide trophic connections among marine and freshwater biomes in northern Australia. *Mar Ecol Prog Ser* 565:181–196.

Falk-Petersen S, Hagen W, Kattner G, Clarke A, Sargent J (2000) Lipids, trophic relationships, and biodiversity in Arctic and Antarctic krill. *Can J Fish Aquat Sci* 57:178–191.

Farquhar GD, Ehleringer JR, Hubick KT (1989) Carbon isotope discrimination and photosynthesis. *Annu Rev Plant Biol* 40:503–537.

Ferreira LC, Thums M, Heithaus MR, Barnett A, Abrantes KG, Holmes BJ, Zamora LM, et al. (2017) The trophic role of a large marine predator, the tiger shark *Galeocerdo cuvier. Sci Rep* 7:7641.

Fischer J, Schneider JW, Voigt S, Joachimski MM, Tichomirowa M, Tütken T, Götze, Berner U (2013) Oxygen and strontium isotopes from fossil shark teeth: environmental and ecological implications for Late Palaeozoic European basins. *Chem Geol* 342:44–62.

Fisk AT, Tittlemier SA, Pranschke JL, Norstrom RJ (2002) Using anthropogenic contaminants and stable isotopes to assess the feeding ecology of Greenland sharks. *Ecology* 83:2162–2172.

Fraile I, Arrizabalaga H, Rooker JR (2015) Origin of Atlantic bluefin tuna (*Thunnus thynnus*) in the Bay of Biscay. *ICES J Mar Sci* 72:625–634.

France RL (1995) Carbon-13 enrichment in benthic compared to planktonic algae—foodweb implications. *Mar Ecol Prog Ser* 124:307–312.

Fry B (2006) *Stable Isotope Ecology.* Springer Science+Business Media, New York.

Fry B, Scalan RS, Parker PL (1977) Stable carbon isotope evidence for two sources of organic matter in coastal sediments: seagrasses and plankton. *Geochim Cosmochim Acta* 41:1875–1877.

Fry B, Scalan RS, Winters JK, Parker PL (1982) Sulphur uptake by salt grasses, mangroves, and seagrasses in anaerobic sediments. *Geochim Cosmochim Acta* 46:1121–1124.

Gat JR (2000) Atmospheric water balance—the isotopic perspective. *Hydrol Process* 14:1357–1369.

Germain LR, Koch PL, Harvey J, McCarthy MD (2013) Nitrogen isotope fractionation in amino acids from harbor seals: implications for compound-specific trophic position calculations. *Mar Ecol Prog Ser* 482:265–277.

Goldstein L, Forster R (1971) Osmoregulation and urea metabolism in the little skate *Raja erinacea. Am J Physiol* 220:742–746.

Graham BS, Koch PL, Newsome SD, McMahon KW, Aurioles D (2010) Using isoscapes to trace the movements and foraging behavior of top predators in oceanic ecosystems. In: West J, Bowen G, Dawson T, Tu K (eds) *Isoscapes.* Springer, Dordrecht, pp 299–318.

Hadwen WL, Arthington AH (2007) Food webs of two intermittently open estuaries receiving ^{15}N-enriched sewage effluent. *Estuar Coast Mar Sci* 71:347–358.

Hammerschlag N, Gallagher AJ, Lazarre DM (2011) A review of shark satellite tagging studies. *J Exp Mar Biol Ecol* 398:1–8.

Heithaus MR, Vaudo JJ, Kreicker S, Layman CA, Krützen M, Burkholder DA, Gastrich K, et al. (2013) Apparent resource partitioning and trophic structure of large-bodied marine predators in a relatively pristine seagrass ecosystem. *Mar Ecol Prog Ser* 481:225–237.

Heupel MR, Semmens JM, Hobday AJ (2006) Automated acoustic tracking of aquatic animals: scales, design and deployment of listening station arrays. *Mar Freshwater Res* 57:1–13.

Hobson KA (1999) Tracing origins and migration of wildlife using stable isotopes: a review. *Oecologia* 120:314–326.

Hobson KA (2008) Applying isotopic methods to tracking animal movements. In: Keith AH, Leonard IW (eds) *Terrestrial Ecology,* vol 2. Elsevier, Amsterdam pp 45–78.

Hobson KA, Gloutney ML, Gibbs HL (1997) Preservation of blood and tissue samples for stable-carbon and stable-nitrogen isotope analysis. *Can J Zool* 75:1720–1723.

Hobson KA, Barnett-Johnson R, Cerling T (2010) Using isoscapes to track animal migration. In: West J, Bowen G, Dawson T, Tu K (eds) *Isoscapes.* Springer, Dordrecht, pp 273–298.

Hoefs J (1997) *Stable Isotope Geochemistry.* Springer, Berlin.

Howland MR, Corr LT, Young SMM, Jones V, Jim S, Van Der Merwe NJ, Mitchell AD, Evershed RP (2003) Expression of the dietary isotope signal in the compound-specific $\delta^{13}C$ values of pig bone lipids and amino acids. *Int J Osteoarchaeol* 13:54–65.

Humston R, Doss SS, Wass C, Hollenbeck C, Thorrold SR, Smith S, Bataille CP (2017) Isotope geochemistry reveals ontogeny of dispersal and exchange between main-river and tributary habitats in smallmouth bass *Micropterus dolomieu. J Fish Biol* 90:528–548.

Hussey NE, Brush J, McCarthy ID, Fisk AT (2010) $\delta^{15}N$ and $\delta^{13}C$ diet–tissue discrimination factors for large sharks under semi-controlled conditions. *Comp Biochem Physiol A Mol Integr Physiol* 155:445–453.

Hussey NE, Dudley SFJ, McCarthy ID, Cliff G, Fisk AT (2011) Stable isotope profiles of large marine predators: viable indicators of trophic position, diet, and movement in sharks? *Can J Fish Aquat Sci* 68:2029–2045.

Hussey NE, MacNeil MA, Olin JA, McMeans BC, Kinney MJ, Chapman DD, Fisk AT (2012) Stable isotopes and elasmobranchs: tissue types, methods, applications and assumptions. *J Fish Biol* 80:1449–1484.

Hussey NE, Kessel ST, Aarestrup K, Cooke SJ, Cowley PD, Fisk AT, Harcourt RG (2015) Aquatic animal telemetry: a panoramic window into the underwater world. *Science* 348:1255642.

Hutchinson G (1957) Concluding remarks. In: Whittaker R, Levin S (eds) *Niche: Theory and Application.* Dowden, Hutchinson & Ross, Stroudsburg, PA, pp 387–399.

Iverson SJ (2009) Tracing aquatic food webs using fatty acids: from qualitative indicators to quantitative determination. In: Arts M, Brett M, Kainz M (eds) *Lipids in Aquatic Ecosystems.* Springer, Dordrecht, pp 281–308.

Iverson SJ, Field C, Don Bowen W, Blanchard W (2004) Quantitative fatty acid signature analysis: a new method of estimating predator diets. *Ecol Monogr* 74:211–235.

Jackson AL, Inger R, Parnell AC, Bearhop S (2011) Comparing isotopic niche widths among and within communities: SIBER—Stable Isotope Bayesian Ellipses in R. *J Anim Ecol* 80:595–602.

Jaeger A, Lecomte VJ, Weimerskirch H, Richard P, Cherel Y (2010) Seabird satellite tracking validates the use of latitudinal isoscapes to depict predators' foraging areas in the Southern Ocean. *Rapid Commun Mass Spectrom* 24:3456–3460.

Kaehler S, Pakhomov E (2001) Effects of storage and preservation on the $\delta^{13}C$ and $\delta^{15}N$ signatures of selected marine organisms. *Mar Ecol Prog Ser* 219:299–304.

Kelly JR, Scheibling RE (2011) Fatty acids as dietary tracers in benthic food webs. *Mar Ecol Prog Ser* 446:1–22.

Kennedy BP (2000) Using natural strontium isotopic signatures as fish markers: methodology and application. *Can J Fish Aquat Sci* 57:2280–2292.

Kim SL, Koch PL (2012) Methods to collect, preserve, and prepare elasmobranch tissues for stable isotope analysis. *Environ Biol Fish* 95:53–63.

Kim SL, Casper DR, Galván-Magaña F, Ochoa-Díaz R, Hernández-Aguilar SB, Koch PL (2012) Carbon and nitrogen discrimination factors for elasmobranch soft tissues based on a long-term controlled feeding study. *Environ Biol Fish* 95:37–52.

Kinney M, Hussey N, Fisk A, Tobin A, Simpfendorfer C (2011) Communal or competitive? Stable isotope analysis provides evidence of resource partitioning within a communal shark nursery. *Mar Ecol Prog Ser* 439:263–276.

Krummen M, Hilkert AW, Juchelka D, Duhr A, Schlüter HJ, Pesch R (2004) A new concept for isotope ratio monitoring liquid chromatography/mass spectrometry. *Rapid Commun Mass Spectrom* 18:2260–2266.

Larsen T, Taylor DL, Leigh MB, O'Brien DM (2009) Stable isotope fingerprinting: a novel method for identifying plant, fungal, or bacterial origins of amino acids. *Ecology* 90:3526–3535.

Larsen T, Ventura M, Andersen N, O'Brien DM, Piatkowski U, McCarthy MD (2013) Tracing carbon sources through aquatic and terrestrial food webs using amino acid stable isotope fingerprinting. *PLoS ONE* 8:e73441.

Layman CA, Arrington DA, Montaña CG, Post DM (2007) Can stable isotope ratios provide for community-wide measures of trophic structure? *Ecology* 88:42–48.

Layman CA, Araujo MS, Boucek R, Hammerschlag-Peyer CM, Harrison E, Jud ZR, Matich P, et al. (2012) Applying stable isotopes to examine food-web structure: an overview of analytical tools. *Biol Rev* 87:545–562.

Li Y, Zhang Y, Hussey NE, Dai X (2016) Urea and lipid extraction treatment effects on $\delta^{15}N$ and $\delta^{13}C$ values in pelagic sharks. *Rapid Commun Mass Spectrom* 30:1–8.

Logan J, Lutcavage M (2010) Stable isotope dynamics in elasmobranch fishes. *Hydrobiologia* 644:231–244.

Lorrain A, Graham B, Ménard, Popp B, Bouillon S, Breugel PV, Cherel Y (2009) Nitrogen and carbon isotope values of individual amino acids: a tool to study foraging ecology of penguins in the Southern Ocean. *Mar Ecol Prog Ser* 391:293–306.

MacKenzie K, Longmore C, Preece C, Lucas C, Trueman C (2014) Testing the long-term stability of marine isoscapes in shelf seas using jellyfish tissues. *Biogeochemistry* 121:441–454.

MacNeill MA, Skomal GB, Fisk AT (2005) Stable isotopes from multiple tissues reveal diet switching in sharks. *Mar Ecol Prog Ser* 302:199–206.

Madhavan S, Treichel I, O'Leary MH (1991) Effects of relative humidity on carbon isotope fractionation in plants. *Botanica Acta* 104:292–294.

Magnone L, Bessonart M, Gadea J, Salhi M (2015) Trophic relationships in an estuarine environment: a quantitative fatty acid analysis signature approach. *Estuar Coast Mar Sci* 166:24–33.

Magozzi S, Yool A, Vander Zanden HB, Wunder MB, Trueman CN (2017) Using ocean models to predict spatial and temporal variation in marine carbon isotopes. *Ecosphere* 8:e01763.

Malpica-Cruz L, Herzka Sharon Z, Sosa-Nishizaki O, Lazo JP (2012) Tissue-specific isotope trophic discrimination factors and turnover rates in a marine elasmobranch: empirical and modeling results. *Can J Fish Aquat Sci* 69:551–564.

Malpica-Cruz L, Herzka SZ, Sosa-Nishizaki O, Escobedo-Olvera MA (2013) Tissue-specific stable isotope ratios of shortfin mako (*Isurus oxyrinchus*) and white (*Carcharodon carcharias*) sharks as indicators of size-based differences in foraging habitat and trophic level. *Fish Oceanogr* 22:429–445.

Marcus L, Virtue P, Pethybridge HR, Meekan MG, Thums M, Nichols PD (2016) Intraspecific variability in diet and implied foraging ranges of whale sharks at Ningaloo Reef, Western Australia, from signature fatty acid analysis. *Mar Ecol Prog Ser* 554:115–128.

Marcus L, Virtue P, Nichols PD, Meekan MG, Pethybridge H (2017) Effects of sample treatment on the analysis of stable isotopes of carbon and nitrogen in zooplankton, micronekton and a filter-feeding shark. *Mar Biol* 164:124.

Matich P, Heithaus MR (2014) Multi-tissue stable isotope analysis and acoustic telemetry reveal seasonal variability in the trophic interactions of juvenile bull sharks in a coastal estuary. *J Anim Ecol* 83:199–213.

Matich P, Heithaus MR, Layman CA (2011) Contrasting patterns of individual specialisation and trophic coupling in two marine apex predators. *J Anim Ecol* 80:294–305.

Matich P, Kiszka JJ, Heithaus MR, Mourier J, Planes S (2015) Short-term shifts of stable isotope ($\delta^{13}C$, $\delta^{15}N$) values in juvenile sharks within nursery areas suggest rapid shifts in energy pathways. *J Exp Mar Biol Ecol* 465:83–91.

McCauley DJ, Young HS, Dunbar RB, Estes JA, Semmens BX, Michel F (2012) Assessing the effects of large mobile predators on ecosystem connectivity. *Ecol Appl* 22:1711–1717.

McClelland JW, Montoya JP (2002) Trophic relationships and the nitrogen isotopic composition of amino acids in plankton. *Ecology* 83:2173–2180.

McKinney CR, McCrea JM, Epstein S, Allen H, Urey HC (1950) Improvements in mass spectrometers for the measurement of small differences in isotope abundance ratios. *Rev Sci Instrum* 21:724–730.

McMahon KW, McCarthy MD (2016) Embracing variability in amino acid $\delta^{15}N$ fractionation: mechanisms, implications, and applications for trophic ecology. *Ecosphere* 7(12):e01511.

McMahon KW, Fogel ML, Elsdon TS, Thorrold SR (2010) Carbon isotope fractionation of amino acids in fish muscle reflects biosynthesis and isotopic routing from dietary protein. *J Anim Ecol* 79:1132–1141.

McMahon KW, Hamady LL, Thorrold SR (2013) A review of ecogeochemistry approaches to estimating movements of marine animals. *Limnol Oceanogr* 58:697–714.

McMahon KW, Thorrold SR, Elsdon TS, McCarthy MD (2015) Trophic discrimination of nitrogen stable isotopes in amino acids varies with diet quality in a marine fish. *Limnol Oceanogr* 60:1076–1087.

McMahon KW, Thorrold SR, Houghton LA, Berumen ML (2016) Tracing carbon flow through coral reef food webs using a compound-specific stable isotope approach. *Oecologia* 180:809–821.

McMeans BC, Arts MT, Fisk AT (2012) Similarity between predator and prey fatty acid profiles is tissue dependent in Greenland sharks (*Somniosus microcephalus*): implications for diet reconstruction. *J Exp Mar Biol Ecol* 429:55–63.

McMeans BC, Arts MT, Lydersen C, Kovacs KM, Hop H, Falk-Petersen S, Fisk AT (2013) The role of Greenland sharks (Somniosus microcephalus) in an Arctic ecosystem: assessed via stable isotopes and fatty acids. *Mar Biol* 160(5):1223–1238.

McMillan M, Izzo C, Wade B, Gillanders B (2017) Elements and elasmobranchs: hypotheses, assumptions and limitations of elemental analysis. *J Fish Biol* 90:559–594.

Meyer LC, Pethybridge H, Nichols PD, Beckmann C, Bruce BD, Werry, JM, Huveneers C (2017) Assessing the functional limitations of lipids and fatty acids for diet determination: the importance of tissue type, quantity and quality. *Front Mar Sci* 4:369.

Moore JW, Semmens BX (2008) Incorporating uncertainty and prior information into stable isotope mixing models. *Ecol Lett* 11:470–480.

Mourier J, Maynard J, Parravicini V, Ballesta L, Clua E, Domeier ML, Planes S (2016) Extreme inverted trophic pyramid of reef sharks supported by spawning groupers. *Curr Biol* 26:2011–2016.

Munroe SEM, Simpfendorfer CA, Heupel MR (2014) Defining shark ecological specialisation: concepts, context, and examples. *Rev Fish Biol Fisher* 24:317–331.

Munroe SEM, Heupel MR, Fisk AT, Logan M, Simpfendorfer CA (2015) Regional movement patterns of a small-bodied shark revealed by stable-isotope analysis. *J Fish Biol* 86:1567–1586.

Murphy HM, Jenkins GP (2010) Observational methods used in marine spatial monitoring of fishes and associated habitats: a review. *Mar Freshwater Res* 61:236–252.

Navarro J, López L, Coll M, Barría C, Sáez-Liante R (2014) Short- and long-term importance of small sharks in the diet of the rare deep-sea shark *Dalatias licha*. *Mar Biol* 161:1697–1707.

Newsome SD, Clementz MT, Koch PL (2010) Using stable isotope biogeochemistry to study marine mammal ecology. *Mar Mam Sci* 26:509–572.

Newsome SD, Martinez del Rio C, Bearhop S, Phillips DL (2007) A niche for isotopic ecology. *Front Ecol Environ* 5:429–436.

Nicol JAC, Somiya H (1989) *The Eyes of Fishes*. Oxford University Press, New York.

O'Connell TC (2017) 'Trophic' and 'source' amino acids in trophic estimation: a likely metabolic explanation. *Oecologia* 184:317–326.

Oh BZL, Sequeira AMM, Meekan MG, Ruppert JLW, Meeuwig JJ (2017) Predicting occurrence of juvenile shark habitat to improve conservation planning. *Conserv Biol* 31:635–645.

Olin JA, Hussey NE, Fritts M, Heupel MR, Simpfendorfer CA, Poulakis GR, Fisk AT (2011) Maternal meddling in neonatal sharks: implications for interpreting stable isotopes in young animals. *Rapid Commun Mass Spectrom* 25:1008–1016.

Olin JA, Hussey NE, Grgicak-Mannion A, Fritts MW, Wintner SP, Fisk AT (2013) Variable $\delta^{15}N$ diet–tissue discrimination factors among sharks: implications for trophic position, diet and food web models. *PLoS ONE* 8:e77567.

Olson K (1999) Rectal gland and volume homeostasis. In: Hamlett W (ed) *Sharks, Skates, and Rays*. Johns Hopkins University Press, Baltimore, MD, pp 329–352.

Osako K, Saito H, Hossain MA, Kuwahara K, Okamoto A (2006) Docosahexaenoic acid levels in the lipids of spotted mackerel *Scomber australasicus*. *Lipids* 41:713–720.

Ostrom NE, Macko SA, Deibel D, Thompson RJ (1997) Seasonal variation in the stable carbon and nitrogen isotope biogeochemistry of a coastal cold ocean environment. *Geochim Cosmochim Acta* 61:2929–2942.

Oxtoby L, Budge S, Iken K, Brien DO, Wooller M (2016) Feeding ecologies of key bivalve and polychaete species in the Bering Sea as elucidated by fatty acid and compound-specific stable isotope analyses. *Mar Ecol Prog Ser* 557:161–175.

Papastamatiou YP, Friedlander AM, Caselle JE, Lowe CG (2010) Long-term movement patterns and trophic ecology of black-tip reef sharks (*Carcharhinus melanopterus*) at Palmyra Atoll. *J Exp Mar Biol Ecol* 386:94–102.

Parnell AC, Phillips DL, Bearhop S, Semmens BX, Ward EJ, Moore JW, Jackson AL, et al. (2013) Bayesian stable isotope mixing models. *Environmetrics* 24:387–399.

Parrish CC (2013) Lipids in marine ecosystems. *Oceanography* 2013:1–16.

Peterson BJ, Fry B (1987) Stable isotopes in ecosystem studies. *Annu Rev Ecol Evol Syst* 18:293–320.

Pethybridge H, Daley R, Virtue P, Nichols P (2010) Lipid composition and partitioning of deepwater chondrichthyans: inferences of feeding ecology and distribution. *Mar Biol* 157:1367–1384.

Pethybridge H, Daley RK, Nichols PD (2011) Diet of demersal sharks and chimaeras inferred by fatty acid profiles and stomach content analysis. *J Exp Mar Biol Ecol* 409:290–299.

Pethybridge HR, Parrish CC, Bruce BD, Young JW, Nichols PD (2014) Lipid, fatty acid and energy density profiles of white sharks: insights into the feeding ecology and ecophysiology of a complex top predator. *PLoS ONE* 9:e97877.

Pethybridge HR, Choy A, Polovina JJ, Fulton EA (2018) Improving marine ecosystem models with biochemical tracers. *Ann Rev Mar Sci* 10:199–228.

Pethybridge HR, Parrish CC, Morrongiello J, Young JW, Farley JH, Gunasekera RM, Nichols PD (2015) Spatial patterns and temperature predictions of tuna fatty acids: tracing essential nutrients and changes in primary producers. *PLoS ONE* 10:e0131598.

Phillips DL (2012) Converting isotope values to diet composition: the use of mixing models. *J Mamm* 93:342–352.

Phillips DL, Gregg JW (2003) Source partitioning using stable isotopes: coping with too many sources. *Oecologia* 136:261–269.

Phillips DL, Newsome SD, Gregg JW (2005) Combining sources in stable isotope mixing models: alternative methods. *Oecologia* 144:520–527.

Phillips DL, Inger R, Bearhop S, Jackson AL, Moore JW, Parnell AC, Semmens BX, Ward EJ (2014) Best practices for use of stable isotope mixing models in food-web studies. *Can J Zool* 92:823–835.

Pillans R, Good J, Anderson WG, Hazon N, Franklin C (2005) Freshwater to seawater acclimation of juvenile bull sharks (*Carcharhinus leucas*): plasma osmolytes and Na$^+$/K$^+$-ATPase activity in gill, rectal gland, kidney and intestine. *J Comp Physiol B* 175:37–44.

Poisot T, Bever JD, Nemri A, Thrall PH, Hochberg ME (2011) A conceptual framework for the evolution of ecological specialisation. *Ecol Lett* 14:841–851.

Post DM (2002) Using stable istopes to estimate trophic position: models, methods, and assumptions. *Ecology* 83:703–718.

Post DM, Layman CA, Arrington DA, Takimoto G, Quattrochi J, Montana CG (2007) Getting to the fat of the matter: models, methods and assumptions for dealing with lipids in stable isotope analyses. *Oecologia* 152:179–189.

Quaeck K (2017) Stable Isotope Analysis of Fish Eye Lenses: Reconstruction of Ontogenetic Trends in Spatial and Trophic Ecology of Elasmobranchs and Deep-Water Teleosts, doctoral thesis. University of Southampton, Southampton, UK, 209 pp.

Ramos R, González-Solís J (2012) Trace me if you can: the use of intrinsic biogeochemical markers in marine top predators. *Front Ecol Environ* 10:258–266.

Reeds PJ (2000) Dispensable and indispensable amino acids for humans. *J Nutr* 130:1835S–1840S.

Rohner CA, Couturier LIE, Richardson AJ, Pierce SJ, Prebble CEM, Gibbons MJ, Nichols PD (2013) Diet of whale sharks *Rhincodon typus* inferred from stomach content and signature fatty acid analyses. *Mar Ecol Prog Ser* 493:219–235.

Rosas-Luis R, Navarro J, Loor-Andrade P, Forero MG (2017) Feeding ecology and trophic relationships of pelagic sharks and billfishes coexisting in the central eastern Pacific Ocean. *Mar Ecol Prog Ser* 573:191–201.

Rosenblatt AE, Heithaus MR (2012) Slow isotope turnover rates and low discrimination values in the American alligator: implications for interpretation of ectotherm stable isotope data. *Physiol Biochem Zool* 86:137–148.

Sarakinos HC, Johnson ML, Zanden MJV (2002) A synthesis of tissue-preservation effects on carbon and nitrogen stable isotope signatures. *Can J Zool* 80:381–387.

Sargent JR, Tocher DR, Bell JG (2002) The lipids. *Fish Nutr* 3:181–257.

Savage C (2005) Tracing the influence of sewage nitrogen in a coastal ecosystem using stable nitrogen isotopes. *Ambio* 34:145–150.

Schaufler L, Heintz R, Sigler M, Hulbert L (2005) Fatty acid composition of sleeper shark (*Somniosus pacificus*) liver and muscle reveals nutritional dependence on planktivores. *ICES CM* 5:1–19.

Schlacher TA, Liddell B, Gaston TF, Schlacher-Hoenlinger M (2005) Fish track wastewater pollution to estuaries. *Oecologia* 144:570–584.

Semeniuk CA, Speers-Roesch B, Rothley KD (2007) Using fatty-acid profile analysis as an ecologic indicator in the management of tourist impacts on marine wildlife: a case of stingray-feeding in the Caribbean. *Environ Manag* 40:665–677.

Sessions AL (2006) Isotope-ratio detection for gas chromatography. *J Sep Sci* 29:1946–1961.

Shipley ON (2017) Stable isotopes reveal food web dynamics of a data-poor deep-sea island slope community. *Food Webs* 10:22–25.

Shipley ON, Brooks EJ, Madigan DJ, Sweeting CJ, Grubbs RD (2017) Stable isotope analysis in deep-sea chondrichthyans: recent challenges, ecological insights, and future directions. *Rev Fish Biol Fisher* 27(3):481–497.

Shuttleworth T (1988) Salt and water balance—extrarenal mechanisms. In: Shuttleworth T (ed) *Physiology of Elasmobranch Fishes*. Springer-Verlag, Berlin, pp 171–199.

Simpfendorfer C, Goodreid A, McAuley R (2001) Size, sex and geographic variation in the diet of the tiger shark, *Galeocerdo Cuvier*, from Western Australian waters. *Environ Biol Fish* 61:37–46.

Smith BN (1972) Natural abundance of the stable isotopes of carbon in biological systems. *BioScience* 22:226–231.

Smith JA, Mazumder D, Suthers IM, Taylor MD (2013a) To fit or not to fit: evaluating stable isotope mixing models using simulated mixing polygons. *Methods Ecol Evol* 4:612–618.

Smith WD, Miller JA, Heppell SS (2013b) Elemental markers in elasmobranchs: effects of environmental history and growth on vertebral chemistry. *PLoS ONE* 8:e62423.

Somes CJ, Schmittner A, Galbraith ED, Lehmann MF, Altabet MA, Montoya JP, Letelier RM, et al. (2010) Simulating the global distribution of nitrogen isotopes in the ocean. *Global Biogeochem Cycles* 24:GB4019.

Speed C, Meekan M, Field I, McMahon C, Abrantes K, Bradshaw C (2012) Trophic ecology of reef sharks determined using stable isotopes and telemetry. *Coral Reefs* 31:357–367.

Stewart JD, Rohner CA, Araujo G, Avila J, Fernando D, Forsberg K, Ponzo A, et al. (2017) Trophic overlap in mobulid rays: insights from stable isotope analysis. *Mar Ecol Prog Ser* 580:131–151.

Sweeting CJ, Polunin NVC, Jennings S (2006) Effects of chemical lipid extraction and arithmetic lipid correction on stable isotope ratios of fish tissues. *Rapid Commun in Mass Spectrom* 20:595–601.

Syväranta J, Lensu A, Marjomäki TJ, Oksanen S, Jones RI (2013) An empirical evaluation of the utility of convex hull and standard ellipse areas for assessing population niche widths from stable isotope data. *PLoS ONE* 8:e56094.

Tieszen LL, Boutton TW, Tesdahl KG, Slade NA (1983) Fractionation and turnover of stable carbon isotopes in animal tissues: implications for $\delta^{13}C$ analysis of diet. *Oecologia* 57:32–37.

Tillett BJ, Meekan MG, Parry D, Munksgaard N, Field IC, Thorburn D, Bradshaw CJA (2011) Decoding fingerprints: elemental composition of vertebrae correlates to age-related habitat use in two morphologically similar sharks. *Mar Ecol Prog Ser* 434:133–142.

Tilley A, López-Angarita J, Turner JR (2013) Diet reconstruction and resource partitioning of a Caribbean marine mesopredator using stable isotope Bayesian modelling. *PLoS ONE* 8:e79560.

Torniainen J, Lensu A, Vuorinen PJ, Sonninen E, Keinänen M, Jones RI, Patterson WP, Kiljunen M (2017) Oxygen and carbon isoscapes for the Baltic Sea: testing their applicability in fish migration studies. *Ecol Evol* 7:2255–2267.

Trueman CN, MacKenzie KM, St. John Glew K (2017) Stable isotope-based location in a shelf sea setting: accuracy and precision are comparable to light-based location methods. *Meth Ecol Evol* 8:232–240.

Tuross N, Fogel ML, Hare P (1988) Variability in the preservation of the isotopic composition of collagen from fossil bone. *Geochim Cosmochim Acta* 52:929–935.

Vander Zanden HB, Tucker AD, Hart KM, Lamont MM, Fujisaki I, Addison DS, Mansfield KL, et al. (2015a) Determining origin in a migratory marine vertebrate: a novel method to integrate stable isotopes and satellite tracking. *Ecol Appl* 25:320–335.

Vander Zanden MJ, Clayton MK, Moody EK, Solomon CT, Weidel BC (2015b) Stable isotope turnover and half-life in animal tissues: a literature synthesis. *PLoS ONE* 10:e0116182.

Vaudo J, Heithaus M (2011) Dietary niche overlap in a nearshore elasmobranch mesopredator community. *Mar Ecol Prog Ser* 425:247–260.

Wada E, Hattori A (1976) Natural abundance of ^{15}N in particulate organic matter in the North Pacific Ocean. *Geochim Cosmochim Acta* 40:249–251.

Wai T-C, Leung KM, Sin SY, Cornish A, Dudgeon D, Williams GA (2011) Spatial, seasonal, and ontogenetic variations in the significance of detrital pathways and terrestrial carbon for a benthic shark, *Chiloscyllium plagiosum* (Hemiscylliidae), in a tropical estuary. *Limnol Oceanogr* 56:1035–1053.

Wallace AA, Hollander DJ, Peebles EB (2014) Stable isotopes in fish eye lenses as potential recorders of trophic and geographic history. *PLoS ONE* 9:e108935.

Waugh CA, Nichols PD, Schlabach M, Noad M, Nash SB (2014) Vertical distribution of lipids, fatty acids and organochlorine contaminants in the blubber of southern hemisphere humpback whales (*Megaptera novaeangliae*). *Mar Environ Res* 94:24–31.

West JB, Bowen GJ, Dawson TE, Tu KP (eds) (2010) *Isoscapes: Understanding Movement, Pattern, and Process on Earth Through Isotope Mapping.* Springer, Dordrecht.

West JB, Sobek A, Ehleringer JR (2008) A simplified GIS approach to modeling global leaf water isoscapes. *PLoS ONE* 3:e2447.

White TD, Carlisle AB, Kroodsma DA, Block BA, Casagrandi R, De Leo GA, Gatto M, et al. (2017) Assessing the effectiveness of a large marine protected area for reef shark conservation. *Biol Conserv* 207:64–71.

White WM (2013) *Geochemistry.* Wiley-Blackwell, Chichester, UK.

Wintner S, Dudley S, Kistnasamy N, Everett B (2002) Age and growth estimates for the Zambezi shark, *Carcharhinus leucas*, from the east coast of South Africa. *Mar Freshwater Res* 53:557–566.

Wolf JM, Johnson B, Silver D, Pate W, Christianson K (2016) Freezing and fractionation: effects of preservation on carbon and nitrogen stable isotope ratios of some limnetic organisms. *Rapid Commun Mass Spectrom* 30:562–568.

Wunder MB (2010) Using isoscapes to model probability surfaces for determining geographic origins. In: West JB, Bowen GJ, Dawson TE, Tu KP (eds) *Isoscapes: Understanding Movement, Pattern, and Process on Earth Through Isotope Mapping.* Springer, Dordrecht, pp 251–270.

Wunder MB (2012) Determining geographic patterns of migration and dispersal using stable isotopes in keratins. *J Mamm* 93:360–367.

CHAPTER 2

Size-Based Insights into the Ecosystem Role of Sharks and Rays

Nicholas K. Dulvy
Earth to Ocean Research Group, Simon Fraser University, Burnaby, British Columbia, Canada

Rowan Trebilco
Antarctic Climate and Ecosystems Cooperative Research Centre, University of Tasmania, Hobart, Tasmania, Australia

CONTENTS

2.1   INTRODUCTION

Arguably the most profound truth about fishes (including sharks and rays) is that they grow in length (and weight) throughout their life (Froese, 2006; Hoenig and Gruber, 1990). This is so obvious that we rarely stop to think what this means for the ecology of fishes; however, the implications of this truth for understanding how sharks and rays fit into communities and ecosystems are often ignored, potentially to the detriment of the field. The latest trend in ecology is framing biodiversity value in terms of ecological function (Cernansky, 2017). In this approach, species names are substituted with their traits (mined from databases), and the outliers in ecological functional space are identified through multivariate analysis (Violle et al., 2017). Further, we often describe ecosystems as food webs and develop graphs of species (nodes) connected by a web of feeding relationships or interactions (Paine, 1980). But, does it make sense to assign a single trait value, such as trophic level, to a species such as a tiger shark (*Galeocerdo cuvier*) that might grow two orders of magnitude in length and three orders of magnitude in weight—from 51 cm total length and about 2 to 3 kg at birth to a maximum length of over 550 cm and a weight of 800 kg?

Given the ontogenetic change in size over a lifetime, does it make sense to represent the feeding links or interaction strengths of the species as a whole? It is easy to forget that these ecological interaction patterns are underlain by distributions of abundance and biomass (Cohen et al., 2003) and by ontogenetic change in diet, interaction strengths, trophic levels, movement patterns, and habitat use (Grubbs, 2010; Werner and Gilliam, 1984; Wetherbee et al., 2004). Understanding how diet, trophic level, and other traits such as home range and other movement metrics change ontogenetically is a key issue in vertebrate ecology (Simpfendorfer et al., 2011). This chapter seeks to reveal how important size-based interactions might be for understanding the abundance, biomass, migration, and aggregation of shark and ray populations. This chapter is not a comprehensive review but more of a perspective; indeed, we rely heavily on much more detailed work on diet biomarkers, migration, and movement (see Chapter 1). These are all technical areas with which we are less experienced, but we are more experienced with theories and concepts from a range of marine ecosystems and draw heavily from comparative knowledge gained from other systems to highlight how a size-based perspective may well apply more broadly to understanding shark aggregation and migration.

First, we briefly summarize evidence for changes in diet and trophic level with size, revealing transitions from one prey type to the next and increases in size and trophic level of prey which, in turn, lead to (mostly) increases in trophic level of their predators. We highlight the observation that this appears to be related to allometric changes in gape dimensions and dentition (concurrent with changes in habitat use resulting from wider home ranges).

Second, we consider why understanding and measuring ontogenetic changes in trophic level matter. This is illustrated by a quest to understand the baseline ecosystem state of sharks (and rays) and especially to understand the underlying processes that give rise to ecological patterns. Specifically, we focus on the inverted trophic pyramid of high shark abundance at remote, relatively pristine coral reefs and summarize the hypotheses that might have given rise to this pattern. In short, it turns out that when you recognize that trophic pyramids are actually graphic representations of size spectra theory, it becomes apparent that truly inverted pyramids at the community scale seem unlikely.

Third, we return to ontogenetic, trophic-level relationships, because these can be used to understand a key parameter: the *predator–prey mass ratio* (PPMR). We show how the slope of size spectra and the shape of ecological pyramids vary with PPMR and *transfer efficiency* (TE). It turns out that there is a relatively narrow range of PPMRs and TEs in the real world, and indeed there is strong evidence that they are compensatory, which in turn means that the typical shape of a biomass pyramid is just that—a pyramid. A pyramid can be inverted when the prey (or resources) are larger than their predators (or consumers) or when the ecological pyramid represents an accumulation of energy from beyond the local area (i.e., is subsidized by energy from elsewhere) (Mourier et al., 2016; Simpfendorfer and Heupel, 2016; Trebilco et al., 2016).

2.2 ONTOGENETIC SHIFTS IN DIET AND TROPHIC LEVEL

The widespread availability of diet data from fisheries analyses and the use of stable isotopes as an indicator of food carbon source ($\delta^{13}C$) and trophic level ($\delta^{15}N$) (see Chapter 1)

have provided profound insights into the structuring of (teleost-dominated) fish communities (Jennings et al., 2008a,b; Reum et al., 2015). For example, it is increasingly accepted that trophic level is not an invariant species trait. Much of the pioneering work was undertaken in the North Sea. For this ecosystem, when the trophic level of fishes (indexed by $\delta^{15}N$ of white muscle tissue) is plotted against the maximum reported length or mass of each species, there is no relationship (Jennings et al., 2001). In contrast, when the trophic levels of individuals of those same species are plotted against their individual sizes we see strong allometry, such that, with few exceptions, there are positive relationships, with larger individuals feeding at higher trophic positions (Jennings et al., 2002a; Polunin and Pinnegar, 2002). These contrasting patterns at the individual vs. species level compellingly illustrate how trophic level depends not only on species identity but also on the size of the individual. The pattern for strong relationships between individual size and trophic position is by no means unique to the North Sea (e.g., Barnes et al., 2010; Trebilco et al., 2016) and is apparent even when trophic level is assigned using diet data, especially for taxonomically defined subsets (Cortés, 1999; Ebert and Bizzarro, 2007).

This body-size–trophic-level relationship is likely to be (at least partly) a result of the allometry in increasing gape dimensions as individuals grow (Mittelbach and Persson, 1998; Pinnegar et al., 2003; Romanuk et al., 2011; Wainwright and Richard, 1995). Hence, trophic level is not a species trait *per se*; rather, these among-species differences belie (and mask) intraspecific allometry in trophic level, and trophic level is the interaction of species size and identity (Reecht et al., 2013; Romanuk et al., 2011). Are similar patterns evident in sharks and rays?

We searched the literature for combinations of topic terms, including "ontogenetic," "diet," "trophic level," and "shark," and worked forward from key references (e.g., Grubbs, 2010; Heithaus, 2010; Wetherbee et al., 2004). With few exceptions, it appears that almost every analysis examined for sharks and rays reveals diet shifts generally consistent with an increase in trophic level with individual size. Hence, our working hypothesis is that, as a general rule for sharks and rays, mean prey mass is related to individual predator mass, and generally this is likely to be a positive relationship, particularly for piscivorous species (Jennings and van der Molen, 2015; Revill et al., 2009). We caution that there will be exceptions and that negative relationships do exist, particularly for low-trophic-level sharks that transition from one habitat to another or for wide-ranging sharks feeding across ecosystems with spatially varying isotopic signatures (Revill et al., 2009). However, where ontogenetic changes in trophic level have been systematically evaluated, positive relationships are eight times more prevalent than negative body-size–trophic-level relationships and tend to be strongest in species that are piscivorous at some stage of their life (Jennings and van der Molen, 2015).

2.2.1 Diet Data Examples

Within a location, and hence at a place with a wide range of prey types and sizes, different-sized sharks of the same species are likely to differ intraspecifically in mean trophic level due to differences in diet that arise from allometric changes in gape size and dentition (ontogenetic heterodonty). Broad transitions in diet composition have been documented; for example, juvenile tiger sharks (*Galeocerdo cuvier*) eat fishes, cephalopods, and sea snakes but adults eat more elasmobranchs, turtles, and dugongs (Simpfendorfer et al., 2001). These changes can be abrupt rather than gradual. The white shark (*Carcharodon carcharias*) transitions from consuming fish to consuming predominantly marine mammals at a total length of around 300 cm (Klimley, 1985; Tricas, 1984).

Dietary analyses are usually conducted on individuals captured by trawl or longline. If lethal sampling is chosen, the stomachs are dissected out whole, preserved in alcohol or formalin, and then returned to the laboratory for analysis (Hyslop, 1980). In the laboratory, the stomach contents are emptied into white trays and sorted into recognizable prey body parts, identified using catalogs and field guides, before enumeration and weighing. Stomach eversion or flushing is a suitable nonlethal method for obtaining stomach contents. The shark or ray is everted and placed ventral side up on a fine mesh cloth, both to protect the animal and to retain any food that is inadvertently spilled during the procedure (Bangley et al., 2013; Elston et al., 2016). The stomach can be everted with the careful use of forceps, the contents washed out, and the stomach reinserted. Alternatively, a small-bore tube (say, 1.5 to 3 cm in diameter, depending on the size of the chondrichthyan) can be inserted into the buccal cavity and esophagus until resistance is encountered at the gastro-esophageal sphincter. The hose is gently manipulated into the stomach for a few centimeters, and the stomach is then filled with seawater using the deck hose (Bangley et al., 2013) or a submersible electric (bilge) pump (Barnett et al., 2010; Elston et al., 2016). Throughout the process, the stomach region on the ventral is watched carefully for signs of expansion, whereupon the hose is removed. Gentle massage of the abdomen and elevation of the tail while supporting the animal in a head-down position causes the water to flush out food items. These can be captured in a mesh sample bag for preservation, labeling, and delivery to a laboratory for analysis (Bangley et al., 2013).

One of the best quantified examples comes from Port Jackson shark (*Heterodontus portusjacksoni*) individuals sampled in broadly the same habitat (Powter et al., 2010). The diet of the Port Jackson shark differs by size between juveniles and adults, and there is little difference in diet between the sexes (Figure 2.1A–C). Juveniles and subadults have mainly tricuspidate teeth and consume soft-bodied infaunal and epifaunal invertebrates (Echiura, Decopoda, and Gastropoda). Mouths are narrow in juveniles and subadults and broaden in adults (Figure 2.1D–F) because of faster growth of mouth length (slope = 0.026 ± 0.003 with total length) relative to mouth width (slope = 0.14 ± 0.009), and there is very little difference in jaw morphology between the sexes (Figure 2.1C). Most of the variance in jaw length and width is determined by size allometry rather than sex or maturity stage (Box 2.1; Table 2.1). The adults have larger, broader heads and broader gape (relative to body length) to accommodate posterior molariform teeth that enable them to eat a greater fraction of hard-shelled decapods (Figure 2.1D–F). In addition, they also consume larger, higher trophic level octopus, squids, and flatfishes. Based on the diet classification of Cortés (1999), the relative compositions of five diet classes were converted into trophic level estimates for each of the three maturity stages. Consequently, there is an increase of approximately half a trophic level in mean trophic level of different sized individuals (e.g., juvenile trophic level of 3.66 and adult trophic level of 4.05) (Powter et al., 2010).

2.2.2 Stable Isotope Examples

Stable isotopes of nitrogen ($\delta^{15}N$) and carbon ($\delta^{13}C$) are commonly used to infer trophic position and carbon source, respectively, in two dimensions (C and N) of the focal animal relative to samples from the base of the ecosystem (see Chapter 1). Stable isotope analysis generally requires soft tissue, and, unless blood, muscle, or fin clips are used, the method tends to focus on tissue samples from major organs, so individuals may have to be euthanized. Because different tissues have different turnover rates, the stable isotope signature of each tissue reflects diet integrated over differing windows of time (Pinnegar and Polunin, 1999). Nonlethally sampled blood plasma or lethally sampled liver tissue can be used for stable isotopic analysis of diet and trophic position, but these fast turnover tissues tend to reflect the most recent diet history—say, over recent days or weeks—whereas a tissue with a slow turnover time, such as white or red muscle tissue, integrates the dietary signal over weeks to months (Pinnegar and Polunin, 1999, 2000). White muscle tissue is most often used; a small portion is dissected out, labeled, and packaged (e.g., in an Eppendorf tube). In the laboratory, the sample is oven- or freeze-dried before being ground using a washed pestle and mortar. A small sample (1 mg) is then weighed into a tin capsule for determination of stable isotope signature by mass spectrometry (Post et al., 2007; Shiffman et al., 2012).

As explained in Chapter 1 of this volume, the position of a given consumer along the carbon axis provides an indication of the relative contribution of different carbon sources, whereas position on the nitrogen axis is an index of the relative height of the consumer above the base of the food chain. Converting relative trophic positions from $\delta^{15}N$ signatures to absolute trophic level requires information on

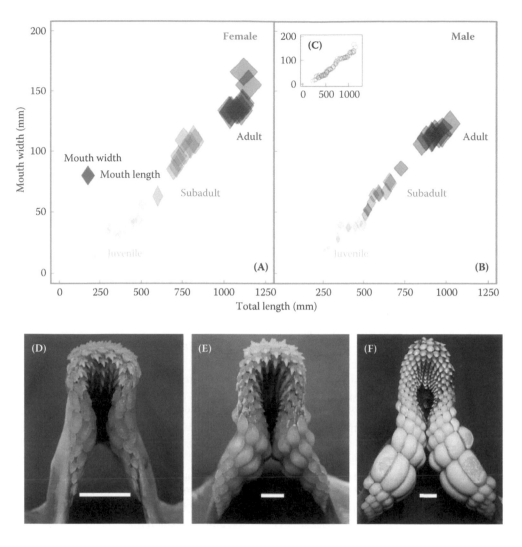

Figure 2.1 Ontogenetic change in gape shape (mouth length [mm] and mouth width [mm]) in the Port Jackson shark (*Heterodontus portusjacksoni*) for (A) females and (B) males, and (C, insert plot) with both sexes together to show the absence of a difference between the sexes. Jaws and teeth of (D) 390-mm female juvenile, (E) 655-mm male subadult, and (F) 905-mm LT male adult *H. portusjacksoni*. Note the reduction in cusps on the anterior teeth and increase in crushing posterior teeth with maturity. Scale bar = 10 mm. (Data and photographs from Powter, D.M. et al., *Mar. Freshw. Res.*, 61, 74–85, 2010.)

both the isotopic baseline of the ecosystem and the rate at which $\delta^{15}N$ becomes enriched with each trophic step (the trophic enrichment factor). To a first approximation, $\delta^{15}N$ is enriched by about 3 to 3.4 parts per thousand (per mill, ‰) per trophic level, and there is growing appreciation that this rate is not fixed (as was widely assumed for almost 30 years) but instead reduces in size for larger values of $\delta^{15}N$ (Hussey et al., 2014; Olive et al., 2003). Critically, it must be borne in mind that $\delta^{15}N$ is not a direct measure of trophic level and can only be used as an indicator of trophic level relative to the $\delta^{15}N$ of basal food sources (e.g., phytoplankton, kelp, primary herbivores such as filter-feeding scallops) (Jennings et al., 2008a; Jennings and Warr, 2003), and isotopic baseline may vary with temperature and salinity (Jennings and van der Molen, 2015) and degree of nitrogen regeneration (Revill et al., 2009). Moreover, if a consumer is supported

by multiple distinct production pathways, each having different isotopic baselines, estimation of absolute trophic level may only be possible using mixing models (see Chapter 1). Finally, spatial (and temporal seasonal) variation in prey composition and isotopic composition may influence a predator isotopic signature (Reum and Essington, 2013), the magnitude of which depends on the lifespan of the consumer and the rate at which changes in the isotopic composition of the diet affect the composition of the tissue being considered (tissue turnover time).

Here, we consider the lessons from two large-scale stable isotope analyses of pelagic fish assemblages including chondrichthyans—one in the western Indian Ocean off eastern Madagascar and the other in the western Pacific Ocean off eastern Australia. Off eastern Madagascar, Kiska et al. (2015) sampled carbon and nitrogen stable isotopes from 92

BOX 2.1   STATISTICAL APPROACHES TO DETECTING ONTOGENETIC SHIFTS IN TROPHIC LEVEL AT LARGE SPATIAL SCALES

The oceanic pelagic studies show that it can be difficult to detect patterns of interest unless spatial scale can be controlled for. There are two ways to overcome this—one is difficult and expensive, and the other is imperfect, free, and quick. The difficult approach is to develop isoscapes of a wide range of prey items and use that to control for the natural variation in the $\delta^{15}N$ or trophic level of potential prey items (Jennings and Warr, 2003a).

This is an exciting but expensive technology-driven solution and one that should be pursued for large ecosystem analyses. But, for graduate students with a more limited time horizon and budget, a simpler statistical approach can make use of the data you have. We recommend that you do not simply rely on correlation or univariate models, as these will tend to give false negatives with small sample sizes. Instead, we suggest using general linear models with geographic covariates (e.g., latitude and longitude) and other oceanographic attributes (e.g., temperature, salinity, ocean color metrics) in your model.

We recommend the R statistical programming language, as it is open access with a wide range of excellent resources (Zuur et al., 2010, 2013), so we offer both mathematical and R code below as examples. For the sake of simplicity, we have only described basic linear models including only "fixed effects" in these examples; however, in many cases hierarchical ("mixed effects") models may be suited to the analysis of these data, which are commonly hierarchical and nested (e.g., Trebilco et al., 2016). We do not delve into these methods here as they fall within the realm of statistics texts, but we encourage readers to consider these approaches.

Note that most stable isotope analyses are univariate:

$$\delta^{15}N = \beta_0 + \beta_1 Mass$$

where β_0 is the intercept term and β_1 is the slope. In R model notation,

$$\delta^{15}N \sim Mass$$

In R code, this would be

```
lm(δ15N ~ Mass, data = yourdata)
```

This approach ignores that part of the variance attributed to mass might actually be accounted for by other variables that have yet to be considered. Following the example of Revill et al. (2009), we suggest using available spatial covariates:

$$\delta^{15}N = \beta_0 + \beta_1 Mass + \beta_2 Longitude + \beta_2 Longitude$$

In R code, this would be

```
lm(δ15N ~ Mass+Latitude+Longitude,
      data = yourdata)
```

This will provide an estimate of the strength of the mass or ontogenetic effect while controlling for location (latitude and longitude). Sometimes ecosystems do not vary along a north–south or east–west axis, so there might be an interaction between latitude and longitude:

$$\delta^{15}N = \beta_0 + \beta_1 Mass + \beta_2 Longitude \times \beta_2 Longitude$$

In R code, this would be

```
lm(δ15N ~ Mass+Latitude*Longitude,
      data = yourdata)
```

In more complex, but not implausible situations, the size distribution of the species of interest may vary across the ocean basin in a manner that might be related to food chain length and primary production, resulting in a three-way interaction between mass and location:

$$\delta^{15}N = \beta_0 + \beta_1 Mass \times \beta_2 Longitude \times \beta_2 Longitude$$

In R code, this would be

```
lm(δ15N ~ Mass*Latitude*Longitude,
      data = yourdata)
```

All hypotheses should be considered and the most plausible evaluated using the Akaike information criterion (AIC); most like the small sample size version (AICc) (e.g., as implemented by the AICc function in R package AICcmodavg::AICc).

How much data are required to detect ontogenetic changes in a trophic level? As a rule of thumb, 10 data points are needed to estimate a parameter, so a total of 20 data points should be adequate to estimate an intercept and a slope parameter for a single explanatory variable. Hence, 30 would be needed for 2 parameters, 40 for 3, etc.

Table 2.1   Variance Is Explained by Size, Sex, and Maturity Stage

	Degrees of Freedom	Sum of Squares	Proportion of Variance Explained	F-Value	p-Value
Total length	1	2485.59	0.896	629.0514	<2.2e-16***
Sex	1	10.62	0.004	2.6878	0.106824
Maturity stage	2	60.65	0.022	7.6741	0.001149**
Residual	55	217.32	0.078	—	—
Total	—	2774.18	—	—	—

Note: This example is drawn from the Port Jackson shark (*Heterodontus portusjacksoni*) jaw morphology data presented in Powter et al. (2010). The procedure is to fit a linear model accounting for total length, sex, and maturity stage—mod1 <-lm(MouthLength ~ TotalLength + Sex + Maturity)—and then examine the ANOVA table to decompose the variance explained by each covariate (anova[mod1]). The variance explained by each covariate is the regression sum of squares (2485.9 + 10.62 + 60.65), and the remaining unexplained variance is the residual sum of squares (217.32). The proportion of variance explained by each covariate (column 4) is simply the sum of squares divided by the total sum of squares; these elements must be calculated by hand. Here, we can see that the overwhelming bulk of the variance (89.6%) is explained by total length; although maturity stage is statistically significant, it is not biologically significant, as it explains around 2% of the variance in these data. Two-way and three-way interactions between and among covariates could be considered within the limitations imposed by data availability. As a rule of thumb, about 10 data points are required to estimate each parameter. Here, we are estimating five parameters: intercept, total length, sex, and two more to account for the three-level factor maturity (adult, juvenile, subadult). Note that R deals with factor levels alphabetically; thus, the parameter estimate of "MaturityJuvenile" and MaturitySubadult" are departures from the "Intercept" parameter, which is, by default, calculated for the "MaturityAdult" stage. More complex general methods of decomposing variance, such as isotopic composition, by season, sex, and location can be found in Reum and Essington (2013). **$p \leq$ 0.01; ***$p \leq$ 0.001.

individuals of 7 species of oceanic pelagic sharks over an area spanning 15° latitude and 25° longitude. They found significant positive relationships between nitrogen stable isotope concentration ($\delta^{15}N$) and fork length for the blue shark (*Prionace glauca*), the species for which they had the greatest number of samples ($n = 31$) and the greatest size range of samples (109 cm, ranging from 160 to 269 cm FL). Obtaining sufficient samples across a broad body size range is not the only problem in recovering the expected pattern; the Kiska team also sampled 29 shortfin mako (*Isurus oxyrinchus*) spanning 182 cm (122 to 304 cm FL) but did not find a significant relationship. There was great heterogeneity in these relationships; for example, for blue sharks with 200 cm FL, observations of $\delta^{15}N$ differed by at least 4 parts per thousand (‰). Assuming that these sharks were feeding on the same food source, this could be interpreted as the trophic level of 200-cm FL blue shark varying by over one trophic level across the southwest Indian Ocean (see central panel in Figure 7 of Kiska et al., 2015). This is implausible, and Kiszka et al. explained that, "The large intraspecies variation in $\delta^{15}N$ ... [is] potentially related to variable regional prey bases and the migratory nature of these animals."

Further insight was revealed by sampling across ocean ecosystem boundaries in eastern Australia. Based on stable isotope analyses of 244 predator and 115 prey samples from 24 species, including 6 species of pelagic shark and ray, Revill et al. (2009) revealed strong positive interspecific fork length and $\delta^{15}N$ relationships. The five largest species were striped marlin, southern bluefin tuna, shortfin mako shark, and blue shark. The shortfin mako had the highest $\delta^{15}N$, with mean values ranging from 9‰ to 16‰—spanning almost two trophic levels if the critical assumption of a common isotopic base signature across these ecosystems were valid. Like the Kiszka et al. (2015) study, the within-species patterns were more heterogeneous than the cross-species patterns; the study was plagued by small sample sizes and a wide latitudinal and longitudinal range in sample locations and ecosystem types (see Box 2.1).

The explanation is revealed once we understand that the eastern Australian pelagic assemblage was sampled across two contrasting ecosystems: the oligotrophic Coral Sea in the north and the nutrient-rich Tasman Sea in the south (Revill et al., 2009). The carbon and nitrogen isotope signatures of the predator assemblage varied such that the nitrogen signature was depleted (lower) in animals captured in the oligotrophic Coral Sea (<28°S; see Figure 7 of Revill et al., 2009). This was not because those animals fed at a lower trophic level, but because the basal $\delta^{15}N$ signature was 4‰ lower (6‰) in the oligotrophic Coral Sea north of the Tasman front compared to 10‰ in the upwelling nutrient-rich waters farther south. The nitrogen isotopic signature of the phytoplankton depends heavily on the form of biologically available nitrogen. The oligotrophic waters are rich in nitrite (NO_3^-) due to nutrient regeneration, and this results in depleted $\delta^{15}N$ values at higher trophic levels compared to nutrient-rich waters (Revill et al., 2009).

In conclusion, available evidence appears to support the null expectation that mean trophic level is positively related to body size within (and across) shark and ray species. When an allometric relationship is not supported by the data it is usually because of one of four reasons: (1) there are too few data; (2) data may not span a sufficiently larger range of body sizes; (3) wide-ranging foraging across different ecosystems and hence $\delta^{15}N$ isoscapes of primary producers may mask any signals of ontogenetic diet and trophic level shifts (Jennings and Warr, 2003; MacKenzie et al., 2014); and (4) species may not be sufficiently piscivorous (Jennings and van der Molen, 2015). Now we set aside ontogenetic diet shifts and transition to summarize a major controversy in our understanding of the structure of ecosystems and specifically the baseline shape of ecological pyramids. However, as will become clear, these fundamental underlying patterns in the relationship between trophic level and body size within and across species have important bearing on expectations for the shape of ecological pyramids.

2.3 ECOSYSTEM BASELINES AND THE SHAPE OF TROPHIC PYRAMIDS

Ecological pyramids have been used to summarize key patterns in abundance or biomass with size or trophic level for nearly a century (Figure 2.2). In a quest to understand the ecosystem effects of fishing, underwater visual censuses (UVCs) of coral reef fish populations revealed high shark abundance and biomass at remote Pacific reefs (DeMartini et al., 2008; Friedlander and DeMartini, 2002; Sandin et al., 2008). The key finding was that top predators comprised the largest fraction of total fish biomass—an estimated 85% at the uninhabited Kingman Atoll but only 19% at the inhabited Kiritimati Island (Sandin et al., 2008). These values equate to 100,000 to 500,000 individuals per km². Specifically, sharks comprised 74% of the top predator biomass (329 g/m²) at Kingman and 57% at Palmyra (97 g/m²). These figures have encountered some controversy in the literature, and one proposed explanation is that the census method was at fault, specifically that non-instantaneous counts may have led to overestimation of large, mobile predatory fish (Ward-Paige et al., 2010).

Indeed, subsequent work using mark–recapture and other survey methodologies that tend to overcount sharks suggests that shark abundance was considerably overestimated in these first studies due to non-instantaneous census estimates (Bradley et al., 2017; Ward-Paige et al., 2010). Mark–recapture estimates of the gray reef shark over a wider range of habitats across Palmyra revealed densities of around 21 individuals per km², compared to UVC estimates of 200 to 1000 individuals per km² implied by the original Sandin et al. (2008) survey of fringing reef habitat (Bradley et al., 2017). From this mark–recapture analysis, the authors go on to estimate that the abundance of top predators is 56% lower than previously reported by Sandin et al. (2008). Nevertheless, the revised estimates suggest that top predators comprise just under half of the censuses biomass, which is still a staggeringly large fraction of standing biomass.

The question remains as to why there are such high abundances of sharks, and relatively inverted pyramids, at relatively uninhabited Pacific atolls and whether inverted pyramids are a plausible expectation for unexploited ecosystems. The explanation offered for the existence of an inverted pyramid was that the turnover time of predators was much lower than that of their prey (Sandin et al., 2008). This is an observation rather than an explanation; as we show next, turnover time is correlated with everything else and is a predictable outcome of fundamental metabolic scaling rather than acting as the underlying driver (Trebilco et al., 2013).

If one sampled an instantaneous snapshot of the ocean, the biomass of predators would be much greater than the biomass of phytoplankton. However, if sampling is integrated across the annual production cycle, then the biomass of phytoplankton and zooplankton (in temperate shelf seas) or the biomass of coral tissue, epiphytic algae, and zooplankton (in coral reefs) will be orders of magnitudes greater than the biomass of the predators. The difference between the inverted instantaneous snapshot and the time-integrated, bottom-heavy pyramid is that production (production-to-biomass ratio, or P/M) or turnover scales as a result of the scaling of abundance and biomass with body size (Banse and Mosher, 1980). Indeed, within a trophic level, the amount of energy available (E) is constant across size (mass, M) classes because of the physical law of the conservation of energy. But, because large things are rare and have long generation spans and hence slow turnover times, there is a strong size structuring in abundance and biomass across mass classes, such that abundance (N) scales with mass class (M) as a power law, where a is a constant and the exponent tends to scale in quarter powers (Brown and Gillooly, 2003; Savage et al., 2004):

$$N = aM^{-3/4 \text{ or } -0.75}$$
(big things are rare)

and biomass (B) scales with mass as

$$B = aM^{1/4 \text{ or } 0.25}$$
(there is a high-standing biomass of large organisms)

Consequently, the production-to-biomass ratio (P/M), or turnover or generation time, scales with mass as

$$P/M = aM^{-1/4 \text{ or } -0.25}$$

Hence, the turnover production-to-biomass ratio is a correlated or emergent phenomenon, rather than the cause of inverted pyramids, *per se.*

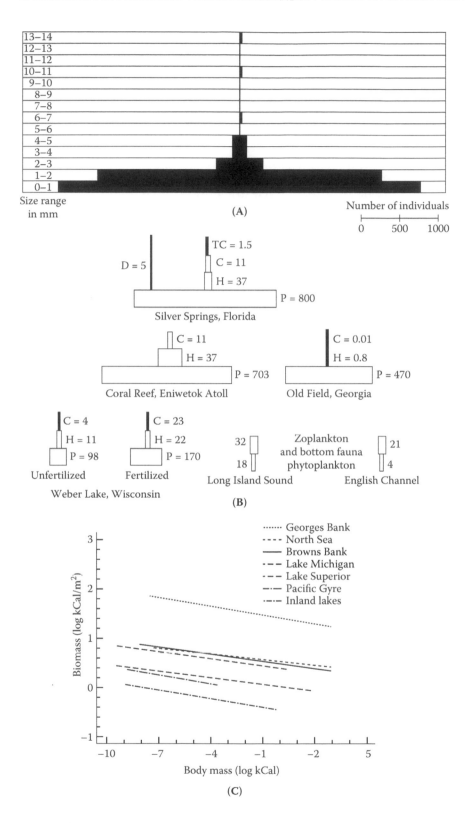

Figure 2.2 Classic examples of ecological pyramids and size spectra: (A) an Eltonian pyramid of numbers of forest-floor fauna of a Panama rainforest (redrawn from Williams, 1941); (B) Biomass pyramids for several ecosystems arranged by trophic level (g/m²; P = producers, H = herbivores, C = carnivores, TC = top carnivores, D = decomposers) (redrawn from Odum and Odum, 1955); (C) biomass spectra for pelagic ecosystems based on summary mean points for phytoplankton, zooplankton, benthos, and fish (redrawn from Boudreau and Dickie, 1992). In particular, note the consistent slopes across a wide range of ecosystems; the height or intercept of each spectrum is related to primary production, such that the highest slopes have the highest primary productivity.

These patterns are true only within a trophic level (Brown and Gillooly, 2003; Jennings and Mackinson, 2003; Trebilco et al., 2013). In order to describe ecosystem energy flow and the shape of ecological pyramids, we need the theory to span more than one trophic level. Specifically, we need to account for the transfer efficiency (*TE*) of energy between trophic levels, which requires a parameter to capture the relative differences in sizes of organisms at different trophic levels: the predator–prey mass ratio (PPMR) (Jennings et al., 2002b).

Now, the scaling of energy (*E*), abundance (*N*), biomass (*B*), and production (*P/M*) can be depreciated by adding a term to convert body size to trophic level (*PPMR*) and then depreciating the amount of energy flowing from one trophic level to the next (*TE*), yielding:

$$E = aM^{\log(TE)/\log(PPMR)}$$

$$N = aM^{-0.75} \times M^{\log(TE)/\log(PPMR)}$$

$$B = aM^{0.25} \times M^{\log(TE)/\log(PPMR)}$$

$$P/M = aM^{-0.25} \times M^{\log(TE)/\log(PPMR)}$$

But, how do these equations relate to ecological pyramids? Charles Elton originally conceived of size-based pyramids with abundance varying with organism size (Elton, 1927). It was only later that Raymond Lindeman simplified this by ignoring size and focusing only on categorical trophic levels (Lindeman, 1942; Trebilco et al., 2013). Nearly every ecology textbook portrays ecological pyramids as numbers at trophic levels; consequently, we have largely forgotten that Elton's view was that size (and the allometry of diet composition) was the ultimate causality for varying abundance and in turn biomass and production, rather than categorical trophic level (Trebilco et al., 2013). The only way we can convert from Lindeman's stepped trophic pyramids back to the more useful Eltonian ecological pyramids is if we know the relationship between trophic level and body size (i.e., PPMR).

A simple, powerful visual translation shows that size spectra are the underlying mathematical basis for ecological pyramids (Figure 2.3). The recipe for converting between the trophic pyramid of biomass or numbers and size spectra is as follows:

1. Left align pyramid of biomass or numbers.
2. Rotate axes counterclockwise by 90° so that size (mass) is on the *x*-axis and numbers (or biomass) are on the *y*-axis.
3. Flip the graph so the *y*-axis is on the left.
4. Log graph to convert the hollow curve of a negative exponential relationship to reveal a linear slope (this linear relationship is also known as *size spectrum*).

Size spectra have intrigued aquatic ecologists for over half a century, but they have largely been overlooked by terrestrial ecologists (Andersen et al., 2016; Blanchard et al., 2017; Trebilco et al., 2013). The profound implication is that

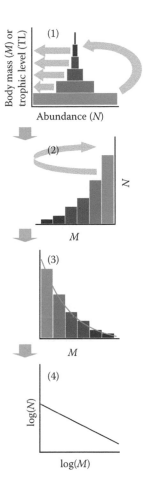

Figure 2.3 Ecological pyramids are a graphical representation of a size spectrum; hence, it follows that the mathematics underlying the size spectrum determines the shape of pyramid shape (Trebilco et al., 2013).

all those terrestrial ecosystems for which we have trophic pyramids have dynamics and interactions that are mostly size based rather than species based. The emerging evidence is beginning to reveal size spectra in a range of terrestrial ecosystems (Hocking et al., 2013; Reuman et al., 2008; Turnbull et al., 2013). Yet, the language and use of metaphors such as "food web" and "trophic pyramid" are species-centric and cause us to overlook critical size-based processes such as the key role of ontogenetic diet change, mean prey size, and the functional role of individuals within species.

2.3.1 Predator–Prey Mass Ratio and the Shape of Ecological Pyramids

The slopes of size spectra and shape of ecological pyramids fundamentally depend on understanding the relative size of prey that predators eat (PPMR) and the transfer efficiency (TE) between trophic levels. Earlier in this chapter, we established the nearly universal generality of ontogenetic shifts in feeding morphology, diet, and trophic level. We then established the importance of understanding and measuring

PPMR in order to understand the shape of ecological pyramids. Here, we get into the technical detail of measuring PPMR and evaluating plausible ranges of PPMR and hence the plausible shapes of ecological pyramids. Later, we ask whether inverted pyramids are likely in the wild and under what conditions they might arise.

Until recently our understanding of trophic level focused on the classical natural history of enumerating the contents of stomachs. There are numerous well-understood issues with stomach content analyses: Most shark stomachs are empty for analysis; if prey are present, stomach contents can be regurgitated on capture; prey are digested at different rates; and a large number of animals is required (Cortés, 1997; Polunin and Pinnegar, 2002). The widespread availability of stable isotope analyses based on $\delta^{15}N$ has expanded our understanding of trophic feeding relationships, but these methods are not without assumptions and issues. Because numerous reviews have summarized their value and utility while flagging their assumptions (e.g., Fry, 2007; Hussey et al., 2012; Jennings et al., 2008a), here we focus instead on estimating the PPMRs of species and whole communities. The sampling is challenging and time consuming but often probably no more challenging than a large diet study; however, one must consider the additional cost of stable isotope analysis.

Measuring species-specific PPMRs requires estimates of trophic level across a size range of individuals of a species (ideally within the same ecosystem). Trophic level can be estimated from the taxonomic composition of diet, notwithstanding some of the challenges of this approach (Cortés, 1997; Polunin and Pinnegar, 2002), or stable isotope analysis. The PPMR can be calculated from the slope coefficient (β) of a regression of trophic level (response) on log body size. Hence, $PPMR = n^{(\beta)}$ if trophic level is calculated based on diet. If trophic level is measured using $\delta^{15}N$, then $PPMR = n^{(\Delta/\beta)}$, where Δ is the fractionation of $\delta^{15}N$ (typically assumed to be 3.4‰), β is the slope estimate, and n is the base of the logarithm; that is, $n = 10$ if the x-axis is $\log_{10}$(body size), and $n = 2$ if a base of $\log_2$ is used (Jennings, 2005; Jennings et al., 2008a).

The principal challenges to estimating community PPMR are twofold. First, an estimate of the relative numbers of individuals of a given size is needed (i.e., one must generate a size spectrum). Second, an estimate of mean trophic level at size is needed. Size spectra are simply the numbers or biomass of animals in a given size class (Figure 2.4A,B). These are most often estimated using trawl surveys but can also be calculated from underwater visual census (UVC) counts taken by scuba divers in coral reefs and kelp forests (Trebilco et al., 2015), typically using $\log_2$ size class bins (2–4 g, 4–8 g, 8–16 g, 16–32 g, ...), as these give more data points along the x-axis than $\log_{10}$ bins (1–10 g, 10–100 g, 100–1000 g, ...) with which to estimate model parameters (Jennings, 2005). With trawl surveys, the individuals in each catch can simply be weighed and sorted into trays and buckets of the corresponding $\log_2$ size range (Figure 2.4A). With UVC, fish cannot be

weighed, but it is straightforward to train even the most inexperienced divers to estimate length to within 5-cm (ideally, 1-cm) length increments using a training set of plastic fish models of known length (Bell et al., 1985; Darwall and Dulvy, 1996; Dulvy et al., 2004; Edgar et al., 2004). Individuals can be trained to a consistent standard within a couple of dives. Individual fish lengths can be converted to mass estimates using species-, genus-, or even family-specific length–weight equations from FishBase and primary publications (Froese, 2006; Letourneur et al., 1998).

Both trawl and UVC approaches have a size window of good catchability (depending on mesh size, tow speeds, etc.) or visual detectability, which may depend on body size, schooling behavior, and human population density (Ackerman and Bellwood, 2000; Kulbicki, 1998; MacNeil et

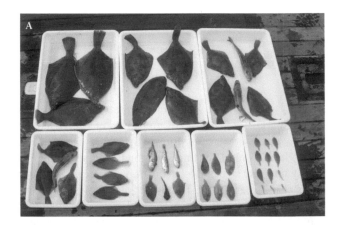

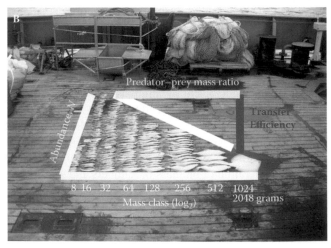

Figure 2.4 A typical size spectrum from the flatfish-dominated shallow southern North Sea: (A) The catch of a 30-minute tow of a 4-m beam trawl would be sorted into trays spanning each $\log_2$ size class from left-to-right top row (256–512 g, 128–256 g, 64–128 g) and from left-to-right bottom row (32–64 g, 16–32 g, 8–16 g, 4–8 g). (B) A nearly complete size spectrum from the same haul; note that the smallest size class was truncated. The slope is illustrated, along with the approximate predator–prey mass ratio and implied transfer efficiency (Jennings and Mackinson, 2003). (Photographs by N.K. Dulvy.)

al., 2008b). The bigger challenge is getting density estimates for large wide-ranging, high-trophic-level species to add onto the right-hand side of the UVC or trawl size spectrum. Density estimates of species poorly selected by sampling gear have been added onto spectra, typically in units of mass per unit area (e.g., g/m^2). For example, fishes are well sampled by otter or large beam trawls, but infauna, epifauna, and small pelagic fishes are better sampled with benthic trawls, cores, dredges, and acoustic sampling to provide appropriate biomass at size estimates (Jennings et al., 2002a,b). Clearly, UVC estimates of sharks have to be used with great caution, and indeed most early coral reef UVC work ignored sharks because of the well-known problems in surveying large-bodied species (Bradley et al., 2017; MacNeil et al., 2008a,b; Ward-Paige et al., 2010). Baited remote underwater videos (BRUVs) can provide relative abundance as the maximum numbers seen per unit time (*maxN*), and at present it is not clear how to convert these values into usable density estimates (g/m^2) (see Chapter 7). Mark–recapture estimates may be useable, and their utility should be considered further if resources allow (Bradley et al., 2017; McCauley et al., 2012a).

The mean trophic level of a community is calculated as mean $\delta^{15}N$ weighted by the biomass of the contribution of each species to each mass class. The simplest way to do this from a community sample (say, a trawl haul) is by sampling tissue from 20 to 25 individuals (or all individuals if fewer than 20 were caught) chosen at random from each size fraction. For each individual fish, white muscle tissue is dissected out as a fixed percentage of the mass of each fish sampled (a typical dissection scheme is provided in Table 1 of Jennings et al., 2001). For each size class, the tissue for all fish can then be homogenized together and frozen before being returned to the laboratory for freeze-drying and encapsulation for stable isotope analysis. If sufficient funds are available for stable isotope analysis, it may be more desirable to separately analyze samples for individual fish, an approach that has the added advantage of enabling the reconstruction of species-level allometries in addition to the community relationship. This process is considerably easier for trawl-caught samples for which both size spectra and $\delta^{15}N$–body size relationships can be calculated for the same sample. For coral reefs and kelp forests, it is possible to calculate species-level $\delta^{15}N$–body size relationships from samples taken by hook and line and spear fishing (Pinnegar and Polunin, 2000). These can be scaled up to provide a community-level PPMR estimate using a hierarchical model such as a mixed effects model that accounts for samples from different species and locations (see Trebilco et al., 2016).

When we have biomass B at body mass class M and corresponding estimates of trophic level from stable isotope data we can calculate the PPMR and TE. Community-wide PPMR can be calculated from the slope of the regression relationship between biomass weighted mean $\delta^{15}N$ and $\log^n$ mass class M, where n is the base of the logarithm of the body mass classes

(solid line in Figure 2.5C). Thus, PPMR = $n^{(\Delta/b)}$, where Δ is the fractionation of $\delta^{15}N$ (typically assumed to be 3.4‰) and b is the slope estimate (Jennings, 2005; Jennings et al., 2008a). Although fractionation has typically been assumed to be fixed in this calculation, to calculate the PPMR we would encourage testing the sensitivity of the estimate of this assumption and/or adopting a scaled-fractionation approach to estimating trophic position and then calculating PPMR as PPMR = $n^{(1/b^*)}$, where b^* is the slope of the relationship between the biomass weighted trophic position (calculated from $\delta^{15}N$, accounting for scaled fractionation) and body mass class (following Reum et al., 2015).

Transfer efficiency is typically 5 to 15% and can be measured as the proportion of prey production converted into predator production: TE = predator production/prey production. The challenge is to convert biomass B of each mass class M into individual biomass production (P/M), which can be achieved using the relationship $P/M = aM^{-0.25}$, where a is a constant. Typical values for a can be derived from published production-at-mass compilations (Banse and Mosher, 1980; Brey, 1999; Brose et al., 2006; Green et al., 2014, 2015; Lorenzen, 1996). However, because of the sensitivity to this scaling, it has been typical to assume plausible TEs (Jennings et al., 2008a). Nevertheless, the collection of more empirical P/M information will be valuable for chondrichthyans.

Now that we have some insight into the technical measurement, what are plausible ecological pyramid shapes given what we know about PPMR and TE? Clearly, estimates of PPMR and TE for whole fish communities are rare but are increasingly available for a broadening range of ecosystems (Reum et al., 2015). Assuming that fractionation of the nitrogen stable isotope ($\Delta^{15}N$) is constant, then community-wide PPMR estimates across four ecosystems range from 111 to 8981; that is, predators are between 100 and 10,000 heavier than their prey, as has broadly been understood for nearly half a century now (Cushing, 1975). This PPMR range narrows considerably to 49 to 312 when the fixed fractionation assumption is relaxed and instead fractionation varies depending on the $\delta^{15}N$ of the prey (P) according to the relationship $\Delta^{15}N = 5.92 - 0.27P$ (Hussey et al., 2014; Reum et al., 2015). Assuming plausible transfer efficiencies and the known range of PPMRs, we can map the corresponding pyramid sizes using the following equation:

$$B = aM^{0.25} \times M^{\log(TE)/\log(PPMR)}$$

In assemblages of the smallest organisms, such as zooplankton, PPMR is low (<1000) and TE is high (~15%). For larger size classes at higher trophic levels, PPMR tends to be greater (>5000) because TE is low (5%) (Barnes et al., 2010). Hence, the smaller sized lower trophic level assemblages tend to exhibit biomass "stacks," where there is a similar biomass in each trophic level (box s in Figure 2.6), whereas at greater body sizes and higher trophic levels (encompassing squids and fishes) the biomass pyramids will always be

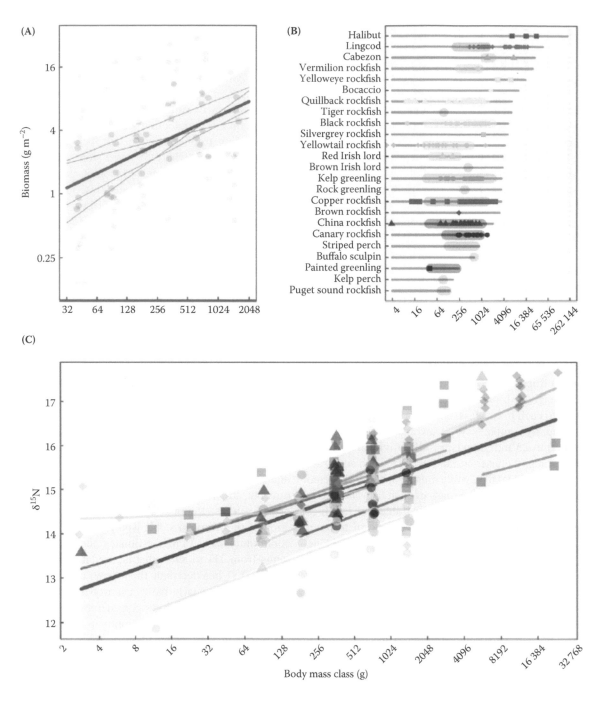

Figure 2.5 (A) The biomass spectrum of a kelp forest in Haida Gwaii, British Columbia, Canada, from underwater visual surveys; (B) size ranges observed for individual species in visual surveys (colored shaded regions) and sampled for isotope analysis (points) relative to λ_{max} (from FishBase, gray bars); and (C) the relationship between $\delta^{15}N$, a proxy for trophic position, and body size for the kelp forest fish community. Dark gray lines in panels (A) and (C) represent the mean mixed-effect model fit for each model (across locations for (A) and across both locations and species for (C)), and gray bands indicate 95% confidence intervals in (A) and Bayesian 95% credible intervals in (C) about the mean fits. Light gray lines in (A) and colored lines in (C) represent random effect fits for location and species, respectively. Colors and shapes of points are used to distinguish among species and are consistent between panels (B) and (C); species that were observed on transects but not sampled for isotope analysis are gray in panel (B). Gray shaded regions in (B) and (C) indicate body sizes outside the range included in (A). (Redrawn from Trebilco, R. et al., *Proc. R. Soc. Lond. B Biol. Sci.*, 283, 20160816, 2016.)

base-larger-than-apex pyramids rather than inverted pyramids (Trebilco et al., 2013, 2016). The range of plausible PPMR and TE is very narrowly bounded and may even be compensatory such that TE = PPMR$^{\beta+0.75}$, where β is the slope of the time-averaged size spectrum (Barnes et al., 2010). The plausible space for an inverted pyramid lies above the solid yellow line, but this space is inconsistent with plausible TE and PPMR estimates (Trebilco et al., 2016). In conclusion, what

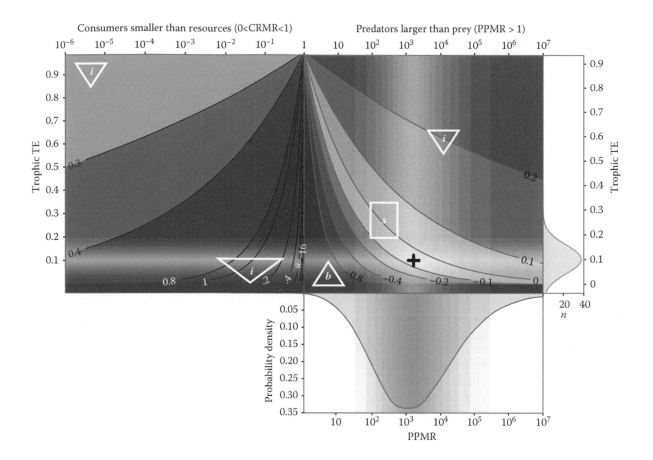

Figure 2.6 Expected biomass spectrum slopes (contours and red/blue shading in top panels) resulting from varying combinations of mean community predator–prey mass ratio (PPMR) and transfer efficiency (TE), shown with reference to the probability density distribution of estimated PPMRs for the reef fish community of Haida Gwaii, British Columbia, Canada (bottom panel). The top right panel shows scenarios with predators larger than prey (PPMR > 1); the top left panel shows scenarios with consumers smaller than resources (0 < CRMR < 1). Positive slopes (red area) correspond to inverted biomass pyramids (represented by triangles labeled i), negative slopes (blue area) correspond to bottom-heavy pyramids (triangle labeled b), and zero slopes imply stacks/columns (rectangle labeled s). Yellow shading lines indicate the range of slopes corresponding to the 95% confidence bounds around the empirically estimated biomass spectrum slope of 0.45 (solid yellow line). Right vertical axis shows TEs derived from marine food web models (*n* = 48, mean = 0.101, s.d. = 0.058). Shaded bands represent 5% quantile increments between 5% and 95% for TE and PPMR, and the black cross-hair indicates the highest probability for both distributions (PPMR = 1650, TE = 0.101). (Redrawn from Trebilco, R. et al., *Proc. R. Soc. Lond. B Biol. Sci.*, 283, 20160816, 2016.)

we know about how TE and PPMR constrain energy flow through communities suggests that if sharks are (1) indeed size-based predators (which the evidence supports) and (2) not operating with a much higher trophic transfer efficiency than is currently considered feasible, then inverted pyramids are energetically impossible in most local marine ecosystems.

But why do we still see high-standing biomass of sharks and other large predators, consistent with inverted pyramids, in remote, relatively unexploited ecosystems, such as the Kingman and Palmyra atolls and French Polynesia? We hypothesize that, in short, these fishes are aggregations of individuals that have skimmed energy off the tops of many nearby local pyramids. They have accumulated biomass either by foraging widely outside the time window encompassed by surveys (Heupel et al., 2014; Trebilco et al., 2013) or by accumulating energy from elsewhere that has come

to them (Mourier et al., 2016; Simpfendorfer and Heupel, 2016). Next, we seek to understand the two natural and well-documented shark phenomena of migration and aggregation through size-based interactions and energy flow.

2.4 MIGRATION AND AGGREGATION THROUGH THE LENS OF SIZE-BASED INTERACTIONS AND ENERGY FLOW

The past two decades of technological development have provided deep insights into the movement patterns of sharks and rays, revealing wide-ranging patterns spanning ecosystems and ocean basins (Heupel et al., 2015; Simpfendorfer and Heupel, 2004; Sims, 2010). Why do they move so far? Simply put, the emerging evidence reveals that there just is not enough energy at the top of any one ecological pyramid to sustain locally viable populations of top predators (Heupel et al., 2014; Hussey et al., 2014, 2015).

Marine food chains are longer than we thought, meaning there must be much less energy at the tips of local pyramids. Recall that transfer efficiencies are such that each successively higher trophic level has only approximately 10% of the biomass of the one below, constraining food chains to be only around four steps. A recent meta-analysis of feeding experiments where predatory fishes are fed prey of known isotopic composition confirms that the fractionation of $\delta^{15}N$ is not fixed at 3.4‰ and instead is negatively related to $\delta^{15}N$ of the prey (P) according to the relationship $\Delta^{15}N = 5.92 - 0.27P$ (Hussey et al., 2014). Hence, fractionation narrows by 0.27 units for each increase in prey $\delta^{15}N$ at higher trophic levels. Consequently, the energy available to sharks and other high-trophic-level species is likely to be 1 to 2 orders of magnitude lower because they are 1 to 2 trophic levels higher (assuming TE is around 10% at high trophic levels). This means that population sizes of apex shark species may be far smaller than we might have previously thought (Hussey et al., 2014). The consequence is that larger individuals and species, which feed at increasingly higher trophic levels, cannot be sustained within an exclusive territory of a size that can be defended and hence must share territory and foraging areas with other individuals (Jetz et al., 2004).

Modeling of mammal home ranges reveals that exclusivity of home range size decreases with animal size; thus, only the smallest animals (feeding at low trophic levels) can defend a territory that is sufficiently large enough to meet their daily energetic requirements (Jetz et al., 2004). As individuals grow larger they must range more widely and have less exclusive use of habitat. They must share their range with increasingly larger numbers of similar-sized animals. Energy availability, more generally, is the principal determinant of vertebrate home ranges (Tamburello et al., 2015). Hence, larger animals must range increasingly widely and skim off higher trophic level prey from an increasing number of ecological pyramids (Grubbs, 2010; Heupel et al., 2014; Tamburello et al., 2015). The key determinants of vertebrate home range size are body mass, locomotion mode (flying, walking, slithering, swimming), and dimensionality (living in two or three dimensions—swimming/flying) (Tamburello et al., 2015). Theory suggests that the predator–prey mass ratio should influence home range (Makarieva et al., 2005), and, although there are fewer data on PPMRs, the addition of this parameter significantly improves a model of home range size such that, all other variables being equal, a larger home range size is necessary when relatively larger prey are consumed (Tamburello et al., 2015). This pattern has been increasingly established for most other vertebrates, but the role of ontogenetic diet shift and trophic level change in driving increasingly large home ranges is only just now being revealed in fishes (Grubbs, 2010; Pittman and McAlpine, 2003; Tamburello et al., 2015).

2.4.1 Predator–Prey Mass Ratio, Ontogenetic Diet Shift, and Home Range Size

Because most non-fish vertebrates can hold down and fragment large prey, the predator–prey mass ratio is small in predatory mammals (median PPMR, 8:1; range, 0.3 to 200:1) and can even be inverted in social mammals (PPMR < 1), where sociality enables pack-hunting lions and wolves to bring down prey much larger than themselves (Jetz et al., 2004). Trophic level has a profound effect on movement. For example, a 1-kg predatory mammal requires a home range 14 times greater than herbivorous mammals of the same size (Tamburello et al., 2015). By comparison to mammals and birds, teleost (and presumably chondrichthyan) fishes have small PPMRs. This is because few fishes—except eels and scavenging or ectoparasitic sharks—can fragment their prey, so generally their prey must be captured and swallowed whole (Romanuk et al., 2011; Tamburello et al., 2015). Prey larger than the gape dimensions have to be discarded and such predation events are probably strongly selected against. But, because fish prey are often so much smaller (maybe 100 to 1000 times smaller) than the predators, they are numerous and highly productive. Hence, a 1-kg fish need not move far to meet its daily needs compared to a similar sized mammal. Indeed, a 1-kg predatory marine fish typically has a home range one order of magnitude smaller than that of a similar sized mammalian predator (Tamburello et al., 2015).

The prediction is that sharks (and rays) are likely to have shallow allometries of home range (and other movement metrics) because they are similarly gape-limited predators like teleost fishes. Extending PPMR movement theory to larger organisms will depend heavily on the development of new metrics of space use (Table 2.2). Because of the reduction in exclusivity of home range use with increasing body size, the utility of enclosed measures of space use will become increasingly irrelevant for the largest, most wide-ranging individuals and species. What, for example, is the home range of a whale shark, white shark, or tiger shark (Grubbs, 2010; Heupel et al., 2015)? Clearly, new metrics are necessary to track the allometry of space use, such as power law relationships of the frequency of movement step lengths (Sims, 2010; Sims et al., 2008).

2.4.2 "Wall of Mouths" and Why Do Sharks Aggregate?

Aggregations of sharks are increasingly documented and the associated dive tourism operations derive significant economic benefit from such aggregations (Cisneros-Montemayor et al., 2013; Gallagher and Hammerschlag, 2011; O'Malley et al., 2013). Indeed, our ability to tag rare, wide-ranging animals relies on this aggregation phenomenon but may yield a biased view of their behavioral ecology (Stewart et al., 2016). Why do aggregations occur? This is clearly a phenomenon worthy of study, yet appears to be

Table 2.2   Research Needs in Ontogenetic Diet, Trophic Level, and Movement Analyses

Research Scale	Research Questions
Within-species	Does dentition change continuously or nonlinearly with ontogeny?
	Does diet, trophic, or stable isotope composition change continuously or nonlinearly with ontogeny?
	Is the change in dentition related to the change in diet, trophic, or stable isotope composition with ontogeny?
	How much variance in diet, stable isotope composition, or trophic level is due to size, sex, season, or location (Reum and Essington, 2013)?
	How prevalent are positive body-size–trophic-level relationships in chondrichthyans (Jennings and van der Molen, 2015)?
	Do only (partially or entirely) piscivorous chondrichthyans exhibit a positive body-size–trophic-level relationship (Jennings and van der Molen, 2015)?
	Do larger species have a steeper body-size–trophic-level relationship and greater PPMR (Barnes et al., 2010)?
	Is steepness of body-size–trophic-level relationship (PPMR) related to increasing home ranges (and other movement metrics) in chondrichthyans (Tamburello et al., 2015)?
Multi-species and community-wide	How much variation is there within vs. among species in the body-size–trophic-level relationship?
	What is the PPMR of fish communities with a significant chondrichthyan biomass? Does that PPMR differ significantly from the narrow reported range (scaled estimate of 49 to 316) (Reum et al., 2015)?
	At what spatial and temporal scales must ecosystems be studied to fully encompass the biomass pyramids of which chondrichthyans are members?
	How significant is the planktonic "wall of mouths" subsidy to a local fish community with a significant chondrichthyan biomass (Hamner et al., 1988)?
	How significant and prevalent is the subsidy to a local fish community with a significant chondrichthyan biomass?

incredibly understudied (Heyman et al., 2001; Wilson et al., 2001). The size-based worldview has revealed that individuals are constrained to range more widely because they are able to eat increasingly larger prey, which, in turn, ranges more widely (Grubbs, 2010). But, teleost prey fishes must eventually overcome the ontogenetic spreading dictated by their gape sizes and dietary needs. At some point, they need to come together at higher than expected densities to reproduce (Sadovy de Mitcheson and Colin, 2011).

The very nature of broadcast spawning in teleosts means that mating is rarely exclusive. The effective population sizes of broadcast spawning fishes suggest that success is rare; it is not unusual for only one-in-a-million such matings to result in a successful contribution to the next generation (Hare et al., 2011; Hutchings and Reynolds, 2004). As a consequence, competition to win this mating-ground lottery is fierce, and high densities of larger numbers of fishes appear routinely in oceanographically favorable locations (Choat, 2012; Molloy et al., 2012). These prey fishes arrive from far and wide, each embodying the energy from a faraway ecological pyramid, to release this energy as a subsidy at the spawning location. This subsidy is released in the form of gametes, eggs, and sperm, but also as their bodies, which are preyed upon by predators including aggregations of sharks and rays (Hartup et al., 2013; Heyman et al., 2001; Mourier et al., 2016). Superficially, these aggregations of sharks might appear to be an inverted pyramid, but the inversion of the pyramid is transient and caused by an energetic subsidy in the form of a superabundance of mating fishes in a spawning aggregation comprised of energy derived from beyond the local pyramid (Mourier et al., 2016).

Energy subsidies resulting from animal migrations and aggregations are widely documented, not least in the salmon- and herring-fueled ecosystems of the Pacific

Northwest. Each year, billions of Pacific salmon return to their natal streams as pink waves surfed upon by migrating and aggregating predators and decomposers (Armstrong et al., 2016). These include bears, wolves, and bald eagles that remove salmon from rivers and transport the carcasses into nearby forests, thus subsidizing the production of maggots (Hocking et al., 2013) and influencing the ecology of woodland birds and the phenology of flowering plants (Hocking and Reynolds, 2011). It is intuitively obvious that Pacific salmon are inverting the ecological pyramids of streams by driving the regular cyclical aggregation of numerous species of top predators, but understanding these cyclical migratory aggregations in the oceans is much more difficult.

We have two vignettes relevant to the role of energy subsidies and ecological pyramids in the oceans. One of the most striking and forgotten analyses of coral reef energetics is the phenomenon known as the "wall of mouths" (Hamner et al., 1988). Much size-based interaction theory has been developed in phytoplankton-fueled shelf sea ecosystems hosting the largest fisheries in the world. There, most of the energy comes from the pelagic phytoplankton pathway (Duarte and Cebrián, 1996), but there is a benthic pathway fueled by detritus or marine snow derived from the pelagic path. This energy is eventually captured by deposit-feeding epifaunal and infaunal organisms that are preyed upon by fishes large enough to derive energy from and hence couple both the "fast" pelagic pathway and the "slow" benthic pathway (Blanchard et al., 2011; Rooney et al., 2006). Reefs are different, in the sense that the large size and slow turnover time (*P/M* ratio) of the primary producers is more akin to terrestrial ecosystems in that the foundation species (corals, seagrasses, mangroves, kelps) are relatively large bodied and long lived compared to phytoplankton (Duarte and Cebrián, 1996).

We are familiar with primary production fueled by these large foundation species, but we seem to overlook that planktivorous fishes form a "wall of mouths" and strip incoming oceanic water of zooplankton, transfering it to and retaining it within the reef ecosystem (Hamner et al., 1988). The numbers are staggering. Approximately 500 individuals from 10 species of reef fish visually inspect each cubic meter of incoming water, stripping out 0.5 kg of zooplankton per linear meter of reef front per day (Hamner et al., 1988). A significant fraction (~86%) of gray reef shark (*Carcharhinus amblyrhynchos*) biomass is supported by off-reef pelagic resources at Palmyra Atoll (McCauley et al., 2012b). Less is known of oceanic subsidies to kelp reefs, but the scale of subsidy from herring spawn is only now being enumerated (Armstrong et al., 2016), and the inputs from zooplankton are also thought to be substantial. Theory suggests that pulsed subsidies, such as from waves of zooplankton, tend to accumulate in the largest size classes, which appear as an inverted pyramid when viewed as an instantaneous snapshot (Pope et al., 1994). For example, it was estimated that the biomass of the 1- to 2-kg size class of kelp forest fishes was four to five times greater than expected given the productivity of the local ecological pyramid (Trebilco et al., 2016). The scale of this subsidy was measured, for the first time, by comparing the observed fish size structure against that expected from stable isotope estimates of community-wide PPMR. These subsidies have long been overlooked, and now that we have a methodology for measuring them and a theoretical–conceptual framework for understanding them we expect they can provide profound insights into shark migration and aggregation.

2.4.3   Subsidies, Sharks, and Inverted Pyramids

Large-scale oceanographic drivers underlie the scale of subsidies to reefs, which in turn influence the expectations of abundance and biomass of top predators. It has long been understood that, on coral reefs, (1) productivity is greater in nutrient-rich rather than nutrient-poor waters, (2) small islands with high reef perimeter-to-area ratios are more productive than larger island reefs, and (3) reefs with rich adjacent communities are more productive (Gove et al., 2016; Polunin, 1996). Large-scale analyses of Pacific reefs are revealing that, in addition to human population density, local productivity (usually indexed by chlorophyll *a* concentration) is a critical determinant of fish biomass, particularly planktivores, secondary consumers, and piscivores (Nadon et al., 2012; Williams et al., 2015). Note the critical role of planktivores, suggesting the need to further explore the "wall of mouths" subsidy pathway (Hamner et al., 1988). It seems that the expected biomass of top predators is particularly high on some of the islands studied by Sandin et al. (2008), such as Kingman and Palmyra, not just because these islands are relatively unfished but because these islands have particularly strong local oceanographic productivity (Williams

et al., 2015). Hence, the relatively inverted pyramid and high biomass of top predators are a result of years of accumulated oceanic subsidy, stored in the abundance and biomass as a result of a "wall of mouths" type of subsidy.

2.5   SUMMARY: SIZE ISN'T EVERYTHING BUT IS CENTRAL TO A NEW UNDERSTANDING OF ECOLOGICAL ROLES

Sharks and rays grow indeterminately throughout their lives, with profound consequences for diet, trophic level, and movement ecology. As they grow, their dentition and gape dimensions change, enabling them to consume larger bodied prey. Consequently, their home ranges expand and their foraging movements extend. These patterns can be neatly encapsulated by the predator–prey mass ratio (PPMR), which can be measured by judicious nitrogen stable isotope analysis of individual species and whole communities. We suggest that the null expectation is for a positive relationship between trophic level and body size, the slope of which is a measure of PPMR. A likely explanation for departures from this expectation are the nuances of stable isotope analyses across widely varying isoscapes, and we have provided statistical suggestions for mitigating this issue (see Tables 2.1 and 2.2). The generality of these findings and hypotheses is far from determined, and a series of key single and multispecies questions must be answered to determine the generality of the patterns and processes suggested in this chapter.

The shape of ecological pyramids is determined by the PPMR, along with transfer efficiency (TE). The few available community-wide measures of PPMR and available knowledge of transfer efficiency suggest that inverted pyramids dominated by a large standing biomass of sharks are implausible. The emerging evidence suggests that inverted pyramids arise from energetic subsidies, either from migration or aggregations. Subsidies can be detected by comparing the ecological pyramid shape (size spectra slope) of community samples with the expectations derived from community-wide measures of PPMR. Hence, PPMR and ontogenetic changes in diet and trophic level have profound effects on the behavior and ecology of individuals, populations, communities, and ecosystems.

ACKNOWLEDGMENTS

We thank David Powter for providing data and photographs and Michelle Heupel, Simon Jennings, John K. Pinnegar, Nicholas V. C. Polunin, Anne K. Salomon, and Colin Simpfendorfer for sharing their knowledge of diet and stable isotope analysis. Nick Dulvy was supported by Discovery and Accelerator grants from the Natural Science and Engineering Research Council and a Canada Research Chair, and Rowan Trebilco by the RJL Hawke Postdoctoral Fellowship.

REFERENCES

Ackerman JL, Bellwood DR (2000) Reef fish assemblages: a re-evaluation using enclosed rotenone stations. *Mar Ecol Ser* 206:227–237.

Andersen KH, Jacobsen NS, Farnsworth KD (2016) The theoretical foundations for size spectrum models of fish communities. *Can J Fish Aquat Sci* 73:575–588.

Armstrong JB, Takimoto G, Schindler DE, Hayes MM, Kauffman MJ (2016) Resource waves: phenological diversity enhances foraging opportunities for mobile consumers. *Ecology* 97:1099–1112.

Bangley CW, Rulifson RA, Overton AS (2013) Evaluating the efficiency of flushed stomach-tube lavage for collecting stomach contents from dogfish sharks. *Southeast Nat* 12:523–533.

Banse K, Mosher S (1980) Adult body mass and annual production/biomass relationships of field populations. *Ecol Monogr* 50:355–379.

Barnes C, Maxwell D, Reuman DC, Jennings S (2010) Global patterns in predator–prey size relationships reveal size dependency of trophic transfer efficiency. *Ecology* 91:222–232.

Barnett A, Redd KS, Frusher SD, Stevens JD, Semmens JM (2010) Non-lethal method to obtain stomach samples from a large marine predator and the use of DNA analysis to improve dietary information. *J Exp Mar Biol Ecol* 393:188–192.

Bell JD, Craik GJS, Pollard DA, Russell BC (1985) Estimating length frequency distributions of large reef fish underwater. *Coral Reefs* 4:41–44.

Blanchard JL, Heneghan RF, Everett JD, Trebilco R, Richardson AJ (2017) From bacteria to whales: using functional size spectra to model marine ecosystems. *Trends Ecol Evol* 32:174–186.

Blanchard JL, Law R, Castle MD, Jennings S (2011) Coupled energy pathways and the resilience of size-structured food webs. *Theor Ecol* 4:289–300.

Boudreau PR, Dickie LM (1992) Biomass spectra of aquatic ecosystems in relation to fisheries yield. *Can J Fish Aquat Sci* 49:1528–1538.

Bradley D, Conklin E, Papastamatiou YP, McCauley DJ, Pollock K, Pollock A, Kendall BE, et al. (2017) Resetting predator baselines in coral reef ecosystems. *Sci Rep* 7:43131.

Brey T (1999) Growth performance and mortality in aquatic macrobenthic invertebrates. *Adv Mar Biol* 35:153–223.

Brose U, Jonsson T, Berlow EL, Warren P, Banasek-Richter C, Bersier LF, Blanchard JL, et al. (2006) Consumer-resource body-size relationships in natural food webs. *Ecology* 87:2411–2417.

Brown JH, Gillooly JF (2003) Ecological food webs: high-quality data facilitate theoretical unification. *Proc Natl Acad Sci USA* 100:1467–1468.

Cernansky R (2017) Biodiversity moves beyond counting species. *Nature* 546:22.

Choat JH (2012) Spawning aggregations in reef fishes; ecological and evolutionary processes. In: Sadovy de Mitcheson Y, Colin PL (eds) *Reef Fish Spawning Aggregations: Biology, Research and Management*. Springer Science+Business Media, London, pp 85–116.

Cisneros-Montemayor AM, Barnes-Mauthe M, Al-Abdulrazzak D, Navarro-Holm E, Sumaila UR (2013) Global economic value of shark ecotourism: implications for conservation. *Oryx* 47:381–388.

Cohen JE, Jonsson T, Carpenter SR (2003) Ecological community description using the food web, species abundance, and body size. *Proc Natl Acad Sci USA* 100:1781–1786.

Cortés E (1997) A critical review of methods of studying fish feeding based on analysis of stomach contents: application to elasmobranch fishes. *Can J Fish Aquat Sci* 54:726–738.

Cortés E (1999) Standardized diet compositions and trophic levels of sharks. *ICES J Mar Sci* 56:707–717.

Cushing DH (1975) *Marine Ecology and Fisheries*. Cambridge University Press, Cambridge, UK.

Darwall WRT, Dulvy NK (1996) An evaluation of the suitability of non-specialist volunteer researchers for coral reef fish surveys. Mafia Island, Tanzania—a case study. *Biol Conserv* 78:223–231.

DeMartini EE, Friedlander AM, Sandin SA, Sala E (2008) Differences in fish-assemblage structure between fished and unfished atolls in the northern Line Islands, central Pacific. *Mar Ecol Prog Ser* 365:199–215.

Duarte CM, Cebrián J (1996) The fate of marine autotrophic production. *Limnol Oceanogr* 41:1758–1766.

Dulvy NK, Polunin NVC, Mill AC, Graham NAJ (2004) Size structural change in lightly exploited coral reef fish communities: evidence for weak indirect effects. *Can J Fish Aquat Sci* 61:466–475.

Ebert DA, Bizzarro JJ (2007) Standardized diet compositions and trophic levels of skates (Chondrichthyes: Rajiformes: Rajoidei). *Environ Biol Fish* 80:221–237.

Edgar GJ, Barrett NS, Morton AJ (2004) Biases associated with the use of underwater visual census techniques to quantify the density and size-structure of fish populations. *J Exp Mar Biol Ecol* 308:269–290.

Elston C, von Brandis RG, Cowley PD (2016) Gastric lavage as a non-lethal method for stingray (Myliobatiformes) diet sampling. *Afr J Mar Sci* 38:415–419.

Elton C (1927) *Animal Ecology*. Macmillan, New York.

Friedlander AM, DeMartini EE (2002) Contrasts in density, size, and biomass of reef fishes between the northwestern and the main Hawaiian islands: the effects of fishing down apex predators. *Mar Ecol Prog Ser* 230:253–264.

Froese R (2006) Cube law, condition factor and weight–length relationships: history, meta-analysis and recommendations. *J Appl Ichthyol* 22:241–253.

Fry B (2007) *Stable Isotope Ecology*. Springer Science+Business Media, London.

Gallagher AJ, Hammerschlag N (2011) Global shark currency: the distribution, frequency, and economic value of shark ecotourism. *Curr Issues Tour* 14:797–812.

Gove JM, McManus MA, Neuheimer AB, Polovina JJ, Drazen JC, Smith CR, Merrifield MA, et al. (2016) Near-island biological hotspots in barren ocean basins. *Nat Commun* 7:10581.

Green SJ, Dulvy NK, Brooks AML, Akins JL, Cooper, Miller S, Côté IM (2014) Linking removal targets to the ecological effects of invaders: a predictive model and field test. *Ecol Appl* 24:1311–1322.

Green SJ, Dulvy NK, Côté IM, Brooks AM, Miller SE, Akins JL, Cooper AB (2015) Response to Valderrama and Fields: effect of temperature on biomass production in models of invasive lionfish control. *Ecol Appl* 25:2048–2050.

Grubbs R (2010) Ontogenetic shifts in movements and habitat use. In: Carrier JF, Musick JA, Heithaus MR (eds) *Sharks and Their Relatives*. II. *Biodiversity, Adaptive Physiology, and Conservation*. CRC Press, Boca Raton, FL, pp 319–350.

Hamner WM, Jones MS, Carleton JH, Hauri IR, Williams DM (1988) Zooplankton, planktivorous fish, and water currents on a windward reef face—Great Barrier Reef, Australia. *Bull Mar Sci* 42:459–479.

Hare MP, Nunney L, Schwartz MK, Ruzzante DE, Burford M, Waples RS, Ruegg K, Palstra F (2011) Understanding and estimating effective population size for practical application in marine species management. *Conserv Biol* 25:438–449.

Hartup JA, Marshell A, Stevens G, Kottermair M, Carlson P (2013) *Manta alfredi* target multispecies surgeonfish spawning aggregations. *Coral Reefs* 32:367.

Heithaus MR (2010) Unraveling the ecological importance of elasmobranchs. In: Carrier JC, Musick JA, Heithaus MR (eds) *Sharks and Their Relatives*. II. *Biodiversity, Adaptive Physiology, and Conservation*. CRC Press, Boca Raton, FL, pp 611–638.

Heupel M, Knip DM, Simpfendorfer CA, Dulvy NK (2014) Sizing up the ecological role of sharks as predators. *Mar Ecol Prog Ser* 495:291–298.

Heupel M, Simpfendorfer C, Espinoza M, Smoothey A, Tobin A, Peddemors V (2015) Conservation challenges of sharks with continental scale migrations. *Front Mar Sci* 2:1–7.

Heyman WD, Graham RT, Kjerfve B, Johannes RE (2001) Whale sharks *Rhincodon typus* aggregate to feed on fish spawn in Belize. *Mar Ecol Prog Ser* 215:275–282.

Hocking MD, Dulvy NK, Reynolds JD, Ring RA, Reimchen TE (2013) Salmon subsidize an escape from a size spectrum. *Proc R Soc Lond Ser B* 280:20122433.

Hocking MD, Reynolds JD (2011) Impacts of salmon on riparian plant diversity. *Science* 331:1609–1612.

Hoenig JM, Gruber SH (1990) Life-history patterns in the elasmobranchs: implications for fisheries management. In: Pratt HL, Gruber SH, Taniuchi T (eds) *Elasmobranchs as Living Resources: Advances in the Biology, Ecology, Systematics, and the Status of the Fisheries*. National Oceanographic and Atmospheric Administration, Silver Spring, MD, pp 1–16.

Hussey NE, MacNeil MA, Olin JA, McMeans BC, Kinney MJ, Chapman DD, Fisk AT (2012) Stable isotopes and elasmobranchs: tissue types, methods, applications and assumptions. *J Fish Biol* 80:1449–1484.

Hussey NE, Macneil MA, McMeans BC, Olin JA, Dudley SF, Cliff G, Wintner SP, et al. (2014) Rescaling the trophic structure of marine food webs. *Ecol Lett* 17:239–250.

Hussey NE, MacNeil MA, Siple MC, Popp BN, Dudley SFJ, Fisk AT (2015) Expanded trophic complexity among large sharks. *Food Webs* 4:1–7.

Hutchings JA, Reynolds JD (2004) Marine fish population collapses: consequences for recovery and extinction risk. *Bioscience* 54:297–309.

Hyslop EJ (1980) Stomach contents analysis: a review of methods and their application. *J Fish Biol* 17:411–429.

Jennings S (2005) Size-based analyses of aquatic food webs. In: Belgrano A, Scharler UM, Dunne J, Ulanowicz RE (eds) *Aquatic Food Webs: An Ecosystem Approach*. Oxford University Press, Oxford, UK, pp 86–97.

Jennings S, Mackinson S (2003) Abundance–body mass relationships in size-structured food webs. *Ecol Lett* 6:971–974.

Jennings S, Warr KJ (2003) Environmental correlates of large-scale spatial variation in the $\delta^{15}N$ of marine animals. *Mar Biol* 142:1131–1140.

Jennings S, van der Molen J (2015) Trophic levels of marine consumers from nitrogen stable isotope analysis: estimation and uncertainty. *ICES J Mar Sci* 72(8):2289–2300.

Jennings S, Pinnegar JK, Polunin NVC, Boon T (2001) Weak cross-species relationships between body size and trophic level belie powerful size-based trophic structuring in fish communities. *J Anim Ecol* 70:934–944.

Jennings S, Pinnegar JK, Polunin NVC, Warr K (2002a) Linking size-based and trophic analyses of benthic community structure. *Mar Ecol Prog Ser* 226:77–85.

Jennings S, Warr KJ, Mackinson S (2002b) Use of size-based production and stable isotope analyses to predict trophic transfer efficiencies and predator–prey body mass ratios in food webs. *Mar Ecol Prog Ser* 240:11–20.

Jennings S, Barnes C, Sweeting CJ, Polunin NV (2008a) Application of nitrogen stable isotope analysis in size-based marine food web and macroecological research. *Rapid Commun Mass Spectrom* 22:1673–1680.

Jennings S, Melin F, Blanchard JL, Forster RM, Dulvy NK, Wilson RW (2008b) Global-scale predictions of community and ecosystem properties from simple ecological theory. *Proc R Soc Lond Ser B* 275:1375–1383.

Jetz W, Carbone C, Fulford J, Brown JH (2004) The scaling of animal space use. *Science* 306:266–268.

Kiszka JJ, Aubail A, Hussey NE, Heithaus MR, Caurant F, Bustamante P (2015) Plasticity of trophic interactions among sharks from the oceanic south-western Indian Ocean revealed by stable isotope and mercury analyses. *Deep Sea Res Part 1 Oceanogr Res Pap* 96:49–58.

Klimley A (1985) The areal distribution and autoecology of the white shark, *Carcharodon carcharias*, off the west coast of North America. *Mem S Cal Acad Sci* 9:15–40.

Kulbicki M (1998) How the acquired behaviour of commercial reef fishes may influence the results obtained from visual censuses. *J Exp Mar Biol Ecol* 222:11–30.

Letourneur Y, Kulbicki M, Labrosse P (1998) Length–weight relationships of fish from coral reefs and lagoons of New Caledonia, Southwestern Pacific Ocean. An update. *Naga* 21:39–46.

Lindeman RL (1942) The trophic–dynamic aspect of ecology. *Ecology* 23:399–417.

Lorenzen K (1996) The relationship between body weight and natural mortality in juvenile and adult fish: a comparison of natural ecosystems and aquaculture. *J Fish Biol* 49:627–647.

MacKenzie KM, Longmore C, Preece C, Lucas CH, Trueman CN (2014) Testing the long-term stability of marine isoscapes in shelf seas using jellyfish tissues. *Biogeochemistry* 121:441–454.

MacNeil MA, Graham NAJ, Conroy MJ, Fonnesbeck CJ, Polunin NVC, Rushton SP, Chabanet P, McClanahan TR (2008a) Detection heterogeneity in underwater visual census data. *J Fish Biol* 73:1748–1763.

MacNeil MA, Tyler EHM, Fonnesbeck CJ, Rushton SP, Polunin NVC, Conroy MJ (2008b) Accounting for detectability in reef-fish biodiversity estimates. *Mar Ecol Prog Ser* 367:249–260.

Makarieva AM, Gorshkov VG, Li B-L (2005) Why do population density and inverse home range scale differently with body size? Implications for ecosystem stability. *Ecol Complex* 2:259–271.

McCauley DJ, McLean KA, Bauer J, Young HS, Micheli F (2012a) Evaluating the performance of methods for estimating the abundance of rapidly declining coastal shark populations. *Ecol Appl* 22:385–392.

McCauley DJ, Young HS, Dunbar RB, Estes JA, Semmens BX, Michel F (2012b) Assessing the effects of large mobile predators on ecosystem connectivity. *Ecol Appl* 22:1711–1717.

Mittelbach GG, Persson L (1998) The ontogeny of piscivory and its ecological consequences. *Can J Fish Aquat Sci* 55:1454–1465.

Molloy PP, Côté IM, Reynolds JD (2012) Why spawn in aggregations? In: Sadovy de Mitcheson Y, Colin PL (eds) *Reef Fish Spawning Aggregations: Biology, Research and Management*. Springer Science+Business Media, London, pp 57–83.

Mourier J, Maynard J, Parravicini V, Ballesta L, Clua E, Domeier ML, Planes S (2016) Extreme inverted trophic pyramid of reef sharks supported by spawning groupers. *Curr Biol* 26:2011–2016.

Nadon MO, Baum JK, Williams ID, McPherson JM, Zgliczynski BJ, Richards BL, Schroeder RE, Brainard RE (2012) Re-creating missing population baselines for Pacific reef sharks. *Conserv Biol* 26:493–503.

O'Malley MP, Lee-Brooks K, Medd HB (2013) The global economic impact of manta ray watching tourism. *PLoS ONE* 8:e65051.

Odum HT, Odum EP (1955) Trophic structure and productivity of a windward coral reef community on Eniwetok Atoll. *Ecol Monogr* 25:291–320.

Olive PJW, Pinnegar JK, Polunin NVC, Richards G, Welch R (2003) Isotope trophic-step fractionation: a dynamic equilibrium model. *J Anim Ecol* 72:608–617.

Paine RT (1980) Foodwebs: linkage, interaction strength and community infrastructure. *J Anim Ecol* 49(3):666–685.

Pinnegar JK, Polunin NVC (1999) Differential fractionation of delta C-13 and delta N-15 among fish tissues: implications for the study of trophic interactions. *Funct Ecol* 13:225–231.

Pinnegar JK, Polunin NVC (2000) Contributions of stable-isotope data to elucidating food webs of Mediterranean rocky littoral fishes. *Oecologia* 122:399–409.

Pinnegar JK, Trenkel VM, Tidd AN, Dawson WA, Du Buit MH (2003) Does diet in Celtic Sea fishes reflect prey availability? *J Fish Biol* 63:197–212.

Pittman SJ, McAlpine CA (2003) Movements of marine fish and decapod crustaceans: process, theory and application. In: Southward AJ, Tyler PA, Young CM, Fuiman LA (eds) *Advances in Marine Biology*, vol 44. Academic Press, London, pp 205–294.

Polunin NVC (1996) Trophodynamics of reef fisheries productivity. In: Polunin NVC, Roberts CM (eds) *Reef Fisheries*. Chapman & Hall, London, pp 113–135.

Polunin NVC, Pinnegar JK (2002) Trophic ecology and the structure of marine food webs. In: Hart PJB, Reynolds JD (eds) *Handbook of Fish Biology and Fisheries*, vol 1. Blackwell Science, Malden, MA, pp 301–320.

Pope JG, Shepherd JG, Webb J (1994) Successful surf-riding on size spectra: the secret of survival in the sea. *Philos Trans R Soc Lond B* 343:41–49.

Post DM, Layman CA, Arrington DA, Takimoto G, Quattrochi J, Montana CG (2007) Getting to the fat of the matter: models, methods and assumptions for dealing with lipids in stable isotope analyses. *Oecologia* 152:179–189.

Powter DM, Gladstone W, Platell M (2010) The influence of sex and maturity on the diet, mouth morphology and dentition of the Port Jackson shark, *Heterodontus portusjacksoni*. *Mar Freshwater Res* 61:74–85.

Reecht Y, Rochet MJ, Trenkel VM, Jennings S, Pinnegar JK (2013) Use of morphological characteristics to define functional groups of predatory fishes in the Celtic Sea. *J Fish Biol* 83:355–377.

Reum JCP, Essington TE (2013) Spatial and seasonal variation in delta N-15 and delta C-13 values in a mesopredator shark, *Squalus suckleyi*, revealed through multitissue analyses. *Mar Biol* 160:399–411.

Reum JCP, Jennings S, Hunsicker ME (2015) Implications of scaled $\delta^{15}N$ fractionation for community predator–prey body mass ratio estimates in size-structured food webs. *J Anim Ecol* 84:1618–1627.

Reuman DC, Mulder C, Raffaelli D, Cohen JE (2008) Three allometric relations of population density to body mass: theoretical integration and empirical tests in 149 food webs. *Ecol Lett* 11:1216–1228.

Revill AT, Young JW, Lansdell M (2009) Stable isotopic evidence for trophic groupings and bio-regionalization of predators and their prey in oceanic waters off eastern Australia. *Mar Biol* 156:1241–1253.

Romanuk TN, Hayward A, Hutchings JA (2011) Trophic level scales positively with body size in fishes. *Glob Ecol Biogeogr* 20:231–240.

Rooney N, McCann K, Gellner G, Moore JC (2006) Structural asymmetry and the stability of diverse food webs. *Nature* 442:265–269.

Sadovy de Mitcheson Y, Colin PL (2011) *Reef Fish Spawning Aggregations: Biology, Research and Management*, vol 35. Springer Science+Business Media, London.

Sandin SA, Smith JE, Demartini EE, Dinsdale EA, Donner SD, Friedlander AM, Konotchick T, et al. (2008) Baselines and degradation of coral reefs in the Northern Line Islands. *PloS ONE* 3:e1548.

Savage VM, Gillooly JF, Woodruff WH, West GB, Allen AP, Enquist BJ, Brown JH (2004) The predominance of quarter-power scaling in biology. *Func Ecol* 18:257–282.

Shiffman DS, Gallagher AJ, Boyle MD, Hammerschlag-Peyer CM, Hammerschlag N (2012) Stable isotope analysis as a tool for elasmobranch conservation research: a primer for non-specialists. *Mar Freshwater Res* 63(7):635–643.

Simpfendorfer CA, Heupel MR (2004) Assessing habitat use and movement. In: Carrier J, Musick J, Heithaus M (eds) *Biology of Sharks and Their Relatives*. CRC Press, Boca Raton, FL, pp 553–572.

Simpfendorfer CA, Heupel MR (2016) Ecology: the upside-down world of coral reef predators. *Curr Biol* 26:R708–R710.

Simpfendorfer CA, Goodreid AB, McAuley RB (2001) Size, sex and geographic variation in the diet of the tiger shark, *Galeocerdo cuvier*, from Western Australian waters. *Environ Biol Fishes* 61:37–46.

Simpfendorfer CA, Heupel MR, White WT, Dulvy NK (2011) The importance of research and public opinion to conservation management of sharks and rays: a synthesis. *Mar Freshwater Res* 62:518–527.

Sims DW (2010) Tracking and analysis techniques for understanding free-ranging shark movements and behavior. In: Carrier JC, Musick JA, Heithaus MR (eds) *Sharks and Their Relatives*. II. *Biodiversity, Adaptive Physiology, and Conservation*. CRC Press, Boca Raton, FL, pp 351–392.

Sims DW, Southall EJ, Humphries NE, Hays GC, Bradshaw CJ, Pitchford JW, James A, et al. (2008) Scaling laws of marine predator search behaviour. *Nature* 451:1098–1102.

Stewart JD, Beale CS, Fernando D, Sianipar AB, Burton RS, Semmens BX, Aburto-Oropeza O (2016) Spatial ecology and conservation of *Manta birostris* in the Indo-Pacific. *Biol Conserv* 200:178–183.

Tamburello N, Côté IM, Dulvy NK (2015) Energy and the scaling of animal space use. *Am Nat* 186:196–211.

Trebilco R, Baum JK, Salomon AK, Dulvy N (2013) Ecosystem ecology: size-based constraints on the pyramids of life. *Trends Ecol Evol* 28:423–431.

Trebilco R, Dulvy NK, Stewart H, Salomon AK (2015) The role of habitat complexity in shaping the size structure of a temperate reef fish community. *Mar Ecol Prog Ser* 532:197–211.

Trebilco R, Dulvy NK, Anderson SC, Salomon AK (2016) The paradox of inverted biomass pyramids in kelp forest fish communities. *Proc R Soc Lond B Biol Sci* 283:20160816.

Tricas TCM (1984) Predatory behavior of the white shark (*Carcharodon carcharias*), with notes on its biology. *Proc Cal Acad Sci* 9:123–135.

Turnbull MS, George PBL, Lindo Z (2013) Weighing in: size spectra as a standard tool in soil community analyses. *Soil Biol Biochem* 68:366–372.

Violle C, Thuiller W, Mouquet N, Munoz F, Kraft NJB, Cadotte MW, Livingstone SW, Mouillot D (2017) Functional rarity: the ecology of outliers. *Trends Ecol Evol* 32:356–367.

Wainwright PC, Richard BA (1995) Predicting patterns of prey use from morphology of fishes. *Environ Biol Fishes* 44:97–113.

Ward-Paige C, Mills Flemming J, Lotze HK (2010) Overestimating fish counts by non-instantaneous visual censuses: consequences for population and community descriptions. *PloS One* 5:e11722.

Werner EE, Gilliam JF (1984) The ontogenetic niche and species interactions in size-structured populations. *Annu Rev Ecol Syst* 15:393–425.

Wetherbee BM, Cortés E, Bizzarro JJ (2004) Food consumption and feeding habits. In: Carrier JC, Musick JA, Heithaus MR (eds) *Biology of Sharks and Their Relatives*. CRC Press, Boca Raton, FL, pp 225–246.

Williams EC (1941) An ecological study of the floor fauna of the Panama rain forest. *Bull Chicago Acad Sci* 6(4):63–124.

Williams ID, Baum JK, Heenan A, Hanson KM, Nadon MO, Brainard RE (2015) Human, oceanographic and habitat drivers of central and western pacific coral reef fish assemblages. *PLoS ONE* 10:e0120516.

Wilson SG, Taylor JG, Pearce AF (2001) The seasonal aggregation of whale sharks at Ningaloo reef, western Australia: currents, migrations and the El Nino/Southern oscillation. *Environ Biol Fishes* 61:1–11.

Zuur AF, Ieno EN, Elphick CS (2010) A protocol for data exploration to avoid common statistical problems. *Meth Ecol Evol* 1:3–14.

Zuur AF, Hilbe J, Ieno EN (2013) *A Beginner's Guide to GLM and GLMM with R: A Frequentist and Bayesian Perspective for Ecologists*. Highland Statistics, Newburgh, UK.

CHAPTER 3

Advances in the Application of High-Resolution Biologgers to Elasmobranch Fishes

Nicholas M. Whitney
Anderson Cabot Center for Ocean Life, New England Aquarium, Boston, Massachusetts

Karissa O. Lear
Centre for Aquatic and Ecosystems Research, Harry Butler Institute, Murdoch University, Murdoch, Western Australia, Australia

Adrian C. Gleiss
Centre for Aquatic and Ecosystems Research, Harry Butler Institute, Murdoch University, Murdoch, Western Australia, Australia

Nicholas Payne
School of Natural Sciences, Trinity College Dublin, Dublin, Ireland

Connor F. White
Shark Lab, California State University, Long Beach, California

CONTENTS

3.1 INTRODUCTION

Defined broadly, *biologging* is any collection of data from one or more sensors using an animal-borne tag (Boyd et al., 2004; Hooker et al., 2007). This term can be applied to a wide range of devices that have been used for over a century (Kooyman, 2004), and the past few decades have seen rapid expansion of the availability and application of these tools in the study of wild animals (e.g., Cooke et al., 2004a). Whereas many animal-borne tags transmit their data via satellite or acoustic signals, this chapter focuses primarily on instruments that store information to memory and must therefore be physically recovered in order to download their data. These devices, often referred to as *biologgers*, *data loggers*, or *loggers* for short, have been used for decades (e.g., Eckert and Eckert, 1986; Kooyman, 1965, 1966; Wilson and Bain, 1984a,b) and were originally developed before transmission of data was possible (Kooyman, 2004). Today, biologgers are used to record high-resolution data, which cannot be easily transmitted, to address an extremely diverse set of questions (Payne et al., 2014).

This chapter is not intended to be an exhaustive review of biologging research or devices (see Cooke et al., 2004a; Ropert-Coudert and Wilson, 2005; Wilson et al., 2015b) or their application to elasmobranchs (see Whitney et al., 2012). Instead, our intention is to highlight recent advances in order to provide a guide for investigators to review and select technologies that are appropriate to address questions in their own research. We aim to help readers understand what types of data they can expect to obtain from these devices and what kind of questions these data can be used to address, as well as their limitations, so the chapter serves as a realistic guide to what works.

3.2 SENSORS AND MEASURED PARAMETERS

To understand the various questions that are currently being addressed with biologging devices, a brief overview of sensor technology is useful. Different biologging sensors vary not only in the data they collect but also in their range, precision, minimum required sampling rate, and power and memory requirements; therefore, great care must be taken in selecting a device that is appropriate for the study species and research goals.

Some of the first sensors used in biologgers for marine animals measured pressure and temperature (Kooyman, 2004). These sensors are often combined together into one unit as temperature is used to correct and calibrate the pressure sensor to accurately measure depth. Although depth measurements can be collected at high frequency (at more than 10 times per second, or 10 Hz), a single sample every second or two (i.e., 1 or 0.5 Hz) is usually sufficient to capture the rate of depth change exhibited by most elasmobranchs (Wilson and Vandenabeele, 2012; Wilson et al., 2008). In the late 1990s, researchers began placing motion-sensitive tags on animals (Brown et al., 2013; Davis et al., 1999; Rutz and Hays, 2009; Wilson et al., 2015b; Yoda et al., 1999). These tags are comprised of at least one of three sensors: accelerometers to measure body acceleration, gyroscopes to measure angular velocity, and magnetometers to measure magnetic fields (and hence direction relative to the Earth's magnetic field). In more recent years, aided by their increased use in personal electronics, inertial measurement units have become ubiquitous and often have all three sensors, each measuring along three axes (Noda et al., 2014). To be useful in reconstructing information about movement of the animal, these sensors often need to record at high frequencies, meaning the loggers are relatively power consumptive and thus are usually battery limited; few current loggers are able to log high-resolution (>20 Hz) information from multiple sensors for more than a couple of weeks.

Each of these three types of motion sensors provides a slightly different perspective on animal movement. Accelerometers measure the total body acceleration of an animal, comprised of both static acceleration, which is the constant acceleration due to gravity and can be used to infer animal pitch or posture and dynamic acceleration, which is due to the movements of the animal (e.g., Shepard et al., 2008a). These acceleration data provide a quantitative measure of the kinematics of an animal and have many applications, including classifying behaviors, estimating energy expenditure, and studying biomechanics. Gyroscopes, which measure angular or rotational velocity, can be paired with accelerometers to provide additional detail and accuracy in the estimation of kinematics of individual animals. Unlike accelerometers, these sensors measure only the rotation of an animal and not its posture, and therefore allow for the unbiased estimation of the dynamic acceleration of an animal during unsteady motions. These sensors can be used to help accurately separate dynamic and static acceleration measured by accelerometers, providing a better estimate of attitude and kinematics while also increasing the accuracy of behavioral classification (Noda et al., 2012, 2014). Magnetometers are also commonly combined with accelerometers to provide more detail on animal movements (Williams et al., 2017). Magnetometers provide information on the posture of an animal in reference to the orientation of the Earth's magnetic field. When this information is taken in the context of the pitch and roll of an animal measured by an accelerometer, these concurrent data can produce a compass heading and, when combined with an estimate of velocity, can be used to reconstruct the swimming paths of the animal, generally referred to as a *pseudo-track*.

In addition to these three-dimensional motion sensors, swimming speed sensors have been integrated into many biologging packages. Modern versions of these sensors are comprised of two parts: (1) a freely rotating propeller (Muramoto et al., 2004, Nakamura et al., 2011; Watanabe et al., 2012) or flexible paddle (Shepard et al., 2008b), and usually (2) an optical or magnetic sensor that detects the movement of the propeller or paddle. Swimming speed sensors need to be oriented along the same axis as the swimming direction of the animal and have a clear path for water to flow through and move the propeller. Additionally, these sensors only provide accurate speed measurements across a discrete range of swimming speeds, and they poorly estimate speed when the animal is moving very slowly (e.g., <0.1 m/s) or very fast (e.g., >5 m/s) (Watanabe et al., 2012). Speed sensors are usually coupled with other three-dimensional motion sensors designed for short (days) deployments.

For all of these motion sensors, it is essential for measurements to be sampled at a sufficient frequency to render accurate behavioral data. Generally, the sampling rate should be at least double the frequency of the sinusoidal movements that make up the typical tailbeat or stride frequency of the animal (Nyquist frequency), and sampling at 5 to 10 times that rate is optimal (Ropert-Coudert and Wilson, 2005). For accelerometers, gyroscopes, and magnetometers, many studies have chosen frequencies of 5 to 40 Hz as effective sampling rates for elasmobranchs, with slower sampling rates producing results that may be more difficult to interpret. Some studies have recorded at higher rates (100 to 600 Hz) and have mainly been used to characterize rapid motions typical of foraging (Broell et al., 2013; Noda et al., 2014). However, these high sampling frequencies are unnecessary for many studies and can slow down analytical procedures and decrease tag battery and memory life without producing significantly better results than sampling at 30 Hz (Broell et al., 2013). This is particularly the case for larger bodied animals such as most elasmobranchs, which will generally produce smaller acceleration values and lower tailbeat frequencies than smaller fishes. Swimming speed, depth, and temperature sensors generally sample at a lower frequency (<1 Hz) because most animals are unlikely to undergo dynamic changes in these parameters at more rapid rates, although there may be notable exceptions to this for some pelagic sharks, such as makos (*Isurus* spp.).

These sensors can provide a large amount of information on the behavior and movements of animals, but it can be difficult to accurately determine animal behaviors from sensor data alone, unless ground-truthing experiments are undertaken. To solve this problem, imaging sensors, including both still cameras and animal-borne video cameras (Rutz and Troscianko, 2013), have been increasingly used to provide context to other sensor data (Heithaus et al., 2001, 2002; Marshall et al., 2007; Payne et al., 2016a). Imaging sensors are useful tools but are also power and memory intensive compared to other sensors and of limited use during periods of low light or poor visibility. As video loggers are increasingly well-integrated with motion sensors, the operation of the camera can be linked to the sensor data, selectively recording footage only during periods of interest, thereby avoiding the recording of excessive amounts of video during periods of limited interest (Rutz et al., 2013; Jorgensen, pers. comm; see also Chapter 5 in this volume).

In addition to the wide assortment of sensors that can quantify the behavior and environment in which animals operate, various sensors have also been developed to meet the challenging task of measuring physiological processes, including pH sensors (Papastamatiou and Lowe, 2004; Papastamatiou et al., 2007), impedance sensors (Meyer and Holland, 2012), and motility sensors (Papastamatiou et al., 2007, 2015), to study the various aspects of digestive physiology as proxies of energy intake (reviewed in Whitney et al., 2012). Other sensors that log biopotentials include electrocardiograms (ECGs), electromyograms (EMGs), and electroencephalograms (EEGs) (Cooke et al., 2004b; Scharold et al., 1989; Vyssotski et al., 2006). These sensors can provide invaluable information on muscle function and heart rate, but it is important to note that obtaining clean signals over long periods (greater than hours) requires the sensor electrode to be placed subcutaneously (Cooke et al., 2004b).

Many other types of biologging sensors provide additional context to animal behavior by describing an animal's surrounding environment. These include oceanographic sensors, such as conductivity loggers to measure salinity (Luo et al., 2008), and dissolved oxygen sensors that can be particularly useful in studying species that dive deep and utilize the oxygen minimum layer (Coffey and Holland, 2015). Acoustic transducers acting as animal-borne sonar (Lawson et al., 2015), passive acoustic recorders (Johnson and Tyack 2003; Meyer et al., 2007), or telemetry receivers (Haulsee et al., 2016a) have also recently been incorporated into biologging tags to help us understand more about the biotic environment and how animals interact with prey items or conspecifics. Despite the tremendous potential for the use of these types of sensors, many are still in developmental stages and have yet to achieve wide use in elasmobranchs.

3.3 FINE-SCALE BEHAVIORS

One of the biggest advantages to the use of high-resolution biologgers is that they can provide the opportunity to move beyond locational telemetry and "observe" the *in situ* behaviors of individuals. This is a particularly valuable tool for the study of elasmobranchs, as most species cannot be observed in the wild except when attracted by human-introduced bait or chum. To date, acceleration data loggers (ADLs) have proven to be one of the most useful devices for this purpose because they provide body movement data that can be used to quantify behavior and activity patterns, including fine-scale information on an animal's tailbeat frequency and amplitude and body posture (reviewed in Whitney et al., 2012).

3.3.1 Diel Cycles of Activity

Historically, changes in activity were inferred from either vertical or horizontal displacement through conventional (acoustic or satellite) telemetry (e.g., Cartamil et al., 2010; Gruber et al., 1988; Sims et al., 2010; Sundstrom et al., 2001), but more subtle differences in activity went undetected. ADLs offer an opportunity to resolve such uncertainties due to their ability to clearly distinguish between periods of swimming or resting (Figure 3.1) (Gleiss et al., 2009a; Whitney et al., 2007). Securely attaching ADLs to a shark not only can allow the activity state of the animals to be assessed but can also allow for examination of fine-scale tailbeat frequency and amplitude (Gleiss et al., 2009a), which can be used to distinguish

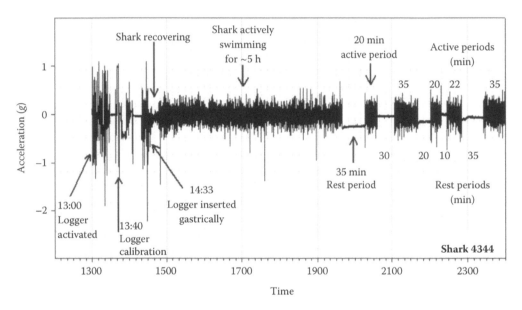

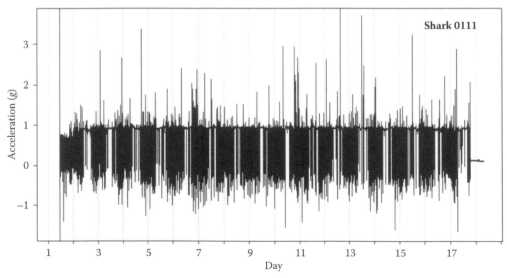

Figure 3.1 (A) Raw, single-axis acceleration data from a gastrically inserted acceleration logger in a whitetip reef shark showing acceleration signature differences between active swimming and resting behavior. (B) Single-axis acceleration data from a 17.3-day deployment of an accelerometer externally tethered to a whitetip reef shark showing consistent nocturnal activity and mid-day resting. Grid lines represent midnight. (From Whitney, N.M. et al., *Aquat. Living Resour.*, 20(4), 299–305, 2007.)

different types of swimming and resting (e.g., resting in a surge zone vs. calm water) (Whitney et al., 2010). Direct comparisons of diel patterns in vertical movements (based on a depth sensor) and activity (based on acceleration data) have revealed that these can be strikingly different (Gleiss et al., 2013, 2017a), cautioning against the use of depth data alone to infer activity rhythms. The incorporation of acceleration sensors into acoustic tags further increases the capacity to infer activity rhythms beyond diel patterns to include both seasonal and annual scales, as the lower resolution of these tags allows for longer deployments (Curtis et al., 2014; Murchie et al., 2011; O'Toole et al., 2011; Papastamatiou et al., 2015).

3.3.2 Foraging and Feeding

Energy gain is of prime importance to organisms, as it closely relates to growth and reproductive fitness. Subsequently, some of the most sought-after questions in ecology relate to where, when, and how individuals forage. Feeding events in some species of elasmobranchs are often associated with bursts of speed to capture prey items, potentially distinctive body postures, and subsequent movements of the jaw and head as the prey item is consumed. Yet, simple identification of periods of high activity in the wild is unlikely to provide sufficient evidence of feeding, as these movements could also represent other behaviors such as predator avoidance or social interactions. Distinctive body movements associated with foraging behavior can sometimes be ground-truthed by feeding tagged animals under laboratory conditions, allowing potential feeding events to be extracted from wild data with greater certainty. This has been done for several teleost species (Broell et al., 2013; Tanoue et al., 2012) and recently for lemon sharks (Brewster et al., 2018). Other studies have even been able to estimate prey type based on the acceleration signature during the capture event (Horie et al., 2017; Kawabata et al., 2014). Acceleration loggers or Hall sensors have also been deployed directly on the mouths of animals to sense the opening and closing movements of the jaw (Iwata et al., 2012; Makiguchi et al., 2012; Metcalfe et al., 2009; Wilson et al., 2002). Although these procedures hold great promise in determining the incidence and extent of foraging behavior, they have yet to be applied for this purpose in elasmobranchs.

In addition to externally placed loggers, the larger body size of many elasmobranchs has permitted multiple loggers to be placed in their stomachs to monitor prey consumption and digestion. These include simple temperature loggers to detect the rapid influx of seawater during feeding in endothermic sharks and the following heat increment of feeding (Jorgensen et al., 2015; Sepulveda et al., 2004), which has been shown to correlate to meal size in a few select species (Whitlock et al., 2013; Wilson et al., 1992).

Other loggers have been designed to monitor the changes in pH and the physical movement (motility) of the stomach during digestion. Immediately following ingestion, buffering of the stomach substantially raises pH. As the prey item is digested, the pH again begins to fall and the impedance within the stomach increases (Meyer and Holland, 2012; Papastamatiou and Lowe, 2004; Papastamatiou et al., 2007, 2015). By monitoring the amount of time it takes for these signals to return to baseline, meal size can be estimated. However, these methods require species-specific laboratory validations in order to understand the baseline variation in the measured parameters and the relationship between meal size and duration of digestion, and there may be issues with classifying prey consumed in rapid succession (Papastamatiou and Lowe, 2005).

3.3.3 Reproductive Behavior and Social Interactions

Biologgers can also provide insight into behaviors that are infrequent or difficult to observe directly, such as mating. Although mating behavior has been observed in relatively few elasmobranch species, there have been some consistently observed postural and movement patterns to this behavior across several species (Pratt and Carrier, 2001). Whitney et al. (2010) used direct observations of tagged animals to validate acceleration data from mating nurse sharks (*Ginglymostoma cirratum*) and suggested aspects of depth and acceleration data that could be used to identify mating in other species even without direct validation.

In addition to identifying individual acts of mating behavior, the development of various animal-borne proximity transmitters or logger–transmitter hybrid tags (Guttridge et al., 2010; Holland et al., 2009) provides promise for understanding such animal social networks (e.g., Jacoby et al., 2012; Mourier et al., 2017; see also Chapter 18 in this volume). Social network data may provide important context for interpreting acceleration data (in cases of mating, for instance). To date, acceleration data have been integrated into network analyses in a single study, which found that lemon sharks (*Negaprion brevirostris*) preferentially associated with conspecifics based on similarity or nonsimilarity of locomotor traits (Wilson et al., 2015a). The ability to surgically implant archival proximity loggers into the abdominal cavity of animals and recover the loggers and data after periods of nearly a year also presents an exceptional opportunity to study long-term associations between tagged conspecifics (Haulsee et al., 2016b).

3.3.4 Identification of Rare or Novel Behaviors

Applying biologging techniques to a new species or system very frequently reveals unanticipated, novel, or rare behaviors. Some unique behaviors recorded by accelerometers and other data-logging sensors are relatively straightforward to identify, such as jumping behavior in a blacktip shark (Figure 3.2); however, even in this case it is difficult to say whether the animal actually broke the surface of the water, although this can be detected with a swim speed

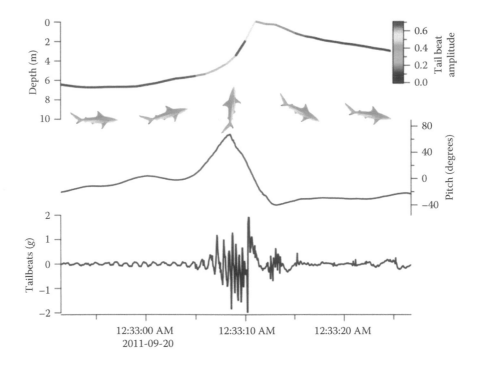

Figure 3.2 Depth, body pitch, and lateral (tailbeat) acceleration from a blacktip shark in Florida outfitted with an acceleration data logger package (Whitmore et al., 2016) indicating putative jumping behavior. Putative jumps were visually identified as periods in which the individual displayed high pitch angles (>70°) and rapid ascents to the surface with high tailbeat frequency and amplitude.

propeller (Watanabe et al., 2008), or whether it was pursuing prey, trying to dislodge its tag, or being pursued itself by a larger predator. Often these behaviors require some type of validation or additional context to confirm the activity the animal is engaged in from direct observation (Whitney et al., 2010), animal-borne cameras (Nakamura et al., 2015; Payne et al., 2016a; see also Chapter 5 in this volume), or extensive trials in a controlled setting (Brewster et al., 2018). As an additional hurdle for these studies, validation techniques and high-resolution biologging in general are often temporally constrained to a period of days or weeks due to limited battery and memory capacity of the tags and thus may fail to detect behavior that occurs infrequently. Although the development of acoustic and satellite tags capable of transmitting acceleration and other sensor data shows tremendous promise for long-term monitoring, their use in detecting fine-scale behaviors is limited by the volume of data that can be transmitted.

Because the physics of data transmission from small platforms is unlikely to change in the near future, a more feasible development is onboard processing that detects and logs specific behaviors and then transmits a summary of events. However, in order for such unsupervised processing to be reliable, it would have to be based on an algorithm that was derived and validated using high-resolution data from a biologger. Thus, even though future datasets may include years' worth of transmitted data reporting the

number of feeding or mating events of an individual shark, those datasets will only be made possible by algorithms built on a foundation of short-term datasets from high-resolution biologgers.

3.4 BIOMECHANICS

Locomotion is among the most energetically costly behaviors vertebrates engage in, especially for the many ram-ventilating fishes that never cease swimming (Schmidt-Nielsen, 1972). This substantial metabolic cost must be balanced by consumption, tightly linking locomotion to vital rates. Thus, understanding the incidence and extent of locomotion represents a crucial aspect of studying the physiological ecology of this group of animals. Quantifying when and how fishes engage in locomotion has historically been difficult, with techniques such as electromyogram telemetry of aerobic or anaerobic swimming muscle (Cooke et al., 2004b), mechanical tailbeat sensors (Lowe et al., 1998), or differential pressure sensors (Webber et al., 2001) finding application in some taxa. None of these techniques has seen broad application, though, largely due to logistical constraints in applying them to a diverse group of animals. Conversely, accelerometers and associated motion sensors have seen rapid proliferation in the study of swimming energetics and kinematics, in both cartilaginous and bony fishes.

3.4.1 Lift and Drag and Its Role in Locomotion

Many elasmobranchs use their lipid-rich, low-density livers to provide substantial hydrostatic forces. Variable liver volumes and densities among different species, sexes, and time of year largely govern animal buoyancy (Bone and Roberts, 1969; Gleiss et al., 2015; Hussey et al., 2010). These differences in hydrostatic forces in turn are expected to impact locomotion, especially in those species that use a broad range of swimming depths, and these impacts can be observed using biologging devices. For example, biologging data from a range of sharks have shown that epipelagic sharks are largely negatively buoyant and often use extended gliding during descents, whereas ascents are characteristic of greater locomotor effort (Figure 3.3) (Gleiss et al., 2011a,b; Nakamura et al., 2011).

This negative buoyancy of the majority of species of sharks has long been postulated to be a potential source of energy savings in elasmobranchs. Weihs (1973) postulated that negative buoyancy should result in animals interspersing gliding with active swimming during vertical movements, in turn reducing the mechanical power requirements of swimming. The central idea in this theory is that undulatory movements of the tail increase drag by approximately two- to threefold compared to a rigid, gliding body, as was recently demonstrated experimentally using digital particle image velocimetry (Floryan et al., 2017). Therefore, if sharks repeatedly dive and use gliding descents with shallow angles and marginally steeper active ascents, the mechanical cost of swimming may be reduced by as much as 50% (Weihs, 1973).

Testing Weihs' model was one of the first applications of biologgers (specifically accelerometers) to study the kinematics of free-ranging sharks. Gleiss et al. (2011b) demonstrated that whale sharks glide almost exclusively on descent, and use diving geometries that minimize the cost of horizontal transport. In further support of Weihs' model, Nakamura et al. (2015b) found that diving kinematics in deep-sea sharks, which are often slightly positively buoyant (Corner et al., 1969), show the reverse pattern. Here,

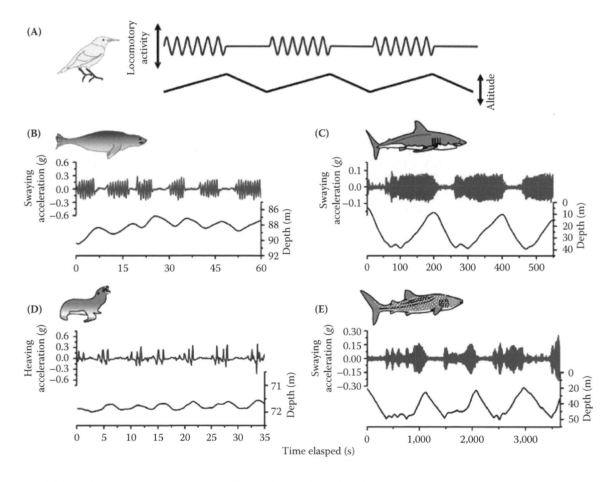

Figure 3.3 Gliding in buoyancy- or gravity-assisted locomotion has been found to be a common feature in vertebrate locomotion, including birds (A), seals (B and D), and sharks (C and E). Although locomotor activity is absent during phases of descent, phases of ascent include active tail or flipper beating. In positively buoyant sharks, the opposite is the case (Nakamura et al., 2015b). This behavior reduces the overall energetic costs of swimming and may be more energetically efficient than horizontal swimming in negatively buoyant fish, assuming that the individual is not neutrally buoyant (Weihs, 1973). (From Gleiss, A.C. et al., *Nat. Commun.*, 2, 352, 2011.)

positive buoyancy allows sharks to glide while ascending, but they have to actively swim during descent. Collectively, these studies have shown that sharks glide while buoyancy assisted, which is a common feature of vertebrate locomotion (Gleiss et al., 2011a; Williams et al., 2000). However, there are some notable exceptions to this trend, as tiger sharks only glide intermittently during descent, failing to support Weihs' model of energy conservation (Iosilevskii et al., 2012). This indicates that other factors, such as prey encounter, may shape the diving kinematics in this species. Recent evidence, however, cautions against the use of short-term biologger deployments to infer swimming kinematics. Whitney et al. (2016) demonstrated that the probability of descents being composed of glides is strongly influenced by time after release from capture. Following stressful capture by hook and line, sharks swam at elevated tailbeat frequencies for a number of hours, and this effect was most noticeable during descents.

Biologging can effectively answer many questions in the field of biomechanics that traditional laboratory methods are unlikely to resolve, but, even more importantly, this suite of techniques also allows for the formulation and testing of novel hypotheses that arise from observations of the swimming kinematics of free-ranging individuals. For example, great hammerhead sharks (*Sphyrna mokarran*) were recently shown to swim for extended periods in a rolled manner, which is not the case for other carcharhiniformes (Figure 3.4) (Payne et al., 2016a). Great hammerheads also feature some of the tallest dorsal fins in sharks, and the authors subsequently tested if rolled swimming would decrease induced drag to counteract negative buoyancy through the use of the dorsal fin as a lift-generating appendage. Wind-tunnel experiments on models of great hammerheads confirmed this prediction, and the authors suggest that the evolution of the tall dorsal fin for the unique maneuverability of hammerheads has resulted in its use for this additional purpose (Payne et al., 2016a). Interestingly, the same sharks would resume swimming in a non-rolled manner once closer to the seabed, presumably as the ground effect provided additional lift, further supporting the notion that rolling increases the lift-to-drag ratio of this species.

3.5 ENERGY EXPENDITURE

Energy is the currency of all life (Brown and Sibly, 2006) and the rate at which animals expend it is a crucial component of biomechanics, physiology, and ecology. As such, quantifying the energetics of free-ranging animals in the wild is a valuable tool for studying elasmobranchs. Most previous methods of measuring field energetics, including doubly labeled water and heart-rate telemetry, are often unsuitable for use in fully aquatic animals (Butler et al., 2004; Thorarenson et al., 1996; Whitney et al., 2012, but see

Clark et al., 2010). Instead of these more traditional techniques, calibrating locomotory activity, such as tailbeat frequency (Lowe, 2001) or swimming speed (Sundström and Gruber, 1998), against oxygen consumption has successfully been used to estimate energy expenditure in sharks.

Building on these initial techniques, accelerometers now offer a commercially available method of measuring animal movement and have been increasingly used in the last decade as a tool to estimate field metabolic rate (FMR) in a variety of both terrestrial and aquatic species (e.g., Halsey et al., 2009, 2011; Wilson et al., 2006). This technique is based on the principle that animal movement is driven by muscle contraction, which requires the hydrolysis of adenosine triphosphate (ATP), the universal carrier of chemical energy, and therefore oxygen consumption (Gleiss et al., 2011b). By running respirometry trials with animals equipped with accelerometers, these simultaneous measurements of metabolic rate and body acceleration—generally measured as overall dynamic body acceleration (ODBA) or partial dynamic body acceleration (PDBA)—can be used to create a calibration between the two metrics. This calibration can then be applied to acceleration data collected from wild fish to estimate energy expenditure. Using accelerometers in this manner also has the added value of concurrently measuring energy use, behavior, and other kinematic parameters, meaning that this technique can provide estimates of context-specific metabolic rates.

Calibrations between oxygen consumption and body acceleration have been conducted for four species of sharks so far: scalloped hammerheads (Gleiss et al., 2010), lemon sharks (Bouyoucos et al., 2017b; Lear et al., 2017), blacktip sharks (*Carcharhinus limbatus*) (Lear et al., 2017), and nurse sharks (*Ginglymostoma cirratum*) (Lear et al., 2017). Each of these studies was successful in creating significant calibrations between oxygen consumption and metabolic rate that can be applied to wild sharks and demonstrated the applicability of using accelerometry as a method of estimating FMRs in elasmobranchs. However, these studies have also identified several caveats in undertaking body acceleration–metabolic rate calibrations, emphasizing the need to carefully consider laboratory protocols and respirometry procedures while conducting these studies in sharks and other large-bodied fish. These caveats are examined below in an effort to discuss best-practice protocols for performing these calibration experiments with elasmobranchs.

3.5.1 Respirometer Choice

The overarching goal of calibrating body acceleration against metabolic rate is to translate laboratory measurements of metabolic rate to wild fish. This means that fish swimming behavior in calibration studies should match free-swimming fish behavior as closely as possible in order to produce body acceleration–metabolism relationships

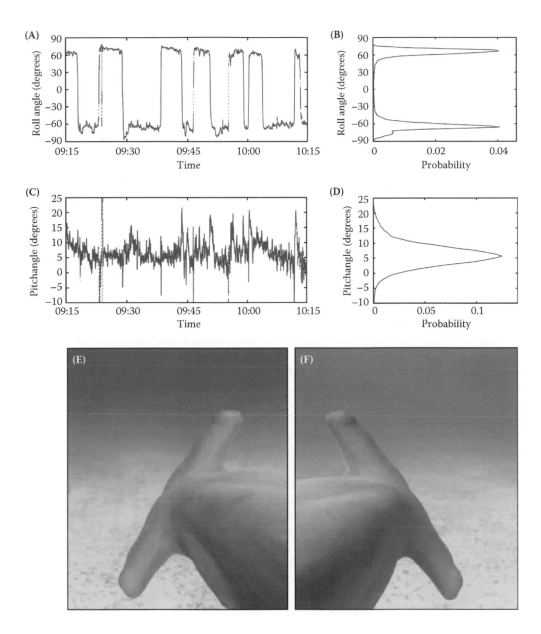

Figure 3.4 While swimming in midwater, accelerometers revealed that great hammerhead sharks (*Sphyrna mokarran*) swim in a rolled manner. For (A) to (D), roll and pitch angles were measured by an acceleration data logger attached to the dorsal fin of a 295-cm shark on the Great Barrier Reef, Australia. A typical hour-long time series for that animal is depicted in (A) and (C), and probability distributions of roll and pitch angles based on the last 15 hours of the monitoring period are shown in (B) and (D). Images in (E) and (F) were taken with a fin-mounted video camera attached to another wild *S. mokarran* (~350 cm) as it rolled to the left and right (respectively) at absolute roll angles of ~60° at South Bimini Island, the Bahamas. (From Payne, N.L. et al., *Nat. Commun.*, 7, 12289, 2016.)

that are applicable to wild individuals. Although simple in concept, this is a difficult goal to achieve in laboratory studies, where captive conditions generally do not equate to the environment experienced by wild fish. This is particularly true of respirometers themselves, which by necessity confine animals to relatively small, sealed chambers where swimming behavior rarely mirrors swimming seen in free-ranging individuals.

The majority of acceleration–metabolic rate calibration studies in fish and elasmobranchs have been conducted in flume or swim-tunnel respirometers, where a propeller and motor push water through a swimming chamber at a controlled speed (e.g., Bouyoucos et al., 2017b; Clark et al., 2010; Gleiss et al., 2010; Mori et al., 2015; Wright et al., 2014; Yasuda et al., 2012). However, many of these studies have stated concerns that the forced swimming conditions in flumes impacted animal behavior or biased their measurements of metabolic rates (Bouyoucos et al., 2017b; Gleiss et al., 2010; Mori et al., 2015; Wright et al., 2014).

3.5.2 Temperature

In addition to swimming behavior, an equally important consideration in the development of laboratory calibrations for FMR estimation is how temperature will affect the relationship between metabolic rate and body acceleration. Metabolic rates scale exponentially with temperature in ectotherms, including most elasmobranch species (Carlson et al., 2004). In fact, temperature is arguably the most influential factor affecting metabolic rate in these animals, even above locomotion (Carlson et al., 2004; Lear et al., 2017). However, few calibration studies in fish have included the effects of temperature (Lowe, 2001; Wright et al., 2014; Lear et al., 2017), making it difficult to apply most calibrations to wild fish, which generally experience substantial temperature variation throughout diel or seasonal cycles. Lear et al. (2017) found that temperature affected the intercept of body acceleration–oxygen consumption calibration curves, but it did not change the slope of these relationships in blacktip, lemon, or nurse sharks. This suggests that if the temperature scaling rate of the standard metabolic rate (SMR) is known, generally expressed as a Q_{10} value, this can be used to correct body acceleration–oxygen consumption curves for changes in temperature in lieu of running full calibration experiments at a range of temperatures. However, even SMR Q_{10} values are known for relatively few elasmobranch species, largely due to the difficulty, expense, and logistical challenge of running respirometry experiments at a large range of temperatures (Carlson et al., 2004; Whitney et al., 2012). Regardless, it is essential to account for temperature effects when estimating metabolic rates in order to accurately apply calibrations to animals under field conditions.

3.5.3 Body Size

Another factor to consider when calibrating body acceleration and metabolism is the body size of individuals used in calibrations. Body size has been shown to be a significant predictor of metabolic rate in a range of species, including fish (Gillooly et al., 2001; Killen et al., 2010; Sims, 1996). However, scaling of metabolic rate with body size has not been investigated in many large-bodied elasmobranchs, and there can be substantial interspecific variability in body size-metabolic rate scaling factors (Killen et al., 2010; Payne et al., 2015). Additionally, body acceleration will also scale with body size (Gleiss et al., 2011b; Whitney et al., 2012), with larger individuals typically having lower ODBA than smaller individuals. This relationship has not been studied extensively, and no calibration study thus far has compared individuals of varying sizes, limiting the application of laboratory calibrations to the same size individuals in the field. Most respirometry studies use relatively small animals because of limitations imposed by respirometer size, making it difficult to assess how the relationship between body acceleration and oxygen consumption fluctuates as fish grow. New respirometry techniques that can accommodate larger animals (e.g., Payne et al., 2015) could allow for the issue of body size to be examined in calibration studies, detangling the effects of simultaneous variation in metabolic rate and body acceleration as size increases. This would greatly expand the potential of this technique to enable the study of field energetics in larger animals, providing information about energy requirements and ecosystem interactions of higher trophic levels.

3.5.4 Application to the Field

Ultimately, the goal of the calibration studies described here is to estimate FMRs by applying the laboratory calibrations to acceleration data collected from wild individuals. With careful consideration of respirometry procedures, these calibrations should be able to render fairly accurate estimates of FMRs based on activity and temperature measurements (Figure 3.5), although there will always be additional factors at play in the field that can alter metabolic rates and which can be difficult to account for in FMR estimates, including, for example, post-exhaustion recovery (anaerobic respiration), circadian rhythms, and specific dynamic action (SDA) (discussed in Whitney et al., 2012). Even with these drawbacks, these calibration experiments allow for the remote measure of absolute energy expenditure with relative accuracy and can provide a substantial amount of information on ecosystem interactions, trophic dynamics, and many other bioenergetics factors. This is particularly important when considering the declines of many shark populations worldwide and assessing the impact of upper trophic level predator removal on these ecosystems.

Direct calibrations of body acceleration and oxygen consumption represent the gold standard in the study of energetics, but performing these calibrations is logistically challenging and even impossible for some species that occur in remote areas, are not amenable to captivity, or are simply too large to fit into even the largest respirometer (but see Payne et al., 2015). Even in such cases, kinematic data can help estimate the energetic cost of locomotion, especially when combined with hydrodynamic models or wind-tunnel experiments. The rationale behind these techniques is that the drag that a shark (or other organism) has to overcome should be proportional to the mechanical power output, which in turn should be proportional to the chemical energy required (Davis and Weihs, 2007). Simple data, such as those on swimming speed or posture, can be used to drive models of varying complexity to arrive at estimates of metabolic power. These can then be used to explore questions ranging

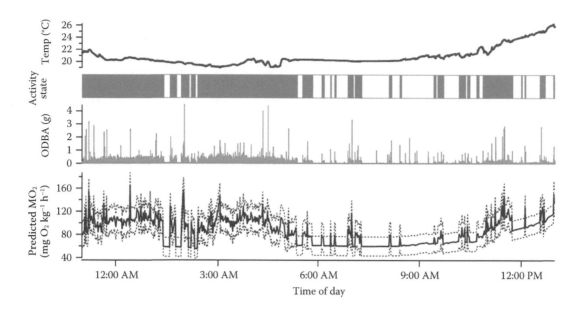

Figure 3.5 Example prediction of the oxygen consumption ($\dot{M}O_2$) of a free-swimming lemon shark in the field based on the laboratory overall dynamic body acceleration (ODBA) calibration. The standard metabolic rate (SMR) of the shark and the intercept of the calibration relationship are determined by the temperature during the deployment (red trace). Activity state is determined as active (gray bars) or resting (white bars), here done using k-means cluster analyses (see Sakamoto et al., 2009; Whitney et al., 2010). Resting intervals are assigned an SMR based on the mean temperature during the interval, and active periods are assigned a routine metabolic rate (RMR) based on ODBA and temperature. The predicted $\dot{M}O_2$ resulting from this method is shown in the bottom panel (blue) ± standard error of the estimate (SEE), with SEE calculated separately for active and resting periods. (From Lear, K.O. et al., *J. Exp. Biol.*, 220(3), 397–407, 2017.)

from optimal swim speeds (Iosilevskii and Papastamatiou, 2016), diving behavior (Iosilevskii et al., 2012), and metabolic physiology (Watanabe et al., 2015) to body composition (Gleiss et al., 2017b).

Additionally, accelerometers can render useful data on the relative energy expenditure of free-ranging sharks even in the absence of laboratory calibrations or wind-tunnel experiments, particularly when used concurrently with behavioral classification. Comparisons of levels of movement (ODBA or PDBA) between different behaviors yield information about the relative energetic cost of each type of behavior. These types of data have many different applications, including assessing the fitness implications of the behavioral decisions that animals make, determining why certain behavioral patterns may be displayed by individuals or populations, and deciphering how environmental changes or anthropogenic effects may impact the energetics and fitness of these animals.

3.6 HUMAN IMPACTS

Of over 1000 species of chondricthyans assessed by the International Union for Conservation of Nature (IUCN), approximately 25% are estimated to be threatened by an elevated risk of extinction, with some experiencing local extinction of adult populations (Dulvy et al., 2014). These

worldwide declines can be largely attributed to anthropogenic impacts, either directly through targeted fishing or bycatch or indirectly as a result of habitat loss and climate change. Although virtually all of the technologies described in this volume can document these effects in one way or another, biologging devices provide a unique ability to quantify fine-scale effects at the individual level and to bridge the gap between laboratory studies and the impacts on animals in their natural environment.

3.6.1 Fishing Effects

Sharks typically have relatively slow growth and low fecundity, making them more vulnerable to fishing pressure compared to many species of teleosts (Cortés, 1999, 2002; Dulvy et al., 2008; Hoenig and Gruber, 1990). While at-vessel mortality rates are relatively easily obtained, quantification of post-release mortality and sublethal behavioral effects are more difficult to measure, as unknown percentages of released animals later succumb to the physical or physiological effects of capture stress (Molina and Cooke, 2012; Skomal, 2007; Skomal and Bernal, 2010). Capture stress and extended recovery after sharks are released from fishing gear can also have more subtle impacts on animal condition and long-term fitness, which can be equally difficult to quantify (Donaldson et al., 2008; Lewin et al., 2006; Olla et al., 1997).

3.6.1.1 *Post-Release Mortality*

Quantifying post-release survival and recovery of sharks following interactions with fishing gear has historically been difficult and expensive. In the last decade, several studies have used acoustic telemetry (Afonso and Hazin, 2014; Kneebone et al., 2013) or satellite tags (e.g., Campana et al., 2009; Gallagher et al., 2014; Musyl et al., 2006; Sepulveda et al., 2015) to quantify diving behavior of pelagic sharks in order to infer mortality and sublethal effects of capture. Although these technologies remain the most feasible for some elasmobranchs, biologgers have recently been shown to be highly effective for several shark species.

Because ADLs provide data on the fine-scale depth and body movements of an animal at subsecond sampling intervals, they can provide clear information on what happens to a shark after capture and release (Whitney et al.,

2016, 2017). Unlike acoustic and satellite tags, which typically assume mortality based on depth information or the absence of detections from a tagged animal, ADLs produce an unambiguous indication of whether an individual lived or died (Figure 3.6) (Whitney et al., 2016). Predation events, another confounding variable in post-release studies, can also be visualized based on acceleration data from the event itself, as well as comparison of the tag position and movement frequency before and after the tag was consumed (Figure 3.7) (Lear and Whitney, 2016).

Another benefit of using ADLs to determine post-release outcomes is that, under the right conditions, these tags are much more cost effective than satellite tags and can generally be used several times if recovered. Lear and Whitney (2016) showed that for the purpose of studying post-release mortality in large coastal sharks in a commercial longline fishery, ADLs cost one-fourth to one-eighth the price of

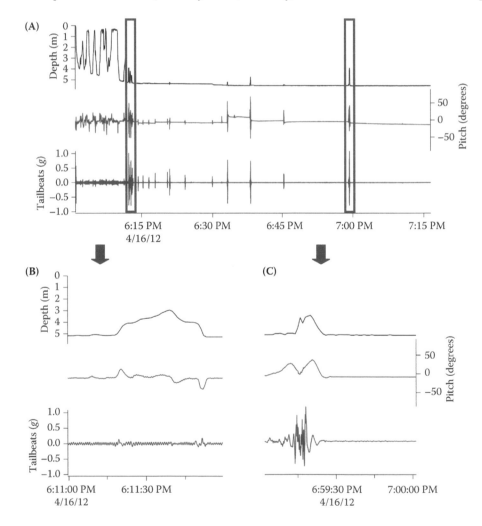

Figure 3.6 Death of a shark: depth, body pitch, and tailbeat acceleration of a dying blacktip shark after release from a recreational fishery. The animal begins swimming normally with an oscillating dive pattern for several minutes (A) before descending to the bottom, where it continues tailbeating and briefly recovers (B). Tailbeats then largely cease for the next several minutes as the animal lies on the bottom, with periodic attempts to recover with a few high-amplitude tailbeats every few minutes (C) until all activity ceases approximately one hour after release. (From Whitney, N.M. et al., *Fish. Res.*, 183, 210–221, 2016.)

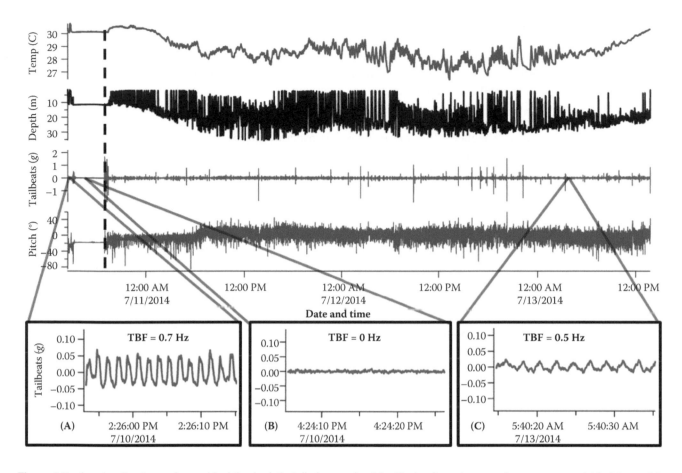

Figure 3.7 Acceleration traces from a blacktip shark that died approximately 40 min after release and was scavenged 4 hr later and the float package ingested (dashed line). (A) Tailbeat acceleration from when the blacktip was alive shows a tailbeat frequency at 0.7 Hz. (B) Tailbeat oscillations stop 40 min after release, showing that the shark died. (C) Tailbeats resume 4 hr later when the package was ingested, but average a lower frequency (0.5 Hz), indicating that the scavenging animal was larger than the blacktip. (From Lear, K.O. and Whitney, N.M., *Anim. Biotelem.*, 4(1), 12, 2016.)

using satellite tags, even after inclusion of tag search-and-recovery efforts. This cost efficiency increases the likelihood of obtaining the large sample sizes (generally >100 animals) thought to be required to achieve high-certainty estimates of post-release mortality rates in many species (Goodyear, 2002; Horodosky and Graves, 2005).

3.6.1.2 Quantifying Recovery Period

Although the effects of capture on post-release survivability have been examined in a few shark species, mortality represents only the most extreme negative effect of capture and release and is therefore an imperfect parameter for quantifying effects of fisheries interactions. Sublethal behavioral effects of capture stress have rarely been examined and then usually only at low resolution (e.g., by inferring recovery time from brief acoustic tracks) (Carey et al., 1990; Gurshin and Szedlmayer, 2004; Holts and Bedford, 1993; Sepulveda et al., 2004; Skomal, 2006), camera deployments (Skomal et al., 2007), or from broad-scale changes in swimming

depth measured from pop-up satellite archival tags (PSATs) (Campana et al., 2009; Moyes et al., 2006; Skomal, 2006). Alternatively, biologgers can provide high-resolution data on fine-scale behavioral changes following capture by calculating a number of different swimming metrics using a combination of kinematic and depth parameters (i.e., tailbeat frequency and acceleration amplitude, body pitch and roll, vertical velocity, and ODBA). Whitney et al. (2016) showed that blacktip sharks resume stable swimming activity after 10.5 ± 3.8 hr following capture. Recovery periods were calculated by comparing acceleration and depth-derived metrics with time post-release and fitting asymptotic nonlinear models to identify the point at which an animal's movements reach a constant level, indicating that it has regained normal swimming behavior. This information is crucial to determining fishery impacts in addition to a simple post-release alive/dead outcome, as sublethal behavioral effects of capture can have substantial impacts on an individual's success in its environment beyond the immediate post-capture period (Lewin et al., 2006).

3.6.1.3 Quantifying Behavior During Capture

In addition to studying post-release behavior, ADLs can also be used to study the behavior of sharks during capture in an effort to understand interspecific differences in susceptibility to capture stress. Bouyoucos et al. (2017c) recently equipped captive juvenile lemon sharks (*Negaprion brevirostris*) with ADLs during simulated longline capture and found a ninefold increase in the frequency of burst swimming events and a 57.7% increase in activity costs compared to unrestrained sharks. Gallagher et al. (2016) attached ADLs directly to fishing gear to show that wild caught blacktip sharks (*Carcharhinus limbatus*) fight more intensely than nurse sharks (*Ginglymostoma cirratum*) and tiger sharks (*Galeocerdo cuvier*) upon being hooked, and this corresponded to higher blood lactate values. Guida et al. (2016, 2017) have used time–depth recorders sampling every 4 sec and attached to gangions to document the behavioral responses of multiple species of shark caught on commercial longlines. This type of information is helpful in determining what drives the physiological disruption and post-release response in certain species and in informing management regulations for specific species and gear types.

3.6.2 Dive Tourism

Just as fisheries capture and release can cause sublethal injuries or stress to surviving elasmobranchs, comparatively innocuous activities such as dive tourism can also have an impact on the long-term health of affected animals (Gallagher et al., 2015; Semeniuk et al., 2009). Although the effects of dive tourism on behavior may be subtle, over time they can produce substantial impacts on the overall health and fitness of animals. Biologgers can be used to measure these subtle changes in behavior, providing information about the behavioral and physiological impacts that such activities can have on elasmobranchs. For example, Barnett et al. (2016) used acceleration transmitters coupled with respirometry to estimate that whitetip reef sharks (*Triaenodon obesus*) showed increased activity and a metabolic rate that was 6.37% higher on days of dive tourist provisioning compared to non-provisioning days. This increase in metabolic demand was not met by the food the animals consumed through provisioning; consequently, dive tour activities may change these animals' energy budgets and allocations. Such effects could potentially have a significant impact on animals involved in ecotourism activities.

3.6.3 Climate Change Effects

In addition to fishing and other direct impacts of human activities, elasmobranchs are also affected by many indirect impacts, the most pressing of which is the suite of environmental shifts driven by climate change. Because the majority of sharks and rays are ectotherms, water temperature regulates an entire spectrum of physiological and behavioral processes in these animals, including metabolic rates, muscle performance, and activity levels (Angilletta et al., 2002; Kingsolver, 2009). As such, the changing temperature regimes associated with climate change are particularly concerning and have the potential to substantially alter many different elements of these animals' physiology and life history, including energy budgets, behavioral strategies, daily activity patterns, and space use. Changing temperatures can drive range shifts in many species (Perry et al., 2005; Stebbing et al., 2002; Sunday et al., 2012), but they can also drive fine-scale changes in day-to-day behavior and activity. This is particularly relevant for species or life stages that are bound to certain habitats and therefore cannot shift their range to seek preferred temperatures (Chin et al., 2010), instead relying on behavioral or physiological plasticity to persist under warmer temperature regimes. Biologgers are well suited to study these more fine-scale effects of climate change.

Many behavioral and physiological processes in ectotherms scale with temperature according to a thermal performance curve, where performance gradually increases with temperature until it reaches a maximum at an animal's optimum temperature before quickly declining as temperatures reach lethal levels (Angilleta, 2006; Angilletta et al., 2002; Huey and Kingsolver, 1989). Many past studies examining thermal performance have done so in the laboratory, but accelerometers offer a tool to study thermal performance in wild fishes, as well. One of the major ways that temperature affects ectotherms is through regulating activity levels, with animals generally becoming more active as temperatures increase (Halsey et al., 2015). By using accelerometers to measure activity at a range of temperatures, generally as ODBA, other kinematic parameters, or percent of time spent active, this wild activity data can be plotted against temperature to form a thermal performance curve. This method has been used to examine thermal performance in several species of teleosts (Gannon et al., 2014; Payne et al., 2016b), as well as one elasmobranch so far, the estuary stingray (*Dasyatis fluviorum*) (Payne et al., 2016b), and has proven to be an effective method of measuring wild performance. These data can help to chart the likely increase in activity, determine how much time animals may spend above their optimum temperatures, evaluate whether certain environments will approach an animal's critical thermal maximum, and determine which types of habitats may become most important for these animals under rising temperature regimes.

Simultaneous logging of temperature and behavioral metrics, including activity and depth use, can also provide insight into how elasmobranchs shift their activity rhythms and behavioral patterns in response to changing temperatures and how climate change may impact these processes. Gleiss et al. (2017a) simultaneously logged acceleration, depth, and temperature in a riverine population of freshwater sawfish (*Pristis pristis*) over a large temperature range and determined that sawfish alter both activity patterns and depth use as temperature increases. In the future, understanding how large-scale changes in temperature modulate activity and behavior over time may help to forecast how increasing temperatures may impact an animal's energy use and requirements, foraging success, and susceptibility to predation, among other factors. It is also important to bear in mind that selection and physiological adaptation may alter thermal reaction norms over long time scales and could shift these measured thermal responses in the future.

3.7 LOGISTICS AND CHALLENGES

Biologgers are clearly powerful tools for addressing a number of questions in elasmobranch studies, but these tools also present a unique set of challenges and are not feasible in all cases. It should be noted that several of the research questions we discuss here can be addressed by transmitters in some cases, and satellite tags often log higher resolution data that can be downloaded if the tag is recovered. This means that the difference between conventional acoustic or satellite transmitters and the high-resolution biologgers discussed in this chapter is more a difference in degree than a difference in kind. However, because biologgers provide unique advantages, it is worth discussing some of their unique disadvantages and ways to overcome them.

3.7.1 Data Acquisition

Data acquisition is one of the biggest challenges in the use of biologgers with high sampling frequencies. Whereas some tags will record high-resolution data and transmit summary or binned values (see Chapter 8 in this volume), high-resolution data loggers store their data to memory and must be physically recovered and downloaded. The necessity of recovering loggers introduces a suite of challenges, particularly when these tags are deployed on large, free-ranging elasmobranchs that are capable of traveling large distances over short periods of time.

3.7.1.1 Logger Package Design, Deployment, and Recovery

Physically recovering a device from a free-swimming fish is very difficult in most cases, and these difficulties are amplified by the fact that most manufacturers of high-resolution biologgers do not provide a mechanism for tag release and recovery. Because logger attachment methods and position on the body are likely to be highly variable depending on the species studied and the questions being addressed, the generation of release-and-recovery mechanisms is highly case specific and therefore their development is usually left to the researcher.

Some biologging tags can be surgically implanted (Haulsee et al., 2016b) or fed to the animal to be later regurgitated (Jorgensen et al., 2015; Papastamatiou and Lowe,

2004), but these methods involve high uncertainty in the timing of tag recovery. For tags that are regurgitated or detach from the animal, several studies have employed negatively buoyant tags coupled with acoustic transmitters that sink to the bottom and are recovered using an underwater hydrophone, but this limits detection range to a few hundred meters (Whitney et al., 2012). In many cases, it is more effective to pair the logger with a float and VHF transmitter or satellite tag that can be detected at the surface from several kilometers away.

Whereas simple floats made of syntactic foam materials have been used to recover tag packages for decades in a variety of taxa (e.g., Blaylock, 1990; Marshall, 1998; Otani et al., 1998; Watanabe and Sato, 2008), few studies have designed logger packages specifically for use on elasmobranchs. Gleiss et al. (2009b) developed a float package made from epoxy and hollow glass microspheres which was used to attach loggers to free-swimming whale sharks using a custom-made, spring-loaded clamp system. This package was attached to the clamp with a galvanic timed release (GTR) made of dissimilar metals that corroded in seawater over time, allowing for the package to release from the animal. Upon release, the package floated to the surface and was detected and recovered via a VHF transmitter that was embedded in the package and could be detected from several kilometers away. This clamp-and-attachment system was later modified and used to attach these tags to free-swimming white sharks (*Carcharodon carcharias*) around cage-diving operations (Chapple et al., 2015).

Similar style packages have also been used to study tiger sharks (*Galeocerdo cuvier*) (Nakamura et al., 2011) and Greenland sharks (*Somniosus microcephalus*) (Watanabe et al., 2012); however, instead of a clamp attachment, these packages were attached through the dorsal fin of sharks using plastic cable ties with a mechanical timed release mechanism built in such that a burn wire was severed by an electric signal at a set time. Whitmore et al. (2016) developed a package for sharks that was designed to carry a G6a acceleration data logger (Cefas Technology Limited; Lowestoft, UK) that recorded acceleration over three axes at subsecond intervals, as well as depth and temperature (Figure 3.8) (Lear and Whitney, 2016). This package also used a VHF transmitter for recovery that was attached to the dorsal fin of sharks using a tether made from zip ties or monofilament; a GTR released the floats. This package was tested on multiple shark species with recovery success rates ranging from 86 to 100%, depending on species (Whitmore et al., 2016). The fine-scale data and reliable recovery rate allowed this package to be used for studies of shark post-release mortality and recovery from fishing (Whitney et al., 2016, 2017).

All of these studies had fairly low sample sizes, partially due to the expense and logistical difficulties in logger deployment and recovery; however, under the right conditions these packages can also be used to economically recover large numbers of tags over large distances.

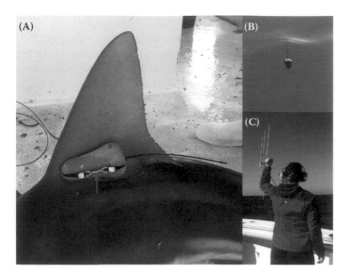

Figure 3.8 Accelerometer–VHF float package designed by Whitmore et al. (2016) and shown attached to the first dorsal fin of a 222-cm TL sandbar shark (*Carcharhinus plumbeus*) (A) using a tether with a built in galvanic timed release (GTR; red arrow). The GTR corrodes after a predetermined amount of time and the float package detaches from the fin, floating upright on the surface (B), so that it can be detected and tracked down using a VHF receiver (C) and recovered. (From Lear, K.O. and Whitney, N.M., *Anim. Biotelem.*, 4(1), 12, 2016.)

Lear and Whitney (2016) tested the same package designed by Whitmore et al. (2016) in a study of acute post-release mortality in longline-caught sharks and achieved an overall recovery rate of 97.4% out of 193 deployed loggers, even with tagged sharks moving an average of 31.3 ± 28.2 km per day, with a maximum displacement of 79 km/day. Although tag package deployment time was relatively short (22.9 ± 22.5 hr), the lower cost of tags and ability to reuse them allowed the work to be done at a fraction of the cost of using conventional satellite tags. It should be noted that the effectiveness of this high-volume approach to logger deployment and recovery is largely dependent on fishing success and works best for projects with high catch per unit effort (CPUE) that aim to tag many animals in a short time period. This allows multiple tags to be deployed at once, making search-and-recovery efforts much more economical.

3.7.1.2 Tag Effects on the Animal

All tagging studies have the potential to alter the movements and behaviors of tagged animals, but high-resolution biologgers often provide a unique opportunity to quantify those effects. Given this capability, biologging tags have shed new light on the effects of equipping animals with tags. First, the act of tagging itself can affect the behavior of the target animals, whether they are caught and tagged while restrained or tagged while free-swimming. Short-term behavioral effects of tag attachment often include increased activity directly following tagging, observed through higher ODBA,

tailbeat frequency, and swimming speeds (Bullock et al., 2015; Gleiss et al., 2009b; Lowe et al., 1998; Sundström and Gruber, 2002; Whitney et al., 2007, 2016). The magnitude of these behavioral effects may be higher with more extreme capture and tagging procedures, such as longline or rod and reel fishing, where animals may take several hours to recover (e.g., Whitney et al., 2016, 2017), whereas tagging free-swimming animals may cause only a slight and brief deviation from normal behavior (e.g., Gleiss et al., 2009b).

Tags can also physically affect their carriers through altering their hydrodynamics and swimming efficiency. Biologging tags are almost exclusively deployed via external attachment to facilitate logger recovery. In addition to potential tissue damage from the attachment method, externally attached tags can impact fish by increasing the amount of drag a fish experiences as it swims or by altering the buoyancy of the fish. Increased drag from externally attached tags has been connected with increased energetic costs and decreased swimming performance in juvenile scalloped hammerheads (Lowe et al., 1998), juvenile lemon sharks (Bouyoucos et al., 2017a), and young of the year cownose rays (Blaylock, 1990; Grusha and Patterson, 2005), including decreases in swimming speed and activity and increases in tailbeat or wingbeat frequency, leading to a substantial increase in the cost of transport.

The impacts of changes in buoyancy due to tagging have not been thoroughly examined for elasmobranchs, despite having the potential to alter swimming dynamics and energetic costs (Gleiss et al., 2015; Grusha and Patterson, 2005; Iosilevskii and Papastamatiou, 2016). Whereas teleosts have been shown to be able to mitigate small changes in buoyancy by altering swim-bladder volume (Lefrancois et al., 2001), elasmobranchs lack a swim bladder and are negatively buoyant. Instead, they generate vertical forces through lift produced by the body and fins, resulting in induced drag (Alexander, 1990). An animal carrying a negatively buoyant tag would generate greater induced drag, and it would therefore have a greater cost of transport, particularly at slow swimming speeds (Alexander, 1990; Gleiss et al., 2015). The opposite would be true for animals carrying a positively buoyant tag, although these tags may also affect an animal's diving behavior and ability to make gliding descents (Grusha and Patterson, 2005).

Effects of tagging, whether due to greater drag or altered buoyancy, are generally proportional to the tag–animal ratio of surface area (for drag) or weight (for buoyancy). Few studies have quantified the increases in drag that a tag imposes on a fish, and there are no widely accepted guidelines for surface area ratios for tagging. Quantifying the impact of changes in buoyancy is more straightforward, and general tagging guidelines recommend that tags remain less than 2% of the body weight of fish to limit impacts of increased weight (Winter, 1983). However, these calculations are often made based on the weight of the tag and animal in air, which can be misleading as it is the underwater buoyancy ratios of tag and animal that provide the physical basis for any impact on the animal's buoyancy. However, most biologging tags are close to neutral buoyancy; therefore, drag-related effects likely play a greater role. Many elasmobranchs are large-bodied animals, where tagging effects are likely limited because of high animal to tag size ratios, although these effects will be much more pronounced when using smaller species or juveniles, which have lower tag–animal size ratios (e.g., Bouyoucos et al., 2017a; Lowe et al., 1998) and must be carefully considered when designing tagging studies.

3.7.1.3 Transmitting vs. Archiving Data

The decision to use a transmitter or biologger must be made on a case-by-case basis and will depend on several factors that include the specific question being addressed; the size, location, and mobility of the study species; and the geography and logistics of the study site. Transmitting sensor data is highly advantageous for long-term studies and is especially feasible for species that (1) are large enough to carry a satellite tag, or (2) are likely to remain within range of an array of acoustic listening stations to which data can be continually transmitted. Even though transmitted acceleration data are extremely coarse compared to logged and unprocessed data (Whitney et al., 2012), they have been useful for long-term studies of post-release survivorship, activity, and metabolic rate in various species of teleosts (e.g., Curtis et al., 2014; Murchie et al., 2011; Payne et al., 2016b) and sharks (Barnett et al., 2016; Papastamatiou et al., 2015; Shipley et al., 2017). The fundamental problem with transmission technologies is that they lack the bandwidth to transmit data at a high enough resolution to adequately address questions regarding things such as fine-scale behaviors or biomechanics. This loss of information can be clearly seen in the degradation of dive information as high-resolution depth data from a biologger are down sampled to simulate the effect of several common transmitter regimes (Figure 3.9). Given the reduction in resolution and number of dives visible in these transmitted data, it is easy to see how similar downsampling of acceleration data would prohibit their use for behavioral studies. Moreover, transmitted data are inadvertently spatially biased because data are only recorded when sharks are near receivers, and imperfect receiver coverage can result in gaps in data, which is not the case with biologging tags. Thus, although transmitted data can be extremely valuable, they cannot be used to address the same suite of questions as high-frequency biologgers (for a comparison of sampling frequencies, see Whitney et al., 2012), and they also lack the resolution that allows logged datasets to be used for multiple studies (Payne et al., 2014).

Pop-up satellite archival tags (PSATs), as their name implies, are data loggers that summarize, compress, and then transmit their logged information. The physical recovery of satellite tags often provides access to much higher resolution archived data (<1-min sampling intervals) than those that are transmitted. Although recoveries of PSATs have typically been serendipitous, their commonality in some studies and the development of receivers to locate and recover tags are beginning to bridge the divide between satellite tag and biologger studies. Jorgensen et al. (2015) used a PSAT to aid in tag recovery, as well as to store depth and temperature information to complement a biologger that logged only acceleration data; however, storing information in memory is power intensive, and PSATs that are designed for long deployments still record at relatively low temporal resolution (>1 sec). Although they may be suitable for depth or temperature measurements, they are not at the scale needed to describe fine-scale movement data (i.e., 5 to 30 Hz acceleration).

3.7.2 Data Handling, Management, and Visualization

Modern biologgers can record multiple channels of information at high frequencies, and the resulting datasets reaching into the millions of data points present difficulties for analysis. Each study will have its own goals that guide how the data are analyzed, but there are common challenges that stretch across most studies involving high-resolution logger data. First, the raw data values provided by loggers are rarely informative in and of themselves, particularly for accelerometers and other three-dimensional movement sensors. Instead, the relevant information is usually encoded in the trends and changes in the sensor values. As such, analyses of motion sensor data largely borrow from the field of signal analysis, and common procedures include using filters or Fourier and wavelet transformations to separate convoluted signals and convert raw movement data into meaningful parameters such as frequency and amplitude (Brown et al., 2013). These techniques can separate out kinematic parameters that are informative for understanding basic swimming behavior, although classifying these data into specific types of behavior is a separate hurdle and often a difficult task. Using machine learning techniques such as k-means clustering, regression trees, and hidden Markov models has allowed researchers to autonomously identify and classify different segments of the data into behaviors (Brewster et al., 2018; Leos-Barajas et al., 2017; Sakamoto et al., 2009).

Additionally, beyond the challenge of learning the specialized techniques necessary to analyze logger data, manipulating these datasets requires the use of appropriate software such as IGOR Pro, MATLAB®, or R and typically involves writing specialized scripts for each program. For extremely large datasets, manipulating and analyzing them can exceed the amount of RAM found on many consumer level computers (<8 GB) and requires splitting datasets, using computers with more memory, or using more advanced software.

The large and multidimensional nature of biologger datasets also presents challenges in effectively visualizing them. Simple methods involve plotting sensor values against time, but only small subsets of data can be effectively displayed

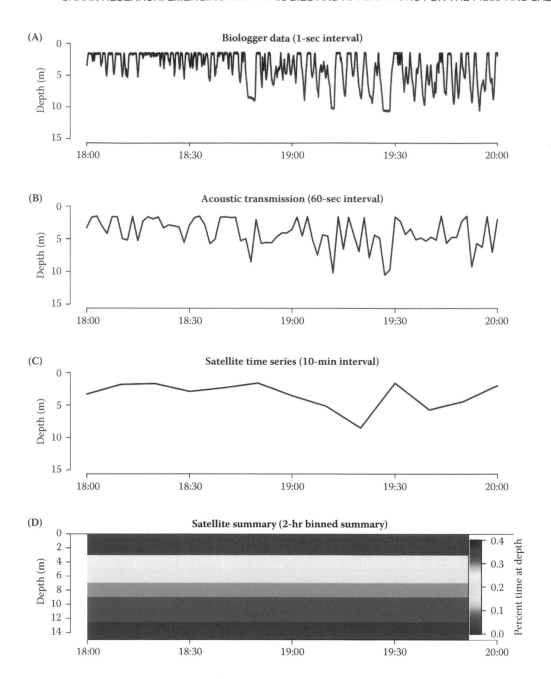

Figure 3.9 Sample depth data logged at 1 Hz from a blacktip shark in Florida over a 2-hr period (data from Whitney et al., 2016). The shark spends time at the surface, with *V* dives to depth (A). When these data are downsampled to simulate data transmitted by an acoustic transmitter (60-sec nominal delay), much of the fine-scale patterns in dive behavior are lost (B). When data are downsampled to simulate transmission via satellite using time series (C) or compressed into 2-hr time bins (D), patterns in vertical habitat use are completely lost. Note that 2-hr time bins represent the finest temporal bin available (0, 2, 4, 6, 8, 10, 15, and 20 hr) for miniPAT transmitters, and many studies use more coarse bin settings.

in this manner. Plotting summarized metrics over time can help to reveal changes in swimming kinematics or behavior over longer periods. Additionally, time series graphs are usually limited to viewing data from a single individual over a relatively small time window and can potentially lead to spurious conclusions (Ropert-Coudert and Wilson, 2004). Three-dimensional (3D) graphs have also been developed for visualizing these data (Grundy et al., 2009) but are often difficult to interpret, and using a two-dimensional graph with a third variable applied as a color scale is often more intuitive (see depth trace in Figure 3.2). Additionally, converting large datasets from XY scatter plots to XY density plots can aid in identifying and visualizing general patterns. This concept has been expanded to 3D spherical plots to aid in identification of general trends in 3D acceleration or magnetic data (Wilson et al., 2016).

From datasets and analyses that are heavily quantitative, results are often presented qualitatively, with authors describing patterns identified visually (Ropert-Coudert and Wilson, 2004). This is partly because each logger deployment contains a wealth of information, often with complex patterns, and generalizing these patterns quantitatively is difficult and often violates assumptions of statistical models. Yet, quantitative analysis is needed to generate commonalities among individuals in order to draw conclusions at the population or species level. To do this, high-resolution time series datasets are often summarized *post hoc* into time bins to form statistical units; however, this is problematic because time bin selection is arbitrary and could be on relatively short (1 min) or long (daily) time scales. This is important because bin size impacts sample size and thus the power of traditional statistical tests. Additionally, these datasets violate assumptions of traditional statistical models, as the data points are repeated measures, and all the measurements from a deployment nested within an individual are non-independent and therefore pseudoreplicated (Bolker et al., 2009). Most biologging studies use statistical models that allow for random effects in order to account for this non-independence. Yet, patterns identified are often nonlinear and have pushed researchers into using generalized additive mixed models (GAMMs) to statistically identify patterns or using nonlinear mixed models (NLMMs) to parameterize coefficients to nonlinear models (Gleiss et al., 2013; Papastamatiou et al., 2015; Whitney et al., 2016).

3.8 SUMMARY AND CONCLUSIONS

Despite their inherent challenges and the need for validation or calibration studies, high-resolution biologgers are increasingly being used to advance our knowledge of shark biology and understand the threats to shark populations. There are already far more sensors available than are typically used in a given biologging study, and as our ability to recover and process those data improves we will undoubtedly see more of these sensors utilized in new and exciting ways.

The transmission bandwidth via acoustic or satellite telemetry is likely to remain limited, but the use of various transmitters to provide the location of, and thus help us find and retrieve, biologging devices has been shown to have great potential even for large, highly mobile species (Lear and Whitney, 2016). The incorporation of Argos transmitters into biologging packages promises to extend the time over which they can be deployed and the distances over which they can be recovered (Jorgensen et al., 2015).

Although these advances are likely to help us apply biologgers to larger, more mobile animals, at the other end of the spectrum we are developing ever-smaller packages to study species previously considered too small to tag. For example, the same biologger technology (logging triaxial acceleration, depth, and temperature) that was included in a float package

122 mm long and 170 g in air by Whitmore et al. (2016) has now been incorporated into a package 80 mm long and 70 g in air that is being attached to blacknose sharks (*Carcharhinus acronotus*) and red drum (*Sciaenops ocellatus*) with no reduction of battery or memory (Whitney, unpublished data).

Further potential to expand this field is provided not only by advances in technology but also by a reduction in the cost of that technology. As these sensors and devices are incorporated into more consumer electronics, their components become commodities that are available at greatly reduced prices. This, in combination with increased use of loggers and a proliferation of manufacturers, has led to a substantial reduction in the cost of these devices over the past several years.

Given the demonstrated ability to retrieve surgically implanted loggers after long periods (Haulsee et al., 2016b), one can also envision future studies in which high-resolution data are collected over extended periods without tag detachment. Instead, inexpensive long-term loggers may be deployed *en masse*, relying on serendipitous recapture or logger recovery to reveal fine-scale information about animal behavior, energetics, or social interactions over periods of many years. This model has been used with identification tags for decades in the National Marine Fisheries Service (NMFS) Cooperative Shark Tagging Program, but has also been used at a smaller scale in biologging studies by researchers studying tuna (Boustany et al., 2010; Domeier et al., 2005), white sea bass (Aalbers and Sepulveda, 2015), cod (Van Der Kooji et al., 2007), and eels (Righton et al., 2016).

The tremendous diversity of size, behavior, and life history of elasmobranchs means that any and all technical advances are likely to be applicable to one or more of these species, and future biologging studies in this group will be extensive.

ACKNOWLEDGMENTS

Much of the work reported in this chapter was supported by grants to ACG and NMW from the National Science Foundation, as well as grants to NMW from the NOAA Cooperative Research Program, the NOAA Bycatch Reduction Program, the Waitt Foundation, and the National Geographic Society. KOL is supported by the Forrest Research Foundation and the Australian Government Research Training Program. ACG is supported by an Australian Research Council Discovery Early Career Research Award (project number 150100321).

REFERENCES

Aalbers SA, Sepulveda CA (2015) Seasonal movement patterns and temperature profiles of adult white seabass (*Atractoscion nobilis*) off California. *Fish Bull* 113(1):1–15.

Afonso AS, Hazin FH (2014) Post-release survival and behavior and exposure to fisheries in juvenile tiger sharks, *Galeocerdo cuvier*, from the South Atlantic. *J Exp Mar Biol Ecol* 454:55–62.

Alexander RM (1990) Size, speed and buoyancy adaptations in aquatic animals. *Am Zool* 30(1):189–196.

Angilletta MJ (2006) Estimating and comparing thermal performance curves. *J Therm Biol* 31(7):541–545.

Angilletta MJ, Niewiarowski PH, Navas CA (2002) The evolution of thermal physiology in ectotherms. *J Therm Biol* 27(4):249–268.

Barnett A, Payne NL, Semmens JM, Fitzpatrick R (2016) Ecotourism increases the field metabolic rate of whitetip reef sharks. *Biol Conserv* 199:132–136.

Blaylock RA (1990) Effects of external biotelemetry transmitters on behavior of the cownose ray *Rhinoptera bonasus* (Mitchill 1815). *J Exp Mar Biol Ecol* 141(2–3):213–220.

Bolker BM, Brooks ME, Clark CJ, Geange SW, Poulsen JR, Stevens MHH, White J-SS (2009) Generalized linear mixed models: a practical guide for ecology and evolution. *Trends Ecol Evol* 24(3):127–135.

Bone Q, Roberts B (1969) The density of elasmobranchs. *J Mar Biol Assoc UK* 49(4):913–937.

Boustany AM, Matteson R, Castleton M, Farwell C, Block BA (2010) Movements of Pacific bluefin tuna (*Thunnus orientalis*) in the Eastern North Pacific revealed with archival tags. *Prog Oceanogr* 86(1):94–104.

Bouyoucos I, Suski C, Mandelman J, Brooks E (2017a) Effect of weight and frontal area of external telemetry packages on the kinematics, activity levels and swimming performance of small-bodied sharks. *J Fish Biol* 90(5):2097–2110.

Bouyoucos IA, Montgomery DW, Brownscombe JW, Cooke SJ, Suski CD, Mandelman JW, Brooks EJ (2017b) Swimming speeds and metabolic rates of semi-captive juvenile lemon sharks (*Negaprion brevirostris*) estimated with acceleration biologgers. *J Exp Mar Biol Ecol* 486:245–254.

Bouyoucos IA, Suski CD, Mandelman JW, Brooks EJ (2017c) The energetic, physiological, and behavioral response of lemon sharks (*Negaprion brevirostris*) to simulated longline capture. *Comp Biochem Physiol Part A Mol Integr Physiol* 207:65–72.

Boyd IL, Kato A, Ropert-Coudert Y (2004) Bio-logging science: sensing beyond the boundaries. *Mem Natl Inst Polar Res* 58:1–14.

Brewster LR, Dale JJ, Guttridge TL, Gruber SH, Elliott M, Cowx IG, Hansell AC, et al. (2018) Development and application of a machine learning algorithm for classification of elasmobranch behavior from accelerometry data. *Mar Biol* 165:62.

Broell F, Noda T, Wright S, Domenici P, Steffensen JF, Auclair JP, Taggart CT (2013) Accelerometer tags: detecting and identifying activities in fish and the effect of sampling frequency. *J Exp Biol* 216(7):1255–1264.

Brown DD, Kays R, Wikelski M, Wilson R, Klimley AP (2013) Observing the unwatchable through acceleration logging of animal behavior. *Anim Biotelem* 1(1):20.

Brown JH, Sibly RM (2006) Life-history evolution under a production constraint. *Proc Natl Acad Sci* 103(47):17595–17599.

Bullock R, Guttridge T, Cowx I, Elliott M, Gruber S (2015) The behaviour and recovery of juvenile lemon sharks *Negaprion brevirostris* in response to external accelerometer tag attachment. *J Fish Biol* 87(6):1342–1354.

Butler PJ, Green JA, Boyd I, Speakman J (2004) Measuring metabolic rate in the field: the pros and cons of the doubly labelled water and heart rate methods. *Funct Ecol* 18(2):168–183.

Campana SE, Joyce W, Manning MJ (2009) Bycatch and discard mortality in commercially caught blue sharks *Prionace glauca* assessed using archival satellite pop-up tags. *Mar Ecol Prog Ser* 387:241–253.

Carey F, Scharold J, Kalmijn AJ (1990) Movements of blue sharks (*Prionace glauca*) in depth and course. *Mar Biol* 106(3):329–342.

Carlson JK, Goldman KJ, Lowe CG (2004) Metabolism, energetic demand, and endothermy. In: Carrier J, Musick J, Heithaus M (eds) *Biology of Sharks and Their Relatives*. CRC Press, Boca Raton, FL, pp 269–286.

Cartamil D, Wegner N, Kacev D, Ben-aderet N, Kohin S, Graham JB (2010) Movement patterns and nursery habitat of the juvenile thresher shark, *Alopias vulpinus*, in the Southern California Bight. *Mar Ecol Prog Ser* 404:249–258.

Chapple TK, Gleiss AC, Jewell OJ, Wikelski M, Block BA (2015) Tracking sharks without teeth: a non-invasive rigid tag attachment for large predatory sharks. *Anim Biotelem* 3(1):14.

Chin A, Kyne PM, Walker TI, McAuley R (2010) An integrated risk assessment for climate change: analysing the vulnerability of sharks and rays on Australia's Great Barrier Reef. *Global Change Biol* 16(7):1936–1953.

Clark TD, Sandblom E, Hinch S, Patterson D, Frappell P, Farrell A (2010) Simultaneous biologging of heart rate and acceleration, and their relationships with energy expenditure in free-swimming sockeye salmon (*Oncorhynchus nerka*). *J Comp Physiol B Biochem Syst Environ Physiol* 180(5):673–684.

Coffey DM, Holland KN (2015) First autonomous recording of *in situ* dissolved oxygen from free-ranging fish. *Anim Biotelem* 3(1):47.

Cooke SJ, Hinch SG, Wikelski M, Andrews RD, Kuchel LJ, Wolcott TG, Butler PJ (2004a) Biotelemetry: a mechanistic approach to ecology. *Trends Ecol Evol* 19(6):334–343.

Cooke SJ, Thorstad EB, Hinch SG (2004b) Activity and energetics of free-swimming fish: insights from electromyogram telemetry. *Fish Fish* 5(1):21–52.

Corner E, Denton E, Forster G (1969) On the buoyancy of some deep-sea sharks. *Proc R Soc Lond B Biol Sci* 171(1025):415–429.

Cortés E (1999) A stochastic stage-based population model of the sandbar shark in the western North Atlantic. In: Musick JA (ed) *Life in the Slow Lane: Ecology and Conservation of Long-Lived Marine Animals*. American Fisheries Society, Bethesda, MD, pp 115–136.

Cortes E (2002) Incorporating uncertainty into demographic modeling: application to shark populations and their conservation. *Conserv Biol* 16(4):1048–1062.

Curtis JM (2014) Discard Mortality, Recruitment, and Connectivity of Red Snapper (*Lutjanus campechanus*) in the Northern Gulf of Mexico, doctoral thesis, Texas A&M University-Corpus Christi.

Davis RW, Weihs D (2007) Locomotion in diving elephant seals: physical and physiological constraints. *Phil Tran R Soc Lond B Biol Sci* 362(1487):2141–2150.

Davis RW, Fuiman LA, Williams TM, Collier SO, et al. (1999) Hunting behavior of a marine mammal beneath the Antarctic fast ice. *Science* 283:993–996.

Domeier ML, Kiefer D, Nasby-Lucas N, Wagschal A, O'Brien F (2005) Tracking Pacific bluefin tuna (*Thunnus thynnus orientalis*) in the northeastern Pacific with an automated algorithm that estimates latitude by matching sea-surface-temperature data from satellites with temperature data from tags on fish. *Fish Bull* 103(2):292–306.

Donaldson MR, Arlinghaus R, Hanson KC, Cooke SJ (2008) Enhancing catch-and-release science with biotelemetry. *Fish Fish* 9(1):79–105.

Dulvy NK, Baum JK, Clarke S, Compagno LJV, Cortés E, Domingo A, Fordham S, et al. (2008) You can swim but you can't hide: the global status and conservation of oceanic pelagic sharks and rays. *Aquat Cons Mar Freshw Ecosyst* 18(5):459–482.

Dulvy NK, Fowler SL, Musick JA, Cavanagh RD, Kyne PM, Harrison LR, John K Carlson JK, et al. (2014) Extinction risk and conservation of the world's sharks and rays. *eLIFE* 3:e00590.

Eckert S, Eckert K (1986) Harnessing leatherbacks. *Mar Turtle Newsl* 37:1–3.

Floryan D, Van Buren T, Smits AJ (2017) Forces and energetics of intermittent swimming. *Acta Mech Sin* 33(4):725–732.

Gallagher A, Serafy J, Cooke S, Hammerschlag N (2014) Physiological stress response, reflex impairment, and survival of five sympatric shark species following experimental capture and release. *Mar Ecol Prog Ser* 496:207–218.

Gallagher AJ, Vianna GM, Papastamatiou YP, Macdonald C, Guttridge TL, Hammerschlag N (2015) Biological effects, conservation potential, and research priorities of shark diving tourism. *Biol Conserv* 184:365–379.

Gallagher AJ, Staaterman ER, Cooke SJ, Hammerschlag N (2016) Behavioural responses to fisheries capture among sharks caught using experimental fishery gear. *Can J Fish Aquat Sci* 74(1):1–7.

Gannon R, Taylor MD, Suthers IM, Gray CA, van der Meulen DE, Smith JA, Payne NL (2014) Thermal limitation of performance and biogeography in a free-ranging ectotherm: insights from accelerometry. *J Exp Biol* 217(17):3033–3037.

Gillooly JF, Brown JH, West GB, Savage VM, Charnov EL (2001) Effects of size and temperature on metabolic rate. *Science* 293(5538):2248–2251.

Gleiss AC, Gruber SH, Wilson RP (2009a) Multi-channel data-logging: towards determination of behaviour and metabolic rate in free-swimming sharks. In: Nielsen JL, Arrizabalaga H, Fragoso N, Hobday A, Lutcavage M, Silbert J (eds) *Tagging and Tracking of Marine Animals with Electronic Devices.* Springer, Dordrecht, pp 211–228.

Gleiss AC, Norman B, Liebsch N, Francis C, Wilson RP (2009b) A new prospect for tagging large free-swimming sharks with motion-sensitive data-loggers. *Fish Res* 97(1):11–16.

Gleiss AC, Dale JJ, Holland KN, Wilson RP (2010) Accelerating estimates of activity-specific metabolic rate in fishes: testing the applicability of acceleration data-loggers. *J Exp Mar Biol Ecol* 385(1):85–91.

Gleiss AC, Norman B, Wilson RP (2011a) Moved by that sinking feeling: variable diving geometry underlies movement strategies in whale sharks. *Funct Ecol* 25(3):595–607.

Gleiss AC, Wilson RP, Shepard EL (2011b) Making overall dynamic body acceleration work: on the theory of acceleration as a proxy for energy expenditure. *Meth Ecol Evol* 2(1):23–33.

Gleiss AC, Jorgensen SJ, Liebsch N, Sala JE, Norman B, Hays GC, Quintana F, et al. (2011c) Convergent evolution in locomotory patterns of flying and swimming animals. *Nat Commun* 2:352.

Gleiss AC, Wright S, Liebsch N, Wilson RP, Norman B (2013) Contrasting diel patterns in vertical movement and locomotor activity of whale sharks at Ningaloo Reef. *Mar Biol* 160(11):2981–2992.

Gleiss AC, Potvin J, Keleher JJ, Whitty JM, Morgan DL, Goldbogen JA (2015) Mechanical challenges to freshwater residency in sharks and rays. *J Exp Biol* 218(7):1099–1110.

Gleiss AC, Morgan DL, Whitty JM, Keleher JJ, Fossette S, Hays GC (2017a) Are vertical migrations driven by circadian behaviour? Decoupling of activity and depth use in a large riverine elasmobranch, the freshwater sawfish (*Pristis pristis*). *Hydrobiologia* 787(1):181–191.

Gleiss AC, Potvin J, Goldbogen JA (2017b) Physical trade-offs shape the evolution of buoyancy control in sharks. *Proc R Soc Lond B Biol Sci* 284:20171345.

Goodyear CP (2002) Factors affecting robust estimates of the catch-and-release mortality using pop-off tag technology. In: Lucy JA, Studholme AL (eds) *Catch and Release in Marine Recreational Fisheries.* American Fisheries Society, Bethesda, MD, pp 172–179.

Gruber SH, Nelson DR, Morrissey JF (1988). Patterns of activity and space utilization of lemon sharks, *Negaprion brevirostris*, in a shallow Bahamian lagoon. *Bull Mar Sci* 43(1):61–76

Grundy E, Jones MW, Laramee RS, Wilson RP, Shepard EL (2009) Visualisation of sensor data from animal movement. *Comput Graph Forum*, 28(3):815–822.

Grusha DS, Patterson MR (2005) Quantification of drag and lift imposed by pop-up satellite archival tags and estimation of the metabolic cost to cownose rays (*Rhinoptera bonasus*). *Fish Bull* 103(1):63–70.

Guida L, Walker TI, Reina RD (2016) Temperature insensitivity and behavioural reduction of the physiological stress response to longline capture by the gummy shark, *Mustelus antarcticus*. *PLoS ONE* 11(2):e0148829.

Guida L, Dapp DR, Huveneers CP, Walker TI, Reina RD (2017) Evaluating time-depth recorders as a tool to measure the behaviour of sharks captured on longlines. *J Exp Mar Biol Ecol* 497:120–126.

Gurshin C, Szedlmayer S (2004) Short-term survival and movements of Atlantic sharpnose sharks captured by hook-and-line in the north-east Gulf of Mexico. *J Fish Biol* 65(4):973–986.

Guttridge TL, Gruber SH, Krause J, Sims DW (2010) Novel acoustic technology for studying free-ranging shark social behaviour by recording individuals' interactions. *PLoS ONE* 5(2):e9324.

Halsey L, Shepard E, Quintana F, Laich AG, Green J, Wilson R (2009) The relationship between oxygen consumption and body acceleration in a range of species. *Comp Biochem Physiol A Mol Integr Physiol* 152(2):197–202.

Halsey LG, Shepard EL, Wilson RP (2011) Assessing the development and application of the accelerometry technique for estimating energy expenditure. *Comp Biochem Physiol A Mol Integr Physiol* 158(3):305–314.

Halsey LG, Matthews P, Rezende E, Chauvaud L, Robson AA (2015) The interactions between temperature and activity levels in driving metabolic rate: theory, with empirical validation from contrasting ectotherms. *Oecologia* 177(4):1117–1129.

Haulsee DE, Fox DA, Breece MW, Brown LM, Kneebone J, Skomal GB, Oliver MJ (2016a) Social network analysis reveals potential fission–fusion behavior in a shark. *Sci Rep* 6:34087.

Haulsee DE, Fox DA, Breece MW, Clauss TM, Oliver MJ (2016b) Implantation and recovery of long-term archival transceivers in a migratory shark with high site fidelity. *PLoS ONE* 11(2):e0148617.

Heithaus MR, Marshall GJ, Buhleier BM, Dill LM (2001) Employing Crittercam to study habitat use and behavior of large sharks. *Mar Ecol Prog Ser* 209:307–310.

Heithaus MR, McLash JJ, Frid A, Dill LM, Marshall GJ (2002) Novel insights into green sea turtle behaviour using animal-borne video cameras. *J Mar Biol Assoc UK* 82(6):1049–1050.

Hoenig JM, Gruber SH (1990) Life-history patterns in the elasmobranchs: implications for fisheries management. In: Pratt HL, Gruber SH, Taniuchi T (eds) *Elasmobranchs as Living Resources: Advances in the Biology, Ecology, Systematics, and the Status of the Fisheries*, NOAA Technical Report 90. National Oceanic and Atmospheric Administration, Washington, DC, pp 1–16.

Holland KN, Meyer CG, Dagorn LC (2009) Inter-animal telemetry: results from first deployment of acoustic 'business card' tags. *Endangered Species Res* 10:287–293.

Holts D, Bedford DW (1993) Horizontal and vertical movements of the shortfin mako shark, *Isurus oxyrinchus*, in the Southern California Bight. *Mar Freshwater Res* 44(6):901–909.

Hooker SK, Biuw M, McConnell BJ, Miller PJ, Sparling CE (2007) Bio-logging science: logging and relaying physical and biological data using animal-attached tags. *Deep Sea Res II Top Stud Oceanogr* 54(3):177–182.

Horie J, Mitamura H, Ina Y, Mashino Y, Noda T, Moriya K, Arai N, Sasakura T (2017) Development of a method for classifying and transmitting high-resolution feeding behavior of fish using an acceleration pinger. *Anim Biotelem* 5(1):12.

Horodysky AZ, Graves JE (2005) Application of pop-up satellite archival tag technology to estimate postrelease survival of white marlin (*Tetrapturus albidus*) caught on circle and straight-shank ('J') hooks in the western North Atlantic recreational fish. *Fish Bull* 103(1):84–96.

Huey RB, Kingsolver JG (1989) Evolution of thermal sensitivity of ectotherm performance. *Trends Ecol Evol* 4(5):131–135.

Hussey NE, Wintner SP, Dudley SF, Cliff G, Cocks DT, MacNeil MA (2010) Maternal investment and size-specific reproductive output in carcharhinid sharks. *J Anim Ecol* 79(1):184–193.

Iosilevskii G, Papastamatiou YP (2016) Relations between morphology, buoyancy and energetics of requiem sharks. *Open Sci* 3(10):160406.

Iosilevskii G, Papastamatiou YP, Meyer CG, Holland KN (2012) Energetics of the yo-yo dives of predatory sharks. *J Theor Biol* 294:172–181.

Iwata T, Sakamoto KQ, Takahashi A, Edwards EWJ, Staniland IJ, Trathan PN, Naito Y (2012) Using a mandible accelerometer to study fine-scale foraging behavior of free-ranging Antarctic fur seals. *Mar Mammal Sci* 28(2):345–357.

Jacoby DM, Brooks EJ, Croft DP, Sims DW (2012) Developing a deeper understanding of animal movements and spatial dynamics through novel application of network analyses. *Meth Ecol Evol* 3(3):574–583.

Johnson MP, Tyack PL (2003) A digital acoustic recording tag for measuring the response of wild marine mammals to sound. *IEEE J Oceanic Eng* 28(1):3–12.

Jorgensen SJ, Gleiss AC, Kanive PE, Chapple TK, Anderson SD, Ezcurra JM, Brandt WT, Block BA (2015) In the belly of the beast: resolving stomach tag data to link temperature, acceleration and feeding in white sharks (*Carcharodon carcharias*). *Anim Biotelem* 1(3):1–10.

Kawabata Y, Noda T, Nakashima Y, Nanami A, Sato T, Takebe T, Mitamura H, et al. (2014) Use of a gyroscope/accelerometer data logger to identify alternative feeding behaviours in fish. *J Exp Biol* 217(18):3204–3208.

Killen SS, Atkinson D, Glazier DS (2010) The intraspecific scaling of metabolic rate with body mass in fishes depends on lifestyle and temperature. *Ecol Lett* 13(2):184–193.

Kingsolver JG (2009) The well-temperatured biologist. *Am Nat* 174(6):755–768.

Kneebone J, Chisholm J, Bernal D, Skomal G (2013) The physiological effects of capture stress, recovery, and post-release survivorship of juvenile sand tigers (*Carcharias taurus*) caught on rod and reel. *Fish Res* 147:103–114.

Kooyman GL (1965) Techniques used in measuring diving capacities of Weddell seals. *Polar Rec* 12(79):391–394.

Kooyman GL (1966) Maximum diving capacities of the Weddell seal, *Leptonychotes weddelli*. *Science* 151(3717):1553–1554.

Kooyman GL (2004) Genesis and evolution of bio-logging devices: 1963–2002. *Mem Natl Inst Polar Res Spec Issue* 58:15–22.

Lawson GL, Hückstädt LA, Lavery AC, Jaffré FM, Wiebe PH, Fincke JR, Crocker DE, Costa DP (2015) Development of an animal-borne "sonar tag" for quantifying prey availability: test deployments on northern elephant seals. *Anim Biotelem* 3(1):22.

Lear KO, Whitney NM (2016) Bringing data to the surface: recovering data loggers for large sample sizes from marine vertebrates. *Anim Biotelem* 4(1):12.

Lear KO, Whitney NM, Brewster LR, Morris JJ, Hueter RE, Gleiss AC (2017) Correlations of metabolic rate and body acceleration in three species of coastal sharks under contrasting temperature regimes. *J Exp Biol* 220(3):397–407.

Lefrancois C, Odion M, Claireaux G (2001) An experimental and theoretical analysis of the effect of added weight on the energetics and hydrostatic function of the swimbladder of European sea bass (*Dicentrarchus labrax*). *Mar Biol* 139(1):13–17.

Leos-Barajas V, Photopoulou T, Langrock R, Patterson TA, Watanabe YY, Murgatroyd M, Papastamatiou YP (2017) Analysis of animal accelerometer data using hidden Markov models. *Meth Ecol Evol* 8(2):161–173.

Lewin W-C, Arlinghaus R, Mehner T (2006) Documented and potential biological impacts of recreational fishing: insights for management and conservation. *Rev Fish Sci* 14(4):305–367.

Lowe C (2001) Metabolic rates of juvenile scalloped hammerhead sharks (*Sphyrna lewini*). *Mar Biol* 139(3):447–453.

Lowe CG, Holland KN, Wolcott TG (1998) A new acoustic tailbeat transmitter for fishes. *Fish Res* 36(2):275–283.

Luo J, Ault JS, Larkin MF, Barbieri LR (2008) Salinity measurements from pop-up archival transmitting (PAT) tags and their application to geolocation estimation for Atlantic tarpon. *Mar Ecol Prog Ser* 357:101–109.

Makiguchi Y, Sugie Y, Kojima T, Naito Y (2012) Detection of feeding behaviour in common carp *Cyprinus carpio* by using an acceleration data logger to identify mandibular movement. *J Fish Biol* 80(6):2345–2356.

Marshall GJ (1998) Crittercam: an animal-borne imaging and data logging system. *Mar Technol Soc J* 32(1):11.

Marshall G, Bakhtiari M, Shepard M, Tweedy III J, Rasch D, Abernathy K, Joliff B, et al. (2007) An advanced solid-state animal-borne video and environmental data-logging device ('Crittercam') for marine research. *Mar Technol Soc J* 41(2):31–38.

Metcalfe J, Fulcher M, Clarke S, Challiss M, Hetherington S (2009) An archival tag for monitoring key behaviours (feeding and spawning) in fish In: Nielsen JL, Arrizabalaga H, Fragoso N, Hobday A, Lutcavage M, Silbert J (eds) *Tagging and Tracking of Marine Animals with Electronic Devices*. Springer, Dordrecht, pp 243–254.

Meyer CG, Holland KN (2012) Autonomous measurement of ingestion and digestion processes in free-swimming sharks. *J Exp Biol* 215(21):3681 3684.

Meyer CG, Burgess WC, Papastamatiou YP, Holland KN (2007) Use of an implanted sound recording device (Bioacoustic Probe) to document the acoustic environment of a blacktip reef shark (*Carcharhinus melanopterus*). *Aquat Living Resour* 20(4):291–298.

Molina JM, Cooke SJ (2012) Trends in shark bycatch research: current status and research needs. *Rev Fish Biol Fish* 22(3):719–737.

Mori T, Miyata N, Aoyama J, Niizuma Y, Sato K (2015) Estimation of metabolic rate from activity measured by recorders deployed on Japanese sea bass *Lateolabrax japonicus*. *Fish Sci* 81(5):871–882.

Mourier J, Bass NC, Guttridge TL, Day J, Brown C (2017) Does detection range matter for inferring social networks in a benthic shark using acoustic telemetry? *R Soc Open Sci* 4(9):170485.

Muramoto H, Ogawa M, Suzuki M, Naito Y (2004) Little Leonardo digital data logger: its past, present and future role in bio-logging science. *Mem Natl Inst Polar Res Spec Issue* 58:196–202.

Murchie KJ, Cooke SJ, Danylchuk AJ, Suski CD (2011) Estimates of field activity and metabolic rates of bonefish (*Albula vulpes*) in coastal marine habitats using acoustic tri-axial accelerometer transmitters and intermittent-flow respirometry. *J Exp Mar Biol Ecol* 396(2):147–155.

Musyl MK, Moyes CD, Fragoso N, Brill RW (2006) Predicting postrelease survival in large pelagic fish. *Trans Am Fish Soc* 135(5):1389–1397.

Nakamura I, Goto Y, Sato K (2015a) Ocean sunfish rewarm at the surface after deep excursions to forage for siphonophores. *J Anim Ecol* 84(3):590–603.

Nakamura I, Meyer CG, Sato K (2015b) Unexpected positive buoyancy in deep sea sharks, *Hexanchus griseus*, and a *Echinorhinus cookei*. *PLoS ONE* 10(6):e0127667.

Nakamura I, Watanabe YY, Papastamatiou YP, Sato K, Meyer CG (2011) Yo-yo vertical movements suggest a foraging strategy for tiger sharks *Galeocerdo cuvier*. *Mar Ecol Prog Ser* 424:237–246.

Noda T, Okuyama J, Koizumi T, Arai N, Kobayashi M (2012) Monitoring attitude and dynamic acceleration of free-moving aquatic animals using a gyroscope. *Aquat Biol* 16(3):265–276.

Noda T, Kawabata Y, Arai N, Mitamura H, Watanabe S (2014) Animal-mounted gyroscope/accelerometer/magnetometer: *in situ* measurement of the movement performance of fast-start behaviour in fish. *J Exp Mar Biol Ecol* 451:55–68.

Olla B, Davis M, Schreck C (1997) Effects of simulated trawling on sablefish and walleye pollock: the role of light intensity, net velocity and towing duration. *J Fish Biol* 50(6):1181–1194.

Otani S, Naito Y, Kawamura A, Kawasaki M, Nishiwaki S, Kato A (1998) Diving behavior and performance of harbor porpoises, *Phocoena phocoena*, in Funka Bay, Hokkaido, Japan. *Mar Mammal Sci* 14(2):209–220.

O'Toole A, Murchie KJ, Pullen C, Hanson KC, Suski CD, Danylchuk AJ, Cooke SJ (2011) Locomotory activity and depth distribution of adult great barracuda (*Sphyraena barracuda*) in Bahamian coastal habitats determined using acceleration and pressure biotelemetry transmitters. *Mar Freshwater Res* 61(12):1446–1456.

Papastamatiou YP, Lowe CG (2004) Postprandial response of gastric pH in leopard sharks (*Triakis semifasciata*) and its use to study foraging ecology. *J Exp Biol* 207(2):225–232.

Papastamatiou YP, Lowe CG (2005) Variations in gastric acid secretion during periods of fasting between two species of shark. *Comp Biochem Physiol A Mol Integr Physiol* 141(2):210–214.

Papastamatiou YP, Purkis SJ, Holland KN (2007) The response of gastric pH and motility to fasting and feeding in free swimming blacktip reef sharks, *Carcharhinus melanopterus*. *J Exp Mar Biol Ecol* 345(2):129–140.

Papastamatiou YP, Watanabe YY, Bradley D, Dee LE, Weng K, Lowe CG, Caselle JE (2015) Drivers of daily routines in an ectothermic marine predator: hunt warm, rest warmer? *PLoS ONE* 10(6):e0127807.

Payne NL, Taylor MD, Watanabe YY, Semmens JM (2014) From physiology to physics: are we recognizing the flexibility of biologging tools? *J Exp Biol* 217(3):317–322.

Payne NL, Snelling EP, Fitzpatrick R, Seymour J, Courtney R, Barnett A, Watanabe YY, et al. (2015) A new method for resolving uncertainty of energy requirements in large water breathers: the 'mega-flume' seagoing swim-tunnel respirometer. *Meth Ecol Evol* 6(6):668–677.

Payne NL, Iosilevskii G, Barnett A, Fischer C, Graham RT, Gleiss AC, Watanabe YY (2016a) Great hammerhead sharks swim on their side to reduce transport costs. *Nat Commun* 7:12289.

Payne NL, Smith JA, van der Meulen DE, Taylor MD, Watanabe YY, Takahashi A, Marzullo TA, et al. (2016b) Temperature dependence of fish performance in the wild: links with species biogeography and physiological thermal tolerance. *Funct Ecol* 30(6):903–912.

Perry AL, Low PJ, Ellis JR, Reynolds JD (2005) Climate change and distribution shifts in marine fishes. *Science* 308(5730):1912–1915.

Pratt Jr HL, Carrier JC (2001) A review of elasmobranch reproductive behavior with a case study on the nurse shark, *Ginglymostoma cirratum*. *Environ Biol Fish* 60:157–188.

Righton D, Westerberg H, Feunteun E, Økland F, Gargan P, Amilhat E, Metcalfe J, et al. (2016) Empirical observations of the spawning migration of European eels: the long and dangerous road to the Sargasso Sea. *Sci Adv* 2(10):e1501694.

Ropert-Coudert Y, Wilson RP (2004) Subjectivity in bio-logging: do logged data mislead? *Mem Natl Inst Polar Res Spec Issue* 58:23–33.

Ropert-Coudert Y, Wilson RP (2005) Trends and perspectives in animal-attached remote sensing. *Front Ecol Environ* 3(8):437–444.

Rutz C, Hays GC (2009) New frontiers in biologging science. *Biol Lett* 5(3):289–292.

Rutz C, Troscianko J (2013) Programmable, miniature video-loggers for deployment on wild birds and other wildlife. *Meth Ecol Evol* 4(2):114–122.

Sakamoto KQ, Sato K, Ishizuka M, Watanuki Y, Takahashi A, Daunt F, Wanless S (2009) Can ethograms be automatically generated using body acceleration data from free-ranging birds? *PLoS ONE* 4(4):e5379.

Scharold J, Lai NC, Lowell W, Graham J (1989) Metabolic rate, heart rate, and tailbeat frequency during sustained swimming in the leopard shark *Triakis semifasciata*. *Exp Biol* 48(4):223–230.

Schmidt-Nielsen K (1972) Locomotion: energy cost of swimming, flying, and running. *Science* 177(4045):222–228.

Semeniuk CA, Haider W, Beardmore B, Rothley KD (2009) A multi-attribute trade-off approach for advancing the management of marine wildlife tourism: a quantitative assessment of heterogeneous visitor preferences. *Aquat Cons Mar Freshw Ecosyst* 19(2):194–208.

Sepulveda CA, Kohin S, Chan C, Vetter R, Graham JB (2004) Movement patterns, depth preferences, and stomach temperatures of free-swimming juvenile mako sharks, *Isurus oxyrinchus*, in the Southern California Bight. *Mar Biol* 145(1):191–199.

Sepulveda CA, Heberer C, Aalbers SA, Spear N, Kinney M, Bernal D, Kohin S (2015) Post-release survivorship studies on common thresher sharks (*Alopias vulpinus*) captured in the southern California recreational fishery. *Fish Res* 161:102–108.

Shepard EL, Wilson RP, Halsey LG, Quintana F, Laich AG, Gleiss AD, Liebsch N, et al. (2008a) Derivation of body motion via appropriate smoothing of acceleration data. *Aquat Biol* 4(3):235–241.

Shepard EL, Wilson RP, Liebsch N, Quintana F, Laich AG, Lucke K (2008b) Flexible paddle sheds new light on speed: a novel method for the remote measurement of swim speed in aquatic animals. *Endangered Species Res* 4(1–2):157–164.

Shipley ON, Brownscombe JW, Danylchuck AJ, Cooke SJ, O'Shea OW, Brooks EJ (2017) Finc-scale movement and activity patterns of Caribbean reef sharks (*Carcharhinus perezi*) in the Bahamas. *Environ Biol Fish* July:1–8.

Sims D (1996) The effect of body size on the standard metabolic rate of the lesser spotted dogfish. *J Fish Biol* 48(3):542–544.

Sims D (2010) Tracking and analysis techniques for understanding free-ranging shark movements and behavior. In: Carrier JF, Musick JA, Heithaus MR (eds) *Sharks and Their Relatives*. II. *Biodiversity, Adaptive Physiology, and Conservation*. CRC Press, Boca Raton, FL, pp 265–289.

Skomal G (2006) The Physiological Effects of Capture Stress on Post-Release Survivorship of Sharks, Tunas, and Marlin, doctoral thesis, Boston University.

Skomal G (2007) Evaluating the physiological and physical consequences of capture on post-release survivorship in large pelagic fishes. *Fish Manage Ecol* 14(2):81–89.

Skomal G, Bernal D (2010) Physiological responses to stress in sharks. In: Carrier JF, Musick JA, Heithaus MR (eds) *Sharks and Their Relatives*. II. *Biodiversity, Adaptive Physiology, and Conservation*. CRC Press, Boca Raton, FL, pp 459–490.

Skomal G, Lobel PS, Marshall G (2007) The use of animal-borne imaging to assess post-release behavior as it relates to capture stress in grey reef sharks, *Carcharhinus amblyrhynchos*. *Mar Technol Soc J* 41(4):44–48.

Stebbing A, Turk S, Wheeler A, Clarke K (2002) Immigration of southern fish species to south-west England linked to warming of the North Atlantic (1960–2001). *J Mar Biol Assoc UK* 82(2):177–180.

Sunday JM, Bates AE, Dulvy NK (2012) Thermal tolerance and the global redistribution of animals. *Nat Climate Change* 2(9):686–690.

Sundström LF, Gruber SH (1998) Using speed-sensing transmitters to construct a bioenergetics model for subadult lemon sharks, *Negaprion brevirostris* (Poey), in the field. *Hydrobiologia* 371:241–247.

Sundström LF, Gruber SH (2002) Effects of capture and transmitter attachments on the swimming speed of large juvenile lemon sharks in the wild. *J Fish Biol* 61(3):834–838.

Sundström LF, Gruber SH, Clermont SM, Correia JPS, de Marignac JRC, Morrissey JF, et al. (2001) Review of elasmobranch behavioral studies using ultrasonic telemetry with special reference to the lemon shark, *Negaprion brevirostris*, around Bimini Islands, Bahamas. *Environ Biol Fish* 60:225–250.

Tanoue H, Komatsu T, Tsujino T, Suzuki I, Watanabe M, Goto H, Miyazaki N (2012) Feeding events of Japanese lates *Lates japonicus* detected by a high-speed video camera and three-axis micro-acceleration data-logger. *Fish Sci* 78(3):533–538.

Thorarensen H, Gallaugher P, Farrell A (1996) The limitations of heart rate as a predictor of metabolic rate in fish. *J Fish Biol* 49(2):226–236.

van der Kooij J, Righton D, Strand E, Michalsen K, Thorsteinsson V, Svedäng H, Neat FC, Neuenfeldt S (2007) Life under pressure: insights from electronic data-storage tags into cod swimbladder function. *ICES J Mar Sci* 64(7):1293–1301.

Vyssotski AL, Serkov AN, Itskov PM, Dell'Omo G, Latanov AV, Wolfer DP, Lipp HP (2006) Miniature neurologgers for flying pigeons: multichannel EEG and action and field potentials in combination with GPS recording. *J Neurophysiol* 95(2):1263–1273.

Watanabe Y, Sato K (2008) Functional dorsoventral symmetry in relation to lift-based swimming in the ocean sunfish *Mola mola*. *PLoS ONE* 3(10):e3446.

Watanabe YY, Lydersen C, Fisk AT, Kovacs KM (2012) The slowest fish: swim speed and tail-beat frequency of Greenland sharks. *J Exp Mar Biol Ecol* 426:5–11.

Watanabe YY, Goldman KJ, Caselle JE, Chapman DD, Papastamatiou YP (2015) Comparative analyses of animal-tracking data reveal ecological significance of endothermy in fishes. *Proc Natl Acad Sci* 112(19):6104–6109.

Webber D, Boutilier R, Kerr S, Smale M (2001) Caudal differential pressure as a predictor of swimming speed of cod (*Gadus morhua*). *J Exp Biol* 204(20):3561–3570.

Weihs D (1973) Hydromechanics of fish schooling. *Nature* 241(5387):290–291.

Whitlock R, Walli A, Cermeño P, Rodriguez L, Farwell C, Block B (2013) Quantifying energy intake in Pacific bluefin tuna (*Thunnus orientalis*) using the heat increment of feeding. *J Exp Biol* 216(21):4109–4123.

Whitmore BM, White CF, Gleiss AC, Whitney NM (2016) A float-release package for recovering data-loggers from wild sharks. *J Exp Mar Biol Ecol* 475:49–53.

Whitney NM, Papastamatiou YP, Holland KN, Lowe CG (2007) Use of an acceleration data logger to measure diel activity patterns in captive whitetip reef sharks, *Triaenodon obesus*. *Aquat Living Resour* 20(4):299–305.

Whitney NM, Pratt Jr HL, Pratt TC, Carrier JC (2010) Identifying shark mating behaviour using three-dimensional acceleration loggers. *Endangered Species Res* 10:71–82.

Whitney NM, Papastamatiou YP, Gleiss AC, Carrier J, Musick J, Heithaus M (2012) Integrative multi-sensor tagging: emerging techniques to link elasmobranch behavior, physiology and ecology. In: Carrier JF, Musick JA, Heithaus MR (eds) *Sharks and Their Relatives. II. Biodiversity, Adaptive Physiology, and Conservation*. CRC Press, Boca Raton, FL, pp 265–289.

Whitney NM, White CF, Gleiss AC, Schwieterman GD, Anderson P, Hueter RE, Skomal GB (2016) A novel method for determining post-release mortality, behavior, and recovery period using acceleration data loggers. *Fish Res* 183:210–221.

Whitney NM, White CF, Anderson P, Hueter RE, Skomal GB (2017) The physiological stress response, postrelease behavior, and mortality of blacktip sharks(*Carcharhinus limbatus*) caught on circle and J-hooks in the Florida recreational fishery. *Fish Bull* 115(4):532–543.

Williams HJ, et al. (2017) Identification of animal movement patterns using tri-axial magnetometry. *Mov Ecol* 5(1):6.

Williams HJ, Holton MD, Shepard ELC, Largey N, Norman B, Ryan PG, Duriez O, et al. (2000) Sink or swim: strategies for cost-efficient diving by marine mammals. *Science* 288(5463):133–136.

Wilson AD, Brownscombe JW, Krause J, Krause S, Gutowsky LFG, Brooks EJ, Cooke SJ (2015a) Integrating network analysis, sensor tags, and observation to understand shark ecology and behavior. *Behav Ecol* 26(6):1577–1586.

Wilson AD, Wikelski M, Wilson RP, Cooke SJ (2015b) Utility of biological sensor tags in animal conservation. *Cons Biol* 29(4):1065–1075.

Wilson RP, Bain CA (1984a) An inexpensive depth gauge for penguins. *J Wildl Manage* 48:1077–1084.

Wilson RP, Bain CA (1984b) An inexpensive speed meter for penguins at sea. *J Wildl Manage* 48:1360–1364.

Wilson RP, Vandenabeele SP (2012) Technological innovation in archival tags used in seabird research. *Mar Ecol Prog Ser* 451:245–262.

Wilson RP, Cooper J, Plötz J (1992) Can we determine when marine endotherms feed? A case study with seabirds. *J Exp Biol* 167:267–275.

Wilson RP, Steinfurth A, Ropert-Coudert Y, Kato A, Kurita M (2002) Lip-reading in remote subjects: an attempt to quantify and separate ingestion, breathing and vocalisation in free-living animals using penguins as a model. *Mar Biol* 140(1):17–27.

Wilson RP, White CR, Quintana F, Halsey LG, Liebsch N, Martin GR, Butler PJ (2006) Moving towards acceleration for estimates of activity-specific metabolic rate in free-living animals: the case of the cormorant. *J Anim Ecol* 75(5):1081–1090.

Wilson RP, Shepard ELC, Liebsch N (2008) Prying into the intimate details of animal lives: use of a daily diary on animals. *Endang Species Res* 4:123–137.

Wilson RP, Holton MD, Walker JS, Shepard ELC, Scantlebury DM, Wilson VL, Wilson GI, et al. (2016) A spherical-plot solution to linking acceleration metrics with animal performance, state, behaviour and lifestyle. *Mov Ecol* 4(1):22.

Winter J (1983) Underwater biotelemetry. In: Nielsen LA and Johnson DL (eds) *Fisheries Techniques*. American Fisheries Society, Bethesda, MD, pp 371–395.

Wright S, Metcalfe JD, Hetherington S, Wilson R (2014) Estimating activity-specific energy expenditure in a teleost fish, using accelerometer loggers. *Mar Ecol Prog Ser* 496:19–32.

Yasuda T, Komeyama K, Kato K, Mitsunaga Y (2012) Use of acceleration loggers in aquaculture to determine net-cage use and field metabolic rates in red sea bream *Pagrus major*. *Fish Sci* 78(2):229–235.

Yoda K, Sato K, Niizuma Y, Kurita M, Bost C, Le Maho Y, Naito Y (1999) Precise monitoring of porpoising behaviour of Adélie penguins determined using acceleration data loggers. *J Exp Biol* 202(22):3121–3126.

CHAPTER 4

Using Aerial Surveys to Investigate the Distribution, Abundance, and Behavior of Sharks and Rays

Jeremy J. Kiszka
Marine Sciences Program, Florida International University, North Miami, Florida

Michael R. Heithaus
Marine Sciences Program, Florida International University, North Miami, Florida

CONTENTS

4.1 INTRODUCTION

Accurately estimating spatiotemporal variation in the abundance and density of marine animals, as well as directly observing behaviors, is crucial for answering many basic questions about their biology and ecology. It is also critical for assessing their ecological importance in marine ecosystems and evaluating the impact of human activities on populations (e.g., fishing, habitat modification, behavioral disturbance, climate change). Yet, it can be extremely challenging to estimate abundances and densities and observe the natural behavior of many taxa, including elasmobranchs.

Indices of relative abundance for elasmobranchs have mostly been generated using captures in fishing gear (e.g., Dudley and Simpfendorfer, 2006; Reid et al., 2011), dedicated surveys using baited hooks (catch per unit of effort)

(e.g., Wirsing et al., 2006), visual surveys by divers and remote (baited and unbaited) cameras (e.g., Bond et al., 2012; Goetze and Fullwood, 2013; Nadon et al., 2012; Rizzari et al., 2014), and boat-based surveys in shallow waters (Vaudo and Heithaus, 2013). Each of these methods has its benefits as well as critical assumptions that limit the types of questions that they can address. Estimating absolute densities using these techniques is difficult or even impossible. Although only appropriate in some situations, aerial surveys are another possible tool for studying sharks and rays. In marine systems, these surveys have been used to study population abundances and distributions over large spatial scales (Marsh and Sinclair, 1989; Martin et al., 2016) for cetaceans (Gilles et al., 2009; Hammond et al., 2013; Laran et al., 2017), sirenians (Bayliss, 1986; Marsh and Sinclair, 1989; Pollock et al., 2006), seabirds (Laran et al., 2017), sea turtles (Cardona et al., 2005; Marsh and Saafeld, 1989), teleosts (Bonhommeau et al., 2010; Fromentin et al., 2003; Lutcavage et al., 1997), and sharks (Burks et al., 2006; Cliff et al., 2007; Rowat et al., 2009; Squire, 1967). For obvious reasons, most aerial surveys have focused on air-breathing species, particularly marine mammals. In the right circumstances, however, studies using aerial methods are appropriate for species that do not surface to breathe. Aerial survey methods have been used to study sharks for at least five decades (e.g., Burks et al., 2006; Jennings, 1985; Laran et al., 2017; Rowat et al., 2009; Squire, 1967) and have contributed to our understanding of the distribution and abundance of elasmobranchs, particularly in coastal habitats or for large epipelagic species in open waters (Table 4.1). Regardless of taxa, surveys have primarily been conducted using manned aircraft with observers recording their observations in real time (e.g., Hodgson et al., 2017). The recent advent of inexpensive unmanned aerial vehicles (UAVs), also known as remotely piloted aerial systems (RPASs), but more commonly referred to as drones, has led to their increasing use in surveys of marine fauna (e.g., Colefax et al., 2017; Hodgson et al., 2013, 2017; Kiszka et al., 2016). However, few published studies are currently available, and methods for employing this new and quickly advancing technology are still being developed. This chapter provides an overview of methods and assumptions that are common to aerial surveys regardless of the type of platform used, describes the aerial platforms and their associated benefits and drawbacks, and highlights past studies using aerial methods. The chapter closes with an eye toward the opportunities ahead as UAV technology improves.

4.2　BENEFITS OF AERIAL SURVEYS

There is no question that aerial methods are not applicable for the majority of elasmobranch species that spend either all or the vast majority of their lives far from the surface or in turbid waters. There are, however, benefits to using aerial methods for those species that are accessible. First, relative to most methods that involve fishing or video surveys, aircraft can cover vast distances in relatively short time periods (Burks et al., 2006; Laran et al., 2017). Second, if flight patterns are appropriate, there is a low probability of double-counting individuals due to the speed of surveys. Photo-identification (see Chapter 12 in this volume) and tagging can also reduce this challenge, but video methods in particular can be susceptible to this bias (see Chapter 7 in this volume). Finally, almost all methods for assessing abundance measure indices of relative abundance, and it is extremely difficult or impossible to link these relative measures to real abundances or densities in habitats. In aerial surveys, if particular assumptions are met, it is possible to sample known areas and to estimate absolute densities.

4.3　METHODOLOGICAL CONSIDERATIONS

4.3.1　Aerial Survey Methods for Assessing Abundance and Density

Regardless of aircraft type, two sampling methods are generally used in aerial surveys aimed at assessing abundances or densities: line and strip (or "belt") transect surveys (Buckland et al., 2001). In both methods, the design is comprised of a set of straight lines spread through a given study area, and surveys are performed by one or several observers. In line transect surveys, it is assumed that the probability of detection decreases with distance from the transect line. The detection function $g(x)$ describes the probability of detection as a function of perpendicular distance from the line x. It is related to the probability density function of the perpendicular distance $f(x)$, which is $g(x)$ rescaled so that the area under the curve equals 1. If $g(0) = 1$, then the density of species D is

$$D = \frac{nf(0)}{2L}$$

Strip transect methods assume that all targeted objects are detected within a predetermined distance from the observer (the "strip") that is surveyed (e.g., Marsh and Sinclair, 1989, Pollock et al., 2006). It is critical, therefore, that the strip width is not so large that the probability of detection decreases at the edges of the strip. In some cases, especially for UAV surveys that record data to video, the strip transect approach is easily applied as videos can be watched multiple times to ensure that all animals are recorded and there is no decrease in detection probability with distance from the transect line (Kiszka et al., 2016).

Critical to analyses of strip transect data, especially when calculating densities of animals, is determining the total area that was surveyed. Calculating the length of strips is fairly straightforward using Global Positioning System (GPS) points for the beginning and end of surveys, but calculating the strip width can be more challenging. Because errors in calculating strip width can have large effects on the estimated area surveyed and, by extension, density estimates, care should be taken in measuring strip width. Because there can be errors associated with reported altitudes of drones and the shooting angles of cameras, ground-truthing these measurements is important. For example, a laser rangefinder can measure drone altitude at the beginning and end of a transect, and the width of a transect can be estimated using markings on the ground or the size of a known-length object in the video frame. Statistical analysis of strip transect data is greatly simplified compared to line transect surveys because distances of detected animals from the transect line are not measured (Buckland and York, 2009); however, both types of sampling must account for availability and perception biases.

Table 4.1 Review of Published Studies Investigating Shark Distribution, Abundance, and Ecology Using Manned Aerial Survey Methods

Species Involved	Aircraft Type	Effort (Study Period)	Study Area	Speed; Altitude	Survey Objective	Refs.
Elasmobranchii	High-wing aircraft (BN2)	89,000 km (December 2009–April 2010)	Southwest Indian Ocean	167 km/hr; 182 m	Marine megafauna diversity abundance and habitats	Laran et al. (2017)
Carcharhinus limbatus	Cessna 172	Biweekly surveys (February 4, 2011–April 17, 2013; January 4–April 1, 2014)	Southeast coast of Florida	150 km/hr; 150 m	Seasonal abundance	Kajiura and Tellman (2016)
Cetorhinus maximus, Prionace glauca, Sphyrna spp.	—	(1978–1982)	Northeast coast of the United States	—	Marine mammal distribution and abundance	Kenney et al. (1985)
Carcharhodon carcharias	Robinson R22 helicopter	43 flights, 2317 minutes of effort (2010–2012)	Algoa Bay, South Africa	90 km/hr; 300 m	Abundance and habitat preferences	Dicken and Booth (2013)
Cetorhinus maximus	—	10,570-km² survey area (1962–1985)	Central and southern California coast	—	Abundance	Squire (1990)
	High-wing, twin-engine Cessna 337	(September 2009; September 2011)	Lower Bay of Fundy, Canada	204 km/hr; 305 m	Density and abundance	Westgate et al. (2014)
	—	(Summer, 1977–2007)	Gulf of St. Lawrence and adjacent waters in Canada	—	Northern right whale abundance	Campana et al. (2008)
Carcharhodon carcharias, Cetorhinus maximus	—	(1948–1950)	Monterey Bay, California	—	Abundance and distribution	Squire (1967)
Negaprion brevirostris	Ultralight aircraft (Wizard J-3)	5 flights (1979–1982)	Bimini, Bahamas	35 km/hr; 50–70 m	Habitats and behavior	Grubber et al. (1988)
	Cessna 172, Beechcraft 35 Bonanza, Piper Pa-28, Piper PA-31-350 Navajo Chieftain	8 flights (2007–2008)	Bimini, Bahamas	185 km/hr; 100 m	Abundance	Kessel et al. (2013)
Rhincodon typus	Cessna 172, 206	(1989–1992)	Ningaloo Reef, Western Australia	370 m	Distribution and abundance	Taylor (1996)
	Top-wing aircraft	8245 km (2001–2005)	Northern KwaZulu-Natal, South Africa	184 km/hr; 305 m	Distribution and abundance	Cliff et al. (2007)
	Cessna 206	61 flights, 105 hours of effort (2005–2009)	Yucatán Peninsula, Mexico	500 m	Abundance and foraging ecology	de la Parra Venegas et al. (2011)
	DeHavilland Twin Otter turbine engine	89,369 km (1989–1998)	Northern Gulf of Mexico	200 km/hr; 229 m	Cetacean abundance	Burks et al. (2006)
	Cessna 182	109 days of survey flights (1996–2002)	Southwestern Gulf of California	330 m	Distribution	Ketchum et al. (2013)
	Cessna 207, ultralight aircraft	70 survey days (2006–2008)	Yucatán Peninsula, Mexico	250 m	Feeding ecology	Motta et al. (2010)
Hammerhead sharks (Sphyrna spp.)	Twin-engine Beechcraft AT-11	2664 km (1980–1981)	East coast of Florida	222 km/hr; 91–228 m	Seasonal abundance and distribution	Jennings (1985)

4.3.2 Availability and Perception Biases

Two major biases associated with aerial surveys that must be taken into account in any study are availability bias and perception bias. Availability bias describes the potential for animals to be unavailable for counting (Marsh and Sinclair, 1989). For example, individuals may be too deep in the water column to be detected, or they may be undetectable hiding in sediment or over substrates where they are cryptic. Environmental conditions such as water visibility, sun glare, and sea conditions can greatly affect availability (e.g., Robbins et al., 2014). It is important, therefore, to have empirical data on the detectability of targeted species at varying depths. Decoy experiments are one way to assess the maximum depth at which an animal can be seen (Westgate et al., 2014). Although correcting for the effects of environmental factors is important to accurately assess abundance, understanding how animal behavior influences availability bias may be even more critical. The vast majority of elasmobranchs will not be found at depths shallow enough to be observed from the air for long periods of time. If the availability of elasmobranchs being targeted for aerial surveys varies spatiotemporally or with variation in environmental conditions (e.g., temperatures), it is possible that differences in abundances measured from the air will not reflect true differences in abundance but rather will reflect behavioral changes in response to environmental variation (e.g., Thomson et al., 2012). Because elasmobranchs modify their habitat use patterns in response to environmental conditions (e.g., Sims 2003) and likely their use of surface and shallow waters, availability bias should be explicitly considered in aerial surveys of sharks and rays. Studies of elasmobranch behavior, such as those using time–depth recorders, animal-borne video (see Chapter 5 in this volume), or active acoustic tracking with depth sensors (see Chapter 8 in this volume) likely will be necessary for the species and location of surveys, as extrapolations from other contexts might not be appropriate or applicable.

Perception bias refers to variation in the probability of detecting targeted objects that are available to be observed. This problem is particularly important to consider during manned aerial surveys that do not record transects to video, because observers have a very limited amount of time to detect animals and estimate the number of individuals before they pass out of view. Methods to estimate perception bias include the use of two independent observers on both sides of the aircraft in combination with a mark–recapture model (Pollock et al., 2006). Increasingly, still photography and video have been employed during aerial (particularly strip transect) surveys (Kajiura and Tellman, 2016), and video is used to document UAV flights. Video recordings can greatly reduce or eliminate perception bias and have the potential to greatly improve size estimates of aggregations, which in turn will improve abundance estimates (Buckland et al., 2012). It is important, however, to have more than a single observer review videos, as individuals can still be missed during annotation. Methods developed to ensure accurate counts during underwater video monitoring data (see Chapter 7 in this volume) can be applied to video data from aerial surveys.

To date, there have been two published studies aimed at assessing detection rates of large sharks during aerial surveys in coastal waters (Robbins et al., 2014; Westgate et al., 2014). In Jervis Bay, Australia, Robbins et al. (2014) investigated the ability of observers to detect 2.5-m shark analogs at known depths (0 to 5 m) from airplanes and helicopters from an altitude of 150 m (Figure 4.1). Analogs could only be detected at shallow depths: 2.5 m and 2.7 m for airplanes and helicopters, respectively (SE = 0.1 in both cases). The deployment of the analogs at shallower depths along a 5-km grid resulted in detection rates of only 25.5% and 17.1% for fixed-wing and helicopter surveys, respectively, highlighting the possibility that aerial beach patrols to prevent shark–human interactions might not be effective (Robbins et al., 2014). An important consideration for analog experiments, however, is how the swimming motion of the shark as well as the specific color relative to the background might influence the probability of detection for a specific set of conditions. Further work is needed on this topic.

A final consideration is that line and strip transect methods assume that animals do not move (or do not respond to the survey platform). If they do, it is assumed that the effect is negligible if the movement is random and animal speed is slow relative to the survey platform (Glennie et al., 2015; Hodgson et al., 2017). Most aerial surveys (manned and unmanned) are carried out at relatively low altitudes and might impact the behavior of focal species, which could affect their availability.

4.4 TYPES OF AERIAL PLATFORMS: STRENGTHS, WEAKNESSES, AND CONSIDERATIONS

Manned aerial vehicles have the advantage of being able to cover vast areas over relatively short periods of time (Buckland et al., 2001); however, their use can be limited by the availability of airfields, and they are costly (e.g., Colefax et al., 2017; Marsh and Sinclair, 1989; Robbins et al., 2014). So far, most marine megafauna surveys (including those dedicated to sharks) have been conducted using manned aircraft (including small planes, helicopters, and ultralights). However, the increasing accessibility of UAVs since the late 2000s has shifted interest to using this technology due to the low cost and availability of smaller platforms. There is also an interest in whether UAVs could outperform manned aircraft in aerial surveys, although limited research has been conducted to test this hypothesis (Colefax et al., 2017). It is unlikely, however, that easily obtainable UAVs will soon be able to replace the

Figure 4.1   Shark analogs used by Robbins et al. (2014) in Jervis Bay, Australia. (Photograph by William D. Robbins.)

ability of manned aircraft to survey extensive areas. The use of military-grade UAVs, though, could equal or exceed the capacities of manned aircraft commonly used in marine megafauna surveys (Hodgson et al., 2013, 2017). The use of digital surveys during both manned and UAV aerial surveys will address issues commonly documented for surveys based on real-time observer data (such as the use of multispectral cameras and other sensors to reduce or alleviate perception biases) (Buckland et al., 2012; Colefax et al., 2017).

An added benefit of UAV surveys is the safety afforded to operators. Although still quite safe given the number of hours researchers have spent in the air, at least five aircraft crashes have resulted in the deaths of 11 marine mammal researchers (Hodgson et al., 2013). Although UAV survey methods pose a limited risk, they are still considered as hazardous for civil aviation, and many countries around the world have recently adopted restrictions that will considerably limit the use of these platforms. For this reason, researchers are spending an increasing amount of time obtaining permits and certifications to operate these platforms (Vincent et al., 2015), which will impact their use in the future.

Finally, manned and unmanned aerial methods generate very different levels of disturbance for wildlife, including elasmobranchs. UAVs generate less noise in the environment than most manned aircraft and thus may have less impact on the behavior of study species. A careful evaluation of the risks of disturbance, however, should be undertaken before carrying out any field research (Hodgson and Koh, 2016). Although only used in a single study on elasmobranchs to date (Nosal et al., 2013), helium-filled balloons tethered to boats can also provide data on sharks and rays. Balloons

offer the benefit of extremely long periods aloft. In situations where observations of sharks are constrained to small areas, a balloon could provide data over long time periods without the need to continually monitor the platform. Tethering the balloon to a boat allows larger areas to be surveyed from the air, although the potential for disturbance from the boat could limit the types of questions that can be addressed with this platform. Furthermore, due to the poor or limited maneuverability of balloons, potential impacts to other aircraft must be considered.

4.5   MANNED AERIAL SURVEYS

Most studies of elasmobranchs that use aerial survey methods have been carried out using small airplanes, ultralight motorized aircrafts, or helicopters (e.g., Cliff et al., 2007; Gruber et al., 1988; Kessel et al., 2013; Rowat et al., 2009). Studies using manned aircraft have generally focused on assessing the distribution and relative abundance of species, including how oceanographic variables influence elasmobranch encounter rates, but research questions and ecosystems sampled have varied considerably (e.g., Burks et al., 2006; Campana et al., 2008; Cliff et al., 2007; Rowat et al., 2009; Westgate et al., 2014). Some studies using aerial methods have also identified critical habitats for species with high tourism value and assessed the potential for developing shark viewing activities (Cliff et al., 2007). Others have used aerial survey methods to monitor the risk of interactions between dangerous shark species and recreational activities, such as surfing and bathing (Lemahieu et al., 2017).

4.5.1 Abundance, Distribution, and Habitat Preferences

Large elasmobranchs are most commonly targeted for aerial surveys (Cliff et al., 2007; Rowat et al., 2009; Westgate et al., 2014), because detecting smaller species (e.g., <2.5-m sharks) at depths greater than approximately 2 m is challenging (Robbins et al., 2014). Thus, it is not surprising that whale sharks (*Rhincodon typus*) and basking sharks (*Cetorhinus maximus*), and to a lesser extent white sharks (*Carcharodon carcharias*), are the species studied most by manned aircraft. A small number of studies, however, have focused on smaller species in habitats where sharks are easily detectable (e.g., shallow and clear waters) (Gruber et al., 1988; Kajiura and Tellman, 2016; Kessel et al., 2013).

The first reference to the use of aircraft surveys to investigate shark distribution and occurrence is Squire (1967), who conducted aerial surveys in Monterey Bay from February 1948 to October 1950, while fish spotters assisted commercial harpooners targeting basking sharks (Table 4.1). During monthly surveys conducted in the bay, white and mako sharks (*Isurus* spp.) were also recorded. These surveys provided data on the occurrence of sharks relative to sea surface temperature (Squire, 1967). Off the north coast of KwaZulu-Natal, South Africa, aircraft surveys between 2001 and 2005 investigated potential development of the whale shark diving industry (Cliff et al., 2007) (Table 4.1). Surveys revealed that whale sharks occur at relatively low densities off KwaZulu-Natal and may have declined in the northern part of the province since the mid-1990s (Cliff et al., 2007). Off Ningaloo Reef in Western Australia, whale shark seasonal distribution has been investigated to delineate their broad-scale occurrence (Taylor, 1996). Seasonal aggregations of whale sharks have since been monitored using a combination of boat-based and aerial surveys, as well as from log sheet data recorded by the tourism industry (Wilson et al., 2001). Along the Caribbean coast of Mexico, off the Yucatán Peninsula, aerial surveys were also conducted to assess the distribution and abundance of whale sharks and other elasmobranchs, including manta and devil rays (*Mobula* spp.) and cownose rays (*Rhinoptera bonasus*) (de la Parra Venegas et al., 2011). When zig-zag surveys were conducted from 2005 to 2009 (May to September) with a Cessna 206 aircraft flying at 500 m, 2295 whale shark sightings were recorded. This study highlights the ability of aerial surveys to provide minimal estimates of the size of whale shark aggregations that are not possible using other survey methods, such as boat-based surveys (de la Parra Venegas et al., 2011). Off southern California, at Bahía de La Paz, the relative abundance of whale shark aggregations has been studied using a combination of aerial and ship surveys (Ketchum et al., 2013) to investigate the drivers of their occurrence, as well as sex and size segregation in relation to their prey (zooplankton). Adults and juveniles appear to be spatially segregated, with juveniles occurring in coastal shallow and more turbid waters and adults feeding offshore. The combination of aerial and ship-based (including prey sampling) surveys advanced our understanding of the effects of habitat conditions and prey composition on whale shark distribution and foraging tactics (Ketchum et al., 2013).

Aerial surveys have also proved to be important for elucidating abundance and population trends of basking sharks. Surveys using fixed-wing aircraft cover areas that are not feasible using ships. In the lower Bay of Fundy in Canada, the conservation status of basking sharks is poorly known. Two day-long line transect aerial surveys were conducted in 2009 and 2011, during which 26 sightings were recorded (Westgate et al., 2014) (see Table 4.1 for details). Corrected density estimates to account for availability biases using dive data were generated in both years (2009: 0.0513 shark per km^2, 95% CI = 0.0188–0.1402; 2011: 0.0598 shark per km^2, 95% CI = 0.0358–0.1001), corresponding to an absolute abundance of 542 sharks in 2009 (95% CI = 198–1482) and 632 in 2011 (95% CI = 377–1058), occupying a 10,570-km^2 area in the lower Bay of Fundy (Westgate et al., 2014).

A number of studies using manned aircraft have investigated the distribution and abundance of smaller shark species, such as white sharks (Dicken and Booth, 2013), hammerhead sharks (*Sphyrna* spp.) (Jennings, 1985), lemon sharks (*Negaprion brevirostris*) (Kessel et al., 2013), and blacktip sharks (*Carcharhinus limbatus*) (Kajiura and Tellman, 2016). Along the southeast coast of the United States, especially off Palm Beach County in Florida, large aggregations of blacktip sharks have been studied using a Cessna 172 aircraft (Kajiura and Tellman, 2016) (Table 4.1). High-definition video and digital still cameras were used to provide a continuous record of transects along the coast (Figure 4.2), allowing perception bias to be minimized. Because surveys were conducted in shallow and relatively clear waters, availability bias was minimal, and it was possible to derive shark density estimates. Blacktip shark abundance peaked from January to March and was inversely correlated with sea surface temperature. Sharks were only observed in waters that were <25°C. This study has generated important baseline data on shark abundance that will allow assessment of how climate change and other human impacts might affect the migration of blacktip sharks along the southeast coast of the United States (Kajiura and Tellman, 2016).

The abundance of lemon sharks in the shallow and clear waters of Bimini, the Bahamas, has been investigated using a variety of manned aircraft (small planes and ultralights) (Gruber et al., 1988; Kessel et al., 2013). The spatial distribution and abundance (mean abundance of 49 ± 8.6 in the study area) of lemon sharks could be linked to time of year and tidal movements, showing that aerial survey methods generate reliable shark abundance estimates in shallow coastal marine habitats (Kessel et al., 2013). In Bimini, the low altitudes and clear water allowed species to be individually

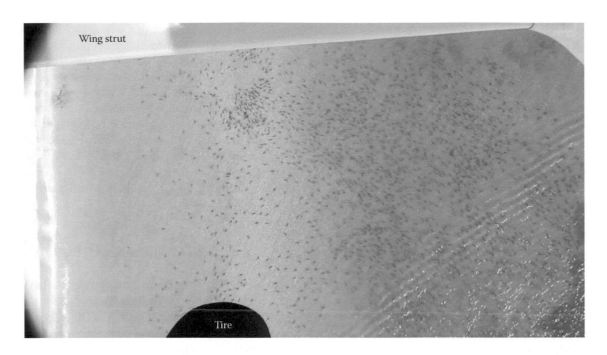

Figure 4.2 Sample frame from a high-definition video from a blacktip shark aggregation off the coast of Florida. The collection of video data during aerial surveys provides precise counts of sharks observed; here, 1678 sharks can be counted. (From Kajiura, S.M. and Tellman, S.L., *PLoS ONE*, 1, e0150911, 2016.)

identified during surveys. Nurse (*Ginglymostoma cirratum*), blacktip, tiger (*Galeocerdo cuvier*), and bull (*Carcharhinus leucas*) sharks were also reliably identified during surveys.

Off Cape Canaveral, Florida, hammerhead shark distribution in relation to the Gulf Stream and time of year was investigated using a twin-engine Beechcraft AT-11 airplane. The presence of hammerheads was linked to sea surface temperature (Jennings, 1985). Species identity, however, could not be determined. In Algoa Bay, South Africa, helicopter surveys over bathing beaches were carried out to investigate the inshore nursery habitats of white sharks (Table 4.1). Observations of juvenile and young-of-the-year white sharks were correlated with barometric pressure and sea surface temperature (Dicken and Booth, 2013).

Observers have regularly collected shark sightings during studies directed at other taxa (e.g., marine mammals, sea turtles), which has enabled opportunistic observations on distribution in areas where information is limited. From October 1978 to January 1982, cetacean and sea turtle surveys were conducted along the Atlantic coast between Cape Hatteras, North Carolina, and Cape Sable, Nova Scotia, from the shoreline to about 9 km offshore (nearly 2000-m isobaths) (Kenney et al., 1985). Nearly 1700 shark sightings from at least three species were collected, including basking, blue (*Prionace glauca*), and hammerhead sharks, highlighting some spatial and seasonal patterns in occurrence of these species (Kenney et al., 1985). Similarly, in the northern Gulf of Mexico, extensive year-round cetacean surveys were conducted between 1989 and 1998 (89,369 km surveyed) in slope waters (100 to 2000 m). A total of 81 whale

shark sightings (119 individuals) were collected (Burks et al., 2006). Whale sharks were more common during the summer than the winter in the eastern Gulf, whereas encounter rates were higher in summer than in winter in the western portion of the Gulf. Moreover, aggregations were only observed in summer and winter, but the number of individuals was significantly higher during the summer (Burks et al., 2006). Along the Atlantic coast of Canada, aerial surveys initially dedicated to assess northern right whale (*Eubalaena glacialis*) abundance in the Bay of Fundy, on the Scotian Shelf, and off Newfoundland provided uncorrected abundance estimates of basking sharks, with more than 10,000 individuals estimated in 2007 (Campana et al., 2008). Similarly, aerial surveys targeting commercial fish species were conducted off the central and southern California coasts from 1962 to 1985 and documented the seasonal distribution and occurrence of basking sharks (Squire, 1990).

Off the coast of Kenya, where information on the distribution and abundance of sharks (and other marine megafauna) is largely nonexistent, a countrywide aerial survey (mostly dedicated to marine mammals) was carried out in November 1994 in coastal waters (Wamukoya et al., 1996). A total of 37 whale sharks, 15 large sharks (unidentified species), and 63 batoids were recorded with concentrations of all species in Ungwana Bay (Wakumoya et al., 1996). Today, these data still represent the only source of fisheries-independent surveys in this country (for a review, see Kiszka and van der Elst, 2015). More recently, large-scale aerial surveys using a high-wing aircraft (BN2) were carried out in the southwestern Indian Ocean, primarily to assess the distribution, abundance, and habitat preferences of cetaceans and seabirds (Laran et al., 2017). These surveys have covered all available habitats present in the region, including shelf, slope, and oceanic waters. Only the whale shark could be identified to the species level during surveys, but mobulid rays (*Mobula* spp., including manta and devil rays) and hammerhead sharks (*Sphyrna* spp.) could be identified at the genus level. Areas with higher encounter rates were the Seychelles and the Mozambique Channel, and whale sharks were only reported in the northern and central Mozambique Channel and east of Madagascar (Laran et al., 2017). This study produced the first regionwide description of the distribution of large elasmobranchs in the southwest Indian Ocean.

Overall, aerial surveys that are not dedicated to elasmobranchs have considerably advanced the knowledge on their distribution and occurrence in data-poor regions. Aerial surveys, such as those in the southwest Indian Ocean, have been conducted in many regions around the world, but few studies have been published to date. Because of the need for even basic data in poorly studied regions, analyzing these types of data should be a priority.

4.5.2 Behavior

Manned aerial surveys have provided insights into the feeding and social interactions of sharks. Although unique and undocumented behaviors have been inferred from manned aerial surveys, limited quantitative data are available. Off Bahía de La Paz, in the Gulf of California, a combination of ship and aerial-based surveys provided insights into the effects of prey composition on whale shark feeding tactics (Ketchum et al., 2013). Adult whale sharks fed by ram-filtering on diffuse patches of euphausiids, whereas juveniles (<9 m in total length) fed in coastal waters on copepods by performing stationary suction feeding (Ketchum et al., 2013). At Cabo Catoche, on the Yucatán Peninsula in Mexico, the largest whale shark aggregation ever reported was documented during aerial surveys (de la Parra Venegas et al., 2011). Off the island of Bimini in the Bahamas, ultralight aircraft were used to study the social behavior and group formation of lemon and bonnethead (*Sphyrna tiburo*) sharks (Gruber et al., 1988). In the Gulf of Maine, aerial photographs were used to document schooling patterns (e.g., echelon, cartwheel) of basking sharks and their possible function for courtship (Wilson, 2004). Similar schooling behaviors have also been documented during opportunistic aerial observations from a helicopter off Nova Scotia (Harvey-Clark et al., 1999).

4.5.3 Other Applications

Sharks have been surveyed from manned aircraft to facilitate fishing operations (Squire, 1967) and to patrol beaches. Low detection rates of sharks off beaches in New South

Wales, Australia, suggest that aerial patrols might not be an effective early-warning system to prevent shark–human interactions (Robbins et al., 2014). In the southwestern Indian Ocean, off Réunion Island, aerial surveys were carried out to assess the distribution of ocean users (surfers, swimmers, snorkelers, and paddle boarders) across inshore waters and were correlated with acoustic tagging data from bull and tiger sharks to identify spatial and temporal overlap (Lemahieu et al., 2016). Three coastal hotspots were identified as having high probabilities of interaction with sharks, and these areas coincided with locations of shark bites on people.

4.6 UNMANNED AERIAL SURVEYS

Rapid developments in commercially available UAVs, including miniaturization, enhanced image quality, improved flight times, and better automation of flight, combined with decreasing costs have dramatically increased their use by researchers, including marine biologists (Anderson and Gaston, 2013; Christie et al., 2016). In addition to off-the-shelf UAVs, researchers have also been able to gain access to more advanced UAVs developed for military and commercial purposes (Hogdson et al., 2013, 2017). UAVs have been used to study the distribution, abundance, and behavior of terrestrial and marine wildlife and how human activities can impact their populations (Anderson and Gaston, 2013; Christie et al., 2016; Hodgson et al., 2013; Kiszka et al., 2016). UAVs are generally equipped with high-resolution video cameras for continuous recording of data, which enables postprocessing videos to enhance detection rates, improve species identification, and more accurately quantify group sizes (Hodgson et al., 2016; Kiszka et al., 2016). Today, UAVs are user friendly, compact, and relatively inexpensive and can be launched from field sites without the need for major infrastructure.

There is a variety of UAV designs, but their main distinction is related to their size, power, operating altitude, and range. Large UAVs (e.g., NASA's Ikhana unmanned aircraft system) operate over large distances (about 500 km) for up to two days and reach an altitude of 20 km, whereas medium platforms can operate for about 10 hours at a maximum altitude of 4 km but at a similar range as large drones (Anderson and Gaston, 2013). However, the most commonly used platforms (either fixed-wing or rotor-based UAVs) are small and operate within a 10-km range at a low altitude (less than 1000 m). They can fly for periods of less than 2 hours but most commonly less than an hour (for a detailed review of the various UAV platforms, see Anderson and Gaston, 2013).

In comparison to manned aerial surveys, the use of UAVs is relatively new. At the time of the writing of this chapter, only one published study focusing on elasmobranchs

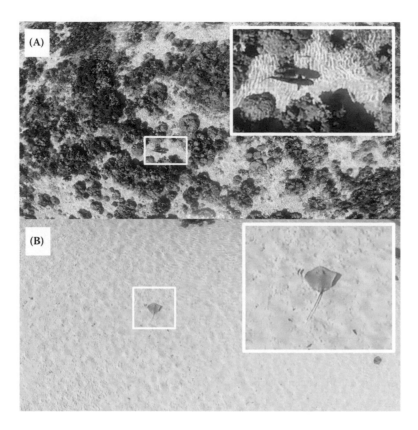

Figure 4.3   Screen shots of (A) a blacktip reef shark (*Carcharhinus melanopterus*) and (B) a pink whipray (*Himantura fai*) on the reefs and sandflats of Moorea (French Polynesia) during unmanned aerial vehicle (UAV) surveys (Kiszka et al., 2016).

was available (Kiszka et al., 2016), although there are a number of UAV-based studies on marine mammal distribution, reproductive biology, and population dynamics (Christiansen et al., 2016; Hodgson et al., 2013, 2017; Koski et al., 2009, 2015). In the single elasmobranch-focused UAV study, transect surveys were conducted using a micro-UAV (DJI Phantom II®) over the reefs of Moorea in the Society Islands, French Polynesia. Surveys were flown at an altitude of 12 m, and surveys were conducted under Beaufort 1 wind conditions to avoid perception biases. Visibility allowed the sea floor to be viewed, making availability bias close to zero, so the authors assumed every targeted object was detected during the surveys. These surveys provided data on fine-scale variation in densities of blacktip reef sharks (*Carcharhinus melanopterus*) and pink whiprays (*Himantura fai*) in relation to habitat type, shark tourism, and provisioning activities (Kiszka et al., 2016) (Figure 4.3). Provisioning activities for tourism had a significant impact on the densities of blacktip reef sharks and pink whiprays, but these density increases were confined to relatively small areas (Kiszka et al., 2016). Small rotor-based UAVs have great potential for studies of shark and ray densities and behavior in coastal and shallow ecosystems, whereas larger platforms provide some of the benefits of manned aircraft

over larger spatial scales and further offshore. There, remains, however, a need to quantify availability biases to correctly estimate densities, as elasmobranchs spend most of their time too deep to be detected. Rotor-based platforms also have a great potential to record data on the behavior and distribution of sharks and rays (Figure 4.4), including their feeding tactics and social behavior when they occur close to the surface (e.g., large planktivorous species such as whale sharks and mobulid rays, basking and white sharks, reef sharks in shallow sandy habitats and patch reefs), because they can remain stationary or slowly travel with animals. In inshore and clear waters, the social interactions of elasmobranchs and their interaction with other species such as teleosts could be easily documented without impact on studied animals. Because UAVs can be operated at a low cost compared to manned aircrafts, they also have a greater potential to monitor potentially dangerous shark species where recreational activities, such as bathing and surfing, could interact with these animals. Finally, because UAVs are so accessible and can be operated easily, they also provide opportunities to generate educational videos on natural systems and wildlife, including sharks and rays in their natural habitats, and to raise awareness on marine conservation in general.

Figure 4.4   Whale shark aggregation in the Gulf of Mexico from a micro-UAV survey. (Photograph by Simon J. Pierce.)

4.7   OTHER PLATFORMS

Aerial photography using helium-filled balloons tethered to boats has been used to study the spatial behavior and social interactions of various species, including coastal bottlenose dolphins (*Tursiops truncatus*) (Lewis et al., 2011). The spatial distribution, abundance, and orientation of leopard sharks (*Triakis semifasciata*) was studied off the southern Californian coast using a 1.8-m diameter helium-filled balloon tethered 45 m above the surface. Aerial photographs taken every 1.25 seconds using a digital camera were used to assess the behavior of this species near the head of a submarine canyon (Nosal et al., 2013). Although limited in the situations in which they can be deployed, tethered balloons offer potential to gather data over long time periods in the right situations.

4.8   IMPLICATIONS FOR CONSERVATION AND MANAGEMENT

When appropriate corrections are applied, aerial surveys represent a unique source of spatially explicit data to identify habitats for elasmobranchs, which potentially includes a number of globally and regionally endangered species. Past studies have identified critical habitats and the location of aggregations of sharks in coastal environments (e.g., de la Parra Venegas et al., 2011). In shallow coastal or reef-associated habitats, manned and unmanned aerial surveys can provide a unique opportunity to document the effects of human disturbance on the distribution and possibly absolute densities of sharks and rays (e.g., Kiszka et al., 2016). Overall, aerial survey methods represent a major opportunity to generate important information on the distribution, abundance, and behavior of elasmobranchs in a variety of ecosystems and contexts. This type of information is critical to detecting spatial and temporal changes in distribution, habitat, and abundance in relation to ongoing impacts, including fisheries, disturbance, and climate change.

4.9   SUMMARY AND CONCLUSION

In comparison to other methods, such as fishing, diver, or video camera surveys, aerial surveys have not been commonly used to study elasmobranchs; however, they represent a unique and independent source of data on epipelagic and/or shallow-water species. They are particularly useful for studies of large species such as mobulid rays and whale sharks, as well as species occurring in shallow waters or close to the surface during migration events. Because of the inherent limitations of aerial surveys and observations of species that spend their entire lives underwater, most applications of aerial technology will require, or be greatly enhanced by, the incorporation of other methods (e.g., boat-based surveys, active tagging and tracking). These approaches should be encouraged, as aerial surveys can cover extended areas, but details on the behavior of sharks and the composition and patch structure of their prey are lacking.

REFERENCES

Anderson K, Gaston, KJ (2013) Lightweight unmanned aerial vehicles will revolutionize spatial ecology. *Front Ecol Environ* 11:138–146.

Bayliss P (1986) Factors affecting aerial surveys of marine fauna, and their relationship to a census of dugongs in the coastal waters of the Northern Territory. *Wildl Res* 13:27–37.

Bond ME, Babcock EA, Pikitch EK, Abercrombie DL, Lamb NF, Chapman DD (2012) Reef sharks exhibit site fidelity and higher relative abundance in marine reserves on the Mesoamerican Barrier Reef. *PLoS ONE* 7:e32983.

Bonhommeau S, Farrugio H, Poisson F, Fromentin JM (2010) Aerial surveys of bluefin tuna in the western Mediterranean Sea: retrospective, prospective, perspective. *Collect Vol Sci Pap ICCAT* 65:801–811.

Buckland ST, York AE (2009) Abundance estimation. In: Perrin WF, Würsig B, Thewissen GM (eds) *Encyclopedia of Marine Mammals*, 2nd ed. Academic Press, San Diego, CA, pp 1–6.

Buckland ST, Anderson DR, Burnham KP, Laake JL, Borchers DL, Thomas L (2001) *Introduction to Distance Sampling: Estimating Abundance of Biological Populations.* Oxford University Press, Oxford, UK.

Buckland ST, Burt, ML, Rexstad EA, Mellor M, Williams AE, Woodward R (2012) Aerial surveys of seabirds: the advent of digital methods. *J Appl Ecol* 49:960–967.

Burks CM, Driggers III WB, Mullin KD (2006) Abundance and distribution of whale sharks (*Rhincodon typus*) in the northern Gulf of Mexico. *Fish Bull* 104579–104585.

Campana SE, Shelton PA, Simpson M, Lawson J (2008) *Status of Basking Sharks in Atlantic Canada.* Fisheries and Oceans Canada, Quebec (http://uni.hi.is/scampana/files/2016/01/basking-shark-Res-Doc-2008_004_e.pdf).

Cardona L, Revelles M, Carreras C, San Felix M, Gazo M, Aguilar A (2005) Western Mediterranean immature loggerhead turtles: habitat use in spring and summer assessed through satellite tracking and aerial surveys. *Mar Biol* 147:583–591.

Christiansen F, Dujon AM, Sprogis KR, Arnould JP, Bejder L (2016) Noninvasive unmanned aerial vehicle provides estimates of the energetic cost of reproduction in humpback whales. *Ecosphere* 7:e011468.

Christie KS, Gilbert SL, Brown CL, Hatfield M, Hanson L (2016) Unmanned aircraft systems in wildlife research: current and future applications of a transformative technology. *Front Ecol Environ* 14:241–251.

Cliff G, Anderson-Reade MD, Aitken AP, Charter GE, Peddemors VM (2007) Aerial census of whale sharks (*Rhincodon typus*) on the northern KwaZulu-Natal coast, South Africa. *Fish Res* 84:41–46.

Colefax AP, Butcher PA, Kelaher BP (2017) The potential for unmanned aerial vehicles (UAVs) to conduct marine fauna surveys in place of manned aircraft. *ICES J Mar Sci* 75(1):1–8.

de la Parra Venegas R, Hueter R, Cano JG, Tyminski J, Remolina JG, Maslanka M, Ormos A, Weigt L, Carlson B, Dove A (2011) An unprecedented aggregation of whale sharks, *Rhincodon typus*, in Mexican coastal waters of the Caribbean Sea. *PLoS ONE* 6:e18994.

Dicken ML, Booth AJ (2013) Surveys of white sharks (*Carcharodon carcharias*) off bathing beaches in Algoa Bay, South Africa. *Mar Freshwater Res* 64:530–539.

Dicken ML, Booth AJ (2013) Surveys of white sharks (*Carcharodon carcharias*) off bathing beaches in Algoa Bay, South Africa. *Mar Freshwater Res* 64:530–539.

Dudley SF, Simpfendorfer CA (2006) Population status of 14 shark species caught in the protective gillnets off KwaZulu-Natal beaches, South Africa, 1978–2003. *Mar Freshwater Res* 57:225–240.

Fromentin JM, Farrugio H, Deflorio M, De Metrio G (2003) Preliminary results of aerial surveys of bluefin tuna in the western Mediterranean Sea. *Collect Vol Sci Pap ICCAT* 55:1019–1027.

Gilles A, Scheidat M, Siebert U (2009) Seasonal distribution of harbour porpoises and possible interference of offshore wind farms in the German North Sea. *Mar Ecol Prog Ser* 383:295–307.

Glennie R, Buckland ST, Thomas L (2015) The effect of animal movement on line transect estimates of abundance. *PLoS ONE*, 10:e0121333.

Goetze JS, Fullwood LAF (2013) Fiji's largest marine reserve benefits reef sharks. *Coral Reefs* 32:21–125.

Gruber SH, Nelson DR, Morrissey JF (1988) Patterns of activity and space utilization of lemon sharks, *Negaprion brevirostris*, in a shallow Bahamian lagoon. *Bull Mar Sci* 43:61–76.

Hammond PS, Macleod K, Berggren P, Borchers DL, Burt L, Cañadas A, Desportes G, et al. (2013) Cetacean abundance and distribution in European Atlantic shelf waters to inform conservation and management. *Biol Cons* 164:107–122.

Harvey-Clark CJ, Stobo WT, Helle E, Mattson M (1999) Putative mating behavior in basking sharks off the Nova Scotia coast. *Copeia* 3:780–782.

Hodgson A, Kelly N, Peel D (2013) Unmanned aerial vehicles (UAVs) for surveying marine fauna: a dugong case study. *PLoS ONE*, 8:e79556.

Hodgson A, Peel D, Kelly N (2017) Unmanned aerial vehicles for surveying marine fauna: assessing detection probability. *Ecol Appl* 27:1253–1267.

Hodgson JC, Koh LP (2016) Best practice for minimising unmanned aerial vehicle disturbance to wildlife in biological field research. *Curr Biol* 26:R404–R405.

Jennings RD (1985) Seasonal abundance of hammerhead sharks off Cape Canaveral, Florida. *Copeia* 1985:223–225.

Kajiura SM, Tellman SL (2016) Quantification of massive seasonal aggregations of blacktip sharks (*Carcharhinus limbatus*) in Southeast Florida. *PLoS ONE* 1:e0150911.

Kenney RD, Owen RE, Winn HE (1985) Shark distributions off the Northeast United States from marine mammal surveys. *Copeia* 1985:220–223.

Kessel ST, Gruber SH, Gledhill KS, Bond ME, Perkins RG (2013) Aerial survey as a tool to estimate abundance and describe distribution of a carcharhinid species, the lemon shark, *Negaprion brevirostris. J Mar Biol* 2013:597383.

Ketchum JT, Galván-Magaña F, Klimley AP (2013) Segregation and foraging ecology of whale sharks, *Rhincodon typus*, in the southwestern Gulf of California. *Environ Biol Fish* 96:779–795.

Kiszka J, van der Elst RP (2015) Elasmobranchs (sharks and rays): a review of status, distribution and interaction with fisheries in the Southwest Indian Ocean. In: Van der Elst RP, Everett BI (eds) *Offshore Fisheries of the Southwest Indian Ocean: Their Status and the Impact on Vulnerable Species.* Oceanographic Research Institute, Durban, South Africa, pp 365–389.

Kiszka JJ, Mourier J, Gastrich K, Heithaus MR (2016) Using unmanned aerial vehicles (UAVs) to investigate shark and ray densities in a shallow coral lagoon. *Mar Ecol Progr Ser* 560:237–242.

Koski WR, Allen T, Ireland D, Buck G, Smith PR, Macrander AM, Halick MA, et al. (2009) Evaluation of an unmanned airborne system for monitoring marine mammals. *Aquat Mamm* 35:347–357.

Koski WR, et al. (2015) Evaluation of UAS for photographic re-identification of bowhead whales, *Balaena mysticetus. J Unmanned Veh Syst* 3:22–29.

Laran S, Authier M, Van Canneyt O, Dorémus G, Watremez P, Ridoux V (2017) A comprehensive survey of pelagic megafauna: their distribution, densities and taxonomic richness in the tropical Southwest Indian Ocean. *Front Mar Sci* 4:139.

Lemahieu A, Blaison A, Crochelet E, Bertrand G, Pennober G, Soria M (2017) Human–shark interactions: the case study of Reunion island in the south-west Indian Ocean. *Ocean Coast Manage* 136:73–82.

Lewis JS, Wartzok D, Heithaus MR (2011) Highly dynamic fission–fusion species can exhibit leadership when traveling. *Behav Ecol Sociobiol* 65:1061–1069.

Lutcavage M, Kraus S, Hoggard W (1997) Aerial survey of giant bluefin tuna, *Thunnus thynnus*, in the Great Bahama Bank, Straits of Florida, 1995. *Fish Bull* 95:300–310.

Marsh H, Saalfeld WK (1989) Distribution and abundance of dugongs in the northern Great Barrier Reef Marine Park. *Wildl Res* 16:429–440.

Marsh H, Sinclair DF (1989) An experimental evaluation of dugong and sea turtle aerial survey techniques. *Wildl Res* 16:639–650.

Motta PJ, Maslanka M, Hueter RE, Davis RL, de la Parra R, Mulvany SL, Habegger ML, et al. (2010) Feeding anatomy, filter-feeding rate, and diet of whale sharks *Rhincodon typus* during surface ram filter feeding off the Yucatán Peninsula, Mexico. *Zoology* 113:199–212.

Nadon MO, Baum JK, Williams ID, Mcpherson JM, Zgliczynski BJ, Richards BL, Schroeder RE, Brainard RE (2012) Re-creating missing population baselines for Pacific reef sharks. *Cons Biol* 26:493–503.

Nosal AP, Cartamil DC, Long JW, Lührmann M, Wegner NC, Graham JB (2013) Demography and movement patterns of leopard sharks (*Triakis semifasciata*) aggregating near the head of a submarine canyon along the open coast of southern California, USA. *Environ Biol Fish* 96:865–878.

Pollock K, Marsh HD, Lawler IR, Alldredge MW (2006). Estimating animal abundance in heterogeneous environments: an application to aerial surveys for dugongs. *J Wildl Manage* 70:255–262.

Reid DD, Robbins WD, Peddemors VM (2011) Decadal trends in shark catches and effort from the New South Wales, Australia, Shark Meshing Program 1950–2010. *Mar Freshwater Res* 62:676–693.

Rizzari JR, Frisch AJ, Connolly SR (2014) How robust are estimates of coral reef shark depletion? *Biol Cons* 176:39–47.

Robbins WD, Peddemors VM, Kennelly SJ, Ives MC (2014) Experimental evaluation of shark detection rates by aerial observers. *PLoS ONE* 9:e83456.

Rowat D, Gore M, Meekan MG, Lawler IR, Bradshaw CJ (2009) Aerial survey as a tool to estimate whale shark abundance trends. *J Exp Mar Biol Ecol* 368:1–8.

Sims DW (2003) Tractable models for testing theories about natural strategies: foraging behaviour and habitat selection of free-ranging sharks. *J Fish Biol* 63:53–73.

Squire JL (1967) Observations of basking sharks and great white sharks in Monterey Bay, 1948–50. *Copeia* 1967:247–250.

Squire Jr JL (1990) Distribution and apparent abundance of the basking shark, *Cetorhinus maximus*, off the central and southern California coast, 1962–85. *Mar Fish Rev* 52:8–11.

Taylor JG (1996) Seasonal occurrence, distribution and movements of the whale shark, *Rhincodon typus*, at Ningaloo Reef, Western Australia. *Mar Freshwater Res* 47:637–642.

Thomson JA, Burkholder DA, Cooper AB, Heithaus MR, Dill LM (2012) Heterogeneous patterns of availability for detection during visual surveys: spatiotemporal variation in sea turtle dive–surfacing behaviour on a feeding ground. *Meth Ecol Evol* 3:378–387.

Vaudo JJ, Heithaus MR (2013) Microhabitat selection by marine mesoconsumers in a thermally heterogeneous habitat: behavioral thermoregulation or avoiding predation risk? *PLoS ONE* 8:e61907.

Vincent JB, Werden LK, Ditmer MA (2015) Barriers to adding UAVs to the ecologist's toolbox. *Front Ecol Environ* 13:74–75.

Wamukoya GM, Mirangi JM, Ottichillo WK, Cockcroft V, Salm R (1996) *Report on the Marine Aerial Survey of Marine Mammals, Sea Turtles, Sharks and Rays*. Kenya Wildlife Service, Nairobi.

Westgate AJ, Koopman HN, Siders ZA, Wong SN, Ronconi RA (2014) Population density and abundance of basking sharks *Cetorhinus maximus* in the lower Bay of Fundy, Canada. *Endanger Spec Res* 23:177–185.

Wilson SG (2004) Basking sharks (*Cetorhinus maximus*) schooling in the southern Gulf of Maine. *Fish Oceanogr* 13:283–286.

Wilson SG, Taylor JG, Pearce AF (2001) The seasonal aggregation of whale sharks at Ningaloo Reef, Western Australia: currents, migrations and the El Nino/Southern Oscillation. *Environ Biol Fish* 61:1–11.

Wirsing AJ, Heithaus MR, Dill LM (2006) Tiger shark (*Galeocerdo cuvier*) abundance and growth in a subtropical embayment: evidence from 7 years of standardized fishing effort. *Mar Biol* 149:961–968.

CHAPTER 5

Animal-Borne Video Cameras and Their Use to Study Shark Ecology and Conservation

Yannis P. Papastamatiou
Marine Sciences Program, Florida International University, North Miami, Florida

Carl G. Meyer
Hawaii Institute of Marine Biology, University of Hawaii at Manoa, Kaneohe, Hawaii

Yuuki Y. Watanabe
National Institute of Polar Research, Tokyo, Japan

Michael R. Heithaus
Marine Sciences Program, Florida International University, North Miami, Florida

CONTENTS

5.1 INTRODUCTION

Animal-borne cameras first appeared over 100 years ago, when pigeons fitted with camera collars captured black-and-white images of buildings and streets during flight, and were even used as spies in World War II. Since those early days, camera technology has evolved tremendously, allowing these devices to shrink drastically yet attain the capacity to record hours of color video (Figure 5.1). These technological advancements have made miniature animal-borne cameras attractive tools for documenting animal behaviors that are difficult or impossible to observe directly. These cameras are often integrated into multi-sensor packages collectively referred to as animal-borne video and environmental data collection systems (AVEDs), which are providing new understandings of the foraging ecology, habitat selection, and social behaviors of animals in terrestrial, aquatic, and aerial environments (Moll et al., 2007). Arguably, some of the greatest insight provided by AVEDs has come from deployments on marine taxa because these animals reside in

Figure 5.1 Miniaturization of AVEDs since the 1990s: (A) The original Little Leonardo tag and (B) its attachment to a Weddell seal. (C) The next smaller generation of still-picture cameras in 2003 and attachment to a smaller Baikal seal. (D,E) By 2010, the camera and other sensors were small enough to attach to penguins. (F) A tiger shark fitted with a 1990s-era National Geographic Crittercam. (G) A 1.5-m gray reef shark fitted with a Little Leonardo camera in 2013.

a highly concealing environment and are almost impossible to observe via other methods. Despite recent advancements, the use of AVED technology is still relatively rare in wildlife studies, and most species have never been equipped with any kind of animal-borne camera; hence, the majority of studies are still descriptive and not hypothesis driven (Moll et al., 2007; but see Heaslip et al., 2014).

The challenges of device recovery from wide-ranging animals that do not return to set locations initially hampered the use of AVEDs in shark research (Heithaus et al., 2001). However, recent improvements in recovery methods, especially the use of satellite telemetry to locate tags that have been displaced large distances, have spurred burgeoning use of these devices in shark research. In this chapter, we draw on the limited number of published shark AVED studies and examples from other taxa to highlight the diverse array of ecological questions that can be addressed using this technology. We discuss important considerations for performing studies using AVED and recommend areas of future research.

5.2 RESEARCH AREAS

5.2.1 Habitat Selection

Habitat selection in sharks can be measured using traditional sampling methods (e.g., fishing surveys) or telemetry (e.g., active tracking, satellite tracking) (Simpfendorfer and Heupel, 2004). However, these methods generally lack

sufficient resolution to quantify microhabitat use, which is often important for testing hypotheses about drivers of habitat use. AVEDs allow biologists to observe the exact habitat being used and to see how animals behave within these habitats. The latter is particularly important for studies investigating the role of prey availability in driving habitat use decisions, as theoretical models that form the foundation for testing predictions are based on habitat use during foraging (e.g., Heithaus and Vaudo, 2012). An additional benefit of AVEDs, when combined with active acoustic tracking (see Chapter 8 in this volume), is the ability to more accurately pinpoint the exact habitat in which tracked sharks occur. Currently, habitat use is determined from the spatial location of the animal, which may have, at best, errors of meters (i.e., active tracking may have errors of tens of meters), which may lead to inaccuracies when calculating habitat selection metrics (Heithaus et al., 2001). For example, AVED-equipped tiger sharks (*Galeocerdo cuvier*) in an Australian bay displayed preference for seagrass habitats, particularly along bank edges, and appeared to be foraging in these locations, which is consistent with several models of foraging habitat use (Heithaus et al., 2002, 2006). Active tracking alone did not provide the resolution to identify bank edge use. AVEDs also revealed that blacktip reef sharks (*Carcharhinus melanopterus*) at a Pacific atoll used very shallow back-reef habitats during daytime low tides, behavior that likely increases their body temperatures (Papastamatiou et al., 2015). Although depth and temperature sensors revealed blacktip reef shark position in the water column (i.e., that they were swimming very close to the surface), the cameras were needed to show that surface swimming occurred in very shallow back-reef habitats (Papastamatiou et al., 2015). Telemetry and computer simulations were used to show that reef sharks at this same atoll partition space, both between and within species, and that competition likely plays a role in helping to drive this spatial partitioning (Papastamatiou et al., 2018a). However, sharks could share space with untagged individuals, which would not be apparent from telemetry results. AVEDs were also attached to gray and blacktip reef sharks, and no individual of the opposite species was ever seen in footage, providing further evidence that these species do not share space (Papastamatiou et al., 2018a). Finally, cameras on sixgill sharks showed that during the day, while at depths of 700 m, sharks were often associated with rocky outcrops, although the importance of these deep-sea habitats is unknown (Nakamura et al., 2015a; Meyer, unpublished data).

Studies of habitat use are greatly enhanced by knowledge of prey abundance and encounter rates, but these data are notoriously difficult to obtain for sharks. AVEDs offer the opportunity to measure the prey fields of predators in real time (Hooker et al., 2002). This provides a powerful tool for testing optimal foraging predictions by comparing and contrasting available prey abundance and encounter rates with predator behavior (see next section). Cameras were able to quantify prey encounter rates of tiger sharks in a subtropical bay (Heithaus et al., 2002) and of oceanic whitetip sharks (*Carcharhinus longimanus*) in the open

ocean (Papastamatiou et al., 2018b) (Figure 5.2). Although cameras can provide information on immediate prey fields, combining AVED studies with broader scale surveys of prey, such as using active acoustics (Moursund et al., 2003) or transect surveys (Heithaus et al., 2002), will add to the suite of questions that AVED-based studies can address.

5.2.2 Foraging

Animal-borne video and environmental data collection systems have tremendous potential as tools for studying foraging ecology. Direct validation of foraging in marine species is extremely challenging, so researchers often resort to analytical methods to infer foraging from patterns of movement; for example, travel speed, path tortuosity, or diving can be used as proxies for foraging activity. However, empirical studies using stomach temperature measurements from free-ranging tuna, seabirds, and elephant seals suggest that foraging does not always occur within area-restricted search zones (i.e., where movements are tortuous and often assumed to represent foraging), emphasizing the need for proper validation of mathematically inferred foraging events (Bestley et al., 2010; Kuhn et al., 2009; Weimerskirch et al., 2007). AVEDs allow the direct observation of feeding events and have already been used to validate foraging activity, characterize hunting strategies, and confirm feeding activity in a variety of taxa. One of the best recent examples was the use of AVEDs to confirm ocean sunfish (*Mola mola*) foraging on invertebrates during deep dives (Nakamura et al.,

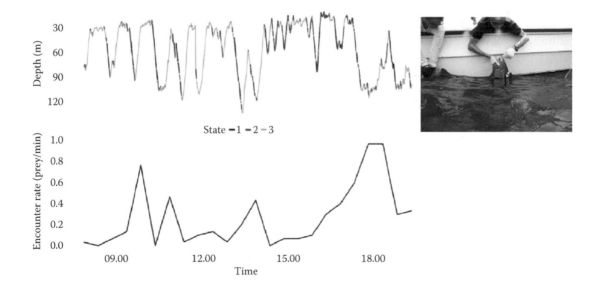

Figure 5.2 Multisensor data loggers with cameras provided insight into the function of diving behavior in an oceanic whitetip shark (*Carcharhinus longimanus*) tagged off Cat Island, the Bahamas. A hidden Markov model was used to analyze acceleration data and assign behavioral states (state 1, low activity; state 2, medium activity; state 3, high activity) throughout the dive (for details, see Leos-Barajas et al., 2016). The video camera was used to estimate encounter rates with potential prey items (i.e., fish and invertebrates other than scyphozoans and pilot fish). Sharks were in a low activity state during dive descent, but the apex of the dives coincided with spikes in prey encounter rates and sharks switching to high activity. These observations suggest that the goal of these dives was searching for potential prey and foraging. (Photograph of tag on dorsal fin by Y.P. Papastamatiou. Data also in Papastamatiou et al., 2018b.)

2015b). AVEDs have been used to compare the success of solitary vs. group hunting in seabirds and have revealed that some seabirds track the position of other predators foraging on their prey, rather than the prey themselves (Sutton et al., 2015; Tremblay et al., 2014). By combining cameras with lights, foraging rates of alligators were measured throughout the diel period, demonstrating that the highest feeding success occurred at night and in the morning (Nifong et al., 2014). Tiger sharks were observed chasing several prey species over seagrass in a shallow Australian bay, sometimes using the descent phase of yo-yo swimming to initiate attacks on benthic prey (Heithaus et al., 2002). Off Hawaii, still cameras suggested that tiger sharks yo-yo dive as a foraging tactic (Nakamura et al., 2011).

One major constraint of AVED technology is the limited camera battery life of the necessarily small devices mounted on animals. Short battery life limits the duration of footage recorded (typically 2 to 12 hours with current devices), which reduces the probability of capturing foraging activities. This issue is particularly pertinent to sharks, as these animals may feed infrequently and sporadically compared to marine mammals and seabirds.

As the majority of sharks and rays are ectotherms, they have considerably lower metabolic rates than endothermic birds and mammals for a given body size; thus, they require lower feeding rates in the wild. Furthermore, recent shark studies suggest that there can be considerable variation among individuals in foraging locations, movements, and behaviors (e.g., Matich et al., 2011; Towner et al., 2016), necessitating sampling a large number of individuals. Moreover, answering ecological-scale foraging questions requires measuring foraging over periods of days to weeks, at the very least. These constraints can be overcome by using information from companion sensors, such as accelerometers (see Chapter 3 in this volume), which require less power than cameras and consequently have much longer battery lives (days to months). As an example, video cameras were used to calibrate the acceleration signals produced by penguins foraging under ice, after which accelerometers quantified the feeding rates of these animals throughout the day (Watanabe and Takahashi, 2013). Similarly, video footage was used to calibrate acceleration signals from foraging Hawaiian monk seals and was further combined with GPS tags to map seal foraging activity (Wilson et al., 2017). More recently, a white shark was successfully filmed using AVEDs when it was chasing and attacking a seal (Watanabe et al., in review). The simultaneous accelerometer records showed an intensive burst event during the chase. Based on the confirmed seal chase, other potential seal chase events were extracted from all acceleration data recorded for multiple individuals, and flexible hunting strategies of white sharks were revealed. Similarly, gray reef sharks with AVEDs were observed feeding on reef fish during the day, and other individual sharks were also observed foraging in frame (Papastamatiou et al., in

review). Despite reef sharks being considered nocturnal foragers, these results show that they will also feed during the day. These examples demonstrate the utility of AVEDs for recording foraging events and calibrating signals, which will be especially useful for species that cannot be kept in captivity.

In addition, AVEDs also provide largely untapped potential for testing hypothesis-driven predictions (Moll et al., 2007) in areas such as optimal foraging theory; for example, cameras and accelerometers were used to show that patch use by foraging penguins matched predictions from marginal value theory (Watanabe et al., 2014). Similarly, the dive dynamics of harbor seals appeared to optimize prey encounter rates as predicted by theory (Heaslip et al., 2014). Many species of pelagic shark exhibit diel changes in diving depth, which may optimize prey encounter rates and/or enable thermoregulation. A combination of cameras, accelerometers, thermal sensors, and active acoustics could reveal the definitive reason for this behavior. This combination of sensors (including a muscle-placed thermistor) showed that ocean sunfish (*Mola mola*) were feeding on deep zooplankton during the day and resting at night (Nakamura et al., 2015). Bounce dives during the day were due to the sunfish having to return to the surface to rewarm their bodies. Similar tools could be applied to pelagic sharks such as blue, thresher and whale sharks.

5.2.3 Mating and Social Behaviors

Animal-borne video and environmental data collection systems also have great potential for revealing social associations and interactions between conspecifics and competitors. Measuring social associations in free-ranging animals is very difficult, although recent technological (e.g., proximity tags) (Haulsee et al., 2016) and analytical (Gaussian mixture modeling) (Jacoby et al., 2016) advances now make it possible to build dynamic social networks. However, these tools require considerable infrastructure to quantify associations between animals and do not reveal any details of the interactions. By contrast, AVEDs can function as proximity loggers and also provide details on the interaction itself (Hooker et al., 2015). For example, video cameras placed on monk seals in the Northwestern Hawaiian Islands revealed that seals would often lose their prey to sharks or large teleosts (Parrish et al., 2008). Kleptoparasitism can play a significant role in the foraging behavior of large terrestrial carnivores and may play an important role in the competitive interactions between marine mammals and sharks. The social dynamics of conspecific foraging and even potential cooperative hunting can be determined from AVEDs (Sutton et al., 2015; Takahashi et al., 2004). Oceanic whitetip sharks live in an oligotrophic and seemingly barren environment, but there are locations where sharks appear to aggregate within a small area. Cameras attached to sharks suggested that they forage by themselves, as no other individuals were ever seen in footage (Papastamatiou et al., 2018b). In other locations, oceanic whitetips are known to follow pods of pilot whales, but the function of this behavior is unknown. AVEDs, however, may reveal details of these interactions, which are impossible to verify using other sensors.

Animal-borne video and environmental data collection systems can also play a role in elucidating mating activity in marine animals; for example, male harbor seal mating strategies and the driving factors behind them were revealed using a combination of cameras, tracking, and molecular techniques (Boness et al., 2006). Of course, using these to study mating tactics and reproductive ecology in sharks will be very challenging due to our lack of knowledge of mating locations in sharks. However, AVEDs have been used in known nurse shark aggregation sites (Simpfendorfer and Heupel, 2004). Furthermore, a camera deployed on a male tiger shark in Hawaii during the mating season captured a mating attempt, further highlighting the potential of this tool to reveal mating frequency and habitats (Figure 5.3) (Meyer et al., 2018). The ultimate advancement in shark social dynamics will be the simultaneous deployment of proximity loggers and AVEDs in order to quantify free-ranging

social networks and understand the behaviors behind them. To date, this combination of methods has only been used with New Caledonian crows, where cameras were used to quantify foraging behavior, and inter-animal telemetry measured dynamic networks of associations between individuals (Troscianko and Rutz, 2015).

5.2.4 Conservation and Fisheries

Animal-borne video and environmental data collection systems could also be used to quantify interactions between sharks and anthropogenic activities and structures, including fishing gears, oil rigs, power generators (e.g., wind and tide generators), and electrical cables. Cameras on seabirds have revealed how often and where birds interact with fishing vessels, even identifying the type of vessel they follow (Tremblay et al., 2014; Votier et al., 2013). Sharks are known to associate with offshore fish cages and oil platforms, and they follow fishing vessels (Papastamatiou et al., 2010; Robinson et al., 2013). AVEDs are one of the only methods available for quantifying the frequency and degree of these interactions, and perhaps more importantly they could help

Figure 5.3 Frame grabs from video recovered from a shark-mounted camera deployed on a 436-cm TL male tiger shark captured off Oahu in January 2015. (A) Rear of tiger shark approached by the camera shark shows no evidence of claspers on the pelvic fins (1), indicating that this shark is female. (B) View of dorsal surface showing evidence of mating scars (2) behind the trailing edge of the dorsal fin (3). (C) Profile view of a female tiger shark as the camera shark approaches, showing that the nictitating membrane is retracted (4). (D) As the camera shark makes a closer approach to the female tiger shark, the nictitating membrane can be seen entirely covering the eye (5). (From Meyer, C.G. et al., *Sci. Rep.*, 8(1), 4945, 2018.)

explain why these associations occur. For example, sharks that associate with fish cages or fish aggregating devices do not appear to be feeding, so the functions of these associations are still unknown (Papastamatiou et al., 2010). AVEDs could also play a role in understanding stress effects associated with catch-and-release fishing and how this may relate to mortality. Traditionally, mortality has been estimated using satellite tags or blood chemistry, although recently recoverable accelerometers have proven to be a cost-effective tool (Whitney et al., 2016). AVEDs were used to assess post-release behavior in gray reef sharks, and, although less cost effective than accelerometers, they may provide additional insight into how catch and release may impact survivorship (Skomal et al., 2007).

5.2.5 Sharks as Observation Platforms

It is becoming increasingly common to use marine animals as remote sampling platforms as they move through areas and at depths that are traditionally very difficult to sample (Fedak, 2013). Marine mammals carrying biologgers have been used to measure salinity and temperature profiles of polar oceans and helped validate oceanographic models (Fedak, 2013). Camera tags may also prove useful with sharks for observing species and community composition in similarly difficult to observe habitats; for example, sixgill sharks with camera tags were able to capture pictures of deep-sea invertebrates at a depth of 700 m, which would otherwise require camera surveys and expensive ship time (Nakamura et al., 2015a) (Figure 5.4). Gray reef sharks were able to "survey" mesophotic reefs in the 60- to 80-m range off a remote Pacific atoll, revealing the invertebrate and fish communities at depths below those surveyed by scientific divers. In addition, these images highlighted the abundance of reef fishes that are not considered in estimates from shallow water surveys (Papastamatiou, unpublished data) (Figure 5.4).

5.3 METHODS AND FUTURE CONSIDERATIONS

5.3.1 Attachment and Retrieval Methods

To date, AVEDs have been attached to either the dorsal fin or the pectoral fin of free-swimming sharks (Heithaus et al., 2001; Nakamura et al., 2011, 2015b). The advantage of AVEDs placed on pectoral fins is that they may provide a better view of the mouth and subsequent feeding behavior, but they may also create a disproportionate distribution of drag costs across the shark's body. AVEDs are embedded in floats (e.g., syntactic foam) along with VHF and sometimes satellite transmitters (e.g., Wildlife Computers SPOT tags), and can be attached in one of two ways (Watanabe et al., 2004). They can be attached via zip ties, cables, or wiring with an associated time-release mechanism (e.g., galvanic links, electronic timers), or they can be attached to spring-action clamps that go over the dorsal fin (Chapple et al., 2015; Nakamura et al., 2011; Papastamatiou et al., 2015). The fixed attachment is more secure and less likely to prematurely release, but it is more invasive than the clamp method (e.g., holes must be made in the dorsal fin) and takes longer to attach. The clamp method also allows tags to be attached to sharks swimming past the boat or to be tagged underwater, which is useful for very large species that are difficult to catch (Chapple et al., 2015) (Figure 5.5). Regardless of attachment, the tag eventually detaches from the shark and floats to the surface. The embedded satellite transmitter sends email messages of the tag's general location, which can then be precisely located using the VHF transmitter (Watanabe et al., 2004).

5.3.2 Miniaturization and Drag Reduction

Animal-borne sensors increase drag and energy expenditure by the host animal (Jones et al., 2013), and although these costs may be relatively small for short-duration deployments (a few days) they may become substantial over longer

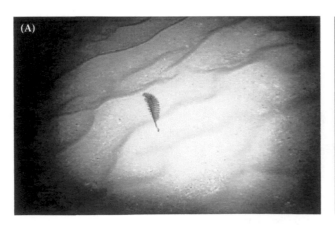

Figure 5.4 Sharks as observation platforms. Sharks carrying AVEDs can survey communities in locations that are difficult to sample. (A) Photograph of a sea pen taken at 700 m off Oahu by a sixgill shark (Meyer, unpublished data); (B) photograph of large schools of reef fish on a mesophotic reef (60 m) off Palmyra Atoll taken by a gray reef shark (Y.P. Papastamatiou, unpublished data).

Figure 5.5 An AVED (Customized Animal Tracking Solutions, CATS, Germany) is attached to a great hammerhead shark by a diver off of Bimini, Bahamas. The AVED is connected to a spring-loaded clamp with a galvanic release. An application pole is used to spring load the clamp underwater. (Photograph by E. Kitsios.)

durations. The greatest component of the increased drag will be related to the cross-sectional area of the tag, highlighting the need to reduce camera area and optimize tag shape (Jones et al., 2013; Ropert-Coudert et al., 2007). Camera tags are now small enough to be applied to medium-sized sharks (minimum 1 m total length), but further miniaturization will allow their application to small and juvenile individuals. The effect of buoyancy from the floats themselves should also be considered, especially for small sharks; however, it is relatively easy to counter flotation by attaching weights to ensure that the sensor package is neutrally buoyant while attached to the shark (Nakamura et al., 2015a).

5.3.3 Energy Efficiency, Recording Time, and Smart Duty Cycling

The second major limitation is battery life. In many cases, the size of AVEDs is determined by a trade-off between tag size and battery life. Most systems can record up to 12 hours at high-definition (HD) resolution. Larger tags or lower resolution recording can extend this to approximately 30 hours for some systems. Longer duration video recording will be particularly beneficial with sharks, as behaviors such as feeding are less frequent than in marine mammals or seabirds and likely vary across habitats and times of day. Effective recording duration can be extended by having the cameras turn on during specific times of the day or can even be based on simultaneous sensor measurements. Acceleration-triggered video cameras (that begin filming when simultaneously recorded acceleration exceeds a threshold and return to the sleep mode after a preset period) have recently been developed and successfully applied to elephant seals (Naito et al., 2017). Elephant seals were revealed to be capturing deep-sea fishes (e.g., ragfish) at depths of about 800 m,

within the oxygen minimum zone. Applying this technique to sharks (especially deep-sea sharks) will greatly enhance our knowledge of their foraging ecology.

5.3.4 Development of New Camera Technologies

A serious limitation of AVED technology is the ability to obtain images of adequate quality in low-light conditions. This is particularly important with sharks because (1) many species show increased activity at night and foraging may be concentrated then (Papastamatiou et al., 2015), and (2) many species occur at depths with little to no light. AVEDs are currently used in conjunction with external light-emitting diodes (LEDs) to record either at night or at great depths (Gilly et al., 2012; Nakamura et al., 2015b). The biggest issue with external LEDs is that they may alter the behavior of conspecifics of prey. For example, Humboldt squid carrying AVEDs with red LEDs were attacked by other squid (Gilly et al., 2012). It is also likely that LEDs could alter the behavior of potential prey in deep-sea habitats where light is used as a lure by multiple taxa. Some AVEDs incorporate image intensifiers (photomultiplier tubes, or PMTs) and make use of infrared lights, but they are expensive and relatively large and will require some modification before entering widespread use. An additional advancement would be to miniaturize 360° cameras so that images can be obtained from all angles. This would greatly improve the ability to infer social associations, when conspecifics may not be directly in front of the animal (e.g., schooling sharks). Finally, it is possible that acoustic cameras might eventually be small enough to obtain acoustic images from the animal. Acoustic cameras have already been shown capable of identifying individual sharks and mobulid rays, although not the species (McCauley et al., 2016), and they can obtain images at night or in zero visibility. An animal-borne echosounder has already been built for use with foraging marine mammals (Lawson et al., 2016), and, although technical challenges remain, future miniaturization of the tags may allow the development of an animal-borne acoustic camera.

5.3.5 Integration with Other Sensors and Data Analysis

As discussed above, AVEDs can be used to calibrate data from other sensors such as accelerometers (see Chapter 3 in this volume). Acceleration sensors have much longer duration of recording than AVEDs and can provide much longer time-series of behaviors such as feeding, if calibrated properly (Watanabe and Takahashi, 2013); however, more quantitative analytical tools can make interpretation and assignment of activity signals less subjective. Machine-learning methods can be used where the algorithm is trained to identify different behaviors using the video footage and then searches for similar records or signals in the activity data (Resheff et al., 2014). A freely available Python-based web application (AcceleRater) can perform a variety of supervised learning methods with accelerometry data (Resheff et al., 2014). Unsupervised methods such as hidden Markov models (HMMs) can also be used to assign the animal to behavioral states based on acceleration signals (Leos-Barajas et al., 2017). HMMs can quantify how the probability of sharks being in a behavioral state (e.g., active vs. inactive) varies with factors such as diel and tidal cycles or abiotic conditions (Leos-Barajas et al., 2017). AVEDs can assign HMM behavioral states to realistic biological interpretations of behavior; for example, if it is active is it foraging? An example of the combination of methods can be seen for data from an oceanic whitetip shark (Figure 5.3). Hidden Markov models were used to assign behavioral states throughout the time-series based on activity derived from the accelerometers (Leos-Barajas et al., 2017). At the same time, AVEDs were used to quantify encounter rates with potential prey throughout the dive. The combination showed that sharks were descending while in a low activity state, but the apex of their dives corresponded with a peak in prey encounter rates and a switch to high activity state. This suggests that the purpose of these dives was searching for prey and foraging.

There are many other combinations of sensors where cameras can validate other information being collected, including sound recorders to study communication in, for example, penguins (Choi et al., 2017) and proximity sensors to look at social associations. As the amount of video data increases, it is getting more difficult for researchers to watch and analyze these data. Unlike other types of digital data (e.g., depth), an overview of video data cannot be easily inspected. Automated image data analysis methods have developed quickly in the private sector, primarily because security cameras became widespread in almost all cities around the world. These technologies could be applied to video data collected by animal-borne cameras and enhance the efficiency and accuracy of data analyses.

5.4 CONCLUSION

Extensive advancements have been made in the technology since the first application of AVEDs to sharks, nearly two decades ago. Tags have become smaller, can record for longer periods, and can be recovered after multiple days or weeks of deployment. Cameras can be combined with other sensors, and novel analytical tools are able to extract as much information from behavioral data as possible. As such, the time should be ripe for a rapid increase in the use of AVEDs to answer ecological, physiological, behavioral, and conservation-focused questions. Despite these advancements, there has been little increase in the use of AVEDs for shark studies and their potential to address important questions. This chapter has provided a framework for the biological questions that can be answered using these methods and how future advancements will allow researchers to approach questions that cannot be answered without a shark's-eye view of their environment.

ACKNOWLEDGMENTS

We thank V. Leos for the hidden Markov analysis used in Figure 5.2 and J. Caselle for support of field work in Palmyra.

REFERENCES

Bestley S, Patterson TA, Hindell MA, Gunn JS (2010) Predicting feeding success in a migratory predator: integrating telemetry, environment, and modeling techniques. *Ecology* 91:2373–2384.

Boness DJ, Bowen WD, Buhleier BM, Marshall GJ (2006) Mating tactics and mating system of an aquatic–mating pinniped: the harbor seal, *Phoca vitulina. Behav Ecol Sociobiol* 61:119–130.

Chapple TK, Gleiss AC, Jewell OJD, Wikelski M, Block BA (2015) Tracking sharks without teeth: a non-invasive rigid tag attachment for large predatory sharks. *Anim Biotelem* 3:14.

Choi N, Kim JH, Kokubun N, Park S, Chung H, Lee WY (2017) Group association and vocal behavior during foraging trips in Gentoo penguins. *Sci Rep* 7:7570.

Fedak MA (2013) The impact of animal platforms on polar ocean observation. *Deep Sea Res II Top Stud Ocean* 88:7–13.

Gilly WF, Zeidberg LD, Booth AT, Stewart JS, Marshall G, Abernathy K, Bell LE (2012) Locomotion and behavior of Humboldt squid, *Dosidicus gigas*, in relation to natural hypoxia in the Gulf of California, Mexico. *J Exp Biol* 215:3175–3190.

Haulsee DE, Fox DA, Breece MW, Brown LM, Kneebone J, Skomal GB, Oliver MJ (2016) Social network analysis reveals potential fission–fusion behavior in a shark. *Sci Rep* 6:34087.

Heaslip SG, Bowen WD, Iverson SJ (2014) Testing predictions of optimal diving theory using animal-borne video from harbor seals (*Phoca vitulina concolor*). *Can J Zool* 92:309–318.

Heithaus MR, Vaudo, JJ (2012) Predator–prey interactions. In: Carrier JC, Musick J, Heithaus MR (eds.) *Biology of Sharks and Their Relatives*, 2nd ed. CRC Press, Boca Raton, FL, pp 505–546.

Heithaus MR, Marshall GJ, Buhleier BM, Dill LM (2001) Employing Crittercam to study habitat use and behavior of large sharks. *Mar Ecol Prog Ser* 209:307–310.

Heithaus MR, Dill LM, Marshall GJ, Buhleier B (2002) Habitat use and foraging behavior of tiger sharks (*Galeocerdo cuvier*) in a seagrass ecosystem. *Mar Biol* 140:237–248.

Heithaus MR, Hamilton IM, Wirsing AJ, Dill LM (2006) Validation of a randomization procedure to assess animal habitat preferences: microhabitat use of tiger sharks in a seagrass ecosystem. *J Anim Ecol* 75:666–676.

Hooker SK, Boyd IL, Jessop M, Cox O, Blackwell J, Boveng PL, Bengston JL (2002) Monitoring the prey-field of marine predators: combining digital imaging with datalogging tags. *Mar Mamm Sci* 18: 680–697.

Hooker SK, Barychka T, Jessopp MJ, Staniland IJ (2015) Images as proximity sensors: the incidence of conspecific foraging in Antarctic fur seals. *Anim Biotelem* 3:37.

Jacoby DMP, Papastamatiou YP, Freeman R (2016) Inferring animal social networks and leadership: applications for passive monitoring arrays. *J R Soc Interface* 13:20160676.

Jones TT, Van Houtan KS, Bostrom BL, Ostafichuk P, Mikkelsen J, Tezcan E, Carey M, Imlach B, Seminoff JA (2013) Calculating the ecological impacts of animal-borne instruments on aquatic organisms. *Meth Ecol Evol* 4:1178–1186.

Kuhn CE, Crocker DE, Tremblay Y, Costa DP (2009) Time to eat: measurements of feeding behaviour in a large marine predator, the northern elephant seal *Mirounga angustirostris. J Anim Ecol* 78:513–523.

Lawson GL, Huckstadt LA, Lavery AC, Jaffre FM, Wiebe PH, Fincke JR, Crocker DE, Costa DP (2016) Development of an animal-borne 'sonar tag' for quantifying prey availability: test deployments on northern elephant seals. *Anim Biotelem* 3:22.

Leos–Barajas V, Photopoulou T, Langrock R, Patterson TA, Watanabe YY, Murgatroyd M, Papastamatiou YP (2017) Analysis of animal accelerometer data using hidden Markov models. *Meth Ecol Evol* 8:161–173.

Matich P, Heithaus MR, Layman CA (2011) Contrasting patterns of individual specialization and trophic coupling in two marine apex predators. *J Anim Ecol* 80:295–304.

McCauley DJ, Salles PA, Young HS, Gardner JPA, Micheli F (2016) Use of high-resolution acoustic camera to study reef shark behavioral ecology. *J Exp Mar Biol Ecol* 482:128–133.

Meyer CG, Anderson JM, Coffey DM, Hutchinson MR, Royer MA, Holland KN (2018). Habitat geography around Hawaii's oceanic islands influences tiger shark (*Galeocerdo cuvier*) spatial behaviour and shark bite risk at ocean recreation sites. *Sci Rep* 8(1):4945.

Moll RJ, Millspaugh JJ, Beringer J, Sartwell J, He Z (2007) A new 'view' of ecology and conservation through animal-borne video systems. *Trend Ecol Evol* 22:660–668.

Moursund RA, Carlson TJ, Peters RD (2003) A fisheries application of a dual–frequency identification sonar acoustic camera. *ICES J Mar Sci* 60:678–683.

Naito Y, Costa DP, Adachi T, Robinson PW, Peterson SH, Mitani Y, Takahashi A (2017) Oxygen minimum zone: an important oceanographic habitat for deep-diving northern elephant seals, *Mirounga angustirostris. Ecol Evol* 7:6259–6270.

Nakamura I, Watanabe YY, Papastamatiou YP, Sato K, Meyer CG (2011) Yo-yo vertical movements suggest a foraging strategy for tiger sharks, *Galeocerdo cuvier. Mar Ecol Prog Ser* 424:237–246.

Nakamura I, Goto Y, Sato K (2015a) Ocean sunfish rewarm at the surface after deep excursions to forage for siphonophores. *J Anim Ecol* 84:590–603.

Nakamura I, Meyer CG, Sato K (2015b) Unexpected positive buoyancy in deep sea sharks, *Hexanchus griseus*, and a *Echinorhinus cookie. PLoS ONE* 10:e0127667.

Nifong JC, Nifong RL, Silliman BR, Lowers RH, Guillette LJ, Ferguson JM, Welsh M, et al. (2014) Animal-borne imaging reveals novel insights into the foraging behavior and diel activity of a large–bodied apex predator, the American alligator (*Alligator mississippiensis*). *PLoS ONE* 9:e83953.

Papastamatiou YP, Itano DG, Dale JJ, Meyer CG, Holland KN (2010) Site fidelity and movements of sharks associated with ocean-farming cages in Hawaii. *Mar Freshwater Res* 61:1366–1375.

Papastamatiou YP, Watanabe YY, Bradley D, Dee LE, Weng K, Lowe CG, Caselle JE (2015) Drivers of daily routines in an ectothermic predator: hunt warm, rest warmer? *PLoS ONE* 10:e0127807.

Papastamatiou YP, Bodey TW, Friedlander AM, Lowe CG, Bradley D, Weng K, Priestley V, Caselle JE. (2018a) Spatial separation without territoriality in shark communities. *Oikos* (in press).

Papastamatiou YP, Iosilevskii G, Leos-Barajas V, Brooks EJ, Howey LA, Chapman DD, Watanabe YY (2018b) Optimal swimming strategies and behavioral plasticity in oceanic whitetip sharks. *Sci Rep* 8:551.

Parrish FA, Marshall GJ, Buhleier B, Antonelis GA (2008) Foraging interaction between monk seals and large predatory fish in the Northwestern Hawaiian Islands. *Endang Species Res* 4:299–308.

Resheff YS, Rotics S, Harel R, Spiegel O, Nathan R (2014) AcceleRater: a web application for supervised learning of behavioral modes from acceleration measurements. *Move Ecol* 2:27.

Robinson DP, Jaidah MY, Jabado RW, Lee-Brooks K, Nour El-Din NM, Al Malki AA, Elmeer K, et al. (2013) Whale sharks, *Rhincodon typus*, aggregate around offshore platforms in Qatari waters of the Arabian Gulf to feed on fish spawn. *PLoS ONE* 8:e58255.

Ropert-Coudert Y, Knott N, Chiaradia A, Kato A (2007) How do different data logger sizes and attachment positions affect the diving behaviour of little penguins? *Deep Sea Res II Top Stud Ocean* 54:415–423.

Simpfendorfer CA, Heupel MR (2004) Assessing habitat use and movement. In: Carrier JC, Musick JA, Heithaus MR (eds.) *Biology of Sharks and Their Relatives.* CRC Press, Boca Raton, FL, pp 553–572.

Skomal G, Lobel PS, Marshall G (2007) The use of animal-borne imaging to assess post-release behavior as it relates to capture stress in grey reef sharks, *Carcharhinus amblyrhynchos. Mar Technol Soc J* 41:44–48.

Sutton GJ, Hoskins AJ, Arnould JPY (2015) Benefits of group foraging depend on prey type in a small marine predator, the little penguin. *PLoS ONE* 10:e0144297.

Takahashi A, Sato K, Naito Y, Dunn MJ, Trathan PN, Crozall JP (2004) Penguin-mounted cameras glimpse underwater group behaviour. *Biol Lett* 271:S281–S282.

Towner AV, Leos-Barajas V, Langrock R, Schick RS, Smale MJ, Kaschke T, Jewell OJD, Papastamatiou YP (2016) Sex-specific and individual preferences for hunting strategies in white sharks. *Funct Ecol* 30:1397–1407.

Tremblay Y, Thiebault A, Mullers R, Pistorius P (2014) Bird-borne video-cameras show that seabird movement patterns relate to previously unrevealed proximate environment, not prey. *PLoS ONE.* 9:e88424.

Troscianko J, Rutz C (2015) Activity profiles and hook-tool use of New Caledonian crows recorded by bird-borne video cameras. *Biol Lett* 11:20150777.

Votier SC, Bicknell A, Cox SL, Scales KL, Patrick SC (2013) A bird's eye view of discard reforms: bird-borne cameras reveal seabird/fishery interactions. *PLoS ONE* 8:e57376.

Watanabe YY, Takahashi A (2013) Linking animal-borne video to accelerometers reveals prey capture variability. *Proc Nat Acad Sci* 110:2199–2204.

Watanabe Y, Baranov EA, Sato K, Naito Y, Miyazaki N (2004) Foraging tactics of Baikal seals differ between day and night. *Mar Ecol Prog Ser* 279:283–289.

Watanabe YY, Ito M, Takahashi A (2014) Testing optimal foraging theory in a penguin–krill system. *Proc R Soc B Biol Sci* 281:20132376.

Weimerskirch H, Pinaud D, Pawlowski F, Bost C (2007) Does prey capture induce area–restricted search? A fine-scale study using GPS in a marine predator, the wandering albatross. *Am Nat* 170:734–743.

Whitney NM, White CF, Gleiss AC, Schwieterman GD, Anderson P, Hueter RE, Skomal GB (2016) A novel method for determining post-release mortality, behavior, and recovery period using acceleration data loggers. *Fish Res* 183:210–221.

Wilson K, Littnan C, Halpin P, Read A (2017) Integrating multiple technologies to understand the foraging behavior of Hawaiian monk seals. *R Soc Open Sci* 4:160703.

CHAPTER 6

Use of Autonomous Vehicles for Tracking and Surveying of Acoustically Tagged Elasmobranchs

Christopher G. Lowe
Shark Lab, California State University, Long Beach, California

Connor F. White
Shark Lab, California State University, Long Beach, California

Christopher M. Clark
Lab for Autonomous and Intelligent Robotics, Harvey Mudd College, Claremont, California

CONTENTS

6.1 INTRODUCTION

Although great strides have been made in understanding shark behavior and ecology due to advances in technology, many technical challenges still exist in quantifying how elasmobranchs behave in response to their environment, conspecifics, prey, and predators. Telemetry (acoustic and satellite), sensor, video, and data logger technology have all provided new insights into how sharks move, select habitat, and interact with other species; however, data collection can be arduous, expensive, time consuming, and spatially and temporally limiting. Satellite telemetry (see Chapter 19 in this volume) can provide valuable data on the water temperature and depths that wide-ranging elasmobranchs move through, but this technology does not provide much in the way of fine-scale geospatial positions or behavioral state. Alternatively, acoustic telemetry (see Chapter 8 in this volume) can provide either short-term, high-resolution position information and corresponding environmental data (e.g., water temperature, depth, activity rate) via active tracking or longer term, lower resolution position information for many animals simultaneously, but over a much more limited spatial scale via passive tracking. Active tracking, in particular, is typically labor intensive and allows for tracking of one individual at a time, whereas passive tracking is much less labor intensive but spatially limited by the size of a stationary array.

Advances in underwater robotics, inertial measurement units (IMUs), global positioning systems (GPSs), and autonomous control systems have greatly enhanced our ability to measure a wide array of oceanographic parameters via autonomous mobile underwater or surface vehicles and simultaneously provide high positional accuracy. Commercially available autonomous underwater vehicles (AUVs) and autonomous surface vehicles (ASVs) have been developed and used primarily for oceanographic purposes; however, recent attempts have been made to incorporate acoustic telemetry hydrophones and receivers into these robots for fully autonomous active tracking of elasmobranchs tagged with acoustic transmitters. The primary benefits of these coupled technologies are they allow for accurate, fine-scale positioning of tagged elasmobranchs but can also simultaneously record environmental conditions, map the benthos, and record biota surrounding focal individuals. The information discussed in this chapter covers the types of autonomous vehicles, previous uses for surveying and tracking of elasmobranchs, and the future of this technology.

6.2 USE OF AUTONOMOUS UNDERWATER VEHICLES IN OCEANOGRAPHY AND MARINE SCIENCE

In 1957, the first autonomous underwater vehicle (AUV) was developed at the University of Washington's Applied Physics Laboratory. The vehicle, named SPURV (Special Purpose Underwater Research Vehicle), had the ability to dive down to 3000 m and measure water temperature and conductivity. Figure 6.1A shows the SPURV being deployed. Its form, consistent with early torpedo designs, was long and thin with a motor-driven propeller located at the rear of the vehicle for propulsion. Four control surfaces at the rear enabled steering for direction control (yaw, pitch, and roll). This general design is still predominant in the AUV industry.

6.2.1 Types of AUVs and Capabilities

Since SPURV was introduced to the research community, a large number of other underwater vehicles with different form factors have been designed, constructed, and deployed (Figure 6.1B,C). Unlike AUVs, which are typically untethered, fully autonomous, and designed to cover longer distances, remotely operated vehicles (ROVs) are tethered to a pilot's control console located above the surface. Figure 6.1B shows an example of a micro-class ROV. Different again are autonomous surface vehicles (ASVs), which are restricted to motion on the surface (Figure 6.1C). These vehicles are generally compared based on their depth rating, speed, level of autonomy (e.g., how much pilot interaction is required during a mission), endurance, sensor payload capacity, maneuverability, and data availability (e.g., whether or not data can be transmitted to the pilot in real time).

6.2.1.1 Remotely Operated Vehicles

Remotely operated vehicles (ROVs) are differentiated from other vehicles by the fact that there is a tether that connects the vehicle to a control station typically located on a dock or boat. Pilots manually control the ROV via a joystick, allowing the pilot to send control signals to the vehicle's actuators (e.g., motors) through a wire bundle inside the tether. Similarly, vehicle sensor data are relayed from the vehicle back to the control station through the tether. These data can take on the form of video, depth, sonar, temperature, etc. and are typically necessary for pilot navigation. Power is also provided through the tether, removing the need for an onboard power supply. The presence of this tether offers several advantages when conducting fish or shark surveys. It provides real-time feedback to a pilot, thus allowing adaption to changing fish behaviors. The video also provides a means for characterizing habitat, prey abundance, etc.; for example, one can survey the bottom to determine if the seafloor would facilitate egg laying for skates and catsharks. The tether, unfortunately, also restricts motion of the vehicle by limiting its range to the length of the tether. Monitoring fish with longer ranges using an ROV is simply not feasible. Because the vehicle is not traveling long distances and power is provided, ROVs are designed not so much for endurance as they are for maneuverability (e.g., their thruster configuration often allows motion in all directions). They can be used for situations requiring up-close, interactive navigation.

Figure 6.1 Examples of underwater robots: (A) Deployment of SPURV, the first autonomous underwater vehicle (AUV) (photograph courtesy of the Applied Physics Laboratory, University of Washington, Seattle). (B) A VideoRay underwater remotely operated vehicle (ROV). (C) This robot is a ClearPath Robotics Kingfisher ASV.

6.2.1.2 Buoyancy-Driven Gliders

Unlike typical AUVs, which are motor or propeller driven, buoyancy-driven gliders provide forward motion by modifying their applied buoyancy force in flight. The gliders are equipped with bladders (often located in the vehicle nose) that can be filled or emptied with water from the surrounding environment. When ballasted correctly, the vehicle will change between being positively and negatively buoyant as the bladder is emptied or filled. Wings affixed to the sides of the vehicle transform these vertical forces into horizontal forces, enabling forward motion. Although this type of motion is slower (e.g., 0.35 m/sec) than propeller-driven AUVs (e.g., 2 m/sec), it requires less energy and hence provides longer endurance; for example, the Slocum glider (Figure 6.2A) can travel a distance of 13,000 km with a maximum deployment time of 18 months. For this reason, the application of such vehicles to fish tracking is limited to conducting more long-range passive surveys than active tracking which requires greater speeds and maneuverability.

6.2.1.3 Wave-Driven Surface Gliders

Similar to buoyancy-driven gliders, wave-driven gliders harness naturally occurring vertical motion and transform it to horizontal motion. As waves carry the vehicle up and

Figure 6.2 Autonomous underwater vehicles: (A) Slocum Glider; (B) underside of Liquid Robotics' Wave Glider (photograph courtesy of Liquid Robotics, Sunnyvale, CA); (C) OceanServer Iver2 AUV; (D) REMUS 100; (E) REMUS 600.

down, a wing system that sits underwater (see Figure 6.2B) provides a force in the forward direction. Also similar to buoyancy-driven gliders, wave-driven gliders have longer endurance but lower maneuverability and speed, making them applicable to long-range passive tracking as opposed to active tracking. Because their primary hull is surface oriented, these vehicles can offer real-time radio connectivity and the use of solar panels to provide the power required to support longer missions.

6.2.1.4 Propeller-Driven AUVs

Propeller-driven AUVs are common, and most take on a form factor similar to the SPURV, in which the body is a long cylinder with a cone-shaped nose and actuation at the tail (e.g., the REMUS vehicles shown in Figure 6.2C–E). The motor-driven propellers located at the AUV tail provide forward and reverse locomotion, and a range of steerable control surfaces act as rudders to set the yaw, pitch, and roll angles of the vehicle. Several propeller-driven AUVs also have configurations with thrusters facing in multiple directions for increased maneuverability. Note that several hybrid buoyancy/propeller-driven AUVs are under development that aim to obtain the advantages of both approaches to locomotion. Examples of off-the-shelf, propeller-driven AUVs include OceanServer's Iver vehicles (Figure 6.2C) and Kongsberg's REMUS vehicles (Figure 6.2D,E). Notably, the REMUS 600 (Figure 6.2E) is larger, has a greater depth rating (up to 1500 m), and has greater endurance (24 hr) than the more typical REMUS 100 (Figure 6.2D), which has a 100-m depth rating and 12-hr battery life at 1.5 knots. In research projects, both the Iver and REMUS vehicles have been outfitted with hydrophone–receiver systems that allow not only passive tracking of tagged fish but also active tracking.

6.2.1.5 Navigation

Autonomous underwater vehicles come equipped with path-following capabilities such that users can simply upload to the vehicle a series of longitude, latitude, and depth points that the vehicle can be deployed to follow autonomously.

Such path following requires a navigation system that iteratively (1) estimates the vehicle's state (three-dimensional position and orientation) in georeferenced coordinates, (2) determines actuator signals, and (3) sends the signals to the actuators to realize path following. A key to this iterative process, which can run at rates of 10 Hz to several 100 Hz, is providing the state estimation step with accurate sensor data. The sensor payload of an AUV used for navigation typically includes a GPS receiver (that only provides data when the vehicle is surfaced), a compass with three degrees of freedom (3DOF), a pressure sensor for depth, and an altimeter that measures the distance from the seafloor. When the vehicle is underwater (and GPS denied), state estimators will often predict position states using additional sensors, such as inertial measurement units (IMUs), which measure accelerations and rotation rates, or Doppler velocity loggers (DVLs), which measure the velocity of the vehicle with respect to the seafloor. Unfortunately, state estimates can "drift" using such techniques, resulting in positioning error. However, when the AUV has surfaced and reacquired a GPS location, the course can be corrected and the preplanned mission path resumed. Notably,

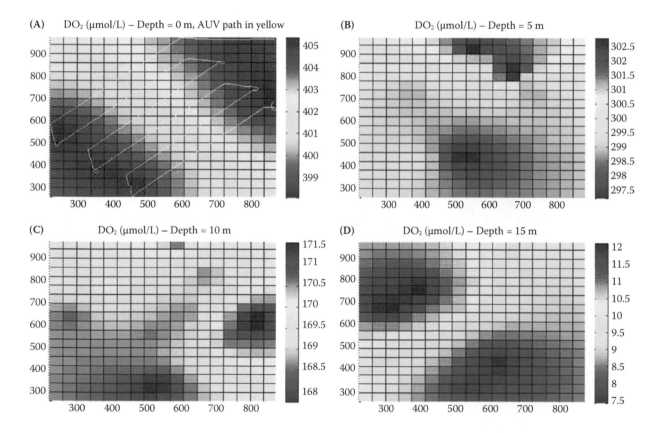

Figure 6.3 Dissolved oxygen concentration map (*x,y*-coordinates) constructed by fusing measurements from an AUV deployment within a Denmark fjord. Consider the AUV's path in yellow at location (200, 100). A jump in the state estimate occurred because the vehicle was drifting off course without realizing it. When the vehicle surfaced at approximately (200, 100), the vehicle obtained a GPS measurement to enable a more accurate position state estimate, thereby allowing the vehicle to make a course correction, turn back to the desired path, and correct the path tracking error.

acoustic positioning systems may also be used for absolute, drift-free underwater position measurements but with additional cost and infrastructure. Improving a vehicle's state estimation improves the ability of the vehicle not only to track paths but also to geolocalize sensor measurements. This is especially challenging when localizing tagged fish relative to a moving AUV. The relative localization problem is difficult due to inaccurate time-of-flight measurements in the variable underwater environment, such as changing water density and sound reflections, and such issues will be compounded by inaccurate AUV state estimation.

6.2.1.6 Sensor Payloads and Applications

Sensors are important for navigation, but most AUVs also have a sensor payload dedicated to sampling the underwater environment. Sensors in this payload section allow for measurements of oceanographic variables including salinity, temperature, chlorophyll, and dissolved oxygen (DO_2) at different depths during a mission (Figure 6.3). Sonar and video camera systems are other typical sensors found on AUVs, and they can be used in downward- and forward-facing orientations. They are often used for characterizing seafloor parameters, such as constructing bathymetry maps, terrain classification (Figure 6.4), or fish school quantification. It is important to note that, although video systems may have higher resolution when compared to sonar systems, they are typically shorter range and require sufficient lighting and water clarity. Although sonar and video data can be used for research data collection purposes, these data streams

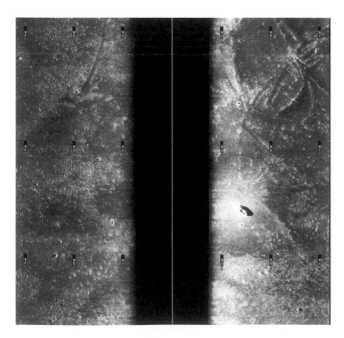

Figure 6.4 Example of side-scan sonar data illustrating a potential shipwreck site (C.M. Clark, unpublished data).

can also be used for navigation and obstacle avoidance. A major advantage of taking environmental sensor measurements when tracking fish is that the fish's behaviors can be tracked while simultaneously characterizing the fish's habitat, providing context for possible state changes in behavior. With the ability to obtain environmental measurements at sampling rates on the order of 1 Hz and geolocalized to within a few meters, AUVs have changed the way passive sampling of the underwater environment is conducted. However, there is potential for AUVs to have even greater impact on the scientific community through active sampling, where the AUVs adapt their trajectories in real time to optimize information gain. For example, some AUVs have been programmed to track steep thermal gradients (Zhang et al., 2010). As described below, this ability of an AUV to modify its behavior in response to live measurements of oceanographic or biological information can be invaluable when tracking marine life that is highly dynamic and can travel long distances.

6.3   USE OF AUTONOMOUS UNDERWATER VEHICLES FOR STUDYING ELASMOBRANCHS

6.3.1   Passive Tracking and Surveying with AUVs and ASVs

A variety of commercially available autonomous underwater and surface vehicles (e.g., Slocum gliders, Wave Gliders, REMUS AUVs, Iver AUVs) have been used to survey oceanographic parameters and map the seafloor of coastal and open-ocean environments over the last 20 years. However, more recently researchers have been equipping these autonomous mobile platforms with a variety of acoustic telemetry receivers to survey for marine animals instrumented with acoustic transmitters (Grothues, 2009; Grothues and Dobarro, 2010; Grothues et al., 2010). For survey applications, AUVs and ASVs are primarily tasked to listen for fish tagged with coded transmitters typically used for passive tracking via stationary receivers, while following a programmed mission path. Because most elasmobranchs are highly mobile, passive tracking methods are constrained by the size of stationary receiver arrays. Mobile acoustic receiver platforms such as AUVs and ASVs provide the ability to autonomously survey a wider range of habitats and areas beyond stationary arrays, and they can simultaneously provide measures of environmental conditions and the benthos in the vicinity (±250 m) where tagged individuals are detected. Because the cost of maintaining large-scale acoustic receiver arrays can be considerable, augmenting area coverage with autonomous mobile acoustic receiver platforms can potentially reduce long-term monitoring costs and provide greater area coverage than expanded stationary receiver arrays. Because of their diving capability, AUVs can provide greater vertical detection capabilities and coverage, as well as oceanographic measurements of the water column, and they achieve better transmitter detection efficiency than surface-oriented platforms alone (Grothues, 2009; Grothues and Dobarro, 2010; Grothues et al., 2010). Autonomous surface vehicles (e.g., wave-driven gliders), however, can provide real-time data via cellular or satellite communication, but they are more limited in terms of their water column characterization and detection efficiency due to sea surface conditions and surface shadows (e.g., acoustic dead spots created when the transmitter and hydrophone are at the surface during rough seascape conditions). Performance of these systems will vary depending on the telemetry system used, habitats and regions surveyed, and duration of deployment.

6.3.1.1 Acoustic Detection Systems

Early attempts to integrate telemetry receiver systems into autonomous mobile platforms varied depending on the acoustic telemetry passive tracking system being used in certain areas and on particular species. In addition, different acoustic telemetry systems may provide different encoding strategies for ID and sensor data. Different transmitter coding schemes have different costs and benefits, often trading off battery life against the number of individuals that can be identified simultaneously. For example, VEMCO coded transmitter systems use a pulse interval coding (PIC) scheme, whereby all transmitters operate at the same frequency, but the code ID is conveyed in a unique pulse train output of varying intervals. This coding scheme only allows a single transmitter to be detected at one time, but it consumes less power than alternative schemes, allowing for longer lived transmitters. The Lotek WHS 3000 series uses a code division multiple access (CDMA) scheme that conveys unique ID and sensor information via frequency spreading and allows for multiple transmitters to be detected simultaneously (Grothues, 2009). This coding scheme is more power consumptive, resulting in a decrease in the battery life of the transmitters.

Typically, coded transmitters have been designed for passive, coarse, spatial-scale tracking or fine-scale tracking using gridded stationary acoustic receiver arrays. Most passive acoustic tracking uses omnidirectional receivers organized in grids and lines (gates) and provides measures of presence/absence within the detection radius of the receiver (Heupel et al., 2006; see also Chapter 8 in this volume). If receiver arrays are organized in a grid formation and close enough that three or more receivers can detect a transmitter as it moves through the array, then a trilaterated position can be obtained by measuring time of arrival to each receiver (Espinoza et al., 2011). These systems include the VEMCO Positioning System (VPS) and VEMCO Radio-Linked Acoustic Positioning (VRAP) system, Lotek MAP

600 and WHS, and HTI Model 290 + 291 systems, which have been deployed to provide fine-scale movements of fishes, some over larger areas (e.g., up to 10 km^2) (Baktoft et al., 2017; Wolfe and Lowe, 2015).

Early integration and comparison of telemetry systems on AUVs were trialed and described by Grothues et al. (2010), who integrated a Lotek WHS 3050 system onto a REMUS 100 AUV to survey for Atlantic sturgeon (*Acipenser oxyrinchus oxyrinchus*) fitted with CDMA-type transmitters in a riverine habitat. They also equipped a REMUS 100 AUV with a VEMCO VR2 receiver to survey another riverine habitat for shortnose sturgeon (*Acipenser brevirostrum*) tagged with VEMCO coded transmitters. Several other project-specific missions were run to survey for and position other fish species using similar systems or setups. Although many of these were proof-of-concept trials, they helped establish the efficacy of using AUV technology coupled with acoustic telemetry systems in quantifying fish movements.

One of the first published studies using an AUV to survey for acoustically tagged elasmobranchs was done by Haulsee et al. (2015), who integrated two VEMCO VR2C acoustic receivers into a Slocum glider. The glider was programmed to survey along the Delaware coastline and listen for migrating sand tiger sharks (*Carcharias taurus*) previously implanted with VEMCO V16 coded transmitters. Stationary acoustic receiver gate arrays deployed across the study area provided additional coverage of this migration route. Using detection data from the AUV, questions about habitat selection could be addressed because the AUV was also programmed to measure water depth, temperature, colored dissolved organic matter (CDOM), and chlorophyll *a* as it moved through the water column. Because the AUV could provide an estimated position of a tagged shark within ±250 m of the AUV at the time of detection, mesoscale habitat characteristics could be measured at those times and compared with locally available conditions to determine habitat selection. The 19-day mission allowed the AUV to cover 337 km, covering depths from 7 to 22 m (Figure 6.5). Using this methodology, Haulsee et al. (2015) concluded that southward migrating sand tiger sharks are likely selecting habitat features that are influenced by distance from shore (water depth), salinity, and CDOM. There are clearly other environmental and biological factors influencing migratory path, but this study demonstrated the efficacy of this technology for offshore research.

Although none is currently published, studies are under way using Wave Glider platforms to autonomously survey for tagged elasmobranchs over larger spatial scales. For example, VEMCO VR2C-equipped Wave Gliders are currently being deployed off central California to survey acoustically tagged adult white sharks while simultaneously gathering oceanographic information along programmed survey paths. Similar missions are being conducted off more

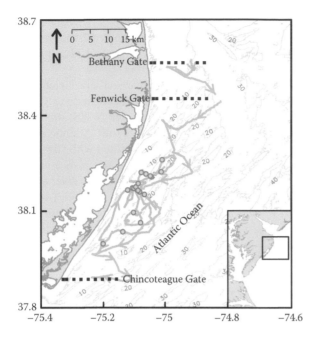

Figure 6.5 Programmed survey path of a Slocum glider in yellow. Location of stationary acoustic receiver gate arrays are shown as black squares, and blue circles indicate locations where tagged sand tiger sharks were detected by the VR2C-equipped glider. (From Haulsee, D. et al., *Mar. Ecol. Prog. Ser.*, 528, 277–288, 2015.)

coastal waters along southern California to survey for acoustically tagged elasmobranchs and food fishes (M. Cimeno, pers. commun.). Because of the greater endurance of Wave Gliders, fewer navigational hazards, and considerable wave energy found in open ocean habitats, these glider missions can run for months and provide periodic or real-time data.

6.3.1.2 Survey Path Planning

As autonomous mobile tracking platform technology becomes more common, developing appropriate surveying methods will be essential for searching for highly mobile or migratory elasmobranchs that have been tagged. For long-endurance gliders (buoyancy-driven or wave-driven), survey paths could be designed for systematic or random surveying across habitats. In Haulsee et al. (2015), the Slocum glider was programmed to move in a sawtooth pattern surveying across depth and habitat contours; however, it was also missioned to return to an area where a tagged shark had been previously detected. This survey plan was necessary to address their particular question, but it might not be the most effective way to survey a large area. Similar paths are often planned for Wave Glider missions, where these platforms are missioned to survey across depth contours and may yield the best opportunity to encounter and detect migrating elasmobranchs that move along coastal shelves. Other considerations for survey path planning vary depending on knowledge of the species tagged, the platform being used,

and degree of boat activity in the areas to be surveyed. For example, buoyancy-driven and wave-driven gliders require a depth of at least 6 m to operate effectively and avoid "dredging," or grounding (Haldeman et al., 2014). In addition, these vehicles are less maneuverable and more likely to be struck by passing vessels operating at high speeds in survey areas. Propeller-driven vehicles can operate in much shallower water (~1 m) and are far more maneuverable if adequately equipped and programmed for obstacle avoidance.

6.3.1.3 Habitats and Species Surveyed

Autonomous survey vehicles may not be appropriate for certain habitats, conditions, or species. Navigational hazards, rapidly changing bathymetry, and strong tides found in estuarine, coastal river, or lagoon habitats would preclude the use of most commercially available buoyancy- and wave-driven gliders for surveying tagged elasmobranchs due to the minimum depth requirements and relatively poor maneuverability in restricted habitats. Smaller propeller-driven AUVs (e.g., Iver2 and Iver3, REMUS 100) or ASVs may offer the best options for surveying in these types of habitats; however, they are much more limited in endurance. Nonetheless, surveying in complex shallow habitats still poses considerable challenges for path planning and successful navigation. Robust obstacle avoidance systems must be implemented to avoid grounding and entanglements with moorings, fishing gear, and kelp/seagrass, as well as negotiating channels with high vessel activity and strong tidal currents. Autonomous vehicle acoustic surveys for tagged elasmobranchs are less problematic for coastal or offshore habitats than inshore waters or embayments due to deeper waters and fewer navigational obstructions. There are also opportunities for much longer missions using less power consumptive buoyancy- and wave-driven vehicles.

6.3.2 Active Tracking Elasmobranchs with AUVs

Acoustic detection data from passive stationary acoustic receiver arrays (see Chapter 8 in this volume) and autonomous mobile acoustic receiver platforms (Section 6.2.1) can be effective in quantifying movements of individuals over large spatial and temporal scales. Yet, these methods often lack sufficient positional accuracy and localization frequency of the same individual to adequately characterize how animals respond to microscale or mesoscale environmental conditions. Thus, limited information on their movements is obtained for animals that move at scales smaller than the typical passive omnidirectional receiver detection ranges (<500 m). Positioning systems such as VEMCO VRAP or VPS, Lotek MAP 600 or WHS, or HTI 3D tracking systems can provide high-resolution estimates of location, but these systems are constrained by the size and location of the stationary arrays of receivers (Hedger et al., 2008; Heupel et al., 2006). This poses major challenges in obtaining high-frequency, high-spatial-resolution measurements for highly mobiles species of elasmobranchs, especially those that show minimal site fidelity to areas (Hussey et al., 2015).

To study aquatic animal movements at higher frequencies and finer spatial scales, researchers have traditionally turned to active tracking. This practice consists of a human tracker in a small vessel using a directional hydrophone to estimate the location of an acoustically tagged animal and following it for 24 to 96 hr (Holland et al., 1992). The amplitude of the signal received by the hydrophone is a function of the direction of the hydrophone relative to the direction and distance to the transmitter; that is, amplitude increases as this relative angle decreases and proximity increases. By rotating the directional hydrophone, the tracker can determine the direction in which the signal is strongest and thus the direction to the animal. The tracker can then navigate the vessel toward the signal source (transmitter), using the position of the vessel as a proxy for the position of the animal (Bass and Rascovich, 1965; Nelson, 1978). However, to obtain high positional accuracy, the tracker must be positioned directly over top of the transmitter (tagged individual), which can affect the animal's behavior.

Active tracking is labor intensive and reliant on the proficiency of the tracker, and it provides variable spatial accuracies (<50 m) (White et al., 2016). Hence, the development of AUV technology has led many researchers to envision AUVs as replacing humans as active trackers. This idea was first conceptualized over 20 years ago with a kayak outfitted with an electric trolling motor and a VEMCO VR60 (Goudey et al., 1998). To replace human trackers, AUVs must be equipped with two key abilities: (1) accurately estimate the location of a tagged animal in real time while at a distance, and (2) incorporate this location state estimate into the AUV's control system and thus realize autonomous tracking and following of the animal, which includes attempting to relocate a transmitter signal temporally lost.

6.3.2.1 Acoustic Positioning System

The acoustic positioning system is the hardware (hydrophone/receiver) and software combination required to provide an estimate of the location of an acoustic transmitter with respect to a georeferenced Cartesian coordinate system (e.g., x, y, z space). A location can be determined when two or more hydrophone/receiver systems are able to derive a distance and bearing to a transmitter, with the tag transmitting information on its depth (z-dimension). This information, when combined with the current position and heading of the AUV, allows for an estimate of the geolocation of the transmitter. Acoustic position systems rely on the fact that sound travels at predictable speeds based on the water density (e.g., temperature and salinity). Thus, accurate measurements of time (microsecond), water density, and temperature allow for precise estimates of location of a transmitter. Two

main strategies have been used for acoustic position systems for the active tracking of animals: short baseline (SBL) and ultrashort baseline (USBL).

Short baseline systems can derive a distance and bearing estimate if at least two hydrophones are fix-mounted at least 2 m apart (Clark et al., 2013; Forney et al., 2012; Lin et al., 2013). When both hydrophones detect a tag transmission, the time difference of arrival (TDOA) between the hydrophones is converted to an angle measurement. The TDOA is multiplied by the speed of sound in water, given current environmental conditions, to form one side of a triangle, and the distance between the hydrophones represents the hypotenuse of the triangle. By taking the arc-cosine of the difference in time of arrival over the distance between the hydrophones, the bearing to the transmitter is calculated. When using transmitters that transmit at fixed intervals, this system can be expanded to incorporate a range measurement by estimating the time of flight (how long it would take the acoustic signal to travel a particular distance through water of known density) (Lin et al., 2013). The AUV can keep track of the estimated time the transmitter should transmit, then, by subtracting the time when the AUV detected the transmission, the time of flight (TOF) can be estimated. The time of flight is multiplied by the speed of sound in seawater based on water density at that location to produce a range measurement. The transmitted depth, estimated bearing, and range can be used to estimate a position of a moving transmitter while the AUV is moving.

For such two-hydrophone configurations, there is a sign ambiguity, and the AUV cannot determine if the tag is on its left or right side from a single detection (Figure 6.6); however, the AUV can be programmed to move in a circular or sigmoidal path until several detections are made, which enables the AUV to differentiate a sign (or side) the transmitter is on relative to its path. To further refine accuracy and sign of position estimates using this time of arrival and TOF approach, a particle filter state estimation algorithm can be applied, which over several successive detections can improve positional

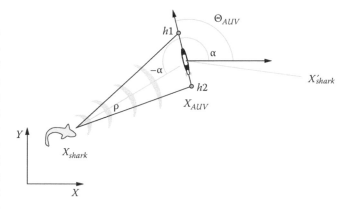

Figure 6.6 Schematic of how a SBL-enabled AUV determines location of a tagged shark. (From Clark, C.M. et al., *J. Field Robot.*, 30(3), 309–322, 2013.)

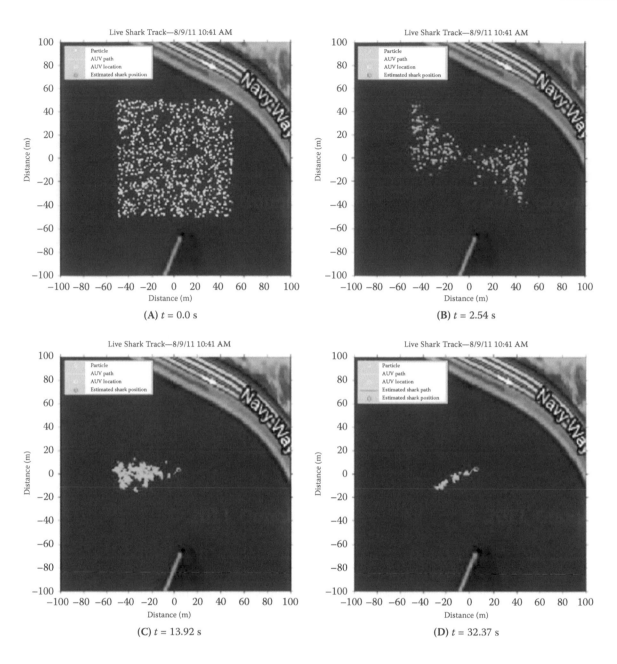

Figure 6.7   Use of a particle filter with a SBL-enabled AUV to refine a position estimate of a moving transmitter over successive detections. (From Forney, C. et al., in *Proceedings—IEEE International Conference on Robotics and Automation*, Institute of Electrical and Electronics Engineers, Piscataway, NJ, 2012, pp. 5315–5321.)

accuracy to ±3 m while both transmitter and AUV are moving simultaneously (Clark et al., 2013; Forney et al., 2012; Lin et al., 2013; Xydes et al., 2013) (Figure 6.7).

The second type of acoustic positioning system used onboard an AUV for active tracking is the ultrashort baseline (USBL) system (Kukulya et al., 2015, 2016; Packard et al., 2013). In this system, the AUV is equipped with a transceiver with a small array of omnidirectional hydrophones, and the animal is tagged with a transponder that can also transmit depth information. The transceiver onboard the AUV sends an acoustic transmission that is received by the animal-borne transponder, which then sends a reply that is received by the AUV. By measuring the time between when the AUV queried the animal's transponder and when it received the transmission in response, the total time of flight to the animal and back to the AUV is calculated. By dividing this by two and multiplying by the speed of sound, a range estimate can be calculated. The AUV is additionally able to compare the received signal phase difference across the multiple hydrophones within the transceiver to generate a bearing estimate. Because the USBL systems were originally designed for tracking large underwater vehicles and require two-way acoustic communication, the transponder tags are large and subsequently powerful, and they also have significantly better range and bearing estimates than the SBL system, which is limited by the accuracy of time recording across the hydrophones as well as the consistency of tag transmissions over time (Kukulya et al., 2015; White et al., 2016). However, USBL system transponder tags are significantly larger (7.6 × 38 cm) and more expensive and power consumptive, thus limiting their application to large sharks (>3 m) capable of carrying larger tag packages. The SBL system, however, was designed to use existing animal acoustic telemetry infrastructure and standard acoustic transmitters from other acoustic biotelemetry systems (i.e., VEMCO, Lotek, Sonotronics), and it allows for much smaller individuals to be tagged (White et al., 2016).

6.3.2.2 AUV Control System

For fully autonomous active tracking, estimating the location of the tagged animal resolves only half of the problem. The AUV must also incorporate the tagged animal's position estimate into the AUV's control system to enable autonomous following of the animal. The AUV control system can be programmed to incorporate a variety of abilities necessary to track a moving animal, which may stop moving, move in a very small discrete space, or travel in nonlinear paths. So, the AUV must be programmed to adapt its movement path to that of the tagged animal while simultaneously avoiding static obstacles, such as docks and pinnacles, and avoiding grounding or colliding with the shoreline or seafloor. In addition, the AUV must be programmed to determine position from a suitable distance to reduce behavioral interference with the tagged individual but be close enough to maintain continuous detection range (Clark et al., 2013; Forney et al., 2012; White et al., 2016).

The simplest control system might be to just have the AUV drive directly toward the position of the shark. Packard et al. (2013) programmed a REMUS AUV with a USBL system to film sharks under water, which required the AUVs to be very close to the tagged animal (<10 m). The shark's path was interpolated 15 sec into the future, and the AUV was programmed to drive directly at that point and pass to the left, right, above, or below the tagged shark (Figure 6.8). After the AUV passed the shark, it was programmed to turn around and perform this "fly by" again. By repeatedly updating the shark's location relative to the AUV, the AUV could continuously update its path to a new location. This control system was refined by Kukulya et al. (2015), who modulated the speed of the AUV proportional to the distance between the AUV and the shark, so that the AUV would travel faster when it was farther away from the tagged individual and match the animal's speed when it was close to the tagged animal.

Such a control system, however, could cause the AUV to influence the tagged animal's behavior by being too close or even colliding with the tagged animal. Because AUVs

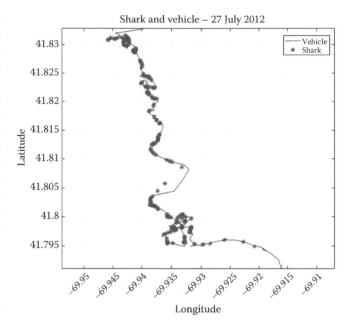

Figure 6.8 The estimated position of a tagged white shark (red asterisk) off Cape Cod using the USBL-enabled system on a REMUS AUV and the path of the AUV (blue line). When frequent detections were obtained, the AUV could accurately predict the shark's path to increase the likelihood of the AUV being close enough to capture video data. (From Packard, G.E. et al., in *MTS/IEEE OCEANS 2013–San Diego*, Institute of Electrical and Electronics Engineers, Piscataway, NJ, 2013, pp. 1–5.)

use propeller-driven propulsion, the vehicles produce low-frequency sounds detectable to the sharks. Sharks have been found to approach or even attack AUVs, following them or staying in close proximity (Skomal et al., 2015; Stanway et al., 2015). A second approach is to have the AUV position itself close enough to the tagged animal to continuously receive acoustic detections but remain far enough away to reduce the likelihood of behavioral interference (>30 m). In this programmed control feature, documented by Lin et al. (2013), an AUV had two different behaviors, depending on the distance to the tagged animal. When the AUV was within detection range, but farther away from the animal, the AUV was programmed to drive directly at the tagged animal. When the AUV was within a user-defined threshold distance of the animal, it switched to nearby circling behavior, such that the AUV circled at a point just outside of the user-defined minimum threshold distance to the tagged animal. Circling allows the AUV to constantly propel itself, which is necessary for stability and to increase the quantity of sensor vantage points, determine bearing and distance to the transmitter, and prevent it from colliding into the tagged animal (Figure 6.9). In addition, if the shark moves toward the AUV and comes within the user-defined minimum threshold distance, the AUV is programmed to move away from the tagged shark. By switching back and forth between point tracking and nearby circle tracking, the AUV can

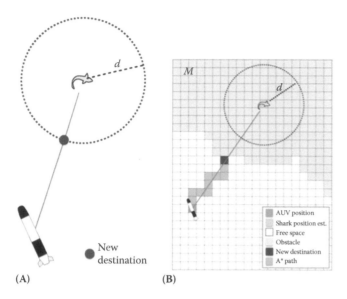

(A)   (B)

Figure 6.9 Programmed control system for SBL-enabled AUV for predicted obstacle avoidance and nearby circling when in user-defined minimum threshold distance to a transmitter. (From Lin, Y. et al., in *Proceedings of the Unmanned Untethered Submersible Technology Conference (UUST)*, Autonomous Undersea Systems Institute, Durham, NH, 2013.)

continuously follow the tagged individual, remaining within detection range. Using this system, the path of the AUV does not closely reflect the path of the tagged individual. It is also worth mentioning that quantifying shark behavior response to AUVs vs. active tracking surface vessels may be useful for determining minimum threshold distances for AUV tracking.

In addition, control systems for active tracking have been modified and adapted to use multiple AUVs for tracking a tagged shark (Kukulya et al., 2016; Lin et al., 2014, 2017). Having both AUVs simultaneously tracking while in communication with each other allows for incorporation of range and bearing measurements from multiple angles, thereby providing further refined position estimates and reducing the likelihood of losing the animal. These systems are programmed to have one leader and could be expanded to incorporate multiple followers. Future swarm programming of smaller, single-hydrophone AUVs could allow for trilateration (position estimates) of multiple cooperative AUVs surrounding a tagged individual, as long as the AUVs were far enough away from the tagged individual to minimize behavioral disturbance.

6.3.2.3 Environmental and Situational Monitoring

Traditional uses of AUVs have been in sampling oceanographic conditions. Enabling these duties can provide environmental context for the focal animal being tracked. Most commercially available AUVs can be equipped with

a wide variety of sensors (e.g., temperature, salinity, DO_2, CDOM, pH, photosynthetically active radiation [PAR], side-scan sonar, image sonar) that can simultaneously sample the environment during tracks within a geospatial context. Measuring these conditions during a track can provide unprecedented resolution of the environment surrounding the animal. This fine-scale environmental information can provide context and insight as to how tracked individuals respond to changing conditions and use gradients as navigational cues.

Environmental sensors can provide valuable information on how tagged individuals respond to changing environmental conditions, but video and sonar imaging can provide insight into how tagged individuals respond when they are in close proximity to other conspecifics, predators, and prey. By outfitting the AUVs with downward- and forward-facing video cameras, or even virtual reality (360° view) video cameras (Figure 6.10), spatially explicit prey and conspecific abundance can be estimated and correlated to changing movement behaviors observed from tracked individuals. Traditional active and passive tracking methods often lack these abilities or require separate teams of researchers to simultaneously collect these data using typical sensor suites, such as sondes; conductivity, temperature, and depth recorders (CTDs); or video camera sleds. Although it is a valuable activity, collecting video data is power consumptive and data intensive and often results in a data processing bottleneck, requiring considerable labor-intensive postprocessing. In addition, underwater video data can be hampered by poor visibility or low light conditions (e.g., nighttime periods), and coastal habitats (e.g., lagoons, bays, estuaries) can often have poor visibility conditions, further limiting the use of video in those ecosystems.

Sonar technology has greatly improved and reduced in size and can now provide low-resolution video image quality at distances of less than 50 m (e.g., DIDSON, BlueView). AUVs can be equipped with forward- and downward-facing image sonar, which is capable of recording and identifying unique targets based on their densities and reflective acoustic properties. These systems can operate in poor visibility and zero light conditions, in addition to providing estimates of distances between objects and the AUV and the size of the targets. Digital target recognition and autonomous tracking of sonar-identified objects has allowed short-term tracking of untagged sharks and schools of fishes (McCauley et al., 2016) and holds great promise for augmenting video for identifying conspecifics and prey items surrounding the tagged animal. However, sonar is much more power consumptive in comparison to video, so its application and duration of use would have to be more conservatively applied in comparison to video.

Coupling transmitters and data loggers with environmental sensors attached to focal individuals and having the AUV sample the habitat around the tagged individual can provide insight into how sharks are selecting habitats

Figure 6.10 REMUS 100 outfitted with six video cameras providing nearly 360° video coverage around the AUV and acoustic transponder tag (inset), used to record white sharks at Guadalupe Island Mexico. (From Skomal, G. et al., *J. Fish Biol.*, 87(6), 1293–1312, 2015.)

and differentiating among environmental gradients. Future AUV control algorithms can incorporate where the AUV has previously sampled and direct the AUV to alternative locations around the moving animals for a more comprehensive characterization of environmental conditions available to the focal animal. This would maximize environmental information gain and provide a more holistic picture of the behavior of the animal.

6.3.2.4 Habitats and Species Tracked

Most of the studies published to date using autonomous active tracking systems have been largely methodological and have focused on the development of the technology and its efficacy. These tracks have taken place on a range of species, including non-obligate ram ventilating species such as leopard sharks (*Triakis semifasciata*) and obligate ram ventilating species such as white and basking sharks (*Cetorhinus maximus*), in shallow coastal areas as well as deeper habitats. Using a modified Iver2 AUV, Clarke et al. (2013) developed an SBL system (Lotek MAP 600) designed for tracking leopard sharks in southern California. Leopard sharks, which are known to aggregate in shallow habitats for thermoregulatory benefits (Hight and Lowe, 2007; Nosal et al., 2014) and exhibit restricted space use during the day, were tracked in embayments ranging from intertidal to 25 m in depth. Tagged leopard sharks were simultaneously actively tracked by a human in a surface vessel and by an SBL-equipped AUV for comparison of positional accuracy and frequency of data acquisition. Comparative tracks (AUV and a human active tracker) indicated that this autonomous system provided similar or better spatial

accuracy than a human active tracker, yet provided significantly more accurate locations than a human (White et al., 2016) (Figure 6.11).

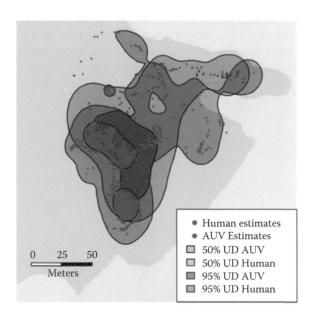

Legend:
- • Human estimates
- • AUV Estimates
- ☐ 50% UD AUV
- ☐ 50% UD Human
- ▨ 95% UD AUV
- ▨ 95% UD Human

0   25   50
Meters

Figure 6.11 Position estimates of a tagged leopard shark tracked in a small cove at Santa Catalina Island. The red dots indicate position estimates made from a human active tracker in a surface vessel using closest estimated range to determine position. The blue dots indicate position estimates from a simultaneous track of the SBL-enabled AUV programmed to remain at least 20 m away from the tagged shark. Colored polygons represent estimated utilization distribution areas of the shark tracked using both methods. (From White, C.F. et al., *J. Exp. Mar. Biol. Ecol.*, 485, 112–118, 2016.)

Individual tracks have lasted longer than 4 hr, and either single or pairs of cooperative AUVs were tested and compared for their accuracy and ability to maintain contact with a tagged leopard shark. These high-resolution spatiotemporal movement data overlaid on high-resolution temperature maps showed leopard sharks spending most of their time in a small temperature range (Bernal and Lowe, 2016). Yet, because the leopard sharks were in shallow (<2 m), rocky, and often narrow habitats, the AUVs would occasionally have to position themselves at a distance (>100 m) from the tagged animal, so detection efficiency and subsequently positional accuracy suffered.

A REMUS 100 outfitted with a USBL system has been used to track basking sharks and white sharks off Cape Cod, Massachusetts (Packard et al., 2013). This study was largely methodological, but the authors were able to show that the tracked white sharks remained closely associated to the benthos while swimming outside seal colonies. Because white sharks were using shallow habitats close to shore, the AUV had difficulties tracking tagged individuals due to periodic loss of transponder detection. White sharks have additionally been tracked off Guadalupe Island, Mexico (Kukulya et al., 2015). Six video cameras mounted to the AUV (forward and reverse) were used to document shark behavior, such as attack events on the AUV (Figure 6.12), as well as shark depth and association with conspecifics (Skomal et al., 2015). White sharks globally are thought to mainly capture prey near the surface by ambushing them, yet the location and the frequency at which these tracking AUVs were attacked by white sharks have led to suggestions that sharks at Guadalupe might be capturing prey at depth (Skomal et al., 2015). Deeper coastal water around Guadalupe Island

allowed the AUVs to encounter few obstacles with less ambient noise, so AUVs were more successful in tracking tagged sharks; however, sharks were able to dive deeper than the depth rating of the AUVs (>100 m). The researchers had a second tracking AUV (REMUS 600) that was capable of diving deeper; however, deploying the AUV in a timely manner before the shark was lost was problematic, as well.

6.4 EXPECTATIONS AND DATA

Like most emerging technologies, autonomous tracking systems have the capability of generating considerably larger and more complex datasets than conventional active and passive tracking due to their ability to gather environmental data while tracking; therefore, there may be considerable need for data postprocessing, synthesis, and error checking. Because these data are spatially explicit and often three dimensional, more sophisticated data visualization methods are needed (Martin et al., 2006; Viswanathan et al., 2017). Not only are these accompanying tools necessary for data visualization and display, but they also will be integral in development of new analytical methods for identifying and quantifying behavioral patterns and changes in behavioral state with changing social and environmental conditions.

6.4.1 *Post Hoc* Analysis of AUV Tracks and Movement Predictions

As demonstrated by Haulsee et al. (2015), using autonomous vehicles to survey for tagged sharks can greatly enhance passive tracking arrays by providing greater area coverage

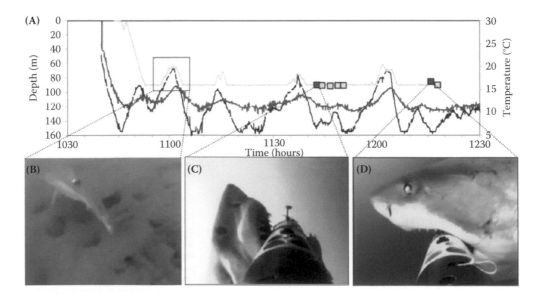

Figure 6.12   The top figure shows the depth of the AUV (green line), and the depth of the shark (black line) recorded onboard a REMUS 100 using an ultra-short base line (USBL) acoustic receiver system. During this span, the shark approached the AUV several times, ultimately biting the AUV twice, while the multiple cameras video recorded the approaches and bites. (From Skomal, G. et al., *J. Fish Biol.*, 87(6), 1293–1312, 2015.)

and simultaneous measurements of environmental conditions. These datasets are quite different from those of conventional stationary acoustic receiver arrays but when analyzed together can provide a much richer understanding of movement and habitat selection than can often be inferred from passive tracking alone. For example, with sufficient sampling and adequate environmental monitoring, habitat suitability models can be developed and used to predict changes in distributions, migrations, and migration routes. Furthermore, these migration routes can potentially be determined from these types of datasets (White, 2016).

One advantage of autonomous vehicle platform active tracking of elasmobranchs is that these systems can provide comparable positional accuracy as human-based active tracking, but at a much higher frequency and while maintaining a distance away from the tagged shark. Using these telemetry systems also provides error measures for each position that can be appropriately filtered to provide more accurate information for assessing habitat association, rate of movement, and fine-scale behaviors.

In addition to obtaining high-resolution position data from autonomous tracking platforms, the ability to overlay these movements over high-resolution, three-dimensional (3D) environmental data (often interpolated from stationary sensors) provides the opportunity to quantify habitat selection and potentially preference. This can allow for development of 3D habitat suitability models, which can be used to make predictions of under what conditions individuals should initiate seasonal migrations, select new habitats, and shift distributions under changing climate and rising ocean temperatures. Coupling these high-resolution movement data with spatially explicit baited remote underwater video (BRUV; see Chapter 7 in this volume) and autonomous camera and image sonar surveys provides the ability to quantitatively evaluate influences such as prey density, intra- and interspecific competition, and presence of mates to predict degree of site fidelity and cues influencing potential emigration. Of course, the utility of autonomous tracking vehicles is still valuable but more limited without corresponding environmental and biological data.

6.4.2 Technical Limitations

Autonomous vehicle and telemetry technology has advanced considerably over the last 20 years, but there are still numerous technical limitations that must be considered before attempting to utilize these tools. Although smaller and less expensive AUVs are commercially available, they are still quite expensive (base models are equivalent to the cost of a new, fully equipped 8-m research vessel), not including the cost of the telemetry system. ASVs are currently the least expensive option for full autonomous tracking platforms, but these systems are limited by battery life, power, and speed. Battery life poses a distinct challenge for either surveying or active tracking. Most smaller class AUVs (e.g., Iver2,

Iver3, REMUS 100) have a depth range of only 100 m and have approximately 12 hr of battery life under low cruising speeds and with conservative use of power-consumptive sensors. Larger, more expensive AUVs (e.g., REMUS 600) can operate at deeper depths (600 m) and offer considerably longer battery life; they are capable of operating at low cruising speeds for up to 24 hr. However, these larger vehicles are heavy, making deployment and retrieval more difficult and requiring larger vessels to support. Continuous multiple-day tracks would require at least two AUVs that could be swapped out during a track while the other is being recharged. This reduces the advantages of autonomy and may only be feasible in protected embayments or for coastal applications. Wave- and buoyancy-driven vehicles have reduced power requirements and therefore can be used for longer deployments, but they may also suffer from shorter battery life if power-consumptive sensors are not used conservatively.

Another technical limitation of AUVs for tracking highly mobile elasmobranchs may be maximum operating speed of the platforms relative to the detection range of the transmitters being applied. Most AUVs have a maximum operating speed of 4.5 knots (2.3 m/sec) and may not be able to stay within the detection range of faster swimming species such as shortfin mako shark (*Isurus oxyrinchus*) or gray reef shark (*Carcharhinus amblyrhynchos*). Even though these highly mobile species of elasmobranchs cannot maintain these speeds for long periods of time, they may be able to lose the AUV by rapidly moving outside the AUV's detection range. In addition, much like while active tracking, acoustic receiver performance decreases with increased speed due to cavitation around the hydrophone. This makes it more difficult for the AUV to maintain contact with the tagged shark while it is traveling at higher speeds. This may also be problematic when tracking smaller elasmobranchs which requires smaller acoustic transmitters that inherently have lower power output and shorter battery lives.

Another major technical limitation required for full autonomous tracking or surveying is the ability to avoid stationary obstacles such as shore lines, seafloors, docks, and reef markers—features for which there are discrete geospatial references. All AUVs have the capabilities to be programmed to remain within coordinate boundaries or to move around a coordinate position; however, their ability to make these movements correctly is based on the accuracy of the geoposition estimates of these features and the AUV's accuracy of its position while moving. The altimeter can help prevent the AUV from running aground, but steep vertical walls can pose navigation challenges; the situation is further complicated because the AUV's ability to estimate its geolocation is reduced the longer it remains underwater. Therefore, operating in complex and enclosed spaces can be challenging for even the most maneuverable AUVs. Moving and small-profile obstacles such as boats, moorings, kelp, and people in the water pose additional challenges for avoidance, as their geopositions are dynamic. To address these challenges, forward-facing sonar or video systems can be used and programmed to enable the AUV to avoid moving and potentially entangling obstacles; however, these systems are power consumptive, which may further reduce tracking duration. Despite these limitations, engineers and programmers are developing new sensors and detection algorithms to improve obstacle avoidance, improve battery technology, develop in-water recharging stations, and control planning for AUVs.

6.4.3 Potential AUV System Configuration and Planning Abilities

The use of AUVs for tracking marine animals is not limited to a single AUV tracking a single animal. Recent research has demonstrated the advantages of using multiple AUVs to track an animal (e.g., Clark et al., 2013; Kukulya et al., 2016; Lin et al., 2013, 2017; Shinzaki et al., 2013; Tang et al., 2014). The combined sensor footprint is increased, and there is an increase in sensor vantage points, which reduces positional errors. There is also the possibility of tracking multiple animals simultaneously. If acoustic CDMA-type transmitters are used (e.g., share the same time slot with a time division multiple access protocol), then even a single AUV can detect and track multiple tagged individuals simultaneously as long as they all remained in the same area. To accomplish this, more sophisticated AUV path planning algorithms may be required that direct vehicles to optimal locations, such as the locations that when visited yield the highest likelihood of the AUV being in range of and detecting all animals. With such applications in mind, algorithms for the autonomous vehicle tracking of multiple targets has been an active area of research that is providing the AUV community with several starting points for such future work.

6.4.4 Future Sensor Integration and Autonomy Capabilities

Autonomous vehicle technology (AUV and ASV) has the potential to completely change the way marine life is surveyed and tracked. Work in the last decade has provided several examples of successful tracking experiments, and the technology has matured to the point where it can be used for systematic tracking deployments. Image sonar offers potentially one of the greatest additions to sensor integration for AUV operations and enhanced tracking capabilities. Forward-facing image sonar could be used not only for AUV obstacle avoidance but also for visualizing the tagged sharks and its associated surroundings. This technology can also be used to measure the size of tagged sharks and any other associated individuals within the sonar window.

Moving from SBL to USBL hydrophone arrays not only improves position estimates but also significantly reduces drag and the power consumption of tracking AUVs and increases the mobility and utility of ASVs. The USBL hydrophone arrays can also be used to detect one-way PIC-type transmitters, which will allow for a greater selection of telemetry systems for integration into autonomous tracking platforms and tracking of much smaller transmitters (Rypkema et al., 2017). AUVs could be equipped with transponder technology and programmed to search out elasmobranchs instrumented with archival transponding tags. Once located, the AUV would be programmed to follow the tagged shark until all the archived data from the shark's transponding tag has been downloaded to the AUV. Because transponding tags are quite large, this technology might only be feasible for larger elasmobranchs in the near future.

The future of this technology will provide greater positioning accuracy, increased temporal resolution, reduced labor intensity, and simultaneous sampling of the environment, as well as access to new datasets that were previously unachievable. Automatic identification of habitat type, quantification of aggregation sizes and densities, and intelligent inference of near-future animal behavior are possible, but these are only a subset of the new capabilities that will provide marine biologists with new data and hence new findings.

6.5 SUMMARY AND CONCLUSIONS

While, historically, autonomous vehicle technology was used and developed for remote open ocean exploration, the need for coastal autonomous oceanographic monitoring has increased, which has driven the development of a range of autonomous vehicle platforms (AUV and ASV) that are smaller, more maneuverable, and more affordable. In addition, acoustic telemetry companies have begun to recognize the growing market for autonomous vehicles and need for adapting their technology for application on these robotic platforms. The expectation is that the development and use of autonomous tracking platforms will expand greatly in the next decade and will require biologists to work more closely with oceanographers, computer scientists, and engineers in order to produce the systems and tools necessary to address more complex questions about the behavior and ecology of elasmobranchs. In addition, there is a growing need for higher resolution positioning and simultaneous environmental monitoring in order to quantify habitat use, migration cues, migration pathways, and drivers of movement strategies of elasmobranch fishes. The continued development of these tools will help address questions regarding habitat recolonization of recovering populations; the effects of climate change on distributions, migrations, and habitat selection; and the evaluation of essential habitat for highly mobile shark and ray species.

ACKNOWLEDGMENTS

Special thanks go to all the students of the CSULB Shark Lab for helping with AUV field trials and to students of the California Polytechnic State University, San Luis Obispo, and Harvey Mudd College LAIR labs for their innovation and determination in making autonomous shark tracking a reality. Funding for shark tracking AUVs was provided by National Science Foundation RI-143620 to CSULB Shark Lab and Harvey Mudd College LAIR lab.

REFERENCES

Baktoft H, Gjelland KØ, Økland F, Thygesen UH (2017) Positioning of aquatic animals based on time–of–arrival and random walk models using YAPS (Yet Another Positioning Solver). *Sci Rep* 7(1):14294.

Bass GA, Rascovich M (1965) A device for the sonic tracking of large fishes. *Zoologica* 50(2):75–82.

Bernal D, Lowe C (2016) Field studies of elasmobranch physiology. In: Shadwick AP, Farrell AP, Brauner CJ (eds) *Physiology of Elasmobranch Fishes: Structure and Interaction with Environment: Fish Physiology*, vol 34A. Elsevier, Amsterdam, pp 311–377.

Clark CM, Forney C. Manii E, Shinzaki D, Gage C, Farris M, Lowe CG, Moline M (2013) Tracking and following a tagged leopard shark with an autonomous underwater vehicle. *J Field Robot* 30(3):309–322.

Espinoza M, Farrugia TJ, Webber DM, Smith F, Lowe CG (2011) Testing a new acoustic telemetry technique to quantify long-term, fine-scale movements of aquatic animals. *Fish Res* 108(2):364–371.

Forney C, Manii E, Farris M, Moline MA, Lowe CG, Clark CM (2012) Tracking of a tagged leopard shark with an AUV: sensor calibration and state estimation. In: *Proceedings—IEEE International Conference on Robotics and Automation*. Institute of Electrical and Electronics Engineers, Piscataway, NJ, pp 5315–5321.

Goudey CA, Consi T, Manley J, Graham M (1998) A robotic boat for autonomous fish tracking. *Mar Technol Soc J* 32(1):47.

Grothues TM (2009) A review of acoustic telemetry technology and a perspective on its diversification relative to coastal tracking arrays. In: Nielsen JL, Arrizabalaga H, Fragoso N, Hobday A, Lutcavage M, Sibert J (eds) *Tagging and Tracking of Marine Animals with Electronic Devices*. Springer, Dordrecht, pp 77–90.

Grothues TM, Dobarro JA (2010) Fish telemetry and positioning from an autonomous underwater vehicle (AUV). *Instrument Viewpoint* (8):78–79.

Grothues TM, Dobarro J, Eiler J (2010) Collecting, interpreting, and merging fish telemetry data from an AUV: remote sensing from an already remote platform. In: *2010 IEEE/OES Autonomous Underwater Vehicles*. Institute of Electrical and Electronics Engineers, Piscataway, NJ, pp 1–9.

Haldeman CD, Aragon D, Roarty H, Kohut J, Glenn S (2014) Enabling shallow water flight on Slocum gliders. In: *MTS/IEEE OCEANS 2014–St John's*. Institute of Electrical and Electronics Engineers, Piscataway, NJ, pp 1–5.

Haulsee D, Breece M, Miller D, Wetherbee BM, Fox D, Oliver M (2015) Habitat selection of a coastal shark species estimated from an autonomous underwater vehicle. *Mar Ecol Prog Ser* 528:277–288.

Hedger RD, Martin F, Dodson JJ, Hatin D, Caron F, Whoriskey FG (2008) The optimized interpolation of fish positions and speeds in an array of fixed acoustic receivers. *ICES J Mar Sci* 65(7):1248–1259.

Heupel M, Semmens JM, Hobday A (2006) Automated acoustic tracking of aquatic animals: scales, design and deployment of listening station arrays. *Mar Freshwater Res* 57(1):1–13.

Hight BV, Lowe CG (2007) Elevated body temperatures of adult female leopard sharks, *Triakis semifasciata*, while aggregating in shallow nearshore embayments: evidence for behavioral thermoregulation? *J Exp Mar Biol Ecol* 352(1):114–128.

Holland K, Lowe C, Peterson J, Gill A (1992) Tracking coastal sharks with small boats: hammerhead shark pups as a case study. *Mar Freshwater Res* 43(1):61–66.

Hussey NE, Kessel ST, Aarestrup K, Cooke SJ, Cowley PD, Fisk AT, Harcourt RG (2015) Aquatic animal telemetry: a panoramic window into the underwater world. *Science* 348(6240):1255642.

Kukulya AL, Stokey R, Littlefield R, Jaffre F, Padilla EMH, Skomal G (2015) 3D real-time tracking, following and imaging of white sharks with an autonomous underwater vehicle. In: *MTS/IEEE OCEANS 2015–Genova: Discovering Sustainable Ocean Energy for a New World*. Institute of Electrical and Electronics Engineers, Piscataway, NJ, pp 1–6.

Kukulya AL, Stokey R, Fiester C, Padilla EMH, Skomal G (2016) Multi-vehicle autonomous tracking and filming of white sharks *Carcharodon carcharias*. In: *2016 IEEE/OES Autonomous Underwater Vehicles*. Institute of Electrical and Electronics Engineers, Piscataway, NJ, pp 423–430.

Lin Y, Kastein H, Peterson T, White C, Lowe C, Clark C (2013) Using time of flight distance calculations for tagged shark localization with an AUV. In: *Proceedings of the Unmanned Untethered Submersible Technology Conference (UUST)*. Autonomous Undersea Systems Institute, Durham, NH.

Lin Y, Kastein H, Peterson T, White C, Lowe CG, Clark CM (2014) A multi-AUV state estimator for determining the 3D position of tagged fish. In: *2014 IEEE/RSJ International Conference on Intelligent Robots and Systems (IROS 2014)*. Institute of Electrical and Electronics Engineers, Piscataway, NJ, pp 3469–3475.

Lin Y, Hsiung J, Piersall R, White C, Lowe CG, Clark CM (2017) A multi-autonomous underwater vehicle system for autonomous tracking of marine life. *J Field Robot* 34(4):757–774.

Martin SC, Whitcomb L, Arsenault R, Plumlee M, Ware C (2006) Advances in real-time spatio-temporal 3D data visualisation for underwater robotic exploration. In: Roberts GN, Sutton R (eds) *Advances in Unmanned Marine Vehicles*. The Institution of Electrical Engineers, Herts, UK, pp 293–310.

McCauley DJ, DeSalles PA, Young HS, Gardner JP, Micheli F (2016) Use of high-resolution acoustic cameras to study reef shark behavioral ecology. *J Exp Mar Biol Ecol* 482:128–133.

Nelson DR (1978) Telemetering techniques for the study of free-ranging sharks. In: Hodgson ES, Mathewson RF (eds) *Sensory Biology of Sharks, Skates, and Rays*. Office of Naval Research, Department of the Navy, Arlington, VA, pp 419–482.

Nosal A, Caillat A, Kisfaludy E, Royer M, Wegner N (2014) Aggregation behavior and seasonal philopatry in male and female leopard sharks *Triakis semifasciata* along the open coast of southern California, USA. *Mar Ecol Prog Ser* 499:157–175.

Packard GE, Kukulya A, Austin T, Dennett M, Littlefield R, Packard G, Purcell M, Stokey R (2013) Continuous autonomous tracking and imaging of white sharks and basking sharks using a REMUS-100 AUV. In: *MTS/IEEE OCEANS 2013–San Diego*. Institute of Electrical and Electronics Engineers, Piscataway, NJ, pp 1–5.

Rypkema NR, Fischell EM, Schmidt H (2017) One-way travel-time inverted ultra-short baseline localization for low-cost autonomous underwater vehicles. In: *IEEE International Conference on Robotics and Automation (ICRA)*. Institute of Electrical and Electronics Engineers, Piscataway, NJ, pp 4920–4926.

Shinzaki D, Gage C, Tang S, Moline M, Wolfe B, Lowe CG, Clark C (2013) A multi-AUV system for cooperative tracking and following of leopard sharks. In: *Proceedings—IEEE International Conference on Robotics and Automation*. Institute of Electrical and Electronics Engineers, Piscataway, NJ, pp 4153–4158.

Skomal G, Hoyos-Padilla E, Kukulya A, Stokey R (2015) Subsurface observations of white shark *Carcharodon carcharias* predatory behaviour using an autonomous underwater vehicle. *J Fish Biol* 87(6):1293–1312.

Stanway MJ, Kieft B, Hoover T, Hobson B, Klimov D, Erickson J, Raanan BY, Ebert DA, Bellingham J (2015) White shark strike on a long-range AUV in Monterey Bay. In: *MTS/IEEE OCEANS 2015–Genova: Discovering Sustainable Ocean Energy for a New World*. Institute of Electrical and Electronics Engineers, Piscataway, NJ, pp 1–7.

Tang S, Shinzaki D, Lowe CG, Clark CM (2014) Multi-robot control for circumnavigation of particle distributions. In: *Proceedings of the 12th International Symposium on Distributed Autonomous Robotic Systems (DARS 2014)*. Springer, Dordrecht, pp 149–162.

Viswanathan VK, Lobo Z, Lupanow J, von Fock SS, Wood Z, Gambin T, Clark C (2017) AUV motion-planning for photogrammetric reconstruction of marine archaeological sites. In: *IEEE International Conference on Robotics and Automation (ICRA)*. Institute of Electrical and Electronics Engineers, Piscataway, NJ, pp 5096–5103.

White C (2016) Quantifying the Habitat Selection of Juvenile White Sharks, *Carcharodon carcharias*, and Predicting Seasonal Shifts in Nursery Habitat use, master's thesis, California State University, Long Beach.

White CF, Lin Y, Clark CM, Lowe CG (2016) Human vs robot: comparing the viability and utility of autonomous underwater vehicles for the acoustic telemetry tracking of marine organisms. *J Exp Mar Biol Ecol* 485:112–118.

Wolfe BW, Lowe CG (2015) Movement patterns, habitat use and site fidelity of the white croaker (*Genyonemus lineatus*) in the Palos Verdes Superfund Site, Los Angeles, California. *Mar Environ Res* 109:69–80.

Xydes A, Moline M, Lowe CG, Farrugia TJ, Clark C (2013) Behavioral characterization and particle filter localization to improve temporal resolution and accuracy while tracking acoustically tagged fishes. *Ocean Eng* 61:1–11.

Zhang Y, Bellingham JG, Godin M, Ryan JP, McEwan RS, Kieft B, Hobson B, Hoover T (2010) Thermocline tracking based on peak–gradient detection by an autonomous underwater vehicle. In: *MTS/IEEE OCEANS 2010–Seattle*. Institute of Electrical and Electronics Engineers, Piscataway, NJ, pp 1–4.

CHAPTER 7

The Use of Stationary Underwater Video for Sampling Sharks

Euan S. Harvey
School of Molecular and Life Sciences, Curtin University, Perth, Western Australia, Australia

Julia Santana-Garcon
Department of Global Change Research, Institut Mediterrani d'Estudis Avançats, Spanish Research Council, Mallorca, Spain

Jordan Goetze
School of Molecular and Life Sciences, Curtin University, Perth, Western Australia, Australia;
Marine Program, Wildlife Conservation Society, Bronx, New York

Benjamin J. Saunders
School of Molecular and Life Sciences, Curtin University, Perth, Western Australia, Australia

Mike Cappo
Australian Institute of Marine Science, Townsville, Queensland, Australia

CONTENTS

7.1 INTRODUCTION

Our understanding of the biology and ecology of sharks and other highly mobile fish species relies largely on fishery-dependent data from commercial and recreational fisheries (Myers and Worm, 2003; Oliver et al., 2015). Catches made by net, hook and line, trawl, or trapping can provide valuable information, especially when biological samples are required (Jaiteh et al., 2014; Santana-Garcon et al., 2014a). However, these methods are often harmful and can result in direct mortality at capture or cryptic mortality after release. Ethically, invasive sampling techniques are becoming less acceptable with growing concerns over the status of some threatened, endangered, and protected species (e.g., sawsharks).

Serial depletion (i.e., the continued removal of individuals when sampling with extractive methods) can bias future sampling and is undesirable for long-term monitoring. Fishery-dependent methods can also introduce sampling biases due to gear selectivity and heterogeneous fishing efforts that discriminate among species, sizes, and habitats (Murphy and Jenkins, 2010; Simpfendorfer et al., 2002). Destructive sampling is prohibited in many "no take" marine reserves and restricted in other management zones (Robbins et al., 2013). This potentially limits the versatility of extractive sampling techniques for evaluating the effectiveness of fisheries management and conservation strategies in these areas.

Non-extractive (i.e., nondestructive) visual methods using scuba divers or cameras have become widely used alternatives to fishing methods when collecting data on marine fishes. The underwater visual census (UVC) technique was developed in 1954 (Brock, 1954) and is now one of the most common non-extractive methods for sampling fishes in shallow waters. Underwater visual censuses of sharks have also been used to survey the population status of carcharhinid reef sharks (Rizzari et al., 2014a,b; Robbins et al., 2006). Many species of sharks are rare, highly mobile, and nocturnally active or vary in their response to divers (MacNeil et al., 2008), so UVC is often an unsuitable survey technique. Previous studies using UVC to sample reef sharks have resulted in overestimation of the density of reef sharks in areas of high abundance (Ward-Paige et al., 2010) or a lack of replication and sampling power in areas of low abundance (McCauley et al., 2012).

Remotely operated camera systems (e.g., "camera traps") have been used to capture images of animals under natural conditions since the late 19th century (Carey, 1926; Kucera and Barrett, 2011). Over the course of the 20th century, rapid technological advancements in imaging systems have increased our ability to explore underwater environments (Mallet and Pelletier, 2014). As a result, marine scientists have increasingly adopted the use of camera technology to sample marine animals since the 1980s (see review by Shortis et al., 2009). The use of video sampling has increased greatly over the last 5 years as the size and cost of video cameras have decreased dramatically, particularly in the consumer market (Struthers et al., 2015). Remote underwater video (RUV), where a camera system is used to make observations rather than a diver, provides an alternative to UVC. Remote video techniques can access depths and habitats inaccessible to divers (Goetze et al., 2011) and have been used to avoid the behavioral biases of fish (Watson and Harvey, 2007) and sharks (McCauley et al., 2012) toward divers. They have also been used to sample for longer time periods through the day and night (Myers et al., 2016). A major advantage of video techniques is that images and video footage can be stored as a permanent digital record. The imagery can be accessed by other users for distribution and repeated examination by an unlimited audience of observers to identify taxa and behaviors in the field of view, verify findings, or garner new information (Cappo et al., 2003).

The objectives of this chapter are to describe the types of stationary video systems that have been used to sample elasmobranchs (sharks, herein), discuss the advantages and disadvantages of these various methods, and to provide recommendations and future directions for the use of stationary video camera systems. This chapter considers only stationary video techniques, not video that is attached to individual animals (see Chapter 5 in this volume) or diver-operated systems used to measure length and behavior of sharks (Marshall and Pierce, 2012; Sequeira et al., 2016; Vianna et al., 2016).

7.1.1 Unbaited Stationary Video Systems

In theory, estimates of shark and ray abundance from unbaited stationary cameras should track true abundance linearly, providing a sample area is calculated (as is possible when using stereo-video technology). However, sharks and rays are generally rare and mobile species, so a large sampling effort is required to survey an adequate sample size that would allow for robust statistical comparisons. For this reason, unbaited stationary cameras have not been used extensively to sample sharks, with a limited number of studies in the literature (Bernard and Götz, 2012; Dunbrack, 2008; Loiseau et al., 2016; McCauley et al., 2012). A common solution is to add some form of bait to the RUV to increase the proportion of predatory fishes, sharks, and rays that are observed in the field of view. This is colloquially known as *baited remote underwater video systems* (BRUVs) (Goetze et al., 2015; Harvey et al., 2007; Mallet and Pelletier, 2014; Santana-Garcon et al., 2014a; Whitmarsh et al., 2017).

7.1.2 Baited Stationary Video Systems

Baited remote underwater video systems (or stations) (Cappo et al., 2000, 2003, 2004, 2006; Harvey et al., 2007) involve the use of video cameras that are deployed underwater with a bait canister in the field of view (Figure 7.1). The initial

Figure 7.1 Baited remote underwater video systems (BRUVs) in action on the seafloor in a coral reef habitat. (Top) BRUVs with a single camera at Scott Reef in northwestern Australia (photograph courtesy of P.Tinkler, AIMS). (Middle) Stereo-BRUVs in the U.S. territory of Guam, Micronesia (photograph courtesy of S. Lindfield, CRRF). (Bottom) New-generation stereo-BRUVs deployed in Papua New Guinea. The recent development of small solid-state cameras has allowed miniaturization of the more modern housings and BRUVs. The larger stereo-BRUVs (middle) measures 145 × 90 × 70 cm (W × D × H) and weighs approximately 35 kg; the modern stereo-BRUVs (bottom) measures 82 × 66 × 50 cm and weighs approximately 7 kg.

bait plume attracts animals into the field of view, as does the "berley effect" of feeding activity on the bait canister (Cappo et al., 2000). The technique also records animals indifferent to the bait attractant but resident in the field of view or passing during normal routines (Cappo et al., 2006; Harvey et al., 2007; Watson et al., 2005). BRUVs are becoming widely adopted due to their minimal impact on the ecosystem, their potential to overcome some of the biases associated with the use of extractive techniques or the presence of scuba divers, and their provision of a permanent digital record (Harvey et al., 2004; Whitmarsh et al., 2017).

The use of bait to attract animals to BRUVs makes the technique particularly suited for sampling large carnivorous species targeted by fisheries and those with special conservation needs (Cappo et al., 2003; Goetze and Fullwood, 2013; Goetze et al., 2015; Langlois et al., 2012), without precluding the sampling of species from other trophic groups (Dorman et al., 2012; Hardinge et al., 2013; Harvey et al., 2007; Watson et al., 2005). BRUVs have become widely adopted and have been used to sample assemblages across a variety of habitats and depths (Figure 7.2). In particular, BRUVs have been used to survey a wide variety of sharks (Bond et al., 2012; Espinoza et al., 2014; Goetze and Fullwood, 2013), offering a non-extractive alternative to longline surveys while providing similar estimates of shark abundance (Brooks et al., 2011; Santana-Garcon et al., 2014a).

Because multiple systems can be deployed simultaneously, this method is particularly useful for covering large spatial scales with high replication. BRUVs have been used to map the patterns of habitat association and relative abundance of common reef sharks across entire biomes (Figure 7.3) (Espinoza et al., 2014). This application demonstrates that BRUVs are a powerful and cost-effective method for assessing spatial and temporal changes in the diversity, relative abundance, and length structure of sharks and fishes. In recent years, the reduction in size and cost of hardware that has occurred with the development of action cameras (e.g., GoPro™) has made the technique relatively cost efficient (Struthers et al., 2015). For these reasons, BRUVs have become the most popular form of video-based sampling method for sharks, rays and fishes (Whitmarsh et al., 2017).

7.2 DEPLOYMENT METHODOLOGY

7.2.1 Demersal or Benthic Baited Remote Underwater Video Systems

Shallow water demersal or benthic BRUVs use video cameras in underwater housings to film the area surrounding a bait canister (Figure 7.4). Bait canisters are generally placed relatively close to the camera system (<1.5 m) to enable identification and counts of individuals in the field of view (Ellis and DeMartini, 1995; Heagney et al., 2007; Willis and Babcock, 2000). Details on bait type and quantity are

Figure 7.2 BRUVs have been used in a range of habitats and depths: (A) Gray reef sharks (*Carcharhinus amblyrhynchos*) over tropical coral reef at Jarvis Island, Pacific Ocean. (B) White-spotted guitarfish (*Rhynchobatus australiae*) over tropical sandy seabed in the Pilbara region of Western Australia. (C) Tiger shark (*Galeocerdo cuvier*) over a tropical macroalgae-covered reef in the Pilbara region of Western Australia. (D) Gray nurse shark (*Carcharias taurus*) over a temperate macroalgae-covered reef off southwestern Australia. (E) Shortfin mako shark (*Isurus oxyrinchus*) over temperate seagrass habitat off southwestern Australia. (F) Port Jackson sharks (*Heterodontus portusjacksoni*) over temperate sandy seabed off southwestern Australia. (G) Great hammerhead shark (*Sphyrna mokarran*) in water 200 m deep off Ningaloo reef in northwestern Australia. (H) Great white shark (*Carcharodon carcharias*) in water 100 m deep off southeastern New Zealand.

discussed in detail in Section 7.3. The camera housings are generally encased within a frame (Figure 7.4) to stabilize the system on the seafloor, protect the camera, and allow the system to sit above the benthos with a clear field of view. A design with a flat bottom that can break away if caught on the benthos (Figure 7.4A) is thought to be the most stable.

These frames are generally deployed from boats onto the seafloor using ropes and floats to enable retrieval. Larger vessels are capable of deploying up to 14 systems or more in sequence to increase efficiency and spatial replication. Cameras are then left to record for a set period of time (see Section 7.5).

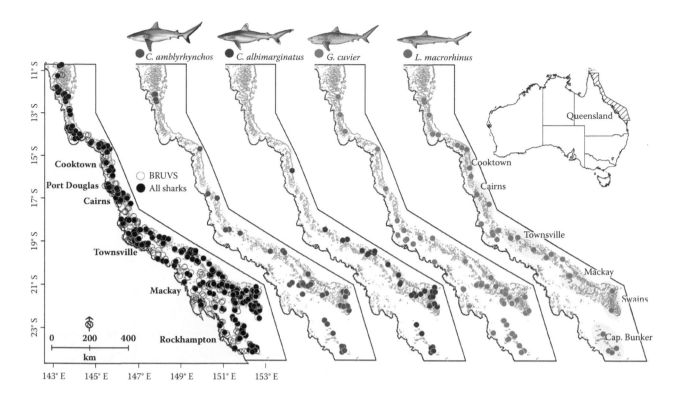

Figure 7.3 Baited video techniques can provide standardized, non-extractive, non-intrusive sampling at any depth or time of day across entire biomes for elasmobranchs. These data have been used to map the patterns of habitat association and relative abundance of common reef sharks in the Great Barrier Reef Marine Park. (From Espinoza, M. et al., *PLoS ONE*, 9, e106885, 2014.)

One of the major differences among the designs of shallow-water benthic BRUVs is the orientation of the cameras to face vertically down at the benthos (i.e., downward-facing) or horizontally across (i.e., forward-facing). Previous studies have shown differences in the fish communities observed between downward- and forward-facing BRUVs (Cundy et al., 2017; Langlois et al., 2006). The advantage of downward-facing BRUVs is that they ensure a standardized field of view, allow for simple measurements of length against a calibrated frame, and are less affected by low water clarity. However, behavioral bias may occur for some species hesitant to enter underneath the camera (Coghlan et al., 2017; Dunlop et al., 2015; Langlois et al., 2006). Downward-facing BRUVs have been used to survey reef fishes (Willis and Babcock, 2000); however, large sharks and rays are rarely recorded as a result of the limited field of view.

Forward-facing BRUVs provide a wide field of view, observe a larger number of species (not just those interested in the bait), and are now the most commonly used configuration (Mallet and Pelletier, 2014). The height and width of the field of view can be standardized when using single cameras, but not the depth of field. The depth of field depends not just on water clarity but also on the size (and hence visibility) of the animals in relation to the camera resolution. For example, a 4-m tiger shark (*Galeocerdo cuvier*) can be seen and identified to species level farther from the cameras than a small shark (<700 mm) in the genera *Loxodon* or *Rhizoprionodon*. Stereo-video systems can be used to standardize the maximum distance and field of view in which individuals are counted and identified (Figure 7.5) (see Section 7.2.5). In contrast to the field of view, which can be calculated, quantifying the actual area from which sharks are attracted to the baited video system is more difficult. Doing so depends on the potential area covered by the bait plume, the swimming speed of different sharks, the local currents, and the sensitivity of different species to detect the bait plume and their behavior when they have detected it (Hannah and Blume, 2016; Martinez et al., 2011; Priede et al., 1990; Westerberg and Westerberg, 2011) (see Section 7.2.4).

Despite some variations in the design of benthic BRUVs, effort has been made to standardize approaches, including system design, bait type and quantity, soak time, and spatial layout (Cappo et al., 2006). Standardization is an important step to enable broadscale synthesis of baited underwater video data. The Global FinPrint project (https://globalfinprint.org/) is the first global assessment of the status of shark and rays across coral reefs and provides a good example of using standardized approaches and protocols for BRUVs. Demersal BRUVs used in shallow water (less than 50 m) vary in shape, size, and weight (7 to 35 kg) depending on the location and habitat. In deeper water, the systems tend to be much heavier (up to several hundred kilograms).

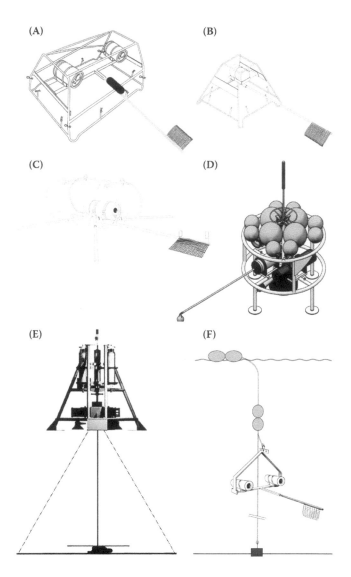

Figure 7.4 System designs used in baited video surveys: (A) Stereo-BRUVs (Harvey et al., 2002b; Watson et al., 2010) (modified from Harvey, E.S. et al., *PLoS ONE*, 8:e80955, 2013). (B) BRUVs (Cappo et al., 2007; Rizzari et al., 2014a) (modified from Harvey, E.S. et al., in *2011 National Workshop*, The University of Western Australia, Perth, 2013, pp. 1–202). (C) BRUVs (Cappo et al., 2004; Ellis and DeMartini, 1995; Harasti et al., 2015; Malcolm et al., 2007) (modified from Harvey, E.S. et al., *PLoS ONE*, 8:e80955, 2013). (D) DeepBRUVs (modified from Marouchos, A. et al., in *MTS/IEEE OCEANS 2011—Santander, Spain*, Institute of Electrical and Electronics Engineers, Piscataway, NJ, 2011, pp. 1–5). (E) Deepwater ROBIO lander (modified from Bailey, D.M. et al., *Mar. Ecol. Prog. Ser.*, 350, 179–192, 2007). (F) Pelagic stereo-BRUVs (Santana-Garcon et al., 2014b) (modified from Santana-Garcon, J. et al., *Meth. Ecol. Evol.*, 5, 824–833, 2014).

7.2.2 Deepwater Stationary Systems

Rapid technological development after World War II and the industrialization of commercial fisheries resulted in increased fishing pressure on the continental shelves of many countries and expansion of fisheries into deeper waters between 200 and 1500 m (Koslow et al., 2000). Many deepwater species have low productivity and fecundity, are late maturing, have slow growth rates, and are long lived, making them vulnerable to overfishing. This is particularly applicable to deep sea sharks such as gulper sharks (*Centrophorus* species), which are believed to have the

lowest reproductive potential of all chondrichthyans (Kyne and Simpfendorfer, 2007). There is a need for non-extractive sampling techniques that can cost-effectively collect data on the relative abundance and size frequency of these deepwater populations, both spatially and temporally (McLean et al., 2015). Baited video systems that are not limited by depth have proven to be a potential solution.

One of the big challenges of sampling deepwater sharks is their rarity. An advantage of baited cameras is that a single deployment can aggregate animals from the surrounding areas (Bailey et al., 2007). However, sampling in deep water (beyond 100 m) with baited cameras can be logistically

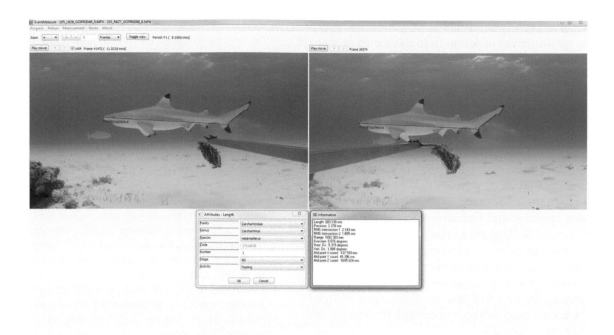

Figure 7.5 Left and right camera images from the same stereo-BRUVs deployment in the *EventMeasure* stereo software. Here, a length measurement is being made of an 88-cm (fork length) blacktip reef shark (*Carcharhinus melanopterus*) at the Cocos (Keeling) Islands..

difficult and costly due to the requirement of artificial lighting (see Section 7.4), underwater housings capable of withstanding greater water pressure, and difficulties in deploying and recovering systems.

Two studies in Australia have used stereo-BRUVs to assess fish and shark populations in deeper waters (300 to 600 m) (McLean et al., 2015; Sih et al., 2017). In New Zealand, similar systems have been deployed in up to 1200 m of water to examine the diversity and composition of fish assemblages (Zintzen et al., 2012, 2017). In both the New Zealand and Australian studies, stereo-BRUVs have been deployed and retrieved using ropes with surface buoys. As a general guide, for every meter of water depth approximately 1.5 m of rope are needed to ensure that there is sufficient surface slack to be able to retrieve the buoys. One of the major challenges is storing, deploying, and retrieving the ropes from the vessel; for example, deployments in 1000 m of water require up to 1500 m of rope, which requires careful management (see supplementary material in Zintzen et al., 2012). Commercial fishers routinely use ropes in these depths to deploy and recover traps for deep sea crabs. However, the use of ropes at depth is a balance between reducing the water drag on the rope by minimizing the diameter of the rope while still maintaining the breaking strain of the rope in order to pull the baited camera system to the surface.

An alternative is to use "ropeless" baited cameras systems that utilize acoustic or mechanical releases to drop a sacrificial weight and float back to the surface. The development of these "landers" was pioneered by Priede and

Bagley (2003) for use in abyssal waters. BotCam, for example, is a fishery-independent tool used to monitor and study the Hawaiian deepwater fish species between 100 and 300 m (Merritt et al., 2011). The system uses two ultra-low-light video cameras for stereo-video and contains a temperature and pressure recorder, a battery pack, and a relay used to trigger a bait release system. BotCam floats ~3 m above sacrificial concrete weights, which are left on the bottom when an acoustic mooring releases the weights. A similar system, DeepBRUVS (Figure 7.4D), developed by CSIRO in Australia, also uses sacrificial weights with acoustic releases and low-light cameras for operating at depths to 1000 m (Marouchos et al., 2011). DeepBRUVS are equipped with a stereo-video system, color balanced white lighting, and a bait release mechanism that permits multiple replicate video samples to be taken. The system has a number of floats and descends at 0.5 m/sec, taking around 30 min to reach its maximum operating depth of 1000 m. When it returns to the surface, it has a 1.5-m reflective mast, a strobe, and a radio range finder to aid relocation. This system has 10-liter storage for a liquefied bait, which can be released in small increments at timed intervals. Some of the deepwater systems, including the DeepBRUVS, have been constructed so they can be deployed for up to 6 months and make multiple recordings over this period (Marouchos et al., 2011).

Baited cameras have been deployed to greater depths, up to 10,000 m (Bailey et al., 2007); for example, chondrichthyans have been recorded at depths of 7703 m (Fujii et al.,

2010). Baited camera systems that are used at these great depths are very different in design from those used in shallower water. One such system, the ROBIO lander (Figure 7.4E), can be deployed on free-falling autonomous vehicles (Bailey et al., 2007). The data collected from these deep baited cameras have made enormous contributions to our understanding of the biology of the deep sea. However, these systems can be large and heavy (weighing up to 250 kg in air) and must be deployed from large vessels that are expensive to run. Given the cost of the systems (e.g., ~$25,000 for BotCam and over $50,000 for DeepBRUVS), most deep-sea sampling programs do not have enough systems to obtain good spatial coverage and replication. Baited cameras are potentially a valuable tool to increase our understanding of deep sea sharks, but there is a need to develop smaller, low-cost systems that can be deployed and recovered easily from vessels of opportunity (e.g., commercial fishing vessels). In turn, decreasing the cost will increase the number of units available for simultaneous deployment and the number of replicate samples obtainable.

7.2.3 Pelagic or Mid-Water Baited Remote Underwater Video Systems

Methods that are used in benthic habitats have been adapted to be applied to mid-water to target pelagic species (Berkenpas et al., 2017; Bouchet and Meeuwig, 2015; Fukuba et al., 2015; Heagney et al., 2007; Santana-Garcon et al., 2014a–c). Variation in design has allowed multiple systems to be deployed simultaneously from both small and large vessels, in order to maximize replication while minimizing field time and costs. For example, the deployment method presented by Santana-Garcon et al. (2014a,b) is moored to the seafloor and has been used at depths from 5 to 120 m in coastal waters and at exposed offshore sites (Figure 7.4F). This method could readily be adapted for deeper or drifting deployments. The design of this system makes the bait arm act as a rudder so that the cameras face downstream of the current, resulting in a clearer view of individuals that approach the camera while following the bait plume. The ballast and subsurface floats minimize the vertical movement of the cameras that may be driven and exacerbated by wave motion (Heagney et al., 2007; Merritt et al., 2011). This has proven to be an advantage over other systems tested in mid-water environments that have failed to effectively stabilize the camera systems (Letessier et al. 2013). The associated costs of building these pelagic stereo-BRUVs and the onboard logistics of using them are similar to those for the commonly used benthic BRUVs (Cappo et al., 2003, 2006).

Current development and trials of drifting deployment methods are expanding the use of baited video systems to open ocean sampling (Berkenpas et al., 2017; Bouchet and Meeuwig, 2015; Fukuba et al., 2015). The main concerns to take into account for drifting systems include the following:

1. Establish cost-effective means to track the position of the systems to enable retrieval at some predetermined time (boats should not remain in the area as this would confound the effects of sampling, such as the potential use of satellite tracking devices by commercial fishers).
2. Explore the differences in sampling area and bait plume dispersal between drifting and moored systems. This is important, as drifting systems have the potential to cover large areas based on current speeds. This will affect the dispersion of the bait plume and thus the area sampled (Heagney et al., 2007).
3. Minimize the effects of both the drifting action and wave motion to ensure stability of the camera systems underwater and thus enhance image quality. For example, attaching stereo-video systems to fish aggregation devices (FADs) such as those used in commercial fisheries could potentially enhance the use of video techniques in the high seas
4. Understand how benthic habitats, water depth, and temperature interact to determine the number of species and individuals sampled by pelagic BRUVs. This will influence the optimum soak time, bait type, and number of replicates necessary to detect meaningful spatial and temporal patterns (Anderson and Santana-Garcon, 2015).

7.2.4 Estimating Density of Sharks and Rays

One of the real challenges with baited stationary underwater video is estimating the density of focal species rather than the relative abundance. Studies using baited stationary underwater video targeting deepwater scavengers (Priede and Merrett, 1996) have modeled the area of attraction using *MaxN* (or n_{peak}) and arrival time, in conjunction with knowledge of current velocities, fish swimming speeds, and models of bait plume behavior and decay to estimate absolute density. Priede and Merrett's density estimation was based on their theory that the number of fish visible at the bait is the result of an equilibrium between arrivals and departures, and the "staying time" or "giving up time" is governed by Charnov's marginal value theorem of optimal foraging, which states that the staying time of an animal at an exhaustible food source is inversely related to the probability of finding an alternative food source (Cappo et al. 2000). Using strict assumptions that all fish are distributed randomly and evenly and that they respond immediately, positively, and independently from one another to the interception of a bait plume, Priede et al. (1990) developed a model of fish density using the "shark fin" curve (Figure 7.4). In a plot of number of fish at time t (N_t) against the soak time (t minutes), an initial fish arrival rate is relatively rapid, rising to a peak (n_{peak}) and declining as fish depart. A curve fitted to the data cloud can be broken up into a steeper arrival curve and a shallower departure curve, which are identical in shape but are separated by a time that corresponds to the mean staying time of fish. The difference between the two curves gives the actual number of individuals of a species present (King et al., 2006).

Theoretical population densities were calculated by Priede et al. (1990) from the time of arrival of the first scavengers to the bait using an inverse square law:

$$N = C/t_{arr}^2$$

where N is the density of fish per square kilometer, and t_{arr} is the time delay between the bait landing on the seafloor and the arrival of the first fish in seconds:

$$C = 0.3848(1/V_f + 1/V_w)^2$$

The constant C depends on the water velocity in (V_w; ms^{-1}) dispersing the bait plume downcurrent and the swimming velocity of the fish toward the bait (V_f; ms^{-1}).

These numerically sophisticated approaches require a number of unverifiable assumptions, and results are not amenable to robust falsification. For example, Priede et al. (1994) found that the n_{peak} of abyssal grenadiers was higher at an oligotrophic location with a low fish population and low food abundance because individuals stayed longer at the bait, whereas in a food-rich area with high population density the arrival rate was high because of the higher population, but n_{peak} was low because individuals gave up trying to gain access to the bait and left within an hour. Bailey and Priede (2002) qualified why such estimates are strongly affected by the assumed foraging behavior of the fish species concerned. They modeled three of the possible foraging strategies (cross-current foraging, sit-and-wait, and passive drifting) of abyssal scavengers, and the likely patterns of fish arrival at bait stations were calculated for the same fish density, swimming and current velocities, and odor plume properties. Each model produced a distinctive pattern of animal arrivals that may be diagnostic of each foraging strategy.

The abyssal scavenger model was tested for Patagonian toothfish by Yau et al. (2001), who noted that, for shallow-water applications, the inverse relationship between abundance and the square of the average arrival time will cause problems. Because abundance is proportional to the reciprocal of the square of the arrival time, a doubling of the arrival time produces a four-fold decline in Priede and Merrett's (1996) abundance estimate. Mean arrival times in shallow deployments occur at the level of seconds to minutes, rather than the tens of minutes to hours in abyssal studies. Shallower deployments can also produce far larger numbers of fish in the field of view. Shallow water studies have therefore neglected Priede and Merrett's theoretical approach to density estimation in favor of informative comparisons of indices of relative abundance among treatments, times, and places.

7.2.5 Stereo-Video

Although single-camera remote underwater video can provide useful measurements of relative abundance, it has limited ability for providing accurate estimates of the length of individuals (Harvey et al., 2002b). As a result, stereo-video systems have become popular in order to facilitate accurate estimates of the length and distance to individuals (Harvey et al., 2001), and they allow for standardization of visibility (Harvey et al., 2004). Stereo-video is being used to make accurate and precise measurements of the lengths of fishes and sharks ranging from small 30-mm damselfishes to very large tiger, white, and even whale sharks up to 9 m in length (Sequeira et al., 2016). Length estimates for stereo-video are made possible through a calibration process (Harvey and Shortis, 1998) and have been shown to be more accurate when compared to those made by a diver (Harvey et al., 2001). In order to achieve this high level of accuracy, it is important to follow a standardized calibration procedure, ideally with a three-dimensional (3D) calibration object (Boutros et al., 2015) and the use of specially designed software such as *CAL* (http://www.seagis.com.au/bundle.html). It is also important to ensure that the camera housings are mounted securely to eliminate any movement of the housings on the base bar which influences the orientation of the housings or the camera separation distance. Finally, it is critical to ensure that, when a camera is removed from the housing to replace memory cards or batteries, it is relocated in exactly the same position to maintain the internal orientation of the cameras and the distance between the lens and the camera port. A very small amount of movement can result in large measurement errors, and it is important to calibrate systems directly before and after each field trip to account for movement that occurs during the transportation and deployments of the systems.

To extract length measurements from stereo-BRUVs, specialist software such as *EventMeasure* (http://www.seagis.com.au/event.html) can be used. *EventMeasure* uses the calibration process and synchronized footage from both cameras to calculate the length and distance to individuals (Figure 7.5). When calibration files have been loaded, the measurement process is relatively simple, such that users simply need to locate the head and the tail for each individual and the software will compute the length and 3D location. Other freeware and two-dimensional calibration objects are available (Delacy et al., 2017), but we strongly recommend the use of specialized software such as *CAL* and *EventMeasure* to ensure preservation of the high level of accuracy that stereo systems are designed for.

An added advantage of extracting the length of sharks using stereo-video technology is the ability to calculate biomass through standard length–weight relationships, such as those available on FishBase (Froese and Pauly, 2015; Harvey et al., 2002a). It is important to obtain both abundance and length/biomass, as ecologically relevant patterns that are not observed in the abundance data are often detected or may be more pronounced in the length/biomass data (Goetze and Fullwood, 2013). Length and biomass are considered the first metrics to respond to protection from fishing (Claudet et al., 2010; Goetze et al., 2017). A recent study by Nash and Graham (2016) suggests that a broad range of metrics is needed to successfully assess the state of fisheries and ecosystem impacts. Although incorporating stereo-video into BRUV systems offers a number of advantages, these benefits should be considered against the extra cost.

7.3 TYPES AND QUANTITY OF BAIT USED

The most common bait used for BRUVs is oily, crushed clupeid baitfish of the genus *Sardinops* (Harvey et al., 2013b; Whitmarsh et al., 2017). Baited camera trials have been undertaken testing different types and quantities of baits (Dorman et al., 2012; Hardinge et al., 2013). These include comparisons of the fish assemblages sampled with tinned cat food, falafel mix (Dorman et al., 2012); urchins, abalone (Wraith et al., 2013); mussels, salmon (Walsh et al., 2017); no bait (Dorman et al., 2012; Hardinge et al., 2013; Harvey et al., 2007; Watson et al., 2005); and sardines. In general, sardines sampled more species and more individuals from targeted and indicator species and had greater statistical power to detect spatial and temporal changes than other bait types. Although researchers using baited underwater video use different volumes of bait, ranging from less than 200 g to 2 kg (Harvey et al., 2013b; Whitmarsh et al., 2017), it is important to ensure that there is enough bait so that it is present throughout the entire deployment, even when animals are attempting to feed from the bait bag. We recommend that a minimum of 500 g be used to ensure persistence throughout the deployment and a maximum of 1.5 kg to limit the total amount of fish used for a sampling method perceived as non-extractive. We also recommend the use of "off cuts" or waste products from fish that have been filleted, as they contain the majority of the oil and align better with conservation work.

In Australia, most researchers targeting multispecies assemblages use *Sardinops sagax neopilchardus*, which is a small pelagic sardine with a high oil content. When crushed in a bait bag and placed in seawater, the oil creates a bait plume, which attracts fish, sharks, and rays to the camera system. The bait type used varies around the world depending on the availability and the sampling objective. For example, in Hawaii (Asher et al., 2017a,b) and Guam (Lindfield et al., 2014, 2016), researchers used Pacific sauri (*Cololabis saira*). In the Mediterranean, Stobart et al. (2007) used sardines and an effervescent bait pellet, and Unsworth et al. (2014) used a commercial fish feed in the Irish Sea containing oily fish meal and fish oils. Although these studies did not target sharks or rays specifically, they did observe them. Baited camera studies that focused on targeting sharks have used tuna (Brooks et al., 2011; Spaet et al., 2016), sardines (Espinoza et al., 2014; Goetze and Fullwood, 2013; Harasti et al., 2016; Rizzari et al., 2014a; White et al., 2013), crushed baitfish (Bond et al., 2012), and mackerel (Clarke et al., 2012). Santana-Garcon et al. (2014a) used chopped sea

mullet (*Mugil cephalus*) to replicate the bait used in scientific longline surveys, which proved effective in attracting sharks to the camera systems. All of these studies have been consistent in the use of oily fish species as bait, but there remains the possibility that there may be differences in the attractant qualities of these baits to sharks. We recommend future research to assess the performance of these varying oily fish baits and the comparability of the data collected.

Despite the use of bait, it is still common to record low numbers of individual sharks due to a combination of their natural rarity, individual foraging patterns, and life histories. As a result, it is important to maximize the statistical power when conducting surveys that target sharks. An alternative to increasing replication may be to try various additions to the bait to maximize the attraction of sharks to BRUVs. This approach has been used, for example, to optimize sampling of pelagic fishes with baited cameras using a combination of visual cues (e.g., reflective materials such as flashes and lures), olfactory cues (e.g., bait such as pilchards), and acoustic cues (e.g., bait fish recordings) (Rees et al., 2015). The use of several simultaneous attractants could increase the effectiveness of stationary underwater video for sampling sharks. This is a potential area for future research; however, it is important that researchers move toward standardization of bait types and quantities to ensure long-term comparability of the data collected worldwide.

7.4 SAMPLING SHARKS AT NIGHT OR IN THE DARK: THE IMPLICATIONS OF LIGHTING

Nocturnal teleosts and sharks represent approximately one-third of the fish within any ecosystem (Helfman, 1978, 1986), but in some habitats (e.g., tropical mangroves) this can be as high as 57% of the species and 75% of the abundance (Ley and Halliday, 2007). Nocturnal and crepuscular movements of fish (Lowry and Suthers, 1998; Sikkel et al., 2017) and sharks (Hammerschlag et al., 2017) occur between habitats, most likely to seek food, to avoid predators, or for reproduction. Extractive sampling (gillnets and longlines) has been a popular technique for examining patterns of movement, habitat use, foraging, and reproduction during dawn and dusk, and at night non-extractive methods using passive and active acoustics and satellite telemetry have become increasingly popular, especially to document movement of individual animals, including habitat use (Hammerschlag et al., 2017). Nocturnal visual sampling targeting sharks and fish (Santana-Garcon et al., 2014a) has been undertaken by scuba divers (Nelson and Johnson, 1970) and remotely using unbaited (Myers et al., 2016) and baited (Harvey et al., 2012a–c) underwater video systems. Sampling when there is little or no natural light requires the use of artificial light, which may damage the eyes of sharks and fish and change their behavior and abundance (Fitzpatrick et al., 2013; Widder et al., 2005). The wavelengths of light to which sharks are sensitive depends on the structure of their eye (Bowmaker, 1990). Nocturnal and deep-sea sharks have developed many structural and physiological adaptations to improve their visual sensitivity in low light levels in comparison to shallow water and diurnal species (Warrant, 2004). Shallow-water sharks see best in light between 496 to 502 nm, midwater sharks (174 to 952 m) see best between 482 and 488 nm, and deepwater sharks see best between 472 and 478 nm (Sillman and Dahlin, 2004). The question of what spectrum of illumination to use for sampling sharks with baited cameras is important, particularly if an objective is to study their behavior. On land, infrared light has been demonstrated to be invisible to animals, but in water the attenuation of infrared is so rapid that it is only possible to illuminate objects at a maximum distance of 1.5 m from the light source (Widder et al., 2005). Research comparing red, white, and blue light revealed differences in the behavior and relative abundance of fishes, with greater numbers of non-target fishes observed under the red light and greater numbers of target fishes under blue light (Fitzpatrick et al., 2013; Harvey et al., 2012a). The choice of lighting is a compromise between the field of view sampled and the potential changes in the natural behavior and abundance recorded. This decision also must take into consideration the potential impact of bait on the behavior of fish and sharks. Does the bait overcome the potential negative effects of the lighting? The use of low-light cameras may overcome some of the potential problems associated with artificial lighting, and such cameras have been applied to BRUVs without artificial lighting in the BotCam systems (Merritt et al., 2011). Highly sensitive intensified CCD (charge-coupled device) and CMOS (complementary metal oxide semiconductor) low-light cameras are available (Graham et al., 2004), but remain expensive. Their use may overcome any possible disturbance from lighting. Future research should investigate the use of light-sensitive cameras illuminated by a light source that is above (Widder et al., 2005) or below (Harvey et al., 2012b) the visual sensitivity of the focal species. Such applications must be offset against the cost and replicability of the baited camera systems (see Section 7.5).

7.5 SAMPLING EFFORT: SOAK TIME AND REPLICATION

The performance of the method used in any study should be assessed under different sampling regimes in order to optimize its precision and effectiveness (Anderson and Santana-Garcon, 2015). Standardizing the soak time (how long the cameras are deployed underwater) is important to facilitate data comparison across time and space. The soak time also must be balanced against optimizing replication. A 1-hour soak time has been the most common and generally accepted time for BRUVs deployments (Whitmarsh et al., 2017). This

was developed out of recommendations by Watson (2006) and De Vos et al. (2014) when sampling fish assemblages. However, some studies suggest that soak times might be reduced or extended depending upon the assemblage and hypotheses to be tested (Gladstone et al., 2012; Harasti et al., 2015; Misa et al., 2016; Stobart et al., 2007). Studies focusing on sharks use soak times that vary between 60 min (De Vos et al., 2015; Goetze and Fullwood, 2013) and 90 min (Bond et al., 2012; Brooks et al., 2011) up to 5 hr (Harasti et al., 2016), but they can be longer when comparing BRUVs to longlines (Santana-Garcon et al., 2014a; Spaet et al., 2016). Therefore, we would recommend BRUVs soak times of at least an hour for surveys targeting sharks, given that this generally provides sufficient sample sizes for studies focusing on mobile species and allows for standardization across studies (Anderson and Santana-Garcon 2015; Santana-Garcon et al., 2014b). It has been suggested that sharks approach the bait in a cautious manner, making a first pass far from the camera system and moving progressively closer over time (Santana-Garcon et al., 2014a; Taylor et al., 2013). This behavior suggests that longer soak times may facilitate species identification and the recognition of individual features, including species, sex, or external markings, in addition to allowing for more accurate length measurements. However, territorial behaviors may prevent other individuals of the same or other species from approaching the cameras, which could affect estimates of species composition and relative abundance (Coghlan et al., 2017; Santana-Garcon et al., 2014a).

The majority of BRUVs studies separate replicate deployments spatially by at least 150 to 250 m (Whitmarsh et al., 2017); however, there are no rigorous studies that have examined the optimal spacing of BRUVs replicates. Most studies take a conservative approach in order to reduce the chances of double-counting wide-ranging individuals. The logic for the separation distance relates to assumptions about the current and bait plume dispersal and the maximum swimming speeds of fish (Cappo et al., 2000, 2003, 2006). Sharks are often wide ranging, and BRUVs studies have used a separation distance of 500 m or greater (Bond et al., 2012; De Vos et al., 2015; Santana-Garcon et al., 2014a; Spaet et al., 2016). Large sharks can swim fast (e.g., burst speeds in excess of 3 m/s) (Ryan et al., 2015), and individuals have been observed to move between up to six replicates, each separated by 250 m, during a 1-hour soak time (Cappo and Goetze, pers. obs.). It is possible to directly measure the range of attraction and the occurrence of repeated visits to "independent" replicates by deploying BRUVs inside an array of acoustic receivers inhabited by sharks tagged with acoustic transmitters (Vanderklift et al., 2014).

Ellis and DeMartini (1995) proposed that at distances of greater than 100 m separation their replicate 10-minute sets of baited videos were independent, because the greatest distance of fish attraction was only 48 to 90 m for a 200-mm fish in a current velocity of 0.1 to 0.2 m/s. This assumed a maximum swimming speed of approximately three body lengths per second for a 200-mm fish ($V_f = 0.6$ m/s). Given a seasonal prevalence of current ($V_c \sim 0.2$ m/s) in Australian studies, 60-minute soaks (S_t) of baited videos may have an effective range of attraction (AR) of ~480 m for fish ~200 to 300 mm in length. This includes 40 minutes of advection of the bait plume downcurrent and 20 minutes of fish swimming time upcurrent to reach the field of view in time to be recorded on the video. This relationship was presented by Cappo et al. (2004) as

$$AR = 60 \times (S_t) \times [(V_f \times V_c) - V_c^2]/V_f$$

in justification of distances of separation of 450 m between BRUVs replicates in biodiversity surveys. Further research is required to determine the optimal spacing of BRUVs for different species of sharks; however, until this has occurred we recommend a precautionary approach with a minimum separation distance of 500 m.

7.6 METHODS FOR IMAGE ANALYSIS: COUNTS AND LENGTH MEASUREMENTS

Converting underwater video into count data can be a time-consuming and challenging process. One of the biggest challenges is avoiding double counts of the same individual that may enter and exit the field of view multiple times. Unless there are distinguishing marks (patterns, distinctive body markings, or damage to body parts), it can be very difficult to tell how many individuals have been observed. A number of different metrics have been proposed to avoid double-counting individuals and to ensure independence of video counts. The maximum number of each species seen at any one time during a video (*MaxN*) was one of the first metrics to be developed (Ellis and DeMartini, 1995; Priede et al., 1994) for video and photographic sampling and is still the most commonly reported metric for stationary underwater video systems (Whitmarsh et al., 2017). *MaxN* counts are recognized as being conservative estimates of abundance and can be used in conjunction with other metrics such as time of first arrival or cumulative increases (Cappo et al., 2003, 2006). Time of first arrival and time to *MaxN* are rarely used but could potentially provide further insight into the status of shark and ray populations. Concerns have been raised that *MaxN* may be nonlinearly related to true abundance, due to saturation at high abundances (Schobernd et al., 2013). This has been confirmed in French Polynesia, where a comparison of single to full spherical cameras revealed that BRUVs with a restricted field of view are limited in their ability to discriminate differences in the relative abundance of sharks at extreme densities (Kilfoil et al., 2017). Despite the potential of full spherical cameras to more accurately capture true abundances at high densities, they are a recent advancement

in technology and are not yet readily available or cost effective. An alternative to *MaxN* that is thought to linearly track true abundance is *MeanCount*, where the mean number of fish is observed over a series of snapshots for a particular viewing interval (Schobernd et al., 2013); however, this method can increase video analysis time and may inflate zero observations (Campbell et al., 2015), making it less desirable for sampling sharks. Further investigation is necessary to assess and compare *MaxN* and *MeanCount* and their relation to true abundance. Regardless, it is important to avoid double-counting on a video by assuming independence of all observations and acknowledge that abundance estimates from stationary video are limited to relative abundance.

Sightings of sharks in pelagic BRUVs have also been converted to catch per unit of effort (CPUE) to facilitate comparisons with data collected from fishing techniques, such as longlines (Brooks et al., 2011; Santana-Garcon et al., 2014a; Spaet et al., 2016). Data have been expressed as the number of sharks (i.e., *MaxN*) per hour of sampling effort (i.e., fishing or video).

The use of generic video players coupled with spreadsheets for data management has become less common in the video analysis process (Whitmarsh et al., 2017), likely due to an increase in analysis time, the potential for operator error (e.g., transcribing), and decreased functionality when compared to specialist software. Specialist software such as the AIMS BRUVS.mdb and *EventMeasure* (www. seaGIS.com.au) (Figure 7.5) has been developed as an alternative to generic programs and relies on pre-populated lists to avoid transcribing/spelling errors and an inclusive video player to increase efficiency. *EventMeasure* has become the most commonly used specialist software for video analysis (Whitmarsh et al., 2017). This software can reduce analysis time by increasing the playback speed of videos, and its advanced functionality allows for the extraction of a broad range of data. Importantly, *EventMeasure* also has an option for stereo-video measurement that facilitates the extraction of the length of individuals, which can later be converted to biomass (see Section 7.2.5). Other software that has been used for annotating fish from video includes Vision Measurement System (Misa et al., 2016), MATLAB® (Hannah and Blume, 2012), and DVLOG (Somerton and Gledhill, 2005).

7.7 STRENGTHS AND LIMITATIONS OF SAMPLING SHARKS USING STATIONARY VIDEO TECHNIQUES

7.7.1 Strengths

The non-extractive nature of stereo-BRUVs allows for deployment in fragile and protected areas across multiple habitats and reduces the negative effects of extractive gears when targeting rare and threatened species, which is particularly relevant when studying sharks. When potentially dangerous or aggressive species are targeted, remote video techniques reduce the risk of having divers in the water or handling distressed sharks aboard vessels. Stationary video techniques can descend to depths inaccessible to divers and provide a permanent record of fish observed, thus allowing for cross-identification, verification, and data extraction after the field work. The use of bait increases the relative abundance and diversity of fish and sharks observed, particularly predatory species that are often targeted by fishers, without precluding the sampling of species from other trophic groups (Cappo et al., 2003; Harvey et al., 2007).

An advantage of using video to sample sharks when compared to extractive fishing techniques is the ability to classify and quantify the benthic habitat in which each individual was observed. Information such as relief (Watson et al., 2005), broad-scale benthic categories (Langlois et al., 2017), and percentage cover (abiotic and biotic) (Cappo et al., 2011) can be extracted from stationary benthic video systems. This information has proven important for the interpretation of habitat-based models of the relative abundance of sharks (Espinoza et al., 2014; White et al., 2013).

Because multiple video systems can be deployed simultaneously, this method is particularly useful for covering large spatial scales with high replication. A reduction in the size and cost of video cameras has made the technique relatively cost efficient. Video recordings also provide the opportunity to record animal behavior in their natural environment (Goetze et al., 2017; Santana-Garcon et al., 2013, 2014a; Zintzen et al., 2012). Timed cameras can be used for sampling around the clock (Myers et al., 2016) and investigating movements that occur with tides, diurnal cycles, and storm events (Munks et al., 2015).

Video imagery collected during remote surveys can be very effective when used for science communication. A number of highlight packages from video surveys are available online and are effective tools for attracting community engagement. For example, when surveying inside and outside of marine reserves, showing a local community excerpts of each video can be more effective than explaining complicated scientific plots or statistics. Similarly, it is possible to involve citizens in the data collection and data analysis process, given that appropriate training and quality analysis/ control procedures are employed. The Global FinPrint project (https://globalfinprint.org/) provides a good example of this approach, as the majority of the videos collected have been analyzed by volunteers and checked by experts.

7.7.2 Limitations

All sampling techniques have some inherent biases and limitations, and video-based methods are no exception (Murphy and Jenkins, 2010). The biggest limitation for the use of BRUVs is the inability to determine the area that is sampled. Although this is not a limitation for unbaited stationary video, baited video is the preferred method for sampling sharks and rays in order to provide sufficient sample sizes across local abundances of most species. Attempts have been made to account for the variation in distance that a bait plume may travel, but sensory threshold will vary across species (Bailey et al., 2007; Farnsworth et al., 2007; Heagney et al., 2007). It may also vary depending on the time of day or the response rates and behavior of different individuals, making estimates of density for baited video methods extremely challenging. As a consequence, we acknowledge that baited video methods can only generate relative estimates of abundance, biomass, or density (see Section 7.2.4). However, provided there is a linear relationship between relative estimates and true abundance, baited video will provide a useful tool for estimating and comparing shark abundance.

The difficulty in identifying all individuals to the species level is a common shortcoming for all video techniques due to the limited resolution of video (Mallet and Pelletier, 2014). This is exacerbated for some shark species because some taxa share similar morphological traits (e.g., coloration, streamlined fusiform body, pointed snouts, forked tails), and some species can only be distinguished by small and inconspicuous features such as tooth formulae and vertebral counts. Notable species pairs in this regard are *Carcharhinus leucas/amboinensis*, *C. tilstoni/limbatus*, and *C. obscurus/brachyurus*. Difficulties associated with species identification are not unique to video techniques but are also reported for other commonly used fishery-independent methods such as hydroacoustic techniques (Lawson et al., 2001). Fishery-dependent data are not immune to species identification issues, either, as fishers may also group species at higher taxonomic levels or under-report the catch of low-value species (Lewison et al., 2004).

Problems with species identification from video can be exacerbated in areas of low visibility and/or high turbidity. Stereo-video techniques can account for variation in visibility to a point by standardizing measurements from all videos in an area or survey to the distance coinciding with the minimum visibility. This has the effect of reducing the field of view used and area surveyed so will reduce the number of individuals seen. In areas with visibility lower than 2 to 3 m, or where visibility does not extend beyond the bait bag, visual sampling with stationary video techniques may not be suitable. These levels of low visibility are common in estuarine waters or coastal areas where there are large tides or a high input of sediment from runoff and rivers. This is problematic, given that some critically endangered sharks such as sawfish (Pristidae) and river sharks (*Glyphis* spp.) inhabit these areas. Extractive methods are also undesirable for sampling these vulnerable species. Emerging technologies such as environmental DNA (Stat et al., 2017; see also Chapter 14 in this volume) or imaging sonar (see Section 7.8.2) may be more suitable in these conditions.

In the above sections we have noted that one of the advantages of stationary underwater video techniques is that they are non-extractive and have minimal impacts on the species sampled or the habitats they are deployed within. However, care still must be taken deploying the systems into fragile habitats such as staghorn or plate corals. In shallow waters, damage to fragile habitats can be mitigated or minimized by using lightweight frames (Figure 7.1), slowing the rate of descent and the impact with the bottom by controlling the speed of deployment with the rope, or using a snorkeler to guide the lowering of the camera system. In deeper water, baited camera systems have been deployed with remotely operated vehicles (ROVs) (Widder et al., 2005), or the deployments have been controlled by rope by viewing the descent of the system on a depth sounder and moving the boat to ensure that the systems come in contact with the bottom in a desirable manner.

Unlike conventional or electronic tags, stationary video techniques do not collect data on the movements of individual animals. It can also be very challenging to consistently sex individuals from video imagery depending on the species and orientation of the shark. Although size estimates can help obtain information about the age structure of a population, they cannot inform full demographic studies. The use of video methods alone might not be appropriate for monitoring or management programs that require such information.

The equipment (cameras, memory cards, housings, frames, base bars for stereo-video systems) and the time required for video analysis can be costly and a major limitation to the uptake and use of stationary video techniques (Goetze et al., 2015). Advances in video processing and the automation of some components of the analysis appear promising with regard to reducing analysis time in the near future (Shortis et al., 2013). Until automation progresses from theoretical to functional software systems that are widely available to ecologists, the cost and time involved in video analysis should not be underestimated when applying for funding or setting up research projects. Another solution to the image analysis problem is the use of volunteers and citizen scientists. This approach has been used in the Global FinPrint project, where a minimum of two volunteers annotate each video and identification is subsequently verified by an expert. Where citizen scientists can simply locate the sharks and make an attempt on identification, an expert is only required to review and confirm each observation, which is more time efficient than watching an entire video. However, this may not always be possible, particularly when information on the entire fish assemblage is required, which requires more advanced identification skills. Understanding the limitations and biases of sampling techniques allows for better use of the method in forthcoming studies and provides direction to future research so that these challenges can be overcome.

7.8 THE FUTURE OF VIDEO METHODS FOR SAMPLING SHARKS

As camera technology develops, the quality and sensitivity of the imaging sensors in video cameras are increasing (particularly in the consumer market), while the size and cost of the video cameras are decreasing (Struthers et al., 2015). Improvements in the quality and resolution of video will result in further increases in the accuracy and precision of stereo-video length measurements. This has been observed as video cameras have developed from Hi8 and MiniDV tape-based recordings to high-definition (Harvey et al., 2010) and more recently 4K video cameras. As the image quality and resolution improve, small morphometric features, movement, shape, and coloration have become more distinguishable, increasing the accuracy and speed of manual identifications of fish and sharks by analysts. A reduction in the size and cost of video cameras has also increased sampling efficiency by allowing for a greater number of systems to be purchased and transported more cost effectively. It is now possible to use smaller boats in the field, further reducing costs and facilitating work in areas where large boats are not available. Despite the development of purpose-built software for viewing and recording events or species and the numbers of individuals seen on BRUVs, the manual detection, counting, and measuring of fish from BRUVs imagery remains time consuming and labor intensive. This can delay the availability of data for researchers and for timely uptake by managers. To facilitate further adoption and make stationary underwater video sampling more cost effective, there is a need for researchers in the fields of pattern recognition, artificial intelligence, and machine learning to focus on automating the analysis of undersea imagery (MacLeod et al., 2010).

7.8.1 Potential for Automatic Image Analysis

Rapid progress has been made in the field of computer vision, pattern recognition, artificial intelligence and machine learning, and many researchers are focused on the recognition of marine life in underwater images. To be comparable to the data produced by trained video analysts a fully automated process will require the detection, measurement, identification, and counting of individuals (Shortis et al., 2013). Various techniques are used to detect objects of interest (often fish) in underwater imagery. Differences between successive video frames (Lines et al., 2001; Spampinato et al., 2008) or histogram thresholds (Khanfar et al., 2010) and the identification of a fish silhouette using edge detectors (Spampinato et al., 2010) have all been used for detection. A shape-based level-sets framework and a Haar classifier have also been used to determine and track the location of the head, snout, and tail of a fish using knowledge of the shape for each species (Ravanbakhsh et al., 2015).

Automated recognition of fish from still and video images is feasible. Initial research used texture- (Rova et al., 2007) and shape-based (Spampinato et al., 2010) methods for identification. The main challenge in using shape-based features for fish classification in natural underwater environments is that the shapes, color, texture, and movement of the plants and animals living on the seabed make segmentation of the fish from the background a very challenging problem (Shafait et al., 2016). Increases in computing power and developments in machine learning are overcoming some of these challenges. Recent studies have achieved an average classification success of over 90% (Salman et al., 2016) on a readily available fish image dataset (http://imageclef.org/). Refining this technique improved identification accuracy to 96.73% on the same dataset and 89% on a more complex unconstrained dataset collected from baited remote underwater video systems (Siddiqui et al., 2017).

Both semiautomated (Shafait et al., 2017) and automated (Muñoz-Benavent et al., 2017) measurements of the lengths of bluefin tuna from stereo-video imagery have been reported with a high degree of accuracy and precision. These protocols can be adapted to sharks and rays. If it is possible to detect, track, and identify animals from video imagery, it is also possible to count them. Research has demonstrated that it is technically feasible to detect, identify, measure, and count fish, but a fully operational and integrated system is not yet available, to the best of our knowledge. Further research that integrates additional information that can be generated from stereo-video systems, such as the swimming motion of organisms using dynamic templates, and the use of measurements of morphometric features of fish and sharks will improve the identification of species and possibly distinguish between individuals (Shortis et al., 2013).

7.8.2 Imaging Sonar to Sample Sharks

In turbid waters (or in dark water without lights) video cameras may not be suitable for detecting and counting sharks. Active acoustic techniques are one possible alternative to video in these environments, and they are regularly used to detect marine fauna in the water column (Simmonds and MacLennan, 2005). These techniques have evolved from single- and split-beam echosounders (SBESs) to multi-beam echosounders (MBESs). High-frequency MBES systems (henceforth called *imaging sonars*) can produce images from reflected acoustic energy of such quality that they have also been referred to as acoustic cameras. The advantage of such systems is that they are able to "see" acoustically in turbid or dark waters. These systems were developed for the oil and gas industry but have found application in ecological studies. For example, imaging sonar systems have been applied for counting and measuring fish (Becker et al., 2011b; Holmes et al., 2006; Parsons et al., 2014; Petreman et

al., 2014), and recent studies have begun to explore the use of imaging sonar to detect and count sharks (Lieber et al., 2014; Parsons et al., 2014).

Active acoustic systems operate by "ensonifying" a sampling volume with an acoustic beam and evaluating the energy reflected back to the system transducer, allowing the sonar to "see" the targets. The amount of energy reflected is dependent on the density contrast between the target and the medium (water) and the frequency at which the imaging sonar operates (Urick, 1983). Lower frequencies travel farther, but higher frequencies provide improved resolution. There is a trade-off between the range of the targets to be detected and the resolution required to discern them. Depending on the system operating frequency, the range can be extended to "see" large targets, such as 3-m sharks, at distances out to 50 m (Parsons et al., 2014); however this increase in range is achieved by decreasing the operating frequency, so it comes at the expense of image resolution and therefore the ability to detect small targets. As a result, the type of sonar used should be tailored to the application; that is, lower frequency systems can detect large sharks at a greater range, but higher frequency systems can provide higher image resolution of smaller sharks at closer range.

As the operating frequency of a system increases, the interface between the target and water plays a greater role in the energy reflected. As a result, although lower frequencies perform well at detecting targets with a high density contrast from water (e.g., >90% of energy reflected by a fish comes from the swim bladder) (Simmonds and MacLennan, 2005), high-frequency systems receive significant reflected energy from the target surface. The level of energy reflected is therefore also highly dependent on the angle of incidence of the target; a surface that is perpendicular to the acoustic beam will reflect more than one that is oblique. The acoustic energy reflected by sharks (in comparison with fish) is increased by their size in relation to the acoustic beam. In short, the acoustic energy reflected back to sonars by sharks is substantial, despite the lack of swim bladder, lungs, or bony structures that would have a high density contrast with water. This is in part due to their size and, at higher frequencies, the amount of surface area available to reflect the energy (Parsons et al., 2014, 2017). Therefore, the reflected energy is also strongly related to shark orientation relative to the transducer. As an extension of this, stingrays can often be difficult to observe as they travel flat along the seabed. Stingrays are often visible on imaging sonar footage as a dark patch that moves through the image. The dark patch is a result of the acoustic signal being reflected away from the sonar rather than back to the transducer due to the low angle of incidence on the flat stingray. However, more orthogonal surfaces of the head around the eye and the fins as they undulate reflect the signal, resulting in a bright image of these parts of the stingray (Figure 7.6A,B). The reflected energy from moving targets is highly variable (Simmonds

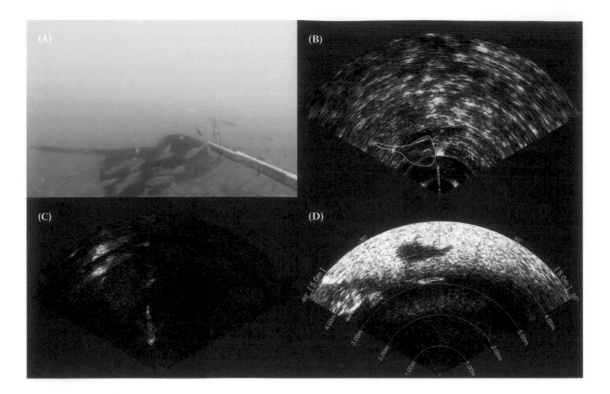

Figure 7.6   (A,B) Smooth stingray (*Bathytoshia brevicaudata*) recorded on a paired stereo-video and imaging sonar BRUVs in Cockburn Sound, southwestern Australia. In (B), the imaging sonar still shows the ray outlined, and the bait bag can be seen as a bright square in the center of the image. (C) Sonar image of a sicklefin lemon shark (*Negaprion acutidens*), and (D) the acoustic shadow of a lemon shark. Note the smaller but bright return signal visible in the center of the image just beyond the 3-m range. Both images were recorded at Shark Bay, Western Australia. Sonar images were recorded using Tritech Gemini 720i imaging sonar.

and MacLennan, 2005) as energy is reflected differently from surfaces dependent on their relative position and orientation. When using imaging sonar to sample sharks or rays, the clearest understanding of the animal comes from observing the pattern of movement through time, using the imaging sonar in a fashion analogous to video.

In addition to the reflected signals (Figure 7.6C), targets can produce acoustic shadows in the same way that an object casts a shadow in the beam of a torch (Parsons et al., 2014). This phenomenon means that with some systems the morphology of the target can be seen in the shadow cast by the system. The acoustic shadows can be useful in detecting sharks on imaging sonars, as these are often larger and more obvious than the reflected signal of the animal itself (Figure 7.6D). However, the size and shape of such acoustic shadows are affected by a variety of factors, including the distance from and orientation of the shark to the system, as well as the bathymetry of the seafloor.

7.8.2.1 *Applications of Imaging Sonar*

Imaging sonar systems can be applied to work in a way similar to a stationary underwater video system. They can be mounted onto a frame and deployed onto the seabed with or without bait, or they can be mounted onto a stationary vessel. They can be positioned so that they are looking horizontally or vertically through the water column. Imaging sonar systems have been used successfully to count and measure the length of fishes in a number of habitats and applications. These include observations of spatially migrating fishes (Crossman et al., 2011; Martignac et al., 2015; Petreman et al., 2014), diel migrations (Becker and Suthers, 2014; Becker et al., 2011a), size structure (Becker et al., 2011b; Grote et al., 2014; Han et al., 2009), and habitat use (Becker and Suthers, 2014). Imaging sonars can provide count and size information, as well as behavioral information such as the frequency of tail beats (Mueller et al., 2010). In applications where the target assemblage is well known it is possible to use size class and behavioral information to draw inferences about the species that might be observed. This information may go some way in helping to overcome the major drawback of imaging sonar systems, which is the identification of the species being observed.

Imaging sonars have recently been applied in conjunction with stereo-BRUVs to improve the detection of fishes in turbid environments (see example in Parsons et al., 2017). The imaging sonar is mounted onto the stereo-BRUVs frame in a central position between the cameras and looking forward horizontal to the seabed. This setup allows synchronization of the video and imaging sonar recordings (Figure 7.6A,B) so that the data can be used to complement one another. The imaging sonar extends the field of view for relative abundance estimates by size class, and the stereo-video footage allows identification of species that approach close to the cameras. The video footage can be used for biodiversity assessment and to identify the species that contribute to a given size class. This technology could be especially applicable in tidal or estuarine environments, where visibility is extremely variable, even over the duration of a single 1-hour BRUVs deployment.

As imaging sonar technology continues to improve, the cost is falling, and as more research is published this tool is likely to become widely valued and applied. We foresee its application in a number of environments and research directions. These might include increased application in turbid waters and deep or dark environments without lights, observing behavioral interactions with artificial lights, and extending the range of detection beyond the 7 to 10 m commonly used with video techniques. This range extension may be especially useful in increasing the detection rates of rare and large-bodied sharks and rays.

7.9 SUMMARY AND CONCLUSION

Stationary underwater video techniques are increasingly being used for sampling the distribution, habitat use, relative abundance, length, biomass, and behavior of sharks. Part of the growing popularity of video techniques is their accessibility, ease of use, and the decreasing cost and size of underwater video cameras. Additionally, improvements in image quality are making stationary underwater video an attractive sampling option. Part of the appeal of video sampling is being able to share imagery with a range of stakeholders to convey messages in a format that most people can understand and relate to. Stationary video techniques (and in particular baited video) collect not only compelling imagery but also robust quantitative data from a range of habitats and to depths not easily sampled by other nonextractive techniques.

As with all techniques, there are limitations and biases to stationary underwater video methods. When using baited videos, we do not know the distance from which sharks are attracted to a bait or how their behavior changes as densities increase artificially around a bait. Counts of sharks from stationary video are conservative because of the strategies used to avoid double counts (i.e., *MaxN*). Often only a small proportion of the data that are potentially available is extracted from the imagery due to the time and cost involved in physically watching videos to record species identifications and the numbers of individuals, to measure lengths, and to record behavioral data.

Standardization of the types and quantity of bait used, where the bait is positioned in front of the cameras and in what sort of container, the soak times, and the orientation of the cameras (vertical or horizontal) is important. The development of standardized protocols and procedures that are used globally creates the opportunity for the synthesis of quantitative datasets at a range of spatial and temporal scales, which will generate meaningful information on the status of shark assemblages. What is really exciting is that people's familiarity with easy-to-use, off-the-shelf action cams means that this sampling technique is accessible to a broad range of stakeholders, from professional scientists to schools and any other groups who might want to collect data on sharks. In the future, fully automated image analysis will facilitate the extraction of a broad range of data shortly after collection. Increasing Internet speeds and improved accessibility, combined with decreasing costs of online storage, will result in the global sharing of metadata, numerical data, and raw imagery. This will increase the pool of knowledge and data available to researchers and managers and facilitate broad-scale collaborations. This information and knowledge will assist in the sustainable management and conservation of shark populations and effective science communication to all stakeholders.

ACKNOWLEDGMENTS

The authors thank M.J.G. Parsons for his valuable input to Section 7.8.2.

REFERENCES

Anderson MJ, Santana-Garcon J (2015) Measures of precision for dissimilarity-based multivariate analysis of ecological communities. *Ecol Lett* 18:66–73.

Asher JM, Williams ID, Harvey ES (2017a) An assessment of mobile predator populations along shallow and mesophotic depth gradients in the Hawaiian archipelago. *Sci Rep* 7:3905.

Asher J, Williams ID, Harvey ES (2017b) Mesophotic depth gradients impact reef fish assemblage composition and functional group partitioning in the main Hawaiian Islands. *Front Mar Sci* 4:98.

Bailey DM, Priede IG (2002) Predicting fish behaviour in response to abyssal food falls. *Mar Biol* 141:831–840.

Bailey DM, King NJ, Priede IG (2007) Cameras and carcasses: historical and current methods for using artificial food falls to study deep-water animals. *Mar Ecol Prog Ser* 350:179–192.

Becker A, Suthers IM (2014) Predator driven diel variation in abundance and behaviour of fish in deep and shallow habitats of an estuary. *Estuar Coast Shelf Sci* 144:82–88.

Becker A, Cowley PD, Whitfield AK, et al. (2011a) Diel fish movements in the littoral zone of a temporarily closed South African estuary. *J Exp Mar Biol Ecol* 406:63–70.

Becker A, Whitfield AK, Cowley PD, et al. (2011b) An assessment of the size structure, distribution and behaviour of
fish populations within a temporarily closed estuary using
dual frequency identification sonar (DIDSON). *J Fish Biol*
79:761–775.

Berkenpas EJ, Henning BS, Shepard CM, et al. (2017) A buoyancy-
controlled Lagrangian camera platform for *in situ* imaging
of marine organisms in midwater scattering layers. *IEEE J
Oceanic Eng* July:1–13.

Bernard A, Götz A (2012) Bait increases the precision in count
data from remote underwater video for most subtidal reef
fish in the warm-temperate Agulhas bioregion. *Mar Ecol
Prog Ser* 471:235–252.

Bond ME, Babcock EA, Pikitch EK, et al. (2012) Reef sharks exhibit
site-fidelity and higher relative abundance in marine reserves
on the Mesoamerican Barrier Reef. *PLoS ONE* 7:e32983.

Bouchet PJ, Meeuwig JJ (2015) Drifting baited stereo-videogra-
phy: a novel sampling tool for surveying pelagic wildlife in
offshore marine reserves. *Ecosphere* 6:1–29.

Boutros N, Shortis MR, Harvey ES (2015) A comparison of calibration methods and system configurations of underwater
stereo-video systems for applications in marine ecology.
Limnol Oceanogr Methods 13:224–236.

Bowmaker JK (1990) Visual pigments of fishes. In: Douglas RH,
Djamgoz, M.B.A. (eds) *The Visual System of Fish*. Springer,
Dordrecht, pp 81–107.

Brock VE (1954) A preliminary report on a method of estimating
reef fish populations. *J Wildl Manage* 18:297–308.

Brooks EJ, Sloman KA, Sims DW, Danylchuk AJ (2011) Validating
the use of baited remote underwater video surveys for assessing the diversity, distribution and abundance of sharks in the
Bahamas. *Endanger Species Res* 13:231–243.

Campbell MD, Pollack AG, Gledhill CT, et al. (2015) Comparison
of relative abundance indices calculated from two methods
of generating video count data. *Fish Res* 170:125–133.

Cappo M, Speare P, Wassenberg T, et al. (2000) The use of
baited remote underwater video stations (BRUVS) to sur-
vey demersal fish stocks—how deep and meaningful?
In: Harvey E, Cappo M (eds) *Direct Sensing of the Size
Frequency and Abundance of Target and Non-Target Fauna
in Australian Fisheries—A National Workshop*. Fisheries
Research Development Corporation, Rottnest Island,
Western Australia, pp 4–7.

Cappo M, Harvey E, Malcolm H, Speare P (2003) Potential of video
techniques to monitor diversity, abundance and size of fish
in studies of marine protected areas. In: Beumer JP, Grant
A, Smith DC (eds) *Aquatic Protected Areas—What Works
Best and How Do We Know? World Congress on Aquatic
Protected Areas Proceedings*. Australian Society for Fish
Biology, North Beach, Western Australia, pp 455–464.

Cappo M, Speare P, De'ath G (2004) Comparison of baited remote
underwater video stations (BRUVS) and prawn (shrimp)
trawls for assessments of fish biodiversity in inter-reefal
areas of the Great Barrier Reef Marine Park. *J Exp Mar Biol
Ecol* 302:123–152.

Cappo M, Harvey E, Shortis M (2006) Counting and measuring
fish with baited video techniques—an overview. In: Furlani
DM, Beaumer JP (eds) *Proceedings of the Australian
Society for Fish Biology Workshop*. Australian Society for
Fish Biology, North Beach, Western Australia, pp 101–114.

Cappo M, De'ath G, Speare P (2007) Shelf-scale patterns in communities of aquatic vertebrates in the inter-reefal waters of the
Great Barrier Reef Marine Park determined by baited remote
underwater video stations. *Mar Ecol Prog Ser* 350:209–221.

Cappo M, Stowar M, Syms C, et al. (2011) Fish-habitat associations in the region offshore from James price point—a rapid
assessment using baited remote underwater video stations
(BRUVS). *J R Soc West Aust* 94:19.

Carey HR (1926) Camera-trapping: a novel device for wild animal
photography. *J Mammal* 7:278–281.

Clarke C, Lea J, Ormond R (2012) Comparative abundance of
reef sharks in the Western Indian Ocean. In: Yellowlees
D, Hughes TP (eds) *Proceedings of the 12th International
Coral Reef Symposium*. James Cook University, Townsville,
Queensland, Australia, pp 9–13.

Claudet J, Osenberg CW, Domenici P, et al. (2010) Marine
reserves: fish life history and ecological traits matter. *Ecol
Appl* 20:830–839.

Coghlan AR, McLean DL, Harvey ES, Langlois TJ (2017) Does
fish behaviour bias abundance and length information collected by baited underwater video? *J Exp Mar Biol Ecol*
497:143–151.

Crossman JA, Martel G, Johnson PN, Bray K (2011) The use of
Dual-frequency IDentification SONar (DIDSON) to docu-
ment white sturgeon activity in the Columbia River, Canada.
J Appl Ichthyol 27:53–57.

Cundy ME, Santana-Garcon J, Ferguson AM, et al. (2017) Baited
remote underwater stereo-video outperforms baited downward-facing single-video for assessments of fish diversity,
abundance and size composition. *J Exp Mar Biol Ecol*
497:19–32.

Delacy CR, Olsen A, Howey LA, et al. (2017) Affordable and
accurate stereo-video system for measuring dimensions
underwater: a case study using oceanic whitetip sharks
Carcharhinus longimanus. *Mar Ecol Prog Ser* 574:75–84.

De Vos L, Götz A, Winker H, Attwood CG (2014) Optimal BRUVs
(baited remote underwater video system) survey design for
reef fish monitoring in the Stilbaai Marine Protected Area.
Afr J Mar Sci 36:1–10.

De Vos L, Watson R, Götz A, Attwood CG (2015) Baited remote
underwater video system (BRUVs) survey of chondrichthyan
diversity in False Bay, South Africa. *Afr J Mar Sci* 37:209–218.

Dorman SR, Harvey ES, Newman SJ (2012) Bait effects in sampling coral reef fish assemblages with stereo-BRUVs. *PLoS
ONE* 7:e41538.

Dunbrack R (2008) Abundance trends for *Hexanchus griseus*,
bluntnose sixgill shark, and *Hydrolagus colliei*, spotted ratfish, counted at an automated underwater observation station
in the strait of Georgia, British Columbia. *Can Field-Nat*
122:124–128.

Dunlop KM, Marian Scott E, Parsons D, Bailey DM (2015) Do
agonistic behaviours bias baited remote underwater video
surveys of fish? *Mar Ecol* 36:810–818.

Ellis DM, DeMartini EE (1995) Evaluation of a video camera
technique for indexing abundances of juvenile pink snapper,
Pristipomoides filamentosus, and other Hawaiian insular
shelf fishes. *Oceanogr Lit Rev* 9:786.

Espinoza M, Cappo M, Heupel MR, et al. (2014) Quantifying
shark distribution patterns and species-habitat associations:
implications of marine park zoning. *PLoS ONE* 9:e106885.

Farnsworth KD, Thygesen UH, Ditlevsen S, King NJ (2007) How to estimate scavenger fish abundance using baited camera data. *Mar Ecol Prog Ser* 350:223–234.

Fitzpatrick C, McLean D, Harvey ES (2013) Using artificial illumination to survey nocturnal reef fish. *Fish Res* 146:41–50.

Froese R, Pauly D (2015) *FishBase*, http://www.fishbase.org/.

Fujii T, Jamieson AJ, Solan M, et al. (2010) A large aggregation of liparids at 7703 meters and a reappraisal of the abundance and diversity of hadal fish. *Bioscience* 60:506–515.

Fukuba T, Miwa T, Watanabe S, et al. (2015) A new drifting underwater camera system for observing spawning Japanese eels in the epipelagic zone along the West Mariana Ridge. *Fish Sci* 81:235–246.

Gladstone W, Lindfield S, Coleman M, Kelaher B (2012) Optimisation of baited remote underwater video sampling designs for estuarine fish assemblages. *J Exp Mar Biol Ecol* 429:28–35.

Goetze JS, Fullwood LAF (2013) Fiji's largest marine reserve benefits reef sharks. *Coral Reefs* 32:121–125.

Goetze JS, Langlois TJ, Egli DP, Harvey ES (2011) Evidence of artisanal fishing impacts and depth refuge in assemblages of Fijian reef fish. *Coral Reefs* 30:507–517.

Goetze JS, Jupiter SD, Langlois TJ, et al. (2015) Diver operated video most accurately detects the impacts of fishing within periodically harvested closures. *J Exp Mar Biol Ecol* 462:74–82.

Goetze JS, Januchowski-Hartley FA, Claudet J, et al. (2017) Fish wariness is a more sensitive indicator to changes in fishing pressure than abundance, length or biomass. *Ecol Appl* 27:1178–1189.

Graham N, Jones EG, Reid DG (2004). Review of technological advances for the study of fish behaviour in relation to demersel fishing trawls. *ICES J Mar Sci* 61:1036–1043.

Grote AB, Bailey MM, Zydlewski J, Hightower JE (2014) Multibeam sonar (DIDSON) assessment of American shad (*Alosa sapidissima*) approaching a hydroelectric dam. *Can J Fish Aquat Sci* 71:545–558.

Hammerschlag N, Skubel RA, Calich H, et al. (2017) Nocturnal and crepuscular behavior in elasmobranchs: a review of movement, habitat use, foraging, and reproduction in the dark. *Bull Mar Sci* 93:355–374.

Han J, Honda N, Asada A, Shibata K (2009) Automated acoustic method for counting and sizing farmed fish during transfer using DIDSON. *Fish Sci* 75:1359.

Hannah RW, Blume MTO (2012) Tests of an experimental unbaited video lander as a marine fish survey tool for high-relief deepwater rocky reefs. *J Exp Mar Biol Ecol* 430:1–9.

Hannah RW, Blume MTO (2016) Variation in the effective range of a stereo-video lander in relation to near-seafloor water clarity, ambient light and fish length. *Mar Coast Fish* 8:62–69.

Harasti D, Malcolm H, Gallen C, et al. (2015) Appropriate set times to represent patterns of rocky reef fishes using baited video. *J Exp Mar Biol Ecol* 463:173–180.

Harasti D, Lee KA, Laird R, et al. (2016) Use of stereo baited remote underwater video systems to estimate the presence and size of white sharks (*Carcharodon carcharias*). *Mar Freshwater Res* 68:1391–1396.

Hardinge J, Harvey ES, Saunders BJ, Newman SJ (2013) A little bait goes a long way: the influence of bait quantity on a temperate fish assemblage sampled using stereo-BRUVs. *J Exp Mar Biol Ecol* 449:250–260.

Harvey ES, Shortis MR (1998) Calibration stability of an underwater stereo-video system: implications for measurement accuracy and precision. *Mar Technol Soc J* 32:3–17.

Harvey E, Fletcher D, Shortis M (2001) Improving the statistical power of length estimates of reef fish: a comparison of estimates determined visually by divers with estimates produced by a stereo-video system. *Fish Bull* 99:72–80.

Harvey E, Fletcher D, Shortis M (2002a) Estimation of reef fish length by divers and by stereo-video: a first comparison of the accuracy and precision in the field on living fish under operational conditions. *Fish Res* 57:255–265.

Harvey ES, Shortis M, Stadler M, Cappo M (2002b) A comparison of the accuracy and precision of measurements from single and stereo-video systems. *Mar Technol Soc J* 36:38–49.

Harvey ES, Fletcher D, Shortis M, Kendrick G (2004) A comparison of underwater visual distance estimates made by scuba divers and a stereo-video system: implications for underwater visual census of reef fish abundance. *Mar Freshwater Res* 55:573–580.

Harvey ES, Cappo M, Butler J, et al. (2007) Bait attraction affects the performance of remote underwater video stations in assessment of demersal fish community structure. *Mar Ecol Prog Ser* 350:245–254.

Harvey ES, Goetze JS, McLaren B, et al. (2010) Influence of range, angle of view, image resolution and image compression on underwater stereo-video measurements: high-definition and broadcast-resolution video cameras compared. *Mar Technol Soc J* 44:75–85.

Harvey ES, Butler JJ, McLean DL, Shand J (2012a) Contrasting habitat use of diurnal and nocturnal fish assemblages in temperate Western Australia. *J Exp Mar Biol Ecol* 426:78–86.

Harvey ES, Dorman SR, Fitzpatrick C, et al. (2012b) Response of diurnal and nocturnal coral reef fish to protection from fishing: an assessment using baited remote underwater video. *Coral Reefs* 31:939–950.

Harvey ES, Newman SJ, McLean DL, et al. (2012c) Comparison of the relative efficiencies of stereo-BRUVs and traps for sampling tropical continental shelf demersal fishes. *Fish Res* 125:108–120.

Harvey ES, Cappo M, Kendrick GA, McLean DL (2013a) Coastal fish assemblages reflect geological and oceanographic gradients within an Australian zootone. *PLoS ONE* 8:e80955.

Harvey ES, McLean DL, Frusher S, et al. (2013b) The use of BRUVs as a tool for assessing marine fisheries and ecosystems: a review of the hurdles and potential. In: *2011 National Workshop*. The University of Western Australia, Perth, pp 1–202.

Heagney EC, Lynch TP, Babcock RC, Suthers IM (2007) Pelagic fish assemblages assessed using mid-water baited video: standardising fish counts using bait plume size. *Mar Ecol Prog Ser* 350:255–266.

Helfman GS (1978) Patterns of community structure in fishes: summary and overview. *Environ Biol Fish* 3:129–148.

Helfman GS (1986) Fish behaviour by day, night and twilight. In: Pitcher TJ (ed) *The Behaviour of Teleost Fishes*. Croon Helm, London, pp 366–387.

Holmes JA, Cronkite GMW, Enzenhofer HJ, Mulligan TJ (2006) Accuracy and precision of fish-count data from a "dual-frequency identification sonar" (DIDSON) imaging system. *ICES J Mar Sci* 63:543–555.

Jaiteh VF, Allen SJ, Meeuwig JJ, Loneragan NR (2014) Combining in-trawl video with observer coverage improves understanding of protected and vulnerable species by-catch in trawl fisheries. *Mar Freshwater Res* 65:830–837.

Khanfar H, Charalampidis D, Ioup G, et al. (2010) *Automated Recognition and Tracking of Fish in Underwater Video*, Final Report, LA Board of Regents Contract NASA (2008)-STENNIS-08.

Kilfoil JP, Wirsing AJ, Campbell MD, et al. (2017) Baited remote underwater video surveys undercount sharks at high densities: insights from full-spherical camera technologies. *Mar Ecol Prog Ser* 585:113–121.

King NJ, Bagley PM, Priede IG (2006) Depth zonation and latitudinal distribution of deep-sea scavenging demersal fishes of the Mid-Atlantic Ridge, 42 to 53°N. *Mar Ecol Prog Ser* 319:263–274.

Koslow JA, Boehlert GW, Gordon JDM, et al. (2000) Continental slope and deep-sea fisheries: implications for a fragile ecosystem. *ICES J Mar Sci* 57:548–557.

Kucera TE, Barrett RH (2011) A history of camera trapping. In: O'Connell AF, Nichols JD, Karanth KU (eds) *Camera Traps in Animal Ecology: Methods and Analyses*. Springer Japan, Tokyo, pp 9–26.

Kyne PM, Simpfendorfer CA (2007) *A Collation and Summarization of Available Data on Deepwater Chondrichthyans: Biodiversity, Life History and Fisheries*, a report prepared by the IUCN SSC Shark Specialist Group for the Marine Conservation Biology Institute. IUCN SSC Shark Specialist Group, Burnaby, British Columbia.

Langlois TJ, Chabanet P, Pelletier D, Harvey E (2006) Baited underwater video for assessing reef fish populations in marine reserves. *SPC Fish Newsl* 118:53–57.

Langlois TJ, Harvey ES, Meeuwig JJ (2012) Strong direct and inconsistent indirect effects of fishing found using stereo-video: testing indicators from fisheries closures. *Ecol Indic* 23:524–534.

Langlois TJ, Bellchambers LM, Fisher R, et al. (2017) Investigating ecosystem processes using targeted fisheries closures: can small-bodied invertivore fish be used as indicators for the effects of western rock lobster fishing? *Mar Freshwater Res* 68:1251–1259.

Lawson GL, Barange M, Fréon P (2001) Species identification of pelagic fish schools on the South African continental shelf using acoustic descriptors and ancillary information. *ICES J Mar Sci* 58:275–287.

Letessier TB, Meeuwig JJ, Gollock M, Groves L, Bouchet PJ, Chapuis L, Vianna G, et al. (2013) Assessing pelagic fish populations: the application of demersal video techniques to the mid-water environment. *Methods Oceanogr* 8:41–55.

Lewison RL, Crowder LB, Read AJ, Freeman SA (2004) Understanding impacts of fisheries bycatch on marine megafauna. *Trends Ecol Evol* 19:598–604.

Ley JA, Halliday IA (2007) Diel variation in mangrove fish abundances and trophic guilds of northeastern Australian estuaries with a proposed trophodynamic model. *Bull Mar Sci* 80:681–720.

Lieber L, Williamson B, Jones CS, et al. (2014) Introducing novel uses of multibeam sonar to study basking sharks in the light of marine renewable energy extraction. In: *Proceedings of the 2nd International Conference on Environmental Interactions of Marine Renewable Energy Technologies (EIMR2014)*. Environmental Interactions of Marine Renewable Energy Technologies.

Lindfield SJ, McIlwain JL, Harvey ES (2014) Depth refuge and the impacts of SCUBA spearfishing on coral reef fishes. *PLoS ONE* 9:e92628.

Lindfield SJ, Harvey ES, Halford AR, McIlwain JL (2016) Mesophotic depths as refuge areas for fishery-targeted species on coral reefs. *Coral Reefs* 35:125–137.

Lines JA, Tillett RD, Ross LG, et al. (2001) An automatic image-based system for estimating the mass of free-swimming fish. *Comput Electron Agric* 31:151–168.

Loiseau N, Kiszka JJ, Bouveroux T, et al. (2016) Using an unbaited stationary video system to investigate the behaviour and interactions of bull sharks *Carcharhinus leucas* under an aquaculture farm. *Afr J Mar Sci* 38:73–79.

Lowry MB, Suthers IM (1998) Home range, activity and distribution patterns of a temperate rocky-reef fish, *Cheilodactylus fuscus. Mar Biol* 132:569–578.

MacLeod N, Benfield M, Culverhouse P (2010) Time to automate identification. *Nature* 467:154–155.

MacNeil MA, Graham N, Conroy MJ, et al. (2008) Detection heterogeneity in underwater visual-census data. *J Fish Biol* 73:1748–1763.

Malcolm HA, Gladstone W, Lindfield S, et al. (2007) Spatial and temporal variation in reef fish assemblages of marine parks in New South Wales, Australia—baited video observations. *Mar Ecol Prog Ser* 350:277–290.

Mallet D, Pelletier D (2014) Underwater video techniques for observing coastal marine biodiversity: A review of sixty years of publications (1952–2012). *Fish Res* 154:44–62.

Marouchos A, Sherlock M, Barker B, Williams A (2011) Development of a stereo deepwater baited remote underwater video system (DeepBRUVS). In: *MTS/IEEE OCEANS 2011—Santander, Spain*. Institute of Electrical and Electronics Engineers, Piscataway, NJ, pp 1–5.

Marshall AD, Pierce SJ (2012) The use and abuse of photographic identification in sharks and rays. *J Fish Biol* 80:1361–1379.

Martignac F, Daroux A, Bagliniere J-L, et al. (2015) The use of acoustic cameras in shallow waters: new hydroacoustic tools for monitoring migratory fish population. A review of DIDSON technology. *Fish Fish* 16:486–510.

Martinez I, Jones EG, Davie SL, et al. (2011) Variability in behaviour of four fish species attracted to baited underwater cameras in the North Sea. *Hydrobiologia* 670:23–34.

McCauley DJ, McLean KA, Bauer J, et al. (2012) Evaluating the performance of methods for estimating the abundance of rapidly declining coastal shark populations. *Ecol Appl* 22:385–392.

McLean DL, Green M, Harvey ES, et al. (2015) Comparison of baited longlines and baited underwater cameras for assessing the composition of continental slope deepwater fish assemblages off southeast Australia. *Deep Sea Res Part I* 98:10–20.

Merritt D, Donovan MK, Kelley C, et al. (2011) BotCam: a baited camera system for nonextractive monitoring of bottomfish species. *Fish Bull* 109:56–67.

Misa WFXE, Richards BL, DiNardo GT, et al. (2016) Evaluating the effect of soak time on bottomfish abundance and length data from stereo-video surveys. *J Exp Mar Biol Ecol* 479:20–34.

Mueller A-M, Burwen DL, Boswell KM, Mulligan T (2010) Tail-beat patterns in dual-frequency identification sonar echograms and their potential use for species identification and bioenergetics studies. *Trans Am Fish Soc* 139:900–910.

Munks LS, Harvey ES, Saunders BJ (2015) Storm-induced changes in environmental conditions are correlated with shifts in temperate reef fish abundance and diversity. *J Exp Mar Biol Ecol* 472:77–88.

Muñoz-Benavent P, Andreu-García G, Valiente-González JM, et al. (2017) Automatic bluefin tuna sizing using a stereoscopic vision system. *ICES J Mar Sci.* 75(1):390–401.

Murphy HM, Jenkins GP (2010) Observational methods used in marine spatial monitoring of fishes and associated habitats: a review. *Mar Freshwater Res* 61:236.

Myers EMV, Harvey ES, Saunders BJ, Travers MJ (2016) Fine-scale patterns in the day, night and crepuscular composition of a temperate reef fish assemblage. *Mar Ecol* 37:668–678.

Myers RA, Worm B (2003) Rapid worldwide depletion of predatory fish communities. *Nature* 423:280–283.

Nash KL, Graham NAJ (2016) Ecological indicators for coral reef fisheries management. *Fish Fish* 17:1029–1054.

Nelson DR, Johnson RH (1970) Diel activity rhythms in the nocturnal, bottom-dwelling sharks, *Heterodontus francisci* and *Cephaloscyllium ventriosum*. *Copeia* 1970:732–739.

Oliver S, Braccini M, Newman SJ, Harvey ES (2015) Global patterns in the bycatch of sharks and rays. *Mar Policy* 54:86–97.

Parsons M, Parnum I, Allen K, et al. (2014) Detection of sharks with the Gemini imaging sonar. *Acoust Aust* 42:185–189.

Parsons MJG, Fenny E, Lucke K, et al. (2017) Imaging marine fauna with a Tritech Gemini 720i sonar. *Acoust Aust* 45:41–49.

Petreman IC, Jones NE, Milne SW (2014) Observer bias and subsampling efficiencies for estimating the number of migrating fish in rivers using Dual-frequency IDentification SONar (DIDSON). *Fish Res* 155:160–167.

Priede IG, Bagley PM (2003) *In situ* studies on deep-sea demersal fishes using autonomous unmanned lander platforms. In: Gibson RN, Barnes M (eds) *Oceanography and Marine Biology: An Annual Review.* CRC Press, Boca Raton, FL, pp 357–392.

Priede IG, Merrett NR (1996) Estimation of abuNdance of abyssal demersal fishes; a comparison of data from trawls and baited cameras. *J Fish Biol* 49:207–216.

Priede IG, Smith KL, Armstrong JD (1990) Foraging behavior of abyssal grenadier fish: inferences from acoustic tagging and tracking in the North Pacific Ocean. *Deep Sea Res A* 37:81–101.

Priede IG, Bagley PM, Smith A, et al. (1994) Scavenging deep demersal fishes of the Porcupine Seabight, north-east Atlantic: observations by baited camera, trap and trawl. *J Mar Biol Assoc UK* 74:481–498.

Ravanbakhsh M, Shortis MR, Shafait F, et al. (2015) Automated fish detection in underwater images using shape-based level sets. *Photogram Rec* 30:46–62.

Rees M, Knott NA, Fenech GV, Davis AR (2015) Rules of attraction: enticing pelagic fish to mid-water remote underwater video systems (RUVS). *Mar Ecol Prog Ser* 529:213–218.

Rizzari JR, Frisch AJ, Connolly SR (2014a) How robust are estimates of coral reef shark depletion? *Biol Conserv* 176:39–47.

Rizzari JR, Frisch AJ, Magnenat KA (2014b) Diversity, abundance, and distribution of reef sharks on outer-shelf reefs of the Great Barrier Reef, Australia. *Mar Biol* 161:2847–2855.

Robbins WD, Hisano M, Connolly SR, Choat JH (2006) Ongoing collapse of coral-reef shark populations. *Curr Biol* 16:2314–2319.

Robbins WD, Peddemors VM, Broadhurst MK, Gray CA (2013) Hooked on fishing? Recreational angling interactions with the critically endangered grey nurse shark *Carcharias taurus* in eastern Australia. *Endanger Spec Res* 21:161–170.

Rova A, Mori G, Dill LM (2007) One fish, two fish, butterfish, trumpeter: recognizing fish in underwater video. In: *Proceedings of the IAPR Conference on Machine Vision Applications (IAPR MVA 2007).* International Association for Pattern Recognition, pp 404–407.

Ryan LA, Meeuwig JJ, Hemmi JM, et al. (2015) It is not just size that matters: shark cruising speeds are species-specific. *Mar Biol* 162:1307–1318.

Salman A, Jalal A, Shafait F, et al. (2016) Fish species classification in unconstrained underwater environments based on deep learning. *Limnol Oceanogr Meth* 14:570–585.

Santana-Garcon J, Leis JM, Newman SJ, Harvey ES (2013) Presettlement schooling behaviour of a priacanthid, the purplespotted bigeye *Priacanthus tayenus* (Priacanthidae: Teleostei). *Env Biol Fish* 97:277–283.

Santana-Garcon J, Braccini M, Langlois TJ, et al. (2014a) Calibration of pelagic stereo-BRUVs and scientific longline surveys for sampling sharks. *Meth Ecol Evol* 5:824–833.

Santana-Garcon J, Newman SJ, Harvey ES (2014b) Development and validation of a mid-water baited stereo-video technique for investigating pelagic fish assemblages. *J Exp Mar Biol Ecol* 452:82–90.

Santana-Garcon J, Newman SJ, Langlois TJ, Harvey ES (2014c) Effects of a spatial closure on highly mobile fish species: an assessment using pelagic stereo-BRUVs. *J Exp Mar Biol Ecol* 460:153–161.

Schobernd ZH, Bacheler NM, Conn PB (2013) Examining the utility of alternative video monitoring metrics for indexing reef fish abundance. *Can J Fish Aquat Sci* 71:464–471.

Sequeira AMM, Thums M, Brooks K, Meekan MG (2016) Error and bias in size estimates of whale sharks: implications for understanding demography. *R Soc Open Sci* 3:150668.

Shafait F, Mian A, Shortis M, et al. (2016) Fish identification from videos captured in uncontrolled underwater environments. *ICES J Mar Sci* 73:2737–2746.

Shafait F, Harvey ES, Shortis MR, et al. (2017) Towards automating underwater measurement of fish length: a comparison of semi-automatic and manual stereo–video measurements. *ICES J Mar Sci* 74:1690–1701.

Shortis M, Harvey E, Abdo D (2009) A review of underwater stereo-image measurement for marine biology and ecology applications. In: Gibson RN, Atkinson JA, Gordon JDM (eds) *Oceanography and Marine Biology: An Annual Review,* vol 47. CRC Press, Boca Raton, FL, pp 257–292.

Shortis MR, Ravanbakskh M, Shaifat F, et al. (2013) A review of techniques for the identification and measurement of fish in underwater stereo-video image sequences. *SPIE Opt Metrol* 8791:87910G.

Siddiqui SA, Salman A, Malik MI, et al. (2017) Automatic fish species classification in underwater videos: exploiting pre-trained deep neural network models to compensate for limited labelled data. *ICES J Mar Sci* 75(1):374–389.

Sih TL, Cappo M, Kingsford M (2017) Deep-reef fish assemblages of the Great Barrier Reef shelf-break (Australia). *Sci Rep* 7:10886.

Sikkel PC, Welicky RL, Artim JM, et al. (2017) Nocturnal migration reduces exposure to micropredation in a coral reef fish. Bull Mar Sci 93:475–489.

Sillman AJ, Dahlin DA (2004) The photoreceptors and visual pigments of sharks and sturgeons. In: von der Emde G, Mogdans J, Kapoor BG (eds) *The Senses of Fish*. Springer, Dordrecht, pp 31–54.

Simmonds J, MacLennan D (2005) *Fisheries Acoustics: Theory and Practice*, 2nd ed. Blackwell Science, Oxford.

Simpfendorfer CA, Hueter RE, Bergman U, Connett SMH (2002) Results of a fishery-independent survey for pelagic sharks in the western North Atlantic, 1977–1994. *Fish Res* 55:175–192.

Somerton DA, Gledhill CT (2005) *Report of the National Marine Fisheries Service Workshop on Underwater Video Analysis*, NOAA Technical Memo NMFS-F/SPO-68. National Oceanic and Atmospheric Administration, Silver Spring, MD.

Spaet JLY, Nanninga GB, Berumen ML (2016) Ongoing decline of shark populations in the Eastern Red Sea. *Biol Conserv* 201:20–28.

Spampinato C, Chen-Burger Y-H, Nadarajan G, Fisher RB (2008) Detecting, tracking and counting fish in low quality unconstrained underwater videos. In: *Proceedings of the Third International Conference on Computer Vision Theory and Applications (VISAPP)*. Science and Technology Publications, Setúbal, Portugal, pp 514–519.

Spampinato C, Giordano D, Di Salvo R, et al. (2010) Automatic fish classification for underwater species behavior understanding. In: *Proceedings of the First ACM International Workshop on Analysis and Retrieval of Tracked Events and Motion in Imagery Streams*. Association for Computing Machinery, New York, pp 45–50.

Stat M, Huggett MJ, Bernasconi R, et al. (2017) Ecosystem biomonitoring with eDNA: metabarcoding across the tree of life in a tropical marine environment. *Sci Rep* 7:12240.

Stobart B, García-Charton JA, Espejo C, et al. (2007) A baited underwater video technique to assess shallow-water Mediterranean fish assemblages: methodological evaluation. *J Exp Mar Biol Ecol* 345:158–174.

Struthers DP, Danylchuk AJ, Wilson ADM, Cooke SJ (2015) Action cameras: bringing aquatic and fisheries research into view. *Fisheries* 40:502–512.

Taylor MD, Baker J, Suthers IM (2013) Tidal currents, sampling effort and baited remote underwater video (BRUV) surveys: are we drawing the right conclusions? *Fish Res* 140:96–104.

Unsworth RKF, Peters JR, McCloskey RM, Hinder SL (2014) Optimising stereo baited underwater video for sampling fish and invertebrates in temperate coastal habitats. *Estuar Coast Shelf Sci* 150:281–287.

Urick RJ (1983) *Principles of Underwater Sound*, 3rd ed. MacGraw- Hill, New York.

Vanderklift MA, Boschetti F, Roubertie C, et al. (2014) Density of reef sharks estimated by applying an agent-based model to video surveys. *Mar Ecol Prog Ser* 508:201–209.

Vianna GMS, Meekan MG, Ruppert JLW, et al. (2016) Indicators of fishing mortality on reef-shark populations in the world's first shark sanctuary: the need for surveillance and enforcement. *Coral Reefs* 35:973–977.

Walsh AT, Barrett N, Hill N (2017) Efficacy of baited remote underwater video systems and bait type in the cool-temperature zone for monitoring "no-take" marine reserves. *Mar Freshwater Res* 68:568–580.

Ward-Paige C, Flemming JM, Lotze HK (2010) Overestimating fish counts by non-instantaneous visual censuses: consequences for population and community descriptions. *PLoS ONE* 5:e11722.

Warrant E (2004) Vision in the dimmest habitats on earth. *J Comp Physiol A Neuroethol Sens Neural Behav Physiol* 190:765–789.

Watson DL (2006) Use of Underwater Stereo–Video to Measure Fish Assemblage Structure, Spatial Distribution of Fishes and Change in Assemblages with Protection from Fishing, doctoral dissertation, School of Plant Biology, University of Western Australia, Perth.

Watson DL, Harvey ES (2007) Behaviour of temperate and subtropical reef fishes towards a stationary SCUBA diver. *Mar Freshw Behav Physiol* 40:85–103.

Watson DL, Harvey ES, Anderson MJ, Kendrick GA (2005) A comparison of temperate reef fish assemblages recorded by three underwater stereo-video techniques. *Mar Biol* 148:415–425.

Watson DL, Harvey ES, Fitzpatrick BM, et al. (2010) Assessing reef fish assemblage structure: how do different stereo-video techniques compare? *Mar Biol* 157:1237–1250.

Westerberg H, Westerberg K (2011) Properties of odour plumes from natural baits. *Fish Res* 110:459–464.

White J, Simpfendorfer CA, Tobin AJ, Heupel MR (2013) Application of baited remote underwater video surveys to quantify spatial distribution of elasmobranchs at an ecosystem scale. *J Exp Mar Biol Ecol* 448:281–288.

Whitmarsh SK, Fairweather PG, Huveneers C (2017) What is Big BRUVver up to? Methods and uses of baited underwater video. *Rev Fish Biol Fish* 27:53–73.

Widder EA, Robison BH, Reisenbichler KR, Haddock SHD (2005) Using red light for *in situ* observations of deep-sea fishes. *Deep Sea Res Part I* 52:2077–2085.

Willis TJ, Babcock RC (2000) A baited underwater video system for the determination of relative density of carnivorous reef fish. *Mar Freshwater Res* 51:755–763.

Wraith J, Lynch T, Minchinton TE, et al. (2013) Bait type affects fish assemblages and feeding guilds observed at baited remote underwater video stations. *Mar Ecol Prog Ser* 477:189–199.

Yau, C, Collins MA, Bagley PM, et al. (2001) Estimating the abundance of Patagonian toothfish *Dissostichus eleginoides* using baited cameras: a preliminary study. *Fish Res* 51: 403-412.

Zintzen V, Anderson MJ, Roberts CD, et al. (2012) Diversity and composition of demersal fishes along a depth gradient assessed by baited remote underwater stereo-video. *PLoS ONE* 7:e48522.

Zintzen V, Anderson MJ, Roberts CD, et al. (2017) Effects of latitude and depth on the beta diversity of New Zealand fish communities. *Sci Rep* 7:8081.

CHAPTER 8

Acoustic Telemetry

Michelle R. Heupel
Australian Institute of Marine Science, Townsville, Queensland, Australia

Steven T. Kessel
Daniel P. Haerther Center for Conservation and Research, John G. Shedd Aquarium, Chicago, Illinois

Jordan K. Matley
Center for Marine and Environmental Studies, University of the Virgin Islands, Charlotte Amalie, St. Thomas, U.S. Virgin Islands

Colin A. Simpfendorfer
Centre for Sustainable Tropical Fisheries and Aquaculture and College of Science and Engineering, James Cook University, Townsville, Queensland, Australia

CONTENTS

8.1　INTRODUCTION

Acoustic telemetry involves the use of sound to convey information relating to the presence of an animal as it moves from one location to another in the aquatic environment. In the context of shark research, this most commonly reflects using acoustic transmitters and receivers to track movement of individuals. Originally, acoustic transmitters simply emitted a pulse that could be detected by a receiving device; researchers followed the sound using a directional hydrophone as the shark swam through the environment and recorded positions every few minutes to represent the movement track of an individual (Holland et al., 1992; Morrissey and Gruber, 1993). As technology developed, information was encoded into acoustic signals by using a series of pings that could be decoded by the receiver. This led to the capacity to provide unique identification codes for an individual tag, which allowed simultaneous tracking of multiple individuals. By combining coded transmitters with data-logging acoustic receivers that could be moored in study sites for long periods, the need for animals to be actively followed was removed. This revolutionized the field of acoustic telemetry by allowing researchers to establish arrays of receivers to detect and track sharks automatically (Heupel et al., 2006). Sensors are also being developed and integrated with transmitters to provide information on the environments that tagged sharks encounter (e.g., depth, temperature) and their behavioral state (e.g., acceleration). Similarly, advances in receiver systems, collaborations, and modes to access stored data provide new ways to examine shark behavior and distribution at broad scales. In this chapter, we discuss the application of acoustic telemetry to track sharks, advances to the technology over time, and the challenges and opportunities this technology has provided to shark research.

8.1.1　Research Applications of Acoustic Telemetry Technology

Acoustic telemetry applications for sharks in their most basic form examine the movement ecology of individuals and species (Donaldson et al., 2014; Heupel and Webber, 2012). This involves determining where individuals move, how fast they move, how long they remain in an area, and if they display any distinctive behaviors such as vertical migrations. These data on animal locations and movements can be further analyzed to provide information on home range and activity space (Heupel et al., 2004), habitat preferences (Davy et al., 2015; Farrugia et al., 2011), and how individuals respond to changes in their environment (Heupel et al., 2003; Schlaff et al., 2017; Udyawer et al., 2013). Numerous studies have examined space use patterns of sharks based on acoustic telemetry; for example, several studies have examined how young sharks use nursery habitats to estimate the home range and space use of naïve individuals over time (Conrath and Musick, 2010; Heupel et

al., 2004; Knip et al., 2011; Rechisky and Wetherbee, 2003) (Figures 8.1 and 8.2). The responses of individuals to changing environmental conditions have been explored in several locations and habitats. Briefly, the response of individuals to changes in salinity levels has been studied on the central coast of Florida (Heupel and Simpfendorfer, 2008; Ortega et al., 2009; Simpfendorfer et al., 2011; Ubeda et al., 2009), the responses to seasonal and acute temperature changes have been explored in multiple species of sharks (Grubbs et al., 2007; Heupel, 2007; Matich and Heithaus, 2012), and movement based on tidal flow and height has also been examined (Ackerman et al., 2000; Carlisle and Starr, 2010; Knip et al., 2011). These studies and many others have improved our understanding of the ecological roles of sharks and the importance of specific habitats by implementing acoustic telemetry technology.

In addition to ecological research questions, acoustic telemetry data are increasingly being applied to fishery and management issues regarding shark populations (Crossin et al., 2017). Management of resources and populations requires scientific input to help inform decision-making. Acoustic telemetry has played a role in several aspects of management; for example, acoustic tracking data have been used to help estimate population size (Dudgeon et al., 2015) and mortality rates of young animals (Heupel and Simpfendorfer, 2002; Heupel and Simpfendorfer, 2011; Knip et al., 2012b). These data have been useful in determining the status of populations and setting fisheries catch quotas. Data from lemon shark (*Negaprion brevirostris*) tracking in Florida were used to support management that prohibited capture of this species in state waters. This provides a clear example of how telemetry data can be used to enhance and influence management decisions (Kessel et al., 2014a). Habitat management is also an area where acoustic telemetry has played a significant role. Designation of critical habitats and marine protected areas (MPAs) is intended to improve the status of key areas and populations by protecting them from habitat loss and fishing. Acoustic telemetry has been used in numerous locations to define the movement patterns and residency of sharks within MPAs (da Silva et al., 2013; Espinoza et al., 2015a; Knip et al., 2012a). Movement data are critical to understanding the level of protection MPAs provide to, or is needed by, mobile species; for example, Espinoza et al. (2015a) showed differing movement patterns among shark species using the same reef. Some species received significant protection from fishing pressure, while others received little (Figure 8.3).

8.1.2　Benefits and Limitations

Like most approaches, acoustic telemetry has both benefits and limitations. One benefit is the capacity to monitor large numbers of individuals simultaneously over long time periods using coded transmitters. Although other technologies also offer this possibility (e.g., satellite telemetry), the low

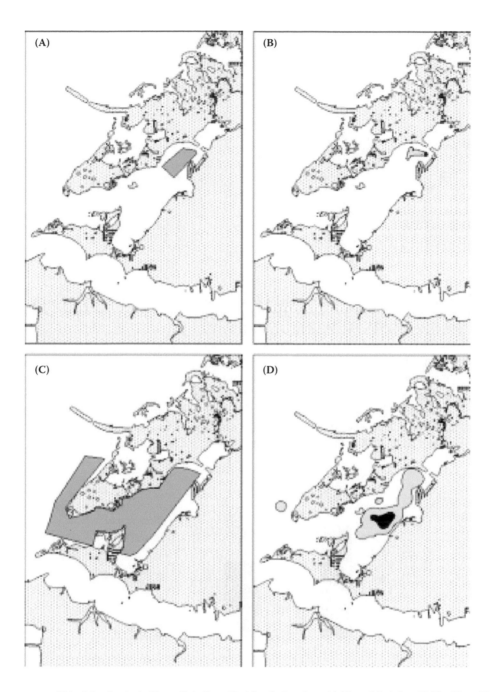

Figure 8.1 Home ranges of blacktip sharks in Terra Ceia Bay, Florida, during June (A,B) and October (C,D). (From Heupel, M.R. et al., *Environ. Biol. Fish.*, 71, 135–142, 2004.)

cost of acoustic telemetry equipment provides an increased opportunity to do so. Acoustic transmitters are also smaller in size than satellite transmitters, thus allowing their application to smaller species (e.g., sharpnose sharks) not suitable for satellite tracking. Directly following an individual via active tracking (Davy et al., 2015) or deployment of a dense data-logging receiver array also produces high-resolution position data for marine species not obtainable with other technologies (Barnett et al., 2010). Acoustic telemetry also provides the capacity to produce high-resolution location data for species that do not surface and more continuous data for species that surface infrequently. This is an advantage over satellite technologies that rely on surfacing to determine animal location.

Acoustic telemetry, however, does have a number of limitations. The most prominent of these is transmitter detection range. Individuals can only be tracked when within the detection range of a hand-held hydrophone or data-logging receiver station. The inability to detect an individual could be the result of the individual having moved completely out of the study site, movement outside of the detection range, or interference causing detections to be masked

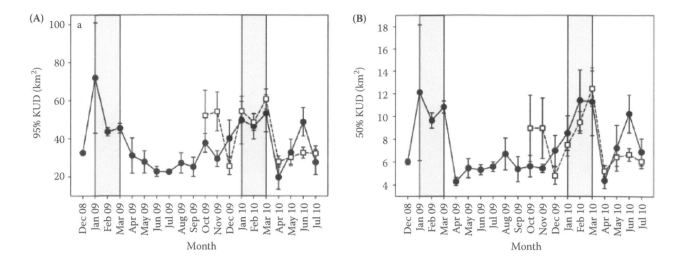

Figure 8.2 Monthly mean and standard error of home ranges of pigeye sharks in Cleveland Bay, Queensland, Australia, for (A) 95% kernel utilization density and (B) 50% kernel utilization density. Shading indicates high rainfall periods. (From Knip, D.M. et al., *Mar. Ecol. Prog. Ser.*, 425, 233–246, 2011.)

(Simpfendorfer et al., 2015). All scenarios produce a similar result and are impossible to disentangle. Thus, the capacity of acoustic telemetry is limited by the detection range of transmitters which is a function of technical specifications and conditions in the study area (Heupel et al., 2008; Kessel et al., 2014b; Simpfendorfer et al., 2008). This means that

the design of acoustic receiver arrays must consider these issues relative to the questions being investigated (Hedger et al., 2008; Heupel et al., 2006). The application of acoustic telemetry to the study of long-range movements is thus diminished in comparison to Global Positioning System (GPS) and satellite technology, which is not restricted by

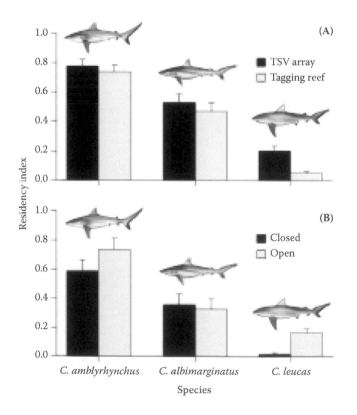

Figure 8.3 Residency index of reef sharks (mean and standard error) to (A) an acoustic array and their capture reef and (B) to reefs open and closed to fishing. (From Espinoza, M. et al., *Mar. Biol.*, 162, 343–358, 2015.)

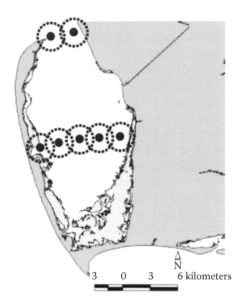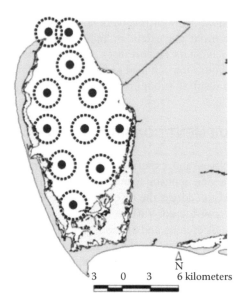

Figure 8.4 Examples of gate or curtain array design (left) vs. grid design (right). (From Heupel, M.R. et al., *Mar. Freshw. Res.*, 57, 1–13, 2006.)

the placement of receivers. However, the ability of disparate research groups, using the same equipment, to share data has enabled long-distance movements to be more easily investigated using acoustic telemetry (Cooke et al., 2011; Heupel et al., 2015).

Another limitation of acoustic telemetry is that as sound travels through water it interacts with physical features and other acoustic signals (both biologically and human produced), which can interfere with signal detection and interpretation (Heupel et al., 2008; Huveneers et al., 2016). The effect on active tracking studies is limited because simple signals are used; however, with acoustic monitoring studies, where automated decoding of complex coded signals is required, it can result in false-positive detections being logged (Simpfendorfer et al., 2015). These false-positive detections can lead to incorrect conclusions if they are not filtered out of datasets, especially those shared by multiple users or made publicly available (Hoenner et al., 2018).

Data management and analysis are also aspects of the limitations of acoustic telemetry. Because acoustic telemetry is typically applied in a customized approach to suit the study species or study site, analysis methods are likewise customized. The lack of a standard set of metrics and reporting of how metrics were applied (e.g., what smoothing factor was applied to kernel utilization distributions) hampers the ability to directly compare results across studies or species. As the number of acoustic telemetry studies continues to increase (Heupel and Webber, 2012; Hussey et al., 2015), the capacity to compare or replicate analyses will be increasingly important. In recognition of this issue and in an attempt to analyze data from multiple species and locations, Udyawer et al. (in review) established a series of standardized metrics and an analysis toolbox for acoustic telemetry

users. This type of approach, if used by the research community, will alleviate many of the issues around comparison of research outputs and improve opportunities to collaborate and share data.

8.1.3 Defining Research Questions

Prior to conducting any acoustic telemetry research, assessing the applicability of this technology to the species or research question is critical. This requires a clearly defined project objective, consideration of whether the study species can reliably be fitted with transmitters, and whether the study site is amenable to acoustic tracking or receiver deployment. Research questions should be well defined and testable (Papastamatiou and Lowe, 2012), which will inform experimental design.

Experimental design is one of the main considerations in acoustic telemetry projects using data-logging receivers. The number of receivers required and their placement are crucial to the success of projects and collection of data suitable for addressing research questions. There is no standard approach to how an acoustic receiver array is designed and deployed; such arrays are customized to suit the purposes of each project (Heupel et al., 2006). A number of factors must be considered when developing an acoustic array. For example, array design will differ if the intent of the project is to determine how long individuals are resident within a particular area vs. defining the activity space of individuals. Determination of residency could be achieved with receivers deployed as gates or curtains at key entry and exit points of a study site (Figure 8.4). Gates and curtains can also be used to determine direction of movement if deployed in parallel lines. In contrast, defining an activity space would require a grid or other spatial array of receivers spread within the study site to determine the extent of movement and any preferential use of habitats (Figure 8.4). Failure to clearly define the research question can lead to flawed experimental design, which will result in suboptimal outcomes.

8.2 EQUIPMENT CONSIDERATIONS

After study objectives and experimental design considerations are complete and acoustic telemetry is determined to be the best method to address the research question, equipment type must be considered. Various equipment manufacturers offer a suite of options and capability. It is important to select the correct equipment for the location to ensure that it performs optimally and delivers the expected results; for example, conditions and equipment needs will vary based on water depth, substrate type, etc.

8.2.1 Equipment Performance

A major consideration in equipment selection is performance. Can the equipment function as designed in the study site? A variety of factors must be considered relative to equipment performance, including the noise level in the study site, depth influences on sound transmission, current speed or tidal effects, etc. Here, we discuss some of the major aspects of acoustic telemetry equipment performance.

8.2.1.1 Detection Efficiency and Influencing Factors

To effectively employ a receiver array and accurately interpret the data it produces, it is necessary to gain an understanding of the performance of the telemetry system and how this varies over time. Telemetry system performance can be generally characterized by detection efficiency (DE), with the specific measure of DE relating to the study question and receiver array design. Detection efficiency can be broadly defined as the probability that a tagged animal will be detected in time and space, and it is most influenced by the effective detection range (Huveneers et al., 2016; Kessel et al., 2014a). Detection range is variable over time and influenced by a variety of factors including, but not limited to, tag power output, environmental conditions, ambient noise, air entrainment, and physical obstructions (Gjelland and Hedger, 2013; Kessel et al., 2014a). Different water body types (e.g., lakes, ocean, freshwater) can, therefore, produce varying effective detection ranges for the same model of receiver. Not accounting for how DE varies throughout a study can result in misinterpretation of data and erroneous conclusions, such as inaccurate estimates of survival/mortality, site fidelity, habitat use, and diel activity (Kessel et al., 2014a; Payne et al., 2010).

Detection efficiency can be broadly categorized into four types based on the study design and focus. These were defined by Melnychuk (2012) as the probability of detecting (1) individual tag transmissions (DE_{single}), (2) tagged animals residing in a given area (DE_{res}), (3) tagged animals moving past a specific location (DE_{mig}), and (4) tagged animals present during a mobile survey (DE_{mobile}). The most common assessment undertaken in acoustic telemetry studies is DE_{single}, which is typically used for detection range tests when calculating the proportion of tag transmissions recorded at specified distances from the receiver. Understanding typical effective detection ranges between tags and receivers can inform study design in terms of most effective and efficient receiver placement. For example, when studies use receiver gates or curtains, the required spacing between receivers is defined by the effective detection range (Heupel et al., 2006; Welch et al., 2008).

It is advisable to conduct some form of detection range testing prior to receiver array design and establishment. Given the variables that can influence detection range profiles over time, the best practice is to continue to assess effective detection range throughout the study and incorporate this into the analysis of the data from tagged animals. Detection range profiles usually show detection of a high proportion of transmissions near the receiver which decrease with increasing distance (Kessel et al., 2014a). Alternatively, in systems using coded transmissions where sheltered or calm waters and hard benthic surfaces are present, detection profiles can be influenced by close proximity detection interference (CPDI) (Kessel et al., 2015). In these cases, transmissions can produce echoes, which interfere with the code sequence, producing a profile with a low proportion of transmissions recorded near and far from the receiver, peaking somewhere in between. In such study sites, the potential for CPDI to exist should be accounted for in the assessment of the detection range.

8.2.1.2 Data Standardization

All digital communication systems, especially those operating in water, will incur a level of error or deviation between what is transmitted and what is recorded (Pincock, 2012). In acoustic telemetry systems using coded transmissions, this can result in the generation of false detections. False detections are generated when the transmissions of two tags of the same frequency overlap and sections of both transmissions form a valid code sequence (Pincock, 2012). In these circumstances, two types of false detection can be generated. Simpfendorfer et al. (2015) categorized these as Type A, an unknown tag ID, or Type B, an ID from a known transmitter from either your study or another study with a data-sharing agreement. In the current climate of data sharing through groups and via contact with the manufacturer, there is increasing potential for Type A false detections to become Type B and, therefore, increasing potential for false detections to cause inaccurate interpretation of data.

Failure to adequately account for potential false detections can lead to false data being erroneously accepted as true detections, with associated consequences. Incorrectly accepting false detections can cause errors associated with, but not limited to, the timing of movements/migrations, mortality estimates, space use, interactions between individuals, and falsely reporting detections of tagged animals from other studies. In most cases with acoustic telemetry, particularly given the large datasets typically generated, the negative effects of accepting a false detection as true far outweigh the opposite—excluding a true detection as false. As such, it is recommended that researchers adopt a conservative approach to false detection filtering.

Several criteria have been used for identifying false detections, but two main criteria have become an industry standard for quality control of acoustic telemetry data: (1) single detections isolated before and after by a given time period (Pincock, 2012), and (2) transitions between receiver stations that would require an unfeasibly fast swimming speed (e.g., as accounted for in Heupel et al., 2010; Kessel et al., 2014b). Isolated detections are usually considered false because they are not validated by a sequence of detection of the same code. In early cases of false detection filtering, isolated detections were usually considered on a receiver-by-receiver basis. More recently, detections are considered isolated relative to the detection records from the entire receiver array. The latter is a more realistic approach for identifying potentially false detections, particularly for more mobile animals, and is the recommended choice. The set time period that defines "isolation" in a given study is variable and typically based on programming of the tag (relative to the delay between transmissions), the known behavior of the animal, and the design of the receiver array. In most cases, either 30 minutes or 1 hour is selected, unless the aforementioned factors (programming, behavior or array design) make an alternative interval more logical. In rare cases where isolated detections, which would typically be considered false through standard filtering practices, have considerable implications for a given study, a more thorough assessment approach can be adopted, including analysis of raw ping data to verify the validity of the detection. A good example of where this would be relevant or necessary and appropriate is provided in Kessel et al. (2017).

Filtering for false detections based on unfeasible swimming speeds between receiver stations is conducted after the first stage of filtering for isolated detections (e.g., Hoenner et al., 2018). It is obvious why these detections would be identified as potentially false and removed, but there are a few considerations to be made when selecting a maximum feasible swimming speed for sharks and when making these calculations. For selecting a maximum feasible swimming speed, it is advised to adopt the maximum sustained swimming speed for your study species. In some cases, this can be obtained directly from the literature. For other species,

however, this information will not be available. In these cases, it can be acceptable to select the speed based on published data for a similar species or otherwise justify your choice of speed based on knowledge of your study species. To calculate transition speeds between receiver sites, the distance between receivers where sequential detections occurred must be calculated and the line-of-sight and physical obstructions must be accounted for. Additionally, receiver detection range must be considered in these calculations, particularly when they are very large, as this can give the impression of a very fast transition when in fact it is just a product of high receiver performance.

When the data have been filtered to remove potentially false detections, a level of standardization is required before further analysis can be conducted. After data filtering, you are left with a record of detections for each tagged shark. These detections, however, do not necessarily hold a specific value as a unit of data. This is because not all transmitted detections, even if within detection range of the receiver, are recorded due to collisions and other forms of acoustic interference. Additionally, over the course of a long-term study, tag programming may change between deployment periods for a number of reasons, creating inconsistency in transmission frequencies between study animals; thus, the number of detections received cannot be directly interpreted as a graduated value of presence. To account for this, it is necessary to standardize detections by some criteria. Commonly this is achieved by converting raw detections to presence or absence based on whether the tagged animal was or was not detected within a given time frame (e.g., an hour, a day, a week). Another approach is to consider all detections within a given time period of each other (say, 30 minutes) as a period of residence between the first and last detection. These approaches eliminate the majority of the biases associated with variable detection potential.

8.2.2 Equipment Deployment and Retrieval

A fundamental consideration for passive tracking is the deployment and retrieval of equipment. Accessing acoustic telemetry equipment in the underwater environment can be costly, labor intensive, and logistically difficult. As a result, the retrieval and downloading of acoustic receivers are typically interspersed by large chunks of time based on factors such as seasonality/weather, receiver battery life, and budget. Logistical and technical foresight is needed to ensure that appropriate equipment is used and reliable procedures are followed. Oversight or lack of rigor can lead to damage or loss of equipment worth thousands of dollars, in addition to the financial costs and time spent overseeing deployment and maintenance. Just as important, the inappropriate choice or application of equipment may lead to ineffective data collection or loss of data. Congruent to many aspects of acoustic telemetry, the type of equipment and approach to deployment and retrieval are dependent on the study site, duration of study, species tagged, sample size, and budget, to name a few. One of the primary considerations is the type and style of mooring used to anchor equipment. The type of anchor or mooring employed will be directly related to water depth, habitat type, influence of currents or other environmental conditions, and ability to access the equipment, among others. In some cases, anchors are screwed into the sediment or chained to habitat to secure tracking equipment in place. In other scenarios, heavy anchor materials such as concrete-filled tires or metal are used to ensure that the mooring remains stationary. Retrieval of equipment for data downloading can be done via several means. The equipment can be lifted to the surface while still on the mooring, divers can access the unit to swap it or attach cables to offload data, or an acoustic release incorporated into the mooring can be used to release the equipment to the surface for collection and downloading. Which type of mooring system is most suitable depends on the scenario as well as the facilities available to the researcher.

8.2.2.1 Surface-Based Downloading

Advances in equipment deployment and retrieval have created greater access to data, often with greater ease. One major advance is the ability to digitally communicate with underwater receivers from the surface via an acoustic modem or cable which enables remote data acquisition. Surface-based downloading reduces the need to put divers in the water which can reduce the amount of time required to service equipment and also reduces risks associated with diving. Surface downloading may also remove the need to use acoustic release devices which can be unreliable if equipment is heavily fouled. There are two main approaches to surface downloading: cabled and modem. The former consists of a receiver connected to surface instrumentation via cable. In this case, data are usually transmitted remotely to a base station via surface-based data transfer networks, such as mobile phones or satellites. This approach allows continuous real-time data access or the unit can be programmed to transmit data at set intervals depending on study requirements. Approaches using acoustic modems require a boat or other platform (e.g., wave glider) to be positioned over the top of the equipment to facilitate communication and data transfer. These systems connect wirelessly between dual-modem transducers (one originating from the surface and one in the receiver underwater) to facilitate downloading at the surface. In both cases, the capacity to download data from the surface reduces the amount of time and effort required to offload data from acoustic receivers. It is worth noting, however, that equipment still requires maintenance such as cleaning off biofouling and replacing batteries, so occasional retrieval to the surface is still necessary.

Although most research applications do not require real-time data on shark movements, in some instances information from acoustically tagged animals is time sensitive, thus requiring greater access to receiver data. For example, bather protection shark monitoring programs that employ acoustic telemetry to track the occurrence of sharks require real-time location updates, especially in beach or surf areas where human–shark interactions typically occur (McAuley et al., 2017). Therefore, immediate transmission of shark detection data via satellite and mobile phone networks is a crucial component of this shark tracking. The ability to quickly access or visualize receiver data may also be advantageous when designing and deploying receiver arrays because initial placement may not be adequate.

8.2.2.2 Underwater Release/Retrieval

When depth or conditions negate the potential to service receivers through diving or the surface recovery of moorings (e.g., dragging a grapnel to snag lines), acoustic releases can be used. Traditionally, industrial grade acoustic release systems, developed primarily for the recovery of deep sea oceanographic equipment, have been incorporated into the mooring design. These units are robust and usually rated to several thousand meters depth, but, as a result, they are also expensive. More recently, telemetry manufacturers have developed acoustic receivers with integrated acoustic release systems, such as the VEMCO VR2AR. Although these units are more expensive than regular acoustic receivers, the offset of the acoustic release cost makes these a more attractive prospect. These receivers are not rated to the same depths as stand-alone acoustic releases, but most receivers that would be used with an acoustic release are not rated to those depths, either. The more compact nature of the integrated receiver releases is also favorable because it reduces the necessary mooring size and design complexity. Both systems function through a surface-based communication system, usually referred to as a "deck box," to communicate with the unit and trigger the release mechanism.

The use of acoustic release systems is a necessity in deep water deployments and challenging or extreme conditions that may negate the possibility of scuba diving or surface recovery. In most other systems, their use may not be necessary and the obvious upfront increase in cost may make them an unattractive option. There are additional benefits, however, associated with the use of acoustic releases, most notably a considerable reduction in recovery time and the ability to range the mooring from the surface to establish the distance from the vessel to the receiver (particularly useful if the station has moved). Recovering a receiver with an acoustic release can take a matter of minutes, whereas diving can take a considerable amount of time for each mooring, in addition to the restrictions associated with the number of dives a researcher can make per day. Similarly, surface snagging mooring lines with a grapnel can often take many attempts, not to mention the associated damage to the substrate or habitat cause by dragging the grapnel. Depending on the spacing between receivers, it is possible to recover large number of receivers in a single field day. This ability can substantially reduce the field time required, which in the long term can offset the increased equipment cost.

As researchers interested in conserving the habitats we study, we have an ethical responsibility to reduce the research footprint we leave behind when our studies are completed. Traditionally, the use of acoustic releases to recover equipment leaves a portion of the mooring remaining in the habitat. Not only is this undesirable, but also in some areas such as marine reserves regulations often prevent this action. As a result of legacy restrictions preventing researchers from deploying integrated acoustic release receivers in the Florida Keys National Park, one manufacturer has developed a fully recoverable deep-water mooring specifically for use in combination with their integrated acoustic release receivers. This solves the issue of legacy burden when using acoustic releases with acoustic telemetry. It is anticipated that similar systems will be developed for use with other acoustic releases.

8.2.2.3 Mobile Platforms

One of the, if not the biggest, shortfalls of passive acoustic telemetry is that you only know where your tagged animal is when it is within detection range of a receiver. It is very difficult, if not impossible, to say where individuals are when they are outside receiver range. This is particularly pertinent for sharks, as many species move beyond acoustic receiver arrays. The unmonitored space between areas of receiver coverage can be assessed using receivers attached to mobile platforms, sometimes referred to as "acoustic fishing," with both directed and opportunistic efforts. Mobile platforms can identify the presence of tagged sharks in unmonitored areas and, where consistent detections occur or a large number of individuals are detected, can identify suitable locations for new moored receiver stations.

The most basic form of directed mobile sampling is the use of *drifters*, which are floating platforms that rely on currents or sometimes the wind for mobility. Drifters can be constructed relatively inexpensively out of basic materials (e.g., PVC, aluminum) and can take several forms. The general design of a drifter consists of central flotation, a structure above the water line for location monitoring and identification to waterway users, and a structure below the water line to catch the current and attach the acoustic receiver (Figure 8.5). For current-driven drifters, the structure above the surface should be streamlined to minimize the influence of wind, and the structure below the water maximized to catch the current. The reverse would be true if wind is the intended form of propulsion. To track the path of the drifter, a GPS logger can be attached to the structure above the surface in a waterproof case. Drifters can be used independently, or

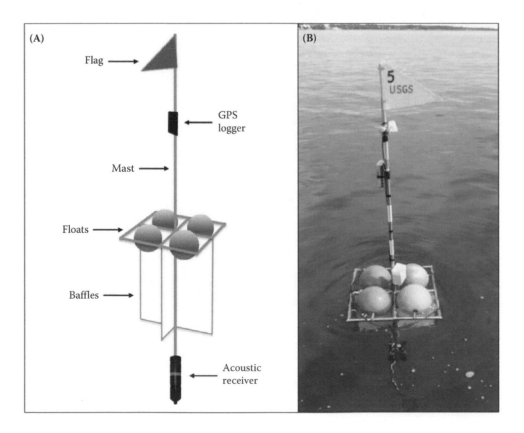

Figure 8.5 Basic setup for acoustic telemetry drifter: (A) Schematic of acoustic telemetry drifter platform, and (B) deployed drifter. (Images courtesy of Christopher Holbrook.)

multiple units can be deployed simultaneously. Where currents are consistent, it can be possible to use several drifters to produce a wide swath of acoustic coverage and maximize the potential for detection (Figure 8.5). Drifters typically have to be tended by a support vessel to reduce the risk of collisions with other water users, re-establish the drift course if the drifter becomes grounded in shallow water, and maintain effective spacing when using multiple drifters in unison.

Slocum and wave gliders provide another option for acoustic receiver mobile platforms (see Chapter 6 in this volume). Unlike drifters, gliders can be programmed or controlled remotely to follow specific routes, allowing for a more structured survey. Both, however, are unable to operate in shallow coastal waters, where the majority of shark telemetry studies are conducted. Slocum gliders operate subsurface, relying on vertical displacement (up and down) for motion, whereas wave gliders, as the name suggests, harness wave energy between the surface and subsurface units for locomotion. As such, wave gliders can operate in shallower environments than Slocum gliders, which typically require at least ~100 m to function, but wave gliders still require at least ~8 m due to their draft. Slocum gliders are battery powered and

can be deployed continuously for about 3 to 4 months. Wave gliders are solar powered with battery banks and can, therefore, be deployed indefinitely; however, they require waves to function and become drifters when conditions are very calm. Both systems are expensive and require constant monitoring, making them labor intensive; however, they can take relatively large payloads with multiple instrumentation packages included. Receivers can, with negotiation and agreement, be piggybacked onto deployments for other purposes, using the gliders as vessels of opportunity (Figure 8.6).

Mobile platforms can also be opportunistically employed as vessels of opportunity. Given the number of vessels that operate on a daily basis, it would seem logical that we could attach receivers to as many as possible to maximize acoustic coverage in the oceans, but this approach has issues. Motor vessels are not ideal platforms for receivers because engine noise interferes with the ability of the acoustic telemetry system to function effectively. Sailing craft provide more suitable platforms, or motor vessels that move from one place to another and then sit on anchor for extended periods. Dive boats, for example, can be suitable platforms for acoustic receivers, and the operators are often interested in

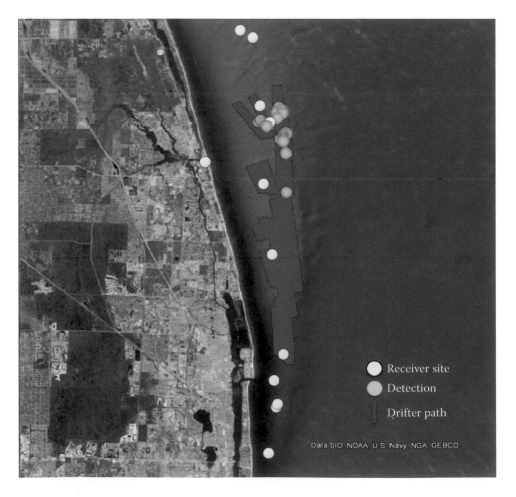

Figure 8.6 Results from acoustic drifter deployments with drifter tracks displayed in red, fixed receiver locations in yellow, and detection of lemon sharks as green circles (S. Kessel, unpublished data).

research participation. Fishing gear, such as drifting long-lines (commercial or survey), also offer potential platforms for receivers and have the added benefit of providing data on the interactions between tagged animals and fishing gear. Finally, drifting buoys or similar can also operate as vessels of opportunity.

8.2.3 Active Tracking

Active acoustic tracking has historically involved following a single individual fitted with an external transmitter for a period of hours to days (Nelson, 1976). Due to the size of transmitters required, active tracking was pioneered in part on sharks; as a result, several keystone telemetry papers were published using active tracking to study the movement patterns of sharks (Carey and Scharold, 1990; Carey et al., 1982; Klimley and Nelson, 1984; McKibben and Nelson, 1986; Nelson, 1977). The development of commercially available transmitters and receivers made this technique available to a wider range of researchers using more established methodology. One of the original papers describing methodology was that of Holland et al. (1992) based on tracking juvenile scalloped hammerhead sharks (*Sphyrna lewini*) in Hawaii. Hammerhead sharks were followed in a small boat and their tracks recorded to determine diel movement patterns in a shallow bay. Active tracking has also been used to define depth use patterns over diel and diurnal scales in a number of species and can be applied to fine-scale objectives such as individual location relative to thermal fronts and plankton migrations (e.g., Nelson, 1977; Sciarrotta and Nelson, 1977). This method has even been applied to little-seen and little-known species such as the megamouth shark (*Megachasma pelagios*) (Nelson et al., 1997); however, due to the labor-intensive nature of active tracking, it usually involves smaller sample sizes and tracking duration than automated approaches.

8.2.3.1 Applications and Recent Advances

Although the overall approach of directly following and recording the movements of an individual animal has not changed, some of the technologies used have advanced significantly. For example, advances in GPS technology have vastly improved the positional accuracy of animal tracks in recent decades. This allows high-resolution location data to be obtained that provides the capacity to relate animal behavior to specific locations or habitat features (Nelson, 1977; Simpfendorfer et al., 2010). Transmitter coding schemes have also allowed multiple individuals to be tracked on a single frequency. In the past, each transmitter was on a separate frequency to avoid collision and the confusion of non-unique transmitter signals. Now, researchers can encounter and record the presence of tagged individuals that are not the subject of the focal track. Improvements in acoustic receiver designs have also allowed the automated recording of signals from multiple individuals across multiple frequencies.

One of the most significant advances in recent years in active tracking is the capacity to remove the human element. The necessity of following an individual in a boat requires a team of researchers to commit to tracking the shark for as long as possible. The cost of having a team and vessel on the water for extended periods, as well as the stamina of the team, can often be limiting factors in active tracking research. Recent studies are trialing the use of autonomous underwater vehicles (AUVs) as a platform for tracking shark movements (see Chapter 6 in this volume). Clark et al. (2013) equipped an AUV with a stereo hydrophone receiver to track stationary and mobile objects. Estimates of position error using the AUV were similar to those produced by humans recording positions via boat-based tracking, indicating that human results are reproducible and the AUV may provide a less labor-intensive option for active tracking. Packard et al. (2013) used a similar approach to provide a three-dimensional track of sharks followed by an AUV. The success of these studies indicates a new path for active acoustic tracking that could revolutionize how we monitor animal movements in the future.

8.3 NEW ADVANCES FOR DEFINING ANIMAL BEHAVIOR

While tracking provides information on where sharks move and spend time, it does not reveal what sharks are doing in these locations. Tracking also does not reveal environmental conditions or whether individuals may be moving or selecting habitat based on conditions. To address these more detailed questions, sensor tags are required to provide additional data.

8.3.1 Advanced Sensor Technology

Standard transmitters allow us to define the movements of sharks in space and time. Sensor tags incorporate additional instrumentation that provide supplementary data on parameters that can increase our understanding of factors influencing movement (Cooke et al., 2004). The additional data provided by sensor tags can allow exploration of complex questions about shark ecology and biology. To date, the most commonly used sensor tags have been pressure (depth) and temperature tags. Pressure sensor tags add a third dimension to the locational data, whereas temperature sensor tags tell us about the environment the animal experiences, either selectively or incidentally. Often, pressure and temperature sensor tags, for example, are combined, with transmitted values alternately switching between the two parameters. As telemetry technology has evolved under the continued miniaturization of instrumentation, vendors have incorporated additional sensors, and found ways to telemeter meaningful data for defining animal behavior.

8.3.1.1 Environmental Sensors

Oxygen (O_2) sensors are the most recent example of instrument miniaturization incorporated into acoustic telemetry tags. To date, no examples exist in the literature of O_2 sensor tags used to study sharks; however, this new technology could be applied to great effect. Laboratory testing of O_2 sensor tags was conducted in a controlled setting by Svendsen et al. (2006), who found them to describe dissolved oxygen levels very accurately at three different temperatures (10, 20, and 30°C). Combined pressure and O_2 tags were used to study striped catfish (*Pangasianodon hypophthalmus*) in a pond in the Mekong Delta, Vietnam, elucidating O_2 levels as a driver for vertical movement (Lefevre et al., 2011). Similarly, for sharks, O_2 tags could be used to investigate factors defining space use in three dimensions and relate available O_2 levels to bioenergetics. This would be particularly useful for sharks inhabiting estuarine areas where conditions can become hypoxic or even anoxic.

In order to investigate the timing and frequency of feeding events in free-ranging sharks, pH sensors have been incorporated into acoustic transmitters. A proof-of-concept study was initially conducted by Papastamatiou and Lowe (2004), who demonstrated the relationship between changes in gastric pH levels and feeding events of leopard sharks (*Triakis semifasciata*). Following this study, Papastamatiou et al. (2007), in partnership with VEMCO, developed and tested an acoustic transmitter with built-in pH and temperature sensors. The transmitter was gastrically inserted into the stomach of captive blacktip reef sharks (*Carcharhinus melanopterus*), and pH levels were periodically telemetered to an acoustic receiver. The test successfully demonstrated that the transmitter could identify feeding events through changes in pH levels. Despite these successful trials, this technology has not been employed on free-ranging sharks. These transmitters are not yet produced commercially, so they would have to be produced on request. This technology has the potential to be used to determine when, where, and how frequently wild sharks are feeding and therefore quantify feeding chronology and daily ration (Papastamatiou et al., 2007).

8.3.1.2 Behavior Sensors

Acceleration sensors measure acceleration on one to three axes and for decades have been used to study bioenergetics, and ecological factors such as behavior, activity, and physiology (Cooke et al., 2016; see also Chapter 3 in this volume). Traditionally, accelerometers had to be recovered (e.g., archival tags) because it was not possible to telemeter the high-resolution data these tags typically collect. Since 2008, however, acceleration sensors have been integrated into acoustic transmitters. Two types of acceleration transmitter have been reported in the literature. The most common of these tags is described in detail in Cooke et al. (2016). With existing technological constraints, it is still not possible to telemeter the high-resolution data that acceleration sensors are able to collect, so summarized values are transmitted that can be interpreted to represent general activity (Cooke et al., 2016). Despite the decreased resolution, acceleration transmitters have several benefits over archival tags. Most desirably, they do not have to be recovered, greatly increasing data acquisition potential. Additionally, archival tags are usually limited to short recording durations (usually up to ~10 days) due to storage capacity constraints, whereas acceleration transmitters remain functional for the duration of the battery life (up to years) because the data are telemetered and not archived. Although compared to archival acceleration tags the resolution of these data is low, it is still possible to quantify the frequency that predefined behavioral events such as foraging occur (e.g., O'Toole et al., 2010). However, acceleration patterns require additional work to validate observed activity patterns with behaviors such as foraging, mating, etc. Failure to validate acceleration data with behaviors may result in inaccurate conclusions. When combined with oxygen consumption rates (achieved through swim tunnel experiments), data from acceleration transmitters can be used to calculate field metabolic rates (e.g., Murchie, 2011). The tailbeat frequency of free-swimming sharks has also been measured to define swimming rate. One of the first studies of this type used custom-built transmitters and active acoustic tracking to measure the tailbeat frequency of juvenile scalloped hammerhead sharks. In conjunction with respirometry data on oxygen consumption, tailbeat frequency was used to estimate metabolic rates for these individuals (Lowe, 2002).

To date, there has been limited use of acceleration transmitters for studies focused on elasmobranch species. Shipley et al. (2017) tagged Caribbean reef sharks (*Carcharhinus perezi*) with acceleration transmitters in Eleuthera, the Bahamas, to investigate their activity patterns. The data obtained from these tags showed the highest level of activity when making trips to the reef shelf. From the combination of the standard locational data and the averaged acceleration values, the authors were able to infer that these sharks mainly occupy deep water but make frequent trips to the reef shelf where high activity levels were likely related to foraging.

Development and testing of acoustic heart rate (electrocardiogram, or ECG) transmitters began in the 1970s (Wardle and Kanwisher, 1973; Young et al., 1972), but ECG transmitters have not been widely used to date. Telemetered ECG systems usually rely on in-tank experiments, but emerging technology is being developed that will expand our ability to study the physiology of wild elasmobranchs through acoustic telemetry. This will allow for improvements in the quantification of metabolic rates and the physiological effects of catch-and-release fishing (Cooke et al., 2016).

Proximity tags provide the ability to explore intra- and interspecific interactions between individuals. Proximity tags to date have been externally attached to the host, and both transmit an ID code and receive and record transmitted IDs from other tagged individuals. To enable receiving and recording functionality, proximity tags are typically quite large (e.g., 9 cm in length), rendering few shark species as suitable candidates for this technology. Using proximity tags, it is possible to quantify the amount of time individuals associate in the wild, addressing questions relating to social behavior in sharks. Proximity tags can also be used to look at the association between elasmobranchs and other species, including symbiotic relationships between pelagic sharks and tuna, for example. Additionally, proximity tags can be used to look at encounter rates between predators and prey. The initial deployment of proximity tags on sharks was conducted by Holland et al. (2009) on Galapagos sharks (*Carcharhinus galapagensis*) around a shark ecotourism operation near Haleiwa, Hawaii. These tags served as proof of concept, providing insights into shark interactions outside fixed receiver range. An alternative proximity transmitter was later produced and field trials conducted by Guttridge et al. (2010). This study focused on the group behavior of juvenile lemon sharks, and one individual was observed associating with nine others on 128 occasions over a 17-day period. Despite the potential of these tags to define interactions among individuals, it must be noted that only interactions among tagged individuals can be recorded. For this reason, this approach does not provide data regarding interactions with individuals not fitted with tags, thus requiring careful consideration of research questions and interpretation of the resulting data.

One limitation of proximity tags has been that they typically have to be recovered to offload the recorded detections, which makes them undesirable for highly mobile sharks, particularly in the pelagic environment. Advances have been made in marine mammal research where proximity tags have been linked via Bluetooth to satellite tags attached to the same animal to allow remote offloading of detections while the animal is at the surface (Baker et al., 2014). This approach requires an additional equipment burden to the study animal; however, large shark species with a high surface association (e.g., tiger or hammerhead sharks) would be good candidates for this emerging technology.

One of the biggest criticisms leveled at acoustic telemetry studies is that it is difficult to know if you are tracking the movements of the fish you tagged … or the bigger fish that ate the fish you tagged. To address this shortfall, a predation tag has been developed to identify predation events. When a tagged animal is consumed, the stomach acid of the predator erodes an external polymer on the tag, which triggers a change in the code transmitted by the tag (Halfyard et al., 2017). This alternative predator code identifies the predation of the tagged animal, removing the previous uncertainty. For many shark species, particularly larger individuals, predation risk is low, reducing the utility of this technology. For smaller species more prone to predation, however, these predation tags offer desirable functionality, particularly if the study aims to address mortality rates. Smaller predation tags could also be very useful for multispecies studies. These tags, for example, would facilitate comprehensive studies exploring dynamics between sharks and their prey.

8.3.1.3 Integration with Other Data

It is rarely appropriate to simply describe observations of movements. Instead, studies that explore the causes and consequences of shark spatial ecology for both the species themselves and the ecosystems they inhabit are of interest. Sensor tags provide a tool for producing more meaningful acoustic telemetry data, but it is also pertinent to combine telemetry data with as much additional data as possible to build the most complete picture. Some data can be acquired from other sources (e.g., weather stations), but focused, simultaneously collected data often provide the clearest picture. Multidisciplinary studies can be conducted through collaborations with other researchers who offer different areas of expertise. The development of *a priori* predictions based on behavioral or ecological theory is an important and recommended component of this approach. Over the past three decades, acoustic telemetry literature has progressively incorporated additional complementary data to explore more complex and comprehensive ecological questions, making these approaches the norm.

Population structure and connectivity are two major questions in ecology, and combining acoustic telemetry and genetic research is a powerful and comprehensive approach to tackling these questions. This is particularly true over small geographical scales where genetic mixing can cloud our ability to differentiate spatially discrete groups of animals that may require separate management considerations. Conversely, on larger geographic scales, genetic data can support or explain population structure. For example, Kessel et al. (2014a) used mitochondrial DNA to investigate the relationship of adult lemon sharks displaying a north–south seasonal migration to juvenile populations found north and south of their winter residency region. The combination of these data indicated that lemon sharks residing off the east coast of the United States are a distinct stock to those residing in the Dry Tortugas, Florida, and do not move between these regions frequently (Figure 8.7).

Nonlethal sampling of sharks for biological materials, such as blood and muscle tissue, at the time of tagging can allow investigation into the behavior and ecological roles of sharks in their ecosystems. Subsequent processing of biological materials via techniques such as stable isotope analysis (see Chapter 1 in this volume) of carbon and nitrogen, fatty acids, and trace elements (e.g., Belicka et al., 2012; Hussey et

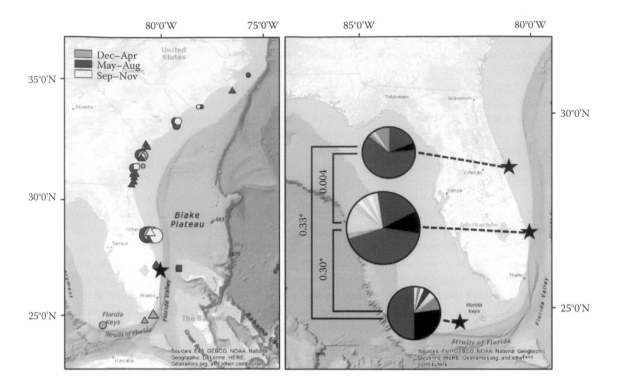

Figure 8.7   Occurrence and movement patterns of lemon sharks in Florida (left) and genetic composition of populations in sampled areas based on mitochondrial DNA (right). (From Hussey, N.E. et al., *Science*, 348(6240), 1255642, 2015.)

al., 2014; McMeans et al., 2007), can be combined with telemetry data to investigate the relationships between feeding and spatial ecology. For example, Matich and Heithaus (2014) used acoustic telemetry and stable isotopes to document a seasonal feeding switch by juvenile bull sharks (*Carcharhinus leucas*) as they moved from freshwater to saltwater environments in the Florida Everglades. Maljkovic and Cote (2011) used stable isotope analysis with acoustic telemetry to investigate the effects of shark-feeding tourism on the spatial and feeding ecology of reef sharks. By biologically sampling sharks at the time of tagging, movement data of tagged individuals can act as a subsample to compare to biological data from the broader population and add a spatial component to strengthen overall data analysis and interpretation. The combination of stable isotope analysis and acoustic tracking can also provide a mechanism for defining patterns of specialization of individuals within a population. The combination of multiple techniques, therefore, has the capacity to provide information about individuals and populations in several ways.

The examples discussed above are only a few of many techniques that can be combined with acoustic telemetry to advance our understanding of shark biology and ecology. The specific combinations to adopt depend on both the question being asked and the technology available at the time the study is conducted. It is recommended that shark researchers consider fields other than their primary expertise and seek collaborative studies where possible to tackle more complex questions important to shark biology and ecology.

8.3.2   Advanced Capacity for Monitoring

Technical advances have greatly increased the capacity for monitoring sharks and are providing new ways to define behavior. Additionally, enhanced data sharing and collaboration among scientists have further expanded this capacity. Many research questions rely on accurate and precise location estimates, and significant progress has been made in developing high-resolution positioning systems to track animals. Concordantly, advances in reducing transmitter signal collisions and in storing and processing datasets have enhanced the capacity to capture and retain data. Miniaturization and other advances in battery technology have increased track duration and the amount of data collected. These technical innovations make tracking animals more accessible than ever before. The establishment of fixed, long-term receiver arrays is now common worldwide. More animals are being tagged every year, and tracks are spanning greater distances thanks to data sharing and collaboration. These broad-scale networks provide spatial receiver coverage over long time scales that are unlikely to be feasible within the confines of a single study. For transient sharks that move beyond the initial study area into other areas and receiver arrays, collaborative efforts are needed or the full scope of movements and associated patterns will go unnoticed. However, this profusion of data requires proper management, and there are still major advances to be made in managing, analyzing, and interpreting large telemetry datasets.

8.3.2.1 Digital Coding Schemes

With the accelerating increase in the number of acoustic telemetry studies in recent years (Hussey et al., 2015), the number of transmitters being used to track sharks and other animals continues to grow. This growth contributes to the advancement of knowledge relating to the ecology and management of relevant species; however, deploying increasing numbers of transmitters raises technical issues that could affect data collection, especially in areas with large numbers of tagged animals. Conventional acoustic telemetry transmitters emit either continuous pings (i.e., active tracking) or coded pings (i.e., passive tracking), depending on the study design and sampling protocol. Because coded pingers typically function at one frequency, signal collisions (more than one transmitter pulse arriving at a receiver simultaneously) often occur within a study site. Similarly limiting, transmitters using a series of pings as part of the coding will only be detected if the full series (consisting of numerous pings for several seconds) is detected without interruption. Thus, detections may not be recorded if signal collisions occur, and a receiver may record erroneous detections when transmissions are corrupted (Simpfendorfer et al., 2015). In this context, if more tags are present in a study area, not only will the occurrence of interrupted and false detections increase but receiver storage and download times will also be impacted. To mediate these issues, new coding schemes are being trialed, including a digital coding scheme in which hundreds of transmitters (with high transmission rates) can be simultaneously detected within the same area. Additionally, signal transmissions that do not rely on lengthy pulse strings have been developed, reducing acoustic collisions and enabling more individuals to be tagged in the same area (see Section 8.3.2.4).

8.3.2.2 National and International Networks

When initiating an acoustic telemetry project, it is pertinent to investigate whether acoustic telemetry network groups operate around your intended study sight. There are many benefits to collaborating in an acoustic telemetry network group for the individual and the group as a whole. The most obvious benefit to the researcher is increased potential to detect your animals outside of the receiver coverage you personally establish and maintain. Because many shark species tend to move large distances, they are particularly good candidates to benefit from network participation. In many cases, it is financially and logistically impossible to establish enough receiver coverage to monitor the full movements of mobile shark species. The proliferation of acoustic telemetry use and the formation of receiver networks over the past few decades promise to provide a level of infrastructure that can begin to represent a suitable level of spatial coverage (Heupel et al., 2015). Participation in these groups facilitates developing the level of data sharing required to reveal large-scale movements and space use.

From a system performance perspective, collaboration with networks operating in a study region will allow researchers to find out what tags may already be deployed in the area and how they are programmed. This is important because it will influence the number of tags that can be deployed in the study and how they can best be programmed to maintain functionality that is not greatly reduced through excessive code collisions (Simpfendorfer et al., 2015). Communication with other telemetry users in the area, facilitated by participation in a network, can maximize the efficiency of receiver deployments and array designs through coordinated efforts to reduce redundancy. Network participation also promotes and facilitates scientific collaboration, creating the potential for more in-depth study questions to be tackled, such as multispecies interactions, as well as the potential to strengthen collaborative grant applications to help continue research projects over longer time periods, desirable for most long lived shark species. Additional benefits of participation in an array group include assistance with data management and data analysis tools and equipment discounts through collective bargaining agreements between the array groups and vendors.

Acoustic telemetry equipment vendors should be able to point researchers in the direction of existing network groups operating in their region. The following is a list (not exhaustive) of current acoustic telemetry network groups and associated websites, at the time of publication: Ocean Tracking Network (oceantrackingnetwork.org), Animal Tracking Network (ioos.noaa.gov/project/atn), Atlantic Cooperative Telemetry and Florida Atlantic Coastal Telemetry network (http://secoora.org/fact/), Australian Integrated Marine Observing System Animal Tracking Facility (imos.org.au/animaltracking.html), Great Lakes Acoustic Telemetry Observing System (data.glos.us/glatos), California Fish Tracking Consortium (californiafishtracking.ucdavis.edu), Acoustic Tracking Array Platform (saiab.ac.za/atap.htm), and Integrated Tracking of Aquatic Animals in the Gulf of Mexico (gcoos2.tamu.edu/itag).

8.3.2.3 Perpetual Tags

The main factor determining study duration when using acoustic telemetry is tag battery life. Most sharks can be issued with tags from the larger end of the available size spectrum, which can usually (depending on tag programming) provide about 10 years of battery life. As such, this is not a big issue for most shark-focused studies, as 10 years is usually sufficient to answer most study questions. When a longer duration is needed (e.g., tracking newborns through to maturation), recently developed perpetual tags could prove useful. These tags are battery free and use a flexible piezoelectric beam to harvest energy from swimming fish as the power source. Currently, only prototypes are available, which are detailed and assessed in Li et al. (2016), but it is anticipated that these will soon be commercially available.

8.3.2.4 *High-Resolution Positioning*

Accurately and reliably identifying the position or location of sharks can expand our understanding of their behavior, which in turn can lead to improved conservation and management. A recurring limitation of acoustic telemetry technology is that the location of an individual tag could be anywhere within the detection range of a single receiver. This creates uncertainty about the location upwards of a kilometer depending on the equipment used (Kessel et al., 2014a). This level of error may not be appropriate for certain study goals. Fine-scale, high-resolution positioning can be applied to address several aspects of shark behavior and will provide greater insight on behavior, particularly habitat use, social interactions, predator–prey interactions, shark-human interactions, socialization, and reproduction.

There are several ways to improve location estimates with acoustic telemetry, each varying in the resolution, equipment, and potential costs involved. Some studies have tried to alleviate this issue by deploying receivers in grids or other array designs with overlapping receiver detection ranges (Heupel et al., 2006). Grid approaches often use a position-averaging algorithm that assigns an average or weighted position based on detection frequency at receivers during a designated period (e.g., Simpfendorfer et al., 2002). Legare et al. (2015) used this method to explore habitat selection and spatial partitioning between juvenile blacktip (*Carcharhinus limbatus*) and lemon (*Negaprion brevirostris*) sharks in nursery areas. Multiple other studies of shark movement have applied the receiver grid and algorithm approach (e.g., Espinoza et al., 2015b; Heupel and Simpfendorfer, 2008; Knip et al., 2012b).

More rigorous positioning alternatives that rely on synchronized time-stamps and proximate receiver locations are increasingly being used. Armansin et al. (2016) used a high-resolution acoustic positioning system to evaluate social networks among spotted wobbegongs (*Orectolobus maculatus*). Similarly, Klimley et al. (2001) tracked the continuous movements of white sharks (*Carcharodon carcharias*) at a foraging site using a system of radio-linked acoustic receivers. These approaches employ conventional geolocation techniques such as time difference of arrival (TDOA) to estimate fine-scale locations within a receiver array with overlapping detection ranges, and they require relatively advanced analytical processing (Smith, 2013). Nevertheless, position accuracy is high and can be within meters (Espinoza et al., 2011; Klimley et al., 2001). The development of transmitters with quick transmission rates (i.e., milliseconds vs. seconds) also has the potential to increase position resolution by producing more transmissions (and animals) within a specific period with which to estimate a position. Guzzo et al. (2018) estimated that positions of stationary and mobile high-residency (HR) transmitters were 3.33 ± 2.27 m and 3.45 ± 3.65 m (mean $\pm$ SD), respectively, from handheld GPS locations at equivalent time periods. They also demonstrated that one-second transmission rates improved estimates of several movement metrics compared to traditional coding with longer transmission delays.

Thus, the capacity for high-resolution position estimates exists and can be used to meet specific study goals. As technology improves, it is likely that high-resolution positioning will become more routine in acoustic telemetry studies. However, acoustic limitations associated with transmission distances continue to restrict high-resolution positioning to dense receiver arrays (e.g., <100-m distance between three or more receivers), which require substantial equipment to cover a large area. Therefore, sharks that are resident to small or specific areas or that move to known areas, such as for feeding or reproduction, are currently best suited for these types of studies.

8.4 DATA CONSIDERATIONS

Although equipment is compatible and data can be shared among users and studies, there are no standardized approaches to data analysis. Most studies apply customized analyses that are difficult or impossible to replicate in other locations. This situation means that much consideration must go into data handling and analysis but also suggests this might be an area of greatest growth and advancement in the future.

8.4.1 Statistical Complications and Requirements

The fundamental concepts and application of acoustics to track sharks have remained relatively consistent over the last few decades (Klimley et al., 1998; McKibben and Nelson, 1986). Yet, the technology and capacity to track sharks have developed significantly (Arnold and Dewar, 2001; Hussey et al., 2015). Currently, scientists are much less restricted by logistical and technical hurdles, leading to a wide array of questions that are feasible to investigate. A result of more advanced technology and probing research is greater accrual and complexity of data. Using acoustic telemetry to study shark movements is no longer limited to presence/absence patterns of few animals on sparsely located receivers. Likewise, interpretation of movement behavior should not be limited to descriptions of broad patterns alone. Descriptive analyses may be insufficient to extract intricacies of the data and may hinder understanding of shark movement and the factors that affect it. Further still, statistical approaches do not have to be rudimentary because the tools are available to overcome issues of computing that existed in the past. Thus, well-informed, statistically derived investigations should be used to explain behaviors and evaluate data at species or ecosystem levels.

With the greater accumulation of data, interpreting acoustic telemetry has increasingly become an analytical playground, but also a potential minefield. Prior to formal analyses, researchers need to become familiar with not only the technical limitations of equipment and the aquatic environment but also statistical biases that are often pervasive in acoustic telemetry data. Identifying these issues and incorporating measures to avoid or reduce their effects are vital to ensure appropriate interpretation of behavior. A major statistical concern prevalent in acoustic telemetry research is spatiotemporal autocorrelation. Briefly, temporal and spatial data (both major aspects of movement) are correlated because consecutive data points are dependent on each other in time and space. Many statistical approaches rely on assumptions that data are independent and not correlated. Often, autocorrelation can lead to underestimates of home range size and inaccurate range utilization (Swihart and Slade, 1985). If model parameters do not sufficiently capture intrinsic or extrinsic sources of autocorrelation, it becomes necessary to take adequate steps to ensure that statistical principles are met. These steps may include adjusting or removing certain data (Rooney et al., 1998), incorporating model parameters that account for autocorrelation (e.g., autoregressive models or weighted functions) (Keitt et al., 2002), or using analytical approaches that are designed not to violate certain statistical assumptions (e.g., Brownian bridge movement models) (Horne et al., 2007). A number of other statistical or related issues should be considered when analyzing acoustic telemetry data of sharks; for more information, see Aarts et al. (2008), Heupel et al. (2006), and Rogers and White (2007).

8.4.2 Data Exploration

Approaches to data exploration will vary depending on study design and research questions. The most fundamental questions are often the best to consider: *What are the limitations/biases with my data? What questions can my data realistically answer? How do I best organize my data for statistical analysis? How will I convey my data and results to colleagues or in print?* Technical aspects of acoustic telemetry data that may confound results include tagging effects on normal behavior, post-release mortality due to infection or improper techniques, natural mortality, low number of detections, environmental variation, and animal behavior, to name a few (Heupel et al., 2006; Kessel et al., 2014a). These issues, and others like them, must be addressed before further analysis is conducted. In many cases, data can be removed prior to analysis to reduce stress behavior (e.g., remove first 24 hr) or if predation or death is evident. Animals deemed to have too few detections to reliably estimate behavior may also have to be excluded from analyses. Nevertheless, sporadic detections are still an important aspect of animal behavior worth investigating (e.g., reproductive movements,

foraging areas, long-range migrations). Data can also be standardized to reduce the effect of environmental, temporal, or behavioral variation; however, the specific way this is done should be evaluated because behavioral information may be lost in the process (De Solla et al., 1999; Payne et al., 2010; Rooney et al., 1998).

Exploratory plots are the most valuable way to identify the issues described above; they also provide a first glance at findings and guide analytical techniques. Because acoustic telemetry provides movement information, it is overwhelmingly associated with the dimensions of space and time. Simply plotting detections on a map will provide basic knowledge of the spatial extent of an individual or population (within the limits of the receiver array). Similarly, plotting pressure sensor values over the course of an individual's detection period will indicate the general patterns of depth use throughout time and highlight spurious detections caused by mortality. Broad habitat designations can be applied to certain receivers, and frequency plots of habitat use can be made at different temporal periods. Standardizing detections by a defined period (e.g., Heupel and Simpfendorfer, 2015; Speed et al., 2016) may further increase reliability of data exploration outputs to help guide future analytical steps.

8.4.3 New Analysis Approaches

Exciting new analyses are continually being developed and explored with regard to tracking aquatic animals. These advanced methods are often statistically relevant to acoustic telemetry data. Compared to analyses conducted at the onset of acoustic telemetry research, these methods are more computationally intensive; however, scientists are now supported by programs intended to process large datasets or employ specifically designed analytical tools. For example, the surfeit of user-friendly packages and functions developed for spatial data that exist in the statistical computing and graphics program R is astounding. Users can choose peer-reviewed statistical or technical approaches to visually or quantitatively explore movement data. Because it is a programming language, R users can adjust and manipulate their data as required. Tutorials and technical support for specific packages are readily available online, as are program-wide learning opportunities. Also, it is a free tool and widely used. Due to its widespread use, functionality, customization, support, and access, both aspiring and established shark researchers should become familiar with R.

In the past, spatial analyses of acoustic telemetry data at local scales (e.g., reef, bay, estuary) have largely focused on establishing the range or focal areas of use either by minimum convex polygons (MCPs) or kernel density estimates (KDEs) (for a review, see Rogers and White, 2007). Both methods do well to visualize and quantify the spatial area in which an animal is present, given presence/absence data and estimated (or perceived) detection ranges. The KDE approach is more dexterous because it uses a probabilistic smoothing factor to estimate space-use size at continuous gradations, known as utilization distributions (UDs), and is less sensitive to outlying locations—an MCP bias that worsens as sample size increases (Burgman and Fox, 2003). However, there is concern that, in some instances, the limitations or biases associated with KDEs are too grievous to reliably apply to acoustic telemetry data; for example, KDEs are sensitive to the areal shape and patchiness of detections, and UD size is dependent on proper selection of smoothing parameters (for a review, see Downs et al., 2011). Further, KDEs assume that each detection (or location) is independent of the others, although consecutive locations are typically linked in space and time (i.e., not independent). Nevertheless, depending on the data and if treated appropriately, many of these limitations can be addressed (Aarts et al., 2008; Kie et al., 2010).

Many alternative or variant approaches exist to quantify space use and movement patterns. Brownian bridge movement models and network analysis (see Chapter 18 in this volume) stand out as useful ways to address the shortcomings of other methods while providing dependable spatial information. Brownian bridge movement models (BBMMs) are similar to KDEs in that both estimate areal use of individuals at varying levels (e.g., 50% and 95% UDs) based on location data within a receiver array. Whereas KDEs incorporate kernel-smoothing techniques with aggregated data points, BBMMs estimate UDs by modeling temporally relevant movement paths based on an animal's mobility (i.e., speed and distance between receivers) and location errors (i.e., receiver detection range) as additional input parameters (Horne et al., 2007). The main advantages of this approach compared to kernel methods is that the subjectivity of smoothing parameter estimates is removed, issues of serial correlation and unequal time intervals between detections are incorporated in model parameters, and spatial estimates are less prone to error because time-dependent movements between locations are modeled (Horne et al., 2007). There is concern that this method does not accurately reflect home ranges (Becker et al., 2016; Fleming et al., 2015); nevertheless, it is increasingly being adopted as a trajectory-based space-use estimator in tracking animal movements (Kie et al., 2010). For sharks, the use of BBMMs has mainly been limited to assessing the influence of habitat and environmental conditions on partial migration (Papastamatiou et al., 2013). As a trajectory-based spatial density estimator, it has significant value for highly mobile species such as sharks.

In the context of acoustic telemetry, network analysis emphasizes the movements between receivers as opposed to estimating common spatial metrics such as activity spaces or utilization distributions (Lédée et al., 2015). Briefly, a network of connections is created in which movements (known as edges) between receivers (known as nodes) is quantified. In graphical displays, the size of edges and nodes and the proximity between nodes (i.e., clustering) typically indicate the relative strength or importance of that movement corridor. Quantitatively, network analysis provides many metrics to compare movement connections among individuals, receivers, and areas/habitats (Jacoby et al., 2012; Lédée et al., 2016). This method is advantageous compared to methods described above, because movement pathways and their interconnectedness are quantified. Also, network analysis is pliable, and physical or environmental attributes can be added to network properties. For sharks, network analysis has demonstrated main pathways and core use areas of different species (Lédée et al., 2015), elucidated species-specific vulnerabilities within and outside of MPAs (Espinoza et al., 2015a), and has helped implement government conservation policy (Lea et al., 2016). Moreover, social network analysis can be used with acoustic telemetry data to investigate behavioral interactions of sharks (e.g., Armansin et al., 2016; Wilson et al., 2015).

Other techniques exist to estimate space-use size and movement trajectories that may be pertinent depending on study design and data demographics. The following approaches should be considered: modified BBMMs (Kranstauber et al., 2012), autocorrelated KDEs (Fleming et al., 2015), time-geographic density estimation (TGDE) (Downs et al., 2011), and state–space modeling (Alós et al., 2016). As analytical approaches continue to be explored and developed, it is imperative to incorporate suitable techniques based on study design, data limitations, and research questions. Many approaches are still in their infancy and require ongoing scrutiny to optimally apply them to acoustic telemetry data. As a result, one of the best ways to understand space use of sharks is to analyze data using multiple methods (e.g., Becker et al., 2016; Lédée et al., 2015). Not only will the advantages and limitations of each approach become apparent but additional behavioral information might also be gained.

8.4.4 Improving Data Sharing and Display

Data exploration, statistical precautions, and advanced analytical techniques are necessary components of acoustic telemetry research. To be effective however, the output generated must be communicated clearly, whether at a conference, to the media, or in a publication (Figure 8.8). As technology and data acquisition expand, new and unique ways to present results have become available. Software that is often freely accessible (e.g., R, Google Earth) offers intricate ways to visualize movement data. Fine-scale activities and long-range migrations can be displayed on maps relative to time to demonstrate temporal behavioral trends. Fine-scale positioning systems that incorporate animal depth are also at the frontier of data visualization because three-dimensional movement patterns can be created and displayed. Not surprisingly, depth is an essential parameter to understanding movements in the aquatic environment that is currently underutilized.

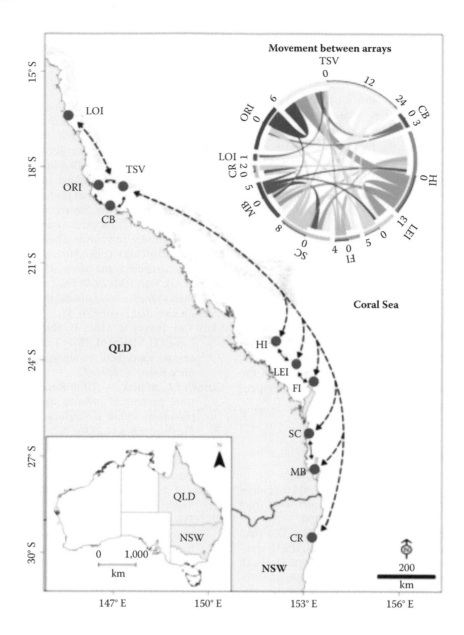

Figure 8.8 Example of visual representations of movement data via a map and a connectivity plot revealing links between locations. (From Espinoza, M. et al., *PLoS ONE*, 11(2), e0147608, 2016.)

The ability to create engaging maps and plots is powerful because it opens the door between scientists and the public, typically via the media (Cooke et al., 2017). This is a vital connection because important themes and concerning trends can be communicated in a manner that is simple to comprehend. However, graphics must be accompanied by clear interpretation and messaging to avoid miscommunication or incorrect information being perpetuated. Of specific concern is the fact that many fish and shark species that are tracked have economic, religious, or cultural significance to humans. Seemingly innocuous information about home ranges, migration patterns, or reproductive sites can inform humans of the timing of a population's location during key

or vulnerable periods (Cooke et al., 2017). If scientific output is used to facilitate overexploitation, the value of the output requires reassessment.

8.5 SUMMARY AND CONCLUSIONS

Future opportunities in acoustic telemetry will likely be based on advances in receiver and transmitter technology. Potential applications of acoustic telemetry to the field of marine science are currently being explored (e.g., Donaldson et al., 2014; Lennox et al., 2017). Researchers and developers are already developing expanded capacity for mobile receivers and remote data uploads from data-logging receivers as well as advanced sensors and perpetual transmitters. These advances along with miniaturization will allow greater numbers of individuals to be tracked and include previously unexamined species and locations. As ocean conditions continue to change, new sensors that reveal the conditions preferred or required by a species will help manage and conserve species into the future. This new technology is likely to require new analytical approaches. Given the vast amounts of data generated from acoustic telemetry systems, there are enormous opportunities for advancement of data handling, management, and analyses that should be pursued. Future advances will also come via networks of people. As use of acoustic telemetry expands, collaboration and data sharing become integral components of research success (Nguyen et al., 2017). National telemetry networks are already established in many locations, but animal movements are not constrained by country. Thus, data sharing and collaboration will have to expand to include opportunities to track individuals across national and jurisdictional boundaries (Heupel et al., 2015). By sharing data and working collaboratively, the field of acoustic telemetry has the capacity to expand and encompass a nearly global perspective of shark movement.

REFERENCES

Aarts G, MacKenzie M, McConnell B, Fedak M, Matthiopoulos J (2008) Estimating space-use and habitat preference from wildlife telemetry data. *Ecography* 31:140–160.

Ackerman JT, Kondratieff MC, Matern SA, Cech JJ (2000) Tidal influence on spatial dynamics of leopard sharks, *Triakis semifasciata*, in Tomales Bay, California. *Env Biol Fish* 58:33–43.

Alós J, Palmer M, Balle S, Arlinghaus R (2016) Bayesian state-space modelling of conventional acoustic tracking provides accurate descriptors of home range behavior in a small-bodied coastal fish species. *PLoS ONE* 11:e0154089.

Armansin NC, Lee KA, Huveneers C, Harcourt RG (2016) Integrating social network analysis and fine-scale positioning to characterize the associations of a benthic shark. *Anim Behav* 115:245–258.

Arnold G, Dewar H (2001) Electronic tags in marine fisheries research: a 30-year perspective. In: Sibert JR, Nielsen JL (eds) *Electronic Tagging and Tracking in Marine Fisheries: Proceedings of the Symposium on Tagging and Tracking Marine Fish with Electronic Devices.* Springer, Dordrecht, pp 7–64.

Baker LL, Jonsen ID, Mills Flemming JE, Lidgard DC, Bowen WD, Iverson SJ, Webber DM (2014) Probability of detecting marine predator–prey and species interactions using novel hybrid acoustic transmitter-receiver tags. *PLoS ONE* 9:e98117.

Barnett A, Abrantes KG, Stevens JD, Bruce BD, Semmens JM (2010) Fine-scale movements of the broadnose sevengill shark and its main prey, the gummy shark. *PLoS ONE* 5(12):e15464.

Becker SL, Finn JT, Danylchuk AJ, Pollock CG, Hillis-Starr Z, Lundgren I, Jordaan A (2016) Influence of detection history and analytic tools on quantifying spatial ecology of a predatory fish in a marine protected area. *Mar Ecol Prog Ser* 562:147–161.

Belicka LL, Matich P, Jaffé R, Heithaus MR (2012) Fatty acids and stable isotopes as indicators of early-life feeding and potential maternal resource dependency in the bull shark *Carcharhinus leucas. Mar Ecol Prog Ser* 455:245–256.

Burgman MA, Fox JC (2003) Bias in species range estimates from minimum convex polygons: implications for conservation and options for improved planning. *Anim Conserv* 6:19–28.

Carey FG, Kanwisher JW, Brazier O, Gabrielson G, Casey JG, Pratt HL (1982) Temperature and activities of a white shark, *Carcharodon carcharias. Copeia* 1982:254–260.

Carey FG, Scharold JV (1990) Movements of blue sharks (*Prionace glauca*) in depth and course. *Mar Biol* 106:329–342.

Carlisle AB, Starr RM (2010) Tidal movements of female leopard sharks (*Triakis semifasciata*) in Elkhorn Slough, California. *Environ Biol Fish* 89:31–45.

Clark CM, Forney C, Manii E, Shinzaki D, Gage C, Farris M, Lowe CG, Moline M (2013) Tracking and following a tagged leopard shark with an autonomous underwater vehicle. *J Field Robot* 30:309–322.

Conrath CL, Musick JA (2010) Residency, space use and movement patterns of juvenile sandbar sharks (*Carcharhinus plumbeus*) within a Virginia summer nursery area. *Mar Freshwater Res* 61:223–235.

Cooke SJ, Hinch SG, Wikelski M, Andrews RD, Kuchel LJ, Wolcott TG, Butler PJ (2004) Biotelemetry: a mechanistic approach to ecology. *Trends Ecol Evol* 19:334–343.

Cooke SJ, Iverson SJ, Stokesbury MJW, Hinch SG, Fisk AT, VanderZwaag DL, Apostle R, Whoriskey F (2011) Ocean Tracking Network Canada: a network approach to addressing critical issues in fisheries and resource management with implications for ocean governance. *Fisheries* 36:583–592.

Cooke SJ, Brownscombe JW, Raby GD, Broell F, Hinch SG, Clark TD, Semmens JM (2016) Remote bioenergetics measurements in wild fish: opportunities and challenges. *Comp Biochem Physiol A Molec Integrat Physiol* 202:23–37.

Cooke SJ, Nguyen VM, Kessel ST, Hussey NE, Young N, Ford AT (2017) Troubling issues at the frontier of animal tracking for conservation and management. *Conserv Biol* 31:1205–1207.

Crossin GT, Heupel MR, Holbrook CM, Hussey NE, Lowerre-Barbieri SK, Nguyen VM, Raby GD, Cooke SJ (2017) Acoustic telemetry and fisheries management. *Ecol Appl* 27:1031–1049.

da Silva C, Kerwath SE, Attwood CG, Thorstad EB, Cowley PD, Okland F, Wilke CG, Naesje TF (2013) Quantifying the degree of protection afforded by a no-take marine reserve on an exploited shark. *Afr J Mar Sci* 35:57–66.

Davy LE, Simpfendorfer CA, Heupel MR (2015) Movement patterns and habitat use of juvenile mangrove whiprays (*Himantura granulata*). *Mar Freshwater Res* 66:481–492.

De Solla SR, Bonduriansky R, Brooks RJ (1999) Eliminating autocorrelation reduces biological relevance of home range estimates. *J Anim Ecol* 68:221–234.

Donaldson MR, Hinch SG, Suski CD, Fisk AT, Heupel MR, Cooke SJ (2014) Making connections in aquatic ecosystems with acoustic telemetry monitoring. *Front Ecol Environ* 12:565–573.

Downs JA, Horner MW, Tucker AD (2011) Time-geographic density estimation for home range analysis. *Ann GIS* 17:163–171.

Dudgeon C, Pollock K, Braccini JM, Semmens J, Barnett A (2015) Integrating acoustic telemetry into mark–recapture models to improve the precision of apparent survival and abundance estimates. *Oecologia* 178:761–772.

Espinoza M, Farrugia TJ, Webber DM, Smith F, Lowe CG (2011) Testing a new acoustic telemetry technique to quantify long-term, fine-scale movements of aquatic animals. *Fish Res* 108(2–3):364–371.

Espinoza M, Ledee EJI, Simpfendorfer CA, Tobin AJ, Heupel MR (2015a) Contrasting movements and connectivity of reef-associated sharks using acoustic telemetry: implications for management. *Ecol Appl* 25:2101–2118.

Espinoza M, Heupel MR, Tobin AJ, Simpfendorfer CA (2015b) Residency patterns and movements of grey reef sharks (*Carcharhinus amblyrhynchos*) in semi-isolated coral reef habitats. *Mar Biol* 162:343–358.

Espinoza M, Heupel MR, Tobin AJ, Simpfendorfer CA (2016) Evidence of partial migration in a large coastal predator: opportunistic foraging and reproduction as key drivers? *PLoS ONE* 11(2):e0147608.

Farrugia TJ, Espinosa M, Lowe CG (2011) Abundance, habitat use and movement patterns of the shovelnose guitarfish (*Rhinobatos productus*) in a restored southern California estuary. *Mar Freshwater Res* 62:648–657.

Fleming CH, Fagan WF, Mueller T, Olson KA, Leimgruber P, Calabrese JM (2015) Rigorous home range estimation with movement data: a new autocorrelated kernel density estimator. *Ecology* 96:1182–1188.

Gjelland KØ, Hedger RD (2013) Environmental influence on transmitter detection probability in biotelemetry: developing a general model of acoustic transmission. *Meth Ecol Evol* 4:665–674.

Grubbs RD, Musick JA, Conrath CL, Romine JG (2007) Long-term movements, migration, and temporal delineation of a summer nursery for juvenile sandbar sharks in the Chesapeake Bay region. In: McCandless CT, Kohler NE, Pratt HL, Jr. (eds) *Shark Nursery Grounds of the Gulf of Mexico and the East Coast Waters of the United States.* American Fisheries Society, Bethesda, MD, pp 87–108.

Guttridge TL, Gruber SH, Krause J, Sims DW (2010) Novel acoustic technology for studying free-ranging shark social behaviour by recording individuals' interactions. *PLoS ONE* 5(2):e9324.

Guzzo MM, Van Leeuwen TE, Hollins J, et al. (2018) Field testing a novel high residence positioning system for monitoring the fine-scale movements of aquatic organisms. *Meth Ecol Evol* doi:10.1111/2041-210X.12993.

Halfyard EA, Webber D, Del Papa J, Leadley T, Kessel ST, Colborne SF, Fisk AT (2017) Evaluation of an acoustic telemetry transmitter designed to identify predation events. *Meth Ecol Evol* 8:1063–1071.

Hedger RD, Martin F, Dodson JJ, Hatin D, Caron F, Whoriskey FG (2008) The optimized interpolation of fish positions and speeds in an array of fixed acoustic receivers. *ICES J Mar Sci* 65:1248–1259.

Heupel MR (2007) Exiting Terra Ceia Bay: an examination of cues stimulating migration from a summer nursery area. In: McCandless CT, Kohler NE, Pratt HL, Jr. (eds) *Shark Nursery Grounds of the Gulf of Mexico and the East Coast Waters of the United States.* American Fisheries Society, Bethesda, MD, pp 265–280.

Heupel MR, Simpfendorfer CA (2002) Estimation of mortality of juvenile blacktip sharks, *Carcharhinus limbatus*, within a nursery area using telemetry data. *Can J Fish Aquat Sci* 59:624–632.

Heupel MR, Simpfendorfer CA (2008) Movement and distribution of young bull sharks *Carcharhinus leucas* in a variable estuarine environment. *Aquat Biol* 1:277–289.

Heupel MR, Simpfendorfer CA (2011) Estuarine nursery areas provide a low-mortality environment for young bull sharks *Carcharhinus leucas*. *Mar Ecol Prog Ser* 433:237–244.

Heupel MR, Simpfendorfer CA (2015) Long-term movement patterns of a coral reef predator. *Coral Reefs* 34:679–691.

Heupel MR, Webber DM (2012) Trends in acoustic tracking: where are the fish going and how will we follow them? In: McKenzie J, Parsons B, Seitz A, Kopf RK, Mesa M, Phelps Q (eds) *Advances in Fish Tagging and Marking Technology.* American Fisheries Society, Bethesda, MD, pp 219–231.

Heupel MR, Simpfendorfer CA, Hueter RE (2003) Running before the storm: blacktip sharks respond to falling barometric pressure associated with Tropical Storm Gabrielle. *J Fish Biol* 63:1357–1363.

Heupel MR, Simpfendorfer CA, Hueter RE (2004) Estimation of shark home ranges using passive monitoring techniques. *Environ Biol Fish* 71:135–142.

Heupel MR, Semmens JM, Hobday AJ (2006) Automated acoustic tracking of aquatic animals: scales, design and deployment of listening station arrays. *Mar Freshwater Res* 57:1–13.

Heupel MR, Reiss KL, Yeiser BG, Simpfendorfer CA (2008) Effects of biofouling on performance of moored data logging acoustic receivers. *Limnol Oceanogr Meth* 6:327–335.

Heupel MR, Simpfendorfer CA, Fitzpatrick R (2010) Large-scale movement and reef fidelity of grey reef sharks. *PLoS ONE* 5(3):e9650.

Heupel MR, Simpfendorfer CA, Espinoza M, Smoothey AF, Tobin A, Peddemors V (2015) Conservation challenges of sharks with continental scale migrations. *Front Mar Sci* 2(12):1–7.

Hoenner X, Huveneers C, Steckenreuter A, et al. (2018). Australia's continental-scale acoustic tracking database and its automated quality control process. *Sci Data* 5:180206.

Holland KN, Lowe CG, Peterson JD, Gill A (1992) Tracking coastal sharks with small boats: hammerhead shark pups as a case study. *Aust J Mar Freshwater Res* 43:61–66.

Holland KN, Meyer CG, Dagorn LC (2009) Inter-animal telemetry: results from first deployment of acoustic 'business card' tags. *Endang Species Res* 10:287–293 doi:10.3354/esr00226.

Horne JS, Garton EO, Krone SM, Lewis JS (2007) Analyzing animal movements using brownian bridges. *Ecology* 88:2354–2363.

Hussey NE, MacNeil MA, McMeans BC, Olin JA, Dudley SF, Cliff G, Wintner SP, Fennessy ST, Fisk AT (2014) Rescaling the trophic structure of marine food webs. *Ecol Lett* 17:239–250.

Hussey NE, Kessel ST, Aarestrup K, Cooke SJ, Cowley PD, Fisk AT, Harcourt RG, Holland KN, Iverson SJ, Kocik JF, Mills Flemming JE, Whoriskey FG (2015) Aquatic animal telemetry: a panoramic window into the underwater world. *Science* 348(6240):1255642.

Huveneers C, Simpfendorfer CA, Kim S, Semmens JM, Hobday AJ, Pederson H, Stieglitz T, Vallee R, Webber D, Heupel MR, Peddemors V, Harcourt RG (2016) The influence of environmental parameters on the performance and detection range of acoustic receivers. *Meth Ecol Evol* 7:825–835.

Jacoby DMP, Brooks EJ, Croft DP, Sims DW (2012) Developing a deeper understanding of animal movements and spatial dynamics through novel application of network analyses. *Meth Ecol Evol* 3:574–583.

Keitt TH, Bjørnstad ON, Dixon PM, Citron-Pousty S (2002) Accounting for spatial pattern when modeling organism-environment interactions. *Ecography* 25:616–625.

Kessel ST, Chapman DD, Franks BR, Gedamke T, Gruber SH, Newman JM, White ER, Perkins RG (2014a) Predictable temperature-regulated residency, movement and migration in a large, highly mobile marine predator (*Negaprion brevirostris*). *Mar Ecol Prog Ser* 514:175–190.

Kessel ST, Cooke SJ, Heupel MR, Hussey NE, Simpfendorfer CA, Vagle S, Fisk AT (2014b) A review of detection range testing in aquatic passive acoustic telemetry studies. *Rev Fish Biol Fish* 24:199–218.

Kessel ST, Hussey NE, Webber DM, Gruber SH, Young JM, Smale MJ, Fisk AT (2015) Close proximity detection interference with acoustic telemetry: the importance of considering tag power output in low ambient noise environments. *Anim Biotel* 3:5.

Kessel ST, Hussey NE, Crawford RE, Yurkowski DJ, Webber DM, Dick TA, Fisk AT (2017) First documented large-scale horizontal movements of individual Arctic cod (*Boreogadus saida*). *Can J Fish Aquat Sci* 74:292–296.

Kie JG, Matthiopoulos J, Fieberg J, Powell RA, Cagnacci F, Mitchell MS, Gaillard J-M, Moorcroft PR (2010) The home-range concept: are traditional estimators still relevant with modern telemetry technology? *Phil Trans Roy Soc Lond B Biol Sci* 365:2221–2231.

Klimley AP, Nelson DR (1984) Diel movement patterns of the scalloped hammerhead shark (*Sphyrna lewini*) in relation to El Bajo Espiritu Santo: a refuging central-position social system. *Behav Ecol Sociol* 15:45–54.

Klimley AP, Voegeli F, Beavers SC, Le Boeuf BJ (1998) Automated listening stations for tagged marine fishes. *Mar Technol Soc J* 32:94–101.

Klimley AP, Le Boeuf BJ, Cantara KM, Richert JE, Davis SF, Van Sommeran S (2001) Radio acoustic positioning as a tool for studying site-specific behavior of the white shark and other large marine species. *Mar Biol* 138:429–446.

Knip DM, Heupel MR, Simpfendorfer CA, Tobin AJ, Moloney J (2011) Ontogenetic shifts in movement and habitat use of juvenile pigeye sharks *Carcharhinus amboinensis* in a tropical nearshore region. *Mar Ecol Prog Ser* 425:233–246.

Knip DM, Heupel MR, Simpfendorfer CA (2012a) Evaluating marine protected areas for the conservation of tropical coastal sharks. *Biol Conserv* 148:200–209.

Knip DM, Heupel MR, Simpfendorfer CA (2012b) Habitat use and spatial segregation of adult spottail sharks *Carcharhinus sorrah* in tropical nearshore waters. *J Fish Biol* 80:767–784.

Knip DM, Heupel MR, Simpfendorfer CA (2012c) Mortality rates for two shark species occupying a shared coastal environment. *Fish Res* 125:184–189.

Kranstauber B, Kays R, LaPoint SD, Wikelski M, Safi K (2012) A dynamic Brownian bridge movement model to estimate utilization distributions for heterogeneous animal movement. *J Anim Ecol* 81:738–746.

Lea JSE, Humphries NE, von Brandis RG, Clarke CR, Sims DW (2016) Acoustic telemetry and network analysis reveal the space use of multiple reef predators and enhance marine protected area design. *Proc R Soc Lond B Biol Sci* 283(1834):20160717.

Lédée EJI, Heupel MR, Tobin AJ, Knip DM, Simpfendorfer CA (2015) A comparison between traditional kernel-based methods and network analysis: an example from two nearshore shark species. *Anim Behav* 103:17–28.

Lédée EJI, Heupel MR, Tobin AJ, Mapleston A, Simpfendorfer CA (2016) Movement patterns of two carangid species in inshore habitats characterised using network analysis. *Mar Ecol Prog Ser* 553:219–232.

Lefevre S, Huong DTT, Ha NTK, Wang T, Phuong NT, Bayley M (2011) A telemetry study of swimming depth and oxygen level in a Pangasius pond in the Mekong Delta. *Aquaculture* 315:410–413.

Legare B, Kneebone J, DeAngelis B, Skomal G (2015) The spatio-temporal dynamics of habitat use by blacktip (*Carcharhinus limbatus*) and lemon (*Negaprion brevirostris*) sharks in nurseries of St. John, United States Virgin Islands. *Mar Biol* 162:699–716.

Lennox RJ, Aarestrup K, Cooke SJ, Cowley PD, Deng ZD, Fisk AT, Harcourt RG et al. (2017) Envisioning the future of aquatic animal tracking: technology, science, and application. *BioScience* 67(10):884–896.

Li H, Tian C, Lu J, Myjak MJ, Martinez JJ, Brown RS, Deng ZD (2016) An energy harvesting underwater acoustic transmitter for aquatic animals. *Sci Rep* 6:33804.

Lowe, CG (2002) Bioenergetics of free-ranging juvenile scalloped hammerhead sharks (*Sphyrna lewini*) in Kāne'ohe Bay, Ōahu. *J Exp Mar Biol Ecol* 278:141–156.

Maljkovic A, Cote IM (2011) Effects of tourism-related provisioning on the trophic signatures and movement patterns of an apex predator, the Caribbean reef shark. *Biol Conserv* 144:859–865.

Matich P, Heithaus MR (2012) Effects of an extreme temperature event on the behavior and age structure of an estuarine top predator, *Carcharhinus leucas*. *Mar Ecol Prog Ser* 447:165–178.

Matich P, Heithaus MR (2014) Multi-tissue stable isotope analysis and acoustic telemetry reveal seasonal variability in the trophic interactions of juvenile bull sharks in a coastal estuary. *J Anim Ecol* 83:199–213.

McAuley RB, Bruce BD, Keay IS, Mountford S, Pinnell T, Whoriskey FG (2017) Broad-scale coastal movements of white sharks off Western Australia described by passive acoustic telemetry data. *Mar Freshwater Res* 68:1518–1531.

McKibben JN, Nelson DR (1986) Patterns of movement and grouping of grey reef sharks, *Carcharhinus amblyrhynchos*, at Enewetak, Marshall Islands. *Bull Mar Sci* 38:89–110.

McMeans BC, Borgå K, Bechtol WR, Higginbotham D, Fisk AT (2007) Essential and non-essential element concentrations in two sleeper shark species collected in arctic waters. *Environ Pollut* 148: 281–290.

Melnychuk MC (2012) Detection efficiency in telemetry studies: definitions and evaluation methods. In: Adams NS, Beeman JW, Eiler JH (eds) *Telemetry Techniques: A User Guide for Fisheries Research*. American Fisheries Society, Bethesda, MD, pp 339–357.

Morrissey JF, Gruber SH (1993) Home range of juvenile lemon sharks, *Negaprion brevirostris*. *Copeia* 1993:425–434.

Murchie KJ, Cooke SJ, Danylchuk AJ, Danylchuk SE, Goldberg TL, Suski CD, Philipp DP (2011) Thermal biology of bonefish (*Albula vulpes*) in Bahamian coastal waters and tidal creeks: an integrated laboratory and field study. *J Therm Biol* 36:38–48.

Nelson DR (1976) Ultrasonic telemetry of shark behavior. *J Acoust Soc Am* 59:1004–1007.

Nelson DR (1977) On the field-study of shark behavior. *Am Zool* 17:501–507.

Nelson DR, McKibben JN, Strong WR, Lowe CG, Sisneros JA, Schroeder DM, Lavenberg RJ (1997) An acoustic tracking of a megamouth shark, *Megachasma pelagios*: a crepuscular vertical migrator. *Environ Biol Fish* 49:389–399.

Nguyen VM, Brooks JL, Young N, Lennox RJ, Haddaway N, Whoriskey FG, Harcourt R, Cooke SJ (2017) To share or not to share in the emerging era of big data: perspectives from fish telemetry researchers on data sharing. *Can J Fish Aquat Sci* 74:1260–1274.

O'Toole AC, Murchie KJ, Pullen C, Hanson KC, Suski CD, Danylchuk AJ, Cooke SJ (2010) Locomotory activity and depth distribution of adult great barracuda (*Sphyraena barracuda*) in Bahamian coastal habitats determined using acceleration and pressure biotelemetry transmitters. *Mar Freshwater Res* 61:1446–1456.

Ortega LA, Heupel MR, Van Beynen P, Motta PJ (2009) Movement patterns and water quality preferences of juvenile bull sharks (*Carcharhinus leucas*) in a Florida estuary. *Environ Biol Fish* 84:361–373.

Packard GE, Kukulya A, Austin T, Dennett M, Littlefield R, Packard G, Purcell M, Stokey R, Skomal G (2013) Continuous autonomous tracking and imaging of white sharks and basking sharks using a REMUS-100 AUV. In: *MTS/IEEE OCEANS 2013–San Diego*. Institute of Electrical and Electronics Engineers, Piscataway, NJ, pp 1–5.

Papastamatiou YP, Lowe CG (2004) Postprandial response of gastric pH in leopard sharks (*Triakis semifasciata*) and its use to study foraging ecology. *J Exp Biol* 207:225–232.

Papastamatiou YP, Lowe CG (2012) An analytical and hypothesis-driven approach to elasmobranch movement studies. *J Fish Biol* 80:1342–1360.

Papastamatiou YP, Meyer CG, Holland KN (2007) A new acoustic pH transmitter for studying the feeding habits of free-ranging sharks. *Aquat Living Resour* 20:287–290.

Papastamatiou YP, Meyer CG, Carvalho F, Dale J, Hutchinson M, Holland K (2013) Telemetry and random walk models reveal complex patterns of partial migration in a large marine predator. *Ecology* 94(11):2595–2606.

Payne NL, Gillanders BM, Webber DM, Semmens JM (2010) Interpreting diel activity patterns from acoustic telemetry: the need for controls. *Mar Ecol Prog Ser* 419:295–301.

Pincock DG (2012) *False Detections: What They Are and How to Remove Them from Detection Data*, Amirix Document DOC-004691, version 03. Vemco, Halifax, Nova Scotia (http://www.vemco.com/pdf/false_detections.pdf).

Rechisky EL, Wetherbee BM (2003) Short-term movements of juvenile and neonate sandbar sharks, *Carcharhinus plumbeus*, on their nursery grounds in Delaware Bay. *Environ Biol Fish* 68:113–128.

Rogers KB, White GC (2007) Analysis of movement and habitat use from telemetry data. In: Guy CS, Brown ML (eds) *Analysis and Interpretation of Freshwater Fisheries Data*. Americna Fisheries Society, Bethesda, MD, pp 625–676.

Rooney SM, Wolfe A, Hayden TJ (1998) Autocorrelated data in telemetry studies: time to independence and the problem of behavioural effects. *Mammal Rev* 28:89–98.

Schlaff AM, Heupel MR, Udyawer V, Simpfendorfer CA (2017) Biological and environmental effects on activity space of a common reef shark on an inshore reef. *Mar Ecol Prog Ser* 571:169–181.

Sciarrotta TC, Nelson DR (1977) Diel behavior of blue shark, *Prionace glauca*, near Santa Catalina Island, California. *Fish Bull* 75:519–528.

Shipley ON, Brownscombe JW, Danylchuk AJ, Cooke SJ, O'Shea OR, Brooks EJ (2017) Fine-scale movement and activity patterns of Caribbean reef sharks (*Carcharhinus perezi*) in the Bahamas. *Environ Biol Fish* July:1–8 (https://doi.org/10.1007/s10641-017-0656-4).

Simpfendorfer CA, Heupel MR, Hueter RE (2002) Estimation of short-term centers of activity from an array of omnidirectional hydrophones and its use in studying animal movements. *Can J Fish Aquat Sci* 59:23–32.

Simpfendorfer CA, Heupel MR, Collins AB (2008) Variation in the performance of acoustic receivers and its implication for positioning algorithms in a riverine setting. *Can J Fish Aquat Sci* 65:482–492.

Simpfendorfer CA, Wiley TR, Yeiser BG (2010) Improving conservation planning for an endangered sawfish using data from acoustic telemetry. *Biol Conserv* 143:1460–1469.

Simpfendorfer CA, Yeiser BG, Wiley TR, Poulakis GR, Stevens PW, Heupel MR (2011) Environmental influences on the spatial ecology of juvenile smalltooth sawfish (*Pristis pectinata*): results from acoustic monitoring. *PLoS ONE* 6:e16918.

Simpfendorfer CA, Huveneers C, Steckenreuter A, Tattersall K, Hoenner X, Harcourt R, Heupel MR (2015) Ghosts in the data: false detections in VEMCO pulse position modulation acoustic telemetry monitoring equipment. *Anim Biotel* 3:55.

Smith F (2013) *Understanding HPE in the Vemco Positioning System (VPS)*. Vemco, Halifax, Nova Scotia (https://vemco.com/wp-content/uploads/2013/09/understanding-hpe-vps.pdf).

Speed CW, Meekan MG, Field IC, McMahon CR, Harcourt RG, Stevens JD, Babcock RC, Pillans RD, Bradshaw CJA (2016) Reef shark movements relative to a coastal marine protected area. *Reg Stud Mar Sci* 3:58–66.

Svendsen JC, Aarestrup K, Steffensen JF, Herskin J (2006) A novel acoustic dissolved oxygen transmitter for fish telemetry. *Mar Technol Soc J* 40:103–108.

Swihart RK, Slade NA (1985) Influence of sampling interval on estimates of home-range size. *J Wildl Manage* 49:1019–1025.

Ubeda AJ, Simpfendorfer CA, Heupel MR (2009) Movements of bonnetheads, *Sphyrna tiburo*, as a response to salinity change in a Florida estuary. *Environ Biol Fish* 84:293–303.

Udyawer V, Chin A, Knip DM, Simpfendorfer CA, Heupel MR (2013) Variable response of coastal sharks to severe tropical storms: environmental cues and changes in space use. *Mar Ecol Prog Ser* 480:171–183.

Wardle CS, Kanwisher JW (1973) The significance of heart rate in free swimming cod, *Gadus morhua*: some observations with ultra-sonic tags. *Mar Behav Physiol* 2:311–324.

Welch DW, Rechisky EL, Melnychuk MC, Porter AD, Walters CJ, Clements S, Clemens BJ, McKinley RS, Schreck C (2008) Survival of migrating salmon smolts in large rivers with and without dams. *PLOS Biol* 6:e265.

Wilson ADM, Brownscombe JW, Krause J, Krause S, Gutowsky LFG, Brooks EJ, Cooke SJ (2015) Integrating network analysis, sensor tags, and observation to understand shark ecology and behavior. *Behav Ecol* 26:1577–1586.

Young AH, Tytler P, Holliday FGT, MacFarlane A (1972) A small sonic tag for measurement of locomotor behaviour in fish. *J Fish Biol* 4:57–65.

CHAPTER 9

Imaging Technologies in the Field and Laboratory

Kara E. Yopak
Department of Biology and Marine Biology, UNCW Center for Marine Science,
University of North Carolina Wilmington, Wilmington, North Carolina

Jeffrey C. Carrier
Department of Biology, Albion College, Albion, Michigan

Adam P. Summers
Department of Biology, SAFS, Friday Harbor Laboratories, University of Washington, Seattle, Washington

CONTENTS

9.1   INTRODUCTION

9.1.1   Purpose of Three-Dimensional Visualization

Gross dissections have historically been vital for our understanding of the internal anatomy of elasmobranchs, and histological techniques have proven invaluable for microstructural analysis; however, these methodologies are highly invasive and thus inappropriate for anatomical investigations *in vivo*. Further, destructive techniques can typically distort the precise relative positions of tissue structures and are often impractical for the study of rare or valuable specimens. Bioimaging techniques, such as ultrasonography, endoscopy, computed tomography (CT), and magnetic resonance imaging (MRI), are unique in their ability to nonsurgically (and often non-invasively) acquire high-resolution, digital, three-dimensional (3D) data, sometimes in real time. These techniques facilitate highly accurate and efficient analysis and visualization of internal anatomy, although not all techniques are useful for *in vivo* examinations. Further, they allow for "digital dissection," which greatly extends the capabilities of researchers to incorporate quantitative anatomical measurements of live specimens, rare samples, and museum collections into the study of organismal biology.

9.1.2   Methods We Will Cover and Why

The purpose of this chapter is to introduce the reader to a range of bioimaging techniques (ultrasonography, endoscopy, CT, and MRI), including their historical uses, guidelines for their application, and methodological limitations. We hope to provide realistic expectations for the values and drawbacks associated with each of these techniques and how they can advance our understanding of shark biology.

9.2   ULTRASONOGRAPHY

9.2.1   Historical Use of Ultrasound in Elasmobranchs

Ultrasonic imaging, or sonography, is a non-invasive imaging technique that relies on interpreting reflected sound waves emitted by a transducer and detected by a receiver that converts the wave patterns into an image. Where radiography excels in visualization of hard tissues such as skeletal elements, ultrasound is more suited for rendering of soft tissues.

9.2.1.1   *Injury and Health Evaluation*

The use of ultrasonic imaging for sharks and rays has been a technique primarily used by veterinarians in captive environments for assessing general questions of elasmobranch health and well-being, as well as diagnosing injury and pathology *in vivo*. Its ability to assess soft-tissue damage and inspect internal organs for injury makes it ideally suited for initial evaluation of injured animals or animals experiencing a range of health issues, as an adjunct to evaluation for possible surgery. As such, it is a common and valuable resource for the clinical setting and is a common tool for marine veterinarians. Routine uses for ultrasound include cardiac monitoring during anesthesia (Stetter, 2004; Walsh et al., 1993), diagnostic applications that include following changes in thyroid size over time (Crow et al., 1998), and evaluating gastrointestinal tract, liver (Grant et al., 2013), gall bladder, spleen and pancreas, urogenital system, and coelom (Mylinczenko et al., 2017) for overall health assessment and progress of disease state and healing.

9.2.1.2 Assessment of Reproductive State

One of the most common applications of ultrasound for non-clinical uses is for the evaluation and monitoring of reproduction and pregnancy. It has been used to determine the presence or absence of eggs and embryos and size as a measurement of developmental stage in many species of sharks and rays, including nurse sharks (*Ginglymostoma cirratum*) (Carrier et al., 2003), small-spotted catsharks (*Scyliorhinus canicula*), thornback rays (*Raja clavata*) (Whittamore et al., 2010), draughtboard sharks (*Cephaloscyllium laticeps*) (Awruch et al., 2008), sand tiger sharks (*Carcharias taurus*), southern stingrays (*Dasyatis americana*) (George et al., 2017), ribbontail stingrays (*Taeniura lymma*) (Peteira et al., 2017), broadnose sevengill sharks (*Notorynchus cepedianus*) (Daly et al., 2007), and tiger sharks (*Galeocerdo cuvier*) (Sulikowski et al., 2016).

Although the majority of studies are restricted to captive facilities with controlled conditions, an increasing number of studies describe field applications, where assessment of reproductive condition is vital for determination of reproductive cycles as well as determination of mating and nursery grounds. Carrier et al. (2003) captured female nurse sharks following mating attempts to assess the likelihood of eventual pregnancy. Animals were examined with ultrasound, and those carrying eggs (and therefore deemed likely to become pregnant) were transported to captive facilities to monitor their pregnancy (Figure 9.1). Continuing ultrasound on captive animals revealed hatchlings *in utero* immediately prior to eventual birth, although survival rates were low. Prior to the use of ultrasound in the field, the selection of females was based solely on observations of mating and were less successful than those evaluated with ultrasound

who eventually gave birth in captivity. The extent to which capture stress and captivity may have interfered with pregnancy is unknown.

Sulikowski et al. (2016) similarly applied ultrasound in the field on tiger sharks (*Galeocerdo cuvier*) which revealed females carrying late-term embryos and aided in identifying the study site as a gestation habitat in this species. Their work demonstrates the value of sonography beyond simply the assessment of animal health and reproductive status. In locations where residency or philopatry can be established by other, more traditional means, ultrasound may be useful to assess and evaluate issues of habitat management and conservation concerns for mating and nursery grounds.

9.2.2 Requirements and Limitations of Instrumentation

9.2.2.1 Portability

Instrumentation size is less of an issue for captive studies, but boat size, availability of power for non-battery-operated devices, and the need for a protective environment for sensitive instruments might restrict widespread use in the field. Where those conditions can be met, new and novel uses might broaden the value of ultrasound to the field biologist. Portable units can range from $1000 to $10,000, depending on the options selected, and specialized probes may be in the same price range.

9.2.2.2 Probe Design and Wave Penetration

Ultrasound relies on the propagation of sound waves through soft tissues and reflection of the waves generated to imaging circuitry. The dermal denticles of shark skin interferes to some degree with wave propagation and resolution. Species with thick scales produce images with reduced quality (Stetter, 2004). Modern probes have been designed to minimize such interference and produce a more faithful image. Probe sizes vary, as do the frequency of the emitted ultrasonic signal and the depth to which the probe will scan. Determination of the optimal size and frequency is likely to be species dependent and may be influenced by skin thickness and body size. For example, an adult female nurse shark (*Ginglymostoma cirratum*), with its extremely thick skin and heavy body, may present more difficulty in evaluating uterine contents compared to a relatively smaller, thin-skinned shark, such as a mature bonnethead shark (*Sphyrna tiburo*). Additionally, the extent to which the probe can tolerate submersion in a seawater environment may also be a determinant. Although some probes are water resistant, others have been enclosed in plastic bags, with and without a conductive gel, to both protect the probe and maximize probe efficiency.

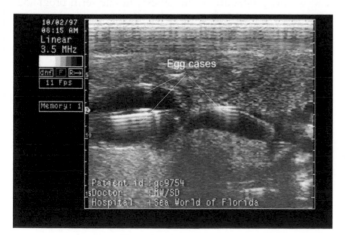

Figure 9.1 Ultrasound image of multiple egg cases from a nurse shark (*Ginglymostoma cirratum*). (Reprinted with permission from Carrier, J.C., et al., *Zoo Biol.*, 22, 179–187, 2003.)

9.2.2.3 Role of Sedation in Field and Laboratory Studies

Unless animals can be maintained in a prolonged state of tonic immobility, either beside the boat or poolside, or immobilized in some other fashion with appropriate gill irrigation (Sulikowski et al., 2016), sedation is required to minimize movement and effectively scan the target region of the animal. Methane tricaine sulfonate (MS-222) has historically been the anesthesia of choice. Animals sedated in this manner can be easily scanned and resuscitated following the procedure of Carrier et al. (2003).

9.2.2.4 Interpretation of Acquired Imagery

Interpretation of an ultrasound image requires some level of training and expertise. For example, depending on factors that may interfere with a clear signal, images are not always resolved to the highest level of image resolution, and interpretation of the resulting imagery may be difficult for new users. Modern instruments include digital measurement scales that can be superimposed onto the image and facilitate measurements where known values exist from prior experience or from previous necropsies (Figure 9.2). Additionally, some internal or external memory for storage and printing of scans for later interpretation is a requirement. Wherever possible, utilizing simultaneous measurements or visualization techniques for the purposes of validation should be considered. Most instruments allow for the capture of single-frame images in common file formats (e.g., JPG, TIFF), as well as video files for later analysis.

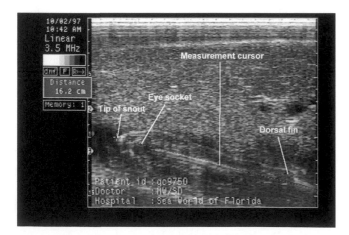

Figure 9.2 Sonogram of a hatchling nurse shark (*Ginglymostoma cirratum*) taken at the same time as the endoscopic procedure depicted in Figure 9.3. Topographical points are labeled for orientation and reference. (Reprinted with permission from Carrier, J.C., et al., *Zoo Biol.*, 22, 179–187, 2003.)

9.2.2.5 Validation of Imagery: Collateral Techniques

In studies with gravid nurse sharks, ultrasound was conducted simultaneously with endoscopy (see Section 9.3 for details on this technique). The solid core of the endoscope was easily visualized with the ultrasound, so that the precise location of both the probe and the resulting ultrasound image could easily be determined. Although endoscopy might not be readily available in a field setting, biopsy cores using a biopsy needle could very easily be inserted to establish a known reference point to validate the location of the sonogram.

9.2.2.6 Side Effects and Unintended Consequences of Ultrasound

The use of ultrasound carries some risk. Candidate animals must be immobilized either by inducing tonic immobility or with the use of sedation. Dosages and recovery must be monitored, and the potential effects of anesthesia on developing embryos is unknown. Perhaps the best documented evidence of potential side effect of ultrasound and, by extension, endoscopy was the work of Carrier et al. (2003) with captive nurse sharks. Animals were monitored with ultrasound every month during gestation. At 2-month intervals during their pregnancy, simultaneous endoscopic examinations of both uteri were conducted. All examinations were conducted under light sedation. The final ultrasound and endoscopy was followed shortly thereafter by spontaneous abortions of most embryos. Few survived to birth; the study revealed a gestation period of 4.5 to 5 months. Unbeknownst to the investigators at the time of the final ultrasound and endoscopic procedure, the final examination was performed near term. It is unknown whether handing, the intrusive nature of the endoscopic procedure, or anesthesia in late-term pregnancy may have resulted in the expulsion of embryos.

9.2.3 Expectations, Limitations, and Data

Ultrasound is a valuable technique for internal, non-invasive examination of living elasmobranchs. It can serve a diagnostic role in health assessment and injury evaluation. It has the added value of allowing some insight into reproductive status and condition. Image resolution is generally not sharp and not always intuitively obvious. Accompanying descriptions are usually required to lead the viewer to the interpretation offered by the investigator.

9.3 ENDOSCOPY

Endoscopy is a nonsurgical procedure that involves insertion of a rigid or flexible tube equipped with a light source and image recording capabilities for internal visualization of body cavities. Best known examples in human systems include examinations of the digestive tract, including upper regions, and examination of the colon. One of its primary benefits is the absence of any need for surgical incision. In this way, it differs from laparoscopy, where small surgical incisions are made in the body wall to facilitate insertion of a more rigid tube for examination of the exterior surface of organs and organ systems, often to be accompanied by surgical intervention. In some circumstances, where only visual examination is the goal, the techniques may be interchangeable.

9.3.1 Historical Use of Endoscopy in Elasmobranchs

9.3.1.1 Injury and Health Evaluation

Murray's (2010) review of endoscopy indicates that the earliest uses of endoscopy in fishes may have occurred in the early 1980s to examine viscera. Its initial application was largely for diagnostic purposes and was used to examine oral cavities, esophagus, stomach, and duodenum, the latter for the appearance of ulcers (Mylinczenko et al., 2017). Eventually, endoscopy allowed sampling of tissues through biopsy or needle aspiration as well as removal of necrotic tissues and parasites while still permitting nondiagnostic examinations.

9.3.1.2 Assessment of Reproductive State

Endoscopy has been used to evaluate the reproductive state of elasmobranchs. Insertion of an endoscope into the uteri of pregnant elasmobranchs can reveal details of the developmental progress of gestation (Carrier et al., 2003). Developing eggs, recently emerged embryos in aplacental viviparous species, empty egg cases, and embryological development of viviparous species can be viewed throughout the course of gestation (Figure 9.3). Samples of uterine fluid may also be removed for assay, and tissue samples may be removed from the uterine wall or developing embryos. In addition, high-resolution imagery from the probe, both still and video, is easily interpreted and does not require the training and expertise necessary to interpret radiographs or ultrasound images.

9.3.2 Requirements and Limitations of Instrumentation

9.3.2.1 Portability, Endoscope Size, and Probe Design

In addition to the electronics that provide illumination and image recording, suitable probes are required to provide entry and access to organs or organ systems. Earlier endoscopy generally required separate instruments for light and recording. Miniaturization and advances in digital technology have allowed the development of units that incorporate light-emitting diode (LED) lighting and recording capabilities into

Figure 9.3 Image obtained from endoscopy of a pregnant nurse shark (*Ginglymostoma cirratum*) approximately 124 days post-mating. The eye and dermal denticles are plainly visible, and the folds of the uterine wall can be seen in the upper right. (Reprinted with permission from Carrier, J.C., et al., *Zoo Biol.*, 22, 179–187, 2003.)

probes that may be battery operated. Such devices have their greatest application in emergency medicine for humans, but they may also be ideally suited for field and poolside applications in elasmobranch biology. These devices are generally built for use on humans, but they might be used for intrauterine examinations of presumed gravid elasmobranchs in the field. Units are available with sizes similar to small laptop computers. The selection of probes depends on the animal and system to be examined. Smaller probes in both length and diameter might be ideal for visualizing parasitic infestations on gills or for use on smaller sharks and batoids, but they may be insufficient for insertion into the uterus for evaluation of the reproductive state of larger species.

9.3.2.2 Role of Sedation in Field and Laboratory Studies

The use of rigid endoscopes and laparoscopes requires some degree of immobilization to prevent damage to both the instrument and soft internal tissues that may be proximal to the tip of the probe. Violent movements by animals not sedated during the procedure risk internal damage. Murray (2010) noted that, even in species where tonic immobility can be induced, ventilatory support (gill irrigation) and restraint are advisable to minimize injury should the animal struggle during the procedure. This requirement may necessitate removal of animals from the water onto a partially submerged stage where restraint can be accomplished or transfer to a small holding tank where restraint, ventilation, and anesthesia such as MS-222 can be administered. Such requirements may potentially limit the use of endoscopy for field use.

9.3.2.3 Interpretation of Acquired Imagery

Imagery acquired from endoscopy includes both video and still photography. Because the images are acquired real-time on an accompanying display device, there is little question what is being observed. The investigator has the added advantage of being able to manipulate the probe to obtain a precise image that can be examined in finer detail at a later date. Unlike sonography, the images are intuitive to decipher because they are color photographs or video rather than being images interpreted from reflected sound waves. Hence, interpretation does not require the same level of training as sonography.

9.3.2.4 Validation of Imagery: Collateral Techniques

Endoscopy may be enhanced by use of radiography or ultrasound (see Section 9.1.2.5) if precise location of the probe during imaging is necessary. Many endoscopes have markings on the exterior surface so that the precise distance of insertion can be determined to facilitate later diagnosis or surgical intervention if such follow-up procedures are warranted.

9.3.2.5 Side Effects and Unintended Consequences of Endoscopy

Endoscopy requires some advanced knowledge and understanding of elasmobranch anatomy to facilitate the placement of the probe and interpretation of the subsequent images. Manipulating the probe to successfully navigate within the body cavities or through passages leading to internal organs requires manual dexterity and is best accomplished following training and experience. In addition, it may be necessary to expand the organ by infusion of saline or carbon dioxide to permit unimpeded movement of the probe. Management of these advanced techniques may necessitate training that is generally not available to an average researcher. Cooperative studies with marine veterinarians specifically skilled in these procedures may eliminate procedural injuries due to absent or inadequate training and experience.

9.3.3 Expectations, Limitations, and Data

The extent to which invasive technologies such as endoscopy disrupt normal biological function, beyond the potential for injury, is unknown. In a study described above with captive nurse sharks (Carrier et al., 2003), spontaneous abortions occurred soon after an endoscopic procedure. It was not clear whether the procedure, restraint, anesthesia, confinement stress, or some combination of these was responsible for termination of the pregnancies. Although endoscopy as a routine procedure for nonclinical examination of elasmobranchs is in its relative infancy, the development of smaller and more portable probes suggests that endoscopy as a technique for examination of elasmobranch species may become valuable.

9.4 X-RAY COMPUTED TOMOGRAPHY

9.4.1 CT Scanning

Radiographic computed tomography (CT) is an imaging technique analogous to magnetic resonance imaging (MRI) (see Section 9.5), in that slice data can be used to create volumes and surfaces. The principle difference is that the images are a reflection of the radio opacity of the tissues rather than their water content. What the end user sees, cross-sectional slice data, is a derived dataset that requires substantial computation on the original shadow images. Cartilaginous fishes are not well mineralized and it is worth investing some time to understand parameters that affect image quality.

A useful analogy to the process of CT is to imagine a shadow cast on a screen by a bright, white light. If the shadow is made by a cunningly configured hand, the viewer might see a crocodile, a horse, or a lizard. However, if the fingers are held in place and the hand is smoothly rotated through 180 degrees, the viewer not only will no longer be fooled by the shadow animal but will also realize exactly how the fingers are placed to do the trick. The brain accomplishes this trick with ease, but in CT scanning these projections are turned into slices by a computer. Instead of a pure black shadow, x-rays cast a grayscale shadow that shows the density of the tissue. From these grayscale images, we can reconstruct anything that casts a shadow. The projections are images taken parallel to the axis of rotation of the scanner (or specimen), and they are used to generate cross-sections that are orthogonal to this axis of rotation.

Two common types of CT scanners are used to image cartilaginous fishes. Since the 1990s, medical scanners have been used on large animals, from lemon sharks to eagle rays (Motta and Wilga, 1995; Summers, 2000). In these scanners, the specimen is placed on a bed while the x-ray source and the imager whirl around the bed at a fixed distance. Modern medical scanners are very fast and can gather data from more than a meter of specimen in a minute. The resolution of these scanners is rarely better than 0.25 mm and is most often 1 to 3 mm. Some very high-quality images of cartilaginous fishes have been gathered using these machines (Dean and Motta, 2004; Dean et al., 2007).

The alternative scanner configuration holds the x-ray source and the imaging system still while the specimen rotates. Movement of the specimen during a scan is a major source of artifact, so the axis of specimen rotation is typically vertical to minimize the effect of gravity. The two advantages of the fixed imaging geometry are that higher precision is possible with the source and imager held perfectly still and, by moving the specimen away from the middle of the chamber, it is possible to achieve different magnifications.

This second advantage bears a closer look. In the shadow puppet analogy, moving the hand closer to the projector makes the shadow larger and moving it farther away makes it smaller. The concept directly translates to the CT scanner, where the rotating specimen table can be moved closer to or farther away from the x-ray source. This allows a tradeoff between the size of the object being scanned and the resolution. Scanners may quote a specimen size of 160 mm in diameter and 5-μm resolution, but these are not available at the same time. With a very small specimen, the maximum resolution will be available by moving the stage very close to the imager. Though this may seem like a weakness, there is no circumstance where you would want high resolution of a large specimen, as the data files would be too large to open.

In both medical style systems, including horizontal scanning beds and the industrial systems with vertical specimen tables, there are parameters that will affect the image quality. The total energy of the x-ray system is determined by the voltage across the source, measured in kilovolts, and the amperage measured in microamps. The energy of medical systems is low relative to industrial systems because the continued viability of the scanned specimen is considered an important design parameter. Industrial systems may have higher source power where fossil cartilaginous fishes can be imaged through rock. Power settings are analogous to the exposure of a camera: They set the white and black points and the dynamic range of the images. The number of images used to reconstruct the slices is determined by the rotation angle between successive images. For most purposes, three images per degree of rotation will be sufficient. Each image has a certain unavoidable amount of noise, and, to reduce this, the scanner averages the images. Averaging more than four images removes more noise, but each extra image costs time. Scan time matters little in the medical context, because the scanners are optimized for low-resolution, high-speed scans, but industrial scanners may take 3 to 24 hours to perform a single scan.

The projection images may be processed directly and automatically in medical systems, but industrial scanners usually do this step separately and with substantial input from the user. Fortunately, this processing can be done on proprietary software from the scanner manufacturer on a computer unconnected to the CT scanner. Typically, the reconstruction software will perform the calculations necessary to reconstruct a single representative slice. The user will then change reconstruction parameters to optimize image quality on that slice and use those parameters to automatically reconstruct the entire stack of slices. For specimens that exceed the height of the imager, there will be several stacks of images to be registered with each other during reconstruction. Beam-hardening artifacts due to low-energy, x-ray photons being preferentially attenuated by a radio-dense material are corrected at this step. Relative to the density of bone, the cartilaginous skeleton's tesserae and areolar mineralization is transparent to x-rays, and the beam hardening is usually negligible. The rest of the artifact correction is the same regardless of the specimen type.

9.4.2 Why Use CT in Elasmobranch Studies?

There are three reasons to use CT in the study of elasmobranch anatomy. The first is that it renders a three-dimensional understanding of the skeleton. Second, the data are collected without having to dissect the specimen or damage it in any way. Third, the resulting anatomical data can be processed into surfaces and volumes that allow quantification and mathematical modeling. Garman's *Plagiostomia* (1913) has dozens of plates with grayscale images of internal and external anatomy. Why, then, should we not simply use a sharp knife and keen eye to gather anatomical data rather than use an expensive, complex, and abstruse CT scanner? The answer is apparent to anyone who has looked hard at one of Garman's figures and really wanted to know what was just under that structure over there. No matter how many views are represented in a paper, there are always other views that would be helpful. A CT scan, properly published, does not have this limitation (Davies et al., 2017). The anatomy rendered is truly three dimensional. The object can be rotated, sliced, and digitally dissected to reveal things that the originator of the data had not yet imagined might be useful. John Maisey, an early adopter of the CT scanner to reveal morphology, published images of a famous Devonian braincase rendered in three dimensions and a good deal of obscuring matrix removed (Maisey, 2001a).

That CT does not damage the specimen is important, because it allows museum specimens to remain intact. After scanning, the specimen can be returned to the collection and examined with other modalities that will complement CT data. Many cartilaginous fishes are known from a few specimens, and even more are rare in collections. This puts a premium on gathering data in a way that does not harm the specimens. CT scanning can be done with alcohol-preserved specimens that remain wrapped in damp cheesecloth, so a drying artifact is minimal. In some cases, it has even been possible to age sharks and rays without physically sectioning vertebrae, although the vertebrae were isolated and air dried before scanning (Geraghty et al., 2012; Parsons et al., 2017).

The slice data processed from the raw projection images are usually visualized and analyzed as either a volume or a surface (Figure 9.4). The differences are at once subtle and important, and they have bearing on the quantitative analyses that can be done on the data. A surface rendering uses ray tracing techniques to bounce virtual light from a source off of the data and to a camera. The edges of the imaged structure are not distinct, and the density of the material below the surface is preserved. A surface can be computed from a volume by connecting all of the volume elements (voxels) that exceed a certain threshold into a set of vertices and faces. This surface encloses a volume, but the volume is empty of data. The first CT scans of cartilaginous fishes used surface renderings to show skeletal anatomy. These were followed by the use of shaded volume scans for the

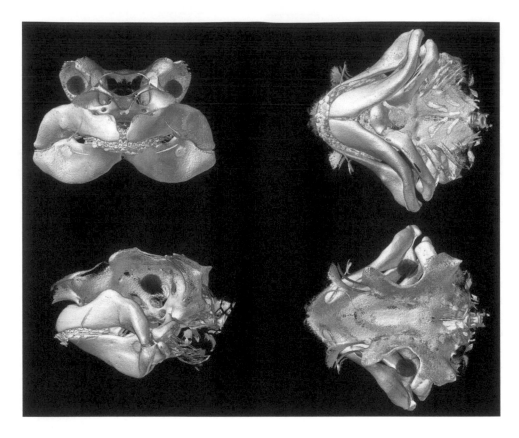

Figure 9.4   Three-dimensional volume rendered images of the skeletal morphology of *Heterodontus ramalhiera*, obtained from a CT scan of the entire head.

same purpose. Volume scans were also used to calculate the second moment of area of skeletal elements, a measure of how well they resist bending (Summers et al., 2004). This simple analysis was succeeded by a far more complex treatment that again used a surface reconstruction—the production of a finite element model of an entire shark head to calculate bite forces (Wroe et al., 2008).

The most exciting trend in CT scanning has been the production of large numbers of datasets that are being released with open access (Kamminga et al., 2017). The study of Kamminga et al. (2017) contains the crania of over 100 species of sharks, and the oVert project at http://www.MorphoSource.org has batoids and other sharks. These data are free for use in any scientific or educational endeavor which promises to both democratize the use of CT scan data and make large-scale meta analyses possible.

9.4.3 Limitations

Computed tomography has substantial drawbacks from a scientific standpoint. The first is the issue of time. The fastest scans are on the order of minutes, and high-quality scans require 30 minutes to 12 hours. Such scans cannot be done on live cartilaginous fishes. There are some technical hurdles to developing a submerged system because of the radio-opacity of water, and currently no aquatic

systems are in development. CT is, therefore, restricted to the study of postmortem anatomy, unlike ultrasonography and endoscopy.

Material will attenuate x-rays in proportion to two factors: density and atomic number. Soft tissues are neither dense nor composed of material with high atomic numbers, being composed primarily of water and organic material. Skeletons are moderately dense by virtue of calcium and phosphate, but many cartilaginous fishes have very poorly mineralized skeletons. This means that for some (e.g., Greenland shark) the soft tissues cannot be imaged and neither can the majority of the rest of the skeleton. Teeth, dermal denticles, vertebrae, and the tesserae that form the rind on the outer surface of many cartilaginous skeletal elements can, however, be imaged. Soft tissues, with their variable amounts of water, are ideal for imaging using MRI (see Section 9.5). It is possible to combine MRI and CT datasets, but in practice it is not often done because the datasets are at such different resolutions and the two types of scanners are rarely deployed on the same specimen. Rygg et al. (2017), however, combined these two methods to visualize the position and anatomy of the nasal rosette in a hammerhead shark.

It is possible to obtain CT images on soft tissue in cartilaginous fishes using a staining technique (Gignac et al., 2016). The only results so far from sharks and rays come from specimens stained with Lugol's solution (Figure 9.5)

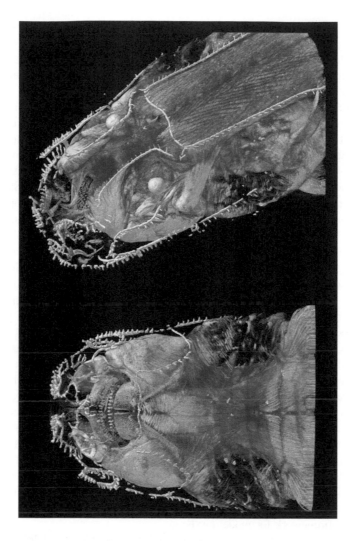

Figure 9.5 Oblique and ventral view of an iodine-stained *Chiloscyllium punctatum* (brownbanded bamboo shark) showing muscle and lateral line canals.

(Kolmann et al., 2017). This aqueous potassium iodide solution is absorbed differentially by tissues and leads to excellent contrast in nervous tissue and muscle tissue with little apparent distortion. The technique is new, and the lack of distortion must be verified, but it holds promise because it works well in other vertebrate groups. The main drawback is that the entire specimen becomes so infused with the dense iodine that the scan times increase and the energy necessary to penetrate the tissue requires the use of a high-end industrial scanner.

9.4.4 Synoptic Survey of Applications

Computed tomography scans of cartilaginous fishes have been used to catalog teeth (Moyer et al., 2015a,b), capture taxonomic diversity (Claeson, 2013), demonstrate the variation in isolated skeletal elements (Claeson, 2008, 2011), and trace the evolutionary and developmental history of dental structures (Rasch et al., 2016; Smith et al., 2015; Underwood

et al., 2015; Welton et al., 2015) and suspensoria (Wilga et al., 2000). There are detailed descriptions of chondrocrania (Maisey, 2004; Maisey and Springer, 2013), gathered with an eye toward understanding fossils, and CT provides orientation in the description of cranial musculature (Kolmann et al., 2014). More than one study has used CT to examine the early stages of cartilaginous fishes, when mineralization has just begun (Criswell et al., 2017). Though the dogma is that the skeletal material of chondrichthyans does not repair, a study of endophytic masses suggests that they may be a response to damage (Seidel et al., 2016, 2017). Finally, even holocephalans have enough mineral to be imaged (Johanson et al., 2015).

In paleontology, there is little to study but skeletal morphology, and the field has made wonderful use of the CT scanner to reveal the anatomy of extinct cartilaginous fishes. There are descriptions and re-descriptions of fossil sharks and rays (Claeson et al., 2010, 2013; Lane, 2010; Lane and Maisey, 2012; Maisey, 2011), as well as lovely images of extant animals used to show the importance of cranial anatomy in understanding fragmentary fossils (Claeson and Hilger, 2011; Mollen, 2010; Mollen and Jagt, 2012). CT has several advantages when interrogating fossils. In many cases, the matrix and fossil are different enough in density that it is possible to digitally prepare the specimen. Segmenting a fragmented specimen, or even one with good preservation but some distortion, allows a pristine whole to be reconstructed and compared to extant taxa. By reconstructing the fossil ray *Burnhamia daviesi*, Underwood et al. (2017) were able to shed light on the evolution of planktivory in mobulids.

One of the end uses of morphological data is the generation and testing of biomechanical hypotheses. This is certainly true for cartilaginous fishes, where CT scan data have been used to infer fluid flow in shark nares (Rygg et al., 2013), prey capture mechanics in megamouth sharks (Tomita et al., 2011), and the mechanics of feeding in electric rays (Dean and Motta, 2004; Dean et al., 2006). Scans of the wings of rays have led to generalizable rules about the architecture of flapping propulsors used in underwater autonomous vehicles (Fontanella et al., 2013; Russo et al., 2015), although a more complete understanding of energy lost during flexion would be useful (Schaefer and Summers, 2005). The issue of crushing prey is an interesting one because of the seeming paradox of doing this with cartilaginous jaws. Nevertheless, diets composed primarily of hard-shelled invertebrates have evolved many times in chondrichthyans and, because the jaws are well mineralized, CT is useful for understanding the process. The hard-prey-crushing sharks include both bonnetheads, which eat crabs, and horn sharks, which eat echinoderms, bivalves, and gastropods (Huber et al., 2005; Mara et al., 2010). CT scans of the cranium provide a framework for modeling the bite forces of the attached muscles. The relative stiffness of the jaws of both horn sharks and the hard-prey-crushing cownose ray was modeled from second moment of area calculations taken from CT scans (Summers et al., 2004). Kolmann et al. (2015) produced the surprising result that the shape of hard-prey-crushing stingray jaws does not allow some to crush prey more easily than others. This study relied on CT scans to provide a template for physical models that were a critical proxy for investigations of failure mechanics.

Although straight morphology is by far the most common use of CT in investigating cartilaginous fishes, some studies have used it for postmortem pathological investigations—for example, the common scoliosis seen in captive sand tiger sharks (Preziosi et al., 2006). Another unusual use of CT is in the discovery of peculiar dietary habits. The wedgefish, *Rhyncobatus*, has a pebble-like dentition and might be expected to eat hard prey. The presence of many stingray spines in and around the jaws revealed by CT scans suggests they frequently consume smaller rays (Dean et al., 2017). Applying comparative morphology to sensory neurobiology, Maisey used CT to trace the evolution of low-frequency sound reception by examining the labyrinth of extant and fossil sharks (Maisey, 2001b; Maisey and Lane, 2010).

9.5 MAGNETIC RESONANCE IMAGING

9.5.1 Historical Applications of MRI in Elasmobranchs

Magnetic resonance imaging (MRI) is a technique that uses a magnetic field and radiofrequency waves to create high-resolution images of soft tissue structures. Although these cutting-edge technologies and methods are extensively developed for applications in humans and veterinary medicine (e.g., Elliot and Skerritt, 2010; Kiessling et al., 2011), in recent years their applicability has been extended to non-invasive visualization and quantification of internal anatomy in elasmobranchs, including whole-body scans (Berquist et al., 2012; Blackband and Stokkopf, 1990; Ziegler et al., 2011) or smaller subregions of the body (e.g., Perry et al., 2007; Yopak and Frank, 2009; Ziegler et al., 2011). Body tissue of most organisms contains a high number of hydrogen atoms (i.e., protons), particularly in water and fat. The magnetic fields in an MR instrument are used to align the hydrogen atoms in the imaged object, essentially mapping the spatial distribution of water vs. fat in an organism. When placed in the magnetic field, the protons absorb the radiofrequency waves, thus generating a detectable signal that is transmitted to a radiofrequency coil and then processed as a series of sliced images by a computer. The two primary types of MR imaging are functional MRI (fMRI) and anatomical MRI (referred to from this point forward as simply MRI). Although both employ the same instrumentation, they measure very different things in an organism.

9.5.1.1 Anatomical MRI

Magnetic resonance imaging, MR microscopy, and MR spectroscopy allow investigators to capture high-resolution (well under 100 μm) 3D data of soft tissue structures *in situ* in cartilaginous fishes. The wide availability of pulse sequences and the varied pulse sequence parameters within this technique also allow for contrast optimization in a particular tissue of interest (described below) and even assessment of chemical composition of fish tissue (Mathiassen et al., 2011). Although histological techniques can often provide higher spatial resolution than MRI data (up to the subcellular level), these more traditional methods are highly destructive, and distortions due to tearing and shearing are unavoidable (for a comparison of histology vs. MRI in fishes, see Ullmann et al., 2010b). MR technology is rapidly improving its resolution capabilities, with MR microscopy allowing for resolutions down to 10 μm (Tyszka et al., 2005; Ullman et al., 2010a). Further, these methods are not restricted to *in situ* studies and have been demonstrated *in vivo* in fishes, allowing for characterization of physiological parameters such as bulk flow and diffusion (Bock et al., 2002).

Anatomical MRI technology has now been established as an effective investigative tool of the internal anatomy of teleost fishes, including variation in the brain (e.g., Ullmann et al., 2010a,b), assessment of barotrauma (Rogers et al., 2008), and investigation of highly specialized internal structures (e.g., Chakrabarty et al., 2011; Forbes et al., 2006; Graham et al., 2014; Sepulveda et al., 2007). This technology has also been increasingly employed for studies of elasmobranch anatomy, such as studies on pelagic shark red muscle volume (Perry et al., 2007), head anatomy (Waller et al., 1994), and comparative brain morphology (Figure 9.5) (Yopak and Frank, 2009; Yopak et al., 2010, 2016), in addition to the generation of online databases of MR data, such as the Digital Fish Library (http://www.digitalfishlibrary.org) (Berquist et al., 2012), where 46 species of cartilaginous fishes across 10 orders have been imaged to date. Three-dimensional data also open doors to new questions in elasmobranch biology. For example, Yopak et al. (2016) investigated variation in shape and complexity of a region of the brain, the cerebellum, which varies greatly in size and folding (or foliation) across different shark and batoid species (Yopak, 2012), ranging from a smooth surface to a highly complex structure, with numerous invaginations and folds. Although previously quantified using a qualitative visual grading scheme (Yopak et al., 2007), MRI provided an automated mathematical assessment of the 3D shape of this structure.

9.5.1.2 Functional MRI

Although anatomical MRI has been increasingly exploited in the last decade in comparative elasmobranch research, no advancements have been published in the primary literature on functional MRI (fMRI) in sharks (and very little in fishes generally; for a review, see Van der Linden et al., 2007). In clinical research, fMRI is used to understand how different parts of the brain respond to external stimuli. When neural activity increases in a particular brain region, the MR signal is also incrementally increased. Contrary to popular belief, however, neuronal electrical activity is not measured directly. Instead, fMRI measures either changes in the blood-oxygen-level-dependent (BOLD) signal, which in essence measures the ratio of oxygenated hemoglobin to deoxygenated hemoglobin across regions of the brain at various time points, or it measures changes in cerebral blood flow (CBF) or volume (CBV). So, rather than measuring absolute activity, fMRI measures activity differences in brain regions during a task compared to a control task.

Theoretically, assessing *in vivo* neural physiology in cartilaginous fishes is possible, with obvious drawbacks, including irrigation of the gills or whole-body submersion (without disrupting the MR signal), anaesthesia that does not cross the blood–brain barrier (which may affect neural activation), and immobilization during MR protocols, all without benefit of instrumentation that would be compromised by the strong magnetic field. To our knowledge, there is only a single known fMRI study on a live fish species, and it explored changes in brain physiology (particularly the hypothalamus) by measuring changes in CBV and oxygenation of hemoglobin in response to thermal stress in the common carp (*Cyprinus carpio*) (van den Burg et al., 2005). At present, much more work is required before fMRI becomes a viable technique for assessing brain activity in live cartilaginous fishes.

9.5.2 Requirements and Limitations of Instrumentation for Anatomical MRI

Magnetic resonance is an incredibly powerful technique for non-invasive visualization, yet it requires highly technical training, access to instrumentation (which can be cost prohibitive), and optimally prepared specimens for best results. The choice of instruments, radiofrequency coils, scan parameters, and protocols to employ when preparing the tissue will depend heavily on the size and shape of the specimen, level of resolving power required to visualize the tissue or structure of interest, tissue composition (e.g., hard tissue, soft tissue, fluids), and budgetary restrictions. In MRI, spatial resolution, or the ability to resolve two points as distinct, is defined by the size of the imaging voxels, which are the individual volume resolution elements in an MR image. They can be thought of as 3D pixels, with the image slice thickness determining its depth. The size of the voxel (and therefore spatial resolution) is a function of slice thickness, field of view (FOV), and matrix size. Ultimately, the smaller the voxel, the higher the spatial resolution. In addition, an MRI image is a combination of pure signal and noise, so the signal-to-noise ratio (SNR) is used to describe and evaluate the contrast in an image. The SNR can be improved by sampling larger volumes (i.e., increasing FOV and slice thickness), which comes at a cost of spatial resolution, or by increasing the size of the magnetic field. Thus, magnets of higher field strength (see Section 9.5.2.1) ultimately produce images with a higher SNR than lower field strength instruments. They are also able to do so in less time, improving protocol efficiency and reducing the cost of scan time. Because most MR facilities charge hourly rates for instrument use (with additional fees for MR technicians), it is critical to determine the minimum spatial resolution required to visualize the structure of interest and obtain that degree of image quality in the shortest time frame. MR acquisition is often a delicate balance among numerous parameters, including spatial resolution, SNR, and scan time, combined with budgetary constraints. We detail below the current best practices for obtaining high-resolution images of soft tissue structures in cartilaginous fishes. This chapter covers only the basics of MR imaging as it pertains to optimizing images; for comprehensive overviews of MR physics, pulse sequences, radiofrequency coils, and scan parameters, see Kiessling et al. (2011) and Brown et al. (2014).

9.5.2.1 Instrumentation

The first choice that must be made in any bioimaging study is the choice of instrument, which will vary primarily in terms of magnetic field strength, measured in the unit Tesla (T) or gauss. Although the list of scanners, their hardware, and thus their capabilities is quite broad, imaging in cartilaginous fishes will likely involve 1.5-T or 3-T human scanners (and, in some facilities, 7-T human scanners); 7-T, 9.4-T, or 11.7-T small animal scanners; and occasionally, 16.4-T small-bore magnets. Although these instruments can vary widely in the kinds of data they can produce, the primary parameters of concern for the purposes of this chapter are (1) spatial resolving power and (2) usable bore size, which (along with the coil used) determine how large a specimen can be successfully imaged.

Human scanners have the benefit of being more widely available, due to the breadth of clinical research in hospitals and medical schools worldwide. In general, 1.5-T and 3-T human scanners will have a larger usable bore (approximately 70 to 110 cm) but will be limited in their field strength and thus the MR signal, producing images with lower spatial resolution (voxel resolutions of 90 to 100 μm^3) than more high-field magnets. Because these scanners are designed to fit a human body, these instruments are ideal for whole-body elasmobranch imaging. Perry et al (2007) imaged a 13-cm-long midbody segment of the shortfin mako shark (*Isurus oxyrinchus*) using a 1.5-T clinical scanner and a standard knee coil. They also obtained a whole-body image (FL = 81.5 cm) of the salmon shark (*Lamna ditropis*) using a 3-T human scanner; five consecutive segments of the animal were imaged separately and registered in postprocessing. Both scans were of sufficient resolution to differentiate and segment red muscle volumes. Ziegler et al. (2011) imaged a whole leopard shark (*Triakis semifasciata*) on a 3-T human scanner, with reasonable contrast and spatial resolution (320 to 280 μm). Yopak and Frank (2009) imaged the crania of a neonate whale shark *in situ* (60.1 cm TL) using a 3-T human scanner and a custom-built rectangular surface coil. They obtained a spatial resolution of 500 μm, which allowed for visualization of the brain but very little contrast between gray and white matter within the brain (Figure 9.6A).

As the field strength increases, often the usable bore size decreases (thus limiting the size of specimens you can image), but the resolving power and SNR improve, allowing for distinction between finer structures. Small animal scanners and the coils available for them provide enough room to fit samples in the size range of mice or rats. Yopak and Frank (2009) imaged the brains (approximately 11 cm in length) from two juvenile whale sharks (*Rhincodon typus*) on a 7-T small animal magnet with a 20-cm usable bore, producing images with high contrast between gray and white matter and a spatial resolution of 100 to 150 μm (Figure 9.6B). Even finer still, in teleosts Ullmann et al. achieved 30-μm resolution in the brain of the barramundi (*Lates calcarifer*) (2010c) and 10-μm resolution in the brain of the zebrafish (*Danio rerio*) (2010a), which were imaged on a 16.4-T vertical magnet, which has a bore diameter of only 89 mm. At present, MRI is not suitable for specimens smaller than 1 mm from which it cannot gather meaningful anatomical data (Figure 9.7) (Ziegler et al., 2011).

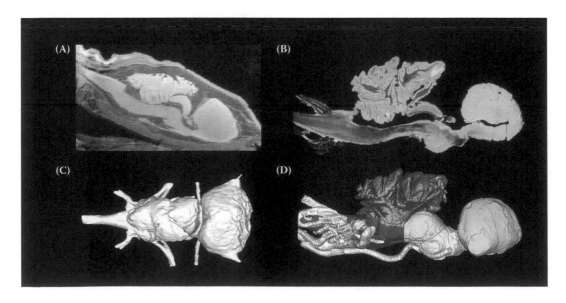

Figure 9.6 Magnetic resonance images of the brain of the whale shark (*Rhincodon typus*). (A) The brain of a neonate specimen in sagittal view, scanned *in situ* using (B) T_1-weighted 3D FSPGR, and (B) digitally segmented. (C) Sagittal slice of MR data of an excised brain from a juvenile specimen, scanned using T_1-weighted 3D FLASH sequence. (D) Digital segmentation of the major structures of the brain and cranial nerves. All digital segmentations were performed using ITK-SNAP (Yushkevich et al., 2006). (Adapted from Yopak, K. and Frank, L.R., *Brain Behav. Evol.*, 74, 121–142, 2009. With permission.)

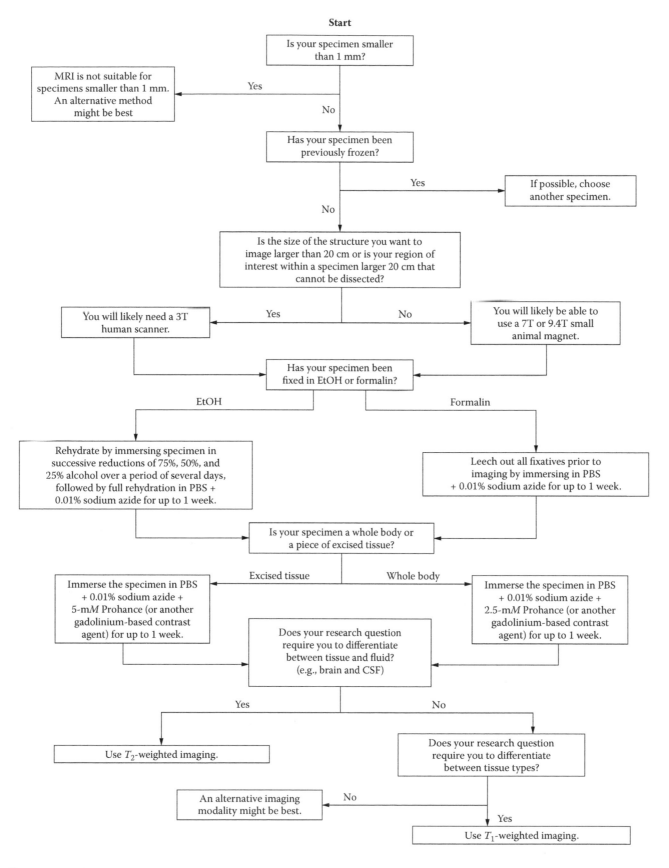

Figure 9.7 A detailed flow chart to guide readers through the process of applying MRI to your research questions, including ideal imaging modality, instrument, specimen preparation, and MR imaging sequences.

9.5.2.2 Specimen Preparation and Optimization

After choosing the magnet, the next step is to properly prepare the specimen for imaging (Figure 9.8). The quality of images obtained is highly dependent on the quality of the specimen, the strength of the magnetic field, and the ability of the scanner to detect the resulting MR signal in specimen tissues. As MRI is extremely sensitive to the structural and chemical nature of tissue, the physical condition of the specimens and the environment to which they have been exposed have significant impacts on imaging results.

Blackband and Stoskopf (1990) were the first to apply MR imaging techniques in fishes *in vivo* (including the elasmobranch *Chiloscyllium arabicum*), and Cloutier et al. (1988) were the first to show selective enhancement of different tissues (including brain, muscles, and cartilage) in a postmortem fish specimen, the coelacanth *Latimeria chalumnae*. Waller et al. (1994) were one of the first to publish nuclear magnetic resonance (NMR) images of the head of an elasmobranch, the bluntnose sixgill shark (*Hexanchus griseus*), which demonstrated the viability of this technique for visualizing cranial tissue and portions of the anterior body. These studies and others now both serve as a platform from which numerous studies since have applied these techniques.

Standardized protocols have now been established, not only for instrumentation and scanning of specimens but also for optimal preparation of tissue (see Figure 9.8). Berquist et al. (2012) wrote a highly comprehensive review outlining best practices for specimen preparation for MRI of fishes, particularly in museum specimens, and Ziegler et al. (2011) determined the viability to apply these techniques across a range of vertebrates, including elasmobranchs. To mitigate adverse effects of suboptimal preparation, most MR studies in fishes recommend the following:

1. Choose the best quality specimens possible. Although this is not always feasible when dealing with rare specimens, when several specimens are available it is best to choose the one that has no external damage and had the shortest postmortem time prior to fixation.
2. Avoid freezing as a storage mechanism where possible (Berquist et al., 2012; Perry et al., 2007). Ideally, imaging fresh tissue provides the best results. Numerous imaging parameters can be affected by freezing, storing, and thawing fish tissue (e.g., Nott et al., 1999a,b). Imaging previously frozen specimens as compared to fresh or fixed tissue will often result in poorer imaging quality, image artifacts, and very little contrast.
3. Avoid the use of alcohol as a preservation medium wherever possible (Figure 9.8A,B), as it can result in tissue shrinkage due to a reduction in the amount of water and lipids in the tissue. This is quite common in museum collections, where specimens are typically fixed in 10% formalin and post-fixed in 50% isopropyl alcohol or 70% ethyl alcohol for long-term storage. Although imaging has been successfully performed on alcohol-preserved fish samples (e.g., Chanet et al., 2009), exposure to alcohol can significantly degrade the tissue MR responses, often does not result in images of high quality, and reduces the SNR (Berquist et al., 2012).
4. To limit the tissue damage described above, Berquist et al. (2012) recommended that specimens preserved in alcohol should be immersed in successive reductions of 75%, 50%, and 25% alcohol over a period of several days, followed by full rehydration in phosphate-buffered saline (PBS) with the addition of 0.01% sodium azide. Although formalin fixation can also result in alteration of image contrast (e.g., Yong-Hing et al., 2005), it is a preferable preservation method compared to the use of alcohol. Regardless of the type fixative used, it is best practice to leech out all fixatives prior to imaging, using PBS + 0.01% sodium azide.

Image quality can be significantly enhanced if specimens undergo some degree of sample preparation prior to scanning. In particular, gadolinium-based contrast agents can enhance the MR responses of tissues and improve their anatomical contrast (e.g., Chakrabarty et al., 2011). For example, prior to imaging the brain of the whale shark (*Rhincodon typus*), Yopak and Frank (2009) transferred the brain tissue, which was previously fixed in 10% formalin, to PBS + 0.01% sodium azide for a week followed by the addition of 5-mM ProHance® (Bracco Diagnostics, Inc.; Princeton, NJ) for 7 days. Equilibrating the tissue in this contrast agent achieved increases in SNR efficiency of the data acquisition and high contrast within the brain (Figure 9.6B). Berquist et al. (2012) similarly found this to be a highly effective method for improving image contrast in small whole-body specimens at 2.5-mM ProHance® (Figure 9.8C,D), although this may be cost prohibitive for larger specimens (e.g., Perry et al., 2007).

The technique for securing specimens while inside of the scanner also seems to be dependent on specimen size. Waller et al. (1994) were able to optimize scanning of *Hexanchus griseus* on a 4.7-T magnet by wrapping the specimen and mounting it directly in a saddle-shaped radiofrequency (RF) coil. Similarly, Berquist et al. (2012) imaged whole-body specimens after wrapping them in sheeting plastic and taping them onto the scanning bed of a 3-T human magnet. In contrast, for small-bodied specimens (<15 cm), suspension in a perfluoropolyether (PFPE) fluid is common, as it is proton free (and thus MR invisible), reduces susceptibility to artifacts at air–tissue boundaries, and helps to avoid tissue dehydration. Further, a vessel in which the sample is suspended can be vacuum sealed, which reduces or removes air bubbles in the chamber, further ensuring optimal images without air-bubble artifacts. The literature details the use of the PFPEs Galden® (Solvay Solexis, Houston, TX) (Berquist et al., 2012) and Fomblin® (Ausimont, Morristown, NJ) (Ullman et al, 2010a,b) in fishes, although numerous brands are available. In some instances, specimens can be scanned in other media (including distilled water, formalin, or even 70% ethanol (Ziegler and Mueller, 2011; Ziegler et al., 2011); however,

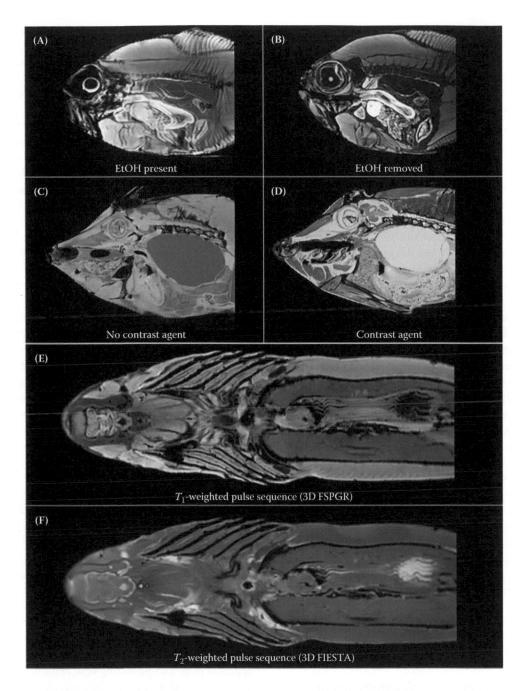

Figure 9.8 Optimization of MR data acquisition. (A) The red bream (*Beryx decadactylus*) preserved in EtOH, imaged with a T_1-weighted FSPGR pulse sequence on the 3-T scanner prior to rehydration, and (B) reimaged following rehydration, resulting in an enhanced image quality. (C) The fantail filefish (*Pervagor spilosoma*) initially imaged on the 7-T scanner using a T_1-weighted FLASH pulse sequence without exposure to the contrast agent, ProHance®. (D) Subsequent reimaging after exposure to 2.5-mM ProHance®, resulting in a significantly brighter MR signal and enhanced visual contrast among tissues. (E) Comparison of T_1-weighted 3D FSPGR and (F) T_2-weighted 3D FIESTA MRI pulse sequences acquired on the 3-T scanner in the smooth hammerhead (*Sphyrna lewini*). (Adapted from Berquist, R., et al., *PLoS ONE*, 7(4), e34499, 2012. With permission.)

Waller et al. (1994) found that scanning *H. griseus* in ethanol provided highly unsatisfactory results. Regardless, it is vital that the specimen not move during imaging, as movement artifacts are highly problematic (for a review, see Ziegler et al., 2011). Securing the specimen to a platform with tape or low-melting agarose can often solve these issues.

Finally, the presence of a strong magnetic field requires that all metal objects be removed from the subject as well as any investigator entering the instrument room. In the case of whole-body elasmobranch imaging, this requires checking the mouth and body cavity for hooks or other metal objects that may inadvertently have been left in the animal.

9.5.2.3   MR Data Acquisition

Magnetic resonance imaging has a unique ability to depict soft tissues because of its sensitivity to the state of tissue water. The signal is sensitive to the local water content (i.e., proton density in a voxel), recovery of the MR signal (longitudinal relaxation time, T_1), and decay of the MRI signal (transverse relaxation time, T_2) (Callaghan, 1991). These key parameters depend on the local tissue environment (including tissue microstructure and protein content) and can differ widely between soft tissue structures; thus, generation of images with high contrast between tissues with different relaxation times is the most common method of anatomical imaging.

Emphasizing T_1 variation (termed T_1-weighted imaging) is particularly useful when distinguishing between tissue types (Frank, 2002), as it emphasizes contrast between microstructures, such that water and dense bone appear dark and fats appear bright (Figure 9.8E). For example, T_1-weighted imaging is the standard for acquiring images with gray and white matter contrast in the brain (e.g., Yopak and Frank, 2009; Yopak et al., 2016); distinguishing among cranial structures such as eye, cartilage, and stomach (e.g., Waller et al., 1994); and differentiating muscle (Perry et al., 2007) in cartilaginous fishes. In contrast, emphasizing T_2 variation (termed T_2-weighted imaging) is generally ideal when distinguishing between tissue and fluid, where water is bright and air and fats appear dark (Figure 9.8F). Although this technique is less commonly used in scanning of non-human subjects, Ullman et al. (2010a) used T_2-weighted imaging sequences to develop an MR-based 3D atlas of the zebrafish brain, which increased the SNR per unit time.

In addition, there are seemingly limitless scan parameters to adjust, including slice thickness, echo time (T_E), repetition time (T_R), flip angle, and number of averages, which will all vary depending on the specimen, instrument, available radiofrequency (RF) coils, and the structure of interest. Detailing them all is outside of the scope of this chapter, but we recommend Berquist et al. (2012) and Ziegler et al. (2011) for excellent guides on standard pulse sequence parameters for a range of fish species imaged in a range of instruments.

9.5.2.4   Data Processing, 3D Segmentation, and Visualization

Almost all MRI scanners can output data into the standard medical imaging DICOM formats (DCM files). These files can also be converted into other formats that may be required by different imaging analysis software. There is, then, a large number of options for software that allows the researcher to visualize, process, and segment MRI data, including the freeware ImageJ (http://rsbweb.nih.gov/ij/), OsiriX (http:// www.osirix-viewer.com)), and ITK-SNAP (http://www.itksnap.org/pmwiki/pmwiki.php) (Yushkevich et al., 2006), as well as licensed software such as Amira™ and Avizo® (https://www.fei.com/software/). From the 3D

data, numerous features of internal anatomy can be quantified, including gross descriptions, as well as linear dimensions, surface area, and volumes of structures.

9.5.3   Expectations, Limitations, and Data

Magnetic resonance imaging is a highly versatile technique that provides very high resolution (down to 10 μm) images of soft tissue structures. Unlike CT, MRI does not use any ionizing radiation and is generally favored over CT when either modality could yield the same information. It is ideal for imaging soft tissue structures, as it provides excellent contrast between tissue types, generating up to 250 shades of gray in optimized samples. For neuroanatomical studies, it also has functional capabilities, although these have not been fully explored in marine organisms. Because many images are taken milliseconds apart, it shows how the brain responds to different stimuli (i.e., fMRI) and can measure physiological parameters such as bulk flow and diffusion. However, MRI has several limitations, as well. This technique is more time consuming and expensive than other digital imaging methods such as x-ray and CT, and it requires highly specialized safety and operator training. There are safety concerns when working with such high magnetic fields, and researchers with pacemakers or implanted metal objects are restricted from working with these instruments. Further, live animal work is impacted, as metallic instruments of any kind are not allowed within the room, hindering experiments that may require special equipment.

9.6   SUMMARY AND CONCLUSIONS

The ideal method for anatomical assessment is non-invasive with infinite resolution and complete discrimination among tissue types, and the imaging can be acquired instantaneously. No current methods to date service all of these needs. However, technological advancements in bioimaging, such as ultrasonography, endoscopy, CT, and MRI, are providing remarkable advancements in resolution, enabling new avenues for longitudinal studies, facilitating rapid digital data dissemination, and providing new opportunities for teaching and outreach. Many of these methods also create a permanent 3D digital record of these samples. Digitization of specimens serves a critical role in preserving biological material in perpetuity, and curating and maintaining an online, accessible database facilitates worldwide access to biological collections. Recent advances in the last decade have allowed researchers access to online repositories of digital material, ranging from MR images of invertebrates, such as echinoids (Ziegler et al., 2008; http://www.morphdbase.de), to fishes (e.g., Berquist et al., 2012; http://www.digitalfishlibrary.org), as well as CT of a wide range of specimens, including living and extinct vertebrates (e.g., www.digimorph.org) and a range of chordates (http://morphosource.org), resulting in digital datasets of high taxonomic breadth. In addition to maintaining the

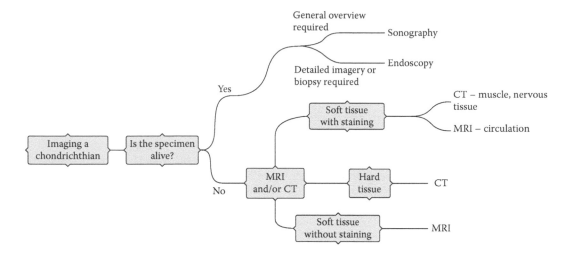

Figure 9.9 A broad overview of the techniques explored in this chapter and the primary specimens and tissue types to which they apply.

physical integrity of specimens, particularly for rare specimens and holotypes, such access provides researchers with a wealth of morphological data, allowing new questions to be asked without the need to collect additional animals. Not all methodologies are ideal for all questions (Figure 9.9). We hope the content provided in this chapter will allow readers to explore the optimal technique for their research, avoid common challenges and pitfalls, and advance our collective understanding of elasmobranch biology.

ACKNOWLEDGMENTS

JCC thanks Dr. Mike Walsh, Dr. Sam Dover, and the veterinary surgical staff and facilities of SeaWorld Adventure Parks in Orlando, Florida, for the use of their surgical facilities and instrumentation for the ultrasound and endoscopy described in this chapter. KEY thanks members of her lab who have contributed feedback to this chapter, particularly E. Peele, and those who assisted with the contribution of figures, including L. Frank and R. Berquist of the Center for Scientific Computation in Imaging (CSCI) at the University of California, San Diego. APS thanks Dr. Matthew Kolmann and Dr. Kyle Newton for iodine contrast staining cartilaginous fishes. APS is funded by NSF (DEB-1701665, IOS-1306718) and the Seaver Institute. Mostly, thank you to anyone who has pushed the field of shark biology, thought beyond the tried and true, and used novel techniques to pave the way for the future of elasmobranch research.

REFERENCES

Awruch CA, Frusher SD, Pankhurst NW, Stevens JD (2008) Non-lethal assessment of reproductive characteristics for management and conservation of sharks. *Mar Ecol Prog Ser* 355:277–285.

Berquist R, Gledhill K, Peterson M, Doan A, Baxter G, Yopak K, Kang N, Walker H, Hastings P, Frank L (2012) The digital fish library: digitizing, databasing, and documenting the morphological diversity of fishes with MRI. *PLoS ONE* 7(4):e34499.

Blackband SJ, Stoskopf MK (1990) *In vivo* nuclear-magnetic-resonance imaging and spectroscopy of aquatic organisms. *Magn Reson Imaging* 8 (2):191–198.

Bock C, Sartoris F, Portner H (2002) *In vivo* MR spectroscopy and MR imaging on non-anaesthetized marine fish: techniques and first results. *Magn Reson Imaging* 20:165–172.

Brown RW, Cheng Y-CN, Haacke EM, Thompson MR, Venkatesan R (eds) (2014) *Magnetic Resonance Imaging: Physical Principles and Sequence Design*, 2nd ed. John Wiley & Sons, New York.

Callaghan P (1991) *Principles of Nuclear Magnetic Resonance Microscopy.* Oxford University Press, Oxford, UK.

Carrier JC, Murru FL, Walsh MT, Pratt HL (2003) Assessing reproductive potential and gestation in nurse sharks (*Ginglymostoma cirratum*) using ultrasonography and endoscopy: an example of bridging the gap between field research and captive studies. *Zoo Biol* 22:179–187.

Chakrabarty P, Davis M, Smith W, Berquist R, Gledhill K, Frank LR, Sparks J (2011) Evolution of the light organ system in ponyfishes (Teleostei: Leiognathidae). *J Morphol* 272:704–721.

Chanet B, Fusellier M, Baudet J, Madec S, Guintard C (2009) No need to open the jar: a comparative study of magnetic resonance imaging results on fresh and alcohol preserved common carps (*Cyprinus carpio* (L. 1758), Cyprinidae, Teleostei). *C R Biol* 332:413–419.

Claeson KM (2008) Variation of the synarcual in the California ray, *Raja inornata* (Elasmobranchii : Rajidae). *Acta Geol Pol* 58:121–126.

Claeson KM (2011) The synarcual cartilage of batoids with emphasis on the synarcual of Rajidae. *J Morphol* 272:1444–1463.

Claeson KM (2013) The impacts of comparative anatomy of electric rays (Batoidea: Torpediniformes) on their systematic hypotheses. *J Morphol* 275:597–612.

Claeson KM, Hilger A (2011) Morphology of the anterior vertebral region in elasmobranchs: special focus, Squatiniformes. *Fossil Rec* 14:129–140.

Claeson KM, Ward DJ, Underwood CJ (2010) 3D digital imaging of a concretion-preserved batoid (Chondrichthyes, Elasmobranchii) from the Turonian (Upper Cretaceous) of Morocco. *C R Palevol* 9:283–287.

Claeson KM, Underwood CJ, Ward DJ (2013) *Tingitanius tenuimandibulus*, a new platyrhinid batoid from the Turonian (Cretaceous) of Morocco and the Cretaceous radiation of the Platyrhinidae. *J Vert Paleontol* 33:1019–1036.

Cloutier R, Schultze H, Wiley E, Musick J, Daimler J, Brown MA, Dwyer III SJ, Cook LT, Laws RL (1988) Recent radiologic imaging techniques for morphological studies of *Latimeria chalumnae*. *Environ Biol Fish* 23:281–282.

Criswell KE, Coates MI, Gillis A (2017) Embryonic development of the axial column in the little skate, *Leucoraja erinacea*. *J Morphol* 278:300–320.

Crow GL, Atkinson MJ, Ron B, Atkinson S, Skillman ADK, Wong GTF (1998) Relationship of water chemistry to serum thyroid hormones in captive sharks with goitres. *Aquat Geochem* 4:469–480.

Daly J, Gunn I, Kirby N, Jones R, Galloway D (2007) Ultrasound examination and behavior scoring of captive broadnose sevengill sharks, *Notorynchus cepedianus* (Peron, 1807). *Zoo Biol* 26:383–395.

Davies TG, Rahman IA, Lautenschlager S, Cunningham JA, Asher RJ, Barrett PM, Bates KT, et al. (2017) Open data and digital morphology. *Proc Biol Sci* 284:20170194.

Dean MN, Motta PJ (2004) Anatomy and functional morphology of the feeding apparatus of the lesser electric ray, *Narcine brasiliensis* (Elasmobranchii : Batoidea). *J Morphol* 262:462–483.

Dean MN, Huber DR, Nance HA (2006) Functional morphology of jaw trabeculation in the lesser electric ray *Narcine brasiliensis*, with comments on the evolution of structural support in the Batoidea. *J Morphol* 267:1137–1146.

Dean MN, Bizzarro JJ, Summers AP (2007) The evolution of cranial design, diet, and feeding mechanisms in batoid fishes. *Int Comp Biol* 47:70–81.

Dean MN, Bizzarro JJ, Clark B, Underwood CJ, Johanson Z (2017) Large batoid fishes frequently consume stingrays despite skeletal damage. *Open Sci* 4:170674.

Elliott I, Skerritt G (2010) *Handbook of Small Animal MRI*. Wiley-Blackwell, Oxford, UK.

Fontanella JE, Fish FE, Barchi EI, Campbell-Malone R, Nicholsa RH, DiNenno NK, Beneskia JT (2013) Two- and three-dimensional geometries of batoids in relation to locomotor mode. *J Exp Mar Biol Ecol* 446:273–281.

Forbes JG, Morris HD, Wang K (2006) Multimodal imaging of the sonic organ of *Porichthys notatus*, the singing midshipman fish. *Magn Reson Imaging* 24:321–331.

Frank L (2002) Magnetic resonance imaging: basic principals. In: Resnick D (ed) *Diagnosis of Bone and Joint Disorders*. WB Saunders, Philadelphia, PA, pp 72–109.

Galloro G, Sivero L, Magno L, Diamantis G, Pastore A, Karagiannopulos P, Inzirillo M, et al. (2007) New technique for endoscopic removal of intragastric balloon placed for treatment of morbid obesity. *Obes Surg* 17:658–662.

Garman S (1913) *The Plagiostomia (Sharks, Skates, and Rays)*, vol XXXVI. Harvard University, Cambridge, MA.

George RH, Steeil J, Baine K (2017) Diagnosis and treatment of common reproductive problems in elasmobranchs. In: Smith M, Warmolts D, Thoney D, Hueter R, Murray M, Ezcurra J (eds) *The Elasmobranch Husbandry Manual II: Recent Advances in the Care of Sharks, Rays and Their Relatives*. Ohio Biological Survey, Columbus, OH, pp 357–362.

Geraghty PT, Jones AS, Stewart J, Macbeth WG (2012) Micro-computed tomography: an alternative method for shark ageing. *J Fish Biol* 80:1292–1299.

Gignac PM, Wegner NC, Miller LA, Jew CJ, Lai NC, Berquist RM, Frank LR, Long JA (2016) Diffusible iodine-based contrast-enhanced computed tomography (diceCT): an emerging tool for rapid, high-resolution, 3-D imaging of metazoan soft tissues. *J Anat* 228:889–909.

Graham JB, Wegner NC, Miller LA, Jew CJ, Lai NC, Berquist RM, Frank LR, Long JA (2014) Spiracular air breathing in polypterid fishes and its implications for aerial respiration in stem tetrapods. *Nat Comm* 5:3022.

Grant KR, Campbell TW, Silver TI, Olea-Popelka FJ (2013) Validation of an ultrasound-guided technique to establish a liver-to-coelom ratio and a comparative analysis of the ratios among acclimated and recently wild-caught southern stingrays, *Dasyatis americana*. *Zoo Biol* 32(1):104–111.

Huber DR, Eason TG, Hueter RE, Motta PJ (2005) Analysis of the bite force and mechanical design of the feeding mechanism of the durophagous horn shark *Heterodontus francisci*. *J Exp Biol* 2008:3553–3571.

Johanson Z, Boisvert C, Maksimenko A, Currie P, Trinajstic K (2005) Development of the synarcual in the elephant sharks (Holocephali; Chondrichthyes): implications for vertebral formation and fusion. *PLoS ONE* 10:e0135138.

Kamminga P, De Bruin PW, Geleijns J, Brazeau MD (2017) X-ray computed tomography library of shark anatomy and lower jaw surface models. *Sci Data* 4:170047.

Kiessling F, Pichler B, Hauff P (2011) *Small Animal Imaging: Basics and Practical Guide*. Springer, Berlin.

Kolmann MA, Huber DR, Dean MN, Grubbs RD (2014) Myological variability in a decoupled skeletal system: batoid cranial anatomy. *J Morphol* 275:862–881.

Kolmann M, Crofts S, Dean M, Summers A, Lovejoy N (2015) Morphology does not predict performance: jaw curvature and prey crushing in durophagous stingrays. *J Exp Biol* 218:3941–3949.

Kolmann MA, Newton K, Summers A (2017) Diffusible Iodine Contrast Enhanced Micro-CT Scanning as a Method to Visualize Soft Tissue Anatomy in Elasmobranch Fishes, paper presented at the Joint Meeting of Herpetologists and Ichthologists, July 12–16, Austin, TX.

Lane JA (2010) Morphology of the braincase in the Cretaceous hybodont shark *Tribodus limae* (Chondrichthyes: Elasmobranchii), based on CT scanning. *Am Mus Novit* 3681:1–70.

Lane JA, Maisey JG (2012) The visceral skeleton and jaw suspension in the durophagous hybodontid shark *Tribodus limae* from the lower Cretaceous of Brazil. *J Paleontol* 86:886–905.

Maisey JG (2001a) CT-scan reveals new cranial features in Devonian chondrichthyan "*Cladodus*" *wildungensis*. *J Vert Paleontol* 21:807–810.

Maisey JG (2001b) Remarks on the inner ear of elasmobranchs and its interpretation from skeletal labyrinth morphology. *J Morphol* 250:236–264.

Maisey JG (2004) Morphology of the braincase in the broadnose sevengill shark *Notorynchus* (Elasmobranchii, Hexanchiformes), based on CT scanning. *Am Mus Novit* 3429:1–52.

Maisey JG (2011) The braincase of the Middle Triassic shark *Acronemus tuberculatus* (Bassani, 1886). *Palaeontology* 54:417–428.

Maisey JG, Lane JA (2010) Labyrinth morphology and the evolution of low-frequency phonoreception in elasmobranchs. *CR Palevol* 9:289–309.

Maisey JG, Springer VG (2013) Chondrocranial morphology of the salmon shark, *Lamna ditropis*, and the porbeagle, *L. nasus* (Lamnidae). *Copeia* 2013:378–389.

Mara KR, Motta PJ, Huber DR (2010) Bite force and performance in the durophagous bonnethead shark, *Sphyrna tiburo*. *J Exp Zool A Ecol Genet Physiol* 313:95–105.

Mathiassen J, Misimi E, Bondo M, Veliyulin E, Ostvik S (2011) Trends in application of imaging technologies to inspection of fish products. *Trends Food Sci Technol* 22:257–275.

Mollen FH (2010) A partial rostrum of the porbeagle shark *Lamna nasus* (Lamniformes, Lamnidae) from the Miocene of the North Sea Basin and the taxonomic importance of rostral morphology in extinct sharks. *Geol Belgica* 13:61–75.

Mollen FH, Jagt JWM (2012) The taxonomic value of rostral nodes of extinct sharks, with comments on previous records of the genus *Lamna* (Lamniformes, Lamnidae) from the Pliocene of Lee Creek Mine, North Carolina (USA). *Acta Geol Pol* 62:117–127.

Motta PJ, Wilga C (1995) Anatomy of the feeding apparatus of the lemon shark, *Negaprion brevirostris*. *J Morphol* 226:309–329.

Moyer JK, Hamilton ND, Seeley RH, Riccio ML, Bemis WE (2015a) Identification of shark teeth (Elasmobranchii: Lamnidae) from a historic fishing station on Smuttynose Island, Maine, using computed tomography imaging. *Northeast Nat* 22:585–597.

Moyer JK, Riccio ML, Bemis WE (2015b) Development and microstructure of tooth histotypes in the blue shark, *Prionace glauca* (Carcharhiniformes: Carcharhinidae) and the great white shark, *Carcharodon carcharias* (Lamniformes: Lamnidae). *J Morphol* 276:797–817.

Murray MI (2010) Endoscopy in sharks. *Vet Clin N Am Exot Anim Pract* 13(2):301–313.

Mylinczenko N, Culpepper EE, Clauss T (2017) Diagnostic imaging of elasmobranchs: updates and case examples. In: Smith M, Warmolts D, Thoney D, Hueter R, Murray M, Ezcurra J (eds) *The Elasmobranch Husbandry Manual II: Recent Advances in the Care of Sharks, Rays and their Relatives*. Ohio Biological Survey, Columbus, OH, pp 303–324.

Nott KP, Evans SD, Hall LD (1999a) The effect of freeze-thawing on the magnetic. resonance imaging parameters of cod and mackerel. *LWT Food Sci Technol* 32:261–268.

Nott KP, Evans SD, Hall LD (1999b) Quantitative magnetic resonance imaging of fresh and frozen-thawed trout. *Magn Reson Imaging* 17:445–455.

Parsons KT, Maisano J, Gregg J, Cotton CF, Latour RJ (2017) Age and growth assessment of western North Atlantic spiny butterfly ray *Gymnura altavela* (L. 1758) using computed tomography of vertebral centra. *Environ Biol Fish* 101(1):137–151.

Pereira N, Batista H, Baylina N (2017) Ultrasound assessment of pregnant ribbontail stingrays, *Taeniura lymma* (Forsskal, 1775). In: Smith M, Warmolts D, Thoney D, Hueter R, Murray M, Ezcurra J (eds) *The Elasmobranch Husbandry Manual II: Recent Advances in the Care of Sharks, Rays and their Relatives*. Ohio Biological Survey, Columbus, OH, pp 325–330.

Perry CN, Cartamil DC, Bernal D, Sepulveda CA, Theilmann RJ, Graham JB, Frank LR (2007) Quantification of red myotomal muscle volume and geometry in the shortfin mako shark (*Isurus oxyrinchus*) and the salmon shark (*Lamna ditropis*), using T1-weighted magnetic resonance imaging. *J Morphol* 268(4):284–292.

Preziosi R, Gridelli S, Borghetti P, Diana A, Parmeggiani A, Fioravanti ML, Marcer F, et al. (2006) Spinal deformity in a sandtiger shark, *Carcharias taurus* Rafinesque: a clinical-pathological study. *J Fish Dis* 29:49–60.

Rasch LJ, Martin KJ, Cooper RL, Metscher BD, Underwood CJ, Fraser GJ (2016) An ancient dental gene set governs development and continuous regeneration of teeth in sharks. *Dev Biol* 415:347–370.

Rogers BL, Lowe CG, Fernández-Juricic E, Frank LR (2008) Utilizing magnetic resonance imaging (MRI) to assess the effects of angling-induced barotrauma on rockfish (*Sebastes*). *Can J Fish Aquat Sci* 65:1245–1249.

Russo RS, Blemker SS, Fish FE, Bart-Smith H (2015) Biomechanical model of batoid (skates and rays) pectoral fins predicts the influence of skeletal structure on fin kinematics: implications for bio-inspired design. *Bioinspir Biomim* 10:046002.

Rygg AD, Cox JPL, Abel R, Webb AG, Smith NB, Craven BA (2013) A computational study of the hydrodynamics in the nasal region of a hammerhead shark (*Sphyrna tudes*): implications for olfaction. *PLoS ONE* 8:e59783.

Schaefer JT, Summers AP (2005) Batoid wing skeletal structure: novel morphologies, mechanical implications, and phylogenetic patterns. *J Morphol* 264:298–313.

Seidel R, Lyons K, Blumer M, Zaslansky P, Fratzl P, Weaver JC, Dean MN (2016) Ultrastructural and developmental features of the tessellated endoskeleton of elasmobranchs (sharks and rays). *J Anat* 229:681–702.

Seidel R, Blumer M, Zaslansky P, Knötel D, Huber DR, Weaver JC, Fratzl P, Omelon S, Bertinetti L, Dean MN (2017) Ultrastructural, material and crystallographic description of endophytic masses—a possible damage response in shark and ray tessellated calcified cartilage. *J Struct Biol* 198:5–18.

Sepulveda CA, Dickson K, Frank LR, Graham JB (2007) Cranial endothermy and a putative brain heater in the most basal tuna species, *Allothunnus fallai*. *J Fish Biol* 70:1720–1733.

Smith MM, Riley A, Fraser GJ, Underwood C, Welten M, Kriwet J, Pfaff C, Johanson Z (2015) Early development of rostrum saw-teeth in a fossil ray tests classical theories of the evolution of vertebrate dentitions. *Proc R Soc Lond B* 282:20151628.

Stetter MD (2004) Diagnostic imaging of elasmobranchs. In: Smith M, Walmolts D, Thoney D, Hueter R (eds) *The Elamosbranch Husbandry Manual: Captive Care of Sharks, Rays, and Their Relatives*. Ohio Biological Survey, Columbus, OH, pp 297–306.

Sulikowski JA, Wheeler CR, Gallagher AJ, Prohaska BK, Langan JA, Hammerschlag N (2016) Seasonal and life-stage variation in the reproductive ecology of a marine apex predator, the tiger shark *Galeocerdo cuvier*, at a protected female-dominated site. *Aquat Biol* 24:175–184.

Summers A (2000) Stiffening the stingray skeleton-an investigation of durophagy in myliobatid stingrays (Chondrichthyes, Batoidea, Myliobatidae). *J Morphol* 243:113–126.

Summers AP, Ketcham RA, Rowe T (2004) Structure and function of the horn shark (*Heterodontus francisci*) cranium through ontogeny: development of a hard prey specialist. *J Morphol* 260:1–12.

Tomita T, Sato K, Suda K, Kawauchi J, Nakaya K (2011) Feeding of the megamouth shark (Pisces: Lamniformes: Megachasmidae) predicted by its hyoid arch: a biomechanical approach. *J Morphol* 272:513–524.

Tyszka J, Fraser S, Jacobs R (2005) Magnetic resonance microscopy: recent advances and applications. *Curr Opin Biotechnol* 16:93–99.

Ullmann JFP, Corwin G, Kurniawan ND, Collin SP (2010a) A three-dimensional digital atlas of the zebrafish brain. *Neuroimage* 51(1):76–82.

Ullmann JFP, Cowin G, Collin SP (2010b) Quantitative assessment of brain volumes in fish: comparison of methodologies. *Brain Behav Evol* 76:261–270.

Ullmann JFP, Cowin G, Collin SP (2010c) Magnetic resonance microscopy of the barramundi (*Lates calcarifer*) brain. *J Morphol* 271:1446–1456.

Underwood C, Johanson Z, Smith MM (2016) Cutting blade dentitions in squaliform sharks form by modification of inherited alternate tooth ordering patterns. *Open Sci* 3:160385.

Underwood CJ, Johanson Z, Welten M, Metscher B, Rasch LJ, Fraser GJ, Smith MM (2015) Development and evolution of dentition pattern and tooth order in the skates and rays (Batoidea; Chondrichthyes). *PLoS ONE* 10:e0122553.

van den Burg E, Peeters R, Verhoye M, Meek J, Flik G, Van der Linden A (2005) Brain responses to ambient temperature fluctuations in fish: reduction of blood volume and initiation of a whole-body stress response. *J Neurophysiol* 93:2849–2855.

Van der Linden A, Van Camp N, Ramos-Cabrer P, Hoehn M (2007) Current status of functional MRI on small animals: application to physiology, pathophysiology, and cognition. *NMR Biomed* 20:522–545.

Waller GNH, Williams SCR, Cookson MJ, Kaldoudi E (1994) Preliminary analysis of elasmobranch tissue using magnetic resonance imaging. *Magn Reson Imaging* 12:535–539.

Walsh MT, Pipers FS, Brendemuehl CA, Murru FL (1993) Ultrasonography as a diagnostic-tool in shark species. *Vet Radiol Ultrasound* 34:213–218.

Welten M, Smith MM, Underwood C, Johanson Z (2015) Evolutionary origins and development of saw-teeth on the sawfish and sawshark rostrum (Elasmobranchii; Chondrichthyes). *Open Sci* 2:150189.

Whittamore JM, Bloomer C, Hanna GM, McCarthy ID (2010) Evaluating ultrasonography as a non-lethal method for the assessment of maturity in oviparous elasmobranchs. *Mar Biol* 157:2613–2624.

Wilga CD, Wainwright PC, Motta PJ (2000) Evolution of jaw depression mechanics in aquatic vertebrates: insights from Chondrichthyes. *Biol J Linnean Soc* 71:165–185.

Wroe S Huber DR, Lowry M, McHenry C, Moreno K, Clausen P, Ferrara TL, Cunningham E, Dean MN, Summers AP (2008) Three-dimensional computer analysis of white shark jaw mechanics: how hard can a great white bite? *J Zool* 276:336–342.

Yong-Hing C, Obenaus A, Stryker R, Tong K, Sarty G (2005) Magnetic resonance imaging and mathematical modeling of progressive formalin fixation of the human brain. *Magn Reson Med* 54:324–332.

Yopak K (2012) Neuroecology in cartilaginous fishes: the functional implications of brain scaling. *J Fish Biol* 80:1968–2023.

Yopak K, Frank LR (2009) Brain size and brain organization of the whale shark, *Rhincodon typus*, using magnetic resonance imaging. *Brain Behav Evol* 74:121–142.

Yopak K, Lisney TJ, Collin SP, Montgomery JC (2007) Variation in brain organization and cerebellar foliation in chondrichthyans: Sharks and holocephalans. *Brain Behav Evol* 69:280–300.

Yopak K, Ainsley SM, Ebert DA, Frank LR (2010) Exploring adaptive evolution in the brains of bathyal skates (Family: Rajidae): phylogenetic and ecological perspectives. *Brain Behav Evol* 75:316.

Yopak K, Galinsky V, Berquist R, Frank L (2016) Quantitative classification of cerebellar foliation in cartilaginous fishes (class: Chondrichthyes) using 3D shape analysis and its implications for evolutionary biology. *Brain Behav Evol* 87:252–264.

Yushkevich PA, Piven J, Cody Hazlett H, Gimpel Smith R, Ho S, Gee JC, Gerig G (2006) User-guided 3D active contour segmentation of anatomical structures: significantly improved efficiency and reliability. *Neuroimage* 31:1116–11128.

Ziegler A, Mueller S (2011) Analysis of freshly fixed and museum invertebrate specimens using high-resolution, high-throughput MRI. In: Schroder L, Faber C (eds) *In Vivo NMR Imaging Methods in Molecular Biology*. Humana Press, New York, pp 633–651.

Ziegler A, Faber C, Mueller S, Bartolomaeus, T (2008). Systematic comparison and reconstruction of sea urchin (Echinoidea) internal anatomy: a novel approach using magnetic resonance imaging. *BMC Biol* 6:33.

Ziegler A, Kunth M, Mueller S, Bock C, Pohmann R, Schröder L, Faber C, Giribet G (2011) Application of magnetic resonance imaging in zoology. *Zoomorphology* 130:227–254.

CHAPTER **10**

History and Mystery of Age and Growth Studies in Elasmobranchs
Common Methods and Room for Improvement

Lisa J. Natanson
Apex Predators Program, National Marine Fisheries Service, Narragansett, Rhode Island

Allen H. Andrews
Pacific Islands Fisheries Science Center, National Marine Fisheries Service, Honolulu, Hawaii

Michelle S. Passerotti
Department of Biological Sciences, University of South Carolina, Columbia, South Carolina

Sabine P. Wintner
KwaZulu-Natal Sharks Board, Umhlanga Rocks, South Africa; Biomedical Resource Unit,
University of KwaZulu-Natal, Durban, South Africa

CONTENTS

10.1 INTRODUCTION

Age determination is a fundamental field of study that is of critical importance to management and conservation of a species (Cortés, 2002; Harry, 2017). Determining population demographics can only be accomplished with the input of age data; yet, despite years of age determination efforts, it continues to be one of the most challenging and nuanced aspects of life history investigation in elasmobranchs. A variety of methods have been used to determine ages in elasmobranchs, some successful and some not. New techniques to obtain accurate and precise ages are being developed, and it is important to recognize the value of both old and new methodologies.

Because elasmobranchs lack otoliths—a well conserved ear stone that is commonly used for teleost age and growth studies—vertebral centra have become the most commonly used hard structure for age determination and in many ways have become somewhat synonymous with age and growth estimation in elasmobranchs. Typically, mineralized band pairs, which are often defined as consisting of one opaque and one translucent band, are counted to determine age (Figure 10.1) (Cailliet, 1990; Cailliet et al., 1986). In the early history, vertebral age reading focused on quantifying the band pairs that were assumed to be annual, although the periodicity of band pair formation has been validated in only a limited number of studies and, in those, a limited portion of the species lifespan (for a summary, see Cailliet, 1990; Cailliet and Goldman, 2004; Goldman, 2005). Vertebral centra cannot be used in some species due to a lack of calcified band pairs or a lack of correspondence between band pairs and time (Hamady et al., 2014; Holden and Meadows, 1962; Kaganovskaia, 1933; Natanson and Cailliet, 1990; Natanson et al., 2008). Other hard structures, such as dorsal fin spines, caudal thorns, and neural arches, have been used with varying success and can provide some level of verification between structures (Cailliet and Goldman, 2004). Additional methods, not related to band pairs or hard structures, have also been used to obtain age and growth data, such as length–frequency analysis and growth data from tagged and recaptured fish (Andrews et al., 2011b; Casey and Natanson, 1992; Dureuil and Worm, 2015; Natanson et al., 2002; Wells et al., 2016).

Extensive reviews have already covered the need for age and growth studies, techniques for enhancing band pairs in vertebral centra, and age-validated species compilations (Cailliet, 1990, 2015; Cailliet and Goldman, 2004; Cailliet et al., 1986, 2006; Goldman et al., 2012). This chapter focuses on the pros and cons of current methods of age determination and validation, explores emerging technologies that go beyond traditional vertebral band pair counting, and more fully examines issues of age estimation using other structures (such as dorsal spines and caudal thorns) and techniques.

10.2 SUBJECTIVITY AND VALIDATION OF AGEING STRUCTURES

The use of a hard part to age a species requires that the structure increase in size in proportion to the body and that it deposit band pairs consistently with increasing age (Lagler, 1952). Because elasmobranchs lack otoliths, other structures

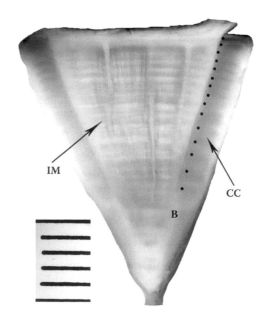

Figure 10.1 Section of a porbeagle (*Lamna nasus*) centra with 5-mm scale and band pairs marked. *Abbreviations:* CC, corpus calcareum; IM, intermedialia; B, birth band.

have been used to estimate age. All of the techniques associated with these structures have positive and negative considerations for their use and each will be discussed separately. However, there are general concerns about counting band pairs in any hard part that must be treated prior to initiating a study. These include the subjectivity of band pair counts (preparation methods, consistency of samples, defined, repeatable criteria, and ager bias), as well as knowledge of the temporal periodicity of the band pairs (validation).

10.2.1 Subjectivity of Band Pair Counts

Assuming that band pair counts on any structure can be used, counts must first be repeatable using a well-defined protocol on both preparation and enumeration criteria to minimize subjectivity. Band pair counts on any growth structure are subject to discrepancies due to structure, preparation methods, and visualization and interpretation of the band pairs themselves. Even with well-defined criteria, their interpretation is based on the subjective eye of the reader, often leading to disparate age estimates for a given specimen. Splits in the intermedialia, the corpus calcareum, or both, as well as double banding patterns, are often confusing and difficult to discern, even with established criteria (Figure 10.2). Most authors use measures to ensure repeatability (e.g., bias analyses, intercalibration, references sets) (Campana, 2001; Goldman et al., 2012), although band pair interpretation is contingent on several factors that include microscope quality and configuration, computer hardware

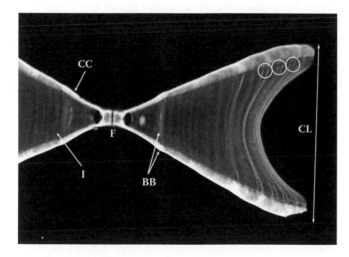

Figure 10.2 X-ray of portion of bow-tie section from a 146-cm female blue shark (*Prionace glauca*). The white circles identify the splits discussed in the text. *Abbreviations:* F, focus; CC, corpus calcareum; I, intermedialia; BB, birth band; CL, centrum length. (Reproduced with permission from Francis, M.P. and Maolagáin, C.O., *Size, Maturity and Length Composition of Blue Sharks Observed in New Zealand Tuna Longline Fisheries*, New Zealand Fisheries Assessment Report 2016/60, Ministry for Primary Industries, Wellington, New Zealand, 2016.)

and image analysis programs, and the variable quality of the centra themselves, with emphasis on reader experience with a particular species. Officer et al. (1996), for example, demonstrated that experienced age readers have higher precision and less bias. Certain species, including the white shark (*Carcharodon carcharias*) and the common thresher shark (*Alopias vulpinus*), are particularly difficult to maintain reader agreement and consistency of the age reading criteria (L.J. Natanson, unpublished data). In the South Pacific, two banding patterns were observed in blue sharks (*Prionace glauca*), and researchers were unable to effectively assign ages to the specimens (Francis and Maolagáin, 2016). This problem affects not only band pair counts but also marginal increment analysis (MIA), a method commonly applied to verify band pair periodicity. MIA relies on precise interpretation of the distal band pair and affects the ability to compare age reading results among studies because the differences may be related to subjectivity rather than real differences in population growth rates (Tanaka, 1990).

Generally, in an ageing study, criteria for counting a band pair are defined prior to initiation of the study. Intercalibration studies and reference sets have become more common tools to help assess and improve the repeatability of both individual counts of one reader (intra-) and counts between two readers (inter-). Reference sets consist of a subsample of specimens of representative sizes whose ages have been agreed upon that readers can recount prior to rereading the entire set to refresh themselves on criteria and maintain quality control (Campana, 2001). Tanaka (1990) was one of the first to address problems with reader bias and precision by exchanging samples between laboratories (intercalibration) and suggested that real differences of growth parameters between populations may be obscured by method and interpretation variability for the blue shark. In addition to intercalibration studies, strict definition of interpretation criteria and examinations of bias and precision have become standard in ageing studies (for details, see Campana, 2001; Goldman et al., 2012; McBride, 2015), although lack of agreement of band pair counts still exists (Francis and Maolagáin, 2016; L.J. Natanson, unpublished data).

10.2.2 Validation

Perhaps the most difficult aspect of age determination is the need for age validation, generally defined as obtaining the absolute age of a fish (Beamish and McFarlane, 1983; Campana, 2001). With respect to elasmobranchs, it is typically the periodicity of band pair formation that is validated for a species; it can then be extended for use as validated criteria for age reading of other conspecifics. However, this must be accomplished over the entire size range of a species because only then can the age of a given fish be considered absolute (Beamish and McFarlane, 1983; Campana, 2001). Although this has been a principle of ageing studies since their initiation, it is often overlooked and rarely accomplished

(Beamish and McFarlane, 1983; Cailliet, 1990, 2015; Cailliet and Goldman, 2004; Cailliet et al., 1986, 2006; Goldman et al., 2012). Due to the increasing numbers of studies showing a limited or lack of an age relationship to vertebral band pair counts, it is important to directly validate every life stage, structure used, and population of a given species. Age validation studies using chemical marking and bomb radiocarbon dating have been variably successful at validating age for some portion of lifespan, but they require careful consideration when applied in conjunction with age determination studies. This is particularly important due to the increase in evidence that age underestimation is occurring in at least 30% of studies that have used band pair counts on vertebral centra (Harry, 2017).

10.2.2.1 Validation Concepts and Missing Time

In 2001, Kalish and Johnston applied bomb radiocarbon (^{14}C) dating to elasmobranchs to validate band pairs of the school shark (*Galeorhinus galeus*); since then, this technique has been used as an age validation tool across a wide range of elasmobranch species and has led to advances in understanding carbon uptake in elasmobranchs (e.g., Andrews et al., 2011b; Kerr et al., 2006; Kneebone et al., 2008; Passerotti et al., 2010). The use of bomb ^{14}C dating has aided validation investigations, and, although several bomb ^{14}C studies have validated annual periodicity throughout life (Ardizzone et al., 2006; Campana et al., 2002; Kneebone et al., 2008), studies are now showing that band pair deposition rate is annual for only a portion of the lifespan in many species. One important result of this technique is that it has highlighted the prevalence of age underestimation in the field. Harry (2017) showed that age underestimation occurred in at least 9 of 29 elasmobranch genera, including 50% of those validated with bomb ^{14}C dating. This has led to a major shift in the mindset of age determination in elasmobranchs. Age underestimation can partly be explained by the concept of "missing time," which is years of life not recorded due to discontinuous band pair deposition (Passerotti et al., 2014).

Irregularity of the band pair deposition could be attributed to variation between or within species, specific life stage events, or a combination of these factors; for example, many species seem to exhibit annual band pair deposition only up to age at maturity (e.g., Andrews et al., 2011b; Natanson et al., 2015). The shortfin mako (*Isurus oxyrinchus*) was validated using bomb ^{14}C dating in the western North Atlantic Ocean (Ardizzone et al., 2006), but oxytetracycline (OTC) validation studies from the Pacific showed an ontogenetic shift from two band pairs per year in early life to one band pair per year later in life (Kinney et al., 2016; Wells et al., 2013). Suggestions for band pair deposition cues include changes in temperature, salinity, photoperiod, season, and annual migrations. None of these factors,

however, has been proven, and in some cases they have even been disproven, such as temperature and salinity in the little skate (*Leucoraja erinacea*) (Natanson, 1993; Sagarese and Frisk, 2010). Little work has been published on the possible triggers for band pair deposition, and it is an area that should be further explored. The cessation of somatic growth (leading to a similar cessation of vertebral growth) has been suggested to account for missing time in the visual age assessments from elasmobranch vertebrae (Francis et al., 2007). Although phase lags observed in the sample chronology of some species can be explained by habitat and dietary sources that are depleted in ^{14}C (e.g., Campana et al., 2002; Kneebone et al., 2008), several well-executed studies have revealed evidence for discontinuous accretion of vertebral band structure. Passerotti et al. (2014) demonstrated that up to 18 years of missing time provided an explanation for an observed offset in the ^{14}C record from the vertebrae of sand tiger sharks (*Carcharias taurus*), but the study also validated annual growth and band pair deposition up to an age of roughly 12 years. This phenomenon was first observed in New Zealand porbeagle (*Lamna nasus*) (Francis et al., 2007) and has been evident in several other shark studies (Andrews and Kerr, 2015; Andrews et al., 2011b; Hamady et al., 2014; Natanson et al., 2014, 2015).

In early growth, the band pairs may be deposited at a rate that coincides with annual band pair deposition, as has been found in many species using OTC validation (Kinney et al., 2016; Natanson et al., 2002; Wells et al., 2013, 2016). As growth slows toward maximum size and the structural requirements of the vertebral centra change, the periodicity of band pair deposition is also bound to change (L.J. Natanson et al., unpublished data). These results corroborate the phenomenon of missing time with increasing age at maximum sizes and could explain both the observed irregularities in band pair deposition and lack of temporal correlation after a certain age or life stage.

10.2.2.2 Validation Methodology

When initiating a validation study, it is important to note that age validation must not be confused with age verification, which is simply the corroboration of indeterminate age estimates for an individual that were produced from different structures or methods—it is simply a confirmation of age estimate consistency and not accuracy (Cailliet, 1990). Validation methods have improved from indirect methods, which include tag–recapture and length–frequency analyses for use in refining age estimation, and MIA for temporal periodicity. These methods are basic comparisons between techniques that have some degree of time specificity or temporal calibration but do not directly assign a year of formation to specific band pairs. Direct age validation is accomplished when a date can be ascribed from a method that is directly correlated to time (Cailliet, 2015; Cailliet and Goldman, 2004; Cailliet et al., 2006; Campana, 2001; Goldman et al., 2012). Methods particularly useful for direct validation include mark–recapture with external tags and chemical markers and, more recently, the use of a time-specific marker of the marine environment (i.e., bomb radiocarbon dating).

10.2.2.3 Chemical Marking

Age validation studies using chemical marking and bomb radiocarbon dating have been variably successful at validating age for some portion of the lifespan but require careful consideration when applied in conjunction with age determination studies. Marking of vertebral centra with fluorochrome markers was first used in elasmobranchs by Holden and Vince (1973), who adapted the technique from teleost studies and tagged, injected with OTC, and released 16 *Raja clavata* into the wild. In addition to OTC, calcein and alizarin red S have been applied to elasmobranchs (Officer et al., 1997). Tetracycline is the most common fluorochrome marker used and has proven to be a successful validation tool in at least 22 studies (for studies up to 2003, see Cailliet and Goldman, 2004; Kinney et al., 2016; Natanson et al., 2006; Wells et al., 2013, 2016). These studies have shown results including annual periodicity, annual periodicity in juveniles changing to biannual in adults, and no temporal periodicity (Kinney et al., 2016; Natanson and Cailliet, 1990; Natanson et al., 2002; Wells et al., 2013, 2016). Recent examples of successful OTC validation include the porbeagle, shortfin mako, and blue shark (Kinney et al., 2016; Natanson et al., 2002; Wells et al., 2013, 2016). Calcein has been used successfully in at least six studies (Gelsclechter et al., 1997; Gutterridge et al., 2013; Harry et al., 2013; McAuley et al., 2006; Officer et al., 1997; Pierce and Bennett, 2009), and alizarin red S use is less common (Officer et al., 1997).

Injection (or immersion) of fishes with a chemical is a standard technique used to mark sites of active calcification in hard parts, such as otoliths in teleost fishes and vertebral centra in elasmobranchs (Campana, 2001; Gelsleichter et al., 1997; Kinney et al., 2016; Natanson et al., 2002; Officer et al., 1997). Dosage of the chemical depends on the marker, although 25 mg/kg has been used for OTC and 5 to 25 mg/kg has been used for calcein (McFarlane and Beamish, 1987; Officer et al., 1997; Pierce and Bennett, 2009). These chemicals permanently bind to the calcium in the growth structure at, or close to, the time of injection to create a temporal marker (Officer et al., 1997). This marker can then be used at some time in the future, when the fish is sacrificed, to provide a time reference to the growth that is past the chemical mark (Figure 10.3). One benefit of calcein is that the mark can be viewed directly, whereas long-wave ultraviolet (UV) light is required to view the OTC mark. This type of study can be conducted in a live-rearing tank or used in conjunction with a tag–recapture program. Live rearing provides

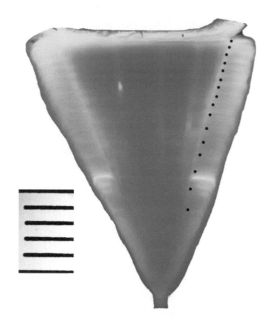

Figure 10.3 The section of a porbeagle (*Lamna nasus*) centra shown in Figure 10.1 with 5-mm scale and band pairs marked taken under ultraviolet light to show the yellow oxytetracycline mark.

control over the fish's environment and the time period between injection and sacrifice, which allows for regulated experimental techniques, but aquarium growth may not accurately reflect growth in the wild. Tag–recapture results represent natural growth, but large numbers of tagged individuals are required to provide few returns.

10.2.2.4 Bomb Radiocarbon

Bomb radiocarbon dating has become a widely adopted standard for validating other age estimation methods in elasmobranchs. The method relies on an anthropogenic rise in atmospheric ^{14}C that resulted from the testing of thermonuclear devices in the 1950s and 1960s; the ^{14}C that diffused into the marine system can be found in the tissues of various marine organisms (e.g., Broecker and Peng, 1982; Druffel and Linick, 1978). The rapid increase in ^{14}C during this period was virtually synchronous for organisms inhabiting shallow marine environments (Campana, 1999; Druffel, 2002; Grottoli and Eakin, 2007), producing a record in known-age tissues that can be used as a reference for validating estimates of age (Andrews et al., 2011a; Kalish 1993). As a result, a number of studies have used this method to validate—and in some cases invalidate—estimates of age, growth, and longevity for elasmobranchs (e.g., Andrews and Kerr, 2015; Campana et al., 2002; Passerotti et al., 2014; for a comprehensive list to date, see Harry, 2017). At present, its application is largely limited to vertebrae, but the use of material extracted from other conserved structures is being explored.

10.2.2.4.1 Sample Requirements

An age estimate and known collection date for vertebral samples is of primary importance when attempting bomb radiocarbon dating because of the potential constraints it places on the dates of formation of a band pair within a given shark vertebrae. Although a single shark vertebrae could be analyzed (via a series of extracted samples from oldest to youngest growth), a better scenario is a series of larger adults whose ages have been estimated to include birth through the ^{14}C rise period; the most informative period is typically from the mid-1950s through the 1960s. The best-case scenario is to add known-age juvenile specimens (or at least younger sharks where age is far less in question) to the adult sample series that cover the most informative period for verification of adult measurement alignment (e.g., Passerotti et al., 2014). Under these circumstances, only archived material will provide a basis for such a study. In cases where underestimated age is suspected for potentially long-lived sharks, more recent collection dates can be explored to determine if the informative period was captured during the formation of the vertebrae (Andrews et al., 2011b; Hamady et al., 2014; Natanson et al., 2014).

For each specimen, it is important to consider the number of growth bands sampled for ^{14}C analyses to increase the chances of success. Ideally, a series of samples from within each shark vertebrae provides individual ^{14}C chronologies to guide the temporal alignment with regional ^{14}C references. This results in numerous samples per individual, perhaps one every few years of estimated growth. Alternatively, a series of specimens from a range of collection years (ranging back several decades) can provide an alignment if the assigned ages are well constrained (e.g., a series of adults collected over several decades with known collection years). In all of these scenarios, the goal is to locate a measurement or series of measurements in vertebral material that can be correlated to a regional ^{14}C reference record in order to provide a dated reference for comparison of estimated age.

Often unstated in the literature is the requirement that vertebral samples analyzed for ^{14}C must be free of exposure to ^{14}C-tracer levels or operations (Zermeño et al., 2004). Because samples can be contaminated by instruments or analysis facilities (e.g., samples should not be handled or stored in laboratories that investigate primary productivity using a ^{14}C tracer), it is important to know the history and handling environment of the tissues to be analyzed. Every accelerator mass spectrometry (AMS) facility offers swipe test sample analyses to determine if the potential exists, and it is essential for any ^{14}C analysis to use this technique.

Vertebral samples are best stored frozen in a standard freezer after extraction from the shark. Samples are then later cleaned and dried for long-term storage. Storage in ethanol is generally acceptable but may cause unknown changes to the original sample matrix over long periods of time.

10.2.2.4.2 Reference Chronologies

Another requirement for successful bomb ^{14}C dating is a regional reference chronology that may or may not be geographically specific for comparison of material measurements. Because the atmospheric ^{14}C produced from nuclear testing was circulated and absorbed differently in various marine environments (e.g., rapid diffusion into surface layers vs. delayed diffusion into deeper waters), the timing and magnitude of ^{14}C in reference chronologies will vary due to location, water depth and mixing, and pre-bomb levels (Druffel, 2002). Hence, when selecting an appropriate ^{14}C reference, one needs to consider habitat through ontogeny of the shark species in question and how it may be similar or different from the reference. In addition, the uptake and sources of the regional reference must be considered. In many cases, hermatypic coral records have been used for species of tropical and subtropical waters. These colonies exhibit rapid uptake of dissolved inorganic carbon (DIC) directly from the water column, whereas the primary source of carbon to the vertebrae of elasmobranchs is dietary carbon from prey items (Fry, 1988). Thus, the cumulative effect of shark habitat and the primary habitat of its prey will determine how closely the timing and magnitude of the bomb ^{14}C record in the vertebrae will be to the reference record over time. For example, great hammerhead sharks (*Sphyrna mokarran*) and sand tiger sharks—both of which feed predominantly on prey from well-mixed, shallow-water environments—exhibited vertebral ^{14}C chronologies that closely mirrored both the timing and magnitude of coral ^{14}C reference chronologies (Passerotti et al., 2010, 2014). Conversely, species undergoing long migrations, extensive vertical movements, or both (e.g., porbeagle, white shark) will almost certainly be consuming prey from a mix of water depths. This kind of uptake would lead to variable or attenuated vertebral ^{14}C levels because of an uptake of carbon from prey that grows in the ^{14}C-depleted depths. As a result, the ^{14}C uptake pattern through ontogeny may be phase shifted, attenuated, or both relative to the timing and magnitude of mixed layer references (e.g., coral or known-age otolith material) (Campana et al., 2002; Hamady et al., 2014; Kerr et al., 2006; McPhie and Campana, 2009). In cases like these, it may be most informative to create a species-specific reference chronology using samples from younger, known-age sharks spanning the period of initial rise in marine ^{14}C, as exemplified in studies of the porbeagle (Campana et al., 2002) and sand tiger shark (Passerotti et al., 2014). In other circumstances, where no good regional reference is available and the life history of the species is highly variable, assumptions must be made that can place constraints on age for what may be a more accurate age estimate compared to other methods that have no temporal constraints (e.g., counting growth bands) (Andrews and Kerr, 2015). Overall, the availability and choice of a regional ^{14}C reference and its applicability to the elasmobranch in question will ultimately determine the resolution and conclusiveness of results (e.g., King et al., 2017; McPhie and Campana, 2009), and they must be considered in concert with sample collection dates and potential birth years when designing a bomb ^{14}C study.

10.2.2.4.3 Sample Extraction

Methods for extracting material from shark vertebrae have advantages and disadvantages relative to the use of otoliths for teleost fishes. Otoliths are typically limited to an extraction of the core material (earliest growth of a few months or up to perhaps a year of carbonate accretion) because the mass growth is spread over a surface area that is increasing with age, such that the annual ring width decreases significantly toward the edge of the otolith and usually precludes sampling through ontogeny. Shark vertebrae—when sectioned to reveal the inner structure of the corpus calcareum (Figure 10.1 and 10.4)—also exhibit an attenuation in the thickness of the material accreted for each successive year, but the amount of material overall is greater and provides an opportunity to measure the uptake of ^{14}C over the life of the shark. With a scalpel, a small block of the centra is usually removed from an area of the corpus calcareum that may hold a particular date in the time series. Usually, several dates are selected to cover the estimated lifespan which are then used to describe the uptake of ^{14}C during this time. A similar situation can happen with a micromill where a block is extracted by drilling around the area of interest (Figure 10.4). In each case, the amount of carbon must be calculated to assess the ability of AMS to detect the ^{14}C signal; for example, a minimum of 2 μmol carbon is required for combustion of small samples at the National Ocean Sciences AMS facility at the Woods Hole Oceanographic Institution. Within this sampling design, early growth can be sampled with greater temporal resolution because the band width is greater than in the older growth; in most cases, it has been necessary to sample several years of growth to acquire enough material for the AMS.

10.2.2.4.4 Sample Processing

The method used to process extracted vertebral material has differed among studies. Early work explored pretreatment of samples because it was thought that the organic component had to be isolated from inorganic carbon. The hypothesis was that the inorganic component would be more mobile or may have been accreted later in life to strengthen the structure and the material would not be time specific, whereas the organic material (collagen) would be a conserved organic structure representing carbon uptake at the time of formation for that part of the vertebra (Andrews et al., 2011b). However, because the principle inorganic component (hydroxyapatite [$Ca_{10}(PO_4)_6(OH)_2$]) does not contain carbon, there should be no difference between treated and

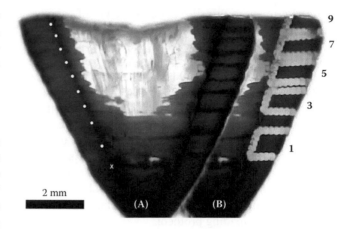

Figure 10.4 Example of an aged vertebrae section from a sand tiger shark (*Carcharias taurus*) showing the corpus calcareum (CC) of a 9-year-old shark. (A) The initial section, prior to extraction with the micromill, provides the band pair counts (white dots on left) and birth mark (X) along with the region of the section to be removed delineated with pencil lines (right side). (B) Post-extraction on the right side provides a visual alignment with the initial section image shown in (A), with the small blocks ready for removal after the micromill traced out the age-specific sample areas. Denoted are ages 1, 3, 5, 7, and 9 years, which were later analyzed for ^{14}C in Passerotti et al. (2014). Within this sample series was a record of ^{14}C from 1957 to 1965, a diagnostic period for the region of study.

untreated samples. A preliminary test between such samples, however, revealed there may be an effect (Kerr et al., 2006). The samples were not true replicates but a split of the extracted block from the same region of the vertebrae. No further work has been performed in this regard, except that the liberation of CO_2 from vertebral material as a result of acidification was too low to acquire measurable ^{14}C levels (S. Campana, University of Iceland, pers. comm.), providing an indication that the carbonate component is negligible.

The primary approach used is to extract small samples from whole vertebrae and submit them for combustion AMS analysis, where carbon is released from the matrix as CO_2. No results from other species have shown problems that could be attributed to some level of isotopic fractionation between materials, but the uncertainty warrants further investigation. Hence, frontier science using laser ablation AMS on carbonates may lead to developing a better way to analyze vertebrae in an extraction process that is a continuous flow of CO_2 to the AMS, as opposed to a series of separate samples and the multistep graphitization process (Welte et al., 2016).

10.2.2.4.5 Analysis

The standards and methods associated with the determination of ^{14}C measurements from AMS analyses is beyond the scope of this chapter and is described in detail elsewhere (e.g., Stuiver and Polach, 1977; Passerotti et al.,

2014). Pertinent to this discussion is the use of appropriate reference records and how to correlate the ^{14}C values with time using the ^{14}C references. Typically, the measured ^{14}C value is given a year of formation (YOF) based on a temporal alignment. Radiocarbon values are typically reported as D^{14}C, which is a time-corrected value, but use of F^{14}C is increasing in age validation of fishes because there is no need to correct for differences in time relative to 1950 (Reimer et al., 2004). Regardless of the system used, care must be taken when assigning the YOF to the samples. In some cases, it is simply a plot of measured ^{14}C value relative to the YOF from the assigned age based on band counting. This is typically the initial scenario because the aim is to validate an age reading protocol.

Reconstruction of ^{14}C sample chronologies is nuanced and must account for regional variations in the dispersal of atmospheric ^{14}C, which will affect results, reinforcing the need for a proximate reference chronology. Current patterns, upwelling, and other waterbody characteristics are also issues that must be considered when analyzing results. Any source of depleted ^{14}C introduced via factors such as migratory patterns, depth profiles, and corresponding variations in diet of the study species can cause ^{14}C values to vary widely. Likewise, any ontogenetic changes to these influences will add variation to the ^{14}C chronology. In cases where the expected relationship of the sample chronology to the reference is unknown (i.e., potential phase lag and attenuation), considerations must be given to the uptake carbon sources for the species throughout time. In some instances, the porbeagle reference from Campana et al. (2002) has been used as a rough proxy for what the ^{14}C reference may look like for the species in question (Ardizzone et al., 2006; Francis et al., 2007; Kerr et al., 2006). Hence, the uncertainty between a timely ^{14}C rise (rapid rise documented from regional carbonate records, such as hermatypic coral, from the mixed layer) and an attenuated and possibly phase-lagged record must be considered and may lead to undesirable age uncertainty (e.g., Andrews and Kerr, 2015). The experimental design is important in minimizing these uncertainties and should include the incorporation of multiple specimens with well-constrained ages. However, it is likely that there will always be some necessary assumptions that will lead to low-resolution constraints on age, as opposed to the more desirable validated age with high precision that is typical of teleost fish studies (e.g., Andrews et al., 2016).

10.2.2.4.6 Limitations Associated with Bomb Radiocarbon Dating

Although bomb radiocarbon dating is a powerful age validation tool and can establish previously unknown estimates of longevity, there are limits to its applicability. In most cases, locating temporally appropriate archival vertebrae can be very difficult or nearly impossible, especially for species that were not historically subject to large-scale fishery or management activity. In some cases, finding one or two vertebrae collected during the informative period is considered fortunate, much less multiple vertebrae from a range of ages and locations for each species. In this regard, it is the time-specific rising ^{14}C signal (late 1950s to perhaps early 1970s, at the latest) that places the greatest limit on a study of this kind. In some studies of tropical teleost fishes, the declining ^{14}C signal has provided an opportunity to validate the age of younger and more recently collected fish (e.g., bluespine unicornfish, *Naso unicornis*) (Andrews et al., 2016). In contrast, carbon sources to the otolith of teleost fishes are from DIC and are more time related than that of vertebrae formed from dietary sources. Hence, it is likely that the decline period—typically a lower slope and post-1980 in the tropics—would not be a clean signal in the vertebrae due to a formational and trophic level mixing of the ^{14}C signal. Some studies provide some evidence for this complication, but there may be other temporal constraints that could provide estimates of age (e.g., hawksbill sea turtle, *Eretmochelys imbricata*) (Van Houtan et al., 2016).

In terms of cost, use of organic sample combustion for AMS analysis at dedicated laboratories can run into the hundreds of dollars per sample (arrangements vary throughout the world), which does not include the time and expense for specimen collection, growth band milling, and the preparation and shipment of the samples for processing (NOSAMS, 2018). Given the number of samples typically necessary for an informative study, use of bomb radiocarbon dating can become cost prohibitive.

10.3 AGEING STRUCTURES

10.3.1 Vertebral Centra

By far the most common structure used in estimating elasmobranch age is the vertebral centrum. Ages for many species have been successfully determined using vertebral band pair counts, as confirmed with chemical and/or bomb carbon validation (for comprehensive discussions of these studies, see Cailliet, 2015; Harry, 2017). Annual band pair periodicity has been confirmed for some species such as the tiger shark (*Galeocerdo cuvier*) and North Atlantic populations of shortfin mako and porbeagle (Ardizzone et al., 2006; Campana et al., 2002; Kneebone et al., 2008). Some species change from a predictable periodicity to a nonpredictable periodicity such as the common thresher, white, and sand tiger sharks (Natanson et al., 2015; Passerotti et al., 2014), whereas in the Pacific angel shark (*Squatina californica*) no relationship has been found between band pair deposition and time (Natanson and Cailliet, 1990). Additionally, band pair deposition among populations of the same species has been shown to differ (Campana et al., 2002; Francis et al., 2007; Natanson et al., 2002). All of these results support the oft-repeated call for validation (Beamish and McFarlane,

1983). Numerous guides to vertebral preparation, age reading, and data analysis have been produced over the years (e.g., Cailliet and Goldman, 2004; Goldman et al., 2012), and the basic processing and analysis methods have not varied significantly over time. Thus, we limit the remainder of discussion of centra to new findings that have called the use of this structure into question.

10.3.1.1 Validation of Vertebral Centra

Recent improvements to validation methodologies are calling into question ages derived from vertebral band counts (see Section 10.2.2.2); however, the concept of validation for each structure and species through ontogeny has been well known (Beamish and McFarlane, 1983). The need for validation in elasmobranchs was directly demonstrated in the early 1990s when two OTC validation studies showed that Pacific angel sharks and several wobbegong (*Orectolobus* spp.) species deposited band pairs relative to somatic growth rather than time, thus confirming that band pairs were not always related to age (Natanson and Cailliet, 1990; Tanaka, 1990). Further studies on a variety of species, including the gummy shark (*Mustelus antarcticus*), school shark, and basking shark (*Cetorhinus maximus*), have shown the relationship of band pairs to somatic growth rather than time (Chidlow et al., 2007; Huveneers et al., 2013; Natanson et al., 2008; Officer et al., 1996). In other species, a shift of the bomb ^{14}C results in time to match the reference chronology using bomb radiocarbon validation revealed underestimation of age for large adults (Andrews et al., 2011b; Francis et al., 2007; Natanson et al., 2014; Passerotti et al., 2014).

10.3.1.2 Limitations Associated with Vertebral Age Estimation

Several well-conducted studies with validation have shown annual band pair deposition in elasmobranch species, but others have shown discontinuous or non-time-related deposition rates. Research on a variety of species has shown that band pair counts vary along the vertebral column (Chidlow et al., 2007; Huveneers et al., 2013; Natanson and Cailliet, 1990; Natanson et al., 2008; Officer et al., 1996; Tribuzio et al., 2018). If these differences are real—not due to reading error or difficulty in reading sections from head or tail centra—then it begs the question of which of the band pair counts are annual, if any? New research suggests that the band pairs may be related to the structure of the vertebral column and the growth in girth rather than age (Natanson, et al., in press). If band pairs are linked to structural support and body girth, which changes along the column and through ontogeny, then growth bands would be deposited relative to their function and may only coincide in relation to time; although this can achieve the same result, it must be kept in mind that the deposition will change with growth rate and thus the deposition rate will change (Natanson et

al., in press). Results of some bomb ^{14}C studies appear to corroborate the change in band pair deposition rate with ontogeny. Given these caveats, the further use of vertebral band pair counts for age determination is questionable and must be undertaken with diligence. Additionally, researchers in age determination need to explore other nonstructural hard parts together with novel age estimation and validation techniques, as well as the processes of band pair formation in all structures.

10.3.2 Dorsal Spines

The use of fin spines in age estimation of sharks has not been well explored in past reviews of age reading methods. Because not all elasmobranchs have dorsal spines, these studies are limited to species in the orders Squaliformes, Heterodontiformes, and Chimaeriformes. In spine-bearing species, such as the family Squalidae (common name dogfishes), the spines are often the only hard part that can be used for band pair counting because vertebral centra are poorly calcified (Holden and Meadows, 1962; Kaganovskaia, 1933; Ketchen, 1975). In cases where both vertebrae and spines are suitable for age reading, the use of spines offers a unique ability to corroborate counts in both structures. Given the emphasis of this chapter on alternative structures and methods for age reading, it is important to expand on the use of spines to age these poorly understood species. Here, we use standardized terminology for fin spine age reading (Clarke and Irvine, 2006). In spiny dogfish (*Squalus acanthias, S. suckleyi*), the second dorsal spine is used most often for age determination because the spine is generally wider and has more evenly spaced bands, and there is usually less wear (loss of the tip from abrasion) than in the first dorsal spine (Holden and Meadows, 1962; Ketchen, 1975). Other species of spine-bearing families also have poorly calcified vertebral centra, so spine studies have expanded to include species of chimera, such as *Centrophorus, Centroselachus,* and *Etmopterus* (Barnett, 2008; Calis et al., 2005; Freer and Griffiths, 1993; Irvine et al., 2006a,b; Tanaka, 1990).

10.3.2.1 Dorsal Spine Methodology

Methods used for the examination of band pairs in spines vary based on species because of structural differences, and either whole spines or sections (cross or longitudinal) are used. Whole spines can be wet sanded and polished to enhance viewing of band pairs (Campana et al., 2009; Tribuzio et al., 2016). In holocephalans, cross-sectioning the spine must be done with care to ensure that the sections encompass the most informative areas of the spine (i.e., where band pairs are most numerous due to differences in growth of internal and external bands). Sullivan (1977) showed that band pairs in the elephant fish (*Callorhinchus milii*) form on the innermost area of the spine; thus, the upper portion (10 mm from the tip) of the spine would contain the most band pairs, and they are thought to be a complete life history record. Other studies have refined these regions, and new studies must include a determination of the best place for cross-sectioning based on the species being aged (Barnett, 2008; Calis et al., 2005).

10.3.2.2 Validation of Dorsal Spines

Traditional methods of age validation are difficult or impossible to perform on spines of many of these species—particularly those that inhabit the deep sea—and it has been suggested that alternative methods should be investigated (Barnett, 2008; Irvine et al., 2006a,b). Validation of annual band pair periodicity for spines has been accomplished for the spiny dogfish using both OTC injection and bomb ^{14}C methodologies (Beamish and McFarlane, 1983; Campana et al., 2006; McFarlane and King, 2006). Barnett (2008) attempted to use OTC injection on the spotted ratfish (*Hydrolagus colliei*), but there was no detectible uptake of the chemical into the spine. Additionally, it was not possible to section at the same point on each spine relative to growth zones due to individual variation, thus eliminating any kind of edge analysis such as MIA (Barnett, 2008). The problem is that the direction of growth in spines is analogous to stacked cones, where the period of formation varies as the location of the section changes along the growth axis. Radiometric dating—methods that rely on the decay of radioactive elements such as lead or radium—are not applicable due to the organic composition of the spine itself which leads to too many potential violations of the necessary uptake and structural conservation assumptions (Cotton et al., 2014). Spines are not expected to be a closed system for these radionuclides (e.g., radium-226), and constant uptake is unlikely. In addition, exogenous lead-210 was found to be high and too variable for lead-210 dating (a form of lead–radium dating) (Cotton et al., 2014). Given these difficulties, OTC and bomb ^{14}C seem the most viable options for validating the annual periodicity in spines, but other methods should continue to be sought for these poorly understood elasmobranchs. In species other than the spiny dogfish, where spine ages have been validated and vertebral ages have not, care needs to be taken in selecting the most appropriate structure to obtain age estimates. Additionally, validation studies are necessary to determine whether internal or external spine counts are more accurate.

10.3.2.3 Limitations Associated with Dorsal Spine Age Estimation

Beyond the fact that most sharks lack fin spines, there are additional caveats to their use as primary age reading structures. These include loss of material due to fin spine wear and differences between internal and external band pair counts. Studies on a variety of species show that external spines wear away over time, leading to a loss of band pairs

and necessitating correction factors that may or may not be accurate (Ketchen, 1975; Sullivan, 1977). Additional studies have compared internal and external band counts and demonstrated lower band counts internally vs. externally, as well as suggesting that internal and external bands are independent and differ in method of formation (Irvine et al., 2006a,b). Recent improvements in vertebral band pair enhancement techniques have allowed researchers to compare band pair deposition between spines and vertebral centra (Bubley et al., 2012; Tribuzio et al., 2018). Bubley et al. (2012) used both structures to age the spiny dogfish (*Squalus acanthias*) in the Gulf of Maine region of the western North Atlantic (WNA) and found that vertebrae were more consistent and provided more biologically realistic age estimates than fin spines. In contrast, Campana et al. (2006) validated age in spiny dogfish (*S. acanthias*) spines from the Canadian region of the WNA using bomb radiocarbon and determined that the fin spine estimates were annual up to at least 25 years. Further, a study comparing the two structures on spiny dogfish (*S. suckleyi*) in the Pacific Ocean found fewer band pairs present in the centra (Tribuzio et al., 2016), indicating that vertebral counts underestimated age, which was supported by validated fin spine ages up to 52 years using bomb radiocarbon dating (Figure 10.5) (Campana et al., 2006). Tribuzio et al. (2016) also noticed a difference in band pair count along the vertebral column and suggested that, because spines had been validated with bomb ^{14}C and did not agree with the centra estimates, the latter was not an appropriate substitute for spines. The inconsistent results between studies species and locations make it difficult to draw broad conclusions on these structures. Although the lack of spines in most species is limiting and there are unresolved issue with this structure, spine age estimation should not be dismissed as a promising tool.

10.3.3 Caudal Thorns

Although the growth of placoid scales is not unlimited (Meyer and Seegers, 2012), a derivative structure, the caudal thorn, has been used as a novel approach to age determination in an attempt to better understand batoid life history (Cailliet and Goldman, 2004). In general, thorns are found in a number of batoids of the order Rajiformes and in some families in the order Rhinopristiformes (Ebert and Stehmann, 2013; Last et al., 2016). Thorns can be broadly categorized into alar (paired patches of thorns on the outer disc of most mature male skates), malar (patches of thorns beside the eyes of many mature male skates), and generalized or caudal thorns (along the back and over the vertebral column) (Meyer and Seegers, 2012). Of the three thorn types, only caudal thorns can be used for age estimation. Caudal thorns have a curved, pyramidal shape with a basal plate, a middle section called the peduncle, and the apex or crown (Kemp, 1999). Bands were shown to form in a stacked fashion under the protothorn, where the base of each successive band becomes an

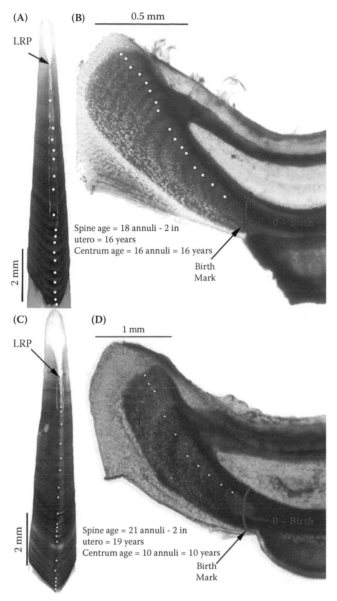

Spine age = 18 annuli - 2 in utero = 16 years
Centrum age = 16 annuli = 16 years
Birth Mark

Spine age = 21 annuli - 2 in utero = 19 years
Centrum age = 10 annuli = 10 years
Birth Mark

Figure 10.5 Comparison of the spine and vertebral methods of ageing on the same specimen of a spiny dogfish (*Squalus suckleyi*), highlighting the different band pair counts between the two structures. (Photographs courtesy of Cindy Tribuzio.)

Caudal thorn tip

Protothorn →

Figure 10.6 Image of a whole caudal thorn from a sandpaper skate (*Bathraja kincaidii*) showing the alternating opaque and translucent bands below the protothorn and tip. Inset is a top view of a smaller thorn revealing another way to see the growth bands. Despite the promising appearance of the thorns for ageing this species, the counts from vertebral sections indicated that age may be underestimated using thorns (Perez et al., 2011). (Photographs courtesy of Colleena Perez-Brazen.)

the presumed nonlethal acquisition of the material and that extraction does not decrease the commercial value of the specimen (Moura et al., 2007). In addition, thorns can be removed at tagging and recapture for age validation or when information is required for depleted stocks (Gallagher et al., 2006). The potential short- and long-term effects of thorn removal, however, should be studied in more detail (Matta and Gunderson, 2007).

10.3.3.1 Caudal Thorn Methodology

Most thorn preparations start by using trypsin for cleaning purposes with varying concentrations and soak times (Gallagher and Nolan, 1999). Additionally, brushes, scalpels, forceps, etc. have been used to remove tissue from the caudal thorns (Ebert et al., 2007; Francis and Maolagáin, 2005; James, 2011); cleaned structures have been stored dry or in solution (Henderson et al., 2004; Gallagher et al., 2006; Moura et al., 2007). Methods to enhance the resolution of the banding pattern include etching in ethylenediaminetetraacetic acid (EDTA) in combination with silver nitrate staining, crystal violet staining, or x-radiography (Ebert et al., 2007; Francis and Maolagáin, 2001; Gallagher and Nolan, 1999; Henderson et al., 2004). Caudal thorn band pairs are counted under a dissecting scope, viewing the thorn from below the base of the protothorn (embryonic thorn), which is situated at the apex of the thorn (Gallagher and Nolan, 1999). The protothorn base is considered the birthmark

external ridge (Figure 10.6). Hence, the most internal band is the most recent (Gallagher, 2000; Gallagher et al., 2002). Gallagher and Nolan (1999) were the first to report the suitability of thorns for age estimation in four *Bathyraja* species and to date at least 24 studies have used caudal thorns, most of them in conjunction with age estimation from vertebrae and various methods were successful (Table 10.1). Thorns have a number of advantages over vertebral centra in their potential use as an ageing tool; for example, thorns are easier to collect in the field and require less storage space (Matta and Gunderson, 2007). The major novel aspect of thorns is

Table 10.1   Summary of Thorn Ageing Studies to Date

Species	Common Name	Vertebrae	Suitability of Thorn Method	Problem	Edge Analysis	Validation	Refs.

Order Rajiformes, Family Rajidae, Subfamily Rajinae

Species	Common Name	Vertebrae	Suitability of Thorn Method	Problem	Edge Analysis	Validation	Refs.
Genus Raja							
R. clavata	Thornback ray	X	U	Faint, variable band pattern	—	—	Gallagher (2000)
		X	S	Some thorns not suitable	X	—	Serra-Pereira et al. (2008)
		X	S	Some thorns not suitable	—	—	Serra-Pereira et al. (2005)
R. stellulata	Starry skate	X	U	No band pattern	—	—	James et al. (2014)
		—	U	No band pattern	—	—	James (2011)
R. undulata	Undulate ray	X	S	Some thorns not suitable	—	—	Moura et al. (2007)
R. montagui	Spotted ray	X	U	No band pattern	—	—	Gallagher (2000)
R. brachyura	Blonde ray	X	U	No band pattern	—	—	Gallagher (2000)
Genus Leucoraja							
L. naevus	Cuckoo ray	X	U	No band pattern	—	—	Gallagher (2000)
Genus Amblyraja							
A. georgiana	Antarctic skate	X	S	First band difficult; OC	—	—	Francis and Maolagáin (2001)
		X	S	First band difficult; poor precision	—	—	Francis and Maolagáin (2005)
		—	S	Novel counting method	—	—	Francis and Gallagher (2009)
A. radiata	Thorny skate	X	S	Slight OC	—	—	Gallagher et al. (2006)

Subfamily Arhynchobatinae

Species	Common Name	Vertebrae	Suitability of Thorn Method	Problem	Edge Analysis	Validation	Refs.
Genus Bathyraja							
B. albomaculata	White-spotted skate	X	S	OC	X	—	Gallagher and Nolan (1999)
		—	S	—	X	X[a]	Gallagher (2000); Henderson et al. (2004)
B. aleutica	Aleutian skate	X	S	Poor precision; OC	—	—	Ebert et al. (2007)
B. brachyurops	Broadnose skate	X	S	OC males	X	X[a]	Gallagher and Nolan (1999)
		—	S	—	X	X[a]	Gallagher (2000)
		X	S	—	—	X[a]	Gallagher et al. (2002)
		—	S	—	—	X[a]	Gallagher et al. (2005)
B. eatonii	Eaton's skate	X	S	First band difficult	—	—	Francis and Maolagáin (2001)
B. griseocauda	Graytail skate	X	S	OC males	X	X[a]	Gallagher and Nolan (1999)
		—	S	—	X	X[a]	Gallagher (2000)
B. interrupta	Bering skate	X	U	UC; inconstistent band pattern	—	—	Ainsley (2009)
		X	S	OC	—	—	Ebert et al. (2007)
B. kincaidii	Sandpaper skate	X	U	UC and OC; poor precision	—	—	Perez et al. (2011)
B. lindbergi	Commander skate	X	U	UC	—	—	Maurer (2009)
		X	S	UC	—	—	Ebert et al. (2009)
B. maculata	Whiteblotched skate	X	U	UC	—	—	Maurer (2009)
		X	S	UC	—	—	Ebert et al. (2009)

Table 10.1 (continued)   Summary of Thorn Ageing Studies to Date

Species	Common Name	Vertebrae	Suitability of Thorn Method	Problem	Edge Analysis	Validation	Refs.
Genus Bathyraja (continued)							
B. minispinosa	Whitebrow skate	X	U	UC	—	—	Ainsley et al. (2011)
		X	U	UC	—	—	Ebert et al. (2009)
B. parmifera	Alaska skate	X	S	—	X	—	Matta and Gunderson (2007)
B. scaphiops	Cuphead skate	X	S	OC	X	X[a]	Gallagher and Nolan (1999)
		—	S	—	X	X[a]	Gallagher (2000)
B. taranetzi	Mud skate	X	S	—	—	—	Ebert et al. (2009)
B. trachura	Roughtail skate	X	U	Poor precision; variable band pattern; UC	—	—	Davis et al. (2007)
		X	U	Poor precision; UC	—	—	Winton (2011)
Genus Atlantoraja							
A. castelnaui	Spotback skate	—	—	No ageing but bands visible	—	—	Rangel et al. (2016)
A. cyclophora	Eyespot skate	—	—	No ageing but bands visible	—	—	Rangel et al. (2016)

Note: X, the technique was used; S, the method was suitable; U, the method was unsuitable; UC, undercounting; OC, overcounting.

[a] Annual periodiciy assumed based on OTC validation of one specimen of a similar species.

(Figure 10.6), which has been confirmed for two species by examination of thorns from hatchling skates (Gallagher et al., 2006; Henderson et al., 2004). Band pair definitions range from concentric ridges separated by an associated broad growth zone on the surface of the thorn (Ebert et al., 2007; Gallagher et al., 2006) to the traditional vertebral translucent and opaque band pair (Matta and Gunderson, 2007; Serra-Pereira et al., 2008). In the latter case, a band pair has been defined as a translucent band with a ridge-like appearance, extending around the thorn and accompanied by an opaque band (Matta and Gunderson, 2007).

10.3.3.2   Validation of Caudal Thorns

Although several studies have used edge analysis in an attempt to validate thorn band periodicity, some were hampered by low sample size or the availability of sufficient samples from across the year. Few studies have successfully determined thorn band periodicity (Table 10.1) (Matta and Gunderson, 2007; Serra-Pereira et al., 2008). Data on thorn ageing have proven variable, ranging from presumably annual patterns to no pattern. Edge analysis and the OTC position on the centra of one recaptured *Bathyraja* spp. individual indicated that the narrow ridge formation occurs during the slower growth phase in caudal thorns (Gallagher and Nolan, 1999). Further research showed that "broad surface bands formed annually during periods of rapid somatic growth" and surface ridges represented a "near stasis in somatic growth formed at the periphery of each cone" (Gallagher et al., 2002, 2005). In

contrast, no banding pattern was observed on three Irish skate species (Gallagher, 2000), which was attributed to the fact that somatic growth cessation, which causes the surface ridges, is more pronounced in slower growing, deepwater species (Gallagher et al., 2005).

10.3.3.3   Limitations Associated with Caudal Thorn Age Estimation

Thorns present challenges similar to those encountered with other hard structures, including variation (sometimes non-linear) with body size, position on the body, and time of formation relative to birth. Additional challenges come from resolution of banding patterns in such small, fine structures and the need for validation. Certain thorns are not suitable for age estimation despite high count agreement and precision between age readers (Moura et al., 2007; Serra-Pereira et al., 2005). Additionally, the cellular development of the different thorn types has been shown to be asynchronous, and the appearance of caudal thorns from different caudal regions vary according to specimen size; thus, there is a need to verify the adequacy of the different thorn types as age reading structures (Moura et al., 2007). Few studies investigated if the thorn was present at the time of hatching (Moura et al., 2007) or explicitly state this assumption, together with the assumption that they are not replaced during the lifespan (Ebert et al., 2009; Maurer, 2009). Furthermore, allowance should be made for the possibility of bands being formed inside the egg case (Francis and Maolagáin, 2001).

Some studies have chosen thorns over vertebrae (Serra-Pereira et al., 2008), but others have dismissed thorns for age estimation (Table 10.1) (Gallagher, 2000; James et al., 2014). Both thorns and vertebral centra have been considered options in some species (Table 10.1) (Moura et al., 2007). Variation has also been seen in the count precision of thorns and vertebral centra, with some studies showing similar precision (Francis and Maolagáin, 2001; Gallagher and Nolan, 1999; Matta and Gunderson, 2007) and some showing higher precision for thorns (Gallagher et al., 2006; Moura et al., 2007) depending on species (Table 10.1). Although other studies reported poor count precision for thorns (Davis et al., 2007; Francis and Maolagáin, 2005; Perez et al., 2011), others reported overcounting in thorns when compared to vertebrae, but in most cases this was attributed to a possible undercounting of vertebra (Ebert et al., 2007; Francis and Maolagáin, 2001; Gallagher and Nolan, 1999). Other studies report undercounting, but, irrespective of age, reading precision and verification are necessary for proper validation of estimates from either structure.

There can also be uncertainty within the same species depending on the study. For example, Ebert et al. (2007) did not rule out the possibility of thorn use for reliable age reading in the Bering skate (*Bathyraja interrupta*), but later Ainsley (2009) reported that using thorns did not appear to be reliable. Similarly, Gallagher (2000) was unsuccessful with estimating age by finding faint and variable thorn band patterns for the Irish thornback ray (*Raja clavata*); yet, Serra-Pereira et al. (2008) were successful using thorns in the same species from Portugal (Table 10.1). All of these limitations vary within and between species, suggesting that caution should be used when initiating thorn sampling on a new species.

Additional issues with thorn use concern the relationship of the thorn dimensions and batoid dimensions, as some studies reported nonlinear relationships (Ebert et al., 2007, 2009; Matta and Gunderson, 2007). Others have shown that thorn growth may not be related to somatic growth (Perez et al., 2011), that it slows in relation to somatic growth (Gallagher et al., 2005), and that thorns are subject to wear and tear that would affect the banding pattern (Ainsley et al., 2011; Maurer, 2009).

Band pair counting challenges and subjectivity issues are the same as in any hard structure and include deciding where to begin counting, determining criteria for band pairs, and encountering wear-related or split bands and poor thorn surface sculpture (Francis and Gallagher, 2009; Francis and Maolagáin, 2001, 2005; Gallagher and Nolan, 1999; Gallagher et al., 2002, 2005).

There is clearly a need for more validation studies, both in the laboratory and field. Identifying cues for thorn band formation, especially when there are regional differences, seasonal variation in stable isotope composition, and cellular structure changes with different somatic growth stages, as well as the analysis of banding structure using histology,

electron microscopy, laser ablation, electron microprobe, and x-ray analysis in conjunction with captive rearing trials, are all methods that could be explored on thorns (Francis and Gallagher, 2009; Gallagher et al., 2006). Together with bomb radiocarbon dating—successfully applied to skate vertebrae by McPhie and Campana (2009)—and near-infrared spectroscopy, as discussed elsewhere in this book (see Chapter 11), using thorns for age estimation of batoids should be explored further and effective use of the structure tested on a species-by-species basis.

10.3.4 Other Structures

The possibilities of using other cartilaginous structures for age determination in elasmobranchs has been explored to a limited degree. Many structures have been shown to be calcified, and some—including the jaws, neural and haemal arches, and pectoral and pelvic girdles—have been examined as possible structures for age estimation that may warrant further exploration (e.g., Campbell, 2010; Clement, 1992; McFarlane et al., 2002; Tanaka, 1990).

10.3.4.1 Other Structure Methodology

Tanaka (1990) examined band pairs in the jaw, teeth, and pelvic and pectoral girdles of the Japanese wobbegong (*Orectolobus japonicas*) and the swell shark (*Cephaloscyllium umbratile*). Three clear growth bands were found in the upper jaw of one wobbegong, and although OTC was at the edge of some of the other structures no non-vertebral band pairs were found in the swell shark or Japanese wobbegong samples. More recently, neural arches were examined for use in the bluntnose sixgill shark (*Hexanchus griseus*) because vertebral centra are poorly calcified and unsuitable for age reading (Campbell, 2010; McFarlane et al., 2002). Neural arches were sectioned dorsoventrally and then stained with silver nitrate to examine calcium deposits (McFarlane et al., 2002). Distinct band pairs were visible at regular intervals and band pairs increased with the size of the shark, making this a promising method for age estimation of this species and possibly for others with poorly calcified centra (McFarlane et al., 2002). The pectoral girdle and mesopterygia (the middle of three principal basal fins cartilages) of the bluntnose sixgill were also examined using sections and stains, as well as an elemental analysis (Campbell, 2010). Although these techniques provided limited success, some banding was observed, indicating that future work is warranted (Campbell, 2010).

10.3.4.2 Validation of Other Structures

To date, validation has not been performed on alternative structures and would be required before adopting these structures for age reading (Campbell, 2010; McFarlane et al., 2002).

10.3.4.3 Limitations Associated with Age Estimation in Other Structures

One of the primary problems with the other structures reviewed here is that the pattern of calcification differs from that of the vertebral centra. Much of the calcified tissue in an elasmobranch is mineralized in a mosaic pattern, as opposed to the periodic mineralization unique to the vertebral column (Clement, 1992; Dean and Summers, 2006). Hence, the mosaic pattern is unlikely to be suitable for observing periodic growth in the form of a band. However, as noted in the previous sections, in some cases banding has been observed to a degree, and developing methods of elucidating the band pairs with validation of the periodicity is warranted. Investigations into the use of alternative ageing structures should be intensified considering the increased information indicating that vertebral-based ages are unreliable in some species.

10.4 NON-BAND-PAIR-DEPENDENT METHODS

Several techniques are available to obtain age estimates for elasmobranchs without the use of hard parts, many of which have been used as verification tools. As hard part analysis becomes more questionable, reliance on these independent methods may increase, warranting re-examination and improvement upon current methods.

10.4.1 Tag–Recapture

Tag–recapture methods have been successfully used to generate growth parameters for a variety of shark species, including the sandbar shark (*Carcharhinus plumbeus*), dusky shark (*C. obscurus*), and porbeagle (*Lamna nasus*) (Figure 10.7) (Casey and Natanson, 1992; Dureuil and Worm, 2015;

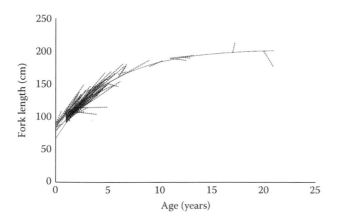

Figure 10.7 Example of a tag and recapture growth curve calculated using the Francis (1988) method (GROTAG). Lines are data used to obtain the curve. Age at tagging is estimated from size based on vertebral growth curve, and age at recapture was calculated using time at liberty.

Natanson et al., 2002; Simpfendorfer, 2000). The different models are relatively simple to use, and their results can be compared to determine which model provides the most biologically meaningful results. There can be uncertainty about which model predicts growth most appropriately due to limited and variable data (Dureuil and Worm, 2015). A new method for evaluating growth from tag–recapture data was developed using a selection procedure to identify the best model for the dataset under consideration, thus obtaining the most applicable growth parameters for the species (Dureuil and Worm, 2015).

Tag–recapture data can be difficult to obtain. The use of tag and recapture data for age estimates requires a large investment in time and money. Specimens must be accurately measured at the time of both tagging and release to be useful, and a large number of individuals must be released to get adequate returns. Large tagging programs report a recapture rate of <5% for over half the species tagged (Kohler and Turner, 2001). Using these data, growth estimates can be generated using a number of methods (Dureuil and Worm, 2015; Fabens, 1965; Francis, 1988; Gulland and Holt, 1959; James, 1991; Simpfendorfer, 2000; Wang, 1998). In conjunction with chemical marking, validation of vertebral centra can also be accomplished using tag–recapture data. Several species have been successfully validated using this technique (e.g., Kinney et al., 2016; Natanson et al., 2002, 2006; Officer et al., 1997; Pierce and Bennett, 2009; Wells et al., 2013).

10.4.2 Laboratory Growth and Captive Rearing

Laboratory growth and captive rearing can provide direct estimates of growth and growth rate, and in conjunction with chemical marking can provide validation of band pair periodicity. Typically, animals are measured at the beginning and end of the study period and sometimes at intervals in between. Often, captive rearing takes place at public aquaria where the primary goal is display and the secondary goal is research. In these cases, the environmental parameters follow protocols for maintaining optimal animal health. In an experimental laboratory rearing situation, care should be taken to mimic natural conditions, if possible, and to avoid unknown effects of uncontrolled environmental parameters on growth. In addition, it must be considered that captivity itself may affect individual growth.

Laboratory growth studies using OTC have been used in several cases to obtain information on band pair periodicity and the causes for deposition. Branstetter (1987) determined annual band pair periodicity using this technique on juvenile Atlantic sharpnose sharks (*Rhizoprionodon terraenovae*), sandbar sharks (*Carcharhinus plumbeus*), and blacktip sharks (*Carcharhinus limbatus*). Tanaka (1990) found mixed results in the Japanese wobbegong, suggesting that band pair deposition had a stronger relationship to centrum growth, as was discovered for the Pacific angel shark

(Natanson and Cailliet, 1990). Natanson (1993) found that two ovulating female little skates did not deposit annual band pairs and suggested this was due to reproductive activity; however, using lab rearing and OTC, it has been found that large specimens of both sexes stop depositing annual band pairs (K.C. James, pers. comm.). Experiments to alter band pair periodicity using temperature and salinity variables failed to change band pair deposition in the little skate (Natanson, 1993; Sagarese and Frisk, 2010). Although care needs to be taken in applying results from captive rearing to wild growth, this kind of study can provide useful information as tools for verification and validation of elasmobranch age and growth.

10.4.3 Length–Frequency Analyses

Length–frequency or size mode analyses are often used to verify ages obtained from band pair counting in vertebral centra or recapture data (Bishop et al., 2006; Natanson et al., 2002; Pratt and Casey, 1983; Wells et al., 2016). The method has been used with varying success and is best suited for juvenile sharks because age modes or classes become indistinguishable due to decreased growth rate of older animals as they approach maximum size (Figure 10.8). Computer programs, however, can assist with identifying all modes in an unbiased manner to provide growth curve estimates (Bishop et al., 2006; Natanson et al., 2002; Wells et al., 2016), which can then be used to verify those obtained from other methods. Recent studies on teleosts have highlighted concerns

with using length–frequency models to assign age without validation of age at size. Deepwater redfish (*Sebastes mentella*), Acadian redfish (*S. fasciatus*), and bluespine unicornfish (*Naso unicornis*) all display growth that is decoupled from age as determined by bomb ^{14}C validation (Andrews et al., 2016; Campana et al., 2016). Campana et al. (2016) found near cessation of growth in both deepwater and Acadian redfish after sexual maturation, resulting in age ranges of 15 to 50 years for fish of 38-cm length of either species. Andrews et al. (2016) found similar challenges with the use of length to predict age because bluespine unicornfish exhibit rapid and variable early growth, with instances of fish near maximum size at 4 years while longevity exceeds 50 years. Similar circumstances are likely for elasmobranchs considering the recent findings of underestimated longevity for large-bodied shark species and other elasmobranchs, such as the sandpaper skate, where fish near maximum length may be ~4 to 20 years old, and the little skate, which ceases band pair deposition over time (Perez et al., 2011; K.C. James, pers. comm.).

10.5 NEW FRONTIERS IN AGE AND GROWTH STUDIES

Due to the issues associated with band pair deposition highlighted by bomb ^{14}C dating and other recent studies, it is evident that new age estimation and validation techniques must be explored and developed, such as use of alternative structures that do not rely on band growth.

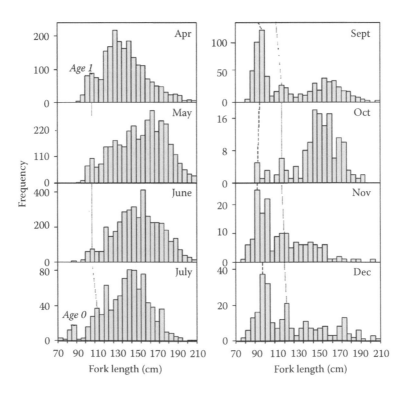

Figure 10.8 Example of length–frequency modes for juvenile porbeagle (*Lamna nasus*).

10.5.1 Eye Lenses in Age Validation

The recent use of bomb and pre-bomb radiocarbon dating in the eye lens of Greenland sharks (*Somniosus microcephalus*) has shown some promise in determining the age of what may be a very long-lived species (Nielsen et al., 2016). Use of radiocarbon in eye lens proteins is likely a robust measure of radiocarbon levels near birth because of a conservation of this tissue over time (e.g., George et al., 1999). For the Greenland sharks studied, isotopic data indicated maternal diet as the primary carbon source for the eye lens nucleus, providing an indication that the material was as old as the individual. Because Greenland sharks lack calcified tissues, no growth-band-derived age estimates were available for comparison with radiocarbon results from the eye lens core. Most of the lenses revealed radiocarbon levels that could be attributed to the pre-bomb period, where ^{14}C levels decrease slowly with time and are not as time specific as bomb ^{14}C levels due to the atmospheric testing of thermonuclear devices in the 1950s and 1960s. Hence, the authors used a correlation of ^{14}C-derived ages to the measured fish lengths and made the assumption of a positive size/age relationship to derive the record longevity of 392 ± 120 years for this species (Nielsen et al., 2016).

A caveat for a study that uses pre-bomb ^{14}C levels (true radiocarbon dating) (Libby, 1955) to age marine organisms is that it is necessary to have a proper understanding of the carbon sources to the material because ^{14}C levels are affected by the age of the carbon sources. Consequently, use of ^{14}C dating requires a correction for reservoir age (ΔR) in marine systems to compensate for the age of the environment, along with knowledge of natural ^{14}C variations due to biological fractionation (Stuiver and Polach, 1977; Stuiver et al., 1986; Wanamaker et al., 2008). These unknowns can be estimated or measured by proxy using other regional records, but the uncertainty and necessary assumptions can lead to a birth year range estimate that spans several centuries, including modern time (post-1950). Hence, the age of an organism that may have been born in the last several hundred years typically cannot be defined by the ^{14}C measurement alone. Evidence for challenges in this regard can be seen with vertebral ^{14}C measurements in porbeagle and sand tiger sharks (Campana et al., 2002; Passerotti et al., 2014). Each of these species exhibits uptake of pre-bomb ^{14}C levels that could be dated to several hundred years BP, but known limitations on growth from other observations preclude these potentially great ages. Ages approaching and even exceeding 200 years have been estimated for cold-water or deepwater teleosts (e.g., Andrews et al., 2009; Cailliet et al., 2001), and it is important to note that these fish were not necessarily the largest, as age was often decoupled from maximum length (e.g., yelloweye rockfish, *Sebastes ruberrimus*) (Andrews et al., 2002). Hence, use of growth modeling that assumes that fish size is proportional to age to provide further constraints on ^{14}C age estimates must be considered with caution.

Eye lenses may also be useful in another type of age analysis. Eye lenses are among the first tissues to form during embryonic development, and they accrete layers throughout the life cycle; the nucleus becomes metabolically inert after formation (Bada et al., 1980). Amino acid molecules exist in tissues as optical stereoisomers, initially formed as L-isomers which then convert to D-isomers at a species-specific rate that is constant and a result of the aging process (Helfman and Bada, 1975). The conversion rate is dependent on a host of external factors, including temperature and pH (Arany and Ohtani, 2010), and most accurately approximates aging in metabolically stable tissues that have not undergone protein synthesis since the time of formation. Thus, measurement of the ratio of D- to L-enantiomers of amino acids in eye lens nuclei, coupled with the conversion rate constant and D/L ratio at birth, can provide an estimate of age. Detection of aspartic acid stereoisomers is traditionally carried out using either gas chromatography or high-performance liquid chromatography (HPLC), but Pleskatch et al. (2016) used tandem mass spectrometry (HPLC–MS/MS) on eye lenses to evaluate beluga whale (*Delphinapterus leucas*) age and found the method provided better results than traditional gas chromatography or HPLC. Use of aspartic acid racemization to estimate age via eye lenses has been demonstrated as effective in several types of marine mammals (Garde et al., 2007, 2010, 2012; George et al., 1999; Nielsen et al., 2013; Rosa et al., 2013; Wetzel et al., 2014) and is under investigation in Greenland sharks (Nielsen and Nielsen, 2012). This is a promising new approach to age estimation in elasmobranchs, but there are limitations to its application. Species-specific racemization rate constants must be calculated from known-age individuals; hence, the ability for D/L ratios to predict age is only as accurate as the calibration curve, and there is no actual measure of time. Best-case scenarios for future study would involve specimens whose ages are independently verified or validated via another method. Because elasmobranch studies tend to be lacking in validated ages, this could prove challenging to accomplish.

10.5.2 Near-Infrared Spectroscopy Technology

Another relatively new method is the use of near-infrared spectroscopy (NIRS) to detect age-related changes in elasmobranch tissues (see Chapter 11 in this volume) via vertebrae, spines, and fin tissue (Rigby et al., 2014, 2016). It is not currently clear, however, what age-related processes are being detected by this method. Some evidence suggests that changes in elemental composition of centra resulting from growth band deposition are the underlying cause of the NIRS/age correlation. Thus far in elasmobranchs, the relationship ostensibly falls off after ~10 years of age for vertebrae (Rigby et al., 2016), which may reflect growth discontinuity or a lack of detectable accretion in older sharks (e.g., Passerotti et al., 2014). The relationship, however, was good to 25 and 31 years of age for dorsal fin spines from two species of spurdogs (Rigby et al., 2014), although ages in these spines are currently not validated. As with acid racemization techniques, the NIRS technique relies upon known ages of tissues to form the correlation model, indicating that validated ages are necessary in order to generate the most accurate results.

10.6 SUMMARY AND CONCLUSIONS

As work on age validation of elasmobranchs has progressed, it has become apparent that for some species vertebral band pairs either are not related to time or are only related to time for a specific period of life (e.g., Andrews and Kerr, 2015; Francis et al., 2007; Harry, 2017; Natanson and Cailliet, 1990; Natanson et al., 2008). In most cases, the conclusions showed distinct underestimation of age for the species studied based on bomb carbon validation (Andrews et al., 2011a,b; Francis et al., 2007; Harry, 2017; Harry et al., 2013; Natanson et al., 2014; Passerotti et al., 2014). Due to the importance of age in determining stock status and management strategies—coupled with the effect underestimated age can have on management decisions (Harry, 2017; Tyler et al., 1989; Yule et al., 2008)—determining the basis of band pair formation in vertebral centra must be part of the next step in the exploration of elasmobranch age and growth. But perhaps of equal importance is the discovery of new methods for determining elasmobranch age that are not related to band pair counting.

ACKNOWLEDGMENTS

We thank all the scientists, technicians, and students who have dedicated time and energy to elasmobranch age and growth research since its inception. We particularly thank C. Brazen (Perez) for providing caudal thorn images, C. Tribuzio for providing dorsal spine and vertebral images, and M. Francis for providing a vertebral image.

REFERENCES

Ainsley SM (2009) Age, Growth and Reproduction of the Bering Skate, *Bathyraja interrupta* (Gill & Townsend, 1897), from the Eastern Bering Sea and Gulf of Alaska, master's thesis, California State University, Monterey Bay.

Ainsley SM, Ebert DA, Cailliet GM (2011) Age, growth, and maturity of the whitebrow skate, *Bathyraja minispinosa*, from the eastern Bering Sea. *ICES J Mar Sci* 68:1426–1434.

Andrews AH, Kerr LA (2015) Estimates of maximum age for white sharks of the northeastern Pacific Ocean: altered perceptions of vertebral growth shed light on complicated bomb $\Delta^{14}C$ results. *Environ Biol Fish* 98: 971–978.

Andrews AH, Cailliet GM, Coale KH, Munk KM, Mahoney MM, O'Connell VM (2002) Radiometric age validation of the yelloweye rockfish (*Sebastes ruberrimus*) from southeastern Alaska. *Mar Freshwater Res* 53:139–146.

Andrews AH, Tracey DM, Dunn MR (2009) Lead-radium dating of orange roughy (*Hoplostethus altanticus*): validation of a centenarian life span. *Can J Fish Aquat Sci* 66:1130–1140.

Andrews AH, Kalish KM, Newman SJ, Johnston JM (2011a) Bomb radiocarbon dating of three important reef-fish species using Indo-Pacific $\Delta^{14}C$ chronologies. *Mar Freshwater Res* 62:1259–1269.

Andrews AH, Natanson LJ, Kerr LA, Burgess GH, Cailliet GM (2011b) Bomb radiocarbon and tag-recapture dating of sandbar shark (*Carcharhinus plumbeus*). *Fish Bull* 109:454–465.

Andrews AH, DeMartini EE, Eble JA, Taylor BM, Lou DC, Humphreys RL (2016) Age and growth of bluespine unicornfish (*Naso unicornis*): a half-century life-span for a keystone browser, with a novel approach to bomb radiocarbon dating in the Hawaiian Islands. *Can J Fish Aquat Sci* 73:1575–1586.

Arany S, Ohtani S (2010) Age estimation by racemization method in teeth: application of aspartic acid, glutamate, and alanine. *J Foren Sci* 55:701–705.

Ardizzone D, Cailliet GM, Natanson LJ, Andrews AH, Kerr LA, Brown TA (2006) Application of bomb radiocarbon chronologies to shortfin mako (*Isurus oxyrinchus*) age validation. *Environ Biol Fish* 25:355–366.

Bada JL, Brown S, Masters PM (1980) Age determination of marine mammals based on aspartic acid racemization in the teeth and lens nucleus. *Rep IWC Special Issue* 3:113–118.

Barnett LAK (2008) Life History, Abundance, and Distribution of the Spotted Ratfish, *Hydrolagus colliei*, master's thesis, California State University, Monterey Bay.

Beamish RJ, MacFarlane GA (1983) The forgotten requirement for age validation in fisheries biology. *Can J Fish Aquat Sci* 38:982–983.

Bishop SDH, Francis MP, Duffy C, Montgomery JC (2006) Age, growth, maturity, longevity and natural mortality of the shortfin mako shark (*Isurus oxyrinchus*) in New Zealand waters. *Mar Freshwater Res* 57:143–154.

Branstetter S (1987) Age and growth validation of newborn sharks held in laboratory aquaria, with comments on the life history of the Atlantic sharpnose shark, *Rhizoprionodon terraenovae*. *Copeia* 1987:291–300.

Broecker WS, Peng T-H (1982) *Tracers in the Sea*. Lamont–Doherty Geological Observatory, Palisades, NY.

Bubley WJ, Kneebone J, Sulikowski JA, Tsang PCW (2012) Reassessment of spiny dogfish *Squalus acanthias* age and growth using vertebrae and dorsal-fin spines. *J Fish Biol* 80:1300-1319.

Cailliet GM (1990) Elasmobranch age determination and verification: an updated review. In: Pratt Jr HL, Gruber SH, Taniuchi T (eds) *Elasmobranchs as Living Resources: Advances in the Biology, Ecology, Systematics, and the Status of the Fisheries*. National Oceanic and Atmospheric Administration, Silver Spring, MD, pp 157–165.

Cailliet GM (2015) Perspectives on elasmobranch life-history studies: a focus on age validation and relevance to fishery management. *J Fish Biol* 87:1271–1292.

Cailliet GM, Goldman KJ (2004) Age determination and validation in chondrichthyan fishes. In: Carrier JC, Musick JA, Heithaus MR (eds) *Biology of Sharks and Their Relatives*. CRC Press, Boca Raton, FL, pp 399–447.

Cailliet GM, Radtke RL, Welden BA (1986) Elasmobranch age determination and verification: a review. In: Uyeno T, Arai R, Taniuchi T, Matsuura K (eds) *Indo-Pacific Fish Biology: Proceedings of the Second International Conference on Indo-Pacific Fishes*. Ichthyological Society of Japan, Tokyo, pp 345–359.

Cailliet GM, Andrews AH, Burton EJ, Watters DL, Kline DE, Ferry-Graham LA (2001) Age determination and validation studies of marine fishes: do deep-dwellers live longer? *Exp Gerontol* 36:739–764.

Cailliet GM, Smith WD, Mollet HF, Goldman KJ (2006) Chondrichthyan growth studies: an updated review, stressing terminology, sample size sufficiency, validation, and curve fitting. In: Carlson JK, Goldman KJ (eds) *Age and Growth of Chondrichthyan Fishes: New Methods, Techniques and Analysis*. Springer, Dordrecht, pp 211–228.

Calis E, Jackson EH, Nolan CP, Jeal F (2005) Preliminary age and growth estimates of the rabbitfish, *Chimaera monstrosa*, with implications for future resource management. *J Northw Atl Fish Sci* 35:15–26.

Campana SE (1999) Chemistry and composition of fish otoliths: pathways, mechanisms and applications. *Mar Ecol Prog Ser* 188:263–297.

Campana SE (2001) Accuracy, precision and quality control in age determination, including a review of the use and abuse of age validation methods. *J Fish Biol* 59:197–242.

Campana SE, Natanson LJ, Myklevoll S (2002) Bomb dating and age determination of large pelagic sharks. *Can J Fish Aquat Sci* 59:450–455.

Campana SE, Jones C, McFarlane GA, Myklevoll S (2006) Bomb dating and age validation using the spines of spiny dogfish (*Squalus acanthias*). *Environ Biol Fish* 77:327–336.

Campana SE, Joyce W, Kulka DW (2009) Growth and reproduction of spiny dogfish off the Eastern coast of Canada, including inferences on stock structure. In: Gallucci VF, McFarlane GA, Bargmann GG (eds) *Biology and Management of Dogfish Sharks*. American Fisheries Society, Bethesda, MD, pp 195–207.

Campana SE, Valentin AE, MacLellan SE, Groot JB (2016) Image-enhanced burnt otoliths, bomb radiocarbon and the growth dynamics of redfish (*Sebastes mentella* and *S. fasciatus*) off the eastern coast of Canada. *Mar Freshwater Res* 67:925–936.

Campbell SJ (2010) Age Determination of the Sixgill Shark from Hard Parts Using a Series of Traditional and Novel Approaches, master's thesis, Western Washington University, Bellingham.

Casey, JG, Natanson LJ (1992) Revised estimates of age and growth of the sandbar shark (*Carcharhinus plumbeus*) from the western North Atlantic. *Can J Fish Aquat Sci* 49:1474–1477.

Chidlow JA, Simpfendorfer CA, Russ GR (2007) Variable growth band deposition leads to age and growth uncertainty in the western wobbegong shark, *Orectolobus hutchinsi*. *Mar Freshwater Res* 58:856–865.

Clarke MW, Irvine SB (2006) Terminology for the ageing of chondrichthyan fish using dorsal-fin spines. *Environ Biol Fish* 77:273–277.

Clement JG (1992) Re-examination of the fine structure of endoskeletal mineralization in chondrichthyans: implications for growth, ageing and calcium homeostasis. *Aust J Mar Freshw Res* 43:157–181.

Cortés E (2002) Incorporating uncertainty into demographic modeling: application to shark population and their conservation. *Cons Biol* 16:1050–1062.

Cotton CF, Andrews AH, Cailliet GM, Dean Grubbs R, Irvine SB, Musick JA (2014) Assessment of radiometric dating for age validation of deep-water dogfish (Order: Squaliformes) finspines. *Fish Res* 151:107–113.

Davis CD, Cailliet GM, Ebert DA (2007) Age and growth of the roughtail skate *Bathyraja trachura* (Gilbert 1892) from the eastern North Pacific. *Environ Biol Fish* 80:325–336.

Dean MN, Summers AP (2006) Mineralized cartilage in the skeleton of chondrichthyan fishes. *Zoology* 109:164–168.

Druffel ERM (2002) Radiocarbon in corals: records of the carbon cycle, surface circulation and climate. *Oceanography* 15:122–7.

Druffel EM, Linick TW (1978) Radiocarbon in annual coral rings of Florida. *Geophys Res Lett* 5:913–916.

Dureuil M, Worm B (2015) Estimating growth from tagging data: an application to north-east Atlantic tope shark *Galeorhinus galeus*. *J Fish Biol* 87:1389–1410.

Ebert DA, Stehmann MFW (2013) *Sharks, Batoids, and Chimaeras of the North Atlantic*. FAO Species Catalogue for Fishery Purposes No. 7. Food and Agriculture Organization of the United Nations, Rome.

Ebert DA, Smith WD, Haas DL, Ainsley SM, Cailliet GM (2007) *Life History and Population Dynamics of Alaskan Skates: Providing Essential Biological Information for Effective Management of Bycatch and Target Species*, North Pacific Research Board Project 510 Final Report. North Pacific Research Board, Anchorage, AK.

Ebert DA, Maurer JR, Ainsley SM, Barnett LAK, Cailliet GM (2009) *Life History and Population Dynamics of Four Endemic Alaskan Skates: Determining Essential Biological Information for Effective Management of Bycatch and Target Species*. North Pacific Research Board Project 715 Final Report. North Pacific Research Board, Anchorage, AK.

Fabens AJ (1965) Properties and fitting of the von Bertalanffy growth curve. *Growth* 29:265–289.

Francis MP, Gallagher MJ (2009) Revised age and growth estimates for Antarctic starry skate (*Amblyraja georgiana*) from the Ross Sea. *CCAMLR Sci* 16:211–220.

Francis MP, Maolagáin CÓ (2001) Age and growth of the Antarctic skates, *Bathyraja eatonii* and *Amblyraja georgiana*. *CCAMLR* 12:183–94.

Francis MP, Maolagáin CÓ (2005) Age and growth of the Antarctic skate, (*Amblyraja georgiana*), in the Ross Sea. *CCAMLR* 12:183–194.

Francis MP, Maolagáin CÓ (2016) *Size, Maturity and Length Composition of Blue Sharks Observed in New Zealand Tuna Longline Fisheries*, New Zealand Fisheries Assessment Report 2016/60. Ministry for Primary Industries, Wellington, New Zealand.

Francis MP, Campana SE, Jones CM (2007) Age under-estimation in New Zealand porbeagle sharks (*Lamna nasus*): is there an upper limit that can be determined from shark vertebrae? *Mar Freshwater Res* 58:10–23.

Francis RICC (1988) Maximum likelihood estimation of growth and growth variability from tagging data. *New Zeal J Mar Fresh* 22:43–51.

Freer DWL, Griffiths CL (1993) Estimation of age and growth in the St Joseph *Callorhinchus capensis* (Dumeril). *So Afr J Mar Sci* 13:1, 75–81.

Fry B (1988) Food web structure on Georges Bank from stable C, N, and S isotopic compositions. *Limnol Oceanogr* 33:1182–1190.

Gallagher MJ (2000) The Fisheries Biology of Commercial Ray Species from Two Geographically Distinct Regions, PhD dissertation, University of Dublin.

Gallagher MJ, Nolan CP (1999) A novel method for the estimation of age and growth in rajids using caudal thorns. *Can J Fish Aquat Sci* 56:1590–1599.

Gallagher MJ, Nolan CP, Jeal F (2002) The Structure and Growth Processes of Caudal Thorns, paper presented at Northwest Atlantic Fisheries Organization Scientific Council Meeting, September 16–20.

Gallagher MJ, Nolan CP, Jeal F (2005) Structure and growth processes of caudal thorns. *J Northw Atl Fish Sci* 35:125–129.

Gallagher MJ, Green MJ, Nolan CP (2006) The potential use of caudal thorns as a non-invasive ageing structure in the thorny skate (*Amblyraja radiata* Donovan, 1808). In: Carlson JK, Goldman KJ (eds) *Special Issue: Age and Growth of Chondrichthyan Fishes: New Methods, Techniques and Analysis*. Springer, Netherlands, pp 265–272.

Garde E, Heide-Jørgensen MP, Hansen SH, Nachman G, Forchhammer MC (2007) Age-specific growth and remarkable longevity in narwhals (*Monodon monoceros*) from west Greenland as estimated by aspartic acid racemization. *J Mamm* 88:49–58.

Garde E, Frie AK, Dunshea G, Hansen SH, Kovacs KM, Lydersen C (2010) Harpseal ageing techniques—teeth, aspartic acid racemization, and telomere sequence analysis. *J Mamm* 91:1365–1374.

Garde E, Heide-Jorgensen MP, Ditlevsen S, Hansen SH (2012) Aspartic acid racemization rate in narwhal (*Monodon monoceros*) eye lens nuclei estimated by counting of growth layers in tusks. *Polar Res* 31:15865.

Gelsleichter J, Cortés E, Manire CA, Hueter RE, Muaick JA (1997) Use of calcein as a fluorescent marker for elasmobranch vertebral cartilage. *Trans Am Fish Soc* 126:862–865.

George JC, Bada J, Zeh J, Scott L, Brown SE, O'Hara T, Suydam R (1999) Age and growth estimates of bowhead whales (*Balaena mysticetus*) via aspartic acid racemization. *Can J Zool* 77:571–580.

Goldman KJ (2005) Age and growth of elasmobranch fishes. In: Musick JA, Bonfil R (eds) *Management Techniques for Elasmobranch Fisheries*, FAO Fisheries Technical Paper 474. Food and Agriculture Organization of the United Nations, Rome, pp 97–132.

Goldman KJ, Cailliet GM, Andrews AH, Natanson LJ (2012) Assessing the age and growth of chondrichthyan fishes. In: Carrier JC, Musick JA, Heithaus MR (eds) *Biology of Sharks and Their Relatives*, 2nd ed. CRC Press, Boca Raton, FL, pp 423–451.

Grottoli AG, Eakin CM (2007) A review of modern coral δ^{18}O and Δ^{14}C proxy records. *Earth-Sci Rev* 81:67–91.

Gulland JA, Holt SJ (1959) Estimation of growth parameters for data at unequal time intervals. *J Cons Int Explor Mer* 25:47–49.

Gutteridge AN, Huveneers C, Marshall LJ, Tibbetts IR, Bennett MB (2013) Life-history traits of a small-bodied coastal shark. *Mar Fresh Res* 64:54–65.

Hamady LL, Natanson LJ, Skomal GB, Thorrold S (2014). Bomb carbon age validation of the white shark, *Carcharodon carcharias*, in the western North Atlantic Ocean. *PLoS ONE* 9(1):e84006.

Harry AV (2017) Evidence for systemic age underestimation in shark and ray ageing studies. *Fish Fish* 19(2):185–200.

Harry AV, Tobin AJ, Simpfendorfer CA (2013) Age, growth and reproductive biology of the spot-tail shark, *Carcharhinus sorrah*, and the Australian blacktip shark, *C. tilstoni*, from the Great Barrier Reef World Heritage Area, north-eastern Australia. *Mar Freshwater Res* 64:277–293.

Helfman PM, Bada JL (1975) Aspartic acid racemization in tooth enamel from living humans. *Proc Nat Acad Sci USA* 72:2891–2894.

Henderson AC, Arkhipkin AI, Chtcherbich JN (2004) Distribution, growth and reproduction of the white-spotted skate *Bathyraja albomaculata* (Norman, 1937) around the Falkland Islands. *J Northw Atl Fish Sci* 35:79–87.

Holden MJ, Meadows P (1962) The structure of the spine of the spur dogfish (*Squalus acanthias* L.) and its use for age determination. *J Mar Biol Assoc UK* 42:179–197.

Holden MJ, Vince MR (1973) Age validation studies on the centra of *Raja clavata* using tetracycline. *J Cons Int Explor Mer* 35:13–17.

Huveneers C, Stead J, Bennett MB, Lee KA, Harcourt RG (2013) Age and growth determination of three sympatric wobbegong sharks: how reliable is growth band periodicity in Orectolobidae? *Fish Res* 147:413–425.

Irvine SB, Stevens JD, Laurenson JB (2006a) Surface bands on deepwater squalid dorsal-fin spines: an alternative method for ageing *Centroselachus crepidater*. *Can J Fish Aquat Sci* 63:617–627.

Irvine SB, Stevens JD, Laurenson JB (2006b) Comparing external and internal dorsal-spine bands to interpret the age and growth of the giant lantern shark, *Etmopterus baxteri* (Squaliformes: Epmpteridae). *Environ Biol Fish* 77:253–264.

James IR (1991) Estimation of von Bertalanffy growth curve parameters from recapture data. *Biometrics* 47:1519–1530.

James KC (2011) Life History Characteristics of the Starry Skate, *Raja stellulata*, from Californian Waters, master's thesis, San Jose State University.

James KC, Ebert DA, Natanson LJ, Cailliet GM (2014) Age and growth characteristics of the starry skate, *Raja stellulata*, with a description of life history and habitat trends of the central California, USA, skate assemblage. *Environ Biol Fish* 97:435–448.

Kaganovskaia S (1933) A method of determining the age and the composition of the catches of the spiny dogfish (*Squalus acanthias* L.). *Vestn Dalnevost Fil Akad Nauk SSSR* 1–3:139–141 (translated from Russian by Fisheries Research Board of Canada, 1960).

Kalish JM (1993) Pre-and post-bomb radiocarbon in fish otoliths. *Earth Planet Sci Lett* 114:549–554.

Kalish JM, Johnston J (2001) Determination of school shark age based on analysis of radiocarbon in vertebral collagen. In: Kalish JM (ed) *Use of the Bomb Radiocarbon Chronometer to Validate Fish Age*. Fisheries Research and Development Corporation, Canberra, Australia, pp 116–122.

Kemp NE (1999) Integumentary system and teeth. In: Hamlett WC (ed) *Sharks, Skates and Rays: The Biology of Elasmobranch Fishes*. The Johns Hopkins University Press, Baltimore, MD, pp 43–68.

Kerr LA, Andrews AH, Cailliet GM, Brown TA, Coale KH (2006) Investigations of Δ^{14}C, δ^{15}N, and δ^{13}C in vertebrae of white shark (*Carcharodon carcharias*) from the eastern Pacific Ocean. *Environ Biol Fish* 77:337–353.

Ketchen KS (1975) Age and growth of dogfish *Squalus acanthias* in British Columbia waters. *J Fish Res Board Can* 32:43–59.

King JR, Helser T, Gburski C, Ebert DA, Cailliet G, Kastelle CR (2017) Bomb radiocarbon analyses validate and inform age determination of longnose skate (*Raja rhina*) and big skate (*Beringraja binoculata*) in the north Pacific Ocean. *Fish Res* 193:195–206.

Kinney MJ, Wells RJD, Kohin S (2016) Oxytetracycline age validation of an adult shortfin mako shark *Isurus oxyrinchus* after 6 years at liberty. *J Fish Biol* 89:1828–1833.

Kneebone J, Natanson LJ, Andrews AH, Howell H (2008) Using bomb radiocarbon analyses to validate age and growth estimates for the tiger shark, *Galeocerdo cuvier*, in the western North Atlantic. *Mar Biol* 154:423–434.

Kohler NE, Turner PA (2001) Shark tagging: a review of conventional methods and studies. *Environ Biol Fish* 60:191–223.

Lagler KF (1952) *Freshwater Fishery Biology*. Brown Co., Dubuque, IA.

Last P, Naylor G, Séret B, White W, Stehmann M, de Carvalho M (2016) *Rays of the World*. Cornell University Press, Ithaca, NY.

Libby WF (1955) *Radiocarbon Dating*, 2nd ed. University of Chicago Press, Chicago, IL.

Matta ME, Gunderson DR (2007) Age, growth, maturity, and mortality of the Alaska skate, *Bathyraja parmifera*, in the eastern Bering Sea. *Environ Biol Fish* 80:309–323.

Maurer JRF (2009) Life History of Two Bering Sea Slope Skates: *Bathyraja lindbergi* and *B. maculata*, master's thesis, California State University, Monterey Bay.

McAuley RB, Simpfendorfer CA, Hyndes GA, Allison RR, Chidlow JA, Newman SJ, Lananton RCJ (2006) Validated age and growth of the sandbar shark, *Carcharhinus plumbeus* (Nardo 1827) in the waters off Western Australia. *Environ Biol Fish* 77:385–400.

McBride R (2015) Diagnosis of paired age agreement: a simulation of accuracy and precision effects. *ICES J Mar Sci* 72:2149–2167.

McFarlane GA, Beamish RJ (1987) Validation of the dorsal spine method of age determination for spiny dogfish. In: Summerfelt RC, Hall GE (eds) *The Age and Growth of Fish*. Iowa State University Press, Ames, pp 287–300.

McFarlane GA, King JR (2006) Age and growth of the big skate (*Raja binoculata*) and longnose skate (*Raja rhina*) in British Columbia waters. *Fish Res* 78:169–178.

McFarlane GA, King JR, Saunders MW (2002) preliminary study on the use of neural arches in the age determination of bluntnose sixgill sharks (*Hexanchus griseus*). *Fish Bull* 100(4):861–864.

McPhie RP, Campana SE (2009) Bomb dating and age determination of skates (family Rajidae) off the eastern coast of Canada. *ICES J Mar Sci* 66:546–560.

Meyer W, Seegers U (2012) Basics of skin structure and function in elasmobranchs: a review. *J Fish Biol* 80:1940–1967.

Moura T, Figueiredo I, Farias I, Serra-Pereira B, Coelho R, Erzini K, Neves A, Gordo LS (2007) The use of caudal thorns for ageing *Raja undulata* from the Portuguese continental shelf, with comments on its reproductive cycle. *Mar Freshwater Res* 58:983–992.

Natanson LJ (1993) Effect of temperature on band deposition in the little skate, *Raja erinacea*. *Copeia* 1993:199–206.

Natanson LJ, Cailliet GM (1990) Vertebral growth zone deposition in Pacific angel sharks. *Copeia*. 1990:1133–1145.

Natanson LJ, Mello JJ, Campana SE (2002) Validated age and growth of the porbeagle shark (*Lamna nasus*) in the western North Atlantic Ocean. *Fish Bull* 100:266–278.

Natanson L, Kohler N, Ardizzone D, Cailliet G, Wintner S, Mollet S (2006) Validated age and growth estimates for the shortfin mako, *Isurus oxyrinchus*, in the North Atlantic Ocean. *Environ Biol Fish* 77:367–383.

Natanson LJ, Wintner S, Johansson F, Piercy AN, Campbell P, De Maddalena A, Gulak SJ, et al. (2008) Ontogenetic vertebral growth patterns in the basking shark, *Cetorhinus maximus*. *Mar Ecol Prog Ser* 361:267–278.

Natanson LJ, Gervelis BJ, Winton MV, Hamady LL, Gulak SJB, Carlson JK (2014) Pre- and post-management growth comparisons for *Carcharhinus obscurus* in the northwestern Atlantic Ocean, with age validation. *Environ Biol Fish* 97:881–896.

Natanson LJ, Hamady LL, Gervelis BJ (2015) Analysis of bomb radiocarbon data for common thresher sharks, *Alopias vulpinus*, in the northwestern Atlantic Ocean with revised growth estimates. *Environ Biol Fish* 99:39–47.

Natanson LJ, Skomal GB, Hoffmann S, Porter M, Goldman KJ, Serra D (in press) Age and growth of sharks: do vertebral band pairs record age? *Mar Freshwater Res*.

Nielsen J, Nielsen M (2012) *The Greenland Shark (Somniosus microcephalus)*. DESCNA, http://www.descna.com/index.php/knowledge-base/the-greenland-shark-somniosus-microcephalus.

Nielsen J, Hedeholm RB, Heinemeier J, Bushnell PG, Christiansen JS, Olsen J, Ramsey CB, et al. (2016) Eye lens radiocarbon reveals centuries of longevity in the Greenland shark (*Somniosus microcephalus*). *Science* 353(6300):702–704.

Nielsen NH, Garde E, Heide-Jørgensen MP, Lockyer CH, Ditlevsen S, Olafsdottir D, Hansen SH (2013) Application of a novel method for age estimation of a baleen whale and a porpoise. *Mar Mamm Sci* 29:E1–E23.

NOSAMS (2018) *Fees for Radiocarbon Analysis*. National Ocean Sciences Accelerator Mass Spectrometry, Woods Hole Oceanographic Institution, www.whoi.edu/website/nosams/fees.

Officer RA, Gason AS, Walker TI, Clement JG (1996) Sources of variation in counts of growth increments in vertebrae from gummy shark, *Mustelus antarcticus*, and school shark, *Galeorhinus galeus*: implications for age determination. *Can J Fish Aquat Sci* 53:1765–1777.

Officer RA, Day RW, Clement JG, Brown LP (1997) Captive gummy sharks, *Mustelus antarcticus*, form hypermineralised bands in their vertebrae during winter. *Can J Fish Aquat Sci* 54:2677–2683.

Passerotti MS, Carlson JK, Piercy AN, Campana SE (2010) Age validation of great hammerhead shark (*Sphyrna mokarran*), determined by bomb radiocarbon analysis. *Fish Bull* 108:346–351.

Passerotti MS, Andrews AH, Carlson JK, Wintner SP, Goldman KJ, Natanson LJ (2014) Maximum age and missing time in the vertebrae of sand tiger shark (*Carcharias taurus*): validated lifespan from bomb radiocarbon dating in the western North Atlantic and southwestern Indian Oceans. *Mar Freshwater Res* 65:1131–1140.

Perez CR, Cailliet GM, Ebert DA (2011) Age and growth of the sandpaper skate, *Bathyraja kincaidii*, using vertebral centra, with an investigation of caudal thorns. *J Mar Biol Assoc UK* 91:1149–1156.

Pierce SJ, Bennett MB (2009) Validated annual band-pair periodicity and growth parameters of blue-spotted maskray *Neotrygon kuhlii* from south-east Queensland, Australia. *J Fish Biol* 75:2490–2508.

Pleskach K, Hoang W, Chu M, Halldorson T, Loseto L, Ferguson SH, Tomy GT (2016) Use of mass spectrometry to measure aspartic acid racemization for ageing beluga whales. *Mar Mamm Sci* 32:1370–1380.

Pratt Jr HL, Casey JG (1983) Age and growth of the shortfin mako, *Isurus oxyrinchus*, using four methods. *Can J Fish Aquat Sci* 40:1944–1957.

Rangel BS, Wosnic N, Magdanelo LR, de Amorim AF, Kfoury JR, Rici REG (2016) Thorns and dermal denticles of skates *Atlantoraja cyclophora* and *A. castelnaui*: microscopic features and functional implications. *Microsc Res Techn* 79(12):1133–1138.

Reimer PJ, Brown TA, Reimer RW (2004) Discussion: reporting and calibration of post-bomb ^{14}C data. *Radiocarbon* 46:1299–1304.

Rigby CL, Wedding BB, Grauf S, Simpfendorfer CA (2014) The utility of near infrared spectroscopy for age estimation of deepwater sharks. *Deep-Sea Res* I 94:184–194.

Rigby CL, Wedding BB, Grauf S, Simpfendorfer CA (2016) Novel method for shark age estimation using near infrared spectroscopy. *Mar Freshwater Res* 67:537–545.

Rosa C, Zeh J, George JC, Botta O, Zauscher M, Bada J, O'Hara TM (2013) Age estimates based on aspartic acid racemization for bowhead whales (*Balaena mysticetus*) harvested in 1998–2000 and the relationship between racemization rate and body temperature. *Mar Mamm Sci* 29:424–445.

Sagarese SR, Frisk MG (2010) An investigation on the effect of photoperiod and temperature on vertebral band deposition in little skate *Leucoraja erinacea*. *J Fish Biol* 77:935–946.

Serra-Pereira B, Figueiredo I, Bordalo-Machado P, Farias I, Moura T, Gordo LS (2005) Age and growth of *Raja clavata* Linnaeus, 1758—evaluation of ageing precision using different types of caudal denticles. *ICES CM* 17:10.

Serra-Pereira B, Figueiredo I, Farias I, Moura T, Gordo LS (2008) Description of dermal denticles from the caudal region of *Raja clavata* and their use for the estimation of age and growth. *ICES J Mar Sci* 65(9):1701–1709.

Simpfendorfer CA (2000) Growth rates of juvenile dusky sharks, *Carcharhinus obscurus* (Lesueur, 1818), from southwestern Australia estimated from tag-recapture data. *Fish Bull* 98:811–822.

Stuiver M, Polach HA (1977) Discussion: reporting of ^{14}C data. *Radiocarbon* 19:355–363.

Stuiver M, Pearson GW, Braziunas T (1986) Radiocarbon age calibration of marine samples back to 9000 cal yr BP. *Radiocarbon* 28:980–1021.

Sullivan KJ (1977) Age and growth of the elephant fish *Callorhinchus milii* (Elasmobranchii: Callorhynchidae). *New Zeal J Mar Fresh* 11:745–753.

Tanaka S (1990) Age and growth studies on the calcified structures of newborn sharks in laboratory aquaria using tetracycline. In: Pratt Jr HL, Gruber SH, Taniuchi T (eds) *Elasmobranchs as Living Resources: Advances in the Biology, Ecology, Systematics, and the Status of the Fisheries.* National Oceanic and Atmospheric Administration, Silver Spring, MD, pp 189–202.

Tribuzio CA, Matta ME, Gburski C, Atkins N, Bubley W (2016) Methods for the preparation of Pacific spiny dogfish, *Squalus suckleyi*, fin spines and vertebrae and an overview of age determination. *Mar Fish Rev* 78(1–2):1–13.

Tribuzio CA, Matta ME, Gburski C, Blood C, Bubley W, Kruse GH (2018) Are Pacific spiny dogfish lying about their age? A comparison of ageing structures for *Squalus suckleyi*. *Mar Freshwater Res* 69(1):37–47.

Tyler AV, Beamish RJ, McFarlane GA (1989) Implications of age determination errors to yield estimates. *Can Spec Pub Fish Aquat Sci* 108:27–35.

Van Houtan KS, Andrews AH, Jones TT, Murakawa SKK, Hagemann ME (2016) Time in tortoiseshell: a bomb radiocarbon-validated chronology in sea turtle scutes. *Proc R Soc Lond B* 283:20152220.

Wanamaker AD, Heinemeier J, Scourse JD, Richardson CA, Butler PG, Eiríksson J, Knudsen KL (2008) Very long-lived mollusks confirm 17th century AD tephra-based radiocarbon reservoir ages for north Icelandic shelf waters. *Radiocarbon* 50:399–412.

Wang YG (1998) An improved Fabens method for estimation of growth parameters in the von Bertalanffy model with individual asymptotes. *Can J Fish Aquat Sci* 55:397–400.

Wells RJD, Smith S, Kohin S, Freund E, Spear N, Ramon DA (2013) Age validation of juvenile shortfin mako (*Isurus oxyrinchus*) tagged and marked with oxytetracycline off southern California. *Fish Bull* 111:147–60.

Wells RJD, Spear N, Kohin S (2016) Age validation of the blue shark (*Prionace glauca*) in the eastern Pacific Ocean. *Mar Fresh Res* 68:1130–1136.

Welte C, Wacker L, Hattendorf B, Christl M, Fohlmeister J, Breitenbach SFM, Robinson LF, et al. (2016) Laser ablation–accelerator mass spectrometry: a novel approach for rapid radiocarbon analyses of carbonate archives at high spatial resolution. *Anal Chem* 88:8570–8576.

Wetzel DL, Reynolds JE, Mercurio P, Givens GH, Pulster EL, George JC (2014) Age Estimation for Bowhead Whales, *Balaena mysticetus*, Using Aspartic Acid Racemization with Enhanced Hydrolysis and Derivatization Procedures, paper presented to the Scientific Committee of the International Whaling Commission.

Winton MV (2011) Age, Growth, and Demography of the Roughtail Skate, *Bathyraja trachura* (Gilbert, 1892), from the Eastern Bering Sea, master's thesis, Moss Landing Marine Labs, California State University Monterey Bay.

Yule DL, Stockwell JD, Black JA, Cullis KI, Cholwek GA, Meyers JT (2008) How systematic age underestimation can impede understanding of fish population dynamics: lessons learned from a Lake Superior cisco stock. *Trans Am Fish Soc* 137:481–495.

Zermeño P, Kurdyla DK, Buchholz BA, Heller SJ, Kashgarian M, Frantz BR (2004) Prevention and removal of elevated radiocarbon contamination in the LLNL/CAMS natural radiocarbon sample preparation laboratory. *Nucl Instr Meth Phys Res B* 223–224:293–297.

CHAPTER 11

Near-Infrared Spectroscopy for Shark Ageing and Biology

Cassandra L. Rigby
Centre for Sustainable Tropical Fisheries and Aquaculture and College of Science and Engineering,
James Cook University, Townsville, Queensland, Australia

William J. Foley
Animal Ecology and Conservation, University of Hamburg, Hamburg, Germany; Research School of Biology,
The Australian National University, Canberra, Australian Capital Territory, Australia

Colin A. Simpfendorfer
Centre for Sustainable Tropical Fisheries and Aquaculture and College of Science and Engineering,
James Cook University, Townsville, Queensland, Australia

CONTENTS

11.1 INTRODUCTION

Accurate and reliable age estimates of sharks are important for informing management that will achieve sustainable outcomes for populations. Age is the foundation of many of the essential parameters, such as growth rate and productivity, that are used in demographic analyses and fisheries assessments (Cailliet et al., 2006; Campana, 2001). Here, "sharks" is used as a general term to refer to sharks, rays, and chimaeras, otherwise known as chondrichthyans. Traditionally, to estimate age in sharks, growth bands are counted in their hard parts. Vertebrae or dorsal fin spines are primarily used, although caudal thorns have also been found suitable for ageing in a few species of skates (Cailliet, 2015; Goldman et al., 2012; Serra-Pereira et al., 2008). As sharks age, calcified material accumulates in these structures and can produce visible band pairs that, when formation periodicity has been validated, enable age determination (Goldman et al., 2012; see also Chapter 10 in this volume).

Counting these band pairs requires experience and time to achieve consistent results, and repeated reads are necessary to ensure precision of the counts (Cailliet et al., 2006). It also can require time-consuming preparation, such as sectioning of the structures and enhancement with stains to improve clarity and readability of the band pairs (Irvine et al., 2006b; Matta et al., 2017). In addition, this approach normally requires the lethal removal of the structures used for ageing from an individual. Given the vulnerability of many shark species to exploitation (Dulvy et al., 2014), nonlethal methods for ageing would be beneficial. These issues prompted investigation of near-infrared spectroscopy (NIRS) as a complementary approach to shark ageing. Although NIRS requires traditional band counts of some age structures, it can greatly reduce the time taken to estimate age from a structure and has the potential to be nonlethal (Rigby et al., 2014, 2016b). This chapter reviews how NIRS works and the application and considerations for use of NIRS in shark ageing.

11.1.1 What Is NIRS and How Does It Work?

Near-infrared spectroscopy uses light in combination with statistical methods to investigate the composition of organic material (Murray and Williams, 1987; Williams, 2008; Wold and Sjöström, 1998). Spectroscopy measures the interaction of organic material with electromagnetic radiation (light). Organic matter is composed of atoms of carbon, hydrogen, oxygen, nitrogen, phosphorus, and sulfur that combine by bonds to form molecules. These bonds vibrate (bend and stretch) at specific electromagnetic frequencies, and when organic material is irradiated with light the bonds absorb energy at different wavelengths. This modifies the light energy that is reflected back from the material. This reflected light is detected and analyzed by NIRS instruments to produce a plot or spectrum of the absorbance at each wavelength that essentially represents the composition of the material (Figure 11.1) (Blanco et al., 1998; Morisseau and Rhodes, 1995; Murray and Williams, 1987).

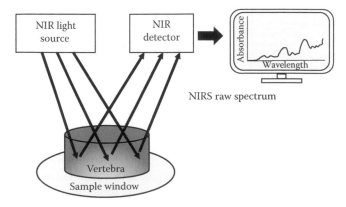

Figure 11.1   Schematic of near-infrared reflectance spectrum collection from shark vertebra.

Different forms of spectroscopy are based on different parts of the electromagnetic spectrum, and each has advantages and disadvantages. The other major types of vibrational spectroscopy are mid-infrared and Raman spectroscopy. Near-infrared spectroscopy uses light in the near-infrared region at the wavelength range of 12,800 to 4000 cm^{-1} (780 to 2500 nm). The NIR spectra detected from irradiated organic material consist of overtones and combination bands derived from stretching and bending of bonds between atoms found in the mid-infrared region of 3300 to 800 cm^{-1} (3000 to 12,500 nm) (see Section 11.2.1).

Near-infrared spectra are typically very complex, which means that unlike some other spectroscopic methods it is not possible to directly assign spectral features to particular components. Therefore, the compositional data must be extracted from the spectra by multivariate statistical approaches (Cozzolino and Murray, 2012; Reich, 2005). However, the major advantage of NIRS over some other spectroscopic methods is that the near-infrared energy penetrates further into the sample so samples can be investigated rapidly, with little preparation and at low cost (Roggo et al., 2007; Williams, 2008).

The need to use statistical approaches to extract compositional data from NIR spectra means that this is an indirect method of analysis. In order to use NIRS for analysis of a trait such as age, a calibration must first be established. This is a statistical model that describes the relationship between the NIR spectra and a set of samples where the values for the trait are known, referred to as the primary reference samples (Foley et al., 1998; Williams, 2008). In the case of shark ageing, these reference samples are vertebrae or dorsal fin spines that have been aged by band counts. When the statistical model has been established, it can be used to rapidly predict the trait (in this case, age) in future samples from the NIR spectra alone.

Near-infrared spectroscopy is widely used in the agriculture, pharmaceutical, and medical industries to measure traits such as protein or fat content in plants and animals, the concentration of active ingredients in pharmaceutical products, and oxygen saturation of tissues (Cozzolino and Murray, 2012; Ferrari et al., 2004; Roggo et al., 2007; Solberg et al., 2003). NIRS has a multitude of very different uses; for example, it has the capacity to detect seafood species substitution and can infer historical changes in penguin population sizes from the proportions of guano in sediment cores (Cozzolino and Murray, 2012; Liu et al., 2011). NIRS is a valuable method for analyzing a wide range of materials of interest to fisheries biologists, as it is capable of providing information on complex traits such as age, species discrimination, and geographic origins (Vance et al., 2016).

In the first study of the application of NIRS for ageing sharks, the spectra of whole vertebrae from 80 great hammerhead sharks (*Sphyrna mokarran*) were collected using a NIR spectrophotometer to produce 80 NIRS spectra (Rigby et al., 2016b). Individuals were also aged by traditional sectioning and band counts (Harry et al., 2011). The statistical relationship between the 80 spectra and the age of each vertebra derived from band counts was established using a calibration model. This calibration model could be used to accurately predict the age of other great hammerheads from vertebrae scanned by NIRS and not aged by traditional band counts (Rigby et al., 2016b).

The development of calibration models requires enough samples to incorporate the variability in the trait being examined (Nicolaï et al., 2007; Reich, 2005; Wedding et al., 2014). In the case of great hammerheads, this was variability in age among great hammerheads of similar length, yet only 80 vertebrae were required to develop a calibration model that was statistically robust (Rigby et al., 2016b) (see Section 11.2). The positioning, scanning, and production of a spectrum for each whole vertebra took less than a minute; therefore, when the initial 80 vertebrae had been scanned and aged by traditional means and a calibration model developed, the NIRS method had the capacity to collect more than 60 spectra from vertebrae per hour. This offers significant time savings in the preparation time because no sectioning or staining of the vertebrae is required, as well as in the time required to estimate the age, because collecting NIR spectra is rapid compared to the time required to do traditional band counts (Rigby et al., 2016b). These time savings translate into cost savings because, when a calibration model has been developed, far less labor is required for ageing as no experience in ageing and no repeated band count reads are required. This cost efficiency is one of the major benefits of NIRS (Roggo et al., 2007; Williams, 2008). The simplicity and speed of the ageing process using NIRS could be beneficial for stock and risk assessments of shark species taken as target or bycatch in commercial fisheries where large numbers of vertebrae must be aged or when stock assessments are conducted annually and regular ageing is required (Matta et al., 2017). Another considerable benefit is the nondestructive nature of NIRS (Luypaert et al., 2007). When a calibration model has been developed, the spectra of whole vertebrae can be collected using NIRS without destroying them because they do not have to be sectioned for age estimation. This can be beneficial for samples that are difficult to obtain or when it is advantageous to retain the samples for other studies (e.g., vertebral elemental analysis to inform stock structure and investigate habitat use) (McMillan et al., 2016).

Near-infrared spectroscopy has also been used successfully to accurately predict the age of teleost fish from whole otoliths (Wedding et al., 2014). The first application of NIRS to ageing teleosts was in saddletail snapper (*Lutjanus malabaricus*), and subsequently the approach was successfully used in two other commercially harvested fish species in Australia: barramundi (*Lates calcarifer*) and snapper (*Pagrus auratus*) (Robins et al., 2015; Wedding et al., 2014). The use of NIRS thus provides the opportunity for significant time and cost savings for fish ageing. With an estimated 60,000 otoliths being used to age fish annually in Australia alone

for stock assessments, adoption of NIRS technology offers the ability to substantially reduce the costs of work that is an essential part of fisheries management (Robins et al., 2015).

The success of NIRS in estimating ages of sharks and teleosts from the spectra of vertebrae or otoliths implies that ageing in these species is accompanied by compositional changes in the vertebrae and otoliths visible as growth band pairs. This is discussed further below (Section 11.2.1), but it is important to note that reliably ageing sharks with NIRS does not require an understanding of these compositional changes. Provided that the traditional age data are accurate and sound statistical procedures are used to establish a robust calibration model for age and spectra, NIRS can be employed as a routine tool.

11.1.2 Practical Aspects of the Application of NIRS for Shark Ageing

A near-infrared spectrophotometer is used to acquire the spectra, and it includes a sample presentation module, a light source, and detectors. Normally, instrument-specific software is provided to enable the spectra to be recorded and exported to different analytical packages (Blanco et al., 1998; Williams, 2008). Comprehensive and detailed treatises on the full range of NIRS analytical procedures are provided in many references on the subject (e.g., McClure, 2003; Reich, 2005; Siesler et al., 2002; Williams and Norris, 1987). The NIRS procedures found to be applicable and successful for shark ageing are summarized here, and more detailed descriptions of each aspect are available in Rigby et al. (2014, 2016b). Most NIRS instruments would be suitable for ageing sharks, although the first studies were performed using a Bruker MPA Multi Purpose FT–NIR Analyzer (Bruker Optik, Ettlingen, Germany). The two studies to date that used NIRS to estimate the age of sharks used 80 to 100 whole dried samples from each of four species: vertebrae from great hammerhead sharks and spot-tail sharks (*Carcharhinus sorrah*) and dorsal fin spines from the piked spurdog (*Squalus megalops*) and Philippine spurdog (*Squalus montalbani*) (Rigby et al., 2014, 2016b). The vertebrae were placed face down and the fin spines in lateral orientation on the round sample window of the spectrophotometer and irradiated with NIR light (Figure 11.2). The orientation of the structure with respect to the sample window should be consistent; for example, all fin spines should be placed in a lateral position rather than a mixture of positions. Orientation may potentially affect the NIR spectra (Rigby et al., 2014; Siesler et al., 2002). The circular nature of vertebrae negates any orientation issues. For species with small vertebrae, a potential alternative to hand positioning each vertebra could be the use of an automated sample presentation method such as a carousel, which is used for many other types of samples. This method was successfully used to collect spectra from saddletail snapper otoliths and can further reduce the time taken for ageing a large number of samples (Wedding et al., 2014).

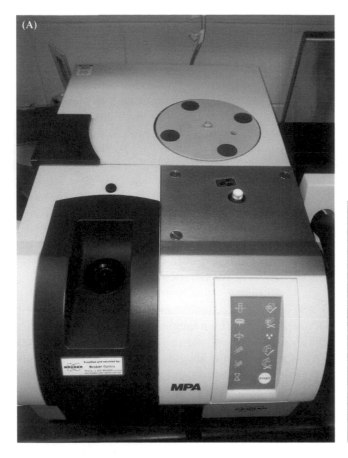

Figure 11.2   Fourier transform NIR spectrophotometer with (A) great hammerhead vertebra face down, and (B) Philippine spurdog dorsal fin spine in lateral orientation.

When light in the NIR region of the electromagnetic spectrum penetrates the sample, some wavelengths are absorbed by the sample and others reflected; these are detected and displayed as a reflectance spectrum. Although the collection of a spectrum is sometimes referred to as a "scan," it is not a linear scan across the vertebra or otolith (i.e., the instrument does not traverse from the inner to the outer edge of a vertebra). The NIR light interacts with the bonds of the sample to provide a holistic view of the organic composition, albeit one that must be interpreted via calibration with samples of known age. As the spectrum of each individual vertebra or fin spine is collected, it can be displayed on a computer linked to the spectrophotometer using associated software to show the absorbance at each wavelength (Figure 11.3) (Williams, 2008, 2013a). As mentioned earlier, it is not possible to correlate particular peaks directly with specific chemical compounds. Similarly, there is no single part of the spectrum that can be referred to as being diagnostic of age or any other trait of interest. However, the degree of absorbance at different wavelengths reflects different stretching and bending of organic bonds, and multivariate statistical procedures can extract combinations of wavelengths from the whole spectra that correlate most strongly with the trait of interest such as age.

11.1.2.1   Development and Validation of Calibration Models

The development of a robust calibration model is critical, as all future analyses are based on the model. Ideally, the available samples should be randomly split into two sets: a calibration set and an independent validation set (Reich, 2005; Robins et al., 2015; Roggo et al., 2007; Williams, 2013b). The calibration set should contain samples that include the full range of the parameter that the calibration model will be used to predict; for sharks, this is the age range. The range of spectral variation with age should be represented and included in development of the calibration model to enable the model to attain an acceptable level of accuracy for prediction (Foley et al., 1998; Wedding et al., 2014). The level of accuracy that is considered acceptable is determined by the objectives and data requirements of the end user (Robins et al., 2015). Although a separate validation set is sometimes omitted in studies with limited samples, the robustness of the calibration model is best tested with an independent validation set. When there are limited sample sizes (that is, less than 100), although not ideal, the samples are not split into two sets but rather all samples are used to develop the calibration model. A process is undertaken during model development to ensure

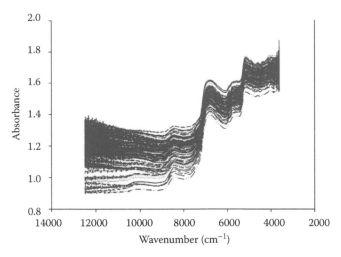

Figure 11.3 Raw near-infrared spectra from 95 Philippine spur-dog dorsal fin spines.

that it accurately predicts the known ages. It is an iterative process, where the calibration model is built on a subset of samples and used to predict the ages of samples not included. This is repeated using successive subsets of samples until all sample ages have been predicted by models that have not included them (Williams, 2013b). This is not a true independent validation method, but it does optimize the model and prevent overfitting of the calibration equation.

Understanding sources of error in any analytical procedure is important. In NIR analysis, the most overlooked source of error is in the determination of the reference values. For estimating the age of sharks, these are the subjective band counts. To reduce the risk of reader bias and provide consistency and precision, quality control methods are incorporated into the age-reading procedures, such as average percent error and age-bias plots (Campana et al., 1995; Goldman et al., 2012). NIRS relies entirely on these reference values, and knowing the error around these values is important as they are incorporated into the calibration model. Consequently, the more precise and accurate the counts of age band pairs, the more accurate the NIRS model age predictions (Robins et al., 2015; Wedding et al., 2014). The level of error and confidence in the age estimation should be reported in association with the NIRS calibration model. When the model has been built, the advantage of NIRS is that further predictions of age are then objective (Foley et al., 1998; Williams, 2008). This avoids any additional age-reading subjectivity that could occur if ages from all animals have be generated from traditional band pair counts.

A calibration model was built for each shark species studied by Rigby et al. (2014, 2016b). For each species, the known ages determined from band counts of the vertebrae and dorsal fin spines were statistically related to the NIRS spectra. For ease of reading, the term "known age" is used for both validated and verified ages, although it is recognized that verified ages are estimates of age. In these studies,

band counts were from sectioned vertebrae (Harry et al., 2011, 2013) and from external bands on the whole dorsal fin spines (Rigby et al., 2016a).

The raw NIRS spectra data are pretreated by mathematical transformation and outliers are eliminated as a first step in the analyses to avoid overfitting of the calibration equation and to remove noise mostly caused by instrument effects (Blanco et al., 1998; Reich, 2005). After pretreatment, partial least squares (PLS) regression is used for each species to establish a relationship between the NIRS spectra and the known ages that identified the combinations of wavelengths that most closely correlated to known ages (Hruschka, 1987; McClure, 2003).

The final calibration model is defined by a combination of statistical criteria that include the highest coefficient of determination (R^2), lowest root mean square error of cross-validation (RMSECV), optimal number of terms, and lowest bias, where bias is the mean difference between known ages and predicted ages and reflects the accuracy of the model (Smith and Flinn, 1991; Williams, 2008). It is important to avoid overfitting of the calibration model which can occur through the inclusion of too many terms in the model. In such a case, the model may fit the calibration data but cannot be successfully applied to unknown spectra. The root mean square of the error of prediction of the model can be plotted against the number of terms to visualize this error and minimize the likelihood of overfitting. Many ecological studies accumulate samples at different times and from different sources. In such cases, it is important to incorporate new sources of variability into the model and rebuild models as the additional samples become available.

11.2 APPLICATION OF NIRS TO SHARK AGEING

Near-infrared spectroscopy can be used as a complementary ageing approach to traditional methods. It has been applied successfully to four species of sharks from different habitats with contrasting body sizes and life histories: the large, semi-oceanic and pelagic great hammerhead, the medium-sized coastal spot-tail shark, and two relatively small deepwater demersal dogfish, the piked spurdog and the Philippine spurdog (Last and Stevens, 2009; Rigby et al., 2014, 2016b). These species all have different growth rates and longevity. The relatively fast-growing spot-tail shark (von Bertalanaffy growth constant $k = 0.34$ year[-1]) lives as long as 14 years, the very slow-growing Philippine spurdog ($k = 0.007$ year[-1]) lives to 31 years, and the great hammerhead has a longevity of 39 years (Harry et al., 2011, 2013; Rigby et al., 2016a). The ability of NIRS to accurately predict the ages of disparate species displaying ecological and biological diversity supports the potentially wide applicability of the NIRS ageing method to sharks. In addition, age was successfully predicted from both of the most common ageing structures: vertebrae and dorsal fin spines.

Consequently, it is likely that the NIRS ageing method would also be suitable for ageing of a wide range of shark taxa, including batoids and chimaerids.

Calibration models have enabled successful prediction of ages across the validated and verified ages of 0.3 to 10 years for great hammerhead sharks (NIRS calibration model $R^2 = 0.89$) and spot-tail sharks ($R^2 = 0.84$) and across the verified and estimated ages of 5 to 25 and 3 to 31 years for the piked spurdog ($R^2 = 0.82$) and Philippine spurdog ($R^2 = 0.73$), respectively (Rigby et al., 2014, 2016b). All models predicted the ages accurately with small mean differences between the known ages and predicted ages in years (bias) of 0.012, –0.005, –0.008, and 0.052 for the great hammerhead shark, spot-tail shark, piked spurdog, and Philippine spurdog, respectively (Rigby et al., 2014, 2016b). The considerable ages of the dogfish successfully predicted attests to the ability of NIRS to be used with older sharks. Thus, with the NIRS approach, ages can be reliably and accurately predicted up to the maximum age successfully calibrated in the NIRS model. To build the calibration models for each of the four species, samples were included across the age ranges predicted. Sharks vary in length with age (Cailliet et al., 1986), and the PLS regression incorporates this individual variation in growth. It is thus preferred to include an even representation of samples from all age classes to be predicted, because this can help improve the calibration model by providing spectral data for each age class (Reich, 2005; Williams, 2008). With sharks, it can be logistically difficult to collect an even representation of samples across the range to be predicted, but a calibration model can still be built with uneven representation. For the piked spurdog, there were proportionally fewer older age classes sampled yet the model was robust and able to predict older ages accurately (Rigby et al., 2014).

The age estimates used in the calibration models should be as accurate as possible. Ideally, the periodicity of the formation of band pairs across the entire age range should be validated; that is, the accuracy of how often band pairs are laid down and absolute age can be proven by a determinate method such as chemical tagging, mark–recapture of known age animals, or bomb radiocarbon dating (Campana, 2001; Campana et al., 2006; Harry et al., 2011; see also Chapter 10 in this volume). Although validation should always be attempted where possible, a far greater number of shark studies have verified rather than validated age estimates due to the difficulties with validation (Cailliet, 2015; Goldman et al., 2012). Verification is another way of evaluating the band pair deposition where the age estimate is corroborated by comparison with other indeterminate methods such as marginal increment analysis and edge analysis (Braccini et al., 2007; Coelho and Erzini, 2008; Goldman et al., 2012).

The NIRS method is not an age validation method, as it requires a set of samples of reliable known ages to develop the calibration model. NIRS cannot determine the age of a structure without prior reference to a known age sample.

It can, however, be considered a verification method, as it provides verification of the band counts through correlation of age with the NIR spectra, allowing comparison of the traditional band count ages with the elemental composition to confirm the band count age estimates.

One of the main reasons for ageing sharks is to generate length-at-age data for growth curves to determine the species growth rates, a key parameter of demographic and fisheries assessments that estimate population status and exploitation risk (Cailliet et al., 2006; Campana, 2001). Ages derived from NIRS for spot-tail and great hammerhead sharks were fitted to three-parameter von Bertalanffy growth models (Rigby et al., 2016b). The growth curves for both species were not significantly different from those generated for known ages up to 10 years which were determined from traditional band pair counts of the vertebrae. This confirmed that the ages derived from the NIRS models are suitable to generate growth curves for species; hence, the NIRS shark ageing approach is feasible for studies of the age and growth of sharks.

11.2.1 Age-Related Changes in Shark Structures

The NIRS method is not dependent on knowing the underlying chemistry of age-related compositional variation in the vertebrae and fin spines (Siesler et al., 2002). The vibrational frequencies of molecular bonds in a sample could be associated with a large number of chemical compounds; although statistical comparison determines the combinations of wavelengths most closely correlated with age, the wavelengths are related to bonds, not any particular chemical compound. NIRS is not an appropriate method to determine which particular chemical component is related to changes in age. Understanding such age-related compositional changes in vertebrae and fin spines requires collaboration with chemists and materials scientists to apply compositional analysis using a range of analytical approaches such as Raman spectroscopy, nuclear magnetic resonance spectroscopy, or soft-ionization mass spectrometry (Jansen van Vuuren et al., 2015; Loch et al., 2014; Schiller and Huster, 2012). Nonetheless, the nature of the age-related compositional changes are of interest to many biologists. As such, some of the features of the spectra that were strongly correlated with age in sharks previously investigated using NIRS are discussed below, along with a description of the composition of vertebrae and fin spines that may be related to ageing.

The spectral areas that correlated most strongly with age in the calibration models described by Rigby et al. (2014, 2016b) were similar for both vertebrae and dorsal fin spines and for all four shark species studied for the application of NIRS to ageing. This suggests that similar compositional changes related to ageing in the vertebrae and dorsal fin spines of the four species were detected by the NIRS method. The three main wavelength areas related to age were in the ranges of 9300 to 8200, 7800 to 6800,

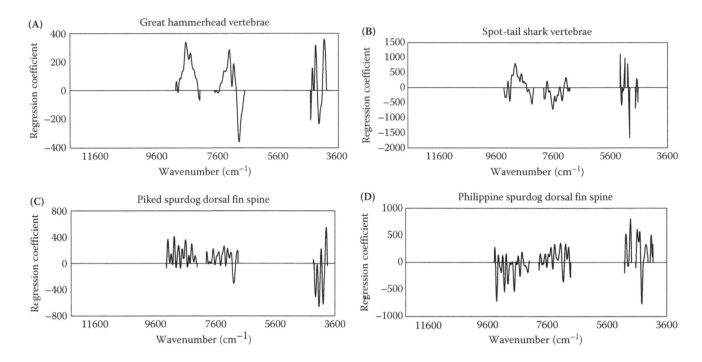

Figure 11.4 Spectral areas of the optimal calibration model plot for vertebrae of (A) great hammerhead and (B) spot-tail shark and dorsal fin spines of (C) piked spurdog and (D) Philippine spurdog. (Parts (C) and (D) from Rigby, C.L. et al., *Deep-Sea Res. Pt. 1,* 94(0), 189, 2014.)

and 5400 to 4000 cm^{-1} (Figure 11.4). Wavelengths in the NIR spectrum are broadly linked to the vibrations in different types of bonds (Siesler et al., 2002). The main spectral areas from the studied shark species that correlated with age correspond to these atomic groups: (1) carbon–carbon alkene, CH (aromatic), –CH$_3$ methyl, and –CH$_2$ methylene, all combination and second overtones; CH, second overtone; –NH$_2$ primary amines, combination and first overtone; –CONH$_2$ primary amide, combination, first, and second overtones; –CONH secondary amides, combination and second overtone; and P–OH and PH, first overtone. The terms "overtone" and "combination" relate to the manner in which energy is absorbed in the NIR region. When energy is absorbed by a molecule, the bonds vibrate and the molecule changes the vibrational energy state. Some molecules jump from a ground state to the next highest energy level, other molecules jump to the second or third energy level, (referred to as the first and second overtones, respectively), and a combination occurs when there are two or more simultaneous jumps (Murray and Williams, 1987).

The primary growth compound in all shark hard parts is the calcium phosphate mineral hydroxyapatite, Ca$_{10}$(PO$_4$)$_6$(OH$_2$), which is present in both vertebral cartilage and dentine and enamel in dorsal fin spines (Kemp, 1999; Maisey, 1979; Walker et al., 1995). Understanding vertebral and fin spine microchemistry might help begin to resolve the compositional changes detected by NIRS that are correlated with age. More detailed information is available for vertebrae than for dorsal fin spines. Growth occurs in these

two structures by the deposition of crystallites of hydroxyapatite within an organic matrix of proteins that are mainly collagen types I and II (which vary in amino acid composition) and proteoglycan (Clement, 1992; Michelacci and Horton; 1989; Porter et al., 2006; Rama and Chandrakasan, 1984). The concentrations of these different components in vertebrae are reported to vary among shark species, with hydroxyapatite accounting for between 39 and 55%; water, 26 to 53%; collagen, 17 to 27%; and proteoglycan, 12 to 28% (Porter et al., 2006). Calcium is reported to account for 40 to 48% of the dried weight of shark vertebrae, with little variation among species (McMillan et al., 2016). Chondroitin sulfate is a component of proteoglycan that provides much of the cartilage resistance to compression and appears to account for 1 to 2% of the dried weight of shark vertebrae (Ge et al., 2016; Takeda et al., 2016). Minor and trace elements (including radioisotopes and stable isotopes) are also known to occur in the vertebrae (Kerr and Campana, 2014; Kerr et al., 2006; Tillett et al., 2011).

The vertebral mineralized accretion is considered to provide mechanical support to the skeleton and occurs in a double cone, but the mechanism of calcium ion (Ca^{2+}) and phosphorus (as both inorganic and organic phosphorus) incorporation into the vertebrae and dorsal fin spines and the relationship with the three-dimensional anatomy are largely unexamined and little understood (Clement, 1992; Dean and Summers, 2006; Dean et al., 2015; McMillan et al., 2016). It is known, though, that in many sharks the accretion of mineral occurs incrementally in concentric rings in vertebrae and at the base and center in dorsal fin spines (Clement, 1992; Dean and Summers, 2006; Irvine et al., 2006a). It often deposits seasonally, with an alternate calcium-rich opaque band due to faster growth in warm months and a calcium-poor translucent band due to slower growth in cool months. This optically distinct band pair is used in traditional age reading and has been validated or verified in a considerable number of shark species to represent one year of growth (Goldman et al., 2012; Hale et al., 2006; Kerr and Campana, 2014; Walker et al., 1995). When the hydroxyapatite and the organic matrix are laid down in the vertebrae or enamel of dorsal fin spines, they are both thought to be metabolically stable, with little to no remodeling, and thus record growth through an animal's lifetime (Campana et al., 2002, 2006; Clement, 1992; Cotton et al., 2014; Koch et al., 1994). The metabolic and temporal stability is not conclusively resolved, but if any remodeling occurs it is likely to be minimal (Campana et al., 2002; Dean et al., 2015; Kerr and Campana, 2014; Kerr et al., 2006). Consequently, the chemical compound changes related to ageing are retained in the age structures throughout an animal's life time and provide a record of ageing.

Yet, at this stage, no specific compounds or potential interactions among several compounds in NIRS-studied shark vertebrae and dorsal fin spines that correlate to the NIRS spectra are known. The most likely candidates are calcium phosphate or a component of the organic matrix. Studies using NIRS in the otoliths of fish found that the spectral regions most correlated with age in the calibration models were 6250 to 4000, 4832 to 4327, and 6160 to 4580 cm[-1] for saddletail snapper, barramundi, and snapper, respectively (Robins et al., 2015; Wedding et al., 2014). These regions overlap most with the lower of the three main spectral areas related to age in the NIRS-studied shark species (i.e., 5400 to 4000 cm[-1]) (Rigby et al., 2014, 2016b). The growth bands in fish are due to the deposition of calcium carbonate rather than calcium phosphate, and although the biochemistry of growth bands in structures used to age sharks are not fully understood they are currently believed to be similar to those of otoliths (Kerr and Campana, 2014). In otoliths of saddletail snapper, carbonates have been suggested as a possible component that is correlated with age (Wedding et al., 2014); however, this possibility has not been investigated further for the other two fish species in which NIRS has been used to predict age. Although we have a more comprehensive understanding of otolith microchemistry than that of vertebrae or dorsal fin spines (Kerr and Campana, 2014), the compositional causes of age correlation are a separate matter from the application of NIRS, and the studies of otoliths that have used NIRS have focused on producing calibration models that can be applied routinely (Robins et al., 2015).

Studies of otoliths have emphasized the importance of the organic matrix as a driving force in otolith formation and that protein accumulation likely affects the opacity of growth bands (Izzo et al., 2016). NIRS has been used successfully in the analysis of chondroitin sulfate because it strongly absorbs NIR radiation (Baykal et al., 2010; Spencer et al., 2009; Zang et al., 2012). Calibration models of chondroitin sulfate in several different studies have identified absorption bands for chondroitin sulfate, including at 4300, 4730, 5400, 5800, and 7000 cm[-1] (Baykal et al., 2010; Spencer et al., 2009; Zang et al., 2012). All of these, except 5800 cm[-1], are within the spectral areas identified in the calibration models developed with shark vertebrae and dorsal fin spines (Rigby et al., 2014, 2016b). Changes in the type of chondroitin sulfate are associated with the ageing process in mammalian cartilage; in young animals, chondroitin-4-sulfate predominates and the relative proportion of chondroitin-6-sulfate increases with age (Bayliss et al., 1999; Mourão, 1988; Shulman and Meyer, 1968). These two variants of chondroitin sulfate are present in shark cartilage (Higashi et al., 2015; Martell-Pelletier et al., 2015; Volpi, 2007). Based on these observations, chondroitin sulfate is a plausible candidate to contribute to age-related compositional variation in shark vertebrae and dorsal fin spines.

11.2.2 Detection of Age in Vertebrae with No Visible Banding

Many deepwater sharks have poorly calcified vertebrae that lack visible growth bands and thus cannot be aged by traditional band counts or other methods, such as tag–recapture, due to logistical difficulties associated with working in the deep sea (Cotton et al., 2014; Goldman et al., 2012). Ageing is possible in some deepwater dogfish that possess dorsal fin spines with discernible growth bands (Irvine et al., 2006b). The piked spurdog is one such dogfish that can be aged by dorsal fin spine bands but that has no discernible bands in its vertebrae. This provided an opportunity to determine if NIRS can detect any compositional changes associated with ageing in vertebrae with no discernible bands.

Spectra from vertebrae of piked dogfish were collected from the same animals aged using band counts from their dorsal fin spine, and a NIRS-based calibration model of the vertebrae was developed using the known ages from the dorsal fin spines (Rigby et al., 2014, 2016a). The calibration model had a good ability to accurately predict ages ($R^2 = 0.89$) (Rigby et al., 2014). This implies that age-related compositional changes occur in the vertebrae despite there being no visible growth bands. A similar result has been reported for vertebrae from another dogfish, the spiny dogfish (*Squalus acanthias*) which ranges from inshore to deepwater (Ebert et al., 2013; Jones and Geen, 1977). Bands were not visible in the vertebrae of the spiny dogfish until recently, when a modified histological staining technique was developed (Bubley et al., 2012), although this technique failed to reveal visible bands in piked dogfish vertebrae (Rigby et al., 2016c). In the earlier spiny dogfish study, x-ray spectrometry revealed a cyclical pattern of concordant calcium and phosphorus peaks and troughs representative of annual seasonal changes in deposition for 21 years (Jones and Geen, 1977).

The nonvisible age-related changes may be due to more subtle changes in the ratios of the mineral hydroxyapatite to the organic matrix in deep-sea sharks; although not visible, they can still be detected by NIRS. Although the deep sea is a relatively stable environment, there are seasonal fluctuations in food availability (Danovaro et al., 2014; Smith et al., 2013), and annual growth band pairs visible in deep-water teleosts validated by radiometric ageing are possibly due to these seasonal food changes (Gordon, 2001; Morales-Nin and Panfili, 2005; Watters et al., 2006). Band pairs are also apparent in the vertebrae of some deepwater skates, but these are difficult to count accurately due to their poor calcification, although some enhancement techniques improve readability (Gennari and Scacco, 2007; Winton et al., 2013). The deposition of annual band pairs in the vertebrae in a few deepwater sharks with visible bands has been validated with bomb radiocarbon (King et al., 2017; McPhie and Campana, 2009) and verified for a species with no visible bands, the longnose velvet dogfish (*Centroselachus crepidater*), for which the radiometric age correlated with visible dorsal fin spine band counts (Fenton, 2001; Irvine et al., 2006b). Alternatively, the detection of age by NIRS is associated with a compositional change that occurs in a manner that is not visibly discernible. This may be the case with changes in the type of chondroitin sulfate with age and would be worth investigation.

11.2.3 Nonlethal Ageing of Sharks

Ageing without the need to lethally remove the structures necessary for ageing would be highly beneficial, given sharks' vulnerability to exploitation, the threatened status of one-quarter of the world's sharks, and the rarity of some species (Cortés et al., 2012; Dulvy et al., 2014; Heupel and Simpfendorfer, 2010). Caudal thorns have been investigated because they can be removed from a live skate, but their reliability as a structure from which to derive age varies among species (Goldman et al., 2012; Matta et al., 2017; see also Chapter 10 in this volume). Nonlethal ageing may be possible for some dogfish and bullhead sharks (*Heterodontus* spp.) that can be aged by band counts of internal sections from the exposed portion of the dorsal fin spine, yet nonlethal removal has not yet been attempted (Irvine et al., 2006b; Tovar-Ávila et al., 2009). NIRS was able to predict age from two shark structures that could be sampled nonlethally: the tip of dorsal fin spines and pectoral fin clips (Rigby et al., 2014).

11.2.3.1 Dorsal Fin Spines

Near-infrared spectra of dorsal fin spines from the piked spurdog and the Philippine spurdog were collected from the anterior tip of the spine rather than the whole spine (Figure 11.3) (Rigby et al., 2014). The anterior region of the fin spine is the part that is exposed in a live animal (Figure 11.5)

Figure 11.5 Philippine spurdog second dorsal fin spine.

(Irvine et al., 2006a); in the NIRS shark study, spectra were collected from this part alone because the available instrument had only a 2-cm-diameter sample window. Alternative sampling setups could be used to collect spectra from the whole fin spine, which can reach lengths of 5 cm. Handheld NIRS instruments for use in the field (see Section 11.3.4) may be suitable for this task and could be used to collect spectra from the exposed portion of the dorsal fin spine on a live shark before returning it to the sea. Either of the two dorsal fin spines present on most squaloid sharks could be used which is beneficial because the tip of one of the two spines is sometimes worn or damaged (Irvine et al., 2006a; Rigby et al., 2014). Establishing a calibration model would require an initial sacrifice of 80 to 150 animals of different ages to obtain dorsal fin spines for ageing and the collection of spectra; the inclusion of slightly more animals than in the shark studies of NIRS to date would enable an independent validation dataset. However, beyond this, further ageing could be based on NIR spectra alone and be rapid and nonlethal. This would be beneficial for stock and risk assessments of the considerable number of species of threatened dogfish sharks (*Squaliformes* spp.) that are commercially targeted or taken as bycatch (Dell'Apa et al., 2015; Ebert et al., 2013; IUCN, 2017).

11.2.3.2 Fin Clips

The use of fin clips for genetic analyses is a widely accepted nonlethal sampling method (Heist, 2004). NIR spectra were obtained from pectoral fin clips of the piked spurdog and a calibration model developed using ages derived from the dorsal fin spines (Rigby et al., 2014, 2016a). The calibration model was able to predict age ($R^2 = 0.76$) (Rigby et al., 2014). This was an interesting outcome, because fin clips have previously not been used in any shark ageing study. A 2-cm square fin clip was removed from the trailing edge of the pectoral fin (Figure 11.6) to match the 2-cm diameter of

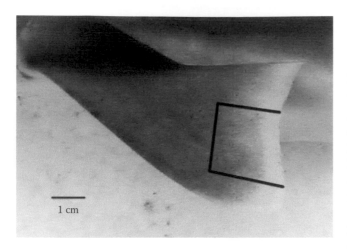

Figure 11.6 Fin clip removed from the piked spurdog left pectoral fin, as shown by the black square. (From Rigby, C.L. et al., *Deep-Sea Res. Pt. 1*, 94(0), 186, 2014.)

the NIR spectrometer window. To reduce the high level of moisture in the fin clips, they were pinned to keep them flat and oven-dried at 50°C for 4 hours before being stored in envelopes (Rigby et al., 2014).

A 2-cm square piece of skin from the dorsal surface between the two dorsal fins was also taken from the same piked spurdog individuals aged by dorsal fin band counts. These were dried and stored in the same manner as the pectoral fin clips, and a similar calibration model was developed to predict age ($R^2 = 0.70$). The main spectral areas in the calibration model of the fin clip were in the ranges of 7800 to 6800 and 5400 to 4000 cm^{-1} (Figure 11.7A), which were similar to the lower two wavelength areas for dorsal fin spine and vertebra calibration models (Section 11.2.1). However, the dorsal skin samples had more spectral areas correlated with age (Figure 11.7B). Some of the skin samples curled and cracked during drying, which may have affected the spectral data, and a different drying method such as a lower oven temperature or air drying in a fume hood may improve the calibration model.

The ability to produce a calibration model suggested that there were compositional changes that were associated with age in the fin clip and, to a lesser extent, in the dorsal skin sample. It is not known what particular part of the fin structure or dorsal skin sample may be related to changes in composition with age. Both the fin clip and dorsal skin are covered in dermal denticles, which are composed of dentine and enamel of a chemical composition similar to that of dorsal fin spines and vertebrae—that is, hydroxyapatite deposited in an organic matrix (Kemp, 1999; Meyer and Seegers, 2012; Miyake et al., 1999). Caudal thorns in skates are derived from dermal denticles that are considered to persist throughout an animal's lifetime with the visible growth band pairs used for ageing, although the banding has not been consistently reliable for ageing for all

skate species (Goldman et al., 2012; Meyer and Seegers, 2012; Serra-Pereira et al., 2008; see also Chapter 10 in this volume). The dermal denticles on shark skin, however, have generally not been considered suitable for ageing because they are not thought to grow continuously throughout the life of the animal; rather, they grow to a definitive size and are discarded and replaced by denticles that can grow larger due to a larger underlying dermal papilla (Kemp, 1999; Maisey, 1979; Meyer and Seegers, 2012). The pectoral fin has other internal components composed of hydroxyapatite and collagen that may be related to age, including cartilaginous platelets, radials, and ceratotrichia (Clement, 1992; Kemp, 1999; Vannuccini, 1999). However, these are not present in shark skin, which consists of an outer epidermis and underlying connective tissue (dermis), neither of which is mineralized, although the dermis does contain chondroitin sulfate (Kemp, 1999; Meyer and Seegers, 2012; Takeda et al., 2016).

Thus, the two components common to both the pectoral fin clips and the dorsal skin samples are dermal denticles and skin. Because dermal denticles are not considered suitable for ageing, it may be chondroitin sulfate in the skin that was correlated with changes in age. As part of a broader chemical investigation of age in sharks, future investigations could include spectra of separate parts of the pectoral fin— skin, dermal denticles, platelets, radials, or ceratotrichia—to determine which is most closely correlated with age.

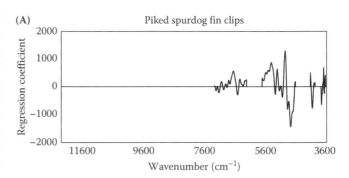

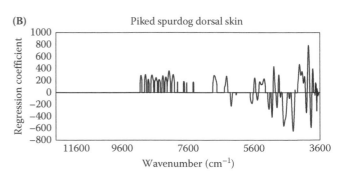

Figure 11.7 Spectral areas of the optimal calibration model plot for piked spurdog (A) pectoral fin clips and (B) dorsal skin samples. (Part (A) from Rigby, C.L. et al., *Deep-Sea Res. Pt. 1*, 94(0), 189, 2014.)

11.3 CONSIDERATIONS IN APPLICATION TO SHARK AGEING

11.3.1 Minimizing Errors in Calibration Models

To reduce any potential confounding factors, it is prudent to keep sample preparation as simple as possible. Dried samples are preferred to be used in NIRS analyses because water is a strong absorber in the NIR region and can dominate and overlap other more informative parts of the NIR spectra (Robins et al., 2015; Williams and Stevensen, 1990). To reduce the potential influence of preservatives or other chemicals on the spectral data, it would be better to use vertebral and dorsal fin samples that have not been stored in ethanol or formalin or stained to enhance band resolution. The influence of ethanol and formalin on the composition of shark vertebrae and dorsal fin spines is unknown, but in teleost otoliths ethanol and formalin can change the concentrations of elements that are not tightly incorporated in the calcium carbonate matrix, and the acidity of formalin can decalcify otoliths (Campana and Neilson, 1985; Kerr and Campana, 2014; McMillan et al., 2016). The cleaning of shark vertebrae for ageing commonly involves immersion in mild household bleach to remove connective tissues. Over short periods of around 40 minutes, this immersion does not alter the chemical composition of the vertebrae, but over longer periods it has been reported to affect concentrations of sodium, magnesium, and manganese (McMillan et al., 2016; Mohan et al., 2017; Tillett et al., 2011). Excessive bleaching would likely affect the hydroxyapatite and organic matrix, as it has been reported to dissolve the vertebrae of some species (Francis and Maolagáin, 2000; Officer et al., 1996). It is probably best to avoid bleaching for more than an hour to maintain both the visibility of band pairs for counts and the integrity of the composition for NIRS analyses.

The possibility that long-term storage may affect the predictive accuracy of ageing using NIRS was considered in an otolith study. Storage up to 8 years was investigated and not thought to affect the ability to predict age because calibration models successfully predicted the age for samples that had been stored for varying lengths of time; however, the findings were not conclusive due to the confounding effect of between-year variability (Robins et al., 2015). No studies of the elemental composition of shark vertebrae have examined the effects of long-term storage, as it has not been considered an issue (McMillan et al., 2016). A collection of older samples is more likely to be from a range of locations collected in different years; this spatial and temporal variability can be accommodated in the NIRS process to produce a single robust calibration model by including the samples from different locations and years in the calibration set (Robins et al., 2015). Thus, long-term storage of dried samples or differences in location and time of collection should not necessarily increase the error in a NIRS calibration model as long

as the variability is included in the calibration samples used to develop the model. As always, ideally a set of independent samples should be used to validate the final model.

11.3.2 Regional Variability and Sexually Dimorphic Growth

Variability in intraspecific age and growth for some species of sharks is apparent across different regions (Cope, 2006; Rigby and Simpfendorfer, 2015). This variability should be considered in the development of shark NIRS age calibration models; if it is significant, then separate NIRS calibration models specific to each region should be developed to provide a more accurate age prediction. For other species of shark, age and growth are more constant across the entire distribution and one calibration model would likely be sufficient (Cailliet and Goldman, 2004). Although the calibration model may be initially developed with samples of such a species from one part of the distribution, the regional scope of the existing calibration model could be increased by later incorporation of samples from a wider geographic area as they become available. The model should then be recalibrated to incorporate any additional spectral variability and validated with an independent dataset to maintain acceptable accuracy in age prediction (Bobelyn et al., 2010; Siesler et al., 2002).

The chemical composition of water can vary with geographic location and may also affect the chemistry of vertebrae and dorsal fin spines through variations in the incorporation of calcium, phosphorus, and other components into the organic matrix (Dean et al., 2015; Kerr and Campana, 2014; McMillan et al., 2016). Diet also contributes to the chemistry of skeletal structures (Dean et al., 2015). Variability in water chemistry and diet and their corresponding NIRS spectral characteristics can be incorporated into the NIRS model by inclusion of samples from different water bodies and life stages in the range of samples collected for development of the calibration model (Robins et al., 2015; Wedding et al., 2014). For example, in the NIRS shark study, great hammerheads were sampled from a large spatial area across the east coast of Australia, which is an area of complex currents (Harry et al., 2011); hence, the NIRS spectra from the great hammerhead encompassed individuals that had resided in different water bodies (Rigby et al., 2016b). Most sharks exhibit diet shifts as they grow from juveniles to adults (Wetherbee et al., 2012); consequently, if juveniles and adults are included in the samples for NIRS, diet changes that result in variability in age structure chemistry would be incorporated into NIRS models. The application of NIRS models to predict age would be most accurate across the regions and sizes that were sampled for each of the species; age predictions of samples outside these regions and sizes should be assessed for accuracy and, if possible, incorporated into the calibration model.

Sexually dimorphic growth is a prevalent trait among sharks, with separate growth curves commonly being generated for males and females where adequate data are available (Cailliet and Goldman, 2004; Cortés, 2000). Where the age and growth are markedly different between the sexes and sufficient samples are available for males and females, the accuracy of NIRS to predict age may potentially be improved by the development of separate calibration models for each sex.

11.3.3　Age, Size, and Older Animals

Ageing structures increase in size as an animal grows, and, as hydroxyapatite and the organic matrix in vertebrae and dorsal fin spines accumulate with size and age, it is logical to assume that NIR spectra correlate with changes in size in addition to age. Size is not suited for age estimation in sharks because of the considerable variation in age-at-length among individuals and the asymptotic nature of their growth (Cailliet et al., 1986). In the NIRS shark study, NIRS spectra related to size for dorsal fin spines, vertebrae, and pectoral fin clips from the piked spurdog (Rigby et al., 2014). Future work to investigate the use of NIRS to predict age and size would benefit from a comparison of NIRS calibration models based on individuals that have reached maximum length and continued to age with adequate sample numbers for both the largest size and older ages from the one species. A suitable candidate could be the North Pacific spiny dogfish (*Squalus suckleyi*), which has been age validated to 52 years by bomb radiocarbon (Campana et al., 2006; Ebert et al., 2010). The size of the species asymptotes at around 100 to 130 cm and about 35 years in the northeast Pacific and then they continue to live for quite a few years without increasing in length (Ketchen, 1975; Orlov et al., 2011; Tribuzio and Kruse, 2012).

The NIRS shark study found that calibration models accurately predicted the older ages of the piked spurdog and Philippine spurdog but not the older ages of the great hammerhead and spot-tail shark (Rigby et al., 2014, 2016b). The older age classes were much better represented in the calibration samples for both dogfish species than for the great hammerhead and spot-tail shark. The lack of proportional representation of the older ages in the latter species may have affected the ability of the calibration models to accurately predict their older ages (Rigby et al., 2016b). Alternatively, it is possible that age underestimation by traditional band counts could have caused poor relationships between the older ages and their corresponding NIR spectra because incorrect older ages would have been used as the reference data (Rigby et al., 2016b). Age underestimation based on vertebral band pair counts has been identified for older individuals of some shark species and is likely caused by either difficulty in discerning and correctly counting the closely spaced band pairs or cessation of band pair deposition (Andrews et al., 2011; Campana et al., 2002; Francis et al., 2007; Harry, 2017; Passerotti et al., 2014).

11.3.4　Instruments

Instruments for NIRS are widely available in industrial and research organizations (Ferrari et al., 2004; Swarbuck, 2014). There are many different configurations and types of instruments and many ways to bring the light to the sample in order to be able to collect the NIR spectrum. A very wide range of instrument configurations have been used in ecological studies, including benchtop instruments and portable handheld instruments, and these can have either a fixed light source or a fiber-flexible optic probe for collecting the spectra (Kirchler et al., 2017; Plans et al., 2013; Vance et al., 2016). Benchtop instruments are the standard in many agricultural research laboratories. These tend to have the advantages of higher signal-to-noise ratios compared to handheld instruments, as well as higher resolution and a wider spectral range. Nonetheless, the performance of many handheld instruments can be good, and the errors and limitations may be acceptable for the intended application (Alcalà et al., 2013; Herberholz et al., 2010).

Different types of instruments do not record identical spectra, and instruments of the exact same model may vary from one another in terms of the spectra that are obtained. Overcoming this issue involves sharing collections of diverse samples and chemometric expertise to ensure that the calibrations are performing identically on different instruments. It would be ideal if these calibrations could be shared among different laboratories all working on the same species of shark. Such approaches are used in very high throughput agricultural analyses (e.g., protein concentration in wheat), but it is not a simple process (Fearn, 2001; Feudale et al., 2002). Larger sample sets would be required for both calibration and validation than have been collected in NIRS studies of sharks to date, and the expertise of instrument specialists would be necessary.

It may be simpler to physically share these shark samples and use them to collect spectra on separate instruments in different laboratories to generate calibration models that can be used locally. Alternatively, a single research center or single instrument configuration may serve as the means where spectra are collected and the spectra files distributed to the original researchers. Given the current status of NIRS in ageing sharks, these latter two options are probably the most feasible way to build expertise and a research community.

11.4　OTHER POTENTIAL USES OF NIRS FOR SHARKS

A recent review of NIRS in wildlife and ecology described a very wide range of applications in insects, amphibians, birds, and mammals (Vance et al., 2016). In all of these studies, the lack of a requirement for extensive sample preparation together with the speed of spectra collection are attractive features. Although most applications have been for simple traits, such as the chemical composition of plant material, the number of applications targeting complex traits is increasing. Compared to some other ecological applications of NIRS, estimating the age of sharks is a complex trait. Based on work in terrestrial systems, there are several other possible traits of sharks that may be amenable to the NIRS approach. These include the reproductive stages of sharks, parasitism, and geographic origins. Many of these applications rely on the ability of NIRS to discriminate individuals and thus provide a means of grouping them into different categories; for example, NIRS has proven to be an excellent aid to identifying cryptic species of insects (Vance et al., 2016). NIRS may be useful for stock structure discrimination in sharks, as principal component analysis of NIR spectra was able to discriminate snapper caught in the northern and southern Gulf St Vincent, South Australia (Robins et al., 2015). Relative to alternatives for species identification, such as an expert taxonomist, molecular analysis, or stable isotope analysis (see Chapter 1 in this volume) for geographic variation, NIRS offers a fast and cost-effective approach. Overall, if there is a trait for which compositional variation between groups is considered possible, then NIRS may provide a means of detecting that variation.

11.5 SUMMARY AND CONCLUSION

Near-infrared spectroscopy is a novel analytical approach to shark ageing which, when employed with validated or verified shark ages, is a cost-effective means of ageing large numbers of sharks. Although it must be used in concert with traditional ageing to develop and validate calibration models, it then has the ability to rapidly, accurately, and reliably predict shark ages from the NIR spectra of vertebrae or fin spines. The method offers the advantages of being nondestructive and low in cost and requiring minimal sample preparation. The NIRS approach offers the potential for nonlethal ageing using dorsal fin spines and fin clips, thus potentially significantly reducing the numbers of sharks lethally sampled. The method accurately predicts the ages of ecologically and biologically diverse shark taxa from both vertebrae and dorsal fin spines and hence is likely to be applicable to a wide range of shark, ray, and chimaerid species.

REFERENCES

Alcalà M, Blanco M, Moyano D, Broad N, O'Brien N, Friedrich D, Pfeifer F, Siesler H (2013) Qualitative and quantitative pharmaceutical analysis with a novel handheld miniature near-infrared spectrometer. *J Near Infrared Spectrosc* 21(6):445–457.

Andrews AH, Natanson LJ, Kerr LA, Burgess GH, Cailliet GM (2011) Bomb radiocarbon and tag–recapture dating of sandbar shark (*Carcharhinus plumbeus*). *Fish Bull* 109(4):454–465.

Baykal D, Irrechukwu O, Ping-Chang L, Fritton K, Spencer RG, Pleshko N (2010) Nondestructive assessment of engineered cartilage constructs using near-infrared spectroscopy. *Appl Spectrosc* 64(10):1160–1166.

Bayliss MT, Osborne D, Woodhouse S, Davidson C (1999) Sulfation of chondroitin sulfate in human articular cartilage. *J Biol Chem* 274(22):15892–15900.

Blanco M, Coello J, Iturriaga H, Maspoch S, de la Pezuela C (1998) Near-infrared spectroscopy in the pharmaceutical industry. *Analyst* 123:135–150.

Bobelyn E, Serban A-S, Nicu M, Lammertyn J, Nicolai BM, Saeys W (2010) Postharvest quality of apple predicted by NIR-spectroscopy: study of the effect of biological variability on spectra and model performance. *Postharvest Biol Technol* 55(3):133–143.

Braccini JM, Gillanders BM, Walker TI, Tovar-Avila J (2007) Comparison of deterministic growth models fitted to length-at-age data of the piked spurdog (*Squalus megalops*) in south-eastern Australia. *Mar Freshwater Res* 58(1):24–33.

Bubley WJ, Kneebone J, Sulikowski JA, Tsang PCW (2012) Reassessment of spiny dogfish *Squalus acanthias* age and growth using vertebrae and dorsal-fin spines. *J Fish Biol* 80(5):1300–1319.

Cailliet GM (2015) Perspectives on elasmobranch life–history studies: a focus on age validation and relevance to fishery management. *J Fish Biol* 87(6):1271–1292.

Cailliet GM, Goldman KJ (2004) Age determination and validation in chondrichthyan fishes. In: Carrier JC, Musick JA, Heithaus MR (eds) *Biology of Sharks and Their Relatives*. CRC Press, Boca Raton, FL, pp 399–447.

Cailliet GM, Radtke RL, Welden BA (1986) Elasmobranch age determination and verification: a review. In: Uyeno T, Arai R, Taniuchi T, Matsura K (eds) *Indo-Pacific Fish Biology: Proceedings of the Second International Conference on Indo-Pacific Fishes*. Ichthyological Society of Japan, Tokyo, pp 345–360.

Cailliet GM, Smith WD, Mollet HF, Goldman KJ (2006) Age and growth studies of chondrichthyan fishes: the need for consistency in terminology, verification, validation, and growth function fitting. *Environ Biol Fish* 77(3):211–228.

Campana SE (2001) Accuracy, precision and quality control in age determination, including a review of the use and abuse of age validation methods. *J Fish Biol* 59(2):197–242.

Campana SE, Neilson JD (1985) Microstructure of fish otoliths. *Can J Fish Aquat Sci* 42(5):1014–1032.

Campana SE, Annand MC, McMillan JI (1995) Graphical and statistical methods for determining the consistency of age determinations. *Trans Am Fish Soc* 124(1):131–138.

Campana SE, Natanson L, Myklevoll S (2002) Bomb dating and age determination of large pelagic sharks. *Can J Fish Aquat Sci* 59:450–455.

Campana SE, Jones C, McFarlane GA, Myklevoll S (2006) Bomb dating and age validation using the spines of spiny dogfish (*Squalus acanthias*). *Environ Biol Fish* 77(3):327–336.

Clement J (1992) Re-examination of the fine structure of endoskeletal mineralization in chondrichthyans: implications for growth, ageing and calcium homeostasis. *Mar Freshwater Res* 43(1):157–181.

Coelho R, Erzini K (2008) Life history of a wide-ranging deepwater lantern shark in the north-east Atlantic, *Etmopterus spinax* (Chondrichthyes: Etmopteridae), with implications for conservation. *J Fish Biol* 73(6):1419–1443.

Cope JM (2006) Exploring intraspecific life history patterns in sharks. *Fish Bull* 104(2):311–320.

Cortés E (2000) Life history patterns and correlations in sharks. *Rev Fish Sci* 8(4):299–344.

Cortés E, Brooks EN, Gedamke T (2012) Population dymanics, demography, and stock assessment. In: Carrier JC, Musick JA, Heithaus MR (eds) *Biology of Sharks and Their Relatives*, 2nd ed. CRC Press, Boca Raton, FL, pp 453–485.

Cotton CF, Andrews AH, Cailliet GM, Grubbs RD, Irvine SB, Musick JA (2014) Assessment of radiometric dating for age validation of deep-water dogfish (Order: Squaliformes) finspines. *Fish Res* 151(0):107–113.

Cozzolino D, Murray I (2012) A review on the application of infrared technologies to determine and monitor composition and other quality characteristics in raw fish, fish products, and seafood. *Appl Spectrosc Rev* 47(3):207–218.

Danovaro R, Snelgrove PVR, Tyler P (2014) Challenging the paradigms of deep-sea ecology. *Trends Ecol Evol* 29(8):465–475.

Dean MN, Summers AP (2006) Mineralized cartilage in the skeleton of chondrichthyan fishes. *Zoology* 109(2):164–168.

Dean MN, Ekstrom L, Monsonego-Ornan E, Ballantyne J, Witten PE, Riley C, Habraken W, Omelon S (2015) Mineral homeostasis and regulation of mineralization processes in the skeletons of sharks, rays and relatives (Elasmobranchii). *Semin Cell Dev Biol* 46:51–67.

Dell'Apa A, Bangley CW, Rulifson RA (2015) Who let the dogfish out? A review of management and socio–economic aspects of spiny dogfish fisheries. *Rev Fish Biol Fish* 25(2):273–295.

Dulvy NK, Fowler SL, Musick JA, Cavanagh RD, Kync PM, Harrison LR, Carlson JK, et al. (2014) Extinction risk and conservation of the world's sharks and rays. *eLife* 3:e00590.

Ebert DA, White WT, Goldman KJ, Compagno LJV, Daly-Engel TS, Ward RD (2010) Resurrection and redescription of *Squalus suckleyi* (Girard, 1854) from the North Pacific, with comments on the *Squalus acanthias* subgroup (Squaliformes: Squalidae). *Zootaxa* 2612(1):22–40.

Ebert DA, Fowler S, Compagno L (2013) *Sharks of the World: A Fully Illustrated Guide*. Wild Nature Press, Plymouth, UK, p 528.

Fearn T (2001) Standardisation and calibration transfer for near infrared instruments: a review. *J Near Infrared Spec* 9(4):229–244.

Fenton GE (2001) *Radiometric Ageing of Sharks*, Project No. 94/021. Fisheries Research Development Corporation and University of Tasmania, Hobart, Australia, p 37.

Ferrari M, Mottola L, Quaresima V (2004) Principles, techniques, and limitations of near infrared spectroscopy. *Can J Appl Physiol* 29(4):463–487.

Feudale RN, Woody NA, Tan H, Myles AJ, Brown SD, Ferré J (2002) Transfer of multivariate calibration models: a review. *Chemometr Intell Lab Syst* 64(2):181–192.

Foley WJ, McIlwee A, Lawler I, Aragones L, Woolnough AP, Berding N (1998) Ecological applications of near infrared reflectance spectroscopy—a tool for rapid, cost-effective prediction of the composition of plant and animal tissues and aspects of animal performance. *Oecologia* 116(3):293–305.

Francis MP, Maolagáin CÓ (2000) Age, growth and maturity of a New Zealand endemic shark (*Mustelus lenticulatus*) estimated from vertebral bands. *Mar Freshwater Res* 51(1):35–42.

Francis MP, Campana SE, Jones CM (2007) Age under-estimation in New Zealand porbeagle sharks (*Lamna nasus*): is there an upper limit to ages that can be determined from shark vertebrae? *Mar Freshwater Res* 58(1):10–23.

Ge D, Higashi K, Ito D, Nagano K, Ishikawa R, Terui Y, Higashi K, Moribe K, Linhardt RJ, Toida T (2016) Poly-ion complex of chondroitin sulfate and spermine and its effect on oral chondroitin sulfate bioavailability. *Chem Pharm Bull* 64(5):390–398.

Gennari E, Scacco U (2007) First age and growth estimates in the deep water shark, *Etmopterus spinax* (Linnaeus, 1758), by deep coned vertebral analysis. *Mar Biol* 152(5):1207–1214.

Goldman KJ, Cailliet GM, Andrews AH, Natanson LJ (2012) Assessing the age and growth of chondrichthyan fishes. In: Carrier JC, Musick JA, Heithaus MR (eds) *Biology of Sharks and Their Relatives*, 2nd ed. CRC Press, Boca Raton, FL, pp 423–451.

Gordon JDM (2001) Deep-sea fishes. In: John HS, Karl KT, Steve AT (eds) *Encyclopedia of Ocean Sciences*, 2nd ed. Academic Press, Oxford, pp 67–72.

Hale LF, Dudgeon JV, Mason AZ, Lowe CG (2006) Elemental signatures in the vertebral cartilage of the round stingray, *Urobatis halleri*, from Seal Beach, California. *Environ Biol Fish* 77:317–325.

Harry AV (2017) Evidence for systemic age underestimation in shark and ray ageing studies. *Fish Fish* 19(2):185–200.

Harry AV, Macbeth WG, Gutteridge AN, Simpfendorfer CA (2011) The life histories of endangered hammerhead sharks (Carcharhiniformes, Sphyrnidae) from the east coast of Australia. *J Fish Biol* 78(7):2026–2051.

Harry AV, Tobin AJ, Simpfendorfer CA (2013) Age, growth and reproductive biology of the spot-tail shark, *Carcharhinus sorrah*, and the Australian blacktip shark, *C. tilstoni*, from the Great Barrier Reef World Heritage Area, north-eastern Australia. *Mar Freshwater Res* 64(4):277–293.

Heist EJ (2004) Genetics: stock identification. In: Musick JA, Bonfil R (eds) *Elasmobranch Fisheries Management Techniques*, FAO Fisheries Technical Paper 474. Food and Agriculture Organization of the United Nations, Rome, pp 79–98.

Herberholz L, Kolomiets O, Siesler H (2010) Quantitative analysis by a portable near infrared spectrometer: can it replace laboratory instrumentation for *in situ* analysis? *NIR News* 21(4):6–8.

Heupel MR, Simpfendorfer CA (2010) Science or slaughter: need for lethal sampling of sharks. *Conserv Biol* 24(5):1212–1218.

Higashi K, Takeuchi Y, Mukuno A, Tomitori H, Miya M, Linhardt RJ, Toida T (2015) Composition of glycosaminoglycans in elasmobranchs including several deep-sea sharks: identification of chondroitin/dermatan sulfate from the dried fins of *Isurus oxyrinchus* and *Prionace glauca*. *PLoS ONE* 10(3):e0120860.

Hruschka WR (1987) Data analysis: wavelength selection methods. In: Williams P, W, Norris K (eds) *Near-Infrared Technology in the Agricultural and Food Industries.* American Association of Cereal Chemists, St. Paul, MN, pp 35–55.

Irvine S, Stevens J, Laurenson L (2006a) Surface bands on deepwater squalid dorsal-fin spines: an alternative method for ageing *Centroselachus crepidater. Can J Fish Aquat Sci* 63:617–627.

Irvine SB, Stevens JD, Laurenson LJB (2006b) Comparing external and internal dorsal-spine bands to interpret the age and growth of the giant lantern shark, *Etmopterus baxteri* (Squaliformes : Etmopteridae). *Environ Biol Fish* 77:253–264.

IUCN (2017) The IUCN Red List of Threatened Species™, www. iucnredlist.org.

Izzo C, Doubleday ZA, Gillanders BM (2016) Where do elements bind within the otoliths of fish? *Mar Freshwater Res* 67(7):1072–1076.

Jansen van Vuuren L, Loch C, Kieser JA, Gordon KC, Fraser SJ (2015) Structure and mechanical properties of normal and anomalous teeth in the sand tiger shark *Carcharias taurus. J Zoo Aqua Res* (2):29–36.

Jones BC, Geen GH (1977) Age determination of an elasmobranch (*Squalis acanthias*) by x-ray spectrometry. *J Fish Res Board Can* 34:44–48.

Kemp NJ (1999) Integumentary system and teeth. In: Hamlett WC (ed) *Sharks, Skates, and Rays: The Biology of Elasmobranch Fishes.* The Johns Hopkins University Press, Baltimore, MD, pp 43–68.

Kerr LA, Campana SE (2014) Chemical composition of fish hard parts as a natural marker of fish stocks. In: Cadrin SX, Kerr LA, Mariani S (eds) *Stock Identification Methods*, 2nd ed. Academic Press, San Diego, CA, pp 205–234.

Kerr LA, Andrews AH, Cailliet GM, Brown TA, Coale KH (2006) Investigations of $\Delta^{14}C$, $\delta^{13}C$, and $\delta^{15}N$ in vertebrae of white shark (*Carcharodon carcharias*) from the eastern North Pacific Ocean. *Environ Biol Fish* 77(3):337–353.

Ketchen KS (1975) Age and growth of dogfish *Squalus acanthias* in British Columbia waters. *J Fish Res Board Can* 32(1):43–59.

King JR, Helser T, Gburski C, Ebert DA, Cailliet G, Kastelle CR (2017) Bomb radiocarbon analyses validate and inform age determination of longnose skate (*Raja rhina*) and big skate (*Beringraja binoculata*) in the north Pacific Ocean. *Fish Res* 193:195–206.

Kirchler CG, Pezzei CK, Bec KB, S. M, Ishigaki M, Ozaki Y, Huck CW (2017) Critical evaluation of spectral information of benchtop vs. portable near-infrared spectrometers: quantum chemistry and two–dimensional correlation spectroscopy for a better understanding of PLS regression models of the rosmarinic acid content in *Rosmarini folium. Analyst* 142:455–464.

Koch PL, Fogel ML, Tuross N (1994) Tracing the diets of fossil animals using stable isotopes. In: Lajtha K, MIchener RH (eds) *Stable Isotopes in Ecology and Environmental Science.* Blackwell Scientific, New York, pp 63–92.

Last PR, Stevens JD (2009) *Sharks and Rays of Australia*, 2nd ed. CSIRO Publishing, Clayton, Victoria, Australia, p 656.

Liu X, Sun J, Sun L, Liu W, Wang Y (2011) Reflectance spectroscopy: a new approach for reconstructing penguin population size from Antarctic ornithogenic sediments. *J Paleolimnol* 45(2):213–222.

Loch C, Swain MV, Fraser SJ, Gordon KC, Kieser JA, Fordyce RE (2014) Elemental and chemical characterization of dolphin enamel and dentine using x-ray and Raman microanalyzers (Cetacea: Delphinoidea and Inioidea). *J Struct Biol* 185(1):58–68.

Luypaert J, Massart DL, Vander Heyden Y (2007) Near-infrared spectroscopy applications in pharmaceutical analysis. *Talanta* 72(3):865–883.

Maisey JG (1979) Finspine morphogenesis in squalid and heterodontid sharks. *Zool J Linn Soc* 66(2):161–183.

Martell-Pelletier JM, Farran A, Montell E, Verges J, Pelletier J (2015) Discrepancies in compostion and biological effects of different formulations of chondroitin sulfate. *Molecules* 20:4227–4289.

Matta ME, Tribuzio CA, Ebert DA, Goldman KJ, Gburski CM (2017) Age and growth of elasmobranchs and applications to fisheries management and conservation in the northeast Pacific Ocean. *Adv Mar Biol* 77:179–220.

McClure WF (2003) 204 years of near infrared technology: 1800–2003. *J Near Infrared Spec* 11:487–518.

McMillan MN, Izzo C, Wade B, Gillanders BM (2016) Elements and elasmobranchs: hypotheses, assumptions and limitations of elemental analysis. *J Fish Biol* 90(2):559–594.

McPhie RP, Campana SE (2009) Bomb dating and age determination of skates (family Rajidae) off the eastern coast of Canada. *ICES J Mar Sci* 66(3):546–560.

Meyer W, Seegers U (2012) Basics of skin structure and function in elasmobranchs: a review. J Fish Biol 80(5):1940–1967. doi:10.1111/j.1095–8649.2011.03207.x

Michelacci YM, Horton DSPQ (1989) Proteoglycans from the cartilage of young hammerhead shark *Sphyrna lewini. Comp Biochem Physiol* 92(4):651–658.

Miyake T, Vaglia JL, Taylor LH, Hall BK (1999) Development of dermal denticles in skates (Chondrichthyes, Batoidea): patterning and cellular differentiation. *J Morphol* 241(1):61–81.

Mohan JA, TinHan TC, Miller NR, Wells RJD (2017) Effects of sample cleaning and storage on the elemental composition of shark vertebrae. *Rapid Commun Mass Spectrom* 31(24):2073–2080.

Morales-Nin B, Panfili J (2005) Seasonality in the deep sea and tropics revisited: what can otoliths tell us? *Mar Freshwater Res* 56(5):585–598.

Morisseau KM, Rhodes CT (1995) Pharmaceutical uses of near-infrared spectroscopy. *Drug Dev Ind Pharm* 21(9):1071–1090.

Mourão PAS (1988) Distribution of chondroitin 4-sulfate and chondroitin 6-sulfate in human articular and growth cartilage. *Arthritis Rheum* 31(8):1028–1033.

Murray I, Williams P (1987) Chemical principles of near-infrared technology. In: Williams P, Norris K (eds) *Near-Infrared Technology in the Agricultural and Food Industries.* American Association of Cereal Chemists, St. Paul, MN, pp 29–31.

Nicolaï BM, Beullens K, Bobelyn E, Peirs A, Saeys W, Theron KI, Lammertyn J (2007) Nondestructive measurement of fruit and vegetable quality by means of NIR spectroscopy: a review. *Postharvest Biol Technol* 46(2):99–118.

Officer RA, Gason AS, Walker TI, Clement JG (1996) Sources of variation in counts of growth increments in vertebrae from gummy shark, *Mustelus antarcticus*, and school shark, *Galeorhinus galeus*: implications for age determination. *Can J Fish Aquat Sci* 53(8):1765–1777.

Orlov A, Kulish E, Mukhametov I, Shubin O (2011) Age and growth of spiny dogfish *Squalus acanthias* (Squalidae, Chondrichthyes) in Pacific waters off the Kuril Islands. *J Ichthyol* 51(1):42–55.

Passerotti MS, Andrews AH, Carlson JK, Wintner SP, Goldman KJ, Natanson LJ (2014) Maximum age and missing time in the vertebrae of sand tiger shark (*Carcharias taurus*): validated lifespan from bomb radiocarbon dating in the western North Atlantic and southwestern Indian Oceans. *Mar Freshwater Res* 65:674–687.

Plans M, Simó J, Casañas F, Sabaté J, Rodriguez-Saona L (2013) Characterization of common beans (*Phaseolus vulgaris* L.) by infrared spectroscopy: comparison of MIR, FT-NIR and dispersive NIR using portable and benchtop instruments. *Food Res Int* 54(2):1643–1651.

Porter ME, Beltrán JL, Koob TJ, Summers AP (2006) Material properties and biochemical composition of mineralized vertebral cartilage in seven elasmobranch species (Chondrichthyes). *J Exp Biol* 209(15):2920–2928.

Rama S, Chandrakasan G (1984) Distribution of different molecular species of collagen in the vertebral cartilage of shark (*Carcharius acutus*). *Connect Tissue Res* 12(2):111–118.

Reich G (2005) Near-infrared spectroscopy and imaging: basic principles and pharmaceutical applications. *Adv Drug Deliv Rev* 57(8):1109–1143.

Rigby C, Simpfendorfer CA (2015) Patterns in life history traits of deep-water chondrichthyans. *Deep-Sea Res Pt I* 115:30–40.

Rigby CL, Wedding BB, Grauf S, Simpfendorfer CA (2014) The utility of near infrared spectroscopy for age estimation of deepwater sharks. *Deep-Sea Res Pt I* 94(0):184–194.

Rigby CL, Daley RK, Simpfendorfer CA (2016a) Comparison of life histories of two deepwater sharks from eastern Australia: the piked spurdog and the Philippine spurdog. *Mar Freshwater Res* 67(10):1546–1561.

Rigby CL, Wedding BB, Grauf S, Simpfendorfer CA (2016b) Novel method for shark age estimation using near infrared spectroscopy. *Mar Freshwater Res* 67(5):537–545.

Rigby CL, White WT, Simpfendorfer CA (2016c) Deepwater chondrichthyan bycatch of the Eastern King Prawn Fishery in the southern Great Barrier Reef, Australia. *PLoS ONE* 11(5):e0156036.

Robins JB, Wedding BB, Wright C, Grauf S, Sellin M, Fowler A, Saunders T, Newman SJ (2015) *Revolutionising Fish Ageing: Using Near Infrared Spectroscopy to Age Fish*, FRDC Project No 2012/11. Department of Agriculture and Fisheries, Brisbane, Queensland, Australia, p 128.

Roggo Y, Chalus P, Maurer L, Lema-Martinez C, Edmond A, Jent N (2007) A review of near infrared spectroscopy and chemometrics in pharmaceutical technologies. *J Pharm Biomed Anal* 44(3):683–700.

Schiller J, Huster D (2012) New methods to study the composition and structure of the extracellular matrix in natural and bioengineered tissues. *Biomatter* 2(3):115–131.

Serra-Pereira B, Figueiredo I, Farias I, Moura T, Gordo LS (2008) Description of dermal denticles from the caudal region of *Raja clavata* and their use for the estimation of age and growth. *ICES J Mar Sci* 65(9):1701–1709.

Shulman HJ, Meyer K (1968) Cellular differentiation and the ageing process in cartilaginous tissues. *J Exp Med* 128(6):1353–1362.

Siesler HW, Ozaki Y, Kawata S, Heise HM (eds) (2002) *Near-Infrared Spectroscopy. Principles, Instruments, Applications*. Wiley-VCH, Weinheim, Germany, p 361.

Smith K, Flinn P (1991) Monitoring the performance of a broad-based calibration for measuring the nutritive value of two independent populations of pasture using near infrared reflectance (NIR) spectroscopy. *Aust J Exp Agric* 31(2):205–210.

Smith KL, Ruhl HA, Kahru M, Huffard CL, Sherman AD (2013) Deep ocean communities impacted by changing climate over 24 y in the abyssal northeast Pacific Ocean. *Proc Natl Acad Sci USA* 110(49):19838–19841.

Solberg C, Saugen E, Swenson L-P, Bruun L, Isaksson T (2003) Determination of fat in live farmed Atlantic salmon using non-invasive NIR techniques. *J Sci Food Agric* 83(7):692–696.

Spencer JA, Kauffman JF, Reepmeyer JC, Gryniewicz CM, Ye W, Toler DY, Buhse LF, Westenberger BJ (2009) Screening of heparin API by near infrared reflectance and Raman spectroscopy. *J Pharm Sci* 98(10):3540–3547.

Swarbuck B (2014) Advances in instrumental technology, industry guidance and data management systems enabling the widespread use of near infrared spectroscopy in the pharmaceutical/biopharmaceutical sector. *J Near Infrared Spec* 22:157–168.

Takeda N, Horai S, Tamura J (2016) Facile analysis of contents and compositions of the chondroitin sulfate/dermatan sulfate hybrid chain in shark and ray tissues. *Carbohydr Res* 424:54–58.

Tillett BJ, Meekan M, Parry D, Munksgaard N, Field I, Thorburn D, Bradshaw C (2011) Decoding fingerprints: elemental composition of vertebrae correlates to age–related habitat use in two morphologically similar sharks. *Mar Ecol Prog Ser* 434:133–142.

Tovar–Ávila J, Izzo C, Walker TI, Braccini JM, Day RW (2009) Assessing growth band counts from vertebrae and dorsal-fin spines for ageing sharks: comparison of four methods applied to *Heterodontus portusjacksoni*. *Mar Freshwater Res* 60(9):898–903.

Tribuzio CA, Kruse GH (2012) Life history characteristics of a lightly exploited stock of *Squalus suckleyi*. *J Fish Biol* 80(5):1159–1180.

Vance CK, Tolleson DR, Rodriguez J, Foley WJ (2016) Near infrared spectroscopy in wildlife and biodiversity. *J Near Infrared Spec* 24:1–25.

Vannuccini S (1999) *Shark Utilization, Marketing and Trade*, FAO Fisheries Technical Paper 389. Food and Agriculture Organization of the United Nations, Rome.

Volpi N (2007) Analytical aspects of pharmaceutical grade chondroitin sulfates. *J Pharm Sci* 96(12):3168–3180.

Walker TI, Officer RA, Clement JG, Brown LP (1995) *Southern Shark Age Validation*. Part 1. *Project Overview, Vertebral Structure and Formation of Growth-Increment Bands Used for Age Determination*, FRDC Project 2001/007. Department of Conservation and Natural Resources, Queenscliff, Victoria, Australia.

Watters DL, Kline DE, Coale KH, Cailliet GM (2006) Radiometric age confirmation and growth of a deep-water marine fish species: the bank rockfish, *Sebastes rufus*. *Fish Res* 81(2–3):251–257.

Wedding BB, Forrest AJ, Wright C, Grauf S, Exley P, Poole SE (2014) A novel method for the age estimation of saddletail snapper (*Lutjanus malabaricus*) using Fourier transform–near infrared (FT–NIR) spectroscopy. *Mar Freshwater Res* 65(10):894–900.

Wetherbee BM, Cortes E, Bizzarro JJ (2012) Food consumption and feeding habits. In: Carrier JC, Musick J, Heithaus MR (eds) *Biology of Sharks and Their Relatives*, 2nd ed. CRC Press, Boca Raton, FL, pp 239–264.

Williams P (2008) Near-infrared technology–getting the best out of light. In: Lawrence S, Warburton P (eds) *A Short Course in the Practical Implementation of Near Infrared Spectroscopy for the User*. PDK Projects, Nanaimo, Canada.

Williams P (2013a) Calibration development and evaluation methods A. Basics. *NIR News* 24(5):24–27.

Williams P (2013b) Calibration development and evaluation methods B. Set-up and evaluation. *NIR News* 24(6):20–24.

Williams P, Norris K (eds) (1987) *Near-Infrared Technology in the Agricultural and Food Industries*. American Association of Cereal Chemists, St. Paul, MN.

Williams P, Stevensen S (1990) Near-infrared reflectance analysis: food industry applications. Trends Food Sci Technol 1:44–48. doi:10.1016/0924–2244(90)90030–3

Winton MV, Natanson LJ, Kneebone J, Cailliet GM, Ebert DA (2013) Life history of *Bathyraja trachura* from the eastern Bering Sea, with evidence of latitudinal variation in a deep–sea skate species. *J Mar Biol Assoc UK* 94(2):411–422.

Wold S, Sjöström M (1998) Chemometrics, present and future success. *Chemometr Intell Lab Syst* 44(1–2):3–14.

Zang H, Li L, Wang F, Yi Q, Dong Q, Sun C, Wang J (2012) A method for identifying the origin of chondroitin sulfate with near infrared spectroscopy. *J Pharm Biomed Anal* 61:224–229.

CHAPTER 12

Photographic Identification of Sharks

Simon J. Pierce
Marine Megafauna Foundation, Truckee, California; Wild Me, Portland, Oregon

Jason Holmberg
Wild Me, Portland, Oregon

Alison A. Kock
South African National Parks, Pretoria, South Africa; South African Institute of Aquatic Biodiversity, Grahamstown, South Africa

Andrea D. Marshall
Marine Megafauna Foundation, Truckee, California; Wild Me, Portland, Oregon

CONTENTS

12.1 INTRODUCTION

The only previous review of photographic identification of sharks (Marshall and Pierce, 2012) considered what was then seen as a relatively new technique that still struggled for acceptance with some peer reviewers and editors. In the years since, the use of photographic identification (henceforth referred to as "photo-ID") has been widely and rapidly adopted, and photo-ID has become a standard method in studies of elasmobranch population ecology, movement, and social behaviors. A small number of large, semiautomated (and, increasingly, fully automated) collaborative online databases are routinely used to facilitate data sharing among research groups. The ubiquity of underwater camera systems has led to a dramatic increase in the volume of visual data posted online. The continuing development of computer vision and machine learning capabilities means that the use of photo-ID will continue to expand, with artificial intelligence systems enhancing, automating, and assuming responsibility for many of the processes and decisions that are currently performed manually.

Here, we define photo-ID as the "recognition of individual fish through their distinctive natural markings, recorded via photographs or video." Photo-ID has been used in elasmobranch studies since at least the early 1970s (e.g., Myrberg and Gruber, 1974). Pigmentation spots, body markings, scars, and fin morphology have all been used as photo-ID characteristics for a variety of shark and ray species. We are not aware of photo-ID techniques being used in chimaeras; many of these species live in deep water, and the more accessible species have not, as yet, been shown to be individually identifiable.

The popularity of photo-ID has been enhanced by the increasing use of waterproof digital cameras by scientists and marine tourists, such as scuba divers and snorkelers, who are being recruited directly as "citizen scientists" to contribute data to broader research efforts or indirectly through data-mining social media. The non-invasive nature of photo-ID (i.e., animals do not have to be touched or restrained) and the inbuilt data validation that it offers, because researchers are able to directly examine original photographs and potentially consult independent computer vision algorithms for confirmation (Bonner and Holmberg, 2013), allow such efforts to be easily applied to studies of elasmobranch species that are popular focal species in marine tourism, such as manta rays (*Mobula* spp.), whale sharks (*Rhincodon typus*), and white sharks (*Carcharodon carcharias*). The routine sharing of images via online social media websites and apps and the ability of researchers to solicit data through such platforms have also played a significant role in expanded participation in these studies (Davies et al., 2012; Robinson et al., 2016). In this chapter, we present a guide for successful photo-ID studies. We consider the current and potential uses of photo-ID as a study method and examine how contemporary developments in computer science are likely to influence and enhance our use of photographs and videos for science.

12.2 ADVANTAGES OF PHOTO-ID

Recognition of individuals within a study species is a fundamental requirement for population biology and demography research. The use of marker tags has a long and illustrious history in elasmobranch studies, and they continue to provide novel information on life-history variables, stock status, reproductive behavior, migrations, and distribution patterns. A review of tagging studies by Kohler and Turner (2001) reported that 64 studies had already used conventional marker tagging in 101 species of shark at that time, and that number has steeply increased since. However, tagging studies have a number of general limitations and associated practical issues. Most challenges stem from the potential for tags to be shed, removed, damaged, or biofouled, thereby limiting their effective lifespan in population studies (Dicken et al., 2006; Graham and Roberts, 2007; Kohler and Turner, 2001; Pierce et al., 2009; Rowat et al., 2009). Reliable re-identification through individual tag numbers or color codes of damaged or heavily fouled tags can also be challenging. Conventional tagging can also be detrimental to individual fitness, may affect natural behavior (Dicken et al., 2006; Feldheim et al., 2002; Fouts and Nelson, 1999; Manire and Gruber, 1991; Wilson and McMahon, 2006), and could lead to increased mortality (Stansbury et al., 2015). The repercussions of these issues depend on the objectives of the research and the focal species. Some studies, such as age validation, require only small numbers of animals to retain tags over time to be successful (Pierce and Bennett, 2009; Smith et al., 2003). In other circumstances, a high rate of tag shedding can mean that re-sighting rates have to be treated as unreliable or used with caution (Rowat et al., 2009).

Photo-ID presents an appealing alternative or supplement to conventional tagging techniques in certain situations. The presence of natural identification marks can eliminate the need for physical marker tags, thus providing a more permanent means of identifying individuals (Dudgeon et al., 2008; Rowat et al., 2009), and marks can be easy to distinguish at a distance. Photo-ID also avoids some of the problems of tags being shed, removed, or fouled and, depending on field practices, can also minimize the risk of inducing stress or behavioral issues. Many shark and ray species are difficult to capture and handle due to their large size or inaccessibility, and photo-ID offers an alternative approach for identification.

Photo-ID has also become an attractive option for researchers seeking to minimize disturbing sensitive populations or threatened elasmobranch species, particularly in cases where the interested public may have a negative view of more invasive research methods. Controversy over shark tagging has rarely been discussed in the scientific literature (but see Hammerschlag et al., 2014; Jewell et al., 2011), although we are aware of a number of additional cases where significant opposition has arisen from community groups due to perceived harm, visual disfigurement, or behavioral avoidance by the target species. These perceptions may have a dubious factual basis, but a lack of local support for research activities can be detrimental to conservation and outreach efforts. The non-invasive nature of photo-ID means that it can be relatively simple to conduct research off platforms of opportunity, such as tourist vessels, and to maximize data collection and outreach potential through the incorporation of citizen science programs into field studies (Davies et al., 2012; Gallagher et al., 2015; Germanov and Marshall, 2014; see also Chapter 16 in this volume).

Although photographic equipment can be an expensive initial outlay, costs tend to remain relatively inexpensive compared to electronic tags, and ongoing expenses of maintenance or service are minor in comparison to satellite tagging fees or maintaining passive acoustic receiver arrays. Because scientists usually use off-the-shelf consumer products in their work, many researchers, students, and volunteers already own suitable cameras and lenses for photo-ID studies. Image capture and processing are also relatively simple and easily trainable, thus broadening the potential pool of study participants.

12.3 WHAT ARE THE RULES FOR SUCCESSFUL PHOTO-ID STUDIES?

Photo-ID is not suitable for all elasmobranch species. The technique has two specific assumptions: (1) individuals can be reliably distinguished from one another, and (2) individuals can be re-identified over time.

12.3.1 Distinguishing Individual Sharks

Many elasmobranchs have natural pigmentation patterns on their skin that act as a unique "fingerprint" for each individual (Figure 12.1). Examples include the spots on the dorsal surfaces of spotted eagle rays (*Aetobatus narinari*) (Corcoran and Gruber, 1999), on the ventral surfaces of reef manta rays (*Mobula alfredi*) (Kitchen-Wheeler, 2010; Marshall et al., 2011), or on the flanks of whale sharks (Taylor, 1994), zebra sharks (*Stegostoma fasciatum*) (Dudgeon et al., 2008), and sand tiger sharks (*Carcharias taurus*) (Bansemer and Bennett, 2008; Van Tienhoven et al., 2007). Other natural patterns, such as the irregular countershading boundaries on white sharks, are also used (Domeier and Nasby-Lucas, 2007). Some species can be distinguished by the color, shape, or notches in their dorsal fins, such as white sharks

(Anderson et al., 2011; Hewitt et al., 2017), blacktip reef sharks (*Carcharhinus melanopterus*) (Porcher, 2005), and basking sharks (*Cetorhinus maximus*) (Gore et al., 2016). Scars, bite marks, fin morphology, and deformities may be useful in the absence of intrinsic patterns (Anderson et al., 2011; Castro and Rosa, 2005; Klimley and Anderson, 1996; Sims et al., 2000) and may also help in secondary confirmation where patterns are present (Marshall et al., 2011).

Most studies employing photo-ID methods have reported 100% of individuals to be recognizable, including whale sharks (Meekan et al., 2006), spotted eagle rays (Corcoran and Gruber, 1999), adult zebra sharks (Dudgeon et al., 2008), white sharks (Domeier and Nasby-Lucas, 2007), whitetip reef sharks (*Triaenodon obesus*) (Whitney et al., 2011), and reef manta rays (Marshall et al., 2011). Some species, however, exhibit lower percentages of identifiable individuals, such as the 54.8% of nurse sharks (*Ginglymostoma cirratum*) reported by Castro and Rosa (2005) and 83% of basking sharks in Scotland (Gore et al., 2016). A photo-ID study on sicklefin lemon sharks (*Negaprion acutidens*) in French Polynesia determined that the uniform color of the focal species, combined with the poor resilience of small spots or color aberrations, resulted in this species being difficult to identify from natural coloration alone (Buray et al., 2009). Furthermore, the assumption of equal individual identifiability across all life stages has not been addressed for any species, and morphometric changes may alter patterning or the applicability of computer vision to analyze it. Although modeling techniques may also be able to compensate, to some extent, for species where a minority of individuals are not identifiable (Hunt et al., 2017), consideration must be given to determining whether photo-ID is preferable to conventional tagging where the percentage of identifiable individuals is low (Pratt and Carrier, 2001).

Even in species with distinct markings, practical considerations may favor tagging over photo-ID. Individuals may be difficult to photograph in their natural environment (e.g., elusive, pelagic, or deepwater species; those living in turbid environments), or large population sizes and low re-sighting rates render the collation and management of a photo-ID library logistically difficult. Photo-ID is most useful—and generally applied—in species that have distinctive markings, that concentrate reliably in areas accessible to observers, and that are reasonably easy to approach and photograph (Marshall et al., 2011).

12.3.2 Re-identifying Individuals Over Time

Individual natural markings must allow for re-identification over time. Animals that have markings caused by fungal infections, or where color is in the mucus layer on the skin rather than the skin itself, have not yet been shown to be reliably identifiable over sufficient time frames. Natural ventral markings, in the form of spots and shading, are present from

Figure 12.1　Examples of photo-identifiable sharks and rays: (A) flank spotting on zebra shark (*Stegostoma fasciatum*); (B) ventral pigmentation and spots on reef manta ray (*Mobula alfredi*); (C) flank spotting on whale shark (*Rhincodon typus*); (D) flank pigmentation gradient and (E) dorsal fin notches on white sharks (*Carcharodon carcharias*). (Photo credits: (A,B) Andrea Marshall; (C) Simon Pierce; (D) Morne Hardenberg; (E) Alison Kock.)

before birth in species such as reef manta rays (Marshall et al., 2008) and are not thought to change over the course of an individual's life span (Couturier et al., 2012). Although the long-term stability of individual markings has been documented in multiple elasmobranch species—to over 10 years in blacktip reef sharks (Mourier et al., 2012; Porcher, 2005), over 20 years in both white sharks and whale sharks (Anderson et al., 2011; Norman and Morgan, 2016), and 30 years in reef manta rays (Couturier et al., 2014)—the longevity of markings is species specific. Therefore, validating their stability across life stages is an important test for photo-ID studies.

Minor pigmentation changes and accumulation of fin damage have been noted in multiple studies, particularly of white sharks, although these were a small minority of individuals in each study population (Domeier and Nasby-Lucas, 2007; Robbins and Fox, 2013; Towner et al., 2013). Scars, wounds, nicks, and scratches may transform or completely heal over time (Anderson et al., 2011; Castro and Rosa, 2005; Domeier and Nasby-Lucas, 2007; Marshall and Bennett, 2010a; Pratt and Carrier, 2001). The intensity of body coloration has also been reported to change over short time periods (Ari, 2014). Depending on the severity of the changes to identifiable markings, photo-ID confirmation may be compromised if individuals can no longer be matched with certainty, an issue equivalent to tag loss in mark–recapture studies using conventional tags. It is important to note, and preferably quantify, the potential for individuals to become less identifiable over time, as this can lead to overestimation of abundance and underestimation of residency (Gubili et al., 2009; Towner et al., 2013).

The stability of natural marks is easiest to assess through some form of double tagging—that is, using an independent feature to confirm individual identification. For example, conventional tags may be applied to a proportion of observed individuals to validate natural pattern stability over the study period (Dudgeon et al., 2008). Alternatively, positive identification can also be achieved by using more than one identifying feature on species where patterns or scarring are reasonably stable, such as using both sides of the animal or using a combination of natural markings and scarring (Domeier and Nasby-Lucas, 2007; Kitchen-Wheeler, 2010; Marshall et al., 2011; Meekan et al., 2006; Norman and Morgan, 2016), or by adding a completely separate analytical technique, such as individual genotyping (Gubili et al., 2009). Validation can alternatively be achieved in partnership with aquariums that house specimens of a desired species by closely monitoring natural coloration or distinctive markings over time, particularly ontogenetic shifts at a certain age or size class (Bansemer and Bennett, 2008). Routine collection of additional metadata, such as sex, maturity status, and size, also provides a useful means of verification.

If ontogenetic shifts in the natural markings of focal species are understood and accounted for, photo-ID may still be successfully employed as long as markings remain stable over the duration of the study (Arzoumanian et al., 2005; Dudgeon et al., 2008). Similarly, identifications are often based largely on scarring patterns in species where distinctive natural markings are absent. Younger animals are unlikely to show as many scars, accumulated bite wounds, or reproductive marks, making identification of these age classes challenging. Removing certain size or age classes from studies, such as by focusing on individuals of over a minimum size, may be effective in cases where the proportion of unidentifiable individuals in the population can be reliably established (Wilson et al., 1999).

12.4 APPLICATIONS OF PHOTO-ID IN SHARK RESEARCH

This overview is not intended to consider all potential applications, but some of the common uses of photo-ID are presented here.

12.4.1 Residency and Movement

Many sharks and rays, even highly mobile species, frequent specific sites (Chapman et al., 2015). This site fidelity has long been evident from telemetry studies and has similarly been documented with photo-ID studies (Anderson and Goldman, 1996; Anderson et al., 2011; Bansemer and Bennett, 2009; Marshall et al., 2011). Individual sharks may use aggregation areas year-round (Cagua et al., 2015), but other sites are characterized by seasonal or aperiodic visitation (Kock et al., 2013; Luiz et al., 2009; Norman and Morgan, 2016). Most elasmobranch population studies using photo-ID have focused on larger, migratory species, but species such as bull sharks (*Carcharhinus leucas*), zebra sharks, and wobbegong sharks (*Orectolobus* spp.) that aggregate in specific locations or have small home ranges have also been investigated (Brunnschweiler and Baensch, 2011; Carraro and Gladstone, 2006; Castro and Rosa, 2005; Dudgeon et al., 2008; Lee et al., 2014).

Photo-ID can be used to evaluate intersite movements; for example, reef manta ray movements were established among Komodo National Park, Nusa Penida, and the Gili Islands in Indonesia (Germanov and Marshall, 2014). Migrations over hundreds of kilometers occur along the eastern coast of Australia among reef manta rays (Couturier et al., 2011, 2014) and sand tiger sharks (Bansemer and Bennett, 2009; Barker and Williamson, 2010). Photo-ID, combined with satellite telemetry, even documented a white shark completing a return migration between South Africa and Western Australia (Bonfil et al., 2005).

Several international photo-database collaborations have provided insights into the broad-scale movements (or lack thereof) in whale sharks. These have used either the online Wildbook for Whale Sharks photo library (www.whale-shark.org) (Arzoumanian et al., 2005) or the downloadable Interactive Individual Identification System (I3S; http://www.reijns.com/i3s) (Speed et al., 2007; Van Tienhoven et al., 2007), both of which semiautomate the photo-matching process. Brooks et al. (2010) and Andrzejaczek et al. (2016) found minimal evidence of population-level interchange among known whale shark feeding areas in the Indian Ocean, although later studies have shown routine movement among countries in the Arabian region (Robinson et al., 2016) and between Mozambique and South Africa (Norman et al., 2017). Regional movements of whale sharks have also been examined in the Western Atlantic, where regular movements were shown between Belize, Honduras, Mexico, and the United States (McKinney et al., 2017). Individual re-sightings occurred between Australia and Indonesia, representing perhaps the longest distance re-sighting to date via photo-ID (~2700 km) (Norman et al., 2017); between Mozambique and Tanzania (~1800 km) (Norman et al., 2017); and between the Philippines and Taiwan (~1600 km) (Araujo et al., 2016). Such collaborative efforts require standardized techniques and often data-exchange among research groups. Several other global photo-library initiatives are underway, such as MantaMatcher (www.manta-matcher.org) (Germanov and Marshall, 2014; Town et al., 2013), ID the Manta (www.mantatrust.org/make-a-difference/id-the-manta), and a white shark database (Hughes and Burghardt, 2017), so it is likely that more regional- to global-scale studies will be developed in the near future.

12.4.2 Population Size and Demographics

When photo-ID data are combined with location and date–time metadata, three fundamental data points underpinning population ecology are established: who, when, and where. These data can then be entered into open or closed capture–mark–recapture (CMR) models or, more specifically, "sight–resight" models, to estimate a variety of population parameters, including abundance, trajectory, sex and size ratios, survival rate, capture probability, and others (Hewitt et al., 2017; Williams et al., 2002), when the assumptions of the underlying analytical models are met. Photo-ID is well-suited for this purpose because individuals can be monitored and non-intrusively re-identified over short or longer time periods. These identified animals can then be used to investigate seasonal or annual population size within the area of interest (Castro and Rosa, 2005; Chapple et al., 2011; Couturier et al., 2014; Deakos et al., 2011; Dudgeon et al., 2008; Holmberg et al., 2009; Marshall et al., 2011; McKinney et al., 2017; Meekan et al., 2006; Rowat et al., 2009). Use of photo-ID may provide an advantage in such studies, as results are less likely to be affected by loss of fitness or mortality resulting from handling or biased through avoidance behavior by the identified individuals, as could be the case with conventionally tagged individuals. Photo-ID also provides access to a potentially greater volume of data for intensive modeling if related tourism is present and can be engaged in concurrent data collection. Care must be taken, however, to understand and account for potential biases in externally sourced data.

Many of the species that have proven suitable for monitoring with photo-ID techniques are also globally threatened. Tracking changes in abundance and survivorship over time can thus provide valuable information on the decline or recovery of these elasmobranchs (Bradshaw et al., 2007; Holmberg et al., 2008, 2009; Hewitt et al., 2017), and photo-ID can also be used to avoid issues with counting the same individual more than once in sighting-based studies (Rohner et al., 2013). That being said, it is possible to inadvertently bias the results of these models in cases where individual sharks vary in their behavior. This issue is discussed further in Section 12.5.

12.4.3 Social Behavior

Photo-ID studies have often focused on areas where sharks aggregate, thereby enabling the study of social interactions between individuals (Jacoby et al., 2012). For example, bonnethead sharks (*Sphyrna tiburo*), maintained in a captive environment and identified using spot patterns, scars, and fin tears, formed size-based dominance hierarchies (Myrberg and Gruber, 1974). Fin morphology of blacktip reef sharks allowed Mourier et al. (2012) to show that these sharks formed stable, long-term social bonds in French Polynesia. On the other hand, white sharks identified by their dorsal fins during chumming activities in South Africa co-occurred at random, displaying no preference or avoidance toward particular individuals, although there was a weak tendency for sharks to co-occur with individuals of similar size and the same sex (Findlay et al., 2016).

12.4.4 Support for Biology and Ecology Studies

Photo-ID can be a useful complement to studies of size at maturity (Acuña-Marrero et al., 2014; Deakos, 2010; Marshall and Bennett, 2010b; Norman and Stevens, 2007; Ramírez-Macías et al., 2012; Rohner et al., 2015), gestation period and reproductive periodicity (Bansemer and Bennett, 2011; Deakos et al., 2011; Marshall and Bennett, 2010b), reproductive behavior (Bansemer and Bennett, 2011; Whitney et al., 2004; Yano et al., 1999), survivorship (Bradshaw et al., 2007; Couturier et al., 2014; Smallegange et al., 2016), growth (Norman and Morgan, 2016; Sims et al., 2000), and longevity (Anderson et al., 2011; Couturier et al., 2014; Norman and Morgan, 2016). One whale shark has been returning to Ningaloo Reef in Australia from 1995 until at least 2016 (Norman and Morgan, 2016), and a male reef manta ray, first sighted when visibly mature in 1982, was re-sighted 30 years later at Lady Elliot Island on the Great Barrier Reef in Australia (Couturier et al., 2014). Where length or other body morphometrics are recorded, photo-ID can be incorporated to investigate individual growth (Graham and Roberts, 2007; Rohner et al., 2015). One of the advantages of photo-ID for such studies is the additional data that can often be collected concurrently to aid in interpretation of results through, for example, assessing how sexual segregation can influence residency and movement patterns (Bansemer and Bennett, 2009; Deakos et al., 2011; Robbins, 2007). Incorporating knowledge of breeding status, such as differences in movement patterns between pregnant and non-pregnant females (Bansemer and Bennett, 2009), can considerably advance understanding of the ecology and management of these animals.

12.4.5 Other Studies

Photo-ID has also been used to examine predator–prey and competitive interactions in elasmobranchs. Marshall and Bennett (2010a) investigated the frequency and effect of shark predation on a population of reef manta rays in southern Mozambique by examining the size, number, and positioning of shark-inflicted bite wounds on individual rays over time. Potential predators were identified through bite mark analysis, and the bite wounds themselves were monitored to track healing rates. Wound healing in whale sharks (Fitzpatrick et al., 2006) and white sharks (Domeier and Nasby-Lucas, 2007; Towner et al., 2012) has also been examined in cases of shark bites, and also for human-induced injuries (Riley et al., 2009). A broader comparison between scarring frequency and origins and their influence on survivorship was conducted among whale shark aggregations in Mozambique, the Seychelles, and Western Australia (Speed et al., 2008). Photo-ID is also a useful addition to studies investigating and quantifying threats such as fishing-related injuries (Bansemer and Bennett, 2008; Riley et al., 2009), net entanglement, and boat strike (Deakos et al., 2011; Speed et al., 2008).

12.5 CHALLENGES ASSOCIATED WITH PHOTO-ID

Like any method, there are drawbacks, limitations, and potential sources of error with photo-ID studies that must be considered to limit bias and ensure robust results. Photo-ID requires both the presence of a photographer, either human or automated, and a relatively close approach by the animal. The presence of boats or divers can have a significant effect on shark behavior, either as an attractant (Bruce and Bradford, 2013) or repellent (Brunnschweiler and Barnett, 2013), particularly where provisioning tourism is a factor (Gallagher et al., 2015; Laroche et al., 2007). Concurrent photo-ID and acoustic telemetry studies have demonstrated that sharks may often be present in the area but not documented via photo-ID (Brunnschweiler and Barnett, 2013; Cagua et al., 2015; Delaney et al., 2012), thereby underestimating residency when using photo-ID alone. This may be overcome in part by using alternative approaches that have less influence on shark behavior, such as rebreather diving systems rather than open-circuit equipment (Lindfield et al., 2014), or by using remote cameras at defined aggregation sites, such as cleaning stations (O'Shea et al., 2010). Passive acoustic receivers also have a far larger range for detection than visual ID approaches, typically ~500 m, and offer continuous sampling coverage. Their disadvantages are the cost of purchase, installation, and maintenance and the normal factors associated with the use of electronic tags (see Section 12.2). Ultimately, the use of complementary methods to assess the bias associated with observer presence can add considerable value to a study and should be implemented where feasible (Brunnschweiler and Barnett, 2013; Cagua et al., 2015; Chapple et al., 2016; Delaney et al., 2012).

Photo-ID is also commonly used in conjunction with mark–recapture models (Section 12.4.2). Individual heterogeneity in shark behavior can, however, cause bias in these analyses (Burgess et al., 2014), and this should be explicitly tested for during modeling studies (Burnham et al., 1987; Holmberg et al., 2009). A re-sighting bias can occur for individuals in which multiple sightings have already been obtained (Holmberg et al., 2009; Van Tienhoven et al., 2007). Individual differences in the sightability of sharks can manifest in a variety of circumstances. On a broad scale, heterogeneity in survivorship estimates has been shown in whale sharks off Western Australia, due to the presence of large numbers of transient individuals (Holmberg et al., 2008), and inferred in whale shark study populations in Belize (Graham and Roberts, 2007) and the Maldives (Riley et al., 2010). Sex- or size-based segregation is also commonly present at aggregation sites (Jacoby et al., 2012). On a finer scale, size-based dominance patterns in white sharks could mean that subordinate sharks are excluded from the area or from the surface, leading to a lower probability of sightings or successful photo-ID (Burgess et al., 2014). If chum is being used to attract sharks, the sharks may also learn to ignore this stimulus when no food reward is offered, also leading to underreporting (Laroche et al., 2007). A failure to detect previously identified individuals that are present would lead to overestimation of population size, whereas a failure to detect unmarked individuals (such as by exclusion through dominance behavior) would cause the population size to be underestimated (Burgess et al., 2014; Irion et al., 2017). Therefore, careful consideration of model assumptions is necessary in mark–recapture studies based on photo-ID data (Holmberg et al., 2008, 2009).

The issue of "tag loss" through changes in natural markings over time has been discussed in Section 12.3.2. It is important to note, however, that although regular effort at a study site may detect fine-scale changes in identifiable characteristics as they occur, allowing individual identification to be continuously updated, long breaks between field work may lead to misidentification of previously identified individuals and consequent overestimation of population abundance (Towner et al., 2013).

Matching errors between photo-identified individuals, through either incorrect assignment of previously identified individuals as "new" or accidental confusion of two different animals with similar markings, can also occur. These can largely be avoided by following the photographic and processing workflow discussed in the following section (Section 12.6), although neither human nor algorithmic matching is 100% accurate (Andreotti et al., 2017; Dureuil et al., 2015;

Speed et al., 2007; Van Tienhoven et al., 2007). Increased automation, particularly when combined with larger photographic databases, is likely to lead to a larger number of incorrect assignments by matching algorithms (in absolute terms). It is therefore important to quantify and incorporate this error rate into analyses using these datasets. This issue is discussed further in Section 12.7.

12.6   PRACTICAL CONSIDERATIONS FOR PHOTO-ID

12.6.1   Photographic Equipment

Photography (either still photography or videography) is generally agreed to be the best method of recording the appearance of natural markings or scars. Photographs can freeze motion and record extremely detailed information, allowing individuals with similar markings or scars to be reliably separated from one another. Photographs also allow a permanent record to be kept for each encounter that can be examined in detail at a later stage and verified by independent observers. Standardized images can also be used in current or future identification software programs to fully or partially automate the image matching process (Arzoumanian et al., 2005; Hughes and Burghardt, 2017; Speed et al., 2007; Van Tienhoven et al., 2007).

It is no coincidence that the popularity of photo-ID as a research technique has grown in conjunction with the increasing use of digital photographic equipment (Markowitz et al., 2003). Digital cameras allow confirmation that suitable pictures have been captured in the field and simplify post-processing workflow, computer-assisted matching, and the storage of images. Decisions on the specific equipment requirements (cameras, lenses) are best made on a case-by-case basis. Digital single-lens reflex (DSLR) cameras, mirrorless cameras, and compact cameras, as well as "action cameras," such as the GoPro™ line and other video cameras, are all in routine use in current studies. Earlier studies have found that video footage was seldom clear enough for successful extraction of photo-IDs (Meekan et al., 2006), but modern video cameras (shooting in high-definition, 4K, or higher resolutions) usually produce acceptable frame grabs in reasonable water visibility (Dureuil et al., 2015). At the time of writing, advanced DSLR cameras still have some autofocus advantages over most mirrorless and compact cameras which may be particularly useful in dorsal fin photo-ID studies, but the gap is rapidly narrowing. Underwater, advanced compact cameras and mirrorless cameras have advantages in reduced size and system cost, although the cost of a complete system, particularly if one or more flash units are required, is still significant. Almost all modern cameras have sufficient resolution for scientific studies. If color correction is required for images, which is routinely the case for underwater images, a camera that shoots

raw images (DNG or the equivalent proprietary format) is advantageous because these contain significantly more data than images processed in-camera. However, they do require more manual processing time, which may not be a worthwhile trade-off in photo-ID studies (Gore et al., 2016).

Large sharks or rays, such as whale sharks or manta rays, are best photographed with rectilinear wide-angle or even fisheye lenses (the latter is the personal preference of authors SJP and ADM). These lenses have an ultrawide field of view that can capture a large portion of the entire animal in a single frame, often capturing a view of additional data such as scars, deformities, or sex. However, compensation for distortion may be necessary with this type of lens to avoid misrepresentation of patterns, particularly if measurements are being obtained from photographs (Bansemer and Bennett, 2008; Deakos, 2010). Where images are significantly distorted, to the extent that computer-based image analysis on natural markings is affected, such distortion could prevent automated individual identification from images. Above the water, a telephoto zoom lens will often be appropriate for photographing dorsal fins as they break the surface. In some cases, artificial lighting (generally through external flash units or video lights) or color-correction filters are useful to capture detailed natural markings properly in filtered underwater light (Marshall et al., 2011).

12.6.2   Standardizing Identification

A standardized area or areas on the animal's body should be chosen for each species. A good reference area will be easy to reliably photograph on a free-swimming shark while also minimizing the influence of the photographer on the animal's natural behavior. For species that have differing patterns or marks on either side of their body it is generally considered best practice to photograph the spot patterns or scarring on one predetermined side of the animal consistently (e.g., the left side) to avoid double-counting individuals, as this could lead to overestimation of population size (Arzoumanian et al., 2005; Dudgeon et al., 2008; Meekan et al., 2006; Van Tienhoven et al., 2007; Whitney et al., 2011), although both sides should be photographed whenever possible (Bonner and Holmberg, 2013; Domeier and Nasby-Lucas, 2007). Taking photographs of multiple standardized areas (e.g., dorsal and ventral surfaces, or both left and right sides of the body) can make re-sighting identification easier and more accurate by allowing for independent confirmation (Dureuil et al., 2015; Robbins and Fox, 2013). Where scars or marks are used as identifying features, assigning a standardized area (e.g., dorsal fin) is also appropriate (Anderson et al., 2011). Reference points (i.e., body parts that can be used to help scale and rotate the image appropriately) may be required by software-matching systems (Speed et al., 2007; Van Tienhoven et al., 2007). Using a reference point, such as the area just behind or between the gill slits or pectoral fins, is also a good way to ensure that the photographed area remains consistent. If a photographic study includes historical images or those donated by the public, using areas that appear consistently in non-specialist photographs is also an important consideration.

12.6.3 Minimizing Sources of Error

Aside from the previously discussed requirements for validation and standardization, a series of other workflow steps should be implemented to ensure accurate matches. In the field, collection of photo-IDs may be influenced by environmental conditions such as wind, swell, glare, visibility, or currents. This variation in detectability is important to account for when standardizing for effort in data analyses, and detailed field logs should be filled out for each photo-ID survey (Evans and Hammond, 2004; Rohner et al., 2013).

Several basic procedures can be implemented to avoid accidental misidentification of individuals in the field. It is useful to take a photograph of a survey sheet or hand signals between each animal, particularly where multiple photos of each individual are taken, so as to associate photos with that individual and distinguish it from any following encounters (Evans and Hammond, 2004). This is particularly the case where a manual assessment of sex and size is made. The camera should be set to the local time and date. Careful downloading, storage, and labeling of images is also necessary to prevent confusion regarding when the photographs were taken and where they were sourced from. Photographs should be cataloged carefully; for example, photographs from a particular survey could be imported into folders arranged in a hierarchical format, such as year, month, and day. Images should be taken perpendicular to the area of interest because the perspective and perception of markings can change with the movement of the animal or the position of the photographer (Bansemer and Bennett, 2008; Van Tienhoven et al., 2007), resulting in increased potential for false-negative matches.

Training programs and reference images are useful for maintaining data quality. This is particularly important when images from citizen scientists are solicited. Although such programs can provide an extremely useful boost to the quantity and geographical extent of data collected, it is important to maintain quality and consistency within the dataset that is actually used for matching (Gubili et al., 2009). It can be useful to develop explicit criteria that must be met for photographs to be included in the matching dataset. Matching itself can also be sped up and made more accurate by categorizing images by metadata such as sex and size (Marshall et al., 2011) or the patterns of coloration or shape that may be relevant to ID classification (Domeier and Nasby-Lucas, 2007; Gore et al., 2016). It is important to note that when more than one person is matching images there exists a potential for bias. Thus, implementing a peer-review system for identifications is a useful means of quality control in photographic datasets (Holmberg et al., 2009).

12.7 COLLABORATIVE DATABASES AND COMPUTER-ASSISTED MATCHING SYSTEMS

Collaboration and increased automation are helping photo-ID studies to achieve greater breadth and depth of coverage for sharks. However, these advances also come with new challenges for humans and wildlife.

12.7.1 Collaborative Databases

Collaboration in photo-ID is a clear pathway to overcoming individual project resource constraints by engaging other research efforts, overlapping tourism activities (e.g., diving, snorkeling) and the public in both data collection and potentially curation, analysis, and publication as well. Collaborative photo-ID projects have reported over 10× increases in data collection through collaboration with tourists, enabling more detailed models for population analysis (Holmberg et al., 2008, 2009), and they have linked shared populations of migratory whale sharks across political, geographic, and individual research project boundaries in the Gulf of Mexico and Caribbean (McKinney et al., 2017). Collaborative platforms and communities provide a powerful foundation for new inquiries among existing researchers, a foundation for the rapid start-up of new field sites, and an open environment for novel, unanticipated, and independent results from disparate participants (Araujo et al., 2016; McKinney et al., 2017; Robinson et al., 2016). Collaboration between scientists and citizen scientists (through data collection and participation in research) can also build relationships between these communities and facilitate buy-in from local stakeholders for the conservation of threatened species (see Chapter 16 in this volume).

Critical to the success of collaborative photo-ID efforts is the ability to scale a project's data collection and curation in an accessible, standardized, equitable, and secure manner. Web-based software such as the Wildbook platform (www.wildbook.org), which originated out of collaborative studies of whale sharks (Arzoumanian et al., 2005; Holmberg et al., 2008), can provide URL-based accessibility to securely engage multiple stakeholders, communities, and projects in shared data collection, management, and analysis. Fundamentally underpinning collaborative databases is a shared information architecture or schema and an accepted study design, allowing for a common understanding of data definitions and types (e.g., date format, study site boundaries, individual identity, GPS coordinate format) and establishing common protocols for data capture and management (e.g., optimum angles and equipment for photographing the species). The Darwin Core biodiversity data model (Wieczorek et al., 2012) provides an excellent foundation for a collaborative photo-ID schema but has required some modification to expressly reflect and store individual identity under photo-ID (Holmberg et al., 2008, 2009). One significant advantage of using an existing data standard, such as the Darwin Core, is the ability to more easily exchange data with other facilities, such as pushing species occurrence data to the Global Biodiversity Information Facility (GBIF, www.gbif.org) and the Ocean Biogeographic Information System (OBIS, www.iobis.org) for long-term storage and third-party analysis.

Although the application of new technology and the *a priori* creation of a good study design can solve many problems in photo-ID and scalability, the potential for conflict among human participants is present in collaborative projects. Shared user agreements among participants (especially agreements addressing publication rights, such as MantaMatcher's User Agreement, http://www.mantamatcher.org/userAgreement.jsp), joint approval of the study design, and nondisclosure agreements can help set expectations early, prevent misunderstandings, and define acceptable methods of resolution where unanticipated disagreements arise. An independent managing authority can also assist in conflict resolution and take responsibility for collaborative database advancement, promotion, maintenance, and sustainability.

12.7.2   Data Security for Humans and Wildlife

Care must be taken in collaborative photo-ID studies to protect any personal information about human participants (e.g., personal information about contributing members of the public) as well as collected photographs and metadata about individually identified animals. Improper exposure of personal data can lead to third-party harassment (e.g., email spamming), and improper exposure of wildlife data (e.g., locations, dates) could potentially be used to better target fishing or tourism activity, exposing the study population to greater impact or even mortality. Authentication, authorization (e.g., function-limiting roles for study participants), and accounting (AAA) software security is recommended for collaborative databases, especially as they grow in scope and participation. Wildbook provides open-source examples of how such security can be flexibly configured and implemented in global-scale research efforts for sharks and rays (e.g., Wildbook for Whale Sharks, www.whaleshark.org; MantaMatcher, www.mantamatcher.org; Spotashark, http://www.spotashark.com). Data security should be addressed at the beginning of photo-ID studies and reflect species- and location-specific threats to personal and wildlife security.

12.7.3   Computer-Assisted Matching of Individuals

Manual (or "by eye") matching of photographs has a declining return on time investment. Dedicated, expert matching can achieve a high proportion of successful matches (Chapple et al., 2009; Gore et al., 2016), but manual processing of identification photographs does not scale well. As Duyck et al. (2015) pointed out, "at 10 seconds per comparison, a 10,000-sized catalog will take approximately 15 person-years to analyze." Although promising new and more in-depth forms of analysis, growth in data from collaboration in photo-ID can conversely slow data curation and introduce additional human bias in collection and curation.

Computer assistance in photograph matching of individual sharks and rays has emerged as a scalable and potentially less biased method that also reduces researcher time and effort. Arzoumanian et al. (2005) introduced computer-assisted matching of whale shark photographs based on the natural spots on their flanks, adapting an algorithm originally developed to match celestial star patterns between photographs (Groth, 1986). Van Tienhoven et al. (2007) introduced a simple, nearest-neighbor-based spot pattern matching algorithm for sand tiger sharks with the I3S software application. Both algorithms are now available for use on species with spot patterns in the open-source Wildbook platform (www.wildbook.org) and are specifically implemented for a global shark research community online in www.whaleshark.org. Other computer-assisted matching applications for sharks and rays have been developed by Hughes and Burghardt (2017) for white sharks (using the natural shape and notches on the trailing edge of the dorsal fin as a unique fingerprint) and Town et al. (2013) for manta rays based on natural, high-contrast markings on their ventral sides. Because multiple areas and forms of individual identification may exist for a single species, advancement of computer assistance has also introduced the need for new research on photographic mark–recapture modeling in the presence of multiple marks (Bonner and Holmberg, 2013).

Successful implementation of one or more computer-assisted algorithms offers a powerful incentive for collaboration, providing a demonstrable savings of time and effort in exchange for collaborative access to data. Such web-based implementations and collaborations (e.g., MantaMatcher) have led to new insights into whale shark abundance (Holmberg et al., 2008, 2009) and movement across borders and research catalogs (McKinney et al., 2017). Implementation online through web-browser access further reduces barriers of accessibility and usability across borders and studies, shifting the computationally intensive matching operations away from disparate desktop systems and resource-constrained users and into more scalable cloud-computing environments. For example, *n* number of virtual computers and CPUs are flexibly and scalably engaged in Amazon Web Services (https://aws.amazon.com/) to quickly match 28,000+ left-side whale shark spot patterns in parallel for www.whaleshark.org, allowing for global access to rapid matching (often completed in less than 4 minutes) through powerful and increasingly inexpensive grid computing.

One important note about current computer-assisted matching systems for sharks and rays: All existing systems either require some amount of human intervention (e.g., manually mapping spots onto whale shark photos before computer-assisted matching) (Arzoumanian et al., 2005; Van

Tienhoven et al., 2007) or can be significantly optimized by an optional manual step, such as cropping images down to a predefined area of the body (Town et al., 2013) or selecting reference points in the image (Hughes and Burghardt, 2016). The required operations are generally less than 2 minutes per photograph and offer significant time savings overall, but none of the systems completely removes the burden of human photographic curation or human analysis of the resulting list of potential matches.

In the near future, evolution of computer-assisted photo-ID for wildlife will significantly engage artificial intelligence (e.g., computer vision trained by deep convolutional neural networks) and remove systematic human involvement in photograph analysis (Menon et al., 2017; Parham et al., 2017). This will enable the extraction of photographic data from disparate data sources, including social media (Menon et al., 2017), and the data mining of video archives such as YouTube.com. This will reduce or remove the role of human input in answering fundamental questions, such as whether the photo-IDs are of new or previously identified individuals or determining how many individuals are present in the study population.

Critical to a fully automated future in photo-ID are a number of required research efforts, including a benchmark of population estimates, biases, and errors from human-curated photo-IDs vs. fully automated computer estimates of the same dataset using only a cloud of photographs and related metadata (e.g., location, date) as inputs. Current population models require fixed-duration "capture" sessions with longer time periods between captures, effectively leaving out data that can be continuously obtained from collaborative activities, such as diving and snorkeling tourism. Research into continuous-time population models that allow for higher volumes of data to be collected at daily intervals could allow for more accurate parameter estimates and detailed ecological insights. More information is also needed on the relative biases involved when studies include multiple modes of data collection, such as data collected from trained researchers, from lightly or untrained tourists, and from social media sources (e.g., YouTube) or which are collected at different spatial and temporal scales.

12.7.4 Leading the Way Forward

Photo-ID studies for sharks and rays have significantly led broader efforts for computer-assisted research on wildlife populations. The datasets acquired and carefully curated over the past two decades (e.g., MantaMatcher, www.mantamatcher.org; Spotashark, http://www.spotashark.com) are likely to provide the foundation for the development of new techniques in computer vision and population analysis for both marine and terrestrial species. Artificial intelligence and computer vision are already increasing data volume (www.whaleshark.org) and reducing the required effort for analysis. This trend is likely to push researchers into new

interactions with machines (computers, drones, etc.) and change roles and responsibilities within research projects, with an increasing focus on understanding and successfully implementing technology. By integrating the cameras of tourists and citizen scientists with research work and augmenting researchers with computer vision and artificial intelligence, we can plausibly imagine a wildlife research and conservation community that is continuously informed about animal population sizes and their individual interactions, movements, and behaviors.

12.8 SUMMARY AND CONCLUSIONS

Photo-ID is a relatively simple research technique, usually requiring only off-the-shelf components and a basic level of training. Its non-invasive nature lends itself to use on opportunistic platforms, such as tourist vessels, and the extension of data collection via citizen science programs, enhancing outreach and public engagement potential. However, the simplicity of photo-ID should not be confused with a lack of power.

Use of photo-ID in shark research continues to increase, and camera sensors and battery life continue to improve. Cameras are getting smaller, sensors are increasing in their resolving power and low-light capabilities, and the number of potential platforms to which they can be affixed is rapidly expanding. It is now possible to extract photo-IDs from autonomous underwater and aerial drones and dedicated research platforms such as animal-mounted tags, baited remote underwater video (BRUV) survey systems (see Chapter 7 in this volume), and remotely operated vehicles (ROVs) (see Chapter 6 in this volume). It is likely that photo-ID will be increasingly used in conjunction with remote cameras placed at such sites as cleaning stations and areas of feeding and reproductive importance, providing improved sampling coverage and standardization over time (Bicknell et al., 2016; Oliver et al., 2011; O'Shea et al., 2010). Taken together, this expansion of use-case scenarios will allow for photo-ID studies of species that may not be adequately surveyed by scientific or recreational divers or that live below normal diving depths.

As well as being an important study methodology in its own right, photo-ID can facilitate or extend studies using complementary techniques, such as telemetry studies, either by bolstering sample size (Guttridge et al., 2017) or by allowing continued long-term monitoring of individuals following tag loss. In the case of white sharks, a combination of satellite tagging and photo-identification allowed return migration from Australian to South African waters to be established (Bonfil et al., 2005), and a whale shark was tracked from the Gulf of Mexico to the mid-Atlantic off Brazil and back (Hueter et al., 2013). Tagging and sighting data can, in fact, be combined within mark–recapture models to improve the precision of results and to help mitigate the lower detectability of photo-ID-only studies (Chapple et al., 2016; Dudgeon et al., 2015, Lee et al., 2014). Incorporating individual IDs into other study methods that may include predictable biases, such as inflation of shark counts in underwater visual census (Ward-Paige et al., 2010) and the underestimation of abundance from BRUV systems (Willis et al., 2000), can also enhance the results obtained from these techniques.

The resolution of modern cameras allows the extraction of considerable information from photographs. This could include details such as parasite loading (Mucientes et al., 2008) and other individual health and fitness information, such as infection or body proportions. Our marine mammal research colleagues are currently well ahead in this area, and it is worth perusing the literature in that field to assess the possibilities (e.g., Hunt et al., 2013, 2015). Elasmobranch photo-ID studies will be enhanced with increasing use of photogrammetric techniques, which can concurrently evaluate the length, body condition, and mass of individuals (Shortis et al., 2009; Waite et al., 2007).

A primary benefit of using photo-ID is the ability to expand data collection through integration with citizen science initiatives, thus enhancing collaboration opportunities due to the ease of matching standardized photographs among research groups. The use of photographs means that it is easy to verify the accuracy of public submissions and could be particularly useful for population studies of elasmobranchs that appear to be at lower than optimal densities for cost-effective dedicated surveys or cryptic species. Many of the larger photo-identifiable species, such as white sharks and basking sharks, routinely traverse political boundaries (Bonfil et al., 2005, 2010; Gore et al., 2008; Skomal et al., 2009). Photo-ID provides a cost-effective means of assessing population-level interchange between discrete areas, the products of which can improve population estimates and stock delineation. To fulfill this potential, there is a need for standardization of species-specific techniques between research groups and increased movement toward routine data sharing. Data collection, processing, and sharing will all be facilitated by computer-assisted data mining and identification, enabling a "big data" approach to shark science.

Photo-ID studies are steadily expanding to new species and sites and asking more ambitious questions. Photo-ID offers a useful alternative or adjunct to conventional tagging where its assumptions and practical constraints are met, and the widespread adoption of this research technique through the scientific community is enhancing opportunities for the public to become directly involved in projects. This can benefit researchers while offering an educational experience for interested participants. As emerging technologies increasingly allow the diverse ecology and behaviors of sharks to be observed first hand, we hope that more and more scientists will bring their cameras along for the journey.

ACKNOWLEDGMENTS

We thank our respective co-authors and collaborators for helping shape our thoughts on this topic. We greatly appreciate comments on this chapter from Chris Rohner, Colin Simpfendorfer, and Jeff Carrier, and we thank Brit Finucci for her advice on the individual identification of chimaeras. Although no specific funding was applied to this work, SJP's research is supported by two private trusts and Aqua-Firma.

REFERENCES

Acuña-Marrero D, Jiménez J, Smith F, Doherty Jr PF, Hearn A, Green JR, Paredes-Jarrín J, Salinas-de-León P (2014) Whale shark (*Rhincodon typus*) seasonal presence, residence time and habitat use at Darwin Island, Galapagos Marine Reserve. *PLoS ONE* 9:e115946.

Anderson SD, Goldman KJ (1996) Photographic evidence of white shark movements in California waters. *Calif Fish Game* 82:182–186.

Anderson SD, Chapple TK, Jorgensen, SJ, Klimley AP, Block BA (2011) Long-term individual identification and site fidelity of white sharks, *Carcharodon carcharias*, off California using dorsal fins. *Mar Biol* 158:1233–1237.

Andreotti S, Holtzhausen P, Rutzen M, Meÿer M, van der Walt S, Herbst B, Matthee CA (2017) Semi-automated software for dorsal fin photographic identification of marine species: application to *Carcharodon carcharias*. *Mar Biodiv* 1–6.

Andrzejaczek S, Meeuwig J, Rowat D, Pierce S, Davies T, Fisher R, Meekan M (2016) The ecological connectivity of whale shark aggregations in the Indian Ocean: a photo-identification approach. *R Soc Open Sci* 3:160455.

Araujo G, Snow S, So CL, Labaja J, Murray R, Colucci A, Ponzo A (2016) Population structure, residency patterns and movements of whale sharks in Southern Leyte, Philippines: results from dedicated photo-ID and citizen science. *Aquat Conserv* 27:237–252.

Ari C (2014) Rapid colouration changes of manta rays (Mobulidae). *Biol J Linnean Soc* 113:180–193.

Arzoumanian Z, Holmberg J, Norman B (2005) An astronomical pattern-matching algorithm for computer-aided identification of whale sharks, *Rhincodon typus*. *J Appl Ecol* 42:999–1011.

Bansemer CS, Bennett MB (2008) Multi-year validation of photographic identification of grey nurse sharks, *Carcharias taurus*, and applications for non-invasive conservation research. *Mar Freshwater Res* 59:322–331.

Bansemer CS, Bennett MB (2009) Reproductive periodicity, localised movements and behavioural segregation of pregnant *Carcharias taurus* at Wolf Rock, southeast Queensland, Australia. *Mar Ecol Prog Ser* 374:215–227.

Bansemer CS, Bennett MB (2011) Sex- and maturity-based differences in movements and migration patterns of grey nurse sharks, *Carcharias taurus*, along the eastern coast of Australia. *Mar Freshwater Res* 62:596–606.

Barker SM, Williamson JE (2010) Collaborative photo-identification and monitoring of grey nurse sharks (*Carcharias taurus*) at key aggregation sites along the eastern coast of Australia. *Mar Freshwater Res* 61:971–979.

Bicknell AWJ, Godley BJ, Sheehan EV, Votier SC, Witt MJ (2016) Camera technology for monitoring marine biodiversity and human impact. *Front Ecol Environ* 14:424–432.

Bonfil R, Meyer M, Scholl MC, Johnson R, O'Brian S, Oosthuizen H, Swanson S, Kotze D, Paterson M (2005) Transoceanic migration, spatial dynamics and population linkages of white sharks. *Science* 310:100–103.

Bonfil R, Francis MP, Duffy C, Manning MJ, O'Brian SO (2010) Large-scale tropical movements and diving behaviour of white sharks *Carcharodon carcharias* tagged off New Zealand. *Aquat Biol* 8: 115–123.

Bonner SJ, Holmberg J (2013) Mark–recapture with multiple, non-invasive marks. *Biometrics* 69:766–775.

Bradshaw CJA, Mollet HF, Meekan MG (2007) Inferring population trends for the world's largest fish from mark–recapture estimates of survival. *J Anim Ecol* 76:480–489.

Brooks K, Rowat D, Pierce SJ, Jouannet D, Vely M (2010) Seeing spots: photo-identification as a regional tool for whale shark identification. *WIO J Mar Sci* 9:185–194.

Bruce BD, Bradford RW (2013) The effects of shark cage-diving operations on the behaviour and movements of white sharks, *Carcharodon carcharias*, at the Neptune Islands, South Australia. *Mar Biol* 160:889–907.

Brunnschweiler JM, Baensch H (2011) Seasonal and long-term changes in relative abundance of bull sharks from a tourist feeding site in Fiji. *PLoS ONE* 6:e16597.

Brunnschweiler JM, Barnett A (2013) Opportunistic visitors: long-term behavioural response of bull sharks to food provisioning in Fiji. *PLoS ONE* 8:e58522.

Buray N, Mourier J, Planes S, Clua E (2009) Underwater photo-identification of sicklefin lemon sharks, *Negaprion acutidens*, at Moorea (French Polynesia). *Cybium* 33:21–27.

Burgess GH, Bruce BD, Cailliet GM, Goldman KJ, Grubbs RD, Lowe CG, MacNeil MA, Mollet HF, Weng KC, O'Sullivan JB (2014) A re-evaluation of the size of the white shark (*Carcharodon carcharias*) population off California, USA. *PLoS ONE* 9:e98078.

Burnham KP, Anderson DR, White GC, Brownie C, Pollock KH (1987) Design and analysis methods for fish survival experiments based on release-recapture. *Am Fish Soc Monogr* 5:1–437.

Cagua EF, Cochran JEM, Rohner CA, Prebble CEM, Sinclair-Taylor TH, Pierce SJ, Berumen ML (2015) Acoustic telemetry reveals cryptic residency of whale sharks. *Biol Lett* 11:20150092.

Carraro R, Gladstone W (2006) Habitat preferences and site fidelity of the ornate wobbegong shark (*Orectolobus ornatus*) on rocky reefs of New South Wales. *Pac Sci* 60:207–223.

Castro ALF, Rosa RS (2005) Use of natural marks on population estimates of the nurse shark, *Ginglymostoma cirratum*, at Atol das Rocas Biological Reserve, Brazil. *Environ Biol Fish* 72:213–221.

Chapman DD, Feldheim KA, Papastamatiou YP, Hueter RE (2015) There and back again: review of residency and return migrations in sharks, with implications for population structure and management. *Annu Rev Mar Sci* 7:547–570.

Chapple TK, Jorgensen SJ, Anderson SD, Kanive PE, Klimley AP, Botsford LW, Block BA (2011) A first estimate of white shark, *Carcharodon carcharias*, abundance off Central California. *Biol Lett* 7:581–583.

Chapple TK, Chambert T, Kanive PE, Jorgensen SJ, Rotella JJ, Anderson SD, Carlisle AB, Block BA (2016) A novel application of multi-event modeling to estimate class segregation in a highly migratory oceanic vertebrate. *Ecology* 97:3494–3502.

Corcoran MJ, Gruber SH (1999) The use of photo-identification to study the social organization of the spotted eagle ray, *Aetobatus narinari. Bahamas J Sci* 11:21–27.

Couturier LIE, Marshall AD, Jaine FRA, Kashiwagi T, Pierce SJ, Townsend KA, Weeks SJ, Bennett MB, Richardson AJ (2012) Biology, ecology and conservation of the Mobulidae. *J Fish Biol* 80:1075–1119.

Couturier LIE, Dudgeon CL, Pollock KH, Jaine FRA, Bennett MB, Townsend KA, Weeks SJ, Richardson AJ (2014) Population dynamics of the reef manta ray *Manta alfredi* in eastern Australia. *Coral Reefs* 33:329–342.

Davies TK, Stevens G, Meekan MG, Struve J, Rowcliffe JM (2012) Can citizen science monitor whale-shark aggregations? Investigating bias in mark–recapture modelling using identification photographs sourced from the public. *Wildlife Res* 39:696–704.

Deakos MH (2010) Paired-laser photogrammetry as a simple and accurate system for measuring the body size of free-ranging manta rays *Manta alfredi. Aquat Biol* 10:1–10.

Deakos MH, Baker JD, Bejder L (2011) Characteristics of a manta ray *Manta alfredi* population off Maui, Hawaii, and implications for management. *Mar Ecol Prog Ser* 429:245–260.

Delaney DG, Johnson R, Bester MN, Gennari E (2012) Accuracy of using visual identification of white sharks to estimate residency patterns. *PLoS ONE* 7:e34753.

Dicken ML, Booth AJ, Smale MJ (2006) Preliminary observations of tag shedding, tag reporting, tag wounds, and tag biofouling for raggedtooth sharks (*Carcharias taurus*) tagged off the east coast of South Africa. *ICES J Mar Sci* 63:1640–1648.

Domeier ML, Nasby-Lucas N (2007) Annual re-sightings of photographically identified white sharks (*Carcharodon carcharias*) at an eastern Pacific aggregation site (Guadalupe Island, Mexico). *Mar Biol* 150:977–984.

Dudgeon CL, Noad MJ, Lanyon JM (2008) Abundance and demography of a seasonal aggregation of zebra sharks *Stegostoma fasciatum. Mar Ecol Prog Ser* 368:269–281.

Dudgeon CL, Pollock KH, Braccini JM, Semmens JM, Barnett A (2015) Integrating acoustic telemetry into mark-recapture models to improve the precision of apparent survival and abundance estimates. *Oecologia* 178:761–772.

Dureuil M, Towner AV, Ciolfi LG, Beck LA (2015) A computer-aided framework for subsurface identification of white shark pigment patterns. *Afr J Mar Sci* 37:363–371.

Duyck J, Finn C, Hutcheon A, Vera P, Salas J, Ravela S (2015) Sloop: a pattern retrieval engine for individual animal identification. *Pattern Recognit* 48:1059–1073.

Evans PGH, Hammond PS (2004) Monitoring cetaceans in European waters. *Mamm Rev* 34:131–156.

Feldheim KA, Gruber SH, Ashley MV (2002) The breeding biology of lemon sharks at a tropical nursery lagoon. *Proc R Soc Lond B* 269:1655–1661.

Findlay R, Gennari E, Cantor M, Tittensor, DP (2016) How solitary are white sharks: social interactions or just spatial proximity? *Behav Ecol Sociobiol* 70:1735–1744.

Fitzpatrick B, Meekan M, Richards A (2006) Shark attacks on a whale shark (*Rhincodon typus*) at Ningaloo Reef, Western Australia. *Bull Mar Sci* 78:397–402.

Fouts WR, Nelson DR (1999) Prey capture by the Pacific angel shark, *Squatina californica*: visually mediated strikes and ambush-site characteristics. *Copeia* 1999:304–312.

Gallagher AJ, Vianna GMS, Papastamatiou YP, Macdonald C, Guttridge TL, Hammerschlag N (2015) Biological effects, conservation potential, and research priorities of shark diving tourism. *Biol Conserv* 184:365–379.

Germanov ES, Marshall AD (2014) Running the gauntlet: regional movement patterns of *Manta alfredi* through a complex of parks and fisheries. *PLoS ONE* 9:e110071.

Gore MA, Rowat D, Hall J, Gell FR, Ormond RF (2008) Transatlantic migration and deep mid-ocean diving by basking shark. *Biol Lett* 4:395–398.

Gore MA, Frey PH, Ormond RF, Allan H, Gilkes G (2016) Use of photo-identification and mark–recapture methodology to assess basking shark (*Cetorhinus maximus*) populations. *PLoS ONE* 11:e0150160.

Graham R, Roberts CM (2007) Assessing the size, growth and structure of a seasonal population of whale sharks (*Rhincodon typus* Smith 1828) using conventional tagging and photo identification. *Fish Res* 84:71–80.

Groth EJ (1986) A pattern-matching algorithm for two-dimensional coordinate lists. *Astron J* 91:1244–1248.

Gubili C, Johnson R, Gennari E, Oosthuizen WH, Kotze D, Meÿer M, Sims DW, Jones CS, Noble LR (2009) Concordance of genetic and fin photo identification in the great white shark, *Carcharodon carcharias*, off Mossel Bay, South Africa. *Mar Biol* 156:2199–2207.

Guttridge TL, Van Zinnicq Bergmann MPM, Bolte C, Howey LA, Finger JS, Kessel ST, Brooks JL, et al. (2017) Philopatry and regional connectivity of the great hammerhead shark, *Sphyrna mokarran* in the U.S. and Bahamas. *Front Mar Sci* 4:3.

Hammerschlag N, Cooke SJ, Gallagher AJ, Godley BJ (2014) Considering the fate of electronic tags: interactions with stakeholders and user responsibility when encountering tagged aquatic animals. *Meth Ecol Evol* 5:1147–1153.

Hewitt, AM, Kock AA, Booth AJ, Griffiths CL (2017) Trends in sightings and population structure of white sharks, *Carcharodon carcharias*, at Seal Island, False Bay, South Africa, and the emigration of subadult female sharks approaching maturity. *Environ Biol Fish* 101:39–54.

Holmberg J, Norman B, Arzoumanian Z (2008) Robust, comparable population metrics through collaborative photo-monitoring of whale sharks *Rhincodon typus*. *Ecol Appl* 18:222–233.

Holmberg J, Norman B, Arzoumanian Z (2009) Estimating population size, structure, and residency time for whale sharks *Rhincodon typus* through collaborative photo identification. *Endangered Species Res* 7:39–53.

Hueter RE, Tyminski JP, de la Parra R (2013) Horizontal movements, migration patterns, and population structure of whale sharks in the Gulf of Mexico and northwestern Caribbean Sea. *PLoS ONE* 8:e71883.

Hughes B, Burghardt T (2017) Automated visual fin identification of individual great white sharks. *Int J Comput Vis* 122:542–557.

Hunt KE, Moore MJ, Rolland RM, Kellar NM, Hall AJ, Kershaw J, Raverty SA, Davis CE, Yeates LC, Fauquier DA, Rowles TK (2013) Overcoming the challenges of studying conservation physiology in large whales: a review of available methods. *Conserv Physiol* 1:cot006.

Hunt KE, Rolland RM, Kraus SD (2015) Conservation physiology of an uncatchable animal: the North Atlantic right whale (*Eubalaena glacialis*). *Integr Comp Biol* 55:577–86.

Hunt TN, Bejder L, Allen SJ, Rankin RW, Hanf DM, Parra GJ (2017) Demographic characteristics of Australian humpback dolphins reveal important habitat toward the southwestern limit of their range. *Endangered Species Res* 32:71–88.

Irion DT, Noble LR, Kock AA, Gennari E, Dicken ML, Hewitt AM, Towner AV, et al. (2017) Pessimistic assessment of white shark population status in South Africa: comment on Andreotti et al. (2016). *Mar Ecol Prog Ser* 577:251–255.

Jacoby DMP, Croft DP, Sims DW (2012) Social behaviour in sharks and rays: analysis, patterns and implications for conservation. *Fish Fish* 13:399–417.

Jewell OJD, Wcisel MA, Gennari E, Towner AV, Bester MN, Johnson RL, Singh S (2011) Effects of smart position only (SPOT) tag deployment on white sharks *Carcharodon carcharias* in South Africa. *PLoS ONE* 6:e27242.

Kitchen-Wheeler AM (2010) Visual identification of individual manta ray (*Manta alfredi*) in the Maldives Islands, Western Indian Ocean. *Mar Biol Res* 6:351–363.

Klimley AP, Anderson SD (1996) Residency patterns of white sharks at the South Farallon Islands, California. In: Klimley AP, Ainley DG (eds) *Great White Sharks: The Biology of Carcharodon carcharias*. Academic Press, San Diego, CA, pp 309–316.

Kock A, O'Riain MJ, Mauff K, Meÿer M, Kotze D, Griffiths C (2013) Residency, habitat use and sexual segregation of white sharks, *Carcharodon carcharias* in False Bay, South Africa. *PLoS ONE* 8:e55048.

Kohler NE, Turner PA (2001) Shark tagging: a review of conventional methods and studies. *Environ Biol Fish* 60:191–223.

Laroche RK, Kock AA, Dill LM, Oosthuizen WH (2007) Effects of provisioning ecotourism activity on the behaviour of white sharks *Carcharodon carcharias*. *Mar Ecol Prog Ser* 338:199–209.

Lee KA, Huveneers C, Gimenez O, Peddemors V, Harcourt RG (2014) To catch or to sight? A comparison of demographic parameter estimates obtained from mark-recapture and mark-resight models. *Biodivers Conserv* 23:2781–2800.

Lindfield SJ, Harvey ES, McIlwain JL, Halford AR (2014) Silent fish surveys: bubble-free diving highlights inaccuracies associated with SCUBA-based surveys in heavily fished areas. *Methods Ecol Evol* 5:1061–1069.

Luiz OJ, Balboni AP, Kodja G, Andrade M, Marum H (2009) Seasonal occurrences of *Manta birostris* (Chondrichthyes: Mobulidae) in southeastern Brazil. *Ichthyol Res* 56:96–99.

Manire CA, Gruber SH (1991) Effect of M-type dart tags on the field growth of juvenile lemon sharks. *Trans Am Fish Soc* 120:776–780.

Markowitz TA, Harlin AD, Wursig B (2003) Digital photography improves efficiency of individual dolphin identification. *Mar Mamm Sci* 19:217–223.

Marshall AD, Bennett MB (2010a) The frequency and effect of shark-inflicted bite injuries to the reef manta ray (*Manta alfredi*). *Afr J Mar Sci* 32:573–580.

Marshall AD, Bennett MB (2010b) Reproductive ecology of the reef manta ray (*Manta alfredi*) in southern Mozambique. *J Fish Biol* 77:169–190.

Marshall AD, Pierce SJ (2012) The use and abuse of photographic identification in sharks and rays. *J Fish Biol* 80:1361–1379.

Marshall AD, Pierce SJ, Bennett MB (2008) Morphological measurements of manta rays (*Manta birostris*) with a description of a foetus from the east coast of Southern Africa. *Zootaxa* 1717:24–30.

Marshall AD, Dudgeon CL, Bennett MB (2011) Size and structure of a photographically identified population of manta rays *Manta alfredi* in southern Mozambique. *Mar Biol* 158:1111–1124.

McKinney JA, Hoffmayer ER, Holmberg J, Graham RT, Driggers WB III, de la Parra-Venegas R, Galván-Pastoriza BE, et al. (2017) Long-term assessment of whale shark population demography and connectivity using photo identification in the Western Atlantic Ocean. *PLoS ONE* 12: e0180495.

Meekan MG, Bradshaw CJA, Press M, McLean C, Richards A, Quasnichka S, Taylor JG (2006) Population size and structure of whale sharks (*Rhincodon typus*) at Ningaloo Reef, Western Australia. *Mar Ecol Prog Ser* 319:275–285.

Menon S, Berger-Wolf TY, Kiciman E, Joppa L, Stewart CV, Parham J, Crall J, Holmberg J, Van Oast J (2017) Animal Population Estimation Using Flickr Images, paper presented at the 2nd International Workshop on the Social Web for Environmental and Ecological Monitoring (SWEEM 2017), June 25, Troy, NY.

Mourier J, Vercelloni J, Planes S (2012) Evidence of social communities in a spatially structured network of a free-ranging shark species. *Anim Behav* 83:389–401.

Mucientes GR, Queiroz N, Pierce SJ, Sazima I, Brunnschweiler JM (2008) Is host ectoparasite load related to echeneid fish presence? *Res Lett Ecol* 2008:107576.

Myrberg Jr AA, Gruber SH (1974) The behavior of the bonnethead shark, *Sphyrna tiburo*. *Copeia* 1974(2):358–374.

Norman BM, Morgan DL (2016) The return of "Stumpy" the whale shark: two decades and counting. *Front Ecol Environ* 14:449–450.

Norman BM, Stevens JD (2007) Size and maturity status of the whale shark (*Rhincodon typus*) at Ningaloo Reef in Western Australia. *Fish Res* 84:81–86.

Norman BM, Holmberg JA, Arzoumanian Z, Reynolds S, Wilson RP, Gleiss AC, Rob D, et al. (2017) Understanding constellations: 'citizen scientists' elucidate the global biology of a threatened marine mega-vertebrate. *BioScience* 67:1029–1043.

Oliver SP, Hussey NE, Turner JR, Beckett AJ (2011) Oceanic sharks clean at coastal seamount. *PLoS ONE* 6:e14755.

O'Shea OR, Kingsford MJ, Seymour J (2010) Tide-related periodicity of manta rays and sharks to cleaning stations on a coral reef. *Mar Freshwater Res* 61:65–73.

Parham J, Crall J, Stewart C, Berger-Wolf T, Rubenstein D (2017) Animal Population Censusing at Scale with Citizen Science and Photographic Identification, paper presented at Association for the Advancement of Artificial Intelligence (AAAI) 2017 Spring Symposium, March 27–29, Palo Alto, CA.

Pierce SJ, Bennett MB (2009) Validated annual band pair periodicity and growth parameters of blue-spotted maskray (*Neotrygon kuhlii*) from southeast Queensland, Australia. *J Fish Biol* 75:2490–2508.

Pierce SJ, Pardo SA, Bennett MB (2009) Reproduction of the blue-spotted maskray, *Neotrygon kuhlii* (Myliobatoidei: Dasyatidae), in south-east Queensland, Australia. *J Fish Biol* 74:1291–1308.

Porcher IF (2005) On the gestation period of the blackfin reef shark, *Carcharhinus melanopterus*, in waters off Moorea, French Polynesia. *Mar Biol* 146:1207–1211.

Pratt Jr HL, Carrier JC (2001) A review of elasmobranch reproductive behaviour with a case study on the nurse shark, *Ginglymostoma cirratum*. *Environ Biol Fish* 60:157–188.

Ramírez-Macías D, Vázquez-Haikin A, Vázquez-Juárez R (2012) Whale shark *Rhincodon typus* populations along the west coast of the Gulf of California and implications for management. *Endangered Species Res* 18:115–128.

Ramírez-Macías D, Queiroz N, Pierce SJ, Humphries NE, Sims DW, Brunnschweiler JM (2017) Oceanic adults, coastal juveniles: tracking the habitat use of whale sharks off the Pacific coast of Mexico. *PeerJ* 5:e3271.

Riley MJ, Harman A, Rees RG (2009) Evidence of continued hunting of whale sharks *Rhincodon typus* in the Maldives. *Environ Biol Fish* 86:371–374.

Riley MJ, Hale MS, Harman A, Rees RG (2010) Analysis of whale shark *Rhincodon typus* aggregations near South Ari Atoll, Maldives Archipelago. *Aquat Biol* 8:145–150.

Robbins RL (2007) Environmental variables affecting the sexual segregation of great white sharks *Carcharodon carcharias* at the Neptune Islands South Australia. *J Fish Biol* 70:1350–1364.

Robbins RL, Fox A (2013) Further evidence of pigmentation change in white sharks, *Carcharodon carcharias*. *Mar Freshwater Res* 63:1215–1217.

Robinson DP, Jaidah MY, Bach S, Lee K, Jabado RW, Rohner CA, March A, Caprodossi S, Henderson AC, Mair JM, Ormond R, Pierce SJ (2016) Population structure, abundance and movement of whale sharks in the Arabian Gulf and Gulf of Oman. *PLoS ONE* 11:e0158593.

Rohner CA, SJ Pierce, AD Marshall, SJ Weeks, MB Bennett, Richardson AJ (2013) Trends in sightings and environmental influences on a coastal aggregation of manta rays and whale sharks. *Mar Ecol Prog Ser* 482:153–168.

Rohner CA, Richardson AJ, Prebble CEM, Marshall AD, Bennett MB, Weeks SJ, Cliff G, Wintner SP, Pierce SJ (2015) Laser photogrammetry improves size and demographic estimates for whale sharks. *PeerJ* 3:e886.

Rowat D, Speed CW, Meekan MG, Gore M (2009) Population abundance and apparent survival of the vulnerable whale shark, *Rhincodon typus*, in the Seychelles aggregation. *Oryx* 43:591–598.

Shortis M, Harvey E, Abdo D (2009) A review of underwater stereo-image measurement for marine biology and ecology applications. *Oceanogr Mar Biol* 47:257–292.

Sims DW, Speedy CD, Fox AM (2000) Movements and growth of a female basking shark re-sighted after a three-year period. *J Mar Biol Assoc UK* 80:1141–1142.

Skomal GB, Zeeman SI, Chrisholm JH, Summers EL, Walsh HJ, McMahon KW, Thorrold SR (2009) Transequatorial migrations by basking sharks in the western Atlantic Ocean. *Curr Biol* 19:1019–1022.

Smallegange IM, van der Ouderaa IBC, Tibiriçá Y (2016) Effects of yearling, juvenile and adult survival on reef manta ray (*Manta alfredi*) demography. *PeerJ* 4:e2370.

Smith SE, Mitchell RA, Fuller D (2003) Age-validation of a leopard shark (*Triakis semifasciata*) recaptured after 20 years. *Fish Bull* 101:194–198.

Speed CW, Meekan MG, Bradshaw JA (2007) Spot the match—wildlife photo-identification using information theory. *Front Zool* 4:1–11.

Speed CW, Meekan MG, Rowat D, Pierce SJ, Marshall AD, Bradshaw CJA (2008) Scarring patterns and relative mortality rates of Indian Ocean whale shark. *J Fish Biol* 72:1488–1503.

Stansbury AL, Götz T, Deecke VB, Janik VM (2015) Grey seals use anthropogenic signals from acoustic tags to locate fish: evidence from a simulated foraging task. *Proc R Soc Lond B* 282:20141595.

Taylor G (1994) *Whale Sharks, the Giants of Ningaloo Reef.* Angus & Robertson, Sydney, Australia.

Town C, Marshall A, Sethasathien N (2013) MantaMatcher: automated photographic identification of manta rays using keypoint features. *Ecol Evol* 3:1902–1914.

Towner A, Smale MJ, Jewell O (2012) Boat strike wound healing in *Carcharodon carcharias*. In: Domeier ML (ed) *Global Perspectives on the Biology and Life History of the White Shark.* CRC Press, Boca Raton, FL, pp 77–84.

Towner AV, Wcisel MA, Reisinger RR, Edwards D, Jewell OJ (2013) Gauging the threat: the first population estimate for white sharks in South Africa using photo identification and automated software. *PLoS ONE* 8:e66035.

Van Tienhoven AM, Den Hartog JE, Reijns RA, Peddemors VM (2007) A computer-aided program for pattern-matching of natural marks on the spotted ragged-tooth shark *Carcharias taurus*. *J Appl Ecol* 44:273–280.

Waite JN, Schrader WJ, Mellish JE, Horning M (2007) Three-dimensional photogrammetry as a tool for estimating morphometrics and body mass of Stellar sea lions (*Eumetopias jubatus*). *Can J Fish Aquat Sci* 64:296–303.

Ward-Paige C, Mills Flemming J, Lotze HK (2010) Overestimating fish counts by non-instantaneous visual censuses: consequences for population and community descriptions. *PLoS ONE* 5:e11722.

Whitney NM, Pratt HL, Carrier JC (2004) Group courtship, mating behaviour, and siphon sac function in the whitetip reef shark, *Triaenodon obesus*. *Anim Behav* 68:1435–1442.

Whitney NM, Pyle RL, Holland KN, Barcz JT (2011) Movements, reproductive seasonality, and fisheries interactions in the whitetip reef shark (*Triaenodon obesus*) from community-contributed photographs. *Environ Biol Fish* 93:121–136.

Wieczorek J, Bloom D, Guralnick R, Blum S, Döring M, Giovanni R, Robertson T, Vieglais D (2012) Darwin Core: an evolving community-developed biodiversity data standard. *PLoS ONE* 7:e29715.

Williams BK, Nichols JD, Conroy MJ (2002) *Analysis and Management of Animal Populations.* Academic Press, San Diego, CA.

Willis TJ, Millar RB, Babcock RC (2000) Detection of spatial variability in relative density of fishes: comparison of visual census, angling, and baited underwater video. *Mar Ecol Prog Ser* 198:249–260.

Wilson B, Hammond PS, Thompson PM (1999) Estimating size and assessing trends in a coastal bottlenose dolphin population. *Ecol Appl* 9:288–300.

Wilson RP, McMahon CR (2006) Measuring devices on wild animals: what constitutes acceptable practice? *Front Ecol Environ* 4:147–154.

Yano K, Sato F, Takahashi T (1999) Observations of the mating behavior of the manta ray, *Manta birostris*, at the Ogasawara Islands, Japan. *Ichthyol Res* 46:289–296.

CHAPTER 13

Genetics and Genomics for Fundamental and Applied Research on Elasmobranchs

Jennifer R. Ovenden
Molecular Fisheries Laboratory, School of Biomedical Sciences, The University of Queensland, St. Lucia, Queensland, Australia

Christine Dudgeon
Molecular Fisheries Laboratory, School of Biomedical Sciences, The University of Queensland, St. Lucia, Queensland, Australia

Pierre Feutry
CSIRO Oceans and Atmosphere, Hobart, Tasmania, Australia

Kevin Feldheim
Pritzker Laboratory for Molecular Systematics and Evolution, Field Museum of Natural History, Chicago, Illinois

Gregory E. Maes
Laboratory for Cytogenetics and Genome Research, Centre for Human Genetics, University of Leuven, Leuven, Belgium

CONTENTS

13.1 INTRODUCTION

For threatened and commercially important species, genetic and genomic tools promise accurate and cost-effective evaluations to contribute to effective conservation strategies and sustainable management (Ovenden et al., 2015; Willette et al., 2014). Genetics and genomic research on elasmobranchs has generally lagged behind studies involving bony fish and other taxa. But, with the increasing interest in elasmobranchs as an important food resource, the corresponding conservation concerns (Dulvy et al., 2017; Simpfendorfer and Dulvy, 2017) provide motivation to harness the power of genetics and genomics (Bernatchez et al., 2017). The earliest genetic studies on population structure in elasmobranchs investigated variants of enzymes coded

for by different alleles in nuclear genomes (allozymes) (e.g., Gardner and Ward, 1998; Smith, 1986). This transitioned into direct examination of the DNA sequence through the advent of the polymerase chain reaction (PCR) (Saiki et al., 1988) combined with nucleotide sequencing (Sanger et al., 1977a,b). These tools allowed elasmobranch researchers to focus on the maternally inherited, haploid mitochondrial DNA (mtDNA) genome. A shift back to the nuclear genome occurred when microsatellite loci were discovered (Powell et al., 1996).

Population genetic diversity and structure can be deduced by comparing alternative forms of microsatellite loci between individuals sampled from the wild (Balloux and Lugon-Moulin, 2002). Furthermore, microsatellite alleles for individuals (genotypes) are used to examine population size (Ovenden et al., 2016), relatedness and relationships (Mourier et al., 2013), and reproductive strategies (Portnoy and Heist, 2012). Genomic technologies are capable of discovering and genotyping not tens or hundreds of loci (as for microsatellites), but tens of thousands of loci (as single nucleotide polymorphisms, or SNPs), in the nuclear genome of almost any species. High-throughput, massively parallel DNA sequencing platforms (da Fonseca et al., 2016) enable rapid and cost-efficient genome-wide SNP marker discovery and genotyping.

Carefully designed studies involving thousands of SNP loci can provide new insight into existing questions of evolution and ecology with increased precision and accuracy (Andrews and Luikart, 2014). Determining causal relationships among genomic variation, phenotypes, and the environment allows a better understanding of the genetic basis of adaptive genetic variation and speciation (Bernatchez, 2016; Nielsen et al., 2009). However, despite the potential for genomics to improve management and conservation practices through improved understanding, it is perceived as difficult to make the transition from theory to practice (McMahon et al., 2014; Shafer et al., 2015). In part, this chapter addresses the transition for elasmobranch species. It begins with the practical aspects of generating genomic data, then focuses on how genomics and existing methods of population genetics are actively being used to address knowledge gaps that are important for conservation and management, such as population structure, population size, and reproductive biology. We review the strengths and weaknesses of genomic and genetic methods to provide insight and realistic expectations for workers involved in theoretical and applied research on elasmobranchs. Details of other applications of DNA technology to elasmobranch species, such as assaying DNA from the environment (eDNA; see Chapter 14 in this volume) and studying hybridization (Marino et al., 2015; Morgan et al., 2012), can be found elsewhere.

13.2 PRACTICAL ASPECTS OF GENOMICS

This section focuses on the use of genomic methods to study the occurrence and frequency of SNPs among individuals, populations, and species. To work with SNPs, it is not necessary to start with comprehensive information about the genome of the study species. Although work is underway on many more, only three chondrichthyan whole nuclear genomes have appeared in the literature to date: (1) the little skate (*Leucoraja erinacea*) (King et al., 2011), (2) the elephant shark (*Callorhinchus milii*) (Venkatesh et al., 2014), and (3) the whale shark (*Rhincodon typus*) (Read et al., 2017). Description (annotation) of gene regions within the whole-genome data is valuable (e.g., for use in studies on selection and adaptation) but challenging due to uncertainties in the genomic data and gene identification in non-model species. Many complete mitochondrial genomes for

elasmobranch species are available on GenBank (Benson et al., 2013) and in the literature (e.g., Vargas-Caro et al., 2016); however, these genomes are not normally the starting point for genomic studies (but see Feutry et al., 2014). Transcriptomes (nuclear gene regions used to produce proteins and peptides) are available for the great white shark (*Carcharias carcharodon*) (Richards et al., 2013), the lesser spotted catshark (*Scyliorhinus canicula*), (Mulley et al., 2014), and a few other species (Marra et al., 2017) that can initiate some types of genomic studies (see below). To bypass the need for full genomes or transcriptomes, several shortcuts have been developed that focus on either pre-selected (targeted approaches) or randomly selected (random approaches, Figure 13.1) gene-regions. After discovery, generally groups of SNPs are selected and examined across a range of individuals depending on project objectives. For example, relatively few SNPs (ten to hundreds) can be used to assign individuals to species or populations (Bylemans et al., 2016; Nielsen et al., 2012), or larger numbers (hundreds to several thousand) can be used to study evolution at the population level (Pazmiño et al., 2017). The genomic methods for SNP discovery summarized here share common problems, however. These include the storage, manipulation and downstream analyses of the massive amounts of data produced by high throughput sequencing, the difficulty of producing sequence data from gene-regions with unusual characteristics, the introduction of artifacts during the preparation of the DNA for sequencing (often associated with the use of the polymerase chain reaction) and the difficulty of working with highly heterozygous gene-regions. The latter issue is of particular importance to elasmobranch researchers who will be dealing with naturally genetically diverse marine organisms. Despite this, genomics in general is moving forward so rapidly that aspects of the methods reviewed below may be soon obsolete. Geneticists working with the conservation and management of wildlife are quick to adopt practical and theoretical advances in medical and agricultural genomics, so a review of the most recent literature is essential before moving forward with research in this area.

High-throughput sequencing of amplicons (Figure 13.1) can facilitate SNP discovery and genomic analyses in a study species. Amplicons (DNA synthesized in the laboratory using PCR) represent gene regions that are preselected by the researcher as being easily targeted or likely to be valuable sources of information (Peñalba et al., 2014). Feutry et al. (2014, 2015) used this approach to collect mitogenome sequences for speartooth shark (*Glyphis glyphis*) collected in northern Australia. The primary disadvantages of this approach are substantial labor and laboratory consumable costs and the need for preexisting knowledge of the DNA sequences for the targeted gene regions. To keep costs down for this and other genomic methods, individuals can be "barcoded" with unique short sequences of artificially synthesized DNA that allows pooling (also called multiplexing)

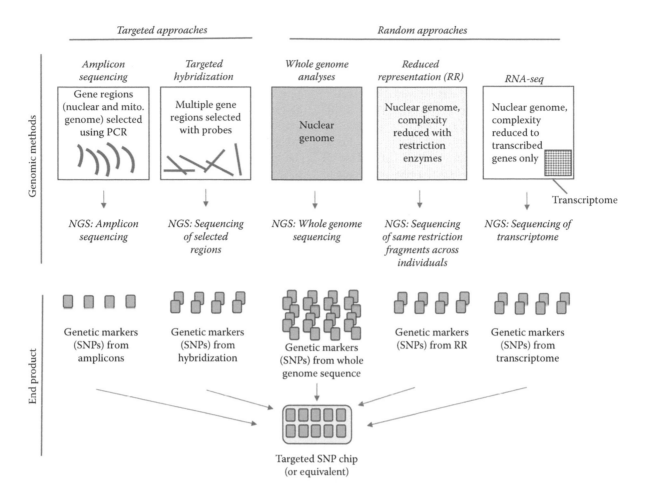

Figure 13.1 Concept map describing potential genomic pathways for the discovery of single nucleotide polymorphisms (SNPs) in non-model species, such as elasmobranchs. See text for full description of next-generation sequencing (NGS).

into one high-throughput sequencing run. Data can be separated back to the individual level for downstream analyses. For example, the Genotyping-in-Thousands by sequencing (GT-seq) method of Campbell et al. (2015) simultaneously sequences multiple amplicons from thousands of individuals to yield data on 50 to 500 SNPs. Genotyping many amplicons of different sizes in multiple barcoded individuals is challenging to optimize and accuracy decreases as the number of amplified loci increases.

A more versatile alternative to amplicon sequencing is targeted sequence capture (targeted hybridization) (Figure 13.1). This method is composed of a diverse suite of technologies designed to selectively capture homologous loci (genes shared among individuals within a species) from multiple individuals for subsequent high-throughput sequencing (Grover et al., 2012; Jones and Good, 2016). The method is an efficient way of identifying SNPs, as it maximizes the number of genes able to be analyzed and the number of individuals (by multiplexing) in a single high-throughput sequencing experiment. Targeted hybridization can also recover common loci between species (orthologous loci) by targeting highly conserved regions of the genome (Lemmon and

Lemmon, 2013; Li et al., 2013). This approach requires some prior genomic data to design probes (artificially synthesized DNA that will bind to the selected gene region, also called baits), and the upfront cost of probes per project can be costly. Probes are often designed from transcriptome sequences to bind to transcribed genes that produce proteins or peptides (also called exome capture) (Bi et al., 2012; Cosart et al., 2011). Exome capture is tolerant of relatively degraded starting DNA, making it ideal for the analysis of historical specimens from museum collections (Bi et al., 2013; Nielsen et al., 2017). Targeted hybridization has been used to assess demographic aspects of the blacktip reef shark (*Carcharhinus melanopterus*) in northern Australia and the Indian Ocean (Delser et al., 2016) and river sharks (*Glyphis* spp.) in northern Australia and southeast Asia (Li et al., 2015). The genus *Manta* was recently demonstrated to be part of the *Mobula* genus using this method (White et al., 2017).

Moving away from targeted approaches, an increasingly popular option due to ever-decreasing sequencing costs is to spread the high-throughput sequencing effort across the whole genome (Figure 13.1). A technique called Pool-seq even further reduces the costs of collecting whole genome sequence data on a population sample (tens to hundreds of individuals). Only one to three sequences per individual (low to shallow coverage) are collected, and DNA from individuals is pooled into a single sample (Schlotterer et al., 2014). Methods such as this have a downside, however, as all information about individuals is lost, making it difficult to control for uneven contributions of individuals to the final dataset and precluding any individual-level analysis. Recently, Therkildsen and Palumbi (2017) optimized an elegant way to apply the power of full genome sequencing to barcoded samples, focusing on coding regions (gene regions translated to proteins or peptides). New developments that produce long stretches of uninterrupted sequences (long-read technology) will make it easier and faster to discover and examine large numbers of SNPs from whole-genome datasets in the future (Rhoads and Au, 2015).

Among the approaches with the most impact in the conservation and wildlife management fields are those that begin with the systematic, experiment-wide breakdown of DNA extracted from individuals (reduced representation; see Figure 13.1). Such approaches are a popular choice for non-model organisms, including elasmobranchs (e.g., Pazmiño et al., 2017), for which few or no genomic data are available. The process of reduced representation (see Figure 1 in Good, 2011) is necessary to deal with the large size and high complexity of whole genomes. Bacterial defense enzymes (restriction enzymes) are commonly used to break down genomic DNA. They cleave DNA into smaller sections in the same way for each individual. The cut ends of the DNA are processed in the laboratory to allow subsequent identification of individuals and to make them compatible with the high-throughput sequencing process; this is the so-called "library" process. Sequence data are collected from the ends of each DNA fragment in the library. The final step to identify and examine SNPs occurs outside the laboratory using custom bioinformatic software (called pipelines) (e.g., Puritz et al., 2014).

Reduced representation methods are ideal for intraspecific studies of populations. They are less suitable for comparisons among species unless they are closely related, such as fiddler rays (*Trygonorrhina* spp.) (Donnellan et al., 2015). Reduced representation significantly reduces the proportion of the nuclear genome that is sequenced and hence reduces the resources required to complete the project. This method is popular due to the streamlined protocols that have been published (Andrews et al., 2016; Davey et al., 2011), their avoidance of difficult-to-sequence regions (e.g., high repetitive DNA sequence), and the ability to simultaneously discover and examine (genotype) thousands of SNPs across tens to hundreds of individuals. Additionally, as these methods work on relatively small amounts of DNA (e.g., 500 to 1000 ng) per individual, the starting material (animal tissue) can be collected using minimally invasive sampling methods; for example, individuals can be released alive after a

tissue biopsy is taken, or remote biopsy methods can be used on free-swimming individuals. This approach is particularly appropriate for studies of rare or endangered species.

Methods generally taking a reduced representation approach include restriction-site-associated DNA sequencing (RAD-seq) and its variants (e.g., Peterson et al., 2012), diversity arrays technology (DArT) (Sansaloni et al., 2011), and genotyping by sequencing (GBS) (Chen et al., 2014). The major drawback to reduced representation approaches is the lack of control over the exact genomic regions that are sequenced. A large proportion of SNPs discovered using this method are typically located outside transcribed sequences, precluding functional analysis of polymorphism patterns. This can be partly addressed during the bioinformatics step if a whole genome sequence for the study species is available (e.g., DiBattista et al., 2017), but as noted above these are rare for elasmobranch species. Using a whole genome sequence, SNPs that are located in functional gene regions can be identified, paving the way for studies focusing on selection and adaptation. Until whole genomes become more common for elasmobranchs, targeted hybridization or RNA sequencing (RNA-seq) (Figure 13.1) methods would be the most useful for studies like these.

High-throughput sequencing of the transcribed portion of a genome (Wang et al., 2009) is commonly used to measure gene expression variation between tissues types (RNA-seq) (Figure 13.1). SNPs in transcribed regions (transcriptome) are useful markers for studies of phenotypic changes in populations in response to local environmental variation (Gagnaire and Gaggiotti, 2016). Transcribed regions include the most conserved (and hence least genetically variable) regions in nuclear genomes, so SNPs in these regions are useful for phylogenetic analyses across taxonomic divides (Karam et al., 2015; White et al., 2017). At present RNA-seq methods are not making a direct contribution to studies of populations of elasmobranchs; however, transcriptomes produced using this method are essential to designing probes (baits) (Nielsen et al., 2017) for targeted hybridization or primers to target individual gene regions (amplicon sequencing). Although RNA-seq potentially makes significant contributions to conservation science, several experimental challenges must be considered. The method begins with the extraction of high-quality RNA that can only be isolated from fresh tissues, which requires individuals to be sacrificed and dissected. This approach is less appropriate for species that have large body size or have high conservation significance and also may be rare. Furthermore, because only a particular portion of the genome is expressed (transcribed) at a given time, RNA must be sampled from the same tissue type from individuals at the same life stage to obtain comparable SNPs. Finally, a general lack of knowledge between transcription of coding regions and their translation into proteins or peptides (Deveson et al., 2017; Ozsolak and Milos, 2011) limits the current usefulness of RNA-seq.

13.3 CURRENT APPLICATIONS TO ELASMOBRANCHS

13.3.1 Population Structure

Molecular tools have been successfully employed to discern the spatial extent of populations (defined here as interbreeding groups of animals of the same species), providing essential information for conservation and management. Most studies examining population genetic structure in elasmobranchs to date have used mitochondrial or microsatellite genes. Although too numerous to summarize in detail (see reviews in Dudgeon et al., 2012; Portnoy and Heist, 2012), some general patterns are emerging. The lack of discrete, physical barriers in the marine environment has often resulted in weak (even if statistically significant) indices of population structure (often measured as F_{ST} or its analogs) (Nei, 1977; Weir, 2012; Weir and Cockerham, 1984; Wright, 1951). For elasmobranchs, the strongest signals of regional population differentiation have come from small-bodied, site-attached species, such as the blacktip reef shark (*Carcharhinus melanopterus*) (Vignaud et al., 2014b). Large, highly mobile species that are mostly pelagic tend to demonstrate broader connectivity, with population structure occurring at ocean basin levels. Examples of these include the scalloped hammerhead shark (*Sphyrna lewini*) (Duncan et al., 2006), the sandbar shark (*Carcharhinus plumbeus*) (Portnoy et al., 2010), the dusky shark (*Carcharhinus obscurus*) (Benavides et al., 2011), the whale shark (*Rhincodon typus*) (Vignaud et al., 2014a), and the tiger shark (*Galeocerdo cuvier*) (Bernard et al., 2016; Holmes et al., 2017). Depending on the species and study system, multiple drivers for observed population structure (or lack thereof) have been proposed. These include biological features such as the potential of the species to move around (vagility) (Ovenden, 2013) and site fidelity (Chapman et al., 2015), as well as modern (Kousteni et al., 2015) and historic (Catarino et al., 2015; Dudgeon et al., 2009; Spaet et al., 2015) environmental features. Different types of genetic markers (e.g., mtDNA, microsatellite loci) do not always give concordant results on the same set of samples, potentially providing novel insights about breeding and mating behavior (i.e., philopatry; see below).

Genetic studies of population differentiation provide information on the appropriate spatial scale for fisheries management and monitoring. When migration between populations is low and significant differentiation is detected, genetics reflects demographic discontinuity (Ovenden, 2013) and thus provides rationale for treating populations as biological stocks (Waples and Gaggiotti, 2006). For example, the presence of significant population genetic structure has led to the management of multiple fisheries stocks in Australian waters for blacktip sharks (*Carcharhinus tilstoni* and *C. limbatus*) (Johnson et al., 2016), spot-tail sharks (*C. sorrah*) (Giles et al., 2014), dusky

sharks (*C. obscurus*) (Braccini et al., 2016; Geraghty et al., 2014b), and sandbar sharks (*C. plumbeus*) (Braccini et al., 2016; Portnoy et al., 2010). For mobile and wide-ranging species, it is challenging to interpret the lack of significant population genetic structure, as this does not always translate into demographic connectivity. Demographically separate stocks may still maintain links that just exceed the threshold required to lead to genetic differentiation (Ovenden, 2013), so a precautionary management approach is often adopted. Genetic studies indicate panmixia in several commercially fished shark species in the North Atlantic Ocean that are treated as single commercial stocks, including the blue shark (*Prionace glauca*) and shortfin mako shark (*Isurus oxyrinchus*) (Anon, 2015; Heist et al., 1996; ICCAT, 2008; ICES, 2016; King et al., 2015; Schrey and Heist, 2003). For some species, tagging and demographic studies support large-scale migration and hence the single stock status deduced from genetic studies—for example, the spiny dogfish (*Squalus acanthias*) (Verissimo et al., 2010) and the tope or school shark (*Galeorhinus galeus*) (Chabot and Allen, 2009; ICES, 2009). However, in the absence of more detailed molecular analysis or auxiliary data, the single stock delineation may be the default based on administrative boundaries or other non-biological criteria, such as single stocks of the Portuguese dogfish (*Centroscymnus coelolepis*) (ICES, 2009; Moura et al., 2008; Verissimo et al., 2011) and the leafscale gulper shark (*Centrophorus squamosus*) (ICES, 2016; Verissimo et al., 2012).

Population structure is also critical for defining conservation management units. Examples of its application are found within International Union for Conservation of Nature (IUCN) Red List assessments, which provide a framework to assess the extinction risks of species (Dulvy et al., 2014). Although this information does not form the basis of a legal instrument, it is widely used by government and nongovernment organizations to inform conservation policy. Within the Red List framework, subpopulations are defined as geographically or otherwise distinct groups that have very little demographic or genetic exchange (IUCN, 2012). Practically, this definition is interchangeable with our definition of population. Population genetic information has recently informed subpopulation delineation for whale sharks (*Rhincodon typus*) (Pierce and Norman, 2016; Vignaud et al., 2014a) and zebra sharks (*Stegostoma fasciatum*) (Dudgeon et al., 2009, 2016). In these examples, both mtDNA and microsatellite markers revealed two major populations (or subpopulations) which were assessed independently. These were then combined and weighted proportionally to the spatial distribution of each population to provide a global assessment. Population genetic structure also has great potential to inform the efficacy of marine protected area management by providing information about the spatial extent of populations and their degree of connectivity. However, designs have primarily incorporated information on connectivity of teleosts and invertebrates with pelagic larval dispersal, while comparatively less consideration has been given to species with high adult mobility, such as elasmobranchs (Momigliano et al., 2015).

Although the utility of genetic data to inform stock structure and management units has been clearly demonstrated, its uptake is still limited across taxa (Bernatchez et al., 2017; Garner et al., 2016; Shafer et al., 2015) and particularly in elasmobranchs. This challenge may be addressed by utilizing genomic tools that give greater coverage of the genomes and markers under differing selection regimes and therefore more power to resolve patterns (Shafer et al., 2015). For example, whole mitogenomes represent a natural improvement for any study that would normally rely on single or multiple mitochondrial gene regions. Previously restricted to model organisms or phylogenetic studies where the number of samples is generally small, mitogenomes can now be sequenced for hundreds of samples for population analyses of non-model species (Meimberg et al., 2016). This is particularly relevant to elasmobranchs given that their mitochondrial DNA is thought to evolve more slowly compared to other taxa (Martin et al., 1992). In combination with bottlenecks, founder events, or pronounced genetic drift, some elasmobranch populations have low or an apparent lack of genetic variation in mtDNA, such as the gray nurse shark (*Carcharias taurus*) (Stow et al., 2006). Thus, for some species, variability uncovered in single mtDNA gene region approaches are simply too low to provide insight into their population structure (Feutry et al., 2014; Wynen et al., 2009). For elasmobranchs, the first investigation of whole mitogenome sequences to analyze population structure was carried out in the speartooth shark (*Glyphis glyphis*) (Feutry et al., 2014). This study demonstrated that analysis of the whole mitogenome data was critical to fully describe population structure present in this threatened shark. Similarly, the use of whole mitogenome sequencing of the largetooth sawfish (*Pristis pristis*) showed that most rivers in northern Australia host a distinct matriarchal lineage (Feutry et al., 2015), adding to the study of Phillips et al. (2011), which used a single mtDNA gene region to show population structure at a much broader scale.

Just as the whole mitochondrial genome is now studied using genomic approaches, similar approaches are facilitating research across more of the nuclear genome. Where microsatellites give a snapshot of tens of markers, genomic approaches enable examination of thousands of markers (i.e., SNPs) (Figure 13.1), potentially increasing the power to test the hypothesis of panmixia (interbreeding) at varying spatial scales to provide information about population structure. Despite considerable interest and research progression, few completed studies have employed nuclear genomic tools for examining population structure in elasmobranchs, although many are planned

and underway. Population genetic analyses using SNPs have been published for at least four shark species, illustrating different outcomes and applications. Portnoy et al. (2015) found regional population structure from SNP data in bonnethead sharks (*Sphyrna tiburo*) between the Gulf of Mexico and Western Atlantic Ocean, but greater population structure was found with mtDNA which they interpreted as arising from male-biased dispersal and female site fidelity. However, even greater population structure was revealed when examining only SNPs that were assumed to be under selection, and these patterns were strongly correlated with latitude. The power of SNPs to further resolve population structure was demonstrated by Pazmiño et al. (2017, 2018). They demonstrated genetic differentiation in the Galapagos shark (*Carcharhinus galapagensis*), a large-bodied, circumglobally distributed species, at local and ocean-basin-wide scales. In contrast, analysis of SNP markers confirmed genetic panmixia in eastern Australian waters for gray reef sharks (*Carcharhinus amblyrhynchos*) (Momigliano et al., 2017). In the studies by Pazmiño et al. (2017) and Momigliano et al. (2017), increased population structure was suggested by examination of outlier SNPs presumed to be under selection. However, these results were prudently not presented due to the absence of an assembled and annotated whole nuclear (reference) genome for these species that would be necessary to ratify the functional role of these gene regions. The fourth study investigated the population structure of juvenile speartooth sharks (*Glyphis glyphis*) in three river systems in northern Australia (Feutry et al., 2017). Population structuring at scales of hundreds of kilometers was revealed by SNP analysis, as well as by whole mitogenome analyses. Feutry et al. (2017) found that the presence of large numbers of siblings within each river system resulted in non-independent sampling events and inflated signals of population structure. The removal of the genetic signal driven by the siblings resulted in non-significant population structuring between the rivers separated by around 150 km.

The ability to identify related animals (such as siblings) adds a new dimension to determining connectivity between populations. The spatial and temporal distribution of kin provides direct insight into individual movements; for example, the spatial distribution of parents and offspring revealed long-term philopatry in lemon sharks (*Negaprion brevirostris*) (Feldheim et al., 2014). Likewise, the spatial distributions of siblings and half-siblings of the speartooth shark (*Glyphis glyphis*) provided information about both juvenile and adult movements without the need to sample any adults (Feutry et al., 2017). The combination of kinship inference from nuclear DNA and the maternally inherited mitochondrial DNA allows adult female and male reproductive movements to be distinguished one from the other (Bravington et al., 2016b), which is important for elasmobranchs, where sex-biased dispersal is often reported (Dudgeon et al., 2012).

13.3.2 Population Size

Estimating population sizes is a huge challenge for elasmobranch species. Population sizes are essential knowledge for elasmobranch conservation and sustainable exploitation. For example, the highest priority action of the Australian government's white shark management plan is to develop and implement a monitoring program for survival, connectivity, fecundity, age-at-maturity, and absolute abundance to assess population trends and dynamics (Commonwealth of Australia, 2013). There are few elasmobranch fisheries worldwide where data are available to estimate these parameters. In some cases, elasmobranch catch is not recorded by species (e.g., Government of Queensland, 2010) or mistakenly identified (Tillett et al., 2012). Population sizes can be inferred from resights of tagged animals. Briefly, in mark–recapture studies, a sample of individuals is marked (generally with physical tags but it can also be genetic tagging or photos identification), and they are then released. Later, more sampling is conducted and the fraction of marked individuals or their individual capture histories can be used to estimate the population size and other demographic parameters (Amstrup et al., 2005; Cormack, 1964; Seber, 1965). However, tagging sufficient numbers of animals is generally expensive and tags can be shed. The interpretation of resights for elasmobranch species, which can be highly mobile, have sophisticated movement behavior, and are long-lived, is challenging. For example, 710 individual basking sharks (*Cetorhinus maximus*) were identified from photographs of dorsal fins, but a low number of resights (41) precluded regionwide, long-term population size estimates (Gore et al., 2016). The imperative for genetic estimates is undeniable.

Genetic estimates of population size have practical value to conservation and management science. They have been used to develop criteria to delist endangered species, such as sea otters (*Enhydra lutris nereis*), from the U.S. Endangered Species Act (Ralls et al., 1996). They have also been used to, for example, estimate absolute population size to implement harvest control rules for southern bluefin tuna (*Thunnus maccoyii*) (Hillary et al., 2016) and to estimate breeding population size to model population size changes for white shark (*Carcharodon carcharias*) in Western Australia (Braccini et al., 2017; Ovenden et al., 2016). Genetic methods rely on obtaining tissue samples from individuals for DNA extraction. Work is also progressing on the extraction of DNA from seawater (Sigsgaard et al., 2016; Simpfendorfer et al., 2016; Weltz et al., 2017; see also Chapter 14 in this volume), but this method is unlikely to replace the need for tissue samples from individuals in the near to medium future for genetic estimates of population size. Unlike many fisheries species that have high fecundity and high natural mortality across life history stages, some elasmobranch species have relatively small population sizes and ideal life-history

characteristics for genetic methods (Ovenden et al., 2016). In addition, elasmobranchs often have low fecundity, low natural mortality, and ecologically and morphologically similar life history stages (Ovenden, 2013). It is, therefore, not surprising that a significant proportion of genetic studies on population size are focusing on elasmobranch species.

The two main approaches to population size estimation using genetics are parameter and individual based. Parameter-based methods relate some genetic characteristic deduced across a sample of individuals to population size (Nomura, 2008; Waples and England, 2011; Waples and Yokota, 2007). Individual-based methods find groups of genetically related individuals and relate them to number of breeding adults (Bravington et al., 2016b; Cope et al., 2014; Creel and Rosenblatt, 2013; Wang, 2009). Both types of approaches rely on a random sample of individuals from the target population or the subsequent stratification of samples based on their biological characteristics. Both approaches rely on the same type of genetic markers from the nuclear genome. Microsatellite loci have been the most common, but SNP markers are becoming more popular. Both methods require the population genetic structure of the target species to be known in advance because both approaches are sensitive to migration (e.g., Waples and England, 2011). Formal comparisons between parameter- and individual-based methods for key elasmobranch species using the same genetic data are around the corner (R.S. Waples, pers. comm.).

A popular parameter-based approach converts empirical estimates of genetic drift to estimates of genetic effective population size (N_e). N_e is formally defined as the size an idealized population experiencing a known amount of genetic drift, such that large populations experience less drift than small populations (Hedrick, 2000). As populations in the real-world rarely meet these conditions, the effect of contravening them has been the focus of a great deal of work. In particular, the presence of overlapping generations has been addressed both in theory and by simulations to the point that the method can now be feasibly applied to iteroparous species (e.g., Jorde and Ryman, 1995), including elasmobranchs. There are numerous ways to measure genetic drift, but two have been closely associated with estimates of N_e. Linkage disequilibrium (LD) is a popular measure of genetic drift, as it can be estimated from a single sample of individuals from the target population at one point in time (Waples and Do, 2010). Genetic drift can also be estimated through time, the so-called temporal method (Waples, 1989). This can be approached in two ways: (1) a population can be sampled twice in time separated by the number of years corresponding to at least three or more elapsed generations (Waples and Do, 2010), or (2) samples that represent cohorts that differ in age by several generations can be taken once from the population (Jorde, 2012). Both approaches are challenging in practice. Generation times (defined as the average age of reproduction) are uncertain for many elasmobranch species.

Estimating the ages of individuals (to assign individuals to age-cohorts) often requires special expertise (Geraghty et al., 2014a; Holmes et al., 2015) and cannot be applied nonlethally, which generally restricts its use to species that are not threatened, endangered, or protected. Although both methods are appropriate, the linkage disequilibrium method is more tractable for elasmobranchs.

Linkage disequilibrium is defined as the correlation (r) for alleles in the population sample at all pairs of loci (see Equation 3 in Jones et al., 2016). It can be either positive or negative, so r^2 is used to measure its magnitude (Waples and England, 2011). Estimates of r^2 are the drivers of inbreeding estimates of N_e. As genetic drift occurs more rapidly in small populations compared to large, values of r^2 will be larger in small populations compared to large. For example, Dudgeon and Ovenden (2015) used the linkage disequilibrium method to explore N_e in a breeding aggregation of zebra sharks (*Stegostoma fasciatum*) in the waters off southeast Queensland. The estimate of N_e from microsatellite genotypes (14 loci, 114 individuals) was 377 (95% CI, 274–584), which was similar to a mark–recapture analysis of the same population (N = 458; 95% CI, 298–618). The N_e/N ratio was 0.82 (SE = 0.27). A similar test of the relationship between N_e and N was performed by Andreotti et al. (2016a). They identified 426 individual white sharks (*Carcharodon carcharias*) from over 4000 photographs of dorsal fins taken in the Gansbaai region of South Africa. Mark–recapture analyses subsequently estimated the population size to be 438 (95% CI, 353–522). Tissue biopsies taken from 233 sharks were genotyped with 14 microsatellite loci. From these data, the LD estimate of N_e was 333 (95% CI, 247–487), yielding a N_e/N ratio of 0.76. The spatial scale to which these estimates apply has been questioned (Irion et al., 2017), but genetic analyses were unable to reject the expectation of panmixia for white sharks sampled from five locations along the South African coastline (Andreotti et al., 2016b). Further analyses are needed to check the assumptions underlying the mark–recapture estimates, but it seems likely that the numbers of white sharks in South Africa are low and a precautionary approach to the conservation of this species is timely (Andreotti et al., 2017).

The similarity of N_e and N in these studies is encouraging for the extension of this method more generally to other species of elasmobranchs; however, there are several issues that must be accounted for. For example, some species may have experienced a large amount of genetic drift leading to naturally low levels of genetic variation (as measured by per-locus heterozygosity and number of alleles). Although this is uncommon in marine species, it has been reported for at least one elasmobranch species, the gray nurse shark (*Carcharias taurus*), on the eastern Australian coast (Stow et al., 2006). Estimation of N_e in these species may require extra effort to find (possibly scarce) variable genetic loci for the measurement of linkage disequilibrium. A genomic approach to the search may be worthwhile. On the other hand, other elasmobranch species, such as the milk shark (*Rhizoprionodon acutus*) (Ovenden et al., 2011) and gummy shark (*Mustelus antarcticus*) (Gardner and Ward, 1998), may have naturally large population sizes (and hence large N_e), meaning that the population may be experiencing low levels of genetic drift. Consequently, estimates of linkage disequilibrium r^2 will be small and difficult to measure with accuracy and precision. Problems like this are exacerbated as N_e is estimated after sampling error ($\approx 1/S$, where S is the number of samples taken from a population) is subtracted from r^2. This can lead to a negative denominator that cannot be converted to finite N_e (see Equation 2 in Jones et al., 2016). Biologically, infinite values have no meaning and practically mean that N_e cannot be computed. When N_e is large (and linkage disequilibrium r^2 is small), empirical geneticists approach this problem by increasing the power of their assays by either increasing sample sizes or increasing the number of genetic loci (or both). The software POWSIM (Ryman and Palm, 2006) is commonly employed in studies of population structure using genetics to calculate necessary sample sizes and loci numbers to estimate low F_{ST} values.

Individual-based approaches for estimating population size commonly use genetics to identify family members among individuals sampled from the population. Some approaches estimate the number of families and thus the likely number of breeding individuals (e.g., Kanno et al., 2011; Ozerov et al., 2015; Wang, 2009). Related individuals are often identified using software such as COLONY (Jones and Wang, 2010). Other approaches use the mark–recapture framework (e.g., Bravington et al., 2016a; Rawding et al., 2014). In close-kin mark–recapture, the recaptures are not the original individual but are recaptures of relatives of the original individual. These relatives or kin are identified genetically using the theory of Mendelian inheritance (Nielsen et al., 2001; Skaug, 2001) and modern genetic approaches. In essence, instead of capturing the same animal twice, genetic markers are used to identify kin pairs such as parent–offspring, siblings, or half-siblings. Therefore, close-kin mark–recapture suppresses the need to capture the same individuals more than once, which greatly expand the scope of applicability compared to mark–recapture. The cost of catching and releasing a substantial fraction of a population for mark–recapture is prohibitive for many species, whereas the sampling for close-kin mark–recapture studies can rely entirely on dead animals collected by industry, although it can be combined with live biopsies. Because the marking happens during reproduction and parents are "recaptures" of their offspring, close-kin mark–recapture provides recent abundance estimates of the parental population, backdated to the birth year of the juveniles, which makes the method suited for the management of wildlife. Importantly, it provides information about adults, not juveniles. Just as mark–recapture has applications beyond the estimation of abundance, close-kin mark–recapture can estimate parameters such as interbirth interval, fecundity-at-age, and mortality rates (Bravington et al., 2016a). Genomic-based approaches have the potential to expand and refine individual-based methods of estimating population size. Increased amounts of empirical data per individual can identify classes of relatives (e.g., half-sibs, which share one parent but not both) that are unable to be reliably discerned with lesser amounts of genetic data. There is one published example of the application of individual-based approaches to elasmobranchs (Hillary et al., 2018). Studies on several other species are underway, including *Glyphis* spp., *Galeorhinus galeus*, *Carcharias taurus*, *Pristis pristis*, and *Raja clavata* (M. Bravington, pers. comm.).

13.4 GENETIC RELATIONSHIP STUDIES TO STUDY MATING SYSTEMS, PHILOPATRY, AND FECUNDITY

Mating system studies on sharks have allowed researchers to address a variety of questions. How often do females give birth? Are females philopatric to nursery areas? How many offspring do individuals have in a given year? Unfortunately, studies of mating in elasmobranchs have lagged behind studies of other vertebrates. The little we do know about mating behavior comes from observational studies that have been performed in shallow water where some shark species congregate to mate. These studies have been bolstered by genetic studies that allow scientists to infer mating behavior. In elasmobranchs, the genetic marker of choice to examine relatedness between individuals has historically been microsatellites due to their ability to assign parentage (Mourier et al., 2013) and genetically tag individuals (Feldheim, 2002). Now that SNPs are becoming more affordable, they will become more prevalent in such studies (e.g., Feutry et al., 2017). Regardless of the genetic marker used, studies of genetic relationships are a vital tool for elasmobranch biologists. In this section, we discuss how genetic tools have enabled scientists to characterize different aspects of elasmobranch reproduction.

Genetic and genomic data can be used to test expectations of multiple paternity without difficulty. In diploid organisms, littermates can have no more than four parental alleles at each nuclear genetic locus (i.e., microsatellite or SNP loci) when the female mates with one male. When five or more parental alleles at a locus are present among littermates, this indicates that the female mated with multiple males. Inference of multiple mating is made easier when the mother of a litter is sampled along with the pups. In this case, the array of genotypes can be used to assess the number of paternal alleles. Given that two of the alleles for a given locus will be contributed to by the mother, three or more paternal alleles in a litter will be indicative of multiple paternity. Recent studies have found that multiple paternity is quite common (Green et al., 2017) rather than the exception in elasmobranchs (Holmes et al., 2018). Given the potential cost of mating to females, such as multiple bite marks from males attempting to mate with a female (Pratt and Carrier, 2001), it is unclear why females mate with multiple males. Only a handful of studies have been able to infer adult mating behavior by examining the genotypes of offspring. In these cases, the genotypes of the adult generation can be reconstructed from the genotypes of the offspring by, for example, employing the software COLONY (Jones and Wang, 2010). The drawback of this method is that intense sampling of offspring is required in order to reconstruct parental genotypes with any certainty. Genetic reconstruction has been used successfully to examine lemon shark mating behavior at Bimini, the Bahamas (Feldheim et al., 2004) and Marquesas Key, Florida (DiBattista et al., 2008a). It has also been used to examine smalltooth sawfish mating habits in Florida nurseries (Feldheim et al., 2017b). Although this latter study found evidence of multiple paternity, the number of pups sampled per female was relatively low, and estimates of multiple paternity could not be made confidently.

Philopatry is defined as the tendency of individuals to remain in or return to their home areas (Mayr, 1963). For sharks and their relatives, Chapman et al. (2015) recognized two types: regional and natal. Regional philopatry occurs when wide-ranging individuals return to their natal region, where females give birth and mating occurs. Natal philopatry describes the phenomenon whereby females return to their exact birthplace to give birth on nursery grounds. Both types of philopatry can lead to genetic differences between broader regions (in the case of regional philopatry) or, at a finer scale, between specific nursery grounds (in the case of natal philopatry). The sampling scheme is important for genetic studies of philopatry. When adults are sampled during parturition and mating seasons, philopatry results in genetic variation between regions in mitochondrial DNA when only the females are philopatric and both mitochondrial and nuclear DNA (e.g., microsatellites) when both sexes are philopatric, such as the white shark (*Carcharodon carcharias*) in Australia (Blower et al., 2012). Outside parturition and mating times, philopatry may not be testable with genetic methods as mature individuals disperse away from these locations; however, sampling sedentary juveniles from nursery areas, regardless of the time of year, can shed light onto philopatric behavior.

Mating between natally philopatric females and wandering males admixes alleles at microsatellite loci in their offspring. Admixture does not occur in the maternally inherited mtDNA, as allelic recombination during reproduction does not occur in this genome. For natal philopatry, this mechanism leads to a situation where there may be contrast in the extent of the mtDNA and microsatellite variation between samples taken in or adjacent to different nursery grounds. Juveniles and females sampled from different nursery grounds may not be distinct when assayed using microsatellite loci, whereas they will be distinct based on their mtDNA. Within limits, this distinction can be used as indirect evidence of natal philopatry, but samples must be taken during breeding times and from breeding locations; otherwise, the natal genetic signal will be obscured by subsequent dispersal of juveniles and females. Practically, deductions about the extent of philopatry using genetic and genomic analyses are strengthened if the sex and age of samples are available. Finally, if the microsatellite loci lack power to detect possible barriers to gene flow between regionally philopatric populations, then natal and regional philopatry may be indistinguishable. Philopatry in elasmobranchs has recently been reviewed in both sharks (Chapman et al., 2015) and skates and rays (Flowers et al., 2016). In sharks, many studies find structure in mtDNA with a lack of differentiation in nuclear markers (e.g., Bernard et al., 2016; Pardini et al., 2001; Portnoy et al., 2010). Population genetic studies have lagged in batoids compared to sharks (Flowers et al., 2016). The few studies that have been published in batoids typically find no contrast between mtDNA and nuclear DNA genetic patterns.

Although philopatry is commonly inferred through indirect genetic methods, a few studies have been able to show philopatry directly using parentage assignment or genetic reconstruction methods, such as the direct evidence obtained for *Glyphis glyphis*, although only juveniles were sampled (Feutry et al., 2017). Mourier and Planes (2013) used parentage analysis in blacktip reef sharks (*Carcharhinus melanopterus*) and sicklefin lemon sharks (*Negaprion acutidens*) to show that some females are philopatric to islands in French Polynesia. Feldheim (2002) and DiBattista et al. (2008b) used genetic reconstruction methods to show that lemon shark females are philopatric to Bimini, the Bahamas, and Marquesas Key, Florida, respectively. Feldheim et al. (2014) further showed that some females exhibit natal philopatry to Bimini. Similar methods have been used to show philopatric behavior of female smalltooth sawfish (*Pristis pectinata*) to nursery sites in Florida (Feldheim et al., 2017b), although it is unknown if these females were born at these Florida sites.

One of the more interesting aspects of reproduction that has been unveiled using genetic markers has been the discovery of parthenogenesis in elasmobranchs. Parthenogenesis occurs when an embryo develops from a female gamete without the contribution of male sperm (Lampert, 2009). Automixis (whereby a polar body fertilizes the egg, thus restoring diploidy) has been the inferred mode of parthenogenesis in elasmobranchs (Dudgeon et al., 2017; Portnoy et al., 2014). This manifests in the offspring as elevated homozygosity compared to the mother. Indeed, in elasmobranchs, most parthenogenetic offspring have been found to be homozygous at all microsatellite loci employed in those studies. Parthenogenesis has been described in six shark species and two batoid species (Table 13.1). These species represent seven families and span the different reproductive modes found in elasmobranchs. Most cases of parthenogenesis have been described from captive animals and occurred when females were isolated from conspecific males. One exception to this is in smalltooth sawfish (*Pristis pectinata*). Fields et al. (2015) found seven juvenile sawfish with elevated homozygosity caught in nursery areas from the west coast of Florida and attributed these individuals to three parthenogenic females. It is unknown what triggers this mode of reproduction, but it has been hypothesized that it is a last-ditch effort for females to pass on their genes when a mate is not available. Switching from sexual to parthenogenetic reproduction has been demonstrated for two captive elasmobranch species, the whitespotted bamboo shark (*Chiloscyllium plagiosum*) (Straube et al., 2016) and the zebra shark (*Stegostoma fasciatum*) (Dudgeon et al., 2017), after the removal of their mates. Smalltooth sawfish populations have declined by more than 95% over the past century (Simpfendorfer, 2000), and it is possible that these females could not find a mate. It is interesting to note that the three sawfish females that gave birth via parthenogenesis had offspring in later years by sexual reproduction (Feldheim et al., 2017b). This suggests that parthenogenesis may play a role as a holding-on strategy, extending the life span of the oocyte through the parthenogenetic offspring until mates become available.

13.5 SUMMARY AND CONCLUSIONS

Know-how adopted from human, plant, and animal genomics combined with the larger and older body of empirical and theoretical population genetics is closing knowledge gaps in the population biology of sharks and rays. This has the potential to significantly improve conservation practices for threatened, exploited, and protected elasmobranch species and the sustainable management of elasmobranch species that are outside this category. New information from genetics and genomics is addressing practical demographic issues such as the extent and drivers of connectivity between populations, numbers of individuals making up those populations, and aspects of reproductive biology and behavior that is rarely observed (Table 13.2). In this chapter, our aim has been to clearly present the most important aspects of this flood of new information from a field that is sometimes regarded as dense and incomprehensible. We will have achieved our goal if conservation plans and harvest strategies for elasmobranchs are updated with information from genetic and genomic studies. Workers on both sides of the interface between the production of new information from research and its application need to sympathetically cooperate to overcome the challenges often associated with the uptake of information from genetics and genomic studies (Bernatchez et al., 2017). Somehow, we need to navigate the discrepancy between the technical expertise of research scientists and the expertise of those who work with social, political, and economic performance criteria.

Table 13.1 Summary of Studies to Date Using Genetics and Genomic Tools for Analyses of Parthenogenesis in Elasmobranch Species

Species	Common Name	Number of Pups	Number of Females	Location of Study Animals	Genetic Markers Used	Mode of Pathenogensis	Mode of Reproduction	Classification	Refs.	Notes
Aetobatus narinari	Spotted eagle ray	2	1	The Seas at Epcot; Orlando, FL	10 microsatellites	Automixis	Ovoviviparity	Myliobatiformes, Myliobatidae	Harmon et al. (2015)	Four pregnancies of this female, the first two sexual; when male was gone, pregnancies 3 and 4 were due to parthenogenesis. Switch to parthenogenesis occurred within a year of no male being present.
Carcharhinus limbatus	Blacktip shark	1	1	Virginia Aquarium and Marine Science Center; Virginia Beach, VA	5 microsatellites	Automixis	Placental viviparity	Carcharhiniformes, Carcharhinidae	Chapman et al. (2008)	Discovered upon necropsy of female
Cephaloscyllium ventriosum	Swell shark	5	1	National Aquarium, Washington, DC	12 microsatellites	Automixis	Oviparity	Carcharhiniformes, Scyliorhinidae	Feldheim et al. (2017a)	Multiple pups
Chiloscyllium plagiosum	White spotted bamboo shark	2	1	Belle Isle Aquarium; Detroit, MI	7 microsatellites and AFLP data	Automixis	Oviparity	Orectolobiformes, Hemiscyllidae	Feldheim et al. (2010)	Four hatched; of the two that were not genotyped, one never initiated feeding and the other jumped out of the tank at 15 months old
Chiloscyllium plagiosum	White spotted bamboo shark	6	1	Vivarium of the State Museum of Natural History; Karlsruhe, Germany	8 microsatellites	Automixis	Oviparity	Orectolobiformes, Hemiscyllidae	Straube et al. (2016)	One of the offspring had external claspers, but upon dissection the internal sexual organs of this specimen were malformed or absent.
Chiloscyllium plagiosum	White spotted bamboo shark	2	1	Vivarium of the State Museum of Natural History; Karlsruhe, Germany	8 microsatellites	Automixis likely (but cannot rule out apomixis)	Oviparity	Orectolobiformes, Hemiscyllidae	Straube et al. (2016)	This one was an offspring of the initial female that gave birth via parthenogenesis.
Pristis pectinata	Small-tooth sawfish	7	3	*Caloosahatchee* and Peace Rivers in Florida	16 microsatellites	Automixis	Ovoviviparous	Pristiformes, Pristidae	Fields et al. (2015)	This was the first known case of normally sexually reproducing species giving birth via parthenogenesis.

(continued)

Table 13.1 (continued) Summary of Studies to Date Using Genetics and Genomic Tools for Analyses of Parthenogenesis in Elasmobranch Species

Species	Common Name	Number of Pups	Number of Females	Location of Study Animals	Genetic Markers Used	Mode of Pathenogensis	Mode of Reproduction	Classification	Refs.	Notes
Sphyrna tiburo	Bonnet-head shark	1	1	Henry Doorly Zoo; Omaha, NE	4 microsatellites	Automixis	Placental viviparity	Carcharhiniformes, Sphyrnidae	Chapman et al. (2007)	Pup was later killed (by stingray?).
Stegostoma fasciatum	Zebra shark	16	1	Burj Al Arab aquarium; Dubai. United Arab Emirates	11 microsatellites	Automixis	Oviparity	Orectolobiformes, Stegostomatidae	Robinson et al. (2011)	One pup exhibited heterozygosity at one locus (and allele was non-maternal); authors attributed it to mutation. As of this writing, four pups were still alive. These were successive pregnancies (four).
Stegostoma fasciatum	Zebra shark	9 (8 from F1; 1 from F2)	2	Reef HQ Great Barrier Reef Aquarium; Townsville City, Queensland, Australia	14 microsatellites	Terminal fusion automixis	Oviparity	Orectolobiformes, Stegostomatidae	Dudgeon et al. (2017)	First female switched from sexual to asexual reproduction
Triaenodon obesus	Whitetip reef shark	1	1	BioPark Aquarium; Albuquerque, NM	24 microsatellites (cross-species)	Developed from unfertilized egg	Oviparity	Carcharhiniformes, Carcharhinidae	Portnoy et al. (2014)	Pup was premature; haploidy was inferred from both microscopy and flow cytometry.

Table 13.2 Summary of Current Issues in Applied and Fundamental Research on Sharks and Their Relatives Addressed Herein, Challenges for Future Research, and Possible Solutions Using Genetics and Genomics

Current Issues and Future Problems	Possible Genetic and Genomic Solutions
Lack of information about population size now and in the past, and connectivity between populations regionally and at an ocean-basin scale	Increase the number of genetic markers to estimate and monitor the effective number of breeders, total population size, and number of migrants with higher precision and accuracy, as well as to identify migrants and estimate the direction of migration. Genomics expands sample sizes and temporal scale through access to DNA from specimens in museums and other historical collections (e.g., vertebrae for ageing, fishers' trophies).
Lack of information about units of conservation (e.g., species, distinct population segments)	Increase the number of genetic markers (under varying levels of selection) to achieve greater resolution to infer population units and phylogenetic relationships, improve detection of cryptic species, and reveal patterns of hybridization and introgression. Incorporate adaptive loci to improve the accuracy and precision to infer fine-scale population structure and identify boundaries between conservation units.
Lack of tools for practical management of elasmobranch species subject to anthropogenic pressure	Increase the number of genetic markers to improve the ability to assign individuals to the population of origin for shark fisheries management, enforcement, migration, bycatch, and mislabeling studies of postprocessed fish products.
Inability to predict the capability of elasmobranch species to adapt to climate change and anthropogenic challenges	Use the unprecedented potential of genomics for understanding adaptive genetic variation which in turn will improve our understanding of the capacity of sharks to adapt to environmental stresses, including rapidly changing temperature, acidity, salinity, and sea levels.
Lack of information about elasmobranch reproductive biology	Increase the number of genetic markers to solve the longstanding debate about whether phylogeographic structure is due to biparental adaptation or female philopatry, as well as identification of family groups using SNPs and inferring movement patterns associated with reproduction.

Source: Adapted from Allendorf, F.W. et al., *Nat. Rev. Genet.*, 11(10), 697–710, 2010.

Many exciting prospects are on the horizon that are made possible by genomics. Despite decades of study, little is known of the genetic and phenotypic nature of local adaptation. For elasmobranchs, like all species in the living world, more details of the genetic architecture and evolution of adaptation may provide predictions of the likely effect of human-induced environmental change. This, in turn, may guide future elasmobranch conservation and fisheries management plans. Genomics also allow us to study the interaction between traits that are directly controlled by genes and those that are controlled by chemical modification to genes (epigenetic changes) within and between generations. Elasmobranch species represent a major slice of marine biodiversity with unique characteristics (longevity, diverse reproductive modes, etc.) that offer the potential for scientific outcomes. For the same reason, comparisons between elasmobranch whole genomes, as they become available, may illuminate the basis of their persistence over geological time and drivers of their species diversity and evolution. The availability of more genomes will boost knowledge gained from future population studies using the range of genomic methods described here and those developed in the future.

ACKNOWLEDGMENTS

Thanks to Colin Simpfendorfer and an anonymous reviewer for comments on previous versions of this chapter. Danielle Davenport collated information on elasmobranch genomic resources.

REFERENCES

Allendorf FW, Hohenlohe PA, Luikart G (2010) Genomics and the future of conservation genetics. *Nat Rev Genet* 11(10):697–710.

Amstrup S, McDonald T, Manly B (2005) *Handbook of Capture–Recapture Analysis.* Princeton University Press, Princeton, NJ, p 313.

Andreotti S, Rutzen M, Van der Walt S, von der Heyden S, Henriques R, Meÿer M, Oosthuizen H, Matthee CA (2016a) An integrated mark-recapture and genetic approach to estimate the population size of white shark in South Africa. *Mar Ecol Prog Ser* 552:241–253.

Andreotti S, von der Heyden S, Henriques R, Rutzen M, Meÿer M, Oosthuizen H, Matthee CA (2016b) New insights into the evolutionary history of white sharks, *Carcharodon carcharias. J Biogeogr* 43(2):328–339.

Andreotti S, von der Heyden S, Henriques R, Rutzen M, Meÿer M, Matthee CA (2017) Erring on the side of caution: reply to Irion et al. (2017). *Mar Ecol Prog Ser* 577:257–262.

Andrews KR, Good JM, Miller MR, Luikart G, Hohenlohe PA (2016) Harnessing the power of RADseq for ecological and evolutionary genomics. *Nat Rev Genet* 17(2):81–92.

Andrews KR, Luikart G (2014) Recent novel approaches for population genomics data analysis. *Mol Ecol* 23(7):1661–1667.

Anon (2015) *Blue Shark Data Preparatory Meeting—Spain 2015* (https://www.iccat.int/Documents/Meetings/Docs/2015-BSH_DATA_PREP_Rep-ENG.pdf).

Balloux F, Lugon-Moulin N (2002) The estimation of population differentiation with microsatellite markers. *Mol Ecol* 11:155–165.

Benavides MT, Horn RL, Feldheim KA, Shivji MS, Clarke SC, Wintner S, Natanson L, et al. (2011) Global phylogeography of the dusky shark *Carcharhinus obscurus*: implications for fisheries management and monitoring the shark fin trade. *Endanger Spec Res* 14:13–22.

Benson DA, Cavanaugh M, Clark K, Karsch-Mizrachi I, Lipman DJ, Ostell J, Sayers EW (2013) GenBank. *Nucleic Acids Res* 41(Database issue):D36–D42.

Bernard AM, Feldheim KA, Heithaus MR, Wintner SP, Wetherbee BM, Shivji MS (2016) Global population genetic dynamics of a highly migratory, apex predator shark. *Mol Ecol* 25(21):5312–5329.

Bernatchez L (2016) On the maintenance of genetic variation and adaptation to environmental change: considerations from population genomics in fishes. *J Fish Biol* 89(6):2519–2556.

Bernatchez L, Wellenreuther M, Araneda C, Ashton D, Barth J, Beacham TD, Maes GE, et al. (2017) Harnessing the power of genomics to secure the future of seafood. *Trends Ecol Evol* 32(9):665–680.

Bi K, Vanderpool D, Singhal S, Linderoth T, Moritz C, Good JM (2012) Transcriptome-based exon capture enables highly cost-effective comparative genomic data collection at moderate evolutionary scales. *BMC Genomics* 13(1):403.

Bi K, Linderoth T, Vanderpool D, Good JM, Nielsen R, Moritz C (2013) Unlocking the vault: next-generation museum population genomics. *Mol Ecol* 22(24):6018–6032.

Blower DC, Pandolfi JM, Gomez-Cabrera MdC, Bruce BD, Ovenden JR (2012) Population genetics of Australian white sharks reveals fine-scale spatial structure, transoceanic dispersal events and low effective population sizes. *Mar Ecol Prog Ser* 455:229–244.

Braccini JM, Johnson G, Rogers P, Hansen S, Peddemors V (2016) *Dusky Whaler Carcharhinus obscurus*. Fisheries Research & Development Corporation, Deakin, Australian Capital Territory, Australia (www.fish.gov.au/report/23–Dusky-Whaler-2016).

Braccini M, Taylor S, Bruce BD, McAuley R (2017) Modelling the population trajectory of West Australian white sharks. *Ecol Model* 360:363–377.

Bravington MV, Grewe PM, Davies CR (2016a) Absolute abundance of southern bluefin tuna estimated by close-kin mark-recapture. *Nat Commun* 7:13162.

Bravington MV, Skaug HJ, Anderson EC (2016b) Close-kin mark–recapture. *Statist Sci* 31:259–274.

Bylemans J, Maes GE, Diopere E, Cariani A, Senn H, Taylor MI, Helyar S, et al. (2016) Evaluating genetic traceability methods for captive-bred marine fish and their applications in fisheries management and wildlife forensics. *Aquacult Env Interac* 8:131–145.

Campbell NR, Harmon SA, Narum SR (2015) Genotyping-in-Thousands by sequencing (GT-seq): a cost effective SNP genotyping method based on custom amplicon sequencing. *Mol Ecol Resour* 15(4):855–867.

Catarino D, Knutsen H, Verissimo A, Olsen EM, Jorde PE, Menezes G, Sannaes H, et al. (2015) The Pillars of Hercules as a bathymetric barrier to gene flow promoting isolation in a global deep-sea shark (*Centroscymnus coelolepis*). *Mol Ecol* 24(24):6061–6079.

Chabot CL, Allen LG (2009) Global population structure of the tope (*Galeorhinus galeus*) inferred by mitochondrial control region sequence data. *Mol Ecol* 18(3):545–552.

Chapman DD, Shivji MS, Louis E, Sommer J, Fletcher H, Prodöhl PA (2007) Virgin birth in a hammerhead shark. *Biol Lett* 3(4):425–427.

Chapman DD, Firchau B, Shivji MS (2008) Parthenogenesis in a large-bodied requiem shark, the blacktip *Carcharhinus limbatus*. *J Fish Biol* 73(6):1473–1477.

Chapman DD, Feldheim KA, Papastamatiou YP, Hueter RE (2015) There and back again: a review of residency and return migrations in sharks, with implications for population structure and management. *Annu Rev Mar Sci* 7(1):547–570.

Chen N, Van Hout CV, Gottipati S, Clark AG (2014) Using Mendelian inheritance to improve high-throughput SNP discovery. *Genetics* 198(3):847–857.

Commonwealth of Australia (2013) *Recovery Plan for the White Shark (Carcharodon carcharias)*. Department of Sustainability, Environment, Water, Population and Communities, Canberra, Australian Capital Territory, Australia.

Cope RC, Lanyon JM, Seddon JM, Pollett PK (2014) Development and testing of a genetic marker-based pedigree reconstruction system 'PR-genie' incorporating size-class data. *Mol Ecol Resour* 14(4):857–870.

Cormack R (1964) Estimates of survival from the sighting of marked animals. *Biometrika* 51(3/4):429–438.

Cosart T, Beja-Pereira A, Chen S, Ng SB, Shendure J, Luikart G (2011) Exome-wide DNA capture and next generation sequencing in domestic and wild species. *BMC Genomics* 12(1):347.

Creel S, Rosenblatt E (2013) Using pedigree reconstruction to estimate population size: genotypes are more than individually unique marks. *Ecol Evol* 3(5):1294–1304.

da Fonseca RR, Albrechtsen A, Themudo GE, Ramos-Madrigal J, Sibbesen JA, Maretty L, Zepeda-Mendoza ML, Campos PF, Heller R, Pereira RJ (2016) Next-generation biology: sequencing and data analysis approaches for non-model organisms. *Marine Genom* 30:3–13.

Davey JW, Hohenlohe PA, Etter PD, Boone JQ, Catchen JM, Blaxter ML (2011) Genome-wide genetic marker discovery and genotyping using next-generation sequencing. *Nat Rev Genet* 12(7):499–510.

Delser PM, Corrigan S, Hale M, Li C, Veuille M, Planes S, Naylor G, Mona S (2016) Population genomics of *C. melanopterus* using target gene capture data: demographic inferences and conservation perspectives. *Sci Rep* 6:33753.

Deveson IW, Hardwick SA, Mercer TR, Mattick JS (2017) The dimensions, dynamics, and relevance of the mammalian noncoding transcriptome. *Trends Genet* 33(7):464–478.

DiBattista JD, Feldheim KA, Gruber SH, Hendry AP (2008a) Are indirect genetic benefits associated with polyandry? Testing predictions in a natural population of lemon sharks. *Mol Ecol* 17(3):783–795.

DiBattista JD, Feldheim KA, Thibert-Plante X, Gruber SH, Hendry AP (2008b) A genetic assessment of polyandry and breeding-site fidelity in lemon sharks. *Mol Ecol* 17(14):3337–3351.

DiBattista JD, Saenz-Agudelo P, Piatek MJ, Wang X, Aranda M, Berumen ML (2017) Using a butterflyfish genome as a general tool for RAD-Seq studies in specialized reef fish. *Mol Ecol Resour* 17(6):1330–1341.

Donnellan SC, Foster R, Junge C, Huveneers C, Rogers P, Kilian A, Bertozzi T (2015) Fiddling with the proof: the magpie fiddler ray is a colour pattern variant of the common southern fiddler ray (Rhinobatidae: *Trygonorrhina*). *Zootaxa* 3981(3):367.

Dudgeon CL, Ovenden JR (2015) The relationship between abundance and genetic effective population size in elasmobranchs: an example from the globally threatened zebra shark *Stegostoma fasciatum* within its protected range. *Conserv Genet* 16:1443–1454.

Dudgeon CL, Broderick D, Ovenden JR (2009) IUCN classification zones concord with, but underestimate, the population genetic structure of the zebra shark *Stegostoma fasciatum* in the Indo-West Pacific. *Mol Ecol* 18(2):248–261.

Dudgeon CL, Blower DC, Broderick D, Giles JL, Holmes BJ, Kashiwagi T, Kruck NC, Morgan JAT, Tillett BJ, Ovenden JR (2012) A review of the application of genetics for fisheries management and conservation of sharks and rays. *J Fish Biol* 80:1789–1843.

Dudgeon CL, Simpfendorfer C, Pillans RD (2016) *Stegostoma fasciatum*. The IUCN Red List of Threatened Species™, http://www.iucnredlist.org/details/41878/0.

Dudgeon CL, Coulton L, Bone R, Ovenden JR, Thomas S (2017) Switch from sexual to parthenogenetic reproduction in a zebra shark. *Sci Rep* 7:40537.

Dulvy NK, Fowler SL, Musick JA, Cavanagh RD, Kyne PM, Harrison LR, Carlson JK, et al. (2014) Extinction risk and conservation of the world's sharks and rays. *eLife* 3(0):e00590.

Dulvy NK, Simpfendorfer CA, Davidson LNK, Fordham SV, Brautigam A, Sant G, Welch DJ (2017) Challenges and priorities in shark and ray conservation. *Curr Biol* 27(11):R565–R572.

Duncan KM, Martin AP, Bowen BW, De Couet HG (2006) Global phylogeography of the scalloped hammerhead shark (*Sphyrna lewini*). *Mol Ecol* 15(8):2239–2251.

Feldheim K (2002) Genetic tagging to determine passive integrated transponder tag loss in lemon sharks. *J Fish Biol* 61(5):1309–1313.

Feldheim K, Gruber SH, Ashley MV (2004) Reconstruction of parental microsatellite genotypes reveals female polyandry and philopatry in the lemon shark, *Negaprion brevirostris*. *Evolution* 58(10):2332–2342.

Feldheim KA, Chapman DD, Sweet D, Fitzpatrick S, Prodohl PA, Shivji MS, Snowden B (2010) Shark virgin birth produces multiple, viable offspring. *J Hered* 101(3):374–377.

Feldheim KA, Gruber SH, Dibattista JD, Babcock EA, Kessel ST, Hendry AP, Pikitch EK, Ashley MV, Chapman DD (2014) Two decades of genetic profiling yields first evidence of natal philopatry and long-term fidelity to parturition sites in sharks. *Mol Ecol* 23(1):110–117.

Feldheim KA, Clews A, Henningsen A, Todorov L, McDermott C, Meyers M, Bradley J, Pulver A, Anderson E, Marshall A (2017a) Multiple births by a captive swellshark *Cephaloscyllium ventriosum* via facultative parthenogenesis. *J Fish Biol* 90(3):1047–1053.

Feldheim KA, Fields AT, Chapman DD, Scharer RM, Poulakis GR (2017b) Insights into reproduction and behavior of the smalltooth sawfish *Pristis pectinata*. *Endang Species Res* 34:463–471.

Feutry P, Kyne PM, Pillans RD, Chen X, Naylor GJP, Grewe PM (2014) Mitogenomics of the speartooth shark challenges ten years of control region sequencing. *BMC Evol Biol* 14:232–240.

Feutry P, Kyne PM, Pillans RD, Chen X, Marthick J, Morgan DL, Grewe PM (2015) Whole mitogenome sequencing refines population structure of the critically endangered sawfish *Pristis pristis*. *Mar Ecol Prog Ser* 533:237–244.

Feutry P, Berry O, Kyne PM, Pillans RD, Hillary RM, Grewe PM, Marthick JR, et al. (2017) Inferring contemporary and historical genetic connectivity from juveniles. *Mol Ecol* 26(2):444–456.

Fields AT, Feldheim KA, Poulakis GR, Chapman DD (2015) Facultative parthenogenesis in a critically endangered wild vertebrate. *Curr Biol* 25(11):R446–R447.

Flowers KI, Ajemian MJ, Bassos-Hull K, Feldheim KA, Hueter RE, Papastamatiou YP, Chapman DD (2016) A review of batoid philopatry, with implications for future research and population management. *Mar Ecol Prog Ser* 562:251–261.

Gagnaire P-A, Gaggiotti OE (2016) Detecting polygenic selection in marine populations by combining population genomics and quantitative genetics approaches. *Curr Zool* 62(6):603–616.

Gardner MG, Ward RD (1998) Population structure of the Australian gummy shark (*Mustelus antarcticus* Günther) inferred from allozymes, mitochondrial DNA and vertebrae counts. *Mar Freshwater Res* 49(7):733–745.

Garner BA, Hand BK, Amish SJ, Bernatchez L, Foster JT, Miller KM, Morin PA, et al. (2016) Genomics in conservation: case studies and bridging the gap between data and application. *Trends Ecol Evol* 31(2):81–83.

Geraghty PT, Macbeth WG, Harry AV, Bell JE, Yerman MN, Williamson JE (2014a) Age and growth parameters for three heavily exploited shark species off temperate eastern Australia. *ICES J Mar Sci* 71(3):559–573.

Geraghty PT, Williamson JE, Macbeth WG, Blower DC, Morgan JAT, Johnson G, Ovenden JR, Gillings MR (2014b) Genetic structure and diversity of two highly vulnerable carcharhinids in Australian waters. *Endanger Spec Res* 24(1):45–60.

Giles JL, Ovenden JR, Dharmadi, AlMojil D, Garvilles E, Khampetch KO, Manjebrayakath H, Riginos C (2014) Extensive genetic population structure in the Indo-West Pacific spot-tail shark, *Carcharhinus sorrah*. *Bull Mar Sci* 90(1):427–454.

Good JM (2011) Reduced representation methods for subgenomic enrichment and next-generation sequencing. In: Orgogozo V, Rockman MV (eds) *Molecular Methods for Evolutionary Genetics*. Humana Press, Totowa, NJ, pp 85–103.

Gore MA, Frey PH, Ormond RF, Allan H, Gilkes G (2016) Use of photo-identification and mark–recapture methodology to assess basking shark (*Cetorhinus maximus*) populations. *PLoS ONE* 11(3):e0150160.

Government of Queensland (2010) *Annual Status Report 2009: East Coast Inshore Fin Fish Fishery*. Department of Employment, Economic Development and Innovation, Brisbane, Queensland, Australia.

Green ME, Appleyard SA, White WT, Tracey SR, Ovenden JR (2017) Variability in multiple paternity rates for grey reef sharks (*Carcharhinus amblyrhynchos*) and scalloped hammerheads (*Sphyrna lewini*). *Sci Rep* 7(1):1528.

Grover CE, Salmon A, Wendel JF (2012) Targeted sequence capture as a powerful tool for evolutionary analysis. *Am J Bot* 99(2):312–319.

Harmon TS, Kamerman TY, Corwin AL, Sellas AB (2015) Consecutive parthenogenetic births in a spotted eagle ray *Aetobatus narinari. J Fish Biol* 88(2):741–745.

Hedrick PW (2000) *Genetics of Populations*, 2nd ed. Jones & Bartlett, Sudbury MA.

Heist EJ, Musick JA, Graves JE (1996) Genetic population structure of the shortfin mako (*Isurus oxyrinchus*) inferred from restriction fragment length polymorphism analysis of mitochondrial DNA. *Can J Fish Aquat Sci* 53(3):583–588.

Hillary RM, Preece AL, Davies CR, Kurota H, Sakai O, Itoh T, Parma AM, et al. (2016) A scientific alternative to moratoria for rebuilding depleted international tuna stocks. *Fish Fish* 17(2):469–482.

Hillary RM, Bravington MV, Patterson TA, Grewe P, Bradford R, Feutry P, Gunasekera R, et al. (2018) Genetic relatedness reveals total population size of white sharks in eastern Australia and New Zealand. *Sci Rep* 8(1):2661.

Holmes BJ, Peddemors VM, Gutteridge AN, Geraghty PT, Chan RWK, Tibbetts IR, Bennett MB (2015) Age and growth of the tiger shark *Galeocerdo cuvier* off the east coast of Australia. *J Fish Biol* 87(2):422–448.

Holmes BJ, Williams SM, Otway NM, Nielsen EE, Maher SL, Bennett MB, Ovenden JR (2017) Population structure and connectivity of tiger sharks (*Galeocerdo cuvier*) across the Indo-Pacific Ocean Basin. *R Soc Open Sci* 4(7):1703099.

Holmes BJ, Pope LC, Williams SM, Tibbetts IR, Bennett MB, Ovenden JR (2018) Lack of multiple paternity in the oceanodromous tiger shark (*Galeocerdo cuvier*). *R Soc Open Sci* 5(1):171385.

ICCAT (2008) *Report of the 2008 Shark Stock Assessments Meeting*, Madrid, Spain, 1–5 September (http://www.iccat. int/Documents/Meetings/Docs/2008_SHK_Report.pdf).

ICES (2009) *Report of the Joint Meeting between ICES Working Group on Elasmobranch Fishes (WGEF) and ICCAT Shark Subgroup*, 22–29 June, 2009, Copenhagen, Denmark.

ICES (2016) *Report of the Working Group on Elasmobranch Fishes (WGEF)*, 15–24 June, 2016, Lisbon, Portugal.

Irion DT, Noble LR, Kock AA, Gennari E, Dicken ML, Hewitt AM, Towner AV, et al. (2017) Pessimistic assessment of white shark population status in South Africa: comment on Andreotti et al. (2016). *Mar Ecol Prog Ser* 577:251–255.

IUCN (2012) *IUCN Red List Categories and Criteria Version 3.1*, 2nd ed. IUCN Species Survival Commission, Gland, Switzerland, and Cambridge, UK (http://www.iucnredlist. org/technical-documents/categories-and-criteria).

Johnson G, Molony B, Jacobsen IP, Peddemors V (2016) *Blacktip Sharks*. Fisheries Research & Development Corporation, Deakin, Australian Capital Territory, Australia (http://www. fish.gov.au/report/11-BLACKTIP-SHARKS-2016).

Jones AT, Ovenden JR, Wang Y-G (2016) Improved confidence intervals for the linkage disequilibrium method for estimating effective population size. *Heredity* 117(4):217–223.

Jones MR, Good JM (2016) Targeted capture in evolutionary and ecological genomics. *Mol Ecol* 25(1):185–202.

Jones OR, Wang J (2010) COLONY: a program for parentage and sibship inference from multilocus genotype data. *Mol Ecol Resour* 10(3):551–555.

Jorde PE (2012) Allele frequency covariance among cohorts and its use in estimating effective size of age-structured populations. *Mol Ecol Resour* 12(3):476–480.

Jorde PE, Ryman N (1995) Temporal allele frequency change and estimation of effective size in populations with overlapping generations. *Genetics* 139(2):1077–1090.

Kanno Y, Vokoun JC, Letcher BH (2011) Sibship reconstruction for inferring mating systems, dispersal and effective population size in headwater brook trout (*Salvelinus fontinalis*) populations. *Conserv Genet* 12(3):619–628.

Karam MJ, Lefèvre F, Dagher-Kharrat MB, Pinosio S, Vendramin GG (2015) Genomic exploration and molecular marker development in a large and complex conifer genome using RADseq and mRNAseq. *Mol Ecol Resour* 15(3):601–612 doi:10.1111/1755-0998.12329.

King BL, Gillis JA, Carlisle HR, Dahn RD (2011) A natural deletion of the HoxC cluster in elasmobranch fishes. *Science* 334(6062):1517.

King JR, Wetklo M, Supernault J, Taguchi M, Yokawa K, Sosa-Nishizaki O, Withler RE (2015) Genetic analysis of stock structure of blue shark (*Prionace glauca*) in the north Pacific ocean. *Fish Res* 172:181–189.

Kousteni V, Kasapidis P, Kotoulas G, Megalofonou P (2015) Strong population genetic structure and contrasting demographic histories for the small-spotted catshark (*Scyliorhinus canicula*) in the Mediterranean Sea. *Heredity* 114(3):333–343.

Lampert KP (2009) Facultative parthenogenesis in vertebrates: reproductive error or chance? *Sex Dev* 2(6):290–301.

Lemmon EM, Lemmon AR (2013) High-throughput genomic data in systematics and phylogenetics. *Annu Rev Ecol Evol Syst* 44(1):99–121.

Li C, Corrigan S, Yang L, Straube N, Harris M, Hofreiter M, White WT, Naylor GJP (2015) DNA capture reveals transoceanic gene flow in endangered river sharks. *Proc Natl Acad Sci USA* 112(43):13302–13307.

Li C, Hofreiter M, Straube N, Corrigan S, Naylor GJ (2013) Capturing protein-coding genes across highly divergent species. *Biotechniques* 54(6):321–326.

Marino IAM, Riginella E, Gristina M, Rasotto MB, Zane L, Mazzoldi C (2015) Multiple paternity and hybridization in two smooth-hound sharks. *Sci Rep* 5:12919.

Marra NJ, Richards VP, Early A, Bogdanowicz SM, Pavinski Bitar PD, Stanhope MJ, Shivji MS (2017) Comparative transcriptomics of elasmobranchs and teleosts highlight important processes in adaptive immunity and regional endothermy. *BMC Genomics* 18(1):87.

Martin AP, Naylor GJP, Palumbi SR (1992) Rates of mitochondrial DNA evolution in sharks are slow compared with mammals. *Nature* 357(6374):153–155.

Mayr E (1963) *Animal Species and Evolution*. Belknap, Cambridge, MA.

McMahon BJ, Teeling EC, Höglund J (2014) How and why should we implement genomics into conservation? *Evol Appl* 7(9):999–1007.

Meimberg H, Schachtler C, Curto M, Husemann M, Habel JC (2016) A new amplicon based approach of whole mitogenome sequencing for phylogenetic and phylogeographic analysis: an example of East African white-eyes (Aves, Zosteropidae). *Mol Phylogenet Evol* 102:74–85.

Momigliano P, Harcourt R, Stow A (2015) Conserving coral reef organisms that lack larval dispersal: are networks of marine protected areas good enough? *Front Mar Sci* 2:16.

Momigliano P, Harcourt R, Robbins WD, Jaiteh V, Mahardika GN, Sembiring A, Stow A (2017) Genetic structure and signatures of selection in grey reef sharks (*Carcharhinus amblyrhynchos*). *Hered (Edinb)* 119(3):142–153.

Morgan J, Harry A, Welch D, Street R, White J, Geraghty PT, Macbeth WG, Broderick D, Tobin A, Simpfendorfer CA, Ovenden JR (2012) Detection of interspecies hybridisation in Chondrichthyes: hybrids and hybrid offspring between Australian (*Carcharhinus tilstoni*) and common (*C. limbatus*) blacktip shark found in an Australian fishery. *Conserv Genet* 13(2):455–463.

Moura T, Figueiredo I, Gordo L (2008) *Analysis of Genetic Structure of the Portuguese Dogfish Centroscymnus coelolepis Caught in the Northeast Atlantic Using Mitochrondrial DNA (Control Region), Preliminary Results*, working document to ICES WGEF meeting.

Mourier J, Planes S (2013) Direct genetic evidence for reproductive philopatry and associated fine-scale migrations in female blacktip reef sharks (*Carcharhinus melanopterus*) in French Polynesia. *Mol Ecol* 22(1):201–214.

Mourier J, Buray N, Schultz JK, Clua E, Planes S (2013) Genetic network and breeding patterns of a sicklefin lemon shark (*Negaprion acutidens*) population in the Society Islands, French Polynesia. *PLoS ONE* 8(8):e73899.

Mulley JF, Hargreaves AD, Hegarty MJ, Heller RS, Swain MT (2014) Transcriptomic analysis of the lesser spotted catshark (*Scyliorhinus canicula*) pancreas, liver and brain reveals molecular level conservation of vertebrate pancreas function. *BMC Genomics* 15:1074.

Nei M (1977) *F*-statistics and analysis of gene diversity in subdivided populations. *Ann Hum Genet* 41(2):225–233.

Nielsen EE, Hemmer-Hansen J, Larsen PF, Bekkevold D (2009) Population genomics of marine fishes: identifying adaptive variation in space and time. *Mol Ecol* 18(15):3128–150.

Nielsen EE, Cariani A, Aoidh EM, Maes GE, Milano I, Ogden R, Taylor M, et al. (2012) Gene-associated markers provide tools for tackling illegal fishing and false eco-certification. *Nat Commun* 3:851.

Nielsen EE, Morgan JAT, Maher SL, Edson J, Gauthier M, Pepperell J, Holmes BJ, et al. (2017) Extracting DNA from 'jaws': high yield and quality from archived tiger shark (*Galeocerdo cuvier*) skeletal material. *Mol Ecol Resour* 17(3):431–442.

Nielsen R, Mattila DK, Clapham PJ, Palsbøll PJ (2001) Statistical approaches to paternity analysis in natural populations and applications to the north Atlantic humpback whale. *Genetics* 157(4):1673–1682.

Nomura T (2008) Estimation of effective number of breeders from molecular coancestry of single cohort sample. *Evol Appl* 1(3):462–474.

Ovenden JR (2013) Crinkles in connectivity: combining genetics and other types of biological data to estimate movement and interbreeding between populations. *Mar Freshwater Res* 64(3):201.

Ovenden JR, Morgan J, Street R, Tobin A, Simpfendorfer CA, Macbeth W, Welch D (2011) Negligible evidence for regional genetic population structure for two shark species (*Rhizoprionodon acutus*, Rüppell, 1837 and *Sphyrna lewini*, Griffith & Smith, 1834) with contrasting biology. *Mar Biol* 158(7):1497–1509.

Ovenden JR, Berry O, Welch DJ, Buckworth RC, Dichmont CM (2015) Ocean's eleven: a critical evaluation of the role of population, evolutionary and molecular genetics in the management of wild fisheries. *Fish Fish* 16:125–159.

Ovenden JR, Leigh GM, Blower DC, Jones AT, Moore A, Bustamante C, Buckworth RC, et al. (2016) Can estimates of genetic effective population size contribute to fisheries stock assessments? *J Fish Biol* 89(6):2505–2518.

Ozerov M, Jurgenstein T, Aykanat T, Vasemagi A (2015) Use of sibling relationship reconstruction to complement traditional monitoring in fisheries management and conservation of brown trout. *Conserv Biol* 29(4):1164–1175.

Ozsolak F, Milos PM (2011) RNA sequencing: advances, challenges and opportunities. *Nat Rev Genet* 12(2):87–98.

Pardini AT, Jones CS, Noble LR, Kreiser B, Malcolm H, Bruce BD, Stevens JD, et al. (2001) Sex-biased dispersal of great white sharks—in some respects, these sharks behave more like whales and dolphins than other fish. *Nature* 412(6843):139–140.

Pazmiño DA, Maes GE, Simpfendorfer CA, Salinas-de-León P, van Herwerden L (2017) Genome-wide SNPs reveal low effective population size within confined management units of the highly vagile Galapagos shark (*Carcharhinus galapagensis*). *Conserv Genet* 18(5):1151–1163.

Pazmiño DA, Maes GE, Simpfendorfer CA, Hoyos-Padilla EM, Duffy CJA, Meyer CG, Kerwath SE, et al. (2017) Strong trans-Pacific break and local conservation units in the Galapagos shark (*Carcharhinus galapagensis*) revealed by genome-wide cytonuclear markers. *Heredity (Edinb)* 120(5):407–421.

Peñalba JV, Smith LL, Tonione MA, Sass C, Hykin SM, Skipwith PL, McGuire JA, et al. (2014) Sequence capture using PCR-generated probes: a cost-effective method of targeted high-throughput sequencing for nonmodel organisms. *Mol Ecol Resour* 14(5):1000–1010.

Peterson BK, Weber JN, Kay EH, Fisher HS, Hoekstra HE (2012) Double digest RADseq: an inexpensive method for *de novo* SNP discovery and genotyping in model and non-model species. *PLoS ONE* 7(5):e37135.

Phillips NM, Chaplin JA, Morgan DL, Peverell SC (2011) Population genetic structure and genetic diversity of three critically endangered *Pristis* sawfishes in Australian waters. *Mar Biol* 158(4):903–915.

Pierce SJ, Norman B (2016) *Rhincodon typus*. The IUCN Red List of Threatened Species™, www.iucnredlist.org/details/19488/0.

Portnoy DS, Heist EJ (2012) Molecular markers: progress and prospects for understanding reproductive ecology in elasmobranchs. *J Fish Biol* 80(5):1120–1140.

Portnoy DS, McDowell JR, Heist EJ, Musick JA, Graves JE (2010) World phylogeography and male-mediated gene flow in the sandbar shark, *Carcharhinus plumbeus*. *Mol Ecol* 19(10):1994–2010.

Portnoy DS, Hollenbeck CM, Johnston JS, Casman HM, Gold JR (2014) Parthenogenesis in a whitetip reef shark *Triaenodon obesus* involves a reduction in ploidy. *J Fish Biol* 85(2):502–508.

Portnoy DS, Puritz JB, Hollenbeck CM, Gelsleichter J, Chapman D, Gold JR (2015) Selection and sex-biased dispersal in a coastal shark: the influence of philopatry on adaptive variation. *Mol Ecol* 24(23):5877–5885.

Powell W, Machray GC, Provan J (1996) Polymorphism revealed by simple sequence repeats. *Trends Plant Sci* 1(7):215–222.

Pratt Jr HL, Carrier JC (2001) A review of elasmobranch reproductive behavior with a case study on the nurse shark, *Ginglymostoma cirratum*. In: Tricas TC, Gruber SH (eds) *The Behavior and Sensory Biology of Elasmobranch Fishes: An Anthology in Memory of Donald Richard Nelson*. Springer, Netherlands, pp 157–188.

Puritz JB, Hollenbeck CM, Gold JR (2014) dDocent: a RADseq, variant-calling pipeline designed for population genomics of non-model organisms. *PeerJ* 2:e431.

Ralls K, Demaster DP, Estes JA (1996) Developing a Criterion for Delisting the Southern Sea Otter under the U.S. Endangered Species Act. *Cons Biol* 10(6):1528–1537.

Rawding DJ, Sharpe CS, Blankenship SM (2014) Genetic-based estimates of adult Chinook salmon spawner abundance from carcass surveys and juvenile out-migrant traps. *Trans Am Fish Soc* 143(1):55–67.

Read TD, Petit III RA, Joseph SJ, Alam MT, Weil MR, Ahmad M, Bhimani R, et al. (2017) Draft sequencing and assembly of the genome of the world's largest fish, the whale shark: *Rhincodon typus* Smith 1828. *BMC Genomics* 18:532.

Rhoads A, Au KF (2015) PacBio sequencing and its applications. *Genom Proteom Bioinform* 13(5):278–289.

Richards VP, Suzuki H, Stanhope MJ, Shivji MS (2013) Characterization of the heart transcriptome of the white shark (*Carcharodon carcharias*). *BMC Genomics* 14:697.

Robinson DP, Baverstock W, Al-Jaru A, Hyland K, Khazanehdari KA (2011) Annually recurring parthenogenesis in a zebra shark *Stegostoma fasciatum*. *J Fish Biol* 79(5):1376–1382.

Ryman N, Palm S (2006) POWSIM: a computer program for assessing statistical power when testing for genetic differentiation. *Mol Ecol Notes* 6(3):600–602.

Saiki RK, Gelfand DH, Stoffel S, Scharf SJ, Higuchi R, Horn GT, Mulluis KB, Erlich HA (1988) Primer-directed enzymatic amplification of DNA with a thermostable DNA polymerase. *Science* 239(4839):487–491.

Sanger F, Air GM, Barrell BG, Brown NL, Coulson AR, Fiddes JC, Hutchison CA, et al. (1977a) Nucleotide sequence of bacteriophage φX174 DNA. *Nature* 265:687–695.

Sanger F, Nicklen S, Coulson AR (1977b) DNA sequencing with chain-terminating inhibitor. *Proc Natl Acad Sci USA* 74:5463–5467.

Sansaloni C, Petroli C, Jaccoud D, Carling J, Detering F, Grattapaglia D, Kilian A (2011) Diversity Arrays Technology (DArT) and next-generation sequencing combined: genome-wide, high throughput, highly informative genotyping for molecular breeding of *Eucalyptus*. *BMC Proc* 5(Suppl 7):P54.

Schlotterer C, Tobler R, Kofler R, Nolte V (2014) Sequencing pools of individuals—mining genome-wide polymorphism data without big funding. *Nat Rev Genet* 15(11):749–763.

Schrey AW, Heist EJ (2003) Microsatellite analysis of population structure in the shortfin mako (*Isurus oxyrinchus*). *Can J Fish Aquat Sci* 60(6):670–675.

Seber GA (1965) A note on the multiple-recapture census. *Biometrika* 52(1/2):249–259.

Shafer ABA, Wolf JBW, Alves PC, Bergström L, Bruford MW, Brännström I, Colling G, et al. (2015) Genomics and the challenging translation into conservation practice. *Trends Ecol Evol* 30(2):78–87.

Sigsgaard EE, Nielsen IB, Bach SS, Lorenzen ED, Robinson DP, Knudsen SW, Pedersen MW, et al. (2016) Population characteristics of a large whale shark aggregation inferred from seawater environmental DNA. *Nat Ecol Evol* 1(1):0004.

Simpfendorfer CA (2000) Predicting population recovery rates for endangered western Atlantic sawfishes using demographic analysis. *Environ Biol Fish* 58(4):371–377.

Simpfendorfer CA, Dulvy NK (2017) Bright spots of sustainable shark fishing. *Curr Biol* 27(3):R97–R98.

Simpfendorfer CA, Kyne PM, Noble TH, Goldsbury J, Basiita RK, Lindsay R, Shields A, Perry C, Jerry DR (2016) Environmental DNA detects critically endangered largetooth sawfish in the wild. *Endang Spec Res* 30:109–116.

Skaug HJ (2001) Allele-sharing methods for estimation of population size. *Biometrics* 57(3):750–756.

Smith PJ (1986) Low genetic variation in sharks (Chondrichthyes). *Copeia* 1986(1):202–207.

Spaet JLY, Jabado RW, Henderson AC, Moore ABM, Berumen ML (2015) Population genetics of four heavily exploited shark species around the Arabian Peninsula. *Ecol Evol* 5(12):2317–2332.

Stow A, Zenger K, Briscoe D, Gillings M, Peddemors V, Otway N, Harcourt R (2006) Isolation and genetic diversity of endangered grey nurse shark (*Carcharias taurus*) populations. *Biol Lett* 2(2):308–311.

Straube N, Lampert KP, Geiger M, Weiß J, Kirchhauser J (2016) First record of second-generation facultative parthenogenesis in a vertebrate species, the whitespotted bambooshark *Chiloscyllium plagiosum*. *J Fish Biol* 88(2):668–675.

Therkildsen NO, Palumbi SR (2017) Practical low-coverage genomewide sequencing of hundreds of individually barcoded samples for population and evolutionary genomics in nonmodel species. *Mol Ecol Resour* 17(2):194–208.

Tillett BJ, Field IC, Johnson G, Buckworth R, Meekan MG, Bradshaw C, Ovenden JR (2012) Accuracy of species identification by fisheries observers in a north Australian shark fishery. *Fish Res* 127–128:109–115.

Vargas-Caro C, Bustamante C, Bennett MB, Ovenden JR (2016) The complete validated mitochondrial genome of the yellownose skate *Zearaja chilensis* (Guichenot 1848) (Rajiformes, Rajidae). *Mitochondrial DNA A DNA Mapp Seq Anal* 27(2):1227–1228.

Venkatesh B, Lee AP, Ravi V, Maurya AK, Lian MM, Swann JB, Ohta Y, et al. (2014) Elephant shark genome provides unique insights into gnathostome evolution. *Nature* 505(7482):174–179.

Verissimo A, McDowell JR, Graves JE (2010) Global population structure of the spiny dogfish *Squalus acanthias*, a temperate shark with an antitropical distribution. *Mol Ecol* 19(8):1651–1662.

Verissimo A, McDowell JR, Graves JE (2011) Population structure of a deep-water squaloid shark, the Portuguese dogfish (*Centroscymnus coelolepis*). *ICES J Mar Sci* 68(3):555–563.

Verissimo A, McDowell JR, Graves JE (2012) Genetic population structure and connectivity in a commercially exploited and wide-ranging deepwater shark, the leafscale gulper (*Centrophorus squamosus*). *Mar Freshwater Res* 63(6):505–512.

Vignaud TM, Maynard JA, Leblois R, Meekan MG, Vazquez-Juarez R, Ramirez-Macias D, Pierce SJ, Rowat D, Berumen ML, Beeravolu C, Baksay S, Planes S (2014a) Genetic structure of populations of whale sharks among ocean basins and evidence for their historic rise and recent decline. *Mol Ecol* 23(10):2590–2601.

Vignaud TM, Mourier J, Maynard JA, Leblois R, Spaet J, Clua E, Neglia V, Planes S (2014b) Blacktip reef sharks, *Carcharhinus melanopterus*, have high genetic structure and varying demographic histories in their Indo-Pacific range. *Mol Ecol* 23(21):5193–5207.

Wang JL (2009) A new method for estimating effective population sizes from a single sample of multilocus genotypes. *Mol Ecol* 18:2148–2164.

Wang Z, Gerstein M, Snyder M (2009) RNA-Seq: a revolutionary tool for transcriptomics. *Nat Rev Genet* 10(1):57–63.

Waples RS (1989) A generalized approach for estimating effective population size from temporal changes in allele frequency. *Genetics* 121(2):379–391.

Waples RS, Do C (2010) Linkage disequilibrium estimates of contemporary Ne using highly variable genetic markers: a largely untapped resource for applied conservation and evolution. *Evol Appl* 3(3):244–262.

Waples RS, England PR (2011) Estimating contemporary effective population size on the basis of linkage disequilibrium in the face of migration. *Genetics* 189(2):633–644.

Waples RS, Gaggiotti O (2006) What is a population? An empirical evaluation of some genetic methods for identifying the number of gene pools and their degree of connectivity. *Mol Ecol* 15:1419–1439.

Waples RS, Yokota M (2007) Temporal estimates of effective population size in species with overlapping generations. *Genetics* 175(1):219–233.

Weir BS (2012) *F*-statistics: a historical view. *Philos Sci* 79(5):637–643.

Weir BS, Cockerham CC (1984) Estimating *F*-statistics for the analysis of population structure. *Evolution* 38:1358–1370.

Weltz K, Lyle JM, Ovenden J, Morgan JAT, Moreno DA, Semmens JM (2017) Application of environmental DNA to detect an endangered marine skate species in the wild. *PLoS ONE* 12(6):e0178124.

White WT, Corrigan S, Yang L, Henderson AC, Bazinet AL, Swofford DL, Naylor G (2017) Phylogeny of the manta and devilrays (Chondrichthyes: mobulidae), with an updated taxonomic arrangement for the family. *Zool J Linn Soc* 182(1):50–75.

Willette DA, Allendorf FW, Barber PH, Barshis DJ, Carpenter KE, Crandall ED, Cresko WA, et al. (2014) So, you want to use next-generation sequencing in marine systems? Insight from the Pan-Pacific Advanced Studies Institute. *Bull Mar Sci* 90(1):79–122.

Wright S (1951) The genetical structure of populations. *Ann Eugenics* 15:323–354.

Wynen L, Larson H, Thorburn D, Peverell S, Morgan D, Field I, Gibb K (2009) Mitochondrial DNA supports the identification of two endangered river sharks (*Glyphis glyphis* and *Glyphis garricki*) across northern Australia. *Mar Freshwater Res* 60(6):554–562.

CHAPTER 14

Environmental DNA (eDNA)
A Valuable Tool for Ecological Inference and Management of Sharks and Their Relatives

Agnes Le Port
Centre for Tropical Water and Aquatic Ecosystem Research (TropWATER) and College of Science and Engineering, James Cook University, Townsville, Queensland, Australia

Judith Bakker
School of Environment and Life Sciences, University of Salford, Salford, United Kingdom

Madalyn K. Cooper
Centre for Tropical Water and Aquatic Ecosystem Research (TropWATER) and College of Science and Engineering, James Cook University, Townsville, Queensland, Australia

Roger Huerlimann
Centre for Tropical Water and Aquatic Ecosystem Research (TropWATER) and College of Science and Engineering, James Cook University, Townsville, Queensland, Australia

Stefano Mariani
School of Environment and Life Sciences, University of Salford, Salford, United Kingdom

CONTENTS

14.1 INTRODUCTION

14.1.1 What Is Environmental DNA?

Knowledge of spatial and temporal variation in abundance is critical for the implementation of effective protective measures for organisms that are both naturally rare and vulnerable to exploitation. The development of management and conservation strategies for elasmobranchs depends on accurate assessment and monitoring of the distribution and abundance of target species in the field, but detecting species occurrences is often even more challenging in the aquatic environment than on land (Webb and Mindel, 2015). Consequently, as is the case for many large, mobile and rare vertebrates, shark detection is inherently difficult.

All organisms continuously leave traces of themselves behind in the environment in the form of shed skin cells, bodily fluids, metabolic waste, gametes, or blood. Any of these materials can contain pieces of the organism's DNA. Environmental DNA (eDNA) analysis is based on the retrieval of this naturally released genetic material from the environment. It generally refers to bulk DNA extracted from an environmental sample such as water but also from soil, sediment, snow, or even from air (Taberlet et al., 2012a). In aquatic systems, macroorganismal-derived eDNA can be present as free DNA, cellular debris, or particle-bound DNA and is mostly present in small fragments, due to rapid degradation (Barnes et al., 2014); however, much of the eDNA is retrieved from cellular material and may therefore contain still relatively undamaged nucleic acid molecules. Nevertheless, eDNA studies focus primarily on the detection of short fragments, as currently available parallel sequencing and qPCR platforms have short-read capabilities limited to a few hundred base pairs. When DNA is present at low concentrations, mitochondrial DNA (mtDNA) is often targeted, as there are substantially more mitochondrial than nuclear DNA copies per cell (Wilcox et al., 2013). Commonly employed mtDNA genes include cytochrome b,

cytochrome c oxidase subunit 1 (COI), 12S rRNA, and 16S rRNA (Kelly et al., 2014; Thomsen et al., 2012b; Valentini et al., 2016), and targeted fragments typically fall within the range of 79 to 285 bp (Ficetola et al., 2008; Minamoto et al., 2012). The level of target specificity is often the main determining factor when choosing or designing primers for eDNA analysis.

Environmental DNA is emerging as a non-invasive method for the detection and identification of rare and elusive species in a wide range of ecosystems, including aquatic environments (Port et al., 2016; Thomsen et al., 2012a; Yamamoto et al., 2017). It is rapidly diffused from its source and degraded under the influence of local environmental conditions such as mechanical forces, ultraviolet (UV) radiation, pH, temperature (Barnes et al., 2014; Jerde et al., 2011; Pilliod et al., 2014), microbial activity (Barnes et al., 2014), and spontaneous chemical reactions such as oxygenation (Lindahl, 1993; Nielsen et al., 2007). This indicates a low probability of long-distance dispersal of eDNA in aquatic ecosystems (Thomsen and Willerslev, 2015); thus, the detection of eDNA from a specific taxon indicates its presence or very recent presence in the environment (Barnes et al., 2014; Jerde et al., 2011; Pilliod et al., 2014). However, there remains much uncertainty with regard to the impact of oceanic currents on the local-scale spatial patterns of trace DNA, especially in open marine systems (O'Donnell et al., 2017).

Due to recent advances in high-throughput sequencing and bioinformatics, the use of eDNA has developed into a cost-effective, rapid, non-invasive method for collecting and analyzing biological samples from large portions of the environment without the necessity of isolating the target species (Hajibabaei et al., 2006; Taberlet et al., 2012b). Using this approach, thousands of species present in any environmental sample can be detected by high-throughput DNA sequencing and identified using molecular taxonomy databases, thus revolutionizing our ability to detect species and conduct genetic analysis for conservation, management, and research of aquatic ecosystems.

14.1.2   A Short History of eDNA

Over the past decade, development of eDNA recovery and sequencing techniques has been substantial and has resulted in an increasing interest in its use as a tool for both targeted species detection and biodiversity assessments (Handelsman, 2004). The term "eDNA" was first used by microbiologists, who have been applying the eDNA method since the mid-1980s to assess the diversity of microorganism communities in ancient marine sediments (Ogram et al., 1987). The general eDNA methods currently used for monitoring aquatic populations arose from this early work (e.g., Willerslev et al., 2003). Subsequently, in the 1990s, eDNA methods were employed to monitor phytoplankton blooms and to assess changes in biomass of bacterial communities (Bailiff and Karl, 1991; Paul et al., 1996; Weirbauer et al., 1993). The use of eDNA has more recently been developed to elucidate macroorganism identity in aquatic environments. However, the nature of eDNA from macroorganisms in environmental samples is different from that of microbial organisms (prokaryotes and microbial eukaryotes) because the former are present only as parts of the organism (cellular remains or free DNA), whereas the latter may be detected by DNA derived from whole, living organisms present in the samples (Thomsen and Willerslev, 2015).

Environmental DNA as a method to assess the diversity of macroorganismal communities was first applied to sediments, revealing DNA from extinct and extant mammals, birds, and plants (Thomsen and Willerslev, 2015; Willerslev et al., 2003). In 2008, the eDNA method was applied for the first time to confirm the presence of an aquatic invasive species, the American bullfrog (*Rana catesbiana*), from water samples in a natural lentic system (Ficetola et al., 2008). Subsequently, the first eDNA study in freshwater lotic systems for the detection of invasive Asian carp was published in 2011 (Jerde et al., 2011). In 2012, eDNA analysis was first applied to the marine environment for the detection of marine mammals (Foote et al., 2012) and for the estimation of marine fish biodiversity (Thomsen et al., 2012a). Environmental DNA has since been applied for the detection of a large range of aquatic species in both freshwater and marine systems (Piaggio et al., 2014; Pilliod et al., 2014; Yamamoto et al., 2017) and, more recently, for the detection of sharks and rays (Bakker et al., 2017; Gargan et al., 2017; Sigsgaard et al., 2016; Simpfendorfer et al., 2016; Weltz et al., 2017).

14.1.3   eDNA vs. Traditional Monitoring Techniques

Currently established survey methods, such as fishing by longlining or gillnetting, acoustic or satellite tagging and monitoring, baited remote underwater video (BRUV), underwater visual census (UVC), ecological knowledge surveys, and fisheries-dependent population surveys, all have associated biases and challenges. These include being potentially resource intensive, selective, and dependent on taxonomic expertise, as well as sometimes being invasive and potentially traumatogenic (Lodge et al., 2012; Simpfendorfer et al., 2016; Wheeler, 2004). Traditional survey methods are also highly susceptible to false negatives, failing to detect rare or cryptic species that are present; therefore, assessment and monitoring of the distribution and abundance of mobile species in aquatic environments remain challenging and would benefit from new, complementary methods of investigation.

Environmental DNA has been shown to be a reliable detection method, matching or even outperforming conventional survey methods (Dejean et al., 2012; Doi et al., 2017b; Hänfling et al., 2016; Sigsgaard et al., 2015; Taberlet et al., 2012a; Takahara et al., 2013; Thomsen et al., 2012a; Valentini et al., 2016). Because eDNA analysis is an inherently non-invasive detection method, it is not necessary for the species of interest (or its habitat) to be either disturbed or caught in order to establish its presence or to acquire a positive taxonomic identification. Because visual detection is not necessary, using eDNA makes it easier to detect rare species (or those species that have juvenile stages that closely resemble other species) (Dejean et al., 2011; Huver et al., 2015; Schmidt et al., 2013).

Species of conservation concern often have low population numbers, making surveys based on eDNA methods particularly suitable for informing applied conservation efforts (Foote et al., 2012; Olson et al., 2012; Thomsen et al., 2012a). Likewise, exotic and invasive species are typically rare at their expanding range margins, requiring highly sensitive detection methods (Davy et al., 2015; Jerde et al., 2011). False negatives in presence/absence data using traditional methods can prevent effective habitat protection for threatened species. A particular case study concerning shark species has recently been described for the New Caledonian archipelago. Here, 2758 UVCs and 385 BRUVs detected 9 shark species. In contrast, with only 22 eDNA samples 13 shark species were detected. Thus, despite two orders of magnitude less sampling effort, with eDNA analysis 44% more shark species were detected compared to UVCs and BRUVs, revealing a greater diversity of sharks than previously thought, thereby indicating the need for large-scale eDNA assessments to improve shark monitoring and conservation efforts (Boussarie et al., 2018). eDNA studies are not immune to false negatives, however, and, similar to traditional survey methods, the false-negative rate in eDNA studies is inversely proportional to the target species' abundance. These "non-detections" are usually due to two sources of error: method error during sample collection or insufficient detection sensitivity of the quantitative polymerase chain reaction (qPCR) or metabarcoding assays (Furlan and Gleeson, 2016b). Contrary to traditional methods, various positive controls can readily be put in place to monitor and exclude potential false negatives at all stages of an eDNA survey (Furlan and Gleeson, 2016b).

In terms of sampling effort, eDNA analysis can offer considerable time and cost benefits (Dejean et al., 2012), especially concerning the distribution of rare and threatened species (Rees et al., 2014a; Valentini et al., 2016). Collecting water samples requires significantly less time and fewer resources compared to traditional survey methods. This is particularly true when target species are found in remote or difficult to access areas (Simpfendorfer et al., 2016). In a study of invasive Asian carp, it took 93 days of person effort to detect one silver carp (*Hypophthalmichthys molitrix*) by electrofishing, whereas eDNA analysis required only 0.174 day of person effort to achieve a positive detection (Jerde et al., 2011; Rees et al., 2014b). Additional advantages of the eDNA method compared to traditional sampling relate to the ease of obtaining permits for the collection and handling of water samples vs. (live) animals, and sampling can often be carried out under more extreme weather conditions.

There are, however, important caveats associated with eDNA detection, and traditional survey methods still have a number of advantages over eDNA methodologies. Foremost among them, when using eDNA analysis, it is not possible to distinguish whether the detected DNA from a certain species has been released by a dead or live animal. Additionally, eDNA methods do not provide information on size, movement patterns, condition, developmental stage (eggs, larvae, juveniles, adults), or sex of the target organism. Moreover, the detection of an individual's DNA, without direct observation, cannot provide information on the exact location of the animal. Furthermore, when using a mitochondrial marker, of which the DNA is mostly maternally inherited (Giles et al., 1980), it will not be possible to distinguish hybrids (which may be the result of breeding between native and invasive species) from their maternal species. Finally, inferring abundance information from eDNA remains challenging and is a key area for further research.

Environmental DNA is becoming a rapid and cost-effective tool for collecting species' presence, distribution, and, with some caveats, (relative) abundance data. Most likely, with continuing developments in the fields of DNA sequencing and bioinformatics, eDNA methods will increasingly complement (rather than completely replace) traditional survey methods.

14.1.4 eDNA Approaches: Species-Specific vs. Metabarcoding

Currently, the use of eDNA can be broadly divided into two main approaches: a single-species approach (eDNA barcoding) or a multispecies approach (eDNA metabarcoding). Environmental DNA barcoding is aimed at detecting a single species in the environment by polymerase chain reaction (PCR) or quantitative PCR (qPCR) to target eDNA sequences belonging to the target species, which is then often confirmed through Sanger sequencing (Eichmiller et al., 2014; Mahon et al., 2013; Sanger et al., 1977). Conventional PCR has previously been used for species-specific eDNA detection (Davison et al., 2016; Dejean et al., 2011; Piaggio et al., 2014; Simpfendorfer et al., 2016). Quantitative PCR, however, offers a distinct advantage over traditional endpoint PCR techniques through the addition of a fluorescent dye (e.g., SYBR™ Green) or a fluorescently labeled reporter probe, which allows the amplification of the target sequence to be monitored in real time by the qPCR platform. Quantification is measured against a standard curve run simultaneously based on samples of a known concentration of reference DNA (Bourlat et al., 2013). Probe-based qPCR increases both detection specificity and sensitivity, as the use of a probe, in combination with forward and reverse primers, ensures that there are three sequences to check against the target template DNA (Herder et al., 2014). However, it is limited to the detection of only one or a few target organisms at a time (Furlan and Gleeson 2016a; Jerde et al., 2011; Mahon and Jerde, 2016; Simpfendorfer et al., 2016; Thomsen and Willerslev, 2015; Uchii et al., 2016; Weltz et al., 2017).

The alternative to traditional DNA Sanger sequencing and eDNA barcoding (which can only sequence specimens individually) for species detection is eDNA metabarcoding (where the prefix "meta" refers to the collection of barcode genes across the taxonomical spectrum of the samples). This multispecies approach simultaneously identifies multiple taxa from an environmental sample without the need for *a priori* knowledge of the species likely to be present (Taberlet et al., 2012b). Metabarcoding offers a tremendously enhanced capability in biodiversity studies because it has the potential to characterize the full community of species present in a set of complex environmental samples (Valentini et al., 2016; Yamamoto et al., 2017). Metabarcoding employs high-throughput sequencing while using more generalized PCR primers in order to mass-amplify a taxonomically informative marker gene and thus can offer a comprehensive view of an ecosystem. This method has the potential to reveal hundreds or thousands of taxa (and potentially their abundances) from a single environmental sample (e.g., Kelly et al., 2017; Leray and Knowlton, 2015; Miya et al., 2015; Yamamoto et al., 2017).

14.2 APPLICATIONS OF eDNA TECHNIQUES IN SHARK BIOLOGY STUDIES: WHAT CAN THEY TELL US ABOUT SHARKS AND THEIR RELATIVES?

One-quarter of all chondrichthyans (sharks, skates, rays, and chimeras) are currently considered threatened (Dulvy et al., 2014). Coastal and continental shelf-dwelling rays and sharks, such as sawfishes and angel sharks, are particularly at risk from overexploitation and other anthropogenic threats

(Dulvy et al., 2016). Traditional survey methods have proven to be useful for determining the presence and distribution of a range of shark and ray species (e.g., Chin 2014; Guttridge et al., 2017; Hansell et al., 2017; Kajiura and Tellman, 2016; Kessel et al., 2016; Vaudo and Heithaus, 2012). However, confirming the presence of a target species relies on locating and/or catching the animals, which can prove challenging and time-consuming for many species due to their rarity, cryptic habits, ecological specialization, and potential occurrence in remote and difficult-to-access locations (Barnes and Turner, 2016). With over half of chondrichthyan species considered data deficient, there is a clear urgency to rapidly increase the knowledge of these species' life histories and current distributional ranges to further conservation and management efforts. Environmental DNA may be the game-changing genetic technique for the study of sharks and their relatives, not only allowing for the time- and cost-effective gathering of crucial species' occurrence and distribution information but also providing much-needed ecosystem-wide species composition and population-level data (see Table 14.1).

14.2.1 Occurrence and Distribution of Rare and Endangered Species

The application of eDNA techniques for the detection of sharks and their relatives has only been described recently. The first study dedicated specifically to the detection of an elasmobranch species successfully detected the critically endangered largetooth sawfish (*Pristis pristis*) in freshwater habitats in northern Australia in locations of both known (based on gillnet surveys and traditional ecological knowledge from local indigenous ranger groups) and unknown sawfish presence (Simpfendorfer et al., 2016). The versatility of using eDNA techniques in elasmobranch species detection has further been demonstrated with the eDNA barcoding approach being successfully applied in two widely different marine habitats: within a coastal embayment for the detection of the endangered Maugean skate (*Zearaja maugeana*) (Weltz et al., 2017) and above the summits of oceanic seamounts for the Chilean devil ray (*Mobula tarapacana*) (Gargan et al., 2017). Positive eDNA detections in water samples, identifying these endangered and critically endangered rays, highlights the value of the method for rare elasmobranch species. Moreover, the detection of oceanic and highly migratory species, such as devil rays, emphasizes that species that are otherwise difficult and rare to encounter can be surveyed expeditiously with eDNA in open-water environments. Although still in its infancy, successful applications of eDNA barcoding for the detection of rare and endangered elasmobranchs in both marine and freshwater environments highlight the potential of this technique for furthering conservation and management outcomes.

14.2.2 Species Composition Using eDNA Metabarcoding

Environmental DNA metabarcoding has the potential to simultaneously identify several taxa from environmental samples (Taberlet et al., 2012b). Shark species inventories and assessment of geographical distributions based on eDNA metabarcoding could be an important tool for rapid environmental monitoring and hence influence conservation management and policy decisions. Although three species of elasmobranch have previously been detected in a large-scale marine eDNA study using a primer set designed for teleosts (bony fish) (Thomsen et al., 2016), other studies have encountered challenges concerning shark-specific detection when applying this multispecific approach in an aquarium-based setting (Kelly et al., 2014; Miya et al., 2015).

The main caveat in using eDNA metabarcoding for the assessment of shark diversity is that sharks are naturally rare compared to most other taxa. Figure 14.1 shows the difference in read abundances between elasmobranchs and teleosts recovered from two marine samples (from an area with relatively high shark abundances) and a sample taken from an aquarium tank, indicating the naturally low abundance of elasmobranch eDNA compared to teleost eDNA. Consequently, when individuals are present in the sampling area, eDNA released by sharks will generally constitute only a very small portion of all the eDNA present in a water sample. This highlights the importance of designing and optimizing protocols specifically geared toward the detection of sharks. This includes sampling relatively large volumes of water (generally >3 liters; see Figure 14.2) per sample and using primers that specifically target sharks while excluding other, non-target taxa.

More recently, eDNA metabarcoding of natural seawater samples was employed to specifically infer shark presence, diversity, and relative abundance in both Atlantic and Pacific tropical ecosystems (Bakker et al., 2017). By using a primer set targeting a 127-bp stretch of the mitochondrial COI region (Fields et al., 2015), 21 different shark species were detected whose geographical patterns of diversity and abundance coincided with geographical differences in levels of anthropogenic pressure and conservation effort in two independent tropical marine systems. Even though issues relating to the taxonomic assignment of closely related species still remain to be resolved, this study demonstrates the potential of the eDNA metabarcoding approach for the detection and monitoring of shark communities.

14.2.3 Population Genetics—From Species Detection to the Analysis of Populations

An additional potential for the use of eDNA—when it stores sufficient population-specific information within the molecular markers used (e.g., mitochondrial haplotypes)—lies in the area of population genetics, with applications

Table 14.1   Summary of eDNA Studies to Date Focusing on Sharks and Their Relatives

Refs.	Study Title	Climate Zone and Habitat	Geographical Location	Study Objective (Genetic Assay)	Species of Interest	Collection Method	Extraction	Target Gene	Successful Detection (No. Positive Detections/Site)
Gargan et al. (2017)	Development of a sensitive detection method to survey pelagic biodiversity using eDNA and quantitative PCR: a case study of devil ray at seamounts	Temperate seamounts	Azores	Species-specific (qPCR)	Chilean devil ray (*Mobula tarapacana*)	3 L collected per site, stored at −20°C before vacuum filtration (0.45-μm nylon filter) and storage in 100% ethanol	DNeasy® Blood & Tissue Kit	COI	5/15
Weltz et al. (2017)	Application of environmental DNA to detect an endangered marine skate species in the wild	Temperate harbor	Macquarie Harbour, Tasmania, Australia	Species-specific (qPCR)	Maugean skate (*Zaeraja maugeana*)	2 × 10-L replicates stored on ice before vacuum filtration (0.45-μm cellulose nitrate filter) and storage at −18°C	DNeasy® PowerWater® DNA Isolation Kit	NADH4	4/4
Bakker et al. (2017)	Environmental DNA reveals tropical shark diversity and abundance in contrasting levels of anthropogenic impact	Tropical marine	Caribbean and New Caledonia; multiple locations	Elasmobranch-specific metabarcoding (Illumina amplicon sequencing)	All elasmobranchs	4-L samples, vacuum filtration (0.45-μm MCE filters); storage in silica beads at −20°C	Mo Bio PowerSoil™ DNA Isolation Kit	COI	50/76
Boussarie et al. (2018)	Environmental DNA illuminates the dark diversity of sharks	Tropical marine	New Caledonia	Elasmobranch-specific metabarcoding (Illumina amplicon sequencing)	All elasmobranchs	4-L samples, vacuum filtration (0.45-μm MCE filters); storage in silica beads at −20°C	Mo Bio PowerSoil™ DNA Isolation Kit	COI	21/22
Sigsgaard et al. (2016)	Population characteristics of a large whale shark aggregation inferred from seawater environmental DNA	Tropical eutrophic, high-saline Gulf	Al Shaheen, Central Arabian Gulf, off Qatar	Species-specific (qPCR and NGS)	Whale shark (*Rhincodon typus*)	3 × 500-mL replicates per site, stored on ice or filtered immediately using Sterivex™ GP filters (0.22-μm) and storage at −18°C	DNeasy® Blood & Tissue Kit	D-loop	13/17

Study	Title	Climate zone and habitat	Location	Genetic assay	Species of interest	Collection and extraction methods	Extraction kit	Target gene	Detection success
Simpfendorfer et al. (2016)	Environmental DNA detects critically endangered largetooth sawfish in the wild	Tropical floodplain waterholes and main-channel river	Daly River, Northern Territory, Australia	Species-specific (PCR)	Largetooth sawfish (*Pristis pristis*)	5 × 2-L replicates per site, stored on ice prior to vacuum filtration (20-µm nylon filter) and storage in ethanol at −20°C	ISOLATE II Genomic DNA Kit	COI	7/11
Thomsen et al. (2016)	Environmental DNA from seawater samples correlate with trawl catches of subarctic, deepwater fishes	Subarctic continental slope	Southwest Greenland	Fish-specific (Illumina amplicon sequencing)	Marine fish	1.5-L samples, vacuum filtration (0.45-µm nylon filters), storage at −20°C	DNeasy® Blood & Tissue Kit	12S rRNA	21/21
Miya et al. (2015)	MiFish, a set of universal PCR primers for metabarcoding environmental DNA from fishes: detection of more than 230 subtropical marine species	Subtropical aquaria	Okinawa, Japan	Metabarcoding (NGS)	Marine fish	10-L replicates subsampled and vacuum filtered (0.7-µm glass fiber) and stored in aluminium foil at −20°C	DNeasy® Blood & Tissue Kit	12S rRNA	17/17 species present in aquaria
Kelly et al. (2014)	Using environmental DNA to census marine fishes in a large mesocosm	Subtropical aquaria	Monterey Bay, California	Metabarcoding (NGS)	Marine fish	20-L replicates subsampled and vacuum filtered (0.22-µm Durapore™ membrane filters) and stored at −80°C	DNeasy® Blood & Tissue Kit	12S rRNA	No detection

Note: Details provided include climate zone and habitat, geographical location of study, genetic assay used, species of interest, collection and extraction methods, target gene, and detection success. COI, cytochrome *c* oxidase subunit 1 (COI); MCE, mixed cellulose ester; NGS, next-generation sequencing.

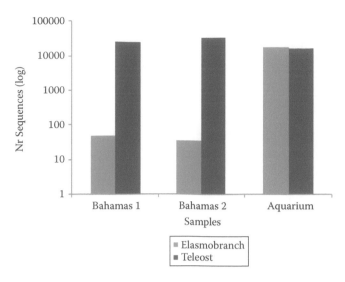

Figure 14.1 Differences in read abundances (on a logarithmic scale) between shark eDNA recovered from natural marine samples and an aquarium sample. Fish-specific primers targeting the cytochrome *b* region were used for eDNA amplification (J. Bakker, unpublished data).

for conservation genetics and phylogeography (Bohmann et al., 2014). To date, only one study applying eDNA to infer population characteristics for shark species has been published (Sigsgaard et al., 2016). Here, samples were collected from areas in the Arabian Gulf, where whale sharks (*Rhincodon typus*) are known to occur. Mitochondrial DNA control region sequences obtained from eDNA samples were compared to sequences from tissue samples collected from the same locality. DNA mutation rate was calculated and female effective population size (N_f) inferred. Subsequently, it proved possible to infer the likely N_f for the entire Indo-Pacific Ocean, with comparable estimates obtained from eDNA and tissue extracted sequences. Moreover, by using eDNA analysis, this study revealed that the whale shark populations in the Indo-Pacific are genetically distinct from those populations occurring in the Atlantic Ocean (Sigsgaard et al., 2016), thus demonstrating for the first time that eDNA methods are capable of using the genetic variation in the DNA fragments isolated from water samples to estimate population sizes, in addition to identifying relatedness between different populations of the same species.

Figure 14.2 Field equipment used in eDNA studies: (Left) Collection of ocean water with a Kemmerer-type water sampler (photograph credit: Diego Camejo). (Right, top) Portable eDNA filtration pump (Grover Scientific eDNA Pump) that can be used to quickly filter samples onsite (photograph credit: Madalyn K. Cooper). (Right, bottom) Extendable pole used in difficult-to-reach areas or to decrease the risk to the sampler of dangerous wildlife (e.g., crocodilians).

14.3 eDNA METHODS

14.3.1 Field Sampling Considerations for Shark eDNA Studies

14.3.1.1 Sampling Methodologies

Effective and accurate detection of organisms in aquatic ecosystems using eDNA is dependent on the development of an appropriate sampling design. There is no single eDNA sampling method that fits all target species and environments (Barnes and Turner, 2016; De Souza et al., 2016), and conducting a pilot study is important before initiating a full study (Furlan and Gleeson 2016b; Goldberg et al., 2016; Kelly et al., 2016). The method of water sample collection is the same for both species-specific and population-level investigations, but there are differences in field sampling design and downstream genetic processing and analyses (Table 14.2). Overall, it is most important to understand the characteristics of eDNA in the context of local environmental conditions, including the influence of biotic and abiotic factors on DNA degradation and dispersal, and factors related to the target species/community, including life history, demographic patterns and ecology. These factors can result in variation in detection sensitivity. Currently, the recommended protocol for each new application should assess detection probabilities for the target species given

the proposed field and laboratory protocols (Goldberg et al., 2016). Preliminary laboratory and aquarium eDNA assays can be applied to test and confirm the sensitivity and specificity of the methodology, and, where possible, controlled tank-based experiments can be conducted to further understand eDNA shedding, degradation, and distribution rates (Turner et al., 2014b; Weltz et al., 2017).

Environmental DNA detection methods are perceived to be highly sensitive, but they are largely contingent on the probability of detecting eDNA where and when it is present in the environment (Dejean et al., 2012; Ficetola et al., 2008; Furlan and Gleeson, 2016b; Goldberg et al., 2013). For sampling approaches that target a single species, estimating the sensitivity (or target species' detection probability) of the assay is crucial for accurately and confidently interpreting results, as it delineates the chances of detection failure (Amberg et al., 2015; Furlan et al., 2016). Detection failure, false positives (incorrect positive detection when the target species is absent), and false negatives (failing to detect the target species when it is present) potentially confound conclusions about species presence or absence and can misinform management. Therefore, the risk of such should be minimized through stringent execution of field and laboratory procedures (Furlan and Gleeson, 2016b).

The field sampling strategy for species-specific eDNA detection should consider the life history, behavior and environment of the target species. Sharks and their relatives have

Table 14.2 Minimum Recommended Reporting for Environmental DNA Studies

Stage	Information
Design	Inferential goal (presence/absence, quantity)
Water collection	Contamination precautions, including negative controls Collection volume, container material, replicates, depth Site descriptions (flow rate, area, etc.)
Sample preservation	Method, temperature, duration Filter type (if applicable), filtering location (e.g., in field)
Extraction process	Contamination precautions (including dedicated laboratory), negative controls Methods, including kit protocol adjustments
Probe-based qPCR	Design and validation methods Primer/probe sequences, amplicon length Positive and negative controls Inhibition detection and handling Reaction concentrations, thermal profile Technical replicates and their interpretation Standard curve preparation and quality
High-throughput sequencing	Library type (shotgun or amplicon) and any enrichment strategy Library preparation protocol or kit Platform, read length, read pairing, expected fragment size Primers, sequencing adapters, sample index tags, exogenous spike-ins Amplicon locus, target taxa, specificity, and bias Read trimming and filtering of artifacts/chimeras Reference database and/or *de novo* OTU generation Taxonomic assignment method and parameters Statistical analysis and rarefaction Positive and negative controls and their interpretation, if applicable Technical replicates and their interpretation Number of raw reads and final reads

Source: Goldberg, C.S. et al., *Meth Ecol Evol.*, 1299–1307, 2016.

diverse life history traits and occur in a vast array of marine, estuarine, and some freshwater systems. Differences in habitat use will influence eDNA concentration and dispersion and impact the likelihood of recovering target DNA from sample locations. Understanding the fine-scale patterns of occurrence and behavior, such as movements and habitat use driven by ontogeny, predator avoidance, environmental tolerances, seasonal change, or fidelity, may allow enhanced detection ability. However, for many species this depth of information is lacking, and data from similar species may render a useful tool to frame the development of an appropriate field sampling strategy. Moreover, where baseline information on patterns of occurrence and distribution does not exist, eDNA methods may be utilized as an exploratory tool to reveal this information.

When initially assessing field sampling strategy effectiveness or when targeting presence/absence information at one point in time, sampling should occur during times and in locations a species is expected to be present (De Souza et al., 2016). For example, by utilizing existing knowledge on occurrence patterns, oceanic, often solitary, deep-swimming elasmobranchs, such as devil rays, can be positively identified in oceanic basins despite the dynamic and turbulent nature of ocean currents and wave action (Gargan et al., 2017). Studies of resident species should account for variations in activity and behavior in response to seasonally variable factors such as temperature or precipitation, which, in turn, may influence eDNA abundance and persistence and thus the probability of detection. Seasonal variations, for example, may influence the timing of reproduction of certain elasmobranch species, which is likely to increase the detectability of eDNA due to the release of reproductive material such as sperm but also neonates (De Souza et al., 2016; Laramie et al., 2015; Spear et al., 2015). One might reasonably expect that discrete habitats used as pupping or nursery grounds by coastal elasmobranchs would contain higher proportions of eDNA as a result of reproductive behaviors. Likewise, increased activity during tidal- or diurnal-driven movements or feeding behavior may also increase eDNA shedding rate. Species-level differences in habitat use and behavior may dictate spatial and temporal considerations for eDNA detection.

14.3.1.2 Water Collection, Filtration, Preservation, and Extraction

Capturing eDNA from an aquatic environment is the crucial first step in the eDNA workflow. Environmental DNA begins to decay immediately after shedding and continues to do so after sample collection (Barnes et al., 2014; Dejean et al., 2012; Pilliod et al., 2014; Sassoubre et al., 2016; Yamanaka et al., 2016). A recent study on eDNA recovery rates following various combinations of eDNA capture, preservation, and extraction methods has indicated that DNA yield (copy number) from stream water samples, prior to filtration,

significantly decreases when stored at room temperature (20°C), refrigerated (4°C), or frozen (−20°C) from day 1 to day 2, regardless of storage temperature (Hinlo et al., 2017). Moreover, in a recent study using decay modeling of Maugean skate eDNA (*Zearaja maugeana*), Weltz et al. (2017) showed that the eDNA concentration in some water samples had fallen below the detection limit of the assay within 4 hours of sampling. For this reason, samples should be filtered and extracted and the eDNA extracts preserved using prescribed protocols as soon as possible after water collection. Precipitation and filtration are the most commonly used methods to recover eDNA from water samples. Other methods include preservation of small volumes of water followed by concentrating the DNA by centrifugation (Klymus et al., 2015).

Generally, precipitation involves the collection of small volumes of water (e.g., 15 mL) (Eichmiller et al., 2016; Ficetola et al., 2008) that are immediately preserved in the field with the addition of sodium acetate and absolute ethanol (salt and ethanol precipitate nucleic acids from water) (Maniatis et al., 1982), prior to storage at −20°C. The precipitation method requires few collection tools (i.e., precipitation solution and collection vials), thus the relative ease of this method is a major benefit for users. To compensate for the inherently lower eDNA yield due to the relatively small water quantities used and to ensure confidence in detection probability, replication effort, preservation, storage, and extraction methods should be thoroughly considered (for examples, see Deiner et al., 2015; Hinlo et al., 2017; Spens et al., 2017. Where the processing of larger volumes of water is required, it is advisable to increase the number of biological replicates or, alternatively, use the filtration method. Filtration is more advantageous when dealing with larger bodies of water such as rivers, estuaries, or marine environments (Hinlo et al., 2017; Turner et al., 2014b).

Filtration requires the passage of water through a membrane that captures the eDNA and generally allows the processing of larger volumes of water (typically 1 to 10 L). Filtration can be carried out onsite with a portable filtration system (Figure 14.2), or water samples can be stored on ice and transported to a laboratory (or equivalent processing facility) for filtration. When not performed in the field, filtration should be undertaken as soon as possible (i.e., within 24 hours) to ensure optimal eDNA recovery (Hinlo et al., 2017; Weltz et al., 2017).

Following filtration, the filter membrane containing the eDNA must be preserved prior to DNA extraction. Depending on field conditions, cold storage of filters wrapped in aluminum foil or contained in sterile microcentrifuge tubes may not be practical; however, this technique is commonly employed in laboratory-based settings or where field locations are close to the laboratory. When field conditions preclude the use of refrigeration, ethanol is the most commonly used alternative for filter preservation and storage. Other ambient-temperature buffers, such as Longmire's solution and cetyl trimethyl ammonium bromide (CTAB), have been used successfully to preserve eDNA contained on filters (Renshaw et al., 2015; Spens et al., 2017; Williams et al., 2016), but they require preparation using several ingredients, and CTAB is a toxic substance. Longmire's buffer can also be used to preserve small volumes of unfiltered water at ambient temperature for up to 56 days prior to DNA extraction (Williams et al., 2016). An alternative method for eDNA preservation is adding silica beads to the vessel containing the filter; the beads function as a desiccator, drying out the filter and preventing the DNA from degrading (Bakker et al., 2017). Long-term eDNA recovery rates from ethanol and other preservatives are currently unclear and further research is necessary; however, eDNA has successfully been recovered after >1 year of storage from filters desiccated with silica beads and stored at –20°C (J. Bakker, unpublished data).

The type of filter membrane used for the separation of eDNA from the environmental samples also varies. Glass fiber, nylon, cellulose nitrate, polycarbonate, polyethersulfone, and cellulose acetate filters have previously been used (Deiner et al., 2015; Goldberg et al., 2016; Renshaw et al., 2015). The inherent properties of the filter material—particles are retained on the surface of depth filters and within the filter matrix vs. surface filters, where particles are trapped only on the filter's surface (Hinlo et al., 2017)—affect the binding affinity of eDNA and, as such, eDNA recovery rates differ, depending on the type of filter used (Liang and Keeley, 2013).

Filter pore size is another important factor to consider when choosing filters for filtration-based eDNA recovery. Intuitively, larger sample volumes will increase eDNA capture success, but there is a trade-off among sample size, pore size, and eDNA particle retention. A smaller pore size captures more eDNA particles but limits sample volume and speed. Conversely, a larger filter pore size allows for a faster flow rate and larger sample volume. This may in turn reduce associated labor costs but may also reduce the amount of eDNA particles captured on the filter. Hence, two important considerations must be taken into account when choosing the correct filter pore size: size distribution of eDNA particles and water turbidity at the sampling location. Knowledge of the size distribution of various intra- or extracellular eDNA particles will assist in informing the trade-off between filter pore size and sample size/volume. Turner et al. (2014a) observed size fractions of common carp (*Cyprinus carpio*) eDNA and concluded that the largest amount of total eDNA recovered was within the 1- to 10-µm size fraction. Comparable studies for sharks and their relatives do not currently exist, and it is unclear whether the aforementioned findings are representative of general size distributions for all eDNA or are taxa or environment specific. With this in mind, small pore sizes should be used where possible to ensure the highest possible eDNA

capture rate; for example, filter pore sizes ranging from 0.45 to 3µm are most commonly used in studies undertaken in less turbid water (Gargan et al., 2017; O'Donnell et al., 2017; Sigsgaard et al., 2016; Weltz et al., 2017). For more turbid water, however, even 3- to 5-µm filters quickly become clogged with suspended particulate matter, necessitating the use of larger pore sizes of up to 20 µm to minimize clogging and maintain an efficient filtration rate (Robson et al., 2016; Simpfendorfer et al., 2016). If filter clogging is a frequent occurrence, multiple filters may be used and eDNA extracts pooled for sample replicates.

Multiple eDNA extraction methods can be applied to isolate eDNA captured by filtration or precipitation but also to remove compounds that can inhibit downstream enzymatic reactions such as PCR (Eichmiller et al., 2016). Inhibitors may range from cellular components to materials in the water, such as humic substances (Wilson, 1997), that are captured together with the eDNA. Both capture methods may be followed by either phenol:chloroform:isoamyl alcohol (PCI) DNA extraction or extraction using a commercial DNA extraction kit (Deiner et al., 2015). DNA extraction kits, such as the commonly used Qiagen DNeasy® Blood & Tissue Kit and the Mo Bio PowerWater® DNA Isolation Kit, PowerSoil™ DNA Isolation Kit, and PowerMax® Soil DNA Isolation Kit, are convenient and simple to use but are more expensive compared to PCI extraction. PCI extraction in turn requires careful preparation and handling of toxic chemicals. Several studies have found that PCI extraction yields more eDNA compared to commercial DNA extraction kits (Deiner et al., 2015; Renshaw et al., 2015; Turner et al., 2015), but another study observed more PCR inhibition in DNeasy-extracted samples compared to PowerWater-extracted samples (Eichmiller et al., 2016), which is likely a result of different additives to alleviate PCR inhibitors. Hence, high eDNA yield does not necessarily accompany increased species detection but is rather dependent on a multitude of factors. Likewise, Deiner et al. (2015) have demonstrated that different combinations of eDNA capture and extraction protocols result in different detection rates of biodiversity.

Environmental characteristics (e.g., water chemistry and temperature), target species, capture method, filter material and pore size, storage, and DNA extraction method interact to produce final detection rates (Deiner et al., 2015; Eichmiller et al., 2016; Goldberg et al., 2016; Renshaw et al., 2015), and no one extraction method is equally beneficial to all taxa or ecosystems for the maximization of eDNA recovery and target species detection. Thus, it is recommended that different combinations of storage, preservation, filter type, and extraction methods be tested and optimized, depending on the research objectives, preference, ease of use, and availability of resources. Finally, detailed information about the field, laboratory, and bioinformatic procedures used in eDNA studies should be reported to enhance the development of the field by increasing communication about techniques and quality control (see Table 14.2).

14.3.2 eDNA in the Laboratory

14.3.2.1 Selecting Gene Regions for Target Organisms or Groups

Methodologically, eDNA detection requires the development of genetic markers specific to the target taxon or taxa. Targeted eDNA fragments may be detected using different molecular methods including Sanger sequencing, qPCR, and metabarcoding. When aiming to detect a single species, primers should be specific to the target species while incorporating as many differences as possible with other sequences of related organisms (Ficetola et al., 2008). Insufficient primer specificity can lead to over- or underestimation of species presence, and, especially when taxa closely related to the target species are present, cross-amplification or interference of amplification can lead to the generation of false positive and negative errors (Wilcox et al., 2013).

Target loci are typically within the mitochondrial genome because of its greater biological abundance and higher level of coverage in genetic databases; however, selecting the correct gene region for a targeted eDNA barcoding approach will ultimately depend on how much intra- and interspecies variability is found for the species of interest at a particular gene. Environmental DNA barcoding studies to date have designed species-specific assays within a wide range of genes, including cytochrome *b* (Hunter et al., 2015; Spear et al., 2015; Wilcox et al., 2013), COI (Brandl et al., 2015; Gargan et al., 2017; Nathan et al., 2014b; Simpfendorfer et al., 2016), nicotine adenine dinucleotide dehydrogenase subunit 4 (NADH4) (Hunter et al., 2015; Weltz et al., 2017), 16S (Robson et al., 2016), and 12S (Furlan and Gleeson, 2016a; Secondi et al., 2016).

In studies where a large number of species co-occur, some of which may be closely related, finding a suitable gene to design a species-specific or even genus-specific primer assay may be challenging. This may also be true for sharks and their relatives. Sharks, and most likely also rays, appear to have slow mutation rates in mtDNA compared to other vertebrates (i.e., mammals and teleost fish) that lead to lower genetic variation (Martin, 1995, 1999; Martin et al., 1992). For example, mitogenomic sequencing in the critically endangered speartooth shark (*Glyphis glyphis*) has revealed one of the lowest known levels of genetic diversity (Feutry et al., 2014). Increasingly, primer assays are being designed using whole mitogenome sequencing to find suitable gene regions (Hunter et al., 2015), as this increases the chances of finding suitably variable gene regions, potentially in less commonly used alternative regions that exhibit useful polymorphisms.

In contrast, when choosing a suitable genetic marker for eDNA metabarcoding, a genomic region with sufficient sequence variability must be targeted in order to be able to distinguish closely-related species. It must be flanked by conserved regions, which act as primer attachment sites.

Moreover, a region with many copies per cell is preferable, as this natural abundance of DNA sequences will facilitate amplification; hence, organelle genomes, such as mitochondrial or chloroplast DNA, or ribosomal RNA clusters are usually preferred targets (Wangensteen et al., 2017; Wilcox et al., 2013). For eDNA applications, the target fragment length must be relatively short, as eDNA released in the environment rapidly degrades into small fragments; thus, the chances of amplifying the full length of the marker from eDNA is inversely proportional to the length of the chosen marker (Wangensteen et al., 2017). For eDNA metabarcoding, additional considerations apply, as the most popular method for eDNA high-throughput sequencing, the Illumina platform, currently has a maximum effective read length of around 500 bp; however, in order to keep sequencing error rates low, smaller fragments are preferred. The ideal length for an eDNA metabarcoding marker should not exceed 350 bp.

The universality or specificity of the primer set is dependent on the breadth of the taxonomic scale of interest. For example, primer sets for the elucidation of elasmobranch (Bakker et al., 2017), teleost (Miya et al., 2015), or arthropod (Zeale et al., 2011) diversity can be used. Conversely, targeting whole eukaryotic community diversity will require a primer set that is as universal as possible in order to be able to attach to the marker flanking sequences in most taxonomic groups, so that all of these groups will be adequately amplified by PCR. As of yet, there is no ideal universal metabarcode that is able to amplify the full taxonomic range of a community for highly variable markers such as COI (Coissac et al., 2012; Deagle et al., 2014; Riaz et al., 2011). Thus, truly universal primers have usually been restricted to markers with more conserved regions such as 18S (Guardiola et al., 2015). However, the development of primers including deoxyinosine (a nucleotide that complements any of the four natural bases) in the fully degenerated sites of the sequence may improve the universality of COI primer sets (Wangensteen et al., 2017).

The use of COI as a metabarcoding marker has previously been criticized, as it has been observed that high rates of sequence variability impair the design of truly universal primers and hamper bioinformatics analysis. Instead, mitochondrial rRNA genes have been recommended for animal identification because they have a similar taxonomic resolution as the COI marker and they present conserved regions that flank variable regions, which allows the design of primers with high resolution power for the target taxonomic group (Deagle et al., 2014). However, it may be argued that COI presents two major advantages over other potential markers. First, the steadily growing international effort headed by the Consortium for the Barcode of Life (CBOL) to develop a public DNA barcoding database with curated taxonomy greatly facilitates taxonomic assignment. Also, the BOLD database (http://www.boldsystems.org/) (Hebert et al., 2003a; Ratnasingham and Hebert, 2007) currently includes >4 million sequences belonging to over 500,000 species, curated and identified by expert taxonomists. Second, the high mutation rate of COI ensures unequivocal identification at the species level, which is crucial for studies aimed at detecting rare or invasive species, such as may be the case for sharks, whereas the highly conserved sequences of other markers, such as 18S, often make it impossible to distinguish at the species or genus levels.

14.3.2.2 Design and Validation of qPCR Assays

The two common types of qPCR assays are based on intercalating dyes (e.g., SYBR green dye) and fluorescent probes (e.g., TaqMan™). The intercalating dye method makes use of two primers that amplify the target region, which is quantified by a fluorescent dye that binds to the double-stranded DNA (dsDNA) of the PCR amplicon. This method is very cost effective and relatively easy to design; however, intercalating dye-based methods tend to be less specific, as the dye can bind nonspecifically to any double-stranded DNA, such as primer dimer and nonspecific products. To minimize this problem, it is advisable to carry out a melting curve analysis. Furthermore, this method is prone to non-specific fluorescence in low-concentration targets. On the other hand, a fluorescent probe assay consists of two parts. The first component is composed of two flanking primers, similar to the primers used in the dye-based method. The second component constitutes of a hydrolysis probe, located close to the 3′ end of either the forward or reverse primer, which consists of a species-specific DNA sequence, a fluorescent dye, a quencher and, most commonly, a minor groove binding (MGB) modification that improves specificity and reduces the required length of the probe. The combination of species-specific primers and a probe increases the specificity of the assay and decreases the amplification of nonspecific products such as primer dimer. This removes the need for a melting curve analysis. Furthermore, the sequence-specific nature of the probe allows for multiplexed assays through the use of multiple probes with differently colored dyes. Disadvantages of fluorescent probe-based assays include increased difficulty in design and higher costs.

Irrespective of the assay type, it is recommended to follow the manufacturer's instructions on primer and probe design (e.g., Taylor et al., 2010). Programs such as Geneious (http://www.geneious.com) (Meintjes et al., 2012) and AlleleID® (Premier BioSoft; Palo Alto, CA) can be used to assist in primer or probe design. The specificity of the primer and probe should be assessed visually for the number of mismatches to exclusion species. Mismatches toward the 3′ end of primers and the middle of a hydrolysis probe are preferred. Furthermore, the online tool Primer-BLAST (www.ncbi.nlm.nih.gov/tools/primer-blast/) can be used to further assess the specificity of the primers against a wider database.

When designing a species-specific qPCR assay it is important to assess the occurrence of false-positive and false-negative results. False positives occur when another, usually closely related, species (i.e., exclusion species) is detected instead of the target species. The likelihood of false-positive detections is determined by the assay specificity to the target species and can be difficult to achieve with closely related species. In contrast, false negatives occur when DNA of the target species is present but not detected. One cause of false negatives is assay sensitivity, which is synonymous with limit of detection and describes the smallest detectable amount of target eDNA. The second possible cause of a false-negative result can occur if the probe or primer sequences do not match the sequence of the local population being tested due to genetic variability between populations and therefore fail to amplify the target. To avoid false-positive and false-negative results, sequence information of the target and co-occurring exclusion species should be collected and assessed *in silico* to design a species-specific assay. When the assay has been designed and *in silico* tested, it is imperative to also carry out laboratory testing. The efficiency of the assay should be tested using either genomic DNA (gDNA) or an artificial oligo of the target sequence. When the optimal qPCR parameters have been determined, the assay should be tested against gDNA of all potential exclusion species, as well as gDNA of the target species collected from across the range if possible. Finally, the designed probes should be validated using tank and field-collected positive and negative eDNA samples.

14.4 CHALLENGES OF eDNA STUDIES

14.4.1 Contamination

One of the main challenges associated with the use of eDNA is dealing with false-positive and false-negative detections (Darling and Mahon, 2011). Due to the high sensitivity of eDNA methods, the most serious stumbling block is the risk of contamination (Goldberg et al., 2016; Thomsen and Willerslev, 2015) and hence the possibility of introducing false-positive results. Contamination of samples may occur anywhere from preparing sampling equipment and collecting the samples in the field (target DNA being carried unintentionally from one locality to another) to every subsequent step of sample preparation, DNA extraction, and analysis in the laboratory. Due to the frequent use of PCR, generating billions of DNA copies, contamination occurring in the laboratory can potentially have serious implications for the resulting dataset, with important downstream repercussions on conservation and management decisions based on these results. Thus, precautions must be taken at all stages by putting strict procedures in place both in the field (establishing clean and consistent field collection protocols) and in the laboratory (implementing strict, clean lab protocols) in order to prevent the occurrence of contamination. This includes the use of disposable gloves and the disinfection/bleaching of sampling devices and all laboratory equipment. Additionally, filtration, DNA extraction, and PCR procedures, as well as pre- and post-PCR procedures, must be separated physically to limit the risk of contamination (Goldberg et al., 2016; Wilson et al., 2015). Moreover, to monitor potential contamination (i.e., to identify the source of contamination when it occurs), the inclusion of field blanks (clean water sampled using the same protocol and equipment and preserved and processed in exactly the same way as the actual field samples), DNA extraction blanks, and PCR blanks is essential (De Barba et al., 2014).

14.4.2 eDNA Detectability

14.4.2.1 Abiotic and Biotic Factors Influencing eDNA Detectability

14.4.2.1.1 eDNA Shedding Rates

The availability of detectable eDNA in environmental samples is reliant on the underlying premise that all organisms shed genetic material. Earlier studies on terrestrial vertebrate eDNA detection in aquatic environments imply that the most probable origin of eDNA is fecal material (Martellini et al., 2005). Although this may be true for a wide range of taxa, the origin of eDNA from aquatic organisms is also linked to species-specific physiological characteristics including skin properties, such as slimy coatings on amphibians (Ficetola et al., 2008) and fish (Jerde et al., 2011); metabolic rates (Klymus et al., 2017); reproductive mode and timing (Bylemans et al., 2017; Spear et al., 2015); feeding rates (Sassoubre et al., 2016); and environmental tolerance (Lacoursiere-Roussel et al., 2016; Robson et al., 2016). The composition of eDNA containing genetic material from these origins remains relatively unclear and particularly difficult to study; however, many complex factors influence eDNA shedding rate, and, as such, interpretation of eDNA detection results benefits from a complete understanding of the ecology of eDNA.

Overall, in marine and freshwater organisms, it is largely understood that eDNA shedding rates are foremost positively related to individual or population biomass (Pilliod et al., 2014; Stoeckle et al., 2017; Thomsen et al., 2016; Weltz et al., 2017). It is this correlation that underpins the use of eDNA concentration in water as a proxy measure of biomass of the focal organisms, which has been applied to both species-specific (Doyle et al., 2017; Sigsgaard et al., 2016) and population-level questions (Kelly, 2016; Leray and Knowlton, 2015; Miya et al., 2015; Yamamoto et al., 2017). In tank-based experiments, eDNA release rates demonstrated positive linear relationships with biomass (Klymus et al., 2015; Sassoubre et al., 2016). However, it is likely that this relationship is more complex, as authors have observed

highly variable eDNA production rates among individuals unrelated to biomass and suggest that this variation may be attributable to animal physiology (Buxton et al., 2017; Maruyama et al., 2014; Pilliod et al., 2014; Wilcox et al., 2016). In juvenile fishes, it was concluded that ontogenetic factors, such as differences in behavior and metabolism, increased eDNA shedding rates per body weight compared to adult conspecifics (Klymus et al., 2015; Maruyama et al., 2014). Moreover, recent work has suggested that stress and feeding behavior can influence eDNA shedding rates (Sassoubre et al., 2016), but these behaviors are intertwined with the physiological tolerances of aquatic organisms (Lacoursiere-Roussel et al., 2016). Additionally, the eDNA contribution from different life stages may vary seasonally. For example, strong temporal increases in eDNA concentration have been observed during months associated with seasonal migration and breeding (Buxton et al., 2017; Doi et al., 2015b; Fukumoto et al., 2015; Spear et al., 2015).

Seasonal migrations and patterns of occurrence associated with specific behaviors are dictated by metabolic function, which, in most sharks, is determined by water temperature. As water temperature increases, mobility and metabolic rate increase until the upper limit of physiological tolerance is reached. Some sharks also perform diel vertical migrations to conserve energy in deeper, cooler waters and search for prey when near the surface in warmer water (Sims et al., 2006). Increased water temperature and digestive function, coupled with movements associated with prey-seeking and feeding behavior, lead to the increased excretion of metabolic waste and the release of epidermal cells containing genetic material. Although some studies have shown no effect of temperature on the accumulation or shedding of eDNA in fishes (Klymus et al., 2015; Takahara et al., 2012), a recent study on the tropical invasive fish species tilapia (*Oreochromis mossambicus*) showed increased eDNA shedding rates at 35°C, a temperature well within their known thermal tolerance (Robson et al., 2016). Moreover, estimates of fish biomass in aquaria samples were better reflected in warmer water as supported by higher eDNA concentration and shedding rates from fish in warm water compared to colder water (Lacoursiere-Roussel et al., 2016).

Other physiological attributes such as skin properties provide clues to possible origins of eDNA-bearing particles. High eDNA detection success rates have been observed for fish that produce slimy coatings (Jerde et al., 2011). Comparably, sharks and their relatives also produce epidermal mucus (Meyer and Seegers, 2012; Tsutsui et al., 2009). The mucus layer on the skin surface of demersal sharks and rays (e.g., angel sharks, *Squatina* spp.) is comparatively thicker than that of pelagic, fast-moving, predatory sharks, but, independent of the variation in mucus production among species, all elasmobranchs have a functional mucus layer covering the skin surface (Meyer and Seegers, 2012). This suggests that a mucus-derived genetic material may contribute to a portion of all elasmobranch eDNA.

14.4.2.1.2   Degradation of eDNA in the Environment

As eDNA possesses limited chemical stability (Lindahl, 1993), once shed from an organism it begins to degrade into small fragments and becomes undetectable within hours to weeks (Dejean et al., 2011; Piaggio et al., 2014; Pilliod et al., 2014). Degradation is the primary mechanism limiting detection of species through eDNA. However, due to the short lifespan of eDNA it is thought to provide approximate real-time data on species' presence in the environment. The persistence of eDNA for aquatic taxa has been estimated at 15 to 30 days for freshwater fishes (Dejean et al., 2011; Takahara et al., 2012) and hours to 7 days for marine fishes (Thomsen et al., 2012a), after which time eDNA concentrations drop below the detection limit. In specific reference to sharks and their relatives, it is accepted that eDNA exponentially decays in aquatic environments and becomes undetectable within hours and up to 5 days (Sigsgaard et al., 2016; Weltz et al., 2017). For example, in controlled degradation experiments, the concentration of whale shark (*Rhincodon typus*) eDNA dropped an order of magnitude within the first 48 hours and was no longer detectable 8 days after sampling (Sigsgaard et al., 2016), and Maugean skate (*Zearaja maugeana*) eDNA remained detectable for up to 5 days (Weltz et al., 2017).

Environmental conditions play an integral role in eDNA persistence and degradation (Barnes et al., 2014). Understanding the interactions of environmental factors controlling degradation is essential to inferring the limits of temporal and spatial inference of eDNA detection results. Drivers of eDNA degradation are classified into three categories: (1) DNA characteristics, including length, conformation, and association with membranous material (Taberlet et al., 2012a; Willerslev et al., 2007); (2) abiotic environment, including temperature, pH, UV radiation, oxygen, and salinity (Barnes et al., 2014; Pilliod et al., 2014; Strickler et al., 2015; Weltz et al., 2017); and (3) biotic environment, including exogenous enzymes and microbial activity (Dejean et al., 2011).

Fragments of DNA in the environment occur in different lengths, sequences, and conformations, thus influencing how eDNA binds to other particles and interacts with microbes in the environment (Lennon, 2007; Ogram et al., 1988) and altering the rate of degradation. Binding to sediment particles can play a role in eDNA preservation (Turner et al., 2015), as does containment within cellular or organelle membranes, by providing protection to external degradative forces. Marine sediment eDNA concentrations have been shown to be three orders of magnitude higher than those of eDNA in seawater (Torti et al., 2015). Moreover, DNA has a stronger affinity for clay particles compared to sand or silt (Romanowski et al., 1992), and, although sediments are not typically sampled in a presence/absence or contemporary occurrence survey because of the longevity of eDNA, consideration should be given to the potential resuspension of sediments in the water column and the subsequent increased probability of detecting this preserved eDNA.

Marine and freshwater tropical environments have high surface temperatures (sometimes above 30°C) and elevated UV radiation at sea level that may increase eDNA degradation rates and reduce its persistence in the water, decreasing the detection probability (Barnes et al., 2014). Higher temperatures can denature DNA molecules, albeit when temperatures are >50°C, and indirectly increase microbial metabolism and exogenous nuclease activity (Fu, 2012; Kreader, 1998). For example, Robson et al. (2016) showed that high water temperatures of up to 35°C did not affect eDNA degradation rates of the invasive Mozambique tilapia (*Oreochromis mossambicus*). Ultraviolet radiation, particularly UVB, can directly damage DNA (but see Andruszkiewicz et al., 2017) and has variable effects on exogenous nuclease production, indirectly inhibiting eDNA persistence.

Acidic, hypersaline, or anoxic environments can influence eDNA stability and increase degradation rates. Deviation from neutral pH may reduce degradation rates, especially considering the pH requirements of extracellular microbial enzymes that are considered to have a large impact on eDNA degradation (Sigsgaard et al., 2015). Highly saline samples may have negative downstream effects, such as inhibition of PCR, but this can be mediated by adding an ethanol wash step to the DNA extraction process in order to remove monovalent Na^+ ions (Foote et al., 2012). The interaction of all biotic and abiotic factors combined is likely to have variable and synergistic effects on the mechanism of eDNA degradation in aquatic systems.

14.4.2.1.3   Variability in eDNA Capture Rate

Understanding the physical movement of eDNA in the environment is essential for correctly inferring the presence of organisms in space and time and, hence, drawing robust conclusions within spatial and temporal boundaries (Barnes and Turner, 2016). Environmental DNA represents a complex mixture of particles ranging in size and composition which behave independently and move freely in aquatic environments. These particles are randomly and heterogeneously distributed in the water column as a result of spatial clumping (Furlan et al., 2016b). The greater the degree of clumping and uneven dispersal of target DNA, the greater the likeliness of false negatives. Consequently, detection sensitivity for a given sampling protocol will vary temporally and spatially, between samples and from site to site, depending on the concentration and dispersion of target eDNA (Furlan et al., 2016b; Weltz et al., 2017). Differences in eDNA detection sensitivity across space and time may also be the result of differences in activity levels or other site-level factors that influence eDNA concentration, including the density or biomass of the target species (see Sections 14.4.2.1.1 and 14.4.2.1.2). The use of hierarchical occupancy and detection sensitivity models that account for the specific survey methods used, both in the field and laboratory, can be applied to optimize capture and detection probabilities

(Furlan et al., 2016; Schmidt et al., 2013). Furthermore, it is recommended that sample replication be increased at individual sites as well as between study areas.

14.4.2.2 Habitat and Ecosystem Effects

14.4.2.2.1 eDNA Transport: Lentic vs. Lotic Systems

Although the high sensitivity of eDNA assays in mesocosms and lentic systems (still waters) is well established (Thomsen et al., 2012b), studies in lotic systems (flowing waters) have more varied results, with potentially important management and conservation implications. This is typically illustrated with eDNA assays that show high detection rates (100%) when tested in ponds but have much lower detection rates when used in the target species' natural lotic environment (54%) (Thomsen et al., 2012b). Environmental DNA occurs at very low concentrations in the aquatic environment and can be heterogeneously distributed. Thus, knowledge of how eDNA distribution is affected by water movement (e.g., currents, eddies, waves) and what additional interacting external drivers may affect its detectability, such as abiotic and biotic factors involved in eDNA persistence in the environment (see Section 14.4.2.1.2) (Barnes et al., 2014; Jane et al., 2015; Strickler et al., 2015), is crucial for the successful detection of species in their environment. This is particularly the case for the detection of rare species, for which eDNA concentrations are likely to be at their lowest (Takahara et al., 2012) and the risk for false-negative errors high.

Long-distance transport of eDNA from hundreds of meters to several kilometers has been reported in river systems and should always be taken into account in eDNA studies in lotic systems (Deiner and Altermatt, 2014; Jane et al., 2015). Although it could easily be expected that eDNA would travel much larger distances in highly dynamic systems such as open oceans or flowing rivers compared to more stagnant systems such as ponds and lakes (Deiner and Altermatt, 2014; Shogren et al., 2016), recent work on a dynamic marine coastline found evidence that eDNA transport was limited to the extent that eDNA metabarcoding methods were able to detect differences among vertebrate communities separated by less than 100 m (Port et al., 2016). Also, Gargan et al. (2017) were able to detect the Chilean devil ray (*Mobula tarapacana*) using a targeted eDNA approach at four out of five remote seamounts that were sampled around the Azores, consistent with visual observation data. However, failure to detect target eDNA at a location where the species had been observed highlights the influence of detection stochasticity and the need for further investigations into how eDNA transport and degradation affect species detection in open ocean environments.

Ultimately, for eDNA to be a useful tool for monitoring species, including low-abundance species (imperiled or invasive), the functionality of an eDNA assay will depend not only on its ability to detect the species in its environment but also on its ability to take into account external factors in eDNA study designs. These include factors such as the physiology and space use of organisms (Goldberg et al., 2016), as well as the environmental conditions that reduce eDNA persistence.

14.4.2.2.2 Freshwater vs. Seawater: How Are Marine Systems Likely to Differ from Freshwater Ones?

For both freshwater and marine ecosystems, eDNA detection is correlated with abundance of the target species and the rate by which DNA is released and degraded by biotic and abiotic factors (Thomsen et al., 2012a,b). A considerable amount of aquatic eDNA research has been focused on freshwater systems (e.g., Dejean et al., 2011; Ficetola et al., 2008; Gustavson et al., 2015; Jerde et al., 2011; Laramie et al., 2015; Takahara et al., 2012). Only more recently have eDNA studies focused on species detection in seawater samples (e.g., Foote et al., 2012; Gargan et al., 2017; Sigsgaard et al., 2016; Thomsen et al., 2012a, 2016; Weltz et al., 2017). Although many of these have been carried out in controlled environments, such as aquarium tanks (Foote et al., 2012; Kelly et al., 2014; Miya et al., 2015), more recently successful eDNA studies have been reported from natural marine environments as varied as coastal waters (Weltz et al., 2017; Yamamoto et al., 2017), open-ocean seamounts (Gargan et al., 2017), offshore oil fields (Saghaï et al., 2015; Sigsgaard et al., 2016), and continental slope depths (Thomsen et al., 2016).

The current lag in eDNA studies and research and development in marine ecosystems may stem from the perception that species detection from seawater samples may be more challenging compared to freshwater due to the larger body of source water and strong tidal and current action potentially diluting and dispersing the eDNA up to 100s of kilometers away (Thomsen et al., 2012a; but see O'Donnell et al., 2017) and perhaps also due to the high salinity of the samples decreasing sample preservation and increasing risks of PCR inhibition (Wilson, 1997). Environmental DNA is subject not only to transport but also to degradation from exposure to various biotic and abiotic stressors (i.e., temperature, salinity, pH, UVB, enzymes; see Sections 14.4.2.1.2, and 14.4.2.4.1) (Goldberg et al., 2016). However, how these combined factors affect the potential of eDNA techniques to detect marine organisms in coastal or open ocean environments has seldom been investigated, and most marine studies in this field have only focused on determining eDNA shedding and decay rates in a handful of species (Andruszkiewicz et al., 2017; Sassoubre et al., 2016; Sigsgaard et al., 2016; Thomsen et al., 2012b). DNA degradation in seawater has previously been suggested to be substantially faster than in freshwater, with an empirical turnover rate as low as 10 hours (Dell'Anno and Corinaldesi, 2004). Although abiotic and biotic stressors in the marine environment are likely to differ from those impacting freshwater systems (Thomsen and Willerslev, 2015), it is not clear which of these are mostly responsible for the increased rate of eDNA degradation. A recent study investigating the impact of sunlight (UVB and UVA+UVB radiation) on the decay of Pacific chub mackerel (*Scomber japonicus*) eDNA in a marine water mesocosm concluded that sunlight was not an important factor in controlling eDNA degradation and suggested that factors other than sunlight, such as bacteria, grazers, and enzymes, are likely to have a more substantial impact (Andruszkiewicz et al., 2017). Moreover, experiments by Weltz et al. (2017) suggest that the time it takes for Maugean skate (*Zearaja maugeana*) eDNA to degrade beyond its detection limit was influenced by the dissolved oxygen (DO) concentration in the eDNA sample.

Environmental DNA in freshwater systems has been shown to degrade beyond the threshold of detectability within a short time frame (days to weeks) (Dejean et al., 2011; Piaggio et al., 2014; Pilliod et al., 2014; Thomsen et al., 2012b), providing a real-time measure of species presence. Conversely, in the marine environment, eDNA may decay below the detection threshold in as little as 4 hours after sampling (Weltz et al., 2017). Other studies have also indicated slightly slower rates of degradation, on a scale of days (Andruszkiewicz et al., 2017; Sigsgaard et al., 2016). As the rate of degradation of eDNA is inherently linked to both the starting concentration and the abiotic factors promoting degradation of the sample (e.g., UV, pH), it is possible that one of the drawbacks of eDNA studies in seawater is the inherently greater dilution of the eDNA signal. This makes the detection of sharks and rays in seawater all the more challenging, although not impossible, as a growing number of studies have shown (Bakker et al., 2017; Gargan et al., 2017; Sigsgaard et al., 2016; Weltz et al., 2017).

Field and laboratory practices for the application of eDNA analysis to seawater may be modified to counter some of these constraints. Larger volumes of water (e.g., 10 L vs. 2 L) and a larger number of field replicates within a study area may be collected to counter for the greater water volume-to-biomass ratio of marine systems. Because eDNA concentrations are expected to be lower in the open ocean than in river systems, this is particularly pertinent when dealing with species that are likely to occur in low numbers or that are sparsely distributed.

Due to the aforementioned challenges and the need for further research, the detection of marine species using environmental DNA should for now be considered only at a local scale. Moreover, caution should be exercised when using eDNA concentrations or metabarcoding reads for the estimation of abundance of fish species in marine systems until further work is carried out to elucidate the persistence of eDNA under the influence of biological, environmental and physical processes and how processes such as eDNA shedding (source), decay (sink), and transport can be integrated into reliable estimates of abundance.

14.4.2.2.3 Challenges Specific to Tropical Ecosystems

Since the introduction of eDNA into mainstream environmental research, the majority of eDNA studies and research and development investigations have been applied to temperate systems (see summary in Table 1 in Thomsen and Willerslev, 2015). Comparatively very little research has been applied to tropical aquatic systems and the applicability and reliability of current eDNA methods for effective species and community detection, and conservation management in these environments is less clear. The tropics present their own set of challenges, with eDNA in marine and freshwater tropical environments exposed to more extreme conditions for longer and more frequent periods of time. Tropical aquatic systems have high surface water temperatures (sometimes >40°C), elevated UV radiation at sea level, and higher levels of microbial activity that may increase eDNA degradation rates and reduce their persistence in the water (Barnes et al., 2014). Furthermore, seasonal precipitations (wet season) lead not only to increased turbidity due to high sedimentation and algal loads but also to increased dilution effects due to high water flow rates (Figure 14.3). The interaction of these factors, specific to tropical systems, is likely to significantly influence the detection of eDNA.

It has been suggested that elevated temperatures may accelerate the rate of eDNA degradation (Strickler et al., 2015). Indeed, Moyer et al. (2014) showed in their trials that for every 1.02°C increase in temperature, the per-liter sample probability of eDNA detection decreased by 1.67 times. However, a similar trend was not reported by Robson et al. (2016), who found that water temperatures of up to 35°C had no detectable effect on invasive Mozambique tilapia (*Oreochromis mossambicus*) eDNA degradation rates. On the other hand, relatively high temperatures (35°C), but well within the range of tropical river systems, have been found to significantly increase fish eDNA shedding rates (Robson et al., 2016), which is likely due to increased metabolism or thermal stress, thus potentially positively affecting the detection probability of eDNA. Similarly, another study has indicated that fish release more eDNA in warm water than in cold water and that eDNA concentration better reflects fish abundance/biomass at high temperatures (Lacoursiere-Roussel et al., 2016).

Exposure to high levels of ultraviolet radiation, particularly ultraviolet B (UVB) light, can photochemically damage DNA, and aquatic environments at higher elevations or closer to the equator are more likely to experience increased effects of UVB radiation on eDNA degradation rate (Strickler et al., 2015). It is likely, however, that it is the interaction of multiple factors (pH, solar radiation, and temperature) either directly or mediated through the biological community that influences the process of eDNA degradation in aquatic systems (Strickler et al., 2015).

Figure 14.3 Tropical aquatic systems present a number of added challenges to eDNA studies, including (A) clogged filters due to (B) high sediment and algal loads; (C) increased water flow and eDNA dilution effects during wet season precipitation events; and (D) dangerous wildlife. (Photograph credits, clockwise: Agnès Le Port, Zoe Bainbridge, Ian McLeod.)

High turbidity levels resulting from increased sedimentation and algal growth occur seasonally in tropical aquatic systems. These conditions present several challenges when collecting eDNA samples. Rapid clogging of filtering apparatus, including filters (Figure 14.3), requires the use of multiple filters per sample or filters with a larger pore size, resulting in either increased filtration times (hours per sample) or decreased capture rates of eDNA molecules. The presence of resuspended sediment in water samples may also affect the temporal scale of the data and lead to downstream PCR inhibition of samples. Although eDNA in water reflects the current state of an ecosystem, eDNA can persist on the order of years in soils and sediments (Pedersen et al., 2015); thus, mixing of contemporary and historic eDNA deposits could lead to misinterpretations as to the actual presence of a rare or invasive species in habitats of pressing environmental concern. Moreover, higher flow rates following wet-season precipitation events may lead to false negatives due to longer than usual downstream transport distances of eDNA.

Several strategies may be adapted to tackle some of the challenges associated with eDNA studies in the tropics. Experimental designs and sampling strategies may be adapted to the conditions encountered. Suggested strategies include avoiding sampling during the wet season or right after heavy precipitation events, avoiding sampling during the hottest part of the day or during summer temperature highs, decreasing the risk of eDNA degradation by targeting areas/habitats that are less exposed (e.g., shaded, still), and filtering and preserving samples as soon as is practically possible. When sampling during heavy precipitation events cannot be avoided, increasing sample replication and sample volumes collected, as well as targeting samples from slow flowing or stagnant areas, is recommended. To avoid rapid clogging of the filtering apparatus and filters, filters with a larger pore size may have to be used. Filtering trials should be performed in order to identify the most optimal compromise between filtering constraints (e.g., filtering time, number of filters per sample) and eDNA capture probability for the species and environment of interest. For example, Robson et al. (2016) found that Mozambique tilapia (*Oreochromis mossambicus*) eDNA detectability decreased from 100% to 57% when using 3-μm and 20-μm filters, respectively. However, the significantly larger pore size decreased the filtering time from 44 minutes to 1.5 minutes per sample. In areas where dangerous wildlife co-occurs (e.g., crocodilians; see Figure 14.3), putting the sampler at risk, an extendable pole (Figure

14.2) or a remotely operated sampling device (e.g., drone) may be used to collect samples safely. Finally, PCR inhibition due to high levels of humic substances (e.g., humic acid, fulvic acid) in water samples is problematic. Strategies such as the use of PCR inhibition removal columns or dilution of samples should be applied in tropical aquatic systems.

14.4.3 Reference Databases

Regardless of whether a species-specific barcoding or a community-based metabarcoding approach is chosen, the reference database is a crucial starting (and end) point. When designing a species-specific qPCR assay, the reference database provides sequence information for target and exclusion species to ensure the specificity of the assay to the target species. In contrast, in eDNA metabarcoding approaches, the reference database is also used for taxonomic identification. If a specific species is present in a sample, but its barcode sequence is missing from the database, it will not be possible to identify it down to species level (but rather to genus or family) in an eDNA metabarcoding analysis. It is essential to have a broad, accurate, and curated sequence database. GenBank® can be used as a starting point, but it may be necessary to collect samples from populations in the target region to build a bespoke reference database appropriate for the sampling region. The Barcode of Life database may be of limited use, as it contains only COI sequences, which may not always be suitable for assay development. Certain gene regions may be too variable, making it very difficult to design universal primers, or, conversely, they may not be variable enough, making it difficult to create species-specific qPCR assays or to distinguish among different taxa in eDNA metabarcoding. See Section 14.3.2.1 for more detail on selecting metabarcoding primers.

14.5 FUTURE ADVANCES IN eDNA METABARCODING

14.5.1 Quantitative Estimates Using eDNA Metabarcoding and Applicability to the Study of Sharks and Their Relatives

14.5.1.1 Quantitative Estimates Using PCR and qPCR

Both field and mesocosm/tank studies have shown that an increase in abundance or density of target species can lead to an increase in either eDNA concentration (Buxton et al., 2017; Klymus et al., 2015; Pilliod et al., 2013; Takahara et al., 2012; Thomsen et al., 2012b) or detectability (Eichmiller et al., 2014; Lacoursiere-Roussel et al., 2016; Mahon et al., 2013). In freshwater systems, the rate of eDNA production has been positively correlated with biomass for several species through the use of PCR and qPCR platforms, including

common carp (*Cyprinus carpio*) in artificial ponds (Takahara et al., 2012), common spadefoot toads (*Pelobatus fuscus*) and great crested newts (*Triturus cristatus*) in natural ponds (Thomsen et al., 2012b), and tailed frog (*Ascaphus montanus*) tadpoles and both giant salamander (*Dicamptodon aterrimus*) larvae and pedomorphic adults (Goldberg et al., 2011; Pilliod et al., 2013). Because this relationship is not a clear-cut one, these studies suggest that eDNA can be used for relative rather than absolute quantification.

The amount of eDNA in the environment depends on both DNA release and degradation rates (which are dependent on a range of biotic and abiotic factors) (Dejean et al., 2011; Strickler et al., 2015), as well as heterogeneous dispersal of eDNA molecules via ecological processes (habitat specificity of target organisms) or the presence of currents (Deiner and Altermatt, 2014) or eddies (particularly in the marine environment). These factors are likely to vary seasonally in response to environmental changes and the life-history stage of the species in question (Barnes et al., 2014; Buxton et al., 2017). For example, great crested newt (*Triturus cristatus*) eDNA concentrations have been shown to increase within the breeding season due to reproductive behavior and egg deposition and, subsequently, with an increase of larval abundance (Buxton et al., 2017; Lacoursiere-Roussel et al., 2016). It may also be associated with an increased eDNA production per biomass in juveniles compared to adults, resulting from increased metabolism during growth (Klymus et al., 2015). Likewise, it has been shown that seasonal variations in stream-dwelling fish eDNA concentration were related to total biomass (associated with breeding season and larval density), rather than abundance or behavior (Doi et al., 2017a). These factors may lead to either over- or underestimation of organism density, and, as such, seasonal changes in eDNA concentrations may have implications for survey strategies, taking into account temporal and spatial patterns to target specific sampling windows, depending on the aim and species of the survey in question. As it relates specifically to sharks, over the course of a year there may, for example, be fluctuations in species densities due to seasonal migrations related to water temperature (Guttridge et al., 2017; Kajiura and Tellman, 2016; Kessel et al., 2016).

14.5.1.2 Quantitative Estimates Using Digital Droplet PCR

Digital droplet PCR (ddPCR), also known as a third-generation PCR, provides a new method of sample analysis allowing for an accurate estimation of low concentrations of DNA. It has been suggested that ddPCR may be better suited for the detection of rare molecules in environmental samples compared to qPCR, as it provides more accurate estimates of the abundance or biomass of a target species (Doi et al., 2015b). Like standard PCR, ddPCR is a direct method that does not use calibration curves (derived from target DNA standards) to estimate target DNA concentration (Vogelstein and Kinzler, 1999), thus decreasing the potential for user error (e.g., pipetting error when preparing standards, the introduction of contamination). Instead of a single measurement, the target DNA in ddPCR is randomly allocated into approximately 20,000 discrete droplets via microfluidics, some of which ideally contain only one or a few copies of the target DNA. The PCR occurs within each droplet, which is subsequently individually screened via fluorescence measurement for the presence of target DNA (Hindson et al., 2011; Pinheiro et al., 2012). Increasing the number of partitions (i.e., droplets) improves precision and therefore enables resolution of small concentration differences between nucleic acid sequences in a sample. With ddPCR it is possible to detect concentration differences between samples as low as 1.25-fold, which is more accurate than qPCR, which only allows for a 2-fold detection difference (Doi et al., 2015b; Hindson et al., 2011). Homogeneous assay chemistries and workflows, similar to those widely used for real-time PCR applications (e.g., TaqMan™) are also used for ddPCR (Hindson et al., 2011).

The ddPCR technique has been used to obtain absolute quantifications from a range of targets including a virus (Hindson et al., 2013), bacteria (Cavé et al., 2016; Kim et al., 2014), fungi (Palumbo et al., 2016), and animal cells (Miotke et al., 2014), and more recently a handful of fish species (Doi et al., 2015b; Jerde et al., 2016; Nathan et al., 2014a). Nathan et al. (2014a) were the first to use ddPCR to quantify the eDNA concentration of a fish species, the invasive round goby (*Neogobius melanostomus*), in mesocosm experiments. Comparing PCR, qPCR, and ddPCR platforms, the authors found that, although both qPCR and ddPCR gave consistent estimates of DNA concentration, smaller variations in estimates were reported for ddPCR (Nathan et al., 2014a). Similarly, ddPCR proved to be more accurate in quantifying common carp (*Cyprinus carpio* L.) eDNA at low concentrations than qPCR, suggesting that this platform is very promising for use in estimating species biomass and/or abundance related to their eDNA concentration in the aquatic environment (Doi et al., 2015b) when the species-specific relationship between biomass and eDNA concentration in the environment has been determined. In lotic systems, detection of eDNA is complicated by continuous dilution of the eDNA signal with simultaneous displacement downstream and/or mixing through physical processes (waves, currents). In a series of experiments in a semi-natural stream setting, Jerde et al. (2016) concluded that at very low eDNA concentrations, there is an advantage to using ddPCR, as was demonstrated by the detection of target eDNA by ddPCR where qPCR failed.

Digital droplet PCR is a promising new eDNA platform for the estimation of fish abundance and biomass in aquatic systems (but for sources of bias, see Hunter et al., 2017). Due to its more accurate quantification of eDNA concentration and better performance in the presence of PCR inhibitors in field samples (Doi et al., 2015a) compared to other platforms (e.g., qPCR), ddPCR may become the next standard method for reliably assessing population abundance in barcoding studies.

14.5.1.3 *Quantitative Estimates Using Metabarcoding*

A remaining controversial issue associated with eDNA metabarcoding is whether it can provide quantitative estimates; that is, are the numbers of reads obtained for each species proportional to the abundance or the biomass of the species present in the original sample? Quantification of eDNA relating to species abundance could provide clues to habitat use, thus identifying spatial conservation priorities such as home ranges and dispersal and migration corridors (Barnes and Turner, 2016). Although amplicon sequencing produces read counts that may contain valuable information about target species abundances (Evans et al., 2016; Port et al., 2016), the interpretation of the results of amplicon studies, in the context of quantitative ecology, is not straightforward and remains difficult (Kelly et al., 2016). This is in part because the precise relationship between amplicon abundance and taxon abundance remains unknown and likely varies among taxa (Evans et al., 2016), as it is argued that PCR products are not fully proportional to real abundances due to the existence of primer bias (Clarke et al., 2014; Elbrecht and Leese, 2015) and instead, some advocate for the use of PCR-free methods (Zhou et al., 2013). Consequently, the number of sequences obtained per taxon may currently not be interpreted as quantitative but rather as semi-quantitative (Kelly et al., 2014; Pompanon et al., 2012). For stream fish in lotic systems, a predictive model incorporating eDNA concentration has been developed to identify detection probabilities and abundance, as well as both eDNA production and degradation rates (Wilcox et al., 2016). Such models that include eDNA production, transport, and decay may improve the ability to infer organism abundance from eDNA quantity (Barnes and Turner, 2016). As the relationships between eDNA and species abundance become clearer, the role of eDNA in estimating species abundance in both freshwater and marine environments is likely to become more valuable, increasing the potential of future eDNA applications in research and conservation.

14.5.2 Increasing Reference Database Coverage and Taxonomic Resolution

Currently, the taxonomic resolution of sequences from eDNA metabarcoding datasets often does not reach the species level. Moreover, taxonomic misidentification poses a significant problem. One of the main causes is the incompleteness of reference databases. Taxonomic resolution may be increased while simultaneously decreasing misidentification by creating and updating a locally curated reference database. Moreover, when using group-specific primers, taxonomic resolution may be improved by complementing the primers with one or several additional primer pairs specifically designed to amplify more discriminately genetic regions for families with many closely related species (Valentini et al., 2016), such as is the case for the elasmobranch family of Carcharhinidae (Bakker et al., 2017). Additionally, broadspectrum primers often amplify non-target groups/species. This may potentially be overcome by the use of blocking primers (Vestheim and Jarman, 2008), a strategy where the amplification of undesired sequences is specifically blocked.

14.5.3 Taking eDNA Analysis into the Field

A current limitation on the range and duration of eDNA field work is the need to keep water samples chilled to prevent DNA degradation and ship them back to a laboratory for processing; however, several recent developments are making it possible to take eDNA assays from the lab into the field. This allows for rapid detection of species or even communities in the field, extends the range and duration of field trips, and removes the need to ship samples to a central wet laboratory. To enable field eDNA analysis, three critical steps must be taken. First, instead of shipping water samples to a central lab for filtration, the development of a mobile pump system allows for the filtration of water as the sample is being collected in the field (e.g., Laramie et al., 2015) (Figure 14.2). Field filtration in itself already simplifies field collection, removes the need for shipping large volumes of water, and improves sample preservation. Filters can be preserved in ethanol (Laramie et al., 2015), silica beads (Bakker et al., 2017), or modified buffers such as Longmire's buffer (Renshaw et al., 2015). Next, magnetic-bead-based extractions (e.g., Tomlinson et al., 2005) or syringe-based extractions (e.g., eDNA Water Filter Sample Prep Kit by Biomeme; Philadelphia, PA) remove the need for an immobile centrifugation step that is restricted to the lab. The final step, analysis of the samples, can be either a qPCR for a species-specific barcoding assay or high-throughput sequencing for an eDNA metabarcoding assay. Improvements in miniaturization and the use of mobile phones as small but powerful computing units have allowed the development of mobile qPCR thermocyclers such as Biomeme's two3™, which can run a qPCR analysis for three samples in parallel using up to two fluorophores (Figure 14.4). However, the increased field capability comes at the cost of reduced throughput. For eDNA metabarcoding, mobile high-throughput sequencing platforms such as Oxford Nanopore Technology's MinION™ can be used in conjunction with a laptop to allow for immediate analysis of the data. They can be used to assess whole communities in the field (Figure 14.4). As these two technologies advance, the throughput of these systems is expected to increase further.

Figure 14.4 (Left) Two3™ mobile qPCR thermocycler (distributed by Biomeme). (Right) MinION™ mobile high-throughput long-read sequencer (Oxford Nanopore Technologies).

14.5.4 Emergence of Autonomous Sampling and Analysis

Applying remote and autonomous sampling techniques for eDNA collection may greatly expand the potential of eDNA applications to inform and improve conservation efforts (Barnes and Turner, 2016). Hydroplane drones such as those employed by Valentini et al. (2016) (Figure 14.5) are a promising way to collect large amounts of surface water with relatively little effort, as they can continuously filter water over an entire body of water and across areas otherwise difficult to sample. The increased sample size may enhance the detection probability of rare species (Hoffmann et al., 2016). Over the past several years, robotic systems, also referred to as *ecogenomic sensors*, for the autonomous collection and molecular analysis of eDNA samples have been under continuous development and are already being used by marine microbiologists to study marine microbial behavior and to

Figure 14.5 Hydroplane drone-assisted water sampling for eDNA metabarcoding. This drone is double hulled, and the outer hull is disposable, which minimizes the risk of water-body cross-contamination. (Photograph credit: Alice Valentini and Tony Dejean/SPYGEN.)

Figure 14.6   Environmental Sample Processor (ESP), a robotic microbiology laboratory that can filter water samples and either preserve the filtrate until recovery or process the filtrate autonomously using a variety of molecular-probe techniques. (Photograph credit: Todd Walsh © MBARI 2017, http://www.mbari.org/.)

detect changes in bacterioplankton communities by utilizing DNA probe (qPCR) and protein arrays to detect target molecules indicative of species and substances they produce (e.g., algal toxins) (Ottesen, 2016; Preston et al., 2011; Scholin, 2010). Additionally, autonomous high-resolution sampling and both *in situ* and *ex situ* molecular essays have been used to study zooplankton distribution (Harvey et al., 2012). Also, a variety of molecular essays, including qPCR, have been applied *in situ* on particulates filtered from seawater from depths up to 4000 m (Ussler et al., 2013). These robotic instruments (Figure 14.6) are designed to autonomously collect, filter, and analyze water samples from surface and subsurface waters and, in near real time, transmit data back to shore. In addition, they collect data on a wide variety of associated environmental parameters, such as currents, turbidity, salinity, and oxygen concentrations (Ottesen, 2016). They may also preserve and archive water samples for laboratory analysis after the instrument is recovered (Breier et al., 2014; Scholin, 2010).

Field portable, molecular analytical techniques such as eDNA metabarcoding are still very challenging to implement in the context of remote instrumentation due to the requirement for multiple wet-chemistry processing steps (including concentration, extraction, and purification of the target DNA, followed by amplification) (Ottesen, 2016). Hence, *post hoc* macrobial eDNA analysis of microbe-motivated samples from these systems likely represents the first step toward broadening their use across taxa and disciplines (Barnes and Turner, 2016). However, the use of ecogenomic sensors, specifically for the large-scale collection and analysis of macrobial eDNA, could bring about significant advances for molecular ecological studies and for the potential of eDNA applications to benefit conservation. As the

range of deployable science instruments increases and their operating costs decrease, ecogenomic sensors will become an increasingly important tool for both oceanographic and ecological research. It will become possible to remotely monitor the presence, biodiversity, and potentially abundance, of any marine species (including sharks and their relatives) through cost-effective, long-term, high-frequency eDNA sampling regimes of the water column, including remote and inaccessible areas, such as the deep sea.

14.5.5   Use of Long-Range PCR for eDNA Applications

As DNA possesses limited chemical stability (Lindahl, 1993), when it has been shed from an organism eDNA will begin to degrade. Hence, the general assumption regarding eDNA presence in an environmental sample is that most eDNA is highly degraded upon capture (Bohmann et al., 2014). Additionally, it has been shown that, in most cases, recently released eDNA becomes undetectable within hours to days (see Section 14.4.2.1.2), underlying its usefulness in providing approximate real-time data on species presence in the environment. Coupled with current sequence length limitations of both qPCR and high-throughput sequencing platforms, most eDNA research has been focused on a short-fragment PCR amplicon sequencing approach to characterize macroorganismal species richness (Deiner et al., 2017; Olds et al., 2016; Valentini et al., 2016). Yet, one of the major drawbacks of targeting short eDNA fragments is that it often limits the utility for species-level assignment (Deiner et al., 2016; Port et al., 2016). This can be particularly hampered when closely related species are concerned (Bakker et al., 2017) due to the highly conserved target sequences between these species.

It has been shown, however, that the largest percentage of common carp (*Cyprinus carpio*) eDNA detected in water samples was from particles that ranged in size from 1 to 10 μm (Turner et al., 2014a). This is consistent with the presence of intact tissues or cells in aquatic environments, indicating that not all eDNA in a water sample is degraded. These findings are corroborated by earlier research on the detection of microbial genetic materials in the environment which recognized that eDNA was present in both intracellular and extracellular forms (Ogram et al., 1987). It is likely that multicellular organisms shed genetic material into their environment first as sloughed tissues and whole cells, which subsequently break down and release DNA into the environment (Barnes and Turner, 2016); consequently, eDNA represents a complex mixture of particles ranging from extracellular DNA molecules up to whole cells and aggregations of cells (Turner et al., 2014a). This suggests that eDNA for species currently occupying a habitat is not primarily free DNA suspended in solution but could also be cellular or membrane-bound DNA in a coiled or circular state, with comparatively more structural resistance to rapid degradation (Deiner et al., 2017; Torti et al., 2015; Turner et al., 2014a).

As opposed to standard PCR amplification, long-range PCR makes it possible to produce a fragment that encompasses an entire mitogenome in a single amplification (Zhang et al., 2013). Hence, a recent study set out to test whether it is possible to amplify and sequence entire mitochondrial fish genomes (mitogenomes) from eDNA isolated from water samples by applying long-range PCR amplification coupled with shotgun sequencing techniques (Deiner et al., 2017). By recovering full-length mitochondrial genes (COI, cytochrome *b*, 12S, and 16S), this study demonstrated that some of the eDNA from macroorganisms currently inhabiting a water body remains intact for a short period, at least at the mitochondrial genome size (Deiner et al., 2017). One drawback of the method used in this study, however, is that mitogenome PCR products require shearing in order to fragment them prior to sequencing (due to the current read-length restrictions of the sequencing technology), and consequently the technique is dependent on short-fragment-based *de novo* assembly or reference mapping (remapping the short reads to a reference sequence). This could still be an obstacle for the identification of closely related species, as conserved regions with high-sequence similarity are difficult to accurately assemble from a complex mixture (Deiner et al., 2017). It is therefore expected that with the continued advancement of single-molecule and long-read technologies (e.g., improved cost-effectiveness, reduced error rates), such as the Oxford Nanopore MinION™ (Laszlo et al., 2014) (see Figure 14.4), it will become possible to couple long-range PCR amplification and sequencing without fragmentation, avoiding problems associated with the use of short fragments.

Being able to sequence whole mitogenomes from eDNA, instead of having to rely on short-fragment PCR amplifications for species identification, could potentially bring about major advances in taxonomic assignment; full-length barcodes, such as the COI region for animals (Hebert et al., 2003b), could be recovered in its entirety and used for species identification and additionally for the investigation of community structure and biodiversity. Future advances in long-read sequencing are expected to further advance eDNA applications into the realm of population and conservation genetics, systematics, and phylogeography (Deiner et al., 2017).

14.6 SUMMARY AND CONCLUSION

Environmental DNA is a tool with a growing list of research and conservation applications in environments as varied as terrestrial, freshwater, and marine. Initially used to gather presence/absence and distribution information for the monitoring of species of interest, other promising lines of inquiry using eDNA tools now include estimates of organism abundance, the study of population genetics, diet characterization, and the description of trophic interactions. However, in order to appropriately apply these emerging applications, as well as refine our current grasp on presence/absence and whole-community eDNA studies, further research is still needed. Indeed, to gain a more complete understanding of the ecology of eDNA and increase our confidence in the interpretation of eDNA results, further investigations into how different environmental conditions affect the production, degradation, and detection of eDNA in a range of taxa are critically needed. This is particularly true for sharks and their relatives, for which this information has yet to be collected. This group's unique biological and physiological traits are likely to impact eDNA persistence and degradation in their environment in different ways from those found in groups studied to date—namely, bony fishes and amphibians. Further, with over 95% of sharks and their relatives found in marine ecosystems (Dulvy et al., 2014) and eDNA research in this environment still in its infancy, future research will have to focus on filling important knowledge gaps in order to improve the reliability of eDNA studies in marine ecosystems. For example, how do environmental conditions in the oceans (e.g., water chemistry, UVB exposure) affect eDNA degradation and, combined with transport forces (e.g., waves, currents, eddies), influence shark eDNA detection?

With a growing number of endangered or critically endangered shark and ray species (e.g., sawfish, scalloped hammerhead shark), there is a clear urgency for cost-effective, non-invasive, and reliable methods to obtain basic distribution, abundance, and biodiversity data to further conservation and management efforts. Environmental DNA sampling can access inhospitable environments, target elusive species, and provide a vast reduction in labor costs. In the future, it may be possible to implement mechanical sampling of (shark) eDNA, similar to that of oil-spill-sampling buoys or military sonobuoys (Bohmann et al., 2014). Although improvements in the methods continue to be implemented, eDNA metabarcoding has great potential for developing into an effective assessment tool for sharks, as it is applicable to a wide range of ecological goals, from mapping diversity gradients in response to environmental variation to monitoring the effectiveness of spatial protection measures (Bakker et al., 2017). Environmental DNA techniques clearly have great potential to aid in conservation, remediation, and restoration efforts of sharks and their relatives. However, although eDNA has shown itself to be a useful tool for identifying habitats of rare and endangered species over the last decade, care must be taken to account for the limitations and levels of uncertainty when applying this new tool.

REFERENCES

Amberg JJ, Grace McCalla S, Monroe E, Lance R, Baerwaldt K, Gaikowski MP (2015) Improving efficiency and reliability of environmental DNA analysis for silver carp. *J Great Lakes Res* 41(2):367–373.

Andruszkiewicz EA, Sassoubre LM, Boehm AB (2017) Persistence of marine fish environmental DNA and the influence of sunlight. *PLoS ONE* 12(9):e0185043.

Bailiff M, Karl D (1991) Dissolved and particulate DNA dynamics during a spring bloom in the Antarctic Peninsula region, 1986–1987. *Deep Sea Res Part 1 Oceanogr Res Pap* 38(8–9):1077–1095.

Bakker J, Wangensteen OS, Chapman DD, Boussarie G, Buddo D, Guttridge TL, Hertler H, et al. (2017) Environmental DNA reveals tropical shark diversity in contrasting levels of anthropogenic impact. *Sci Rep* 7(1):16886.

Barnes MA, Turner CR (2016) The ecology of environmental DNA and implications for conservation genetics. *Conserv Genet* 17(1):1–17.

Barnes MA, Turner CR, Jerde CL, Renshaw MA, Chadderton WL, Lodge DM (2014) Environmental conditions influence eDNA persistence in aquatic systems. *Environ Sci Technol* 48(3):1819–1827.

Bohmann K, Evans A, Gilbert MT, Carvalho GR, Creer S, Knapp M, Yu DW, de Bruyn M (2014) Environmental DNA for wildlife biology and biodiversity monitoring. *Trends Ecol Evol* 29(6):358–367.

Bourlat SJ, Borja A, Gilbert J, Taylor MI, Davies N, Weisberg SB, Griffith JF, et al. (2013) Genomics in marine monitoring: new opportunities for assessing marine health status. *Mar Pollut Bull* 74(1):19–31.

Boussarie G, Bakker J, Wangensteen OS, Mariani S, Bonnin L, Juhel J-B, Kiszka JJ, et al. (2018) Environmental DNA illuminates the dark diversity of sharks. Sci Adv 4(5):eaap9661.

Brandl S, Schumer G, Schreier BM, Conrad JL, May B, Baerwald MR (2015) Ten real-time PCR assays for detection of fish predation at the community level in the San Francisco Estuary–Delta. *Mol Ecol Resour* 15(2):278–284.

Breier JA, Sheik CS, Gomez-Ibanez D, Sayre-McCord RT, Sanger R, Rauch C, Coleman M, et al. (2014) A large volume particulate and water multi-sampler with *in situ* preservation for microbial and biogeochemical studies. *Deep Sea Res Part 1 Oceanogr Res Pap* 94:195–206.

Buxton AS, Groombridge JJ, Zakaria NB, Griffiths RA (2017) Seasonal variation in environmental DNA in relation to population size and environmental factors. *Sci Rep* 7:46294.

Bylemans J, Furlan EM, Hardy CM, McGuffie P, Lintermans M, Gleeson DM (2017) An environmental DNA-based method for monitoring spawning activity: a case study, using the endangered Macquarie perch (*Macquaria australasica*). *Meth Ecol Evol* 8:646–655.

Cavé L, Brothier E, Abrouk D, Bouda PS, Hien E, Nazaret S (2016) Efficiency and sensitivity of the digital droplet PCR for the quantification of antibiotic resistance genes in soils and organic residues. *Appl Microbiol Biotechnol* 100(24):10597–10608.

Chin A (2014) "Hunting porcupines": citizen scientists contribute new knowledge about rare coral reef species. *Pacific Conserv Biol* 20(1):48–53.

Clarke LJ, Soubrier J, Weyrich LS, Cooper A (2014) Environmental metabarcodes for insects: in silico PCR reveals potential for taxonomic bias. *Mol Ecol Resour* 14:1160–1170.

Coissac E, Riaz T, Puillandre N (2012) Bioinformatic challenges for DNA metabarcoding of plants and animals. *Mol Ecol* 21:1834–1847.

Darling JA, Mahon AR (2011) From molecules to management: adopting DNA-based methods for monitoring biological invasions in aquatic environments. *Environ Res* 111(7):978–988.

Davison PI, Créach V, Liang WJ, Andreou D, Britton JR, Copp GH (2016) Laboratory and field validation of a simple method for detecting four species of non-native freshwater fish using eDNA. *J Fish Biol* 89(3):1782–1793.

Davy CM, Kidd AG, Wilson CC (2015) Development and validation of environmental DNA (eDNA) markers for detection of freshwater turtles. *PLoS ONE* 10(7):e0130965.

De Barba M, Miquel C, Boyer F, Mercier C, Rioux D, Coissac E, Taberlet P (2014) DNA metabarcoding multiplexing and validation of data accuracy for diet assessment: application to omnivorous diet. *Mol Ecol Resour* 2:306–323.

De Souza LS, Godwin JC, Renshaw MA, Larson E (2016) Environmental DNA (eDNA) detection probability is influenced by seasonal activity of organisms. *PLoS ONE* 11(10):e0165273.

Deagle BE, Jarman SN, Coissac E, Pompanon F, Taberlet P (2014) DNA metabarcoding and the cytochrome *c* oxidase subunit 1 marker: not a perfect match. *Biol Lett* 10(9):1789–1793.

Deiner K, Altermatt F (2014) Transport distance of invertebrate environmental DNA in a natural river. *PLoS ONE* 9(2):e88786.

Deiner K, Walser JC, Mächler E, Altermatt F (2015) Choice of capture and extraction methods affect detection of freshwater biodiversity from environmental DNA. *Biol Conserv* 183:53–63.

Deiner K, Fronhofer EA, Mächler E, Walser J-C, Altermatt F (2016) Environmental DNA reveals that rivers are conveyer belts of biodiversity information. *Nat Comm* 7(0):12544.

Deiner K, Renshaw MA, Li Y, Olds BP, Lodge DM, Pfrender ME (2017) Long-range PCR allows sequencing of mitochondrial genomes from environmental DNA. *Meth Ecol Evol* 8(12):1888–1898.

Dejean T, Valentini A, Duparc A, Pellier-Cuit S, Pompanon F, Taberlet P, Miaud C (2011) Persistence of environmental DNA in freshwater ecosystems. *PLoS ONE* 6(8):e23398.

Dejean T, Valentini A, Miquel C, Taberlet P, Bellemain E, Miaud C (2012) Improved detection of an alien invasive species through environmental DNA barcoding: the example of the American bullfrog *Lithobates catesbeianus*. *J Appl Ecol* 49(4):953–959.

Dell'Anno A, Corinaldesi C (2004) Degradation and turnover of extracellular DNA in marine sediments: ecological and methodological considerations. *Appl Environ Microbiol* 70:4384–4386.

Doi H, Takahara T, Minamoto T, Matsuhashi S, Uchii K, Yamanaka H (2015a) Droplet digital polymerase chain reaction (PCR) outperforms real-time PCR in the detection of environmental DNA from an invasive fish species. *Environ Sci Technol* 49(9):5601–5608.

Doi H, Uchii K, Takahara T, Matsuhashi S, Yamanaka H, Minamoto T (2015b) Use of droplet digital PCR for estimation of fish abundance and biomass in environmental DNA surveys. *PLoS ONE* 10(3): e0122763.

Doi H, Inui R, Akamatsu Y, Kanno K, Yamanaka H, Takahara T, Minamoto T (2017a) Environmental DNA analysis for estimating the abundance and biomass of stream fish. *Freshwater Biol* 62(1):30–39.

Doi H, Katano I, Sakata Y, Souma R, Kosuge T, Nagano M, Ikeda K, et al. (2017b) Detection of an endangered aquatic heteropteran using environmental DNA in a wetland ecosystem. *R Soc Open Sci* 4(7):170568.

Doyle JR, McKinnon AD, Uthicke S (2017) Quantifying larvae of the coralivorous seastar *Acanthaster* cf. *solaris* on the Great Barrier Reef using qPCR. *Mar Biol* 164(8):176.

Dulvy NK, Fowler SL, Musick JA, Cavanagh RD, Kyne PM, Harrison LR, Carlson JK, et al. (2014) Extinction risk and conservation of the world's sharks and rays. *eLife* 3:e00590.

Dulvy NK, Davidson LNK, Kyne PM, Simpfendorfer CA, Harrison LR, Carson JK, Fordham SV (2016) Ghosts of the coast: global extinction risk and conservation of sawfishes. *Aquat Conserv Mar Freshwater Ecosyst* 26:134–153.

Eichmiller JJ, Bajer PG, Sorensen PW (2014) The relationship between the distribution of common carp and their environmental DNA in a small lake. *PLoS ONE* 9(11):e0112611.

Eichmiller JJ, Miller LM, Sorensen PW (2016) Optimizing techniques to capture and extract environmental DNA for detection and quantification of fish. *Mol Ecol Resour* 16(1):56–68.

Elbrecht V, Leese F (2015) Can DNA-based ecosystem assessments quantify species abundance? Testing primer bias and biomass—sequence relationships with an innovative metabarcoding protocol. *PLoS ONE* 10:e0130324.

Evans NT, Olds BP, Renshaw MA, Turner CR, Li Y, Jerde CL, Mahon AR, et al. (2016) Quantification of mesocosm fish and amphibian species diversity via environmental DNA metabarcoding. *Mol Ecol Resour* 16(1):29–41.

Feutry P, Kyne PM, Pillans RD, Chen X, Naylor GJP, Grewe PM (2014) Mitogenomics of the speartooth shark challenges ten years of control region sequencing. *BMC Evol Biol* 14:232.

Ficetola GF, Miaud C, Pompanon F, Taberlet P (2008) Species detection using environmental DNA from water samples. *Biol Lett* 4(4):423–425.

Fields AT, Abercrombie DL, Eng R, Feldheim K, Chapman DD (2015) A novel mini-DNA barcoding assay to identify processed fins from internationally protected shark species. *PLoS ONE* 10(2):e0114844.

Foote AD, Thomsen PF, Sveegaard S, Wahlberg M, Kielgast J, Kyhn LA, Salling AB, et al. (2012) Investigating the potential use of environmental DNA (eDNA) for genetic monitoring of marine mammals. *PLoS ONE* 7(8):e41781.

Fu XH, Wang L, Le YQ, Hu JJ (2012) Persistence and renaturation efficiency of thermally treated waste recombinant DNA in defined aquatic microcosms. *J Environ Sci Health A Tox Hazard Subst Environ Eng* 47(13):1975–1983.

Fukumoto S, Ushimaru A, Minamoto T (2015) A basin-scale application of environmental DNA assessment for rare endemic species and closely related exotic species in rivers: a case study of giant salamanders in Japan. *J Appl Ecol* 52(2):358–365.

Furlan EM, Gleeson D (2016a) Environmental DNA detection of redfin perch, *Perca fluviatilis*. *Conserv Genet Resour* 8(2):115–118.

Furlan EM, Gleeson D (2016b) Improving reliability in environmental DNA detection surveys through enhanced quality control. *Mar Freshwater Res* 68(2):388–395.

Furlan EM, Gleeson D, Hardy CM, Duncan RP (2016) A framework for estimating the sensitivity of eDNA surveys. *Mol Ecol Resour* 16(3):641–654.

Gargan LM, Morato T, Pham CK, Finarelli JA, Carlsson JEL, Carlsson J (2017) Development of a sensitive detection method to survey pelagic biodiversity using eDNA and quantitative PCR: a case study of devil ray at seamounts. *Mar Biol* 164(5):112.

Giles RE, Blanc H, Cann HM, Wallace DC (1980) Maternal inheritance of human mitochondrial DNA. *Proc Natl Acad Sci USA* 77(11):6715–6719.

Goldberg CS, Pilliod DS, Arkle RS, Waits LP (2011) Molecular detection of vertebrates in stream water: a demonstration using rocky mountain tailed frogs and Idaho giant salamanders. *PLoS ONE* 6(7):e22746.

Goldberg CS, Sepulveda A, Ray A, Baumgardt J, Waits LP (2013) Environmental DNA as a new method for early detection of New Zealand mudsnails (*Potamopyrgus antipodarum*). *Freshw Sci* 32(3):792–800.

Goldberg CS, Turner CR, Deiner K, Klymus KE, Thomsen PF, Murphy MA, Spear SF, et al. (2016) Critical considerations for the application of environmental DNA methods to detect aquatic species. *Meth Ecol Evol* 7(11):1299–1307.

Guardiola M, Uriz MJ, Taberlet P, Coissac E, Wangensteen OS, Turon X (2015) Deep-sea, deep-sequencing: metabarcoding extracellular DNA from sediments of marine canyons. *PLoS ONE* 10(10):e0139633.

Gustavson MS, Collins PC, Finarelli JA, Egan D, Conchúir, RÓ, Wightman GD, King JJ, et al. (2015) An eDNA assay for Irish *Petromyzon marinus* and *Salmo trutta* and field validation in running water. *J Fish Biol* 87(5):1254–1262.

Guttridge TL, Van Zinnicq Bergmann MPM, Bolte C, Howey LA, Finger JS, Kessel ST, Brooks JL, et al. (2017) Philopatry and regional connectivity of the great hammerhead shark, *Sphyrna mokarran*, in the U.S. and Bahamas. *Front Mar Sci* 4(3):1–15.

Hajibabaei M, Smith MA, Janzen DH, Rodriguez JJ, Whitfield JB, Hebert PDN (2006) A minimalist barcode can identify a specimen whose DNA is degraded. *Mol Ecol Notes* 6(4):959–964.

Handelsman J (2004) Metagenomics: application of genomics to uncultured microorganisms. *Microbiol Mol Biol Rev* 68(4):669–685.

Hänfling B, Lawson Handley L, Read DS, Hahn C, Li J, Nichols P, Blackman RC, et al. (2016) Environmental DNA metabarcoding of lake fish communities reflects long-term data from established survey methods. *Mol Ecol* 25(13):3101–3119.

Hansell AC, Kessel ST, Brewslter LR, Cadrin SX, Gruber SH, Skomal GB, Guttridge TL (2017) Local indicators of abundance and demographics for the coastal shark assemblage of Bimini, Bahamas. *Fish Res* 197:34–44.

Harvey JBJ, Ryan JP, Marin R, Preston CM, Alvarado N, Scholin CA, Vrijenhoek RC (2012) Robotic sampling, *in situ* monitoring and molecular detection of marine zooplankton. *J Exp Mar Biol Ecol* 413:60–70.

Hebert PDN, Cywinska A, Ball SL, DeWaard JR (2003a) Biological identifications through DNA barcodes. *Proc R Soc Lond B Biol Sci* 270(1512):313–321.

Hebert PDN, Ratnasingham S, deWaard JR (2003b) Barcoding animal life: cytochrome *c* oxidase subunit 1 divergences among closely related species. *Proc R Soc Lond B Biol Sci* 270(Suppl 1):S96–S99.

Herder J, Valentini A, Bellemain E, Dejean T, van Delft JJCW, Thomsen PF, Taberlet P (2014) *Environmental DNA: A Review of the Possible Applications for the Detection of (Invasive) Species.* Stichting RAVON, Nijmegen, Netherlands.

Hindson BJ, Ness KD, Masquelier DA, Belgrader P, Heredia NJ, Makarewicz AJ, Bright IJ, et al. (2011) High-throughput droplet digital PCR system for absolute quantitation of DNA copy number. *Anal Chem* 83(22):8604–8610.

Hindson CM, Chevillet JR, Briggs HA, Gallichotte EN, Ruf IK, Hindson BJ, Vessella RL, Tewari M (2013) Absolute quantification by droplet digital PCR versus analog real-time PCR. *Nat Meth* 10:1003–1005.

Hinlo R, Gleeson D, Lintermans M, Furlan E (2017) Methods to maximise recovery of environmental DNA from water samples. *PLoS ONE* 12(6):e0179251.

Hoffmann C, Schubert G, Calvignac-Spencer S (2016) Aquatic biodiversity assessment for the lazy. *Mol Ecol* 25(4):846–848.

Hunter ME, Dorazio RM, Butterfield JSS, Meigs-Friend G, Nico LG, Ferrante JA (2017) Detection limits of quantitative and digital PCR assays and their influence in presence–absence surveys of environmental DNA. *Mol Ecol Resour* 17(2):221–229.

Hunter ME, Oyler-McCance SJ, Dorazio RM, Fike JA, Smith BJ, Hunter CT, Reed RN, Hart KM (2015) Environmental DNA (eDNA) sampling improves occurrence and detection estimates of invasive Burmese pythons. *PLoS ONE* 10(4):e0121655.

Huver JR, Koprivnikar J, Johnson PTJ, Whyard S (2015) Development and application of an eDNA method to detect and quantify a pathogenic parasite in aquatic ecosystems. *Ecol Appl* 25(4):991–1002.

Jane SF, Wilcox TM, McKelvey KS, Young MK, Schwartz MK, Lowe WH, Letcher BH, Whiteley AR (2015) Distance, flow and PCR inhibition: EDNA dynamics in two headwater streams. *Mol Ecol Resour* 15(1):216–227.

Jerde CL, Mahon AR, Chadderton WL, Lodge DM (2011) "Sight-unseen" detection of rare aquatic species using environmental DNA. *Conserv Lett* 4(2):150–157.

Jerde CL, Olds BP, Shogren AJ, Andruszkiewicz EA, Mahon AR, Bolster D, Tank JL (2016) Influence of stream bottom substrate on retention and transport of vertebrate environmental DNA. *Environ Sci Technol* 50(16):8770–8779.

Kajiura SM, Tellman SL (2016) Quantification of massive seasonal aggregations of blacktip sharks (*Carcharhinus limbatus*) in southeast Florida. *PLoS ONE* 11(3): e0150911.

Meintjes, P, Duran C, Kearse M, Moir R, Wilson A, Stones-Havas S, Cheung M, et al. (2012) Geneious Basic: an integrated and extendable desktop software platform for the organization and analysis of sequence data. *Bioinformatics* 28(12):1647–1649.

Kelly RP (2016) Making environmental DNA count. *Mol Ecol Resour* 16(1):10–12.

Kelly RP, Port JA, Yamahara KM, Crowder LB (2014) Using environmental DNA to census marine fishes in a large mesocosm. *PLoS ONE* 9(1):e86175.

Kelly RP, O'Donnell JL, Lowell NC, Shelton AO, Samhouri JF, Hennessey SM, Feist BE, Williams GD (2016) Genetic signatures of ecological diversity along an urbanization gradient. *PeerJ* 4:e2444.

Kelly RP, Closek CJ, O'Donnell JL, Kralj JE, Shelton AO, Samhouri JF (2017) Genetic and manual survey methods yield different and complementary views of an ecosystem. *Front Mar Sci* 3:283.

Kessel ST, Hansell AC, Gruber SH, Guttridge TL, Hussey NE, Perkins RG (2016) Three decades of longlining in Bimini, Bahamas, reveals long-term trends in lemon shark *Negaprion brevirostris* (Carcharhinidae) catch per unit effort. *J Fish Biol* 88(6):2144–2156.

Kim TG, Jeong SY, Cho KS (2014) Comparison of droplet digital PCR and quantitative real-time PCR for examining population dynamics of bacteria in soil. *Appl Microbiol Biotechnol* 98:6105–6113.

Klymus KE, Richter Ca, Chapman DC, Paukert C (2015) Quantification of eDNA shedding rates from invasive bighead carp *Hypophthalmichthys nobilis* and silver carp *Hypophthalmichthys molitrix*. *Biol Conserv* 183(SI):77–84.

Klymus KE, Marshall NT, Stepien CA (2017) Environmental DNA (eDNA) metabarcoding assays to detect invasive invertebrate species in the Great Lakes. *PLoS ONE* 12(5):e0177643.

Kreader CA (1998) Persistence of PCR-detectable *Bacteroides distasonis* from human feces in river water. *Appl Environ Microbiol* 64:4103–4105.

Lacoursiere-Roussel A, Rosabal M, Bernatchez L (2016) Estimating fish abundance and biomass from eDNA concentrations: variability among capture methods and environmental conditions. *Mol Ecol Resour* 16:1401–1414.

Laramie MB, Pilliod DS, Goldberg CS (2015) Characterizing the distribution of an endangered salmonid using environmental DNA analysis. *Biol Conserv* 183:29–37.

Laszlo AH, Derrington IM, Ross BC, Brinkerhoff H, Adey A, Nova IC, Craig JM, et al. (2014) Decoding long nanopore sequencing reads of natural DNA. *Nat Biotechnol* 32:829–833.

Lennon JT (2007) Diversity and metabolism of marine bacteria cultivated on dissolved DNA. *Appl Environ Microbiol* 73(9):2799–2805.

Leray M, Knowlton N (2015) DNA barcoding and metabarcoding of standardized samples reveal patterns of marine benthic diversity. *Proc Natl Acad Sci USA* 112(7):2076–2081.

Liang Z, Keeley A (2013) Filtration recovery of extracellular DNA from environmental water samples. *Environ Sci Technol* 47(16):9324–9331.

Lindahl T (1993) Instability and decay of the primary structure of DNA. *Nature* 362:709–715.

Lodge DM, Turner CR, Jerde CL, Barnes MA, Chadderton L, Egan SP, Feder JL, et al. (2012) Conservation in a cup of water: estimating biodiversity and population abundance from environmental DNA. *Mol Ecol* 21(11):2555–2558.

Mahon AR, Jerde CL (2016) Using environmental DNA for invasive species surveillance and monitoring. *Meth Mol Biol* 1452:131–142.

Mahon AR, Jerde CL, Galaska M, Bergner JL, Chadderton WL, Lodge DM, Hunter ME, Nico LG (2013) Validation of eDNA surveillance sensitivity for detection of Asian carps in controlled and field experiments. *PLoS ONE* 8(3):e58316.

Maniatis T, Fritsch EF, Sambrook J (1982) *Molecular Cloning: A Laboratory Manual.* Cold Spring Harbor Laboratory Cold Spring Harbor, NY.

Martellini A, Payment P, Villemur R (2005) Use of eukaryotic mitochondrial DNA to differentiate human, bovine, porcine and ovine sources in fecally contaminated surface water. *Water Res* 39:541–548.

Martin AP (1995) Mitochondrial DNA sequence evolution in sharks: rates, patterns, and phylogenetic inferences. *Mol Biol Evol* 12(6):1114–1123.

Martin AP (1999) Substitution rates of organelle and nuclear genes in sharks: implicating metabolic rate (again). *Mol Biol Evol* 16(7):996–1002.

Martin AP, Naylor GJP, Palumbi SR (1992) Rates of mitochondrial DNA evolution in sharks are slow compared with mammals. *Nature* 357(6374):153–155.

Maruyama A, Nakamura K, Yamanaka H, Kondoh M, Minamoto T (2014) The release rate of environmental DNA from juvenile and adult fish. *PLoS ONE* 9(12):e114639.

Meyer W, Seegers U (2012) Basics of skin structure and function in elasmobranchs: a review. *J Fish Biol* 80(5):1940–1967.

Minamoto T, Yamanaka H, Takahara T, Honjo MN, Kawabata Zi (2012) Surveillance of fish species composition using environmental DNA. *Limnology* 13(2):193–197.

Miotke L, Lau BT, Rumma RT, Ji HP (2014) High sensitivity detection and quantitation of DNA copy number and single nucleotide variants with single color droplet digital PCR. *Anal Chem* 86:2618–2624.

Miya M, Sato Y, Fukunaga T, Sado T, Poulsen JY, Sato K, Minamoto T, et al. (2015) MiFish, a set of universal PCR primers for metabarcoding environmental DNA from fishes: detection of more than 230 subtropical marine species. *R Soc Open Sci* 2(7):150088.

Moyer GR, Díaz-Ferguson E, Hill JE, Shea C (2014) Assessing environmental DNA detection in controlled lentic systems. *PLoS ONE* 9(7):e103767.

Nathan LM, Simmons M, Wegleitner BJ, Jerde CL, Mahon AR (2014a) Quantifying environmental DNA signals for aquatic invasive species across multiple detection platforms. *Environ Sci Technol* 48(21):12800–12806.

Nathan LR, Jerde CL, Budny ML, Mahon AR (2014b) The use of environmental DNA in invasive species surveillance of the Great Lakes commercial bait trade. *Conserv Biol* 29(2):430–439.

Nielsen KM, Johnsen PJ, Bensasson D, Daffonchio D (2007) Release and persistence of extracellular DNA in the environment. *Environ Biosafety Res* 6(1–2):37–53.

O'Donnell JL, Kelly RP, Shelton AO, Samhouri JF, Lowell NC, Williams GD (2017) Spatial distribution of environmental DNA in a nearshore marine habitat. *PeerJ* 5:e3044.

Ogram A, Sayler GS, Barkay T (1987) The extraction and purification of microbial DNA from sediments. *J Microbiol Meth* 7(2–3):57–66.

Ogram A, Sayler GS, Gustln D, Lewis RJ (1988) DNA adsorption to soils and sediments. *Environ Scid Technol* 22(8):982–984.

Olds BP, Jerde CL, Renshaw MA, Li Y, Evans NT, Turner CR, Deiner K, et al. (2016) Estimating species richness using environmental DNA. *Ecol Evol* 6(12):4214–4226.

Olson ZH, Briggler JT, Williams RN (2012) An eDNA approach to detect eastern hellbenders (*Cryptobranchus a. alleganiensis*) using samples of water. *Wildl Res* 39(7):629–636.

Ottesen EA (2016) Probing the living ocean with ecogenomic sensors. *Curr Opin Microbiol* 31:132–139.

Palumbo JD, O'Keeffe TL, Fidelibus MW (2016) Characterization of *Aspergillus* section *Nigri* species populations in vineyard soil using droplet digital PCR. *Lett Appl Microbiol* 63(6):458–465.

Paul J, Kellogg C, Jiang S (1996) Viruses and DNA in marine environments. In: Colwell RR, Simidu U, Ohwada K (eds) *Microbial Diversity in Time and Space.* Plenum Press, New York, pp 115–124.

Pedersen MW, Overballe-Petersen S, Ermini L, Sarkissian CD, Haile J, Hellstrom M, Spens J, et al. (2015) Ancient and modern environmental DNA. *Philos Trans R Soc Lond B Biol Sci* 370(1660):20130383.

Piaggio AJ, Engeman RM, Hopken MW, Humphrey JS, Keacher KL, Bruce WE, Avery ML (2014) Detecting an elusive invasive species: a diagnostic PCR to detect Burmese python in Florida waters and an assessment of persistence of environmental DNA. *Mol Ecol Resour* 14(2):374–380.

Pilliod DS, Goldberg CS, Arkle RS, Waits LP (2013) Estimating occupancy and abundance of stream amphibians using environmental DNA from filtered water samples. *Can J Fish Aquat Sci* 70(8):1123–1130.

Pilliod DS, Goldberg CS, Arkle RS, Waits LP (2014) Factors influencing detection of eDNA from a stream-dwelling amphibian. *Mol Ecol Resour* 14(1):109–116.

Pinheiro LB, Coleman VA, Hindson CM, Herrmann J, Hindson BJ, Bhat S, Emslie KR (2012) Evaluation of a droplet digital polymerase chain reaction format for DNA copy number quantification. *Anal Chem* 84(2):1003–1011.

Pompanon F, Deagle BE, Symondson WO, Brown DS, Jarman SN, Taberlet P (2012) Who is eating what: diet assessment using next generation sequencing. *Mol Ecol* 21(8):1931–1950.

Port JA, O'Donnell JL, Romero-Maraccini OC, Leary PR, Litvin SY, Nickols KJ, Yamahara KM, Kelly RP (2016) Assessing vertebrate biodiversity in a kelp forest ecosystem using environmental DNA. *Mol Ecol* 25(2):527–541.

Preston CM, Harris A, Ryan JP, Roman B, Marin 3rd R, Jensen S, Everlove C, et al. (2011) Underwater application of quantitative PCR on an ocean mooring. *PLoS ONE* 6(8):e22522.

Ratnasingham S, Hebert PDN (2007) The Barcode of Life Data System (http://www.barcodinglife.org/). *Mol Ecol Notes* 7(3):355–364.

Rees HC, Bishop K, Middleditch DJ, Patmore JRM, Maddison BC, Gough KC (2014a) The application of eDNA for monitoring of the great crested newt in the UK. *Ecol Evol* 4(21):4023–4032.

Rees HC, Maddison BC, Middleditch DJ, Patmore JRM, Gough KC (2014b) The detection of aquatic animal species using environmental DNA: a review of eDNA as a survey tool in ecology. *J Appl Ecol* 51(5):1450–1459.

Renshaw MA, Olds BP, Jerde CL, McVeigh MM, Lodge DM (2015) The room temperature preservation of filtered environmental DNA samples and assimilation into a phenol-chloroform-isoamyl alcohol DNA extraction. *Mol Ecol Resour* 15(1):168–176.

Riaz T, Shehzad W, Viari A, Pompanon F, Taberlet P, Coissac E (2011) ecoPrimers: inference of new DNA barcode markers from whole genome sequence analysis. *Nucleic Acids Res* 39(21):e145.

Robson HLA, Noble TH, Saunders RJ, Robson SKA, Burrows DW, Jerry DR (2016) Fine-tuning for the tropics: application of eDNA technology for invasive fish detection in tropical freshwater ecosystems. *Mol Ecol Resour* 16(4):922–932.

Romanowski G, Lorenz MG, Sayler G, Wackernagel W (1992) Persistence of free plasmid DNA in soil monitored by various methods, including a transformation assay. *Appl Environ Microbiol* 58(9):3012–3019.

Saghaï A, Zivanovic Y, Zeyen N, Moreira D, Benzerara K, Deschamps P, Bertolino P, et al. (2015) Metagenome-based diversity analyses suggest a significant contribution of non-cyanobacterial lineages to carbonate precipitation in modern microbialites. *Front Microbiol* 6:797.

Sanger F, Nichlen S, Coulson AR (1977) DNA sequencing with chain-termination inhibitors. *Proc Natl Acad Sci USA* 74(12):5463–5467.

Sassoubre LM, Yamahara KM, Gardner LD, Block BA, Boehm AB (2016) Quantification of environmental DNA (eDNA) shedding and decay rates for three marine fish. *Environ Sci Technol* 50(19):10456–10464.

Schmidt BR, Kéry M, Ursenbacher S, Hyman OJ, Collins JP (2013) Site occupancy models in the analysis of environmental DNA presence/absence surveys: a case study of an emerging amphibian pathogen. *Meth Ecol Evol* 4(7):646–653.

Scholin CA (2010) What are "ecogenomic sensors?" A review and thoughts for the future. *Ocean Sci* 6:51–60.

Secondi J, Dejean T, Valentini A, Audebaud B, Miaud C (2016) Detection of a global aquatic invasive amphibian, *Xenopus laevis*, using environmental DNA. *Amphib Reptil* 37(1):131–136.

Shogren AJ, Tank JL, Andruszkiewicz EA, Olds B, Jerde C, Bolster D (2016) Modelling the transport of environmental DNA through a porous substrate using continuous flow-through column experiments. *J R Soc Interface* 13(119):20160290.

Sigsgaard EE, Carl H, Møller PR, Thomsen PF (2015) Monitoring the near-extinct European weather loach in Denmark based on environmental DNA from water samples. *Biol Conserv* 183:46–52.

Sigsgaard EE, Nielsen IB, Bach SS, Lorenzen ED, Robinson DP, Knudsen SW, Pedersen MW, et al. (2016) Population characteristics of a large whale shark aggregation inferred from seawater environmental DNA. *Nat Ecol Evol* 1(1):4.

Simpfendorfer CA, Kyne PM, Noble TH, Goldsbury J, Basiita RK, Lindsay R, Shields A, et al. (2016) Environmental DNA detects critically endangered largetooth sawfish in the wild. *Endang Spec Res* 30:109–116.

Sims DW, Wearmouth VJ, Southall EJ, Hill JM, Moore P, Rawlinson K, Hutchinson N, et al. (2006) Hunt warm, rest cool: bioenergetic strategy underlying diel vertical migration of a benthic shark. *J Anim Ecol* 75(1):176–190.

Spear SF, Groves JD, Williams LA, Waits LP (2015) Using environmental DNA methods to improve detectability in a hellbender (*Cryptobranchus alleganiensis*) monitoring program. *Biol Conserv* 183:38–45.

Spens J, Evans AR, Halfmaerten D, Knudsen SW, Sengupta ME, Mak SST, Sigsgaard EE, Hellström M (2017) Comparison of capture and storage methods for aqueous macrobial eDNA using an optimized extraction protocol: advantage of enclosed filter. *Meth Ecol Evol* 8(5):635–645.

Stoeckle MY, Soboleva L, Charlop-Powers Z (2017) Aquatic environmental DNA detects seasonal fish abundance and habitat preference in an urban estuary. *PLoS ONE* 12(4):e0175186.

Strickler KM, Fremier AK, Goldberg CS (2015) Quantifying effects of UV-B, temperature, and pH on eDNA degradation in aquatic microcosms. *Biol Conserv* 183:85–92.

Taberlet P, Coissac E, Hajibabaei M, Rieseberg LH (2012a) Environmental DNA. *Mol Ecol* 21:1789–1793.

Taberlet P, Coissac E, Pompanon F, Brochmann C, Willerslev E (2012b) Towards next-generation biodiversity assessment using DNA metabarcoding. *Mol Ecol* 21(8):2045–2050.

Takahara T, Minamoto T, Yamanaka H, Doi H, Kawabata Zi (2012) Estimation of fish biomass using environmental DNA. *PLoS ONE* 7(4):e35868.

Takahara T, Minamoto T, Doi H (2013) Using environmental DNA to estimate the distribution of an invasive fish species in ponds. *PLoS ONE* 8(2):e56584.

Taylor S, Wakem M, Dijkman G, Alsarraj M, Nguyen M (2010) A practical approach to RT-qPCR—publishing data that conform to the MIQE guidelines. *Methods* 50(4):S1–S5.

Thomsen PF, Willerslev E (2015) Environmental DNA—an emerging tool in conservation for monitoring past and present biodiversity. *Biol Conserv* 183:4–18.

Thomsen PF, Kielgast J, Iversen LL, Møller PR, Rasmussen M, Willerslev E (2012a) Detection of a diverse marine fish fauna using environmental DNA from seawater samples. *PLoS ONE* 7(8):e41732.

Thomsen PF, Kielgast J, Iversen LL, Wiuf C, Rasmussen M, Gilbert MTP, Orlando L, Willerslev E (2012b) Monitoring endangered freshwater biodiversity using environmental DNA. *Mol Ecol* 21(11):2565–2573.

Thomsen PF, Møller PR, Sigsgaard EE, Knudsen SW, Jørgensen OA, Willerslev E (2016) Environmental DNA from seawater samples correlate with trawl catches of subarctic, deepwater fishes. *PLoS ONE* 11(11):e0165252.

Tomlinson JA, Boonham N, Hughes KJ, Griffin RL, Barker I (2005) On-site DNA extraction and real-time PCR for detection of *Phytophthora ramorum* in the field. *Appl Environ Microbiol* 71(11):6702–6710.

Torti A, Lever MA, Jørgensen BB (2015) Origin, dynamics, and implications of extracellular DNA pools in marine sediments. *Mar Genom* 24:185–196.

Tsutsui S, Yamaguchi M, Hirasawa A, Nakamura O, Watanabe T (2009) Common skate (*Raja kenojei*) secretes pentraxin into the cutaneous secretion: the first skin mucus lectin in cartilaginous fish. *J Biochem* 146(2):295–306.

Turner CR, Barnes MA, Xu CCY, Jones SE, Jerde CL, Lodge DM (2014a) Particle size distribution and optimal capture of aqueous macrobial eDNA. *Meth Ecol Evol* 5(7):676–684.

Turner CR, Miller DJ, Coyne KJ, Corush J (2014b) Improved methods for capture, extraction, and quantitative assay of environmental DNA from Asian bigheaded carp (*Hypophthalmichthys* spp.). *PLoS ONE* 9(12):e114329.

Turner CR, Uy KL, Everhart RC (2015) Fish environmental DNA is more concentrated in aquatic sediments than surface water. *Biol Conserv* 183:93–102.

Uchii K, Doi H, Minamoto T (2016) A novel environmental DNA approach to quantify the cryptic invasion of non-native genotypes. *Mol Ecol Resour* 16(2):415–422.

Ussler III W, Preston C, Tavormina P, Pargett D, Jensen S, Roman B, Marin III R, et al. (2013) Autonomous application of quantitative PCR in the deep sea: *in situ* surveys of aerobic methanotrophs using the deep-sea environmental sample processor. *Environ Sci Technol* 47(16):9339–9346.

Valentini A, Taberlet P, Miaud C, Civade R, Herder J, Thomsen PF, Bellemain E, et al. (2016) Next-generation monitoring of aquatic biodiversity using environmental DNA metabarcoding. *Mol Ecol* 25(4):929–942.

Vaudo JJ, Heithaus MR (2012) Diel and seasonal variation in the use of a nearshore sandflat by a ray community in a near pristine system. *Mar Freshwater Res* 63(11):1077–1084.

Vestheim H, Jarman SN (2008) Blocking primers to enhance PCR amplification of rare sequences in mixed samples—a case study on prey DNA in Antarctic krill stomachs. *Front Zool* 5:12.

Vogelstein B, Kinzler KW (1999) Digital PCR. *Proc Natl Acad Sci USA* 96:9236–9241.

Wangensteen OS, Palacín C, Guardiola M, Turon X (2017) Metabarcoding shallow marine hard-bottom communities: unexpected diversity and database gaps revealed by two molecular markers. *PeerJ* 5:e3429v1

Webb TJ, Mindel BL (2015) Global patterns of extinction risk in marine and non-marine systems. *Curr Biol* 25(4):506–511.

Weirbauer M, Fuks D, Peduzzi P (1993) Distribution of viruses and dissolved DNA along a coastal trophic gradient in the Northern Adriatic Sea. *Appl Environ Microbiol* 59(12):4074–4082.

Weltz K, Lyle JM, Ovenden J, Morgan JAT, Moreno DA, Semmens JM (2017) Application of environmental DNA to detect an endangered marine skate species in the wild. *PLoS ONE* 12(6):e0178124.

Wheeler QD (2004) Taxonomy: impediment or expedient? *Science* 303(5656):285.

Wilcox TM, McKelvey KS, Young MK, Jane SF, Lowe WH, Whiteley AR, Schwartz MK (2013) Robust detection of rare species using environmental DNA: the importance of primer specificity. *PLoS ONE* 8(3):e59520.

Wilcox TM, McKelvey KS, Young MK, Sepulveda AJ, Shepard BB, Jane SF, Whiteley AR, et al. (2016) Understanding environmental DNA detection probabilities: a case study using a stream-dwelling char *Salvelinus fontinalis*. *Biol Conserv* 194:209–216.

Willerslev E, Hansen AJ, Binladen J, Brand TB, Gilbert MTP, Shapiro B, Bunce M, et al. (2003) Diverse plant and animal genetic records from Holocene and Pleistocene sediments. *Science* 300(5620):791–795.

Willerslev E, Cappellini E, Boomsma W, Nielsen R, Hebsgaard MB, Brand TB, Hofreiter M, et al. (2007) Ancient biomolecules from deep ice cores reveal a forested southern Greenland. *Science* 317(5834):111–114.

Williams KE, Huyvaert KP, Piaggio AJ (2016) No filters, no fridges: a method for preservation of water samples for eDNA analysis. *BMC Res Notes* 9(1):298.

Wilson CC, Wozney KM, Smith CM (2015) Recognizing false positives: synthetic oligonucleotide controls for environmental DNA surveillance. *Meth Ecol Evol* 7(1):23–29.

Wilson IG (1997) Inhibition and facilitation of nucleic acid amplification. *Appl Environ Microbiol* 63:3741.

Yamamoto S, Masuda R, Sato Y, Sado T, Araki H, Kondoh M, Minamoto T, Miya M (2017) Environmental DNA metabarcoding reveals local fish communities in a species-rich coastal sea. *Sci Rep* 7:40368.

Yamanaka H, Motozawa H, Tsuji S, Miyazawa RC, Takahara T, Minamoto T (2016) On-site filtration of water samples for environmental DNA analysis to avoid DNA degradation during transportation. *Ecol Res* 31(6):963–967.

Zeale MR, Butlin RK, Barker GL, Lees DC, Jones G (2011) Taxon-specific PCR for DNA barcoding arthropod prey in bat faeces. *Mol Ecol Resour* 11(2):236–244.

Zhang Y, Snow DD, Parker D, Zhou Z, Li X (2013) Intracellular and extracellular antimicrobial resistance genes in the sludge of livestock waste management structures. *Environ Sci Technol* 47(18):10206–10213.

Zhou X, Li Y, Liu S, Yang Q, Su X, Zhou L, Tang M, et al. (2013) Ultra-deep sequencing enables high-fidelity recovery of biodiversity for bulk arthropod samples without PCR amplification. *GigaScience* 2(1):4.

CHAPTER 15

Shark CSI—The Application of DNA Forensics to Elasmobranch Conservation

Diego Cardeñosa
School of Marine and Atmospheric Science, Stony Brook University, Stony Brook, New York;
Fundación Colombia Azul, Bogotá DC, Colombia

Demian D. Chapman
Department of Biological Sciences, Florida International University, North Miami, Florida

CONTENTS

15.1 INTRODUCTION

Wildlife trafficking, including illegal fisheries and timber harvesting, is the fourth largest global illicit sector after narcotics, counterfeiting, and human trafficking, yet until recently law enforcement agencies have received limited support to address this issue (Agarwal, 2015). Wildlife crime has ravaged formerly recovering populations of charismatic megafauna on land such as elephants (Agarwal, 2015), rhinoceros (Milliken and Shaw, 2012), and big cats (Saif et al., 2016) and also threatens their marine counterparts, such as cetaceans (Taylor et al., 2016), tunas (Pramod et al., 2017), sharks (Chuang et al., 2016), and rays (Asis et al., 2014). There is growing public pressure to rectify this, leading to local, national, and international policy reform to combat wildlife crime with concomitant investment in monitoring and law enforcement (Agarwal, 2015). Wildlife forensics, defined as the application of scientific knowledge to legal issues involving wildlife conservation, is one particular area of increased investment. Here, we review DNA forensics, defined as the use of genetic or genomic approaches to aid in the detection and prosecution of crimes, as it applies to law enforcement involving the exploitation or trade of sharks and rays (elasmobranchs).

Elasmobranchs are increasingly subject to legal protection after severe fisheries and trade-driven population declines have occurred all over the world (Dulvy et al., 2014; Worm et al., 2013). Fisheries and trade supply the demand for a wide range of products, including meat, fins (for luxury soup dishes), liver oil (for health supplements and cosmetics), skins (for leather goods), cartilage (for health supplements), gill rakers (for traditional Chinese medicine), live individuals (for the aquarium trade), and rostra, jaws, and teeth (for curios) (Dent and Clarke, 2015; Wu, 2016). Of more than 1100 species of elasmobranchs potentially exploited for these products, about 25% are currently threatened with extinction (Dulvy et al., 2014), although a few populations, primarily in the United States, Canada, Australia, and New Zealand, are successfully being managed for sustainability (Simpfendorfer and Dulvy, 2017). Some species and populations are only modestly affected by exploitation due to factors including relatively high productivity, inaccessible distribution or habitats, low catchability in typical fishing gear, and low commercial value. In contrast, a number of exploited elasmobranchs have fared poorly under unregulated fishing due to low productivity, high catchability, or deliberate targeting due to high value, resulting in severe reductions in their abundance (Cortés et al., 2009; Dulvy et al., 2014; Worm et al., 2013). These particularly vulnerable species are increasingly targeted for legislative protection. Some elasmobranchs might not have experienced declines in a particular jurisdiction but are locally valued alive for tourism or for cultural reasons, prompting legal measures to protect them from exploitation, including the entire taxon in some cases (e.g., shark sanctuaries) (Ward-Paige, 2017).

The multispecies nature of most elasmobranchs fisheries coupled with low species selectivity of gear often used to catch them (e.g., gillnets, longlines) can result in the capture of protected species alongside legal ones, thus setting up situations whereby fishery participants continue to have the opportunity to conduct illicit trade (Pank et al., 2001). This is enabled by the fact that many elasmobranch products are difficult to identify to the species level after modest processing, which usually takes place early enough in the supply chain to evade easy detection (Pank et al., 2001). Although there are guides to identify whole sharks and rays, dressed carcasses (http://www. sharkid.com/carcassguide.html), gill rakers (https://cites.unia. es/cites/file.php/1/files/pew-manta-ray-gill-plate-id-guide.

pdf), and dried unprocessed fins (Abercrombie et al., 2013), most shark and ray products processed further than this are very difficult to identify to the species level. The challenge for law enforcement lies in resolving whether landed and traded products are derived from illicitly taken individuals. Here, we review how DNA forensics is being applied to meet this challenge on a global scale. We first outline the legal frameworks under which elasmobranchs are protected and the types of situations that then arise requiring law enforcement personnel to turn to DNA forensics to prosecute illicit activity. We then review the DNA approaches that are currently available for sharks and rays and provide an up-to-date summary table of DNA tests available by species. Case studies from the Americas to Asia illustrate how DNA forensics is being applied to law enforcement as it pertains to the elasmobranch trade, and we look at how genetic and genomic approaches will aid in DNA forensic applications for these animals.

15.2 AN OVERVIEW OF LEGAL PROTECTIONS IN PLACE FOR SHARKS AND RAYS AND ENFORCEMENT ISSUES THAT CAN BE ADDRESSED WITH DNA FORENSICS

15.2.1 Protected Species (Local, National, Regional)

A relatively large number of elasmobranch species are protected by local or national legislation in one or more jurisdictions, usually achieved through fisheries or environmental policy. For example, the smalltooth sawfish (*Pristis pectinata*) is protected in federal and state waters in the United States under the Endangered Species Act (NMFS and SSRT, 2009), and landings of a much large number of elasmobranchs are prohibited in federal waters under various fisheries regulations (e.g., Title 50, Code of Federal Regulations), including the white shark (*Carcharodon carcharias*), dusky shark (*Carcharhinus obscurus*), and thorny skate (*Amblyraja radiata*). In the United States, some species are only protected within 3 miles of shore by state legislation (e.g., FFWCC, 2018). Internationally, regional fisheries management organizations (RFMOs) that have been established to manage shared fisheries resources for sustainability include the International Commission for the Conservation of Atlantic Tuna (ICCAT), Western and Central Pacific Fisheries Commission (WCPFC), and Indian Ocean Tuna Commission (IOTC). These organizations have prohibited landings of certain species of elasmobranchs in fisheries under their jurisdiction (Techera and Klein, 2014). The responsibility of enforcement is placed on individual countries, which may or not have harmonized domestic legislation. The enforcement of any type of species protection or retention ban at the state, federal, or RFMO level requires knowing the provenance of the specimen from the perspective of jurisdiction (i.e., was it caught within local or national boundaries?) and confirmation of the species of origin. Although provenance of a specimen pertaining to jurisdiction usually cannot be determined from DNA due to the small geographic distances involved, provenance can be assessed if certain genetically identifiable populations are prohibited but others from the same species are not. Certain populations of scalloped hammerhead (*Sphyrna lewini*), for example, are listed as threatened or endangered under the U.S. Endangered Species Act (50 CFR Parts 223 and 224: Endangered and Threatened Wildlife and Plants; Threatened and Endangered Status for Distinct Population Segments of Scalloped Hammerhead Sharks; Final Rule). Scalloped hammerhead distinct population segments (DPSs) located in the eastern Pacific, eastern Atlantic, and central southwest Atlantic are all prohibited from trade or trans-shipment in the United States. Scalloped hammerhead products such as fins and meat are therefore illegal in the United States if they originate from any of these DPSs. Likewise, some RFMOs have retention bans on silky sharks (*Carcharhinus falciformis*) but others do not (Techera and Klein, 2014). In both of these examples, law enforcement personnel will have to establish the population of origin, and genetics is one of the only tools available to do this.

15.2.2 Elasmobranch Sanctuaries

Sixteen nations have established themselves as shark sanctuaries (Ward-Paige, 2017) and one, Belize, is establishing itself as a ray sanctuary. Although sometimes framed as marine protected areas, they are, in reality, simply national protected species lists that include all sharks, rays, or elasmobranchs, because no legislation actually prohibits all forms of fishing that could affect these taxa within the sanctuary. Because fishing that may take sharks or rays still occurs in most, if not all, of these nations, legal issues arise if incidentally captured individuals are kept and traded instead of being released (Ward-Paige, 2017). Law enforcement personnel, therefore, are faced with the question of whether a certain product has been derived from sharks or rays or from another taxon (e.g., teleost). This issue is not a problem for whole specimens, dressed carcasses, jaws, teeth, or whole fins but is relevant for more processed products, such as meat (especially dried salted meat) or liver oil.

15.2.3 Finning and Fin Bans

Finning is defined as the removal of the fins from a captured shark or ray and discarding the rest of the body (usually at sea), ostensibly to keep the most valuable product (fins) while not wasting vessel freezer space on the less valuable carcass. Most shark fishing nations and RFMOs have anti-finning regulations in place, stipulating that sharks must be landed with their fins naturally attached or that the fin and carcass weights on a vessel must be within a prescribed ratio (Techera and Klein, 2014). Although the former stipulation

is easy to enforce, it is possible to practice high-grade finning when landing fins and carcasses separately by keeping the fins from all or most sharks but only keeping the carcasses of species with more valuable meat, trying to stay within the legal fin-to-carcass ratio. In such cases, law enforcement personnel may need to show that the species present in a collection of landed fins do not correspond to the group of carcasses landed by the same vessel, indicative of finning. Some states in the United States have banned the sale of shark fins and shark fin soup entirely, often due to public perception that all or most shark fins are derived from finning activity. Some of these state fin bans have exemptions for certain species that are caught in local regulated fisheries; for example, the state of New York has prohibited fins except for those derived from smooth dogfish (*Mustelus canis*) and spiny dogfish (*Squalus acanthias*). In these instances, law enforcement personnel must verify that the fins are not derived from prohibited species.

15.2.4 Convention on International Trade of Endangered Species of Wild Flora and Fauna

The Convention on International Trade in Endangered Species of Wild Fauna and Flora (CITES) is an international treaty that aims to encourage sustainable trade of wildlife and eliminate trade of species with high extinction risk (Clarke, 2014; Vincent et al., 2014). CITES is a binding and multilateral environmental agreement (MEA) that operates under three core principles: (1) legality, (2) sustainability, and (3) traceability. Species are proposed for listing by at least one party (nation) at regular meetings (Conference of the Parties, or CoP) and adopted into one of two primary appendices of CITES (Appendix I or II) by vote. Species listed in Appendix I are prohibited from international trade, with rare exceptions allowed through permits issued by both importing and exporting parties. For species listed in Appendix II, CITES exporting parties are required to issue permits that certify that trade is not detrimental to the species' survival in the wild based on a non-detriment finding (NDF) and that traded specimens are legally collected and traceable through the supply chain (Vincent et al., 2014). Over 35,000 animal and plant species have been listed on CITES, and 96% of the species are listed in Appendix II. The first elasmobranch CITES listings were established in 2001 with the inclusion in Appendix II of the whale shark (*Rhincodon typus*) and the basking shark (*Cetorhinus maximus*), followed by the white shark (*Carcharodon carcharias*) in 2004. In 2007, all sawfish (family Pristidae) were listed in Appendix I, making them the only elasmobranchs completely prohibited from international trade. In recent years, an additional nine sharks and eight ray species have been listed on Appendix II, indicating growing momentum for elasmobranch trade management under the auspices of CITES. These include scalloped hammerhead shark (2013), smooth hammerhead shark

(2013), great hammerhead shark (2013), oceanic whitetip shark (2013), porbeagle shark (2013), silky shark (2016), pelagic thresher shark (2016), bigeye thresher shark (2016), common thresher shark (2016), and the family Mobulidae (2013, 2016). CITES enforcement relies entirely on the parties themselves, with exporting nations being responsible for developing systems for traceability, documenting legality, and producing NDFs for listed species. In addition, CITES parties are required to keep track of their trade records and submit an annual report to the CITES Secretariat. Importing parties are required to monitor imports and prosecute cases where listed species cross a border without the proper documentation (Vincent et al., 2014). To enforce CITES Appendix I or II, border control personnel must be able to identify the listed species of origin to determine instances of trade without the correct documentation. Failure to properly implement CITES by a party can lead to international trade sanctions (Vincent et al., 2014).

15.3   DNA FORENSICS APPROACHES

In Section 15.2, we introduced scenarios where law enforcement personnel could use DNA forensics to address illicit activities associated with the exploitation and trade of elasmobranchs. The following section describes the primary DNA forensic approaches available to apply to these scenarios.

15.3.1   Sequence-Based Species Identification

DNA sequence-based approaches (hereafter referred to as "DNA barcoding") are the most commonly used genetic identification methods for most wildlife and can be implemented in most scenarios and at any point in the trade supply chain (Cardeñosa et al., 2017; Naylor et al., 2012; Wong et al., 2009). Successful DNA barcoding depends on DNA sequence variation within species being lower than variance between species, which enables comparison between the sequence derived from a product of uncertain species identity (the "unknown") and a suite of reference sequences to enable inferences to be made about the unknown's identity. DNA barcoding requires several steps. First, genomic DNA must be isolated from a sample of the tissue collected from the specimen or product, followed by a polymerase chain reaction (PCR), an agarose gel run to check for amplification, a PCR product cleaning step, DNA sequencing, sequence editing, and analysis. Typically, the online database GenBank Basic Local Alignment Search Tool (BLAST®; https://blast.ncbi.nlm.nih.gov/Blast.cgi) and the Barcode of Life Data System (BOLD; http://www.boldsystems.org/) are used to compare query sequences to a set of reference sequences from specimens of known species identity. These comparisons result in a percentage of identity (i.e., a 100% sequence identity to a single species constitutes a perfect match) that allows the investigator to identify the unknown sample to the lowest taxonomic category possible (e.g., genus and/or species). If ambiguities occur—for example, the unknown is not a 100% match to a single species in the database or multiple possible species are a match or have the same percentage of identity to the unknown sample—then character-based identification keys can be used that focus on a subset of base positions known to be diagnostic for certain species after extensive, geographically broad sampling efforts (Cardeñosa et al., 2017; Fields et al., 2015; Wong et al., 2009).

All sequence-based approaches require the same basic set of laboratory and analytical steps, but they usually differ in PCR protocol and the type of primers used during the PCR (Cardeñosa et al., 2017; Fields et al., 2015; Naylor et al., 2012; Ward et al., 2005). These primers are usually universal, which means they produce sequences on a wide range of species (e.g., all chondrichthyan species). Elasmobranch DNA barcoding has been successful with the mitochondrial cytochrome *c* oxidase subunit 1 (COI) (~650 bp of sequence) (Cardeñosa et al., 2017; Ward et al., 2005) and the NADH2 gene (~1040 bp) (Naylor et al., 2012). Relatively few sharks and rays cannot be unambiguously identified with either of these two loci, but there are exceptions, as some species pairs are recently diverged or have experienced recent gene flow, such as the dusky and Galapagos shark (*Carcharhinus obscurus* and *C. galapagensis*), blacktip and Australian blacktip shark (*C. limbatus* and *C. tilstoni*), and sandbar and bignose shark (*C. plumbeus* and *C. altimus*) (Naylor et al., 2012; Ward et al., 2005). DNA barcoding approaches that rely on several hundreds of base pairs for identification are most appropriate in the early stages of the supply chain where products are predominantly unprocessed and PCR is likely to yield full-length fragments for good-quality sequencing. However, some products are exposed to more intensive processing practices (e.g., cooking, salting, bleaching) that degrade DNA to the point that DNA barcoding methods requiring large sequences are ineffective because only shorter, degraded DNA remains within the product (Cardeñosa et al., 2017; Fields et al., 2015).

To address this issue, several minibarcode assays (PCR and sequencing of ~150 to 200 bp) have been developed to identify the species or species group (complex, genus) of origin in highly processed products such as processed shark fins, shark fin soup, and cosmetics containing shark liver oil (Cardeñosa et al., 2017; Fields et al., 2015). These minibarcode PCR approaches target a small fragment of the COI gene where most of the interspecific sequence differences occur, allowing for the differentiation of even closely related species with the shorter sequence. However, the use of minibarcodes tends to exacerbate difficulties in identifying unknowns that are from species pairs that have radiated very recently or have experienced recent gene flow. Cardeñosa et al. (2017) presented a multiplex (>2 PCR primers) approach to efficiently obtain two minibarcode COI sequences to help ameliorate this issue.

15.3.2 PCR-Based Species Identification

Polymerase chain reaction-based species identification, also known as species-specific PCR, are PCR tests explicitly designed to identified a target species or group of species (Abercrombie et al., 2005; Caballero et al., 2011; Chapman et al., 2003; Magnussen et al., 2007; Pinhal et al., 2012; Shivji et al., 2002). This is accomplished by designing species-specific PCR primers that bind ("anneal") only to a matching or near-matching target sequence within a particular region; for elasmobranchs, the COI or internal transcribed spacer 2 (ITS2) have been used for species-specific PCR. The primer is designed to match a unique sequence in the target species, thus annealing during PCR and allowing amplification to occur. If the target species' DNA is not present, the primer fails to anneal and PCR amplification will not occur. Usually a species-specific primer is nested within two universal primers, so that the universal primers will amplify all elasmobranch DNA but the species-specific primer will only anneal to the target species and produce the smaller internal fragment. Thus, a positive result for a target species would be the large universal fragment and a smaller species-specific fragment of a diagnostic size, which can be visually identified on an agarose gel (Figure 15.1). A negative result for the target species would consist of just the larger universal fragment, thus providing an internal positive control for the reaction itself (i.e., to avoid false-negative results from reaction failure). Multiple species-specific primers can usually be nested within one set of universal primers and all run together, with each primer being placed in different positions to generate size differences between fragments to enable species identification (e.g., Shivji et al., 2002).

Polymerase chain reaction-based approaches for species identification present several advantages compared to DNA barcoding; for example, they involve fewer steps because they skip the product cleaning step, sequencing, and sequence editing and analysis. Identification is accomplished by analyzing the agarose gel, which requires less time and is

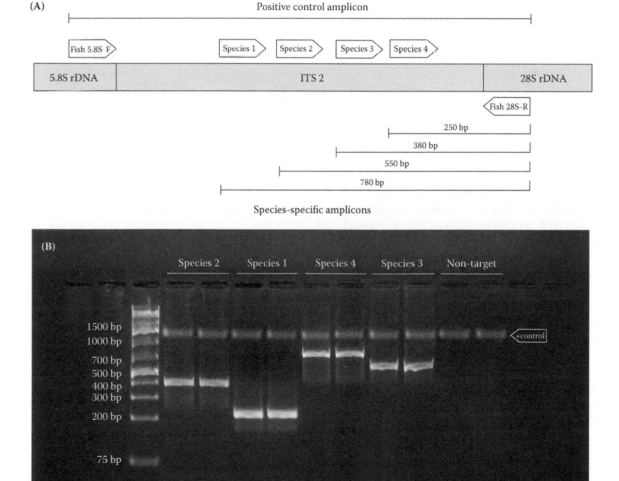

Figure 15.1 (A) Schematic representation of a typical PCR-based species identification technique with relative annealing sites and orientation of each primer. (B) Example of a gel-based identification for several CITES-listed species.

Table 15.1 Published Species-Specific Primers for Shark Species Identification Using PCR-Based Identification Approaches

Species	Primer Sequence (5′–3′)	Amplicon Size (bp)	Refs.
Sphyrna mokarran[a]	AGCAAAGAGCGTGGCTGGGGTTTCGA	782	Abercrombie et al. (2005)
Sphyrna lewini[a]	GGTAAAGGATCCGCTTTGCTGGA	445	Abercrombie et al. (2005)
Sphyrna zygaena[a]	TGAGTGCTGTGAGGGCACGTGGCCT	249	Abercrombie et al. (2005)
Isurus paucus	CCTCAACGACACCCAACGCGTTC	418	Shivji et al. (2002)
Isurus oxyrinchus	AGGTGCCTGTAGTGCTGGTAGACACA	771	Shivji et al. (2002)
Lamna nasus[a]	GTCGTCGGCGCCAGCCTTCTAAC	554	Shivji et al. (2002)
Prionace glauca	AGAAGTGGAGCGACTGTCTTCGCC	929	Shivji et al. (2002)
Carcharhinus falciformis[a]	ACCGTGTGGGCCAGGGTC	1085	Shivji et al. (2002)
Cetorhinus maximus[a]	TCTCGGCCTCCGGGCGAACGAATGAGA	1100	Magnussen et al. (2007)
Cetorhinus maximus[a]	AAGATGCGGCACGCTGTTGGGCACGC	900	Magnussen et al. (2007)
Alopias pelagicus[a]	TAGAAGTGATCCTGGCTGTCCTAA	119	Caballero et al. (2011)
Alopias superciliosus[a]	TAGGAGTGATCCGGGCTGGCCTAA	119	Caballero et al. (2011)
Alopias vulpinus[a]	TAGAAGTGATCCGGGTTGTCCTAA	119	Caballero et al. (2011)
Carcharodon carcharias[a]	GCTGGAGTTCATTCTCCGTGCTG	580	Chapman et al. (2003)
Carcharodon carcharias[a]	AGTCAGAACTAGTATGTTGGCTACAAGAAT	511	Chapman et al. (2003)
Carcharhinus plumbeus	AAAGTGGAGCGACTGTCTGCAGGTC	1018	Pank et al. (2001)
Rhizoprionodon acutus	TTAACGTTCTGTGCGTGTCGAGT	230	Pinhal et al. (2012)
Rhizoprionodon porosus	GCGAGGCACACCTCGGCAC	420	Pinhal et al. (2012)
Rhizoprionodon longurio	GACTTGCTCTGTCCTTGAGCCC	560	Pinhal et al. (2012)
Rhizoprionodon terraenovae	TGTGAATAGGGGCAGCCGACA	720	Pinhal et al. (2012)
Rhizoprionodon oligolinx	TACCGGGAGAGCTCGGAAAACGT	850	Pinhal et al. (2012)
Rhizoprionodon taylori	AACGGTTCGGGTGCTCCGGCA	1150	Pinhal et al. (2012)
Rhizoprionodon lalandii	GGCACGTAGGCACCGCCCGCTAT	1300	Pinhal et al. (2012)

[a] Species listed on CITES Appendix II.

lower in cost in terms of reagents and labor. These species-specific assays have been designed for a number of common and vulnerable species present in large-scale fisheries, including nearly all of the CITES-listed sharks (Table 15.1). One limiting factor is that they require intensive research and development to produce before being adopted. For example, when designing a PCR test one must design the putative species-specific primer, test it against a robust number of individuals of the target species from as much of the species range as possible, confirm that it does not amplify any non-target species, and then integrate it into a multiplex PCR with several other primers.

The other key limitation is that an analyst must pick the correct multiplex to identify an unknown species *a priori* or run the risk of having to test an unknown sample repeatedly with different multiplexes. This rapidly escalates the costs and time spent trying to identify unknowns to the point that they may equal or exceed those of DNA barcoding and may result in ambiguity, as the majority of the species present in the international shark trade, for example, do not have species-specific primers developed for them (Fields et al., 2017). Therefore, PCR-based approaches are most useful when the number of potential species-of-origin is limited, such as when a known group of species occurs in a specific fishery (Caballero et al., 2011) or when there is *a priori* evidence that an unknown sample comes from a certain species (e.g., due

to morphology). It is also useful when there is a narrow suite of species of interest and the others do not need to be identified, such as when you simply need to identify whether or not a product is any of the CITES-listed shark species and do not need to identify it to the species level if it is not one of them.

15.3.3 Population of Origin Assessed Using DNA

One of the most prevalent data gaps in fisheries management is the lack of traceability of elasmobranch products. Advances in fishing technologies, boat capacity, and onboard refrigeration allow fishing boats to go farther and for longer periods of time. Thus, elasmobranch landings at commercial fishing ports or in seafood markets do not necessarily come from a local population or stock. Genetic stock identification (GSI) methods are an approach to assess the stock composition of a fishery or market that could have multiple sources. The global elasmobranch trade focuses mainly on two major commodities (i.e., fins and meat), and GSI of trade hubs, such as Hong Kong for fins and Brazil for meat (Dent and Clarke, 2015), could play an essential role in assessing population-specific exploitation levels (Chapman et al., 2010). These methods could also be applied under particular law enforcement scenarios where certain populations of a species might be protected while others are not (see Sections 15.2.1 and 15.2.2).

To apply these GSI methods the species genetic population structure must be assessed and the geographical resolution determined by the geographic scale of population genetic differentiation. Mitochondrial DNA regions (e.g., mitochondrial control region, or mtCR) that are often used for elasmobranch population genetic studies can exhibit regional population structure in sharks and some rays, thus allowing differentiating populations geographically separated by hundreds to thousands of kilometers (Cardeñosa et al., 2014; Chapman et al., 2010, 2015; Clarke et al., 2015), and they allow the natal source population of a product or individual to be assessed. An example of the protocols and application of these methods is presented in Chapman et al. (2010), who reconstructed the natal source population of origin of 62 scalloped hammerhead shark fins sampled from the Hong Kong shark fin market using GSI methods (Baker et al., 2000; Bowen et al., 2006; Laurent et al., 1998; Waldman et al., 1996). After extracting DNA and running PCRs to amplify a partial sequence of the mtCR, the sequence obtained for each fin (haplotype) was matched against published haplotypes from the western Atlantic, eastern Atlantic, and Indo-Pacific regions (Duncan et al., 2006; Ovenden et al., 2009). Overall, 65% of the fins came from the Indo-Pacific, 21% from the western Atlantic, and 14% from the eastern Atlantic (Chapman et al., 2010). Moreover, the results provided evidence that regional-scale mitochondrial stock structure can be used to accurately reconstruct the contribution of different regional stocks to the international shark fin trade using mixed stock analysis (Chapman et al., 2010).

15.4 CASE STUDIES

Considerable progress is being made on elasmobranch DNA forensics research and development. Yet, despite the widespread availability and diminishing costs of these approaches and the increased international obligation to enforce policies supporting elasmobranch conservation (e.g., CITES listings, RFMO regulations), we have yet to see the widespread adoption of DNA forensics in elasmobranch management. Here, we present case studies of a few bright spots where DNA forensics has been adopted in this context.

15.4.1 The Americas

15.4.1.1 United States

The United States has some of the best-managed shark fisheries in the world, employing assessment-based catch limits with time-area closures and a suite of especially vulnerable species that are prohibited from landings (see https://www.fisheries.noaa.gov/welcome). After years of poor regulation of fisheries resulted in large declines in several species, the adoption of effective management measures resulted in the

beginning stages of recovery for some of them (Peterson et al., 2017). DNA forensics has proven to be a crucial tool in helping to enforce regulations. For the most part, these tools have been used to enforce prohibited species regulations in the federally regulated fisheries, although in some cases they have been used to detect prohibited species for legal cases initially based on fishers being charged with other, more obvious violations. For example, 263 fins seized by law enforcement personnel in seven different finning cases involving Atlantic-based fishing vessels and one New York–based seafood dealer suspected of multiple permit violations in the early 2000s were identified using species-specific PCR tests (Henning, 2005). These analyses revealed that 40% of the fins originated from three prohibited shark species: dusky shark (*Carcharhinus obscurus*), night shark (*C. signatus*), and bignose shark (*C. altimus*) (Henning, 2005). Magnussen (2006) and Magnussen et al. (2007) followed up by reporting that the same dealer also possessed 2 fin sets of the prohibited basking shark (*Cetorhinus maximus*) and 61 fin sets of the prohibited sand tiger shark (*Carcharias taurus*), as determined through the use of PCR tests developed for these species. Arguably, the highest profile application of DNA forensics involved this same seafood dealer, who was found in possession of 21 fin sets from prohibited great white sharks. Shivji et al. (2005) used the PCR test of Chapman et al. (2003) for this species to verify species of origin. Law enforcement personnel investigating the case had *a priori* evidence that these fin sets were derived from great whites because the bag they were contained in was labeled *blanco*, which is Spanish for "white." Today, several federal and academic laboratories are involved with research, development, and application of DNA forensics to elasmobranch management in the United States (see, for example, https://www.nwfsc.noaa.gov/research/divisions/cb/genetics/forensics.cfm and http://cnso.nova.edu/ghri/index.html, https://www.peclabfiu.com).

Together with federal management of sharks, some states also have additional regulations ranging from prohibited species lists for state waters to bans on the trade of fins within the state. New York is an example of the latter, whereby trade in all shark fins was prohibited in 2013 within the state, with the exception of fins from smooth dogfish and spiny dogfish landed by federally managed fisheries. Prior to the adoption of this regulation, seafood dealers in New York City's Chinatown district routinely offered processed shark fins imported from Hong Kong and China for sale. Shortly after the adoption of this regulation, law enforcement personnel detained two shipments of processed shark fins at John F. Kennedy International Airport that were being imported from Hong Kong by local seafood dealers. DNA minibarcoding (using the assay of Fields et al., 2015) was used to verify that these shipments were not derived from the exempt species, instead originating from a range of requiem sharks (genus *Carcharhinus*). Moreover, several fins were derived from great and scalloped hammerhead sharks, which are listed in CITES Appendix II and had been imported without the requisite permits. In 2015, these efforts led to the first successful prosecution of the New York state fin ban (DEC, 2015).

15.4.1.2 Belize

Belize is a relatively small Central American nation bordered by Mexico to the north and Guatemala to the west and south. It has jurisdiction over the world's second largest barrier reef, which supports two of the nation's major industries: tourism and fishing (Cisneros-Montemayor et al., 2013). Sharks constitute a relatively minor part of Belize's total capture production, but shark fishing employs around 75 licensed fishermen who primarily export product (meat and fins) to Guatemala and Mexico (NSWG, 2018). There is evidence that the fishery has depleted some species (Bond et al., 2012), and it is probable that at least two species listed on CITES (scalloped hammerhead and great hammerhead) are part of the catch. Since implementation of CITES Appendix II listing of these hammerheads took place in September 2014 and Belize has yet to produce a NDF, any export of these species from Belize would be illegal. To assess the species composition of the shark fishery, the Belize Department of Fisheries (BFD) and Florida International University (FIU) are collaborating on a project that combines fishery-dependent sampling and DNA forensics. Beginning in January 2017, fishermen renewing their annual shark fishing licensing were informed of a requirement that they needed to collect anal fins, a fin they do not sell, from all landed sharks and make them available to BFD officers conducting sporadic inspections of shark landing sites. The 748 anal fins collected between January and June 2017 were sorted into putative species categories based on morphology and were then genetically tested for species identity using full COI barcoding (Ward et al., 2005). Anal fin morphological types had high concordance with genetically identified species; highly falcate-shaped fins proved to be scalloped or great hammerheads (Figure 15.2). These species collectively made up 5% of all fins submitted by fishermen. Given that there is a very limited local market for the fins and meat of sharks, these findings suggest that illegal export of these CITES-listed species was probably occurring. In October 2017, BFD and FIU held an NDF workshop for CITES-listed sharks involving all major stakeholders and including the fishing industry, and they are currently collaboratively developing a program to track the catch and exports on a species-specific basis based on anal fin sampling and DNA forensics.

15.4.1.3 Brazil

Brazil is one of the largest elasmobranch fishing nations in the world and has also emerged as one of the top importers and consumers of elasmobranch meat (Dent and Clarke,

Figure 15.2 Anal fin from (top) great hammerhead (*Sphyrna mokarran*) and (bottom) Caribbean reef shark (*Carcharhinus perezi*), both taken in the artisanal shark fishery of Belize and identified using DNA barcoding. Species-diagnostic morphological characters were concordant with DNA results for these and other species, enabling a robust monitoring program to be developed.

2015). A number of academic laboratories in Brazil have developed DNA forensics approaches for elasmobranchs (e.g., Pinhal et al., 2012), yet we know relatively little about the species composition of the domestic meat trade in this hub when compared to the species composition of major fin trade hubs in Asia (e.g., Clarke et al., 2006; Fields et al., 2017). It is likely that there are a number of potential avenues for illicit activity in Brazil involving the elasmobranch trade, including importation or exportation of CITES-listed species without permits and landing of species prohibited by ICCAT (Tolotti et al., 2015).

One especially unfortunate case study of the application of DNA forensics in Brazil was a survey of "shark" meat being sold at two fish markets in northern Brazil (Palmeira et al., 2013). Of the 44 meat samples identified using DNA barcoding, 24 originated from a prohibited species: the critically endangered largetooth sawfish (*Pristis pristis*). Similarly, a study of "shark meat" being sold in other parts of Brazil used PCR to show that the protected guitarfish (*Rhinobatos horkelii*) was still being illegally sold (Alexandre de-Franco et al., 2012). The detection of local consumption of meat from these critically endangered, prohibited ray species highlights a need for regular DNA-based monitoring of the meat trade in Brazil and other large meat trade hubs to detect illicit activity. These studies have revealed that the mere prohibition of these species in Brazil without socialization and enforcement is an ineffective strategy for preventing them from being landed and traded.

15.4.2 Asia

15.4.2.1 Philippines

The Philippines lie in the Coral Triangle region of the Indo-Pacific and as such are a hotspot of marine biodiversity. Ecotourism is a major source of revenue, including underwater experiences with manta rays (*Mobula birostris* and *M. alfredi*) (O'Malley et al., 2013). Many rays, including related manta and devil rays (family Mobulidae), are also exploited for meat and gill rakers (Asis et al., 2014; O'Malley et al., 2016). To better protect mantas for the tourism industry, Fisheries Administrative Order No. 193 was established to prohibit all capture and trade of *M. birostris*. In a joint operation between the Bureau of Fisheries and Aquatic Resources, the Bureau of Customs, and the Genetic Fingerprinting Laboratory of the National Fisheries Research and Development Institute, 11 samples of dried shark or ray products suspected to originate from illegal fishing were identified using DNA barcoding. Four (36%) were from mantas, indicating that illegal capture and trade were still occurring, possibly because the mantas were still being captured in a legal fishery for devil rays (Asis et al., 2014).

15.4.2.2 Taiwan

Taiwan is one of the world's largest shark fishing nations and a major exporter of meat and dried fins to global trade hubs for these products, as well as being a domestic consumer of both (Dent and Clarke, 2015). DNA barcoding has recently been used to examine the species composition of the domestic meat and fin trade, port landings, and nine cases of fins detained by border control personnel from 2013 to 2015 that were not declared by importers (a total of 113 fins) (Chuang et al., 2016; Liu et al., 2017). Only one species is domestically protected in Taiwan (whale shark), and neither study found evidence of this species in trade. However, of the detained fins, 8%, were from oceanic whitetips and 14.1% were from thresher sharks (family Alopiidae), which might represent RFMO violations given the retention bans for some populations of these species that existed at the time (Chuang et al., 2016). Unfortunately, the DNA barcoding conducted did not have the spatial resolution to determine the exact geographic origin of these fins. Harmonization between RFMO prohibitions and the domestic landing and possession of these species would help ameliorate this issue.

15.4.2.3 Hong Kong

The People's Republic of China—Hong Kong Special Administrative Region (Hong Kong) is one of the world's largest shark fin trade hubs (Dent and Clarke, 2015) and thus plays a key role in the implementation of CITES regulations.

Until recently, very little was known of the species composition of the contemporary Hong Kong fin trade. From 1999 to 2001, species-specific PCR was combined with importer auction records to assess the contribution of 14 common trade categories that were concordant with particular shark species or species groups; these species, several of which are now listed on CITES, collectively made up ~46% of all of the auctioned fins (Clarke et al., 2006). After a long period with no further species composition data from Hong Kong, a total of 4800 fin trimmings were sampled from randomly selected retail vendors in Hong Kong from 2014 to 2015 (Fields et al., 2017). These trimmings are a byproduct of processed fins for the retail market and are sold inexpensively for use in soup or soup broth. A total of 76 chondrichthyan species were identified, and a Bayesian model was used to assess the relative proportions of some of the more common species in this retail market (Fields et al., 2017). Results revealed that the market is focused on a small subset of globally distributed species, most of which were also important from 1999 to 2001 (Fields et al., 2017). CITES-listed species, such as silky, scalloped, and smooth hammerhead sharks, were some of the top species in the market during this sampling, which largely took place prior to implementation of these regulations in Hong Kong in 2014. The primary insight from this study as it relates to CITES is that the second, fourth, and fifth most common species in the fin trade are now listed, meaning that the import volumes of these species and concomitant monitoring and enforcement obligations are likely to be substantial for trade hubs such as Hong Kong. Since implementing CITES, the Hong Kong Customs Department and Agriculture Fisheries and Conservation Department (AFCD) have seized more than 1.3 tons of shark fins from listed species, primarily using visual identification followed up by DNA barcoding conducted by Hong Kong University (K.H. Shea, Bloom Association, pers. comm.).

15.5 FUTURE DIRECTIONS

A relatively high proportion of elasmobranchs species have experienced population declines mainly due to overfishing and the high demand for products such as shark fins and gill rakers (Dulvy et al., 2014; Worm et al., 2013; Wu, 2016). This has triggered adoption of new policies aimed at reversing these declines (see Section 15.2). Despite this, many fishing and trading nations lack the political will, capacity, or resources to enforce these policies. DNA forensics approaches described in this chapter and new protocols in development can play a major role in the enforcement of legislation at all governance levels. As our cases studies show, these approaches are being used in some areas but mainly by academic laboratories producing snapshots of the species composition of trade or in sporadic government law enforcement operations. In order to successfully transfer and implement these technologies on a much larger scale, including routine inspection of products in trade, collaborative initiatives must be developed among the various stakeholders (e.g., governments, non-governmental organizations, industry, academic units, funding bodies) to ensure the necessary investments in capacity and establishment of secure financing to support these efforts. Although this is especially true in the developing world, there is a widespread perception that DNA forensics is cost prohibitive for routine screening of products even in developed nations and rapidly growing economies such as China. Engagement between nations successfully using these approaches, such as the United States, and others that need to use them would be a substantial step forward to seeing broader uptake.

Several recent analytical advances offer promising avenues for improving the efficiency and resolution of elasmobranch DNA forensics. Genomics is a new approach from the molecular biology field that focuses on the sequencing and analysis of entire genomes instead of a specific set of genes, enabled by next-generation sequencing approaches that allow massive parallel sequencing of many short DNA fragments (see Chapter 13 in this volume). Genomic approaches can be used in DNA barcoding to resolve between species pairs that cannot be resolved with single loci by providing sequence data from many loci at once, potentially revealing many single nucleotide polymorphisms (SNPs) useful for species diagnosis (Liu et al., 2017). Genomic techniques are also increasing our resolution of population genetic differentiation by revealing vast numbers of SNPs that can be used to detect genetic population structure. There is emerging evidence that elasmobranch populations can be structured on much finer geographic scales than evident at single mitochondrial loci when using genome-wide SNPs (Pazmiño et al., 2017). For example, Pazmiño et al. (2017) found SNPs that distinguished two distinct populations of Galapagos sharks (*Carcharhinus galapagensis*) within the Galapagos Islands that were identical in their mitochondrial sequences. Fine-scale structure studies based on genomic analysis may therefore enable GSI to trace some elasmobranch products to country or even island of origin to assess if the specimen was caught within a specific country, shark sanctuary, or marine protected area. While DNA barcoding of a single locus costs from $5 to $10 per sample, next-generation sequencing currently costs closer to $100 per sample, although costs are rapidly diminishing (Liu et al., 2017). Multilocus SNP assays to identify species or populations can be developed using DNA microarrays that are even more efficient than gel- or DNA sequencing-based approaches (Wenne et al., 2016).

Another application of genomic approaches is environmental DNA (eDNA), which uses water samples, taxon-specific primers, and next-generation sequencing (Shokralla et al., 2012) to detect the presence/absence of target species (Ficetola et al., 2008; Takahara et al., 2012; see also Chapter 14 in this volume). In law enforcement scenarios, it is possible that this tool could be used to detect prohibited species that have already been offloaded by a fishing or cargo vessel by sampling the meltwater of freezers or swabs from the cargo hold.

Advances in efficiency and portability of DNA forensic methods are likely to increase uptake. Real-time PCR (rtPCR) techniques incorporate fluorescent dyes into PCR that can be detected by modified thermal cyclers as amplification takes place, eliminating the need for gels and enabling more rapid identification of species through species-specific PCR. We have recently designed new and adapted published ITS2 species-specific shark primers (see Section 15.3.2) to work in a rtPCR format for implementation in major exporting and importing nations (D. Cardeñosa et al., unpublished data). This technique could detect CITES-listed species before they leave the exporting nation or when they enter the importing nation (Figure 15.3). rtPCR approaches can be used outside of the laboratory and are even being adapted for devices with an extremely small footprint to facilitate rapid DNA forensics at the port of entry or in the marketplace. Our reagent costs per sample for a simple SYBR® Green chemistry rtPCR assay are $0.85, and the technique can identify whether a sample is one of 9 CITES-listed sharks in about 5 hours for 95 samples in a field setting (e.g., a port).

15.6 CONCLUSION

DNA forensics will continue to be an important tool to help enforce policies aimed at conserving elasmobranchs and could be deployed at all points of the supply chain and at all governance levels. Substantial research and development have been done for sharks and to a lesser extent rays, and robust online reference sequence databases and protocols are available to support DNA barcoding and PCR tests already published for several commercially important or protected species. Although a much greater research and development effort is needed, population genetic data can be used to look at source population of origin. Despite a few bright spots in fishing and importing nations, uptake and widespread use of these approaches have not occurred. Even where these techniques have been employed they have been used primarily in a reactionary manner (e.g., to confirm identity of products detained based on visual identification or for other infractions) or to produce snapshots of the species composition of the trade. As yet, no country routinely screens elasmobranch products at any point of the supply chain to detect infractions and provide a deterrent to illegal actors. The low risk of detection and modest penalties in some jurisdictions promote illegal trade, as was commonly seen in our case studies, particularly in the domestic meat trade examples. The recent listing of a number of elasmobranch species on CITES now provides an international forum to promote these approaches and offers high potential for capacity building and funding to support much broader uptake. New DNA forensic approaches on the horizon, including real-time PCR and next-generation

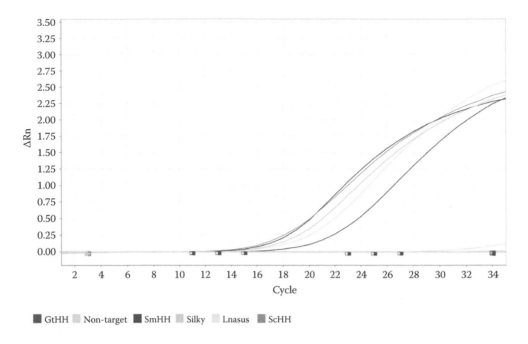

Figure 15.3 Amplification plot showing the successful amplification and detection of hammerhead CITES-listed species using newly developed real time PCR techniques.

sequencing, offer greater resolution and are continually becoming faster, cheaper, and more portable, which should facilitate much greater application of DNA forensics to elasmobranch conservation in the near future.

ACKNOWLEDGMENTS

This work was supported by Paul G. Allen Philanthropies as part of the DNA Tool Kit Project and by the Roe Foundation (DDC). Thanks to COLCIENCIAS–Colombia for supporting DC during his PhD studies.

REFERENCES

Abercrombie DL, Clarke SC, Shivji MS (2005) Global-scale genetic identification of hammerhead sharks: application to assessment of the international fin trade and law enforcement. *Conserv Genet* 6:775–788.

Abercrombie DL, Chapman DD, Gulak SJB, Carlson JK (2013) *Visual Identification of Fins from Common Elasmobranch in the Northwest Atlantic Ocean*, NOAA Technical Memorandum NMFS-SEFSC-643. National Marine Fisheries Service, National Oceanic and Atmospheric Administration, Panama City, FL.

Agarwal P (2015) A global challenge: the illegal wildlife trade chain. *J Commerce Trade* 2:7–14.

Alexandre de-Franco B, Fernandes Mendonça F, Oliveira C, Foresti F (2012) Illegal trade of the guitarfish *Rhinobatos horkelii* on the coasts of central and southern Brazil: genetic identification to aid conservation. *Aquat Conserv Mar Freshw Ecosyst* 22:272–276.

Asis AMJM, Lacsamana JKM, Santos MD (2014) Illegal trade of regulated and protected aquatic species in the Philippines detected by DNA barcoding. *Mitochondrial DNA A DNA Mapp Seq Anal* 27(1):659–666.

Baker CS, Lento GM, Cipriano F, Palumbi SR (2000) Predicted decline of protected whales based on molecular genetic monitoring of Japanese and Korean markets. *Proc R Soc Lond B Biol Sci* 267:1191–1199.

Bond ME, Babcock EA, Pikitch EK, Abercrombie DL, Norlan FL, Chapman DD (2012) Reef sharks exhibit site-fidelity and higher relative abundance in marine reserves on the Mesoamerican Barrier Reef. *PLoS ONE* 7:e32983

Bowen BW, Grant WS, Hillis-Starr Z, Shaver DJ, Bjorndal KA, Bolten AB, Bass AL (2006) Mixed-stock analysis reveals the migrations of juvenile hawksbill turtles (*Eretmochelys imbricata*) in the Caribbean Sea. *Mol Ecol* 16:49–60.

Caballero S, Cardeñosa D, Soler G, Hyde J (2011) Application of multiplex PCR approaches for shark molecular identification: feasibility and applications for fisheries management and conservation in the Eastern Tropical Pacific. *Mol Ecol Resour* 12:233–237.

Cardeñosa D, Hyde J, Caballero S (2014) Genetic diversity and population structure of the pelagic thresher shark (*Alopias pelagicus*) in the Pacific Ocean: evidence for two evolutionarily significant units. *PLoS ONE* 9:e110193.

Cardeñosa D, Fields A, Abercrombie D, Feldheim K, Shea SKH, Chapman DD (2017) A multiplex PCR mini-barcode assay to identify processed shark products in the global trade. *PLoS ONE* 12:e0185368.

Chapman DD, Abercrombie DL, Douady CJ, Pikitch EK, Stanhopen MJ, Shivji MS (2003) A streamlined, bi-organelle, multiplex PCR approach to species identification: application to global conservation and trade monitoring of the great white shark, *Carcharodon carcharias*. *Conserv Genet* 4:415–425.

Chapman DD, Pinhal D, Shivji MS (2010) Tracking the fin trade: genetic stock identification in western Atlantic scalloped hammerhead sharks *Sphyrna lewini*. *Endang Species Res* 9:221–228.

Chapman DD, Feldheim K, Papastamatiou YP, Hueter RE (2015) There and back again: a review of residency and return migrations in sharks, with implications for population structure and management. *Annu Rev Marine Sci* 7:547–570.

Chuang PS, Hung TC, Chang HA, Huang CK, Shiao JC (2016) The species and origin of shark fins in Taiwan's fishing ports, markets, and customs detention: a DNA barcoding analysis. *PLoS ONE* 11:e0147290.

Cisneros-Montemayor AM, Barnes-Mauthe M, Al-Abdulrazzak D, Navarro-Holm E, Sumaila UR (2013) Global economic value of shark ecotourism: implications for conservation. *Oryx* 47:381–388.

Clarke CR, Karl SA, Horn RL, Bernard AM, Lea JS, Hazin FH, Prodöhl PA, Shivji MS (2015) Global mitochondrial DNA phylogeography and population structure of the silky shark, *Carcharhinus falciformis*. *Mar Biol* 162:945–955.

Clarke S (2014) Re-examining the shark trade as a tool for conservation. *SPC Fish Newsl* 145:1–8.

Clarke SC, Magnussen JE, Abercrombie DL, McAllister MK, Shivji MS (2006) Identification of shark species composition and proportion in the Hong Kong shark fin market based on molecular genetics and trade records. *Conserv Biol* 20:201–211.

Cortés E, Arocha F, Beerkircher L, Carvalho F, Domingo A, Heupel M, Holtzhausen H, Santos MN, Ribera M, Simpfendorfer C (2009) Ecological risk assessment of pelagic sharks caught in Atlantic pelagic longline fisheries. *Aquat Living Resour* 23:25–34.

DEC (2015) DEC: First Successful Shark Fin Case Results in Guilty Felony Plea: First Case to Conclude Under NY's New Shark Fin Ban (press release). Department of Environmental Conservation, Albany, NY.

Dent F, Clarke S (2015) *State of the Global Market for Shark Products*, FAO Fisheries and Aquaculture Technical Paper 590. Food and Agriculture Organization of the United Nations, Rome.

Dulvy NK, Fowler SL, Musick JA, Cavanagh RD, Kyne PM, Harrison LR, Carlson JK, et al. (2014) Extinction risk and conservation of the world's sharks and rays. *eLife* 3:e00590.

Duncan KM, Martin AP, Bowen BW, De Couet HG (2006) Global phylogeography of the scalloped hammerhead shark (*Sphyrna lewini*). *Mol Ecol* 15:2239–2251.

FFWCC (2018) *Sharks*. Florida Fish and Wildlife Conservation Commission, Tallahassee (http://myfwc.com/fishing/saltwater/recreational/sharks/).

Ficetola GF, Miaud C, Pompanon F, Taberlet P (2008) Species detection using environmental DNA from water samples. *Biol Lett* 4:423–425.

Fields AT, Abercrombie DL, Eng R, Feldheim K, Chapman DD (2015) A novel mini-DNA barcoding assay to identify processed fins from internationally protected shark species. *PLoS ONE* 10:e0114844.

Fields AT, Fischer GA, Shea SKH, Zhang H, Abercrombie DL, Feldheim KA, Babcock EA, Chapman DD (2017) Species composition of the international shark fin trade assessed through a retail-market survey in Hong Kong. *Conserv Biol* 1–47.

Henning, M (2005) Highly Streamlined PCR-Based Genetic Identification of Carcharhinid Sharks (Family Carcharhinidae) for Use in Wildlife Forensics, Trade Monitoring, and Delineation of Species Distributions, master's thesis, Nova Southeastern University, Fort Lauderdale, FL.

Laurent L, Casale P, Bradai MN, Godley BJ, Gerosa G, Broderick AC, Schroth W, et al. (1998) Molecular resolution of marine turtle stock composition in fishery bycatch: a case study in the Mediterranean. *Mol Ecol* 7:1529–1542.

Liu J, Jiang J, Song S, Tornabene L, Chabarria R, Naylor GJP, Li C (2017) Multilocus DNA barcoding—species identification with multilocus data. *Sci Rep* 7:16601.

Liu SYV, Chan CLC, Lin O, Hu CS, Chen CA (2013) DNA barcoding of shark meats identify species composition and CITES-listed species from the markets in Taiwan. *PLoS ONE* 8:e79373.

Magnussen JE. (2006) DNA Diagnostics for Internationally Protected and Commercially Traded Shark Species, master's thesis, Nova Southeastern University, Ft. Lauderdale, FL (https://nsuworks.nova.edu/occ_stuetd/271).

Magnussen JE, Pikitch EK, Clarke SC, Nicholson C, Hoelzel AR, Shivji MS (2007) Genetic tracking of basking shark products in international trade. *Anim Conserv* 10:199–207.

Milliken T, Shaw J (2012) *The South Africa–Vietnam Rhino Horn Trade Nexus: A Deadly Combination of Institutional Lapses, Corrupt Wildlife Industry Professionals and Asian Crime Syndicates*. TRAFFIC, Johannesburg, South Africa.

Naylor G, Caira JN, Jensen K, Rosana KAM, White WT, Last PR (2012) A DNA sequence-based approach to the identification of shark and ray species and its implications for global elasmobranch diversity and parasitology. *Bull Am Mus Nat Hist* 367:1–262.

NMFS, SSRT (2009) *Smalltooth Sawfish Recovery Plan (Pristis pectinata)*. Smalltooth Sawfish Recovery Team, National Marine Fisheries Service, National Oceanic and Atmospheric Administration, Silver Spring, MD.

NSWG (2018) *National Plan of Action for the Conservation and Management of Sharks (NPOA–Sharks)*. National Shark Working Group, Belize City, Belize.

O'Malley MP, Lee-Brooks K, Medd HB (2013) The global economic impact of manta ray watching tourism. *PLoS ONE* 8:e65051.

O'Malley MP, Townsend KA, Hilton P, Heinrichs S, Stewart JD (2016) Characterization of the trade in manta and devil ray gill plates in China and South-east Asia through trader surveys. *Aquat Conserv Mar Freshw Ecosyst* 27:394–413.

Ovenden JR, Kashiwagi T, Broderick D, Giles J, Salini J (2009) The extent of population genetic subdivision differs among four co-distributed shark species in the Indo-Australian archipelago. *BMC Evol Biol* 9:40.

Palmeira MCA, Rodrigues-Filho LFS, Sales JBL, Vallinoto M, Schneider H, Sampaio I (2013) Commercialization of a critically endangered species (largetooth sawfish, *Pristis perotteti*) in fish markets of northern Brazil: authenticity by DNA analysis. *Food Control* 34:249–252.

Pank M, Stanhope M, Natanson L, Kohler N, Shivji M (2001) Rapid and simultaneous identification of body parts from the morphologically similar sharks *Carcharhinus obscurus* and *Carcharhinus plumbeus* (Carcharhinidae) using multiplex PCR. *Mar Biotechnol* 3:231–240.

Pazmiño DA, Maes GE, Simpfendorfer CA, Salinas-de-León P, van Herwerden L (2017) Genome-wide SNPs reveal low effective population size within confined management units of the highly vagile Galapagos shark (*Carcharhinus galapagensis*). *Conserv Genet* 18:1151–1163.

Peterson CD, Belcher CN, Bethea DM, Driggers WB, III Frazier BS, Latour RJ (2017) Preliminary recovery of coastal sharks in the south-east United States. *Fish Fish* 18(5):845–859.

Pinhal D, Shivji MS, Nachtigall PG, Chapman DD, Martins C (2012) A streamlined DNA tool for global identification of heavily exploited coastal shark species (genus *Rhizoprionodon*). *PLoS ONE* 7:e34797.

Pramod G, Pitcher TJ, Mantha G (2017) Estimates of illegal and unreported seafood imports to Japan. *Mar Policy* 84:42–51.

Saif S, MacMillan DC, Rahman THM (2016) Who is killing the tiger *Panthera tigris* and why? *Oryx* 52(1):46–54.

Shivji M, Clarke S, Pank M, Natanson L, Kohler N, Stanhope M (2002) Genetic identification of pelagic shark body parts for conservation and trade monitoring. *Conserv Biol* 16:1036–1047.

Shokralla S, Spall JL, Gibson JF, Hajibabaei M (2012) Next-generation sequencing technologies for environmental DNA research. *Mol Ecol* 21:1794–1805.

Simpfendorfer CA, Dulvy NK (2017) Bright spots of sustainable shark fishing. *Curr Biol* 27:R97–R98.

Takahara T, Minamoto T, Yamanaka H, Doi H, Kawabata Z (2012) Estimation of fish biomass using environmental DNA. *PLoS ONE* 7:e35868.

Taylor B.L, Rojas-Bracho L, Moore J, Jaramillo-Legorreta A, Ver Hoef JM, Cardenas-Hinojosa G, Nieto-Garcia E, et al. (2016) Extinction is imminent for Mexico's endemic porpoise unless fishery bycatch is eliminated. *Conserv Lett* 10:588–595.

Techera EJ, Klein N (2014) *Sharks: Conservation, Governance and Management*. Routledge, Abingdon, UK.

Tolotti MT, Bach P, Hazin F, Travassos P, Dagorn L (2015) Vulnerability of the oceanic whitetip shark to pelagic longline fisheries. *PLoS ONE* 10:e0141396.

Vincent ACJ, Sadovy de Mitcheson YJ, Fowler SL, Lieberman S (2014) The role of CITES in the conservation of marine fishes subject to international trade. *Fish Fish* 15:563–592.

Waldman JR, Hart JT, Wirgin II (1996) Stock composition of the New York Bight Atlantic sturgeon fishery based on analysis of mitochondrial DNA. *Trans Am Fish Soc* 125:364–371.

Ward RD, Zemlak TS, In es BH, Last PR, Hebert PDN (2005) DNA barcoding Australia's fish species. *Proc R Soc Lond B Biol Sci* 360:1847–1857.

Ward-Paige CA (2017) A global overview of shark sanctuary regulations and their impact on shark fisheries *Mar Policy* 82:87–97.

Wenne R, Drywa A, Kent M, Sundsaasen KK, Lien S (2016) SNP arrays for species identification in salmonids. In Bourlat S (ed) *Marine Genomics. Methods and Protocols*. Humana Press, New York, pp 97–111.

Wong EHK, Shivji MS, Hanner RH (2009) Identifying sharks with DNA barcodes: assessing the utility of a nucleotide diagnostic approach. *Mol Ecol Resour* 9:243–256.

Worm B, Davis B, Kettemer L, Ward-Paige CA, Chapman D, Heithaus MR, Kessel ST, Gruber SH (2013) Global catches, exploitation rates, and rebuilding options for sharks. *Mar Policy* 40:194–204.

Wu J (2016) *Shark Fin and Mobulid Ray Gill Plate Trade*. TRAFFIC, Johannesburg, South Africa.

CHAPTER 16

Citizen Science in Shark and Ray Research and Conservation
Strengths, Opportunities, Considerations, and Pitfalls

Andrew Chin
Centre for Sustainable Tropical Fisheries and Aquaculture, James Cook University, Townsville, Queensland, Australia

Gretta Pecl
Institute for Marine and Antarctic Studies and Centre for Marine Socioecology,
University of Tasmania, Hobart, Tasmania, Australia

CONTENTS

16.1 AN INTRODUCTION TO CITIZEN SCIENCE

16.1.1 Citizen Science Present and Past

Citizen science programs are growing around the world in number, diversity, and prominence, and they are arguably now accepted as a mainstream scientific methodology (Dickinson et al., 2012; Silvertown, 2009). Modern citizen science programs range from observation-based programs, such as recording local sightings of specific species or phenomena, to global efforts to collect "big data," such as the National Aeronautics and Space Administration (NASA) program Global Learning and Observations to Benefit the Environment (GLOBE), where citizen scientists from around the world use a smartphone app to photograph clouds and record data on mosquitoes. In the fields of biology and ecology, citizen scientists are collecting data on changes in species distributions, pollution, invasive species, threatened species, disease, phenology, biodiversity, habitats, and landscapes and are making tangible conservation contributions

(Bonney et al., 2009, 2014; Dickinson et al., 2010; Edgar et al., 2017; Robinson et al., 2015; Silvertown, 2009). In situations where citizen science participants are numerous and widely distributed, these initiatives can help professional scientists greatly expand their spatial and temporal sampling capability, and thus the scope and scale of research questions they can address. In terms of biodiversity monitoring alone, citizen science programs have been estimated to include 1.5 million volunteers that contribute over $2.5 billion worth of in-kind contributions to biodiversity science every year (Theobald et al., 2015).

The reliance of citizen science programs on (often) amateur observers instead of trained scientists to collect data has led to concerns about the accuracy and rigor of the data collected. Indeed, data quality can vary depending on factors such as observer training, experience, and supervision (Buesching et al., 2014; Nerbonne and Nelson, 2008). However, with proper design, implementation, and analysis, citizen science projects can provide reliable and useful data for biological and ecological research and can result in effective conservation outcomes (Bird et al., 2014; Bonney et al., 2014; Pecl et al., 2015).

Moreover, it should be remembered that prior to the emergence of professional, institution-based scientists, science was generally conducted by unpaid, non-professional citizen scientists under the patronage of wealthy benefactors or organizations (Miller-Rushing et al., 2012). Charles Darwin was an unpaid naturalist when he did his iconic work during the 1831 to 1836 voyage of the HMS *Beagle*, and in fact he was attending Cambridge to be trained in the clergy. At Cambridge he was mentored by John Stevens Menslow, who was a "parson-naturalist," a member of the clergy who saw the study of natural history as an extension of their religious profession. These networks of faith-based naturalists were citizen scientists and enabled early ecologists such as Carl Linnaeus to collect specimens "across the known world" (Miller-Rushing et al., 2012). Furthermore, citizen science has existed for thousands of years, and in some cases these records are still providing valuable data for current research. Records of locust outbreaks from China dating back over 1900 years are helping contemporary researchers understand climate change effects on locust outbreaks (Tian et al., 2011). Given the enduring impact of Darwin's work on modern biology and ecology and the continuing value of historical natural history records, it is clear that institutionalized scientific training is not a prerequisite for robust data collection and scientific contribution.

Although citizen science has demonstrated its potential to contribute to contemporary research, the citizen science approach is not a panacea, and researchers interested in exploring citizen science must carefully consider the costs and limitations. It is important to recognize that citizen science is not a data-for-free proposition and that there are transaction costs and investments that scientists must make for citizen science initiatives to deliver on their potential and to provide reciprocity for participants. Furthermore, in some cases, citizen science is inappropriate due to logistical, scientific, and ethical issues. As with any scientific approach or method, citizen science must be fit for purpose.

This chapter provides an overview of citizen science approaches and their potential to contribute to the scientific study, conservation, and management of sharks, rays, and chimaeras. Citizen science on sharks and rays is relatively simple and limited compared to other research fields, but it is growing in the area of chondrichthyan research and has great potential to contribute to our future knowledge and understanding of this important ocean taxon. This chapter aims to

- Introduce readers to a basic understanding of citizen science, what it is, how it works, and the opportunities it provides to researchers.
- Provide real-world examples of citizen science to illustrate its potential as a tool for shark and ray research and conservation.
- Give insights into different citizen science approaches and the potential limitations, challenges, and pitfalls researchers need to consider.
- Address best practices for citizen science projects for sharks and rays including project design and implementation.

16.1.2 Describing Citizen Science

Numerous definitions of *citizen science* can be found in the literature (e.g., McKinley et al., 2017; Miller-Rushing et al., 2012; Riesch and Potter, 2014; Silvertown, 2009), and many of these focus on attributes related to the training and professional status of the observers or whether the observers are being paid. The definition of citizen science is contested, perhaps because so many types of participants and projects could be identified as being citizen science. Indeed, because of the many approaches to citizen science (Bonney et al., 2009; Pecl et al., 2015), crafting an all-encompassing definition is difficult. Rather than a strict definition, we propose that citizen science projects have the following traits:

1. The project involves community-based members, usually acting as unpaid volunteers, who collect data for a specific research question or other specified purpose.
2. The participants usually do not have formal academic qualifications in the specific field of research.
3. The project usually (but not always) involves collaborations with professional scientists.
4. There is some level of formal engagement and outreach to the community.
5. Data collection is usually organized to some degree; that is, observers collect specific types of data in specific ways, record specific parameters, and submit data to identified persons in a prescribed manner.

Independent observers collecting and opportunistically providing data to support research may also be considered citizen scientists; however, they may not be working on a specific citizen science project. Other citizen science contributions may include data or samples that might be provided to scientists as part of a legal requirement or regulatory obligation. For example, commercial fishers may be required to record catch and landings data in catch-and-effort logbooks. Although the fishers are not volunteers, they are members of a community that is systematically collecting data for a specific purpose. Given that unsustainable fishing is one of the key threats facing sharks and rays around the world (Dulvy et al., 2014), this form of citizen science is essential for ongoing conservation and management. Finally, the growth of social media and user-driven web content has made photographs, videos, and other types of data freely available to the scientific community. Although these data were not intended for scientific use, this informal citizen science has helped alert shark scientists about potential new species and range expansions.

16.1.3 Potential Benefits of Citizen Science

A benefit of citizen science is its ability to achieve multiple objectives, including benefits for the participants, broader society, and the research community. Citizen science can provide researchers with data across very large spatial scales, and sometimes temporal scales, that may not be possible through research sampling programs due to costs or logistical limitations. Importantly, widely dispersed citizen scientists are able to collect data over large spatial scales *simultaneously*, meaning that phenomena can be tracked over space through time. Examples of these applications include animal sightings across migration pathways, outbreaks of insects, and the timing of biological events such as flowering. Spatial coverage can even reach global scales where programs access an international network of observers such as for highly migratory birds (Greenwood, 2007) and sharks such as whale sharks (*Rhincodon typus*) (Sequeira et al., 2013). However, greater statistical power to identify these patterns may require participants to also consistently and systematically record zero data, which can be a challenge in some situations (see Section 16.3.3).

Collaborations with volunteer observers can also facilitate or enable research and monitoring programs to continue for long time periods. Since Christmas Day in 1900, citizen scientists in North America have been counting birds between December 14 and January 5 through the National Audubon Society Christmas Bird Count. With 117 years of data, this dataset is probably the longest running citizen science project in existence (Tulloch et al., 2013) and provides invaluable continuous, time-series data that would be cost prohibitive to collect and institutionally difficult to maintain through researcher-based monitoring programs.

The scale of spatial and temporal analyses enabled by citizen science can be especially helpful in research on uncommon and rare species that are difficult (and costly) to sample (Dickinson et al., 2012; Edgar et al., 2017). Local knowledge can help field researchers identify candidate sampling locations, but citizen scientists who are regularly sampling these locations can also collect data on relative abundance and seasonal patterns and potentially collect samples when researchers are absent. For example, citizen scientists have documented new records of rare river sharks (*Glyphis garricki* and *G. glyphis*) in Papua New Guinea and provided the first-ever photographs of adult individuals for researchers and taxonomists (White et al., 2015). Through extended networks, citizen scientists can provide information on a species' wider range, occurrence, and connectivity (Sequeira et al., 2013), and they have contributed knowledge on range extensions of other shark and ray species (Chin, 2014; Meekan et al., 2016) and are helping to populate and validate species checklists in remote areas (Hylton et al., 2017).

Finally, citizen science can also be an excellent means of engaging the community in research and conservation, building social capital between researchers and the public and encouraging the community to engage in conservation (Forrester et al., 2017; McKinley et al., 2017). Well-designed and implemented citizen science programs can democratize science and demystify the research process, helping participants to better understand science and enabling them to participate in the scientific processes of data collection and analysis (Baker, 2016; Resnik et al., 2015). Participants may also gain better knowledge and understanding of the subject area through access to scientific resources and information and immersion in research, and the investigative process may also prompt participants to ask new questions and expand their own knowledge. Such programs can also connect people to the environment and empower and motivate the community to take conservation action (Bonney et al., 2009; Dickinson et al., 2012), such as, for example, advocating mammal conservation (Forrester et al., 2017) or becoming involved in butterfly conservation (Lewandowski and Oberhauser, 2017). These studies report that after participating in projects volunteers are more likely to take action (Lewandowski and Oberhauser, 2017) and to share information in their social circles (Forrester et al., 2017), and they also identified the importance of volunteers feeling connected to each other. This diffusion into the wider community through a participant's own social networks can be very important in engaging some communities such as fishers who place significant value in information sourced from their community and peers (Li, 2016). Effective participation and positive relationships can also build trust between researchers and the community (Martin et al., 2016a), increasing the legitimacy of the project results which in turn may lead to increased acceptance of management actions.

The data collection capabilities of citizen science approaches coupled with community engagement, empowerment, and participation mean that citizen science can be a powerful conservation and management tool. This is perhaps best illustrated in the field of bird conservation and management where volunteer ornithologists have had a significant direct influence in bird research and conservation (Greenwood, 2007). In Australia, data collected by BirdLife Australia's community monitoring programs directly contribute to the Australian Government's periodic State of the Environment reports and the development of policy such as the Action Plan for Australian Birds. Indeed, cryptic citizen science may be making valuable global contributions to assessing the effects of climate change on migratory birds and contributing to the policy addressing global threats (Cooper et al., 2014). Citizen science in the marine environment is also increasing. Projects such as the Reef Life Survey (http://reeflifesurvey.com/) are beginning to provide wide-scale data for reporting on the state of Australia's marine environment, and the Eye on the Reef program (www.gbrmpa.gov.au/managing-the-reef/how-the-reefs-managed/eye-on-the-reef) is an important component of monitoring coral bleaching on the Great Barrier Reef.

16.2 CITIZEN SCIENCE IN ACTION

16.2.1 Scope, Scale, and Diversity of Citizen Science Projects

Citizen science projects are incredibly diverse, and researchers interested in integrating citizen science approaches into their work have many examples from which to choose. Citizen science projects vary in scale and complexity from small, locally focused, species-specific programs, such as

monitoring turtle bycatch in a single fishery (Peckham et al., 2007), to global-scale citizen science monitoring networks and databases used to monitor migrating birds (Cooper et al., 2014). However, citizen science approaches can be categorized according to the type of involvement community members have with the project. Bonney et al. (2009) identified three main types of citizen science projects:

1. *Contributory*—Projects are usually designed by professional scientists, and community members contribute data and/or observations, including analyses of historical citizen science data from journals or other records (e.g., Chin, 2014; Tian et al., 2011).
2. *Collaborative*—Projects are generally designed together by scientists and community members, community members have input into project design and sampling methodology, and community members may help to collect and/or analyze data, review project findings, interpret results, and disseminate project findings (e.g., Whitelaw et al., 2003)
3. *Co-created*—Projects are created and designed by scientists and community members working closely together through all stages of the project. Community members have input into the research question and may be the originators of projects. Such research projects can be conceived and implemented entirely by community members (Miller-Rushing et al., 2012). This type of project may be a form of participatory action research (see Chapter 17 in this volume).

The specific characteristics and differences between these types of citizen science projects are shown in Table 16.1.

The differing levels of engagement and participation between these project types translate to differences in transaction costs, investment, and power dynamics. Contributory projects tend to have lower transaction and investment costs. The researcher selects the question, scope, and scale of the

Table 16.1 Models for Public Participation in Scientific Research

Project Process	Contributory Projects	Collaborative Projects	Co-Created Projects
Define research questions and hypotheses.	—	—	√
Organize funding and marshal resources.	—	√[a]	√
Engage community and stakeholders.	—	√[a]	√
Develop design sampling and quality assurance and quality control protocols.	—	√[a]	√
Collect data/samples.	√	√	√
Analyze data/samples.	√[a]	√[a]	√[a]
Interpret data and draw conclusions.	—	√[a]	√
Disseminate findings to community and stakeholders.	√[a]	√[a]	√
Discuss new questions, further actions.	—	—	√

Source: Adapted from Bonney, R. et al., *Public Participation in Scientific Research: Defining the Field and Assessing Its Potential for Informal Science Education*, a CAISE Inquiry Group Report, Center for Advancement of Informal Science Education (CAISE), Washington, DC, 2009.

Note: Different types of citizen science projects have different levels of participation by community members, from collaborative projects where participants mainly collect data to co-created projects that closely involve the community in every aspect of the project including research question formulation.

[a] Indicates that community members are sometimes involved in this process.

project and then creates the opportunity for community members to participate. The level of investment and ownership by community members may be reduced, as their role is mainly to contribute data when they feel inclined so their expectation of benefits and feedback may also be reduced. In contrast, co-created projects may require greater time and effort by all parties to attend meetings, prepare funding proposals, develop data-sharing agreements, and engage external stakeholders. This increased investment will probably increase participant expectations; thus, significant resources may be required to maintain communication and engagement, jointly develop and implement the project, analyze data, disseminate findings, and, if necessary, resolve conflicts. Researchers should also recognize that in co-created projects intellectual input and ownership of the project and data are shared between participants. In successful projects, this power sharing and co-ownership can deliver powerful benefits gained from much deeper and more meaningful engagement and relationship building between researchers and the community. However, this deeper engagement and shared ownership can introduce additional ethical issues (see Section 16.3.2), and if these relationships fail the lack of progress and loss of trust may derail the entire project, leaving both the researcher and community with little benefit for their efforts. Importantly, such failures can lead to general distrust of research that can compromise future efforts by other researchers to work with these communities. As such, when deciding on a citizen science approach, researchers should carefully consider whether they have the time and resources needed to maintain the necessary level of investment and engagement.

16.2.2 Citizen Science with Sharks and Rays

There are relatively few marine-focused citizen science projects compared to terrestrial based-projects (Cigliano et al., 2015; Theobald et al., 2015). This disparity is probably due to the increased cost and difficulty in working in the marine environment, including additional costs of boating, specialized equipment, and all the associated training and safety considerations of marine operations, as well as the potential impact of adverse weather conditions (Cigliano et al., 2015). Additionally, some marine species can be highly mobile, naturally less abundant, patchily distributed, relatively shy, and difficult to accurately identify, and they may occur in inaccessible or undesirable habitats (e.g., deep ocean, muddy estuaries). Many of these traits apply to sharks and rays, and, indeed, relatively few citizen science projects are focused on sharks and rays. Of these projects, most citizen science projects involving sharks and rays tend to be contributory projects, although collaborative and co-created projects may be emerging.

Some of the best known shark and ray citizen science projects include photo-identification projects of iconic species such as the broadnose sevengill shark (*Notorynchus*

cepedianus) (http://sevengillsharksightings.org/), whale shark (*Rhincodon typus*) (www.whaleshark.org.au/education/citizen-science/), and manta rays (*Mobula birostris*; *M. alfredi*) (www.mantamatcher.org/). These projects tend to be contributory projects where community members are invited to send in photographs to the research team. In some cases, researchers have collaborations with tourist operators whom they may work alongside. Photo-identification is an established technique for identifying individual sharks and rays (Marshall and Pierce, 2012; see also Chapter 12 in this volume). The development of photo-imaging and image-processing software has enabled researchers to analyze photographs contributed by scuba-diving and snorkeling citizen scientists from around the world to study the movement and population dynamics of white sharks (Gubili et al., 2009), whale sharks (Araujo et al., 2017; Arzoumanian et al., 2005; Davies et al., 2012; Sequeira et al., 2013), and manta rays (Jaine et al., 2012; Town et al., 2013). However, photo-identification projects also include less iconic species such as the whitetip reef shark (*Triaenodon obesus*) (Whitney et al., 2012), sicklefin lemon shark (*Negaprion acutidens*) (Buray et al., 2009), and blacktip reef shark (*Carcharhinus melanopterus*) (Mourier et al., 2012; Porcher, 2005).

Iconic species provide the best known examples of shark and ray citizen science, but other innovative projects focus on different species and allow community members to participate without the need to snorkel or scuba dive. Since 2003, the Great Egg Case Hunt (www.sharktrust.org/en/GEH_the_project) run by the Shark Trust in the United Kingdom has mobilized scuba divers, snorkelers, and beachgoers to record the number of spent egg cases they observe to document the relative abundance and distribution of egg-laying sharks and skates. Observers can participate at varying levels—from recording opportunistic finds during beach visits to performing standardized beach surveys according to a formalized sampling protocol. These data can help to identify potential nursery areas for further study and inform potential conservation and management. The project website provides a wide range of training tools and information and, importantly, has an interactive map that displays project results. Participants are also reminded to record zero data, such as visits to beaches where no egg cases were recorded. The data are also separated into verified data, which are records where a photograph or specimen has been checked by a Shark Trust member, or unverified data for records without this quality control check. The project has been very successful and involves over 1000 participants who have collected over 100,000 records, and it has provided quantitative data on skate and catshark egg cases and taxonomy (Gordon et al., 2016). In 2013, the project expanded into the United States and has since expanded into the Netherlands and Portugal and is now receiving records from around the world. This growth into an international network of observers may in time provide the "big data" necessary for global monitoring, but this would not be possible without the large network of volunteers.

Citizen science approaches can also be very useful for documenting the presence or absence of cryptic or rare species and in defining species' ranges and distributions, as the expanded sampling effort across a larger area increases the probability of observations (Dickinson et al., 2012). For example, citizen scientists have provided valuable data about the endangered smalltooth sawfish (*Pristis pectinata*) in the southern United States. Public sightings programs for sawfish have collected observations and photographs that have helped map the species' historic and contemporary range, revealed differences in depth and habitat use between size classes, identified key sawfish habitats (Waters et al., 2014; Wiley and Simpfendorfer, 2010), and identified ongoing threats (Seitz and Poulakis, 2006). Importantly, these public sightings were instrumental in identifying critical habitats that have since been protected to promote sawfish recovery (Norton et al., 2012), and the associated education and outreach programs have increased public understanding of the sawfish's plight and informed fishers of safe handling and release techniques.

In Australia, the Great Porcupine Ray Hunt used diver photographs to record the distribution of the uncommon porcupine ray (*Urogymnus asperrimus*). Citizen science contributions expanded its known southern range, in addition to providing information about its depth range, habitat use, and significance to indigenous communities (see Case Study 1. The Great Porcupine Ray Hunt). Similar photographic evidence approaches are being used in the Shark Search Indo-Pacific project (www.sharksearch-indopacific.org) to create robust checklists of shark and ray diversity in the Indo-Pacific (Hylton et al., 2017). In these cases, photographic evidence is a key component of the project's quality control processes; however, unsolicited photographs collected by community-based observers can also prove valuable in documenting range expansions and rare species. For example, in 2015, scuba divers from the Great Barrier Reef sent scientists photographs of interesting rafting behavior observed in stingrays. Although rafting behavior in rays is not uncommon, the species photographed was a small-eyed stingray (*Dasyatis microps*) and provided the first validated records of that species' occurrence on the east coast of Australia (Meekan et al., 2016).

Citizen science is also being used to collect information on global patterns in shark and ray occurrence. The eShark project (http://eoceans.org/?page_id=424) has collected over 33,000 records of sharks and rays from divers, fishers, snorkelers, boaters, and paddlers around the world. The public is invited to submit opportunistic sightings via an online form that collects information about where sharks and rays are observed, and the aggregated data are used to examine widescale spatial and temporal patterns in shark occurrence (e.g., Ward-Paige et al., 2010b). The Shark Base project (www.shark-base.org) also solicits public sightings of sharks and rays but expands data collection to include records of sharks seen in media and on the Internet. Citizen scientists

may also count sharks and rays as part of broader underwater visual survey projects, such as the Reef Environmental Education Foundation (www.reef.org), which closely collaborates with eShark, or the Reef Life Survey (www.reeflifesurvey.com). These projects generally record data on a large number of taxa and are not specifically designed for surveying sharks and rays. Although these projects can amass a substantial amount of information, sharks and rays can be difficult to monitor through these generalized survey methods as they are typically in much lower abundance than teleost fishes, are highly mobile (including at depths and distances beyond diver visual range), and can be difficult to accurately identity underwater (see Section 16.3.3). However, this does not mean that citizen scientists cannot count sharks. In cases where dives occur at known aggregation sites where sharks are more numerous and potentially acclimatized to diver presence and divers are specifically tasked with counting sharks, comparative studies have shown that citizen scientists can collect high-quality abundance data over time (Vianna et al., 2014).

16.2.3 Technological Advances and Citizen Science

Ongoing improvements in technology have greatly increased the potential for citizen scientists to collect and store large amounts of high-quality data. The increased availability of high-speed Internet has made it possible to share and disseminate large amounts of data over vast networks. The introduction of digital cameras and storage has enabled photographers to take a large number of high-resolution images, to check these images in the field, and then, perhaps most importantly for research applications, easily disseminate these images through email or the Internet. Digital cameras

Figure 16.1 Tailor-made apps for mobile devices, such as the Redmap app for logging climate change range-related species extensions, allow citizen scientists to collect high-quality data while inbuilt data verification and validation processes reduce data errors and data management costs.

have become less expensive and more reliable, and Global Positioning System (GPS)-integrated models enable each image to be geotagged, providing accurate date, time, and location data for each image. For marine researchers, the increasing availability of underwater digital still and video cameras provides citizen scientists with the means to collect and share data on marine species and phenomena. More recently, the introduction of small, rugged action cameras, such as the GoPro®, have allowed divers and snorkelers to collect increasingly large amounts of imagery that can be used raw or processed further to provide novel data (Raoult et al., 2016).

Another major technological advance is the development of the smartphone and other mobile devices. These devices can have considerable computing power, are compact and mobile, and can link to the Internet through cellular communications networks. Furthermore, almost all smartphones have integrated GPS and high-quality cameras. This technology enables a citizen scientist to collect, process, and submit real-time data in a field environment through user-friendly apps. Indeed, numerous free apps are already available for citizen scientists projects, encouraging the community to collect and submit data on a wide range of topics from data on marine debris (NOAA/University of Georgia Marine Debris Tracker) to tracking global patterns in human sexual behavior (Kinsey Reporter). These tailor-made tools allow data to be collected in prescribed formats, with in-built quality control processes that ensure data are entered correctly (Figure 16.1). They can also provide instant support to the citizen scientist in the field by including reference materials such as species identification guides, sampling protocols, and FAQs, as well as means to contact the research team to ask questions and seek advice and participate in other ways (Figure 16.1). Furthermore, when data have been collected, they can be uploaded automatically via cellular networks, reducing time lags between data collecting and data entry and reducing data entry costs. Apps that require users to register also help research teams monitor community engagement and participation and can make it easier to communicate directly with participants. Importantly, intuitive, icon- and image-driven apps can overcome literacy barriers, allowing citizen scientists in developing countries to collect valuable data. For example, the Hapi Fish Hapi Pipol project in the Solomon Islands uses picture- and icon-driven mobile apps that enable community fish monitors to collect fisheries data (www.fisheries.gov.sb/hapi-fis). Given the resource limitations, logistical challenges and conservation pressures in many developing nations, these enabling technologies may be extremely valuable.

It is important to recognize that potential advantages from technological advances extend beyond data collection tools. Image processing and automated pattern recognition software enable researchers to analyze large image libraries to identify individual whale sharks and manta rays by their unique markings, significantly reducing analysis times

(Arzoumanian et al., 2005; Town et al., 2013) and enabling global-scale analysis of movement patterns (Sequeira et al., 2013; see also Chapter 12 in this volume). Data filters can screen for unusual values during data input, alerting users to potentially erroneous data and thus improving data quality (Bonter and Cooper, 2012). Technology is fast reaching the stage where a citizen scientist can take a photograph of a specimen in the field with a mobile device that then processes and analyzes the image to identify the specimen. The device will then send the photograph, tagged with date, time, and location data, directly to the project database which can then trigger such actions as sending an SMS alert to the researcher or, when an urgent response is required (such as with a pollution incident or a new record of an invasive species), directly to managing authorities (Newman et al., 2012). There are also large scientific data repositories (e.g., www.datadryad.org) that will curate data and even integrate data and manuscript submissions for some journals. Furthermore, some online data platforms such as iNaturalist (www.iNaturalist.org) or the Atlas of Living Australia (www.ala.org.au) are specifically designed to accommodate observation-type data commonly collected by citizen science programs. Storing project data within existing data repositories makes project data easier to locate and retrieve and consolidates data, which in turn enables "big data" analyses (Mackechnie et al., 2011; Newman et al., 2012).

Social media also provides unprecedented capabilities to interact in real time with a large audience, enabling researchers to promote projects, advertise opportunities to participate, and quickly disseminate results and project information (Newman et al., 2012). Projects can quickly establish an online presence using pages such as Facebook,® and social media platforms such as Twitter® and Instagram® allow researchers to capitalize on social networks to motivate the community to participate, in addition to engaging the wider, non-participating community in project news and activities. The careful use of hashtags to uniquely identify a project can provide a useful means of archiving project engagement and activities to help the community understand the project and view the results, in addition to helping the research team report on project outcomes. In the case of image-based social media such as Flickr™ and Instagram®, hashtags may even be used to search for particular evidence of species or phenomena among the platform's entire user base and, when such evidence is located, facilitate direct communication with the observer.

The Internet and social media also provide platforms to connect individual observers into global observer networks. For example, the eBird initiative provides an Internet-based platform that enables bird watchers around the world to log, share, and discuss sightings and data and to manage data in a unified database (Sullivan et al., 2009). With millions of data records submitted every year (Bonney et al., 2014), the eBird system enables researchers and conservationists to investigate regional and global scale trends and phenomena that would be impossible to detect otherwise. Web-based projects such as iNaturalist combine mapping, social networking, and data tools into websites where participants can upload data, ask questions, use mapping and query tools to search for sightings, and set up individual profiles. Using the site's social tools, iNaturalist users can identify species for each other, locate other participants nearby, and create local networks and projects. iNaturalist also integrates social networking tools that provide opportunities for participants to interact directly with each other and exchange information. To encourage and maintain participant engagement, some websites build in recognition and badging schemes that reward participants for their efforts by displaying awards on users' profile pages to acknowledge their contributions, expertise, and assistance given to others (Dickinson et al., 2012). Technology is improving at a rapid rate, and modern communications networks allow researchers to directly connect with communities at unprecedented scales and citizen scientists to collect increasingly accurate and detailed information. Researchers interested in harnessing citizen science should carefully consider how to use technology to their advantage in designing, implementing, and evaluating their projects.

16.3 MAKING CITIZEN SCIENCE WORK

16.3.1 Special Considerations for Using Citizen Science

Partnerships between scientists and the community can strengthen research, help address real-world problems, and enhance community awareness of science and conservation issues (see Section 16.1.3). Nevertheless, researchers using citizen science in their work will encounter several challenges. Citizen science has historically met resistance from the research community over doubts about data quality and rigor in analysis and interpretations. In some instances, these concerns remain valid. One of the main concerns involves observer quality—that is, the ability of untrained (usually volunteer) observers to collect accurate data (Dickinson et al., 2010) and the subsequent quality of the data produced (Riesch and Potter, 2014). Numerous studies conducted on citizen science projects have found that multiple factors can affect data quality, including group size, monitoring experience, participant age, training, aptitude, education, and level of involvement by scientists in the data collection (Buesching et al., 2014; Delaney et al., 2008; Nerbonne and Nelson, 2008), but it is also unlikely that there are factors that can be universally applied to predict observer quality (Crall et al., 2011). However, numerous studies also show that with proper project design, participant training, and support, citizen scientists can collect data that are comparable to those collected by professional scientists (Crall et al., 2011; Darwall and Dulvy, 1996; Vianna et al., 2014). Furthermore,

data quality and interpretability can be improved by inserting technological innovations into the project (see Section 16.2.3), and various statistical approaches can account for commonly encountered biases (Bird et al., 2014).

Other challenges and limitations also must be acknowledged. Community members may be more likely to participate in projects that are considered to be easy, fun, and social (Dickinson et al., 2012). This may be relatively simple to accomplish in contributory projects where participants have flexibility over where and when they sample, but this can lead to sampling bias. Participants might collect observations only when the weather is ideal or at easily accessible locations, and they might only record observations they think are interesting. This can lead to under- or overreporting, and results may reflect observer effort rather than actual trends and patterns (Dickinson et al., 2010). Although statistical methods may be able to account for many forms of bias (Bird et al., 2014), in some cases the research team may have to design and implement a standardized sampling protocol for community participants to follow or involve prescriptive sampling at specific times. For example, monitoring certain events (e.g. floods, dust, noise) may require volunteers to collect data in adverse conditions, which could reduce participation rates and even involve health and safety concerns. Furthermore, although data collection can be perceived as fun, citizen scientists may find data entry and management, analysis, reporting, and project administration much less enjoyable and potentially of less interest. Researchers need to carefully consider participant motivations and capabilities to design programs and assign tasks that balance ease, flexibility, and capability with the level of effort, rigor, and detail required. Getting this balance wrong can lead to poor participation, volunteer dissatisfaction, and poor quality data (Martin et al., 2016a,b).

The marine environment introduces additional specific challenges for citizen science due to logistical constraints, environmental conditions, and safety considerations (Cigliano et al., 2015; Theobald et al., 2015). The safety and well-being of volunteers should be a primary consideration. One benefit of contributory projects is that participants contribute information collected during the conduct of their own personal activities, reducing the researchers' responsibility for their health and safety. However, if citizen scientists are collecting data according to a sampling protocol provided by the research team, then supervising researchers need to ensure that participants have the appropriate training, equipment, support, and supervision required to enable them to safely apply the sampling protocol. Additionally, citizen scientists may also have to be made aware of, or even formally trained and authorized in, research ethics and specific protocols if the research project is taking place under specific research permit conditions or ethics approvals. Researchers should also carefully consider liability and insurance issues should a volunteer be injured while conducting a survey or collecting data. Citizen science in the aquatic realm may also exacerbate some of the sampling biases evident in terrestrial studies. Distance and weather conditions may limit the ability to safely access sampling sites and bias observations to nearshore, sheltered sites. In particular, divers and snorkelers may favor sites with good visibility and abundant marine life. They may also be restricted to visiting sites frequented by commercial tour operators who may also select sites based on ease of access, safety, and attractiveness to tourists. Similarly, many citizen scientists may not be able to access sites farther offshore, collect samples from deep water, or sample at night. Community members may also be more interested in sampling iconic or interesting species, which are usually marine megafauna (marine turtles, seabirds, marine mammals, large fishes), leading to underrepresentation of some species.

Given that sampling bias is common in many (if not all) presence-only datasets, new methods are increasingly being developed to account for or correct for such bias when using quantitative analyses. For example, Phillips and Elith (2010) described the presence-only calibration (POC) plot, which uses presence-only data (the type of data often generated by citizen science projects) to determine whether modeled predictions of species distributions are proportional to the conditional probability of species presence. Warton et al. (2013) accounted for observer bias (specifically, pseudoabsence data) by factoring observer biases directly into species distribution modeling approaches. Presence locations are modeled as a function of observer bias variables (e.g., accessibility of different locations), as well as environmental variables, to condition the model and make predictions of species distributions corrected for observer bias (Warton et al., 2013). Researchers should be aware of these statistical tools and their potential to resolve bias issues in citizen science data.

16.3.2 Ethical Considerations

Citizen science provides a powerful means to engage communities in science, but forming these relationships introduces ethical dimensions (Resnik et al., 2015). Unfortunately, these dimensions generally receive less attention but can jeopardize citizen science projects and even the careers of researchers involved (Riesch and Potter, 2014). Researchers need to recognize that participants are donating their time, resources, and often knowledge, so there is a reasonable expectation that researchers will be transparent about the project and use of data and that participants will be treated fairly and gain tangible benefits (Mackechnie et al., 2011; Riesch and Potter, 2014). Researchers should also recognize the potential implicit power imbalance where the project leaders (scientists) assume the role of authority figures that may decide when, where, and how community members participate, gain control over information, and potentially unilaterally decide the fate of the data. This could be seen as exploitative and result in resentment and conflict,

particularly for projects where community members make a great investment of local knowledge, time, and effort. Information about project aims, processes, data ownership, and sharing should be clearly stated in project information sheets, and for co-created projects should be jointly written and agreed upon with community representatives. Many different arrangements exist, but most researchers engaging citizen scientists consider reciprocity (Riesch and Potter, 2014). Appropriate reciprocity could include training in scientific methods and data management; assistance with community-driven activities (e.g., festivals, beach cleanups, social media shoutouts); certificates detailing the training received; project paraphernalia such as t-shirts, posters, and stationary; and making time available to engage with and be responsive to the community (Resnik et al., 2015; Riesch and Potter, 2014).

Intellectual property (IP) can also be a sensitive issue in citizen science projects and has even escalated to costly law suits (Resnik et al., 2015). Data and knowledge can be considered sensitive by data owners, and, depending on the cultural context, different people within communities may feel they should have full access to it or, conversely, that they alone should have control over how it is disseminated and understood (Pulsifer et al., 2014). Consequently, IP and data ownership arrangements must be well thought out and agreed upon by all participants before project commencement. For contributed data such as photographs, details regarding copyright and image use should be clear at the time of image submission. For example, the Shark Search Indo-Pacific project uses contributed photographs to help validate species records, and the terms and conditions by which contributors' photographs are used are relatively simple and clearly stated (www.sharksearch-indopacific.org/get-involved).

The ramifications of IP issues can be greatly magnified when dealing with sensitive or confidential information (Resnik et al., 2015). In some cases, the research team may receive information that should not be shared for conservation reasons. For example, the Chinese cave gecko was harvested nearly to extinction shortly after its discovery was published (Lindenmayer and Scheele, 2017). Social media facilitates the rapid spread of news of rare species or special locations that can result in negative effects on populations before management and protection can be put in place. Scientists receiving sensitive information such as locations of rare species, aggregations, or places of special community interest and value need to carefully consider the potential costs of disseminating this information. For projects that acquire sensitive information, researchers should consider developing data-sharing agreements that clarify how IP and confidentiality issues will be managed, ownership of the data received and the knowledge derived from it, and who authorizes the publication and sharing of information. Researchers should also be careful not to underestimate how long such negotiations may take.

Potential conflict may also arise regarding authorship of project publications (Resnik et al., 2015; Riesch and Potter, 2014). Although it is often inappropriate to list every participant as an author, participants should be formally acknowledged in the paper (Dickinson et al., 2012), although this may not be practical with large numbers of contributors (for example, Redmap has had over 800 contributors). Authorship may be considered for certain participants who make significant contributions to the paper, but it is advisable to have these criteria clearly communicated at the beginning of the project. For more complex projects, these criteria can be appended to data-sharing agreements.

16.3.3 Special Considerations for Shark and Ray Citizen Science Projects

Shark and ray research includes a number of specific considerations for citizen science. First, because most sharks and rays are highly mobile, have complex behavior and movement patterns, and are potentially less abundant than other marine fishes, it is likely that citizen scientists will observe only a few animals unless they are actively fishing for them or scuba diving at a location specifically to see them. This means that some projects may result in large numbers of zero data, which can be difficult to analyze and can also quickly reduce volunteer motivation to maintain survey effort. Second, many sharks can be difficult to accurately identify. For example, many members of the Carcharhinidae are very similar and can be difficult for even experienced shark biologists to identify reliably. These errors may be magnified when surveys take place in low-visibility conditions or when viewing an animal from the surface. Generalized underwater visual surveys by citizen scientists are crucial to monitoring large-scale changes in marine ecosystems, but these limitations mean that data may have to be pooled into species groups (e.g., Ward-Paige et al., 2010b), which compromises the ability to identify species-specific trends and draw robust conclusions on species conservation status.

The adequacy of underwater visual surveys on scuba for surveying mobile marine species such as sharks has been questioned due to concerns over diver effects and depth limitations (Lindfield et al., 2014; Ward-Paige et al., 2010a; Willis et al., 2000). Furthermore, many sharks may live in habitats that are less appealing to community volunteers. Most citizen science projects working with sharks and rays involve scuba divers in relatively clear water on reefs, as divers prefer these conditions and tourism operations visit these locations. Scuba divers and snorkelers are much less likely to be willing to survey muddy coastal waters, and even if they did the surveys would be compromised by low visibility, making these approaches potentially unsuitable for species that are not associated with reefs.

Fishers are not restricted by these issues and thus could sample in these habitats; however, the potential safety issues arising from volunteers handling sharks and rays

must be carefully considered. An additional complication is that some sharks and rays may be more active at night, when fishing and diving activities are likely reduced (Holland et al., 1992; Nelson and Johnson, 1970). Sharks may also change their activity patterns (and thus their detectability) in the presence of divers and boats, especially when provisioning is involved (Brunnschweiler and Barnett, 2013; Fitzpatrick et al., 2011). These behavioral factors are highly variable, difficult to quantify, and thus can skew inferences about local abundances from citizen science projects.

16.3.4 Common Pitfalls in Citizen Science Projects

Numerous papers describe the outcomes of citizen science projects or examine data accuracy and scientific rigor, but far fewer publications describe failures and lessons learned from citizen science projects (Riesch and Potter, 2014). However, our personal experiences and discussions with citizen science practitioners have revealed several common process and planning pitfalls that can compromise citizen science projects. One common pitfall arises when researchers are uncertain about how the data collected by citizen scientists are going to be used. This is not a question about data quality; rather, it is a question about the data being fit for purpose, as different quality data are required for different contexts (Riesch and Potter, 2014). Projects that are ambiguous about how the data will be used risk wasting participants' efforts through inappropriate data collection, which in turn can create volunteer dissatisfaction and distrust. If volunteers perceive that the data they collect are not being used in the ways they thought it would, they are likely to cease participating and may even feel misled by the research team (Ganzevoort et al., 2017).

Other potential pitfalls involve inadequate consideration of the nature of volunteer community members. Volunteers may lack the interest, capability (skills and abilities), or capacity (time and resources) to undertake more intensive and complex tasks. If expectations are set too high, the project risks having low participation rates and high participant dissatisfaction. Where projects require recording of presences and absences, community volunteers may also become frustrated with recording zero data and may not understand the importance of including zero values in a dataset. This issue may be especially pertinent to projects on rare species or phenomena where large amounts of zero data are likely. The risk is that volunteers fail to record zero data or even stop participating because they fail to find what they are searching for. Participant motivation can also vary depending on the species and location. It may be relatively easy to motivate scuba divers to record sightings of charismatic species on coral reefs but much more difficult to motivate volunteers to survey muddy estuaries or mangroves for catsharks or stingrays.

Citizen science programs often cite public education and mobilization as an explicit goal, and there is certainly evidence that participation can inspire community change and action (see Section 16.1.3); however, specific attention is often needed to realize these goals. Claims that citizen science can change community behaviors and attitudes are potentially compromised by the fact that project participants may already be intrinsically motivated toward conservation. For example, scuba divers contributing photos of sharks and rays are already likely to be marine conservation advocates as (1) scuba diving is a pasttime, and (2) they self-select by actively choosing to participate. Thus, citizen science projects may risk preaching to the choir, so to speak. Nevertheless, this may be a valid and worthwhile outcome. Citizen science can provide opportunities for conservation-orientated community members to mobilize, become more organized, and implement local conservation actions. However, if a citizen science project states that one of its objectives is to inform and empower the broader community, project leaders may need to take specific actions to engage community members outside of project participants.

The last pitfall discussed here is that researchers often underestimate the amount of time, effort, and resources required to engage communities, inform them of opportunities to participate, build volunteer capacity, maintain enthusiasm, and report back to communities, as well as the normal tasks required for the research project (Resnik et al., 2015). Citizen science projects often place emphasis on engagement and communication, but traditionally trained scientists may have limited training and experience in these areas. Moreover, there are additional difficulties for projects that require consistent, ongoing engagement and communication over large geographical areas with limited financial resources. Depending on the nature of the project, considerable researcher investment may also be required to prepare ethics applications, project information, and fact sheets; to arrange meetings and travel; to draw up data-sharing agreements; and possibly initiate conflict resolution between project participants or members of the team. Researchers should be very clear about their capacity and capability to build and maintain community engagement efforts and select a citizen science approach (contributory, collaborative, or co-created) and scale that match their research question, resources, and abilities.

16.3.5 Importance of Quality Assurance and Quality Control

Probably the greatest concern scientists and managers express about citizen science is doubt about data quality (Bonter and Cooper, 2012; Dickinson et al., 2010; Riesch and Potter, 2014). Interestingly, although many researchers involved in citizen science may be personally confident about data quality, they may be concerned about how other scientists, reviewers, and journal editors will perceive the data, fearing that these perceptions could make it more difficult to publish the research and maintain credibility (Riesch and Potter, 2014). However, these perceptions may be unfounded. Numerous studies have found that well-designed citizen science programs can provide high-quality data that may be of comparable quality to that collected by professional researchers (Darwall and Dulvy, 1996; Holmberg et al., 2008; Tulloch et al., 2013). Additionally, citizen science research has found that, although data quality is important, many other factors such as project scale and longevity affect publication rates (Theobald et al., 2015); thus, perceptions that citizen science data are more difficult to publish may be unfounded. Researchers might rightly spend less energy on trying to ensure that data are collected to the highest quality attainable and instead focus on making sure that the data are fit for purpose; that is, the data quality matches its intended application (Riesch and Potter, 2014). It follows that researchers who want to demonstrate that the data are fit for purpose need to explicitly describe the quality (fitness) and intended application (purpose) of the data. To do so, the researcher must clearly state the intended purpose of the data and demonstrate how the data meet the quality standard required.

Quality assurance and quality control are two separate but interlinked processes that together depict the quality of a product or output. Quality assurance (QA) describes the steps taken to ensure that tasks and activities are completed to the required standard. This includes providing the training, infrastructure and equipment, resources, and protocols to ensure that the activities producing the product are completed properly. Quality control (QC) processes are the steps taken to make sure that the products or outputs created by these tasks and activities meet the required standard. These include product testing, independent monitoring and evaluation, quality checks, and audits. There is no set minimum standard of QA/QC, as the required standard applied depends on the product's intended purpose. For example, alloys used for kitchen utensils may not have to be produced to the same standard as those used for aircraft components.

Many QA/QC options are available for citizen science projects. Common QA processes focus on the people, enabling the participants to collect data to the required standards. These may include developing training programs to standardize data collection, producing identification guides and sampling protocols, providing access to reference materials, developing standardized data sheets, providing simple equipment and the training to use it properly, and creating the means and opportunities for participants to learn from each other (Bonter and Cooper, 2012; Crall et al., 2010). In contrast, QC processes tend to focus on the data and involve checks to make sure that data are at the required standard. One QC process that should be included involves checking contributed data to ensure correct interpretation (e.g., having expert confirmation of species identification). The terminology used to describe QC processes must be consistent so that

QA/QC protocols can be clearly understood and compared. Unfortunately, the terms *data verification* and *data validation* are sometimes confused. We suggest that researchers adopt the definitions used by the National Biodiversity Network (https://nbn.org.uk/), which has defined data verification as the process of "ensuring the accuracy of the identification of the things being recorded" and data validation as "carrying out standardized, often automated, checks on the 'completeness,' accuracy of transmission and validity of the content of a record." In a biological context, verification is performed by species experts and validation is performed by data experts.

Quality control processes often include expert verification of photographs or emerging trends, automated data validation in databases, or the use of data collection tools to flag records that are incomplete or values that violate preset limits. In marine citizen science projects, QC often involves expert verification of photographs (see case studies, below). Technology can also be used to greatly enhance QA/QC by providing new training tools (e.g., video courses, online tests) and data collection tools that control how data are entered and can check the data in real time during data entry in the field (see Section 16.2.3). QC can also include project protocols—for example, a clear series of steps taken to flag, verify, and assess data anomalies to decide whether to retain or discard these data (Riesch and Potter, 2014).

Only half of marine citizen science projects surveyed between 1983 and 2013 included quality control to ensure the collection of a robust dataset that can be defended (Thiel et al., 2014), despite this being critically important to the credibility of the data (e.g., Crall et al., 2010; Gallo and Waitt, 2011; Hochachka et al., 2012; Sullivan et al., 2014). Furthermore, even if QA/QC processes are in place, the lack of clarity about how they are implemented can also discourage uptake by intended data users (Chin, 2013). In summary, researchers considering citizen science projects should implement QA/QC processes that suit the intended use of the data and ensure that these processes are clearly documented so that third parties can make informed decisions about how to use the data.

Case Study 1. The Great Porcupine Ray Hunt

BACKGROUND

The Great Porcupine Ray Hunt aimed to obtain better information about the occurrence and distribution of the porcupine ray (*Urogymnus asperrimus*), a species assessed as being one of the most at-risk shark and ray species with regard to climate change in the Great Barrier Reef (Chin et al., 2010). However, it was considered a relatively rare species and very little was known about its occurrence, biology, and habitat use. Given the paucity of existing records (Theiss et al., 2010), a citizen

science project was launched to (1) document the distribution, range, and habitat use of the species in Australia; (2) identify any hotspots of occurrence; and (3) document any potentially new information on its biology, ecology, and behavior.

ENGAGING CITIZEN SCIENTISTS

The project focused on the scuba diving community who are more likely to encounter this species than other marine user groups. As a purely contributory project, no training was involved other than showing the community what the species looked like. The project team formed a collaboration with major scuba magazines (*SportDiving Magazine* and *Dive Log Australasia*) and a scuba equipment retailer (Adreno.com) to promote the project and to provide prizes to incentivize scuba divers. A Facebook page (www.facebook.com/The-Great-Porcupine-Ray-Hunt--305641946126804/) was created to enable researchers and the community to interact and post photographs. Project posters were also distributed to dive shops and research stations, and the project was advertised through the Eye on the Reef Program, a large citizen project implemented across the Great Barrier Reef. The project was a relatively small-scale project that was actively promoted and managed between November 2011 and June 2012.

DATA COLLECTION AND VERIFICATION

Citizen scientists were asked to send photographs of porcupine rays encountered in Australian waters and to send details of the date, depth, and location. Project scientists verified the photographs received and asked for additional information, if necessary. The project benefited from the animal's unique appearance (Figure 16.2). Porcupine rays are a relatively large, bold species covered with spikes or thorns, making it easy for snorkelers and divers to approach and identify accurately. Although

Figure 16.2 Porcupine ray photographed on March 18, 2017, at Magnetic Island by Porcupine Ray Hunt contributor Ms. Sanna Persson.

photographs were required for data quality control, none of the photographs received was a misidentification, probably due to the species' unique appearance. Once validated, all data were uploaded to the Atlas of Living Australia (www.ala.org.au), Australia's national database for biodiversity data, for long-term curation. Importantly, the project outcomes were written up in an article in *SportDiving Magazine* to show the scuba community what their efforts had produced.

PROJECT OUTCOMES AND LESSONS LEARNED

The project was very successful. Only 29 new records were received, but these contributions more than doubled existing Australian records of the species in biodiversity and museum databases, and even provided video footage of mating behavior (Chin, 2014). The species range was extended, a depth range determined, and its habitat use expanded to include turbid, coastal habitats. Project findings were disseminated to the scuba diving community and the general public through articles in *SportDiving Magazine* and the media. However, it soon became apparent that the project team incorrectly assumed that the species was restricted to coral-associated habitats and perhaps community engagement should have included a wider range of potential contributors. Following newspaper articles and radio interviews about the project, additional information was provided by other members of the public, including trawler fishers and indigenous communities for whom the species has particular significance and value. Nevertheless, the project did provide valuable information that was used to develop successful funding proposals for additional research on the species. Moreover, contributions from the public are still being received even though the project is no longer being actively promoted, including observations from other countries. All Australian data continue to be uploaded to the Atlas of Living Australia, which makes it freely available to the general public.

Case Study 2. Redmap Australia (Range Extension Database and Mapping Project)

BACKGROUND

Shifts in the geographical distribution of species (i.e., range shifts) are globally some of the most frequently reported impacts of climate change as species alter their distributional limits to keep pace with changing environmental conditions (Burrows et al., 2014). Australia has 60,000 km of coastline, much of which is warming at least twice as fast as the global average (Hobday and Pecl, 2014), and range shifts have already

been detected in a wide range of species (Last et al., 2011; Wernberg et al., 2016). Redmap is a citizen science initiative designed to provide an early indication of which species may be changing their distributions in our coastal marine environments and may therefore require additional research focus or management efforts. The project invites members of the public to submit photographs of anomalous species they observe while undertaking marine activities such as fishing, diving, boating, and beachcombing. Redmap has two main discrete but linked objectives: (1) ecological monitoring for the early detection of species that may be extending their geographic distribution as our climate changes, and (2) engaging with the public on the ecological impacts of climate change, using their own data. Redmap is primarily a contributory citizen science project; however, it intentionally includes elements of a collaborative model. For example, members of the fishing and diving communities were involved in the selection of species being monitored and in the design, testing, and production of various aspects of the website, smartphone application, and major communication outputs such as the Redmap Tasmania Report Card (www.redmap.org.au/article/the-redmap-tasmania-report-card/).

ENGAGING CITIZEN SCIENTISTS

Redmap is an Australia-wide project that engages participants from a range of sectors who may have quite different views on a variety of marine issues. An important challenge for Redmap is maintaining engagement with these individuals and groups and demonstrating that their observations are contributing significant scientific data. Educational messages about critical issues are carefully considered and framed for the target audience, increasing the probability of engagement (Nisbet, 2009). The Redmap team has developed a detailed engagement strategy. This is a dynamic document that is updated over time and is based on a typical engagement framework:

1. Specify goals.
2. Identify who to engage with.
3. Develop engagement strategies and implementation priorities.
4. Monitor progress (project evaluation).

Fishers, divers, and scientists are at the frontline of Australia's changing seas and form Redmap's core audience, and most of Redmap's engagement activities are tailored toward these groups.

Feedback on the project outcomes is delivered at both individual and community levels. A critical aspect of Redmap engagement is the individual and personalized responses that participants receive from scientists for every observation submitted. Redmap is most frequently active on Facebook, encouraging engagement with the audience through competitions, a weekly "What's That Fish?" quiz, and providing the latest news on the marine environment or climate change, links to interesting media on marine research, stories relevant to the interests of fishers and divers, and the latest observations logged around the country. The development of an interactive website and smartphone application has facilitated increased interactivity, better reach, and improved collaboration and innovation. Within the Redmap project, integration of the website, smartphone application, and social media (e.g., Facebook, Instagram, Twitter) has been instrumental in maximizing reach, streamlining workloads and processes, and allowing collaborators to work together.

DATA COLLECTION AND VERIFICATION

Redmap began as a pilot project in 2009 in the island state of Tasmania (population of approximately 500,000 people), which has a well-networked community of fishers and divers and a highly collaborative scientific community across multiple institutions (Frusher et al., 2014). As a small-scale pilot project, Redmap began as a simple web form allowing submission of observations. The pilot was easily manageable by a single person using email to contact relevant local scientific experts to verify species identification and then manually emailing feedback to contributors. In December 2012, Redmap was extended to cover all states of mainland Australia (population 23 million), with the expectation of data collection at the scale of decades. Community members can use region-specific lists of over 200 target species available on the website or smartphone app to help identify which species are unusual to their particular area before logging a sighting, or they can submit photographs of any species they know or consider to be unusual for a given area. Each species listed is linked directly in the database to one of 80 taxonomic experts from over 26 institutes, and sightings are routed automatically to them for verifying. After verification, sightings are displayed on the website and the observer is sent detailed feedback on their observation via email.

PROJECT OUTCOMES AND LESSONS LEARNED

Over 2200 "unusual" observations of a wide range of marine species have been submitted by approximately 800 observers, all of which have been verified by scientists collaborating with the Redmap project. These observations have been used to examine potential range shifts (Robinson et al., 2015), provide detailed examinations of particular observations (Stuart-Smith et al., 2017), including manta rays (Couturier et al., 2015), contribute data to larger studies (e.g. Last et al., 2011; Johnson et al., 2011) and add species distribution data to

the Australian Faunal Directory (www.environment.gov. au/biodiversity/abrs/online-resources/fauna/afd/taxa/ PISCES). The project currently lists 13 species of sharks and rays and has received approximately 35 verified sightings of these, with half of those determined to be outside of known range limits. Observations of another ten shark and ray species have also been submitted. In terms of the engagement objectives of Redmap, approximately 50% of survey respondents said that Redmap had raised their awareness about the importance of marine climate change impacts, and nearly 80% indicated that Redmap had increased their awareness of how marine climate change may lead to shifts in the distribution of marine species (Nursey-Bray et al., 2018). Moreover, 78% said they had discussed Redmap with others (i.e., transmission of knowledge was occurring).

Creating awareness of Redmap and then maintaining communication and engagement over a large and dispersed potential audience have been challenging, particularly on a relatively low budget given the temporal and spatial scales involved. Independent verification of the species identification of the observations submitted to Redmap is considered essential; however, scaling the collection and processing of species observational data to include all of Australia and over a longer duration presented several challenges, including the following:

- Governance of the project in a multijurisdictional environment
- Timely management of potentially larger volumes of observations submitted by the public
- Maximizing system efficiency and minimizing operational costs under financially constrained circumstances
- Identifying species and managing information for species from across a large geographic area encompassing both tropical and temperate zones on a large continent

Many of these challenges were significantly reduced with the introduction of the semi-automated distribution of observations to the large network of verifying scientists.

16.4 DESIGNING CITIZEN SCIENCE PROJECTS FOR SHARK AND RAY RESEARCH

Integrating citizen science into research programs requires careful consideration and planning, especially so for marine-based projects that have particular requirements and limitations. Projects focusing on sharks and rays have additional factors that must be assessed. There are three main areas that require specific consideration: (1) scoping and planning, (2) implementation, and (3) working with communities.

16.4.1 Scoping and Planning

Citizen science projects require thorough planning and consideration. Researchers should

- Clearly identify and communicate the goals and specific research questions of the citizen science project and how the data will be managed, accessed, and used.
- Assess the feasibility of citizen science to deliver data that meet the required spatiotemporal scale and resolution to address project questions (e.g., data are fit for purpose); for example, consider whether the shark and ray species can be reliably identified.
- Match data collection to the intended participants. Sampling tasks that are overly intensive, difficult, technical, or strenuous (e.g., diving under difficult conditions to observe sharks) risk having low participation rates.
- Determine if the research team has the capacity and capability the project may require.
- Identify communities that may have the interest, capacity (amount of time and resources), and capability (aptitude and ability) to participate. Shark and ray research projects will probably be focused on scuba divers, snorkelers, or anglers, so researchers need to understand the motivations of these participants and aligned groups (e.g., PADI Project AWARE, fishing clubs) to design appropriate engagement activities and data collection tasks.
- Identify the appropriate citizen science approach (contributory, collaborative, co-created) and plan for how to engage and maintain participants at the required level.
- Identify whether similar projects and efforts already exist. If so, determine the feasibility of collaboration instead of beginning a separate program. Redundancy may cause confusion among participants and funders and can also create numerous, smaller, patchy datasets that limit the utility of the data (Bonney et al., 2014).
- Plan the project's legacy by identifying its intended time frame, the project fate, and succession if the project is intended to continue. If the project has a specific lifespan, researchers should have an exit strategy to ensure that the community is involved in the transition, especially for projects that have high community investment and participation.

It may be also be advantageous to limit the project to relatively modest goals at the outset to build success (Riesch and Potter, 2014), and then when relationships and understanding are established explore more challenging research questions. If several different types of participants or communities are involved, researchers may also consider scaffolding the project so that participants can self-select different levels of engagement and participation based on their circumstances (Dickinson et al., 2010). Finally, to avoid creating false expectations, community engagement should only proceed when researchers have completed these planning and scoping steps.

16.4.2 Project Implementation

For shark and ray citizen science projects to be successful, researchers need to make sure that the expectations and requirements of the community and intended data users are met. Researchers should ensure that participants remain engaged in collecting quality data that are fit for purpose by

- Clearly communicating the project goals and objectives to participants and intended data users
- Clearly describing the roles, responsibilities, and benefits received for each party in the project and clarifying data ownership and dissemination processes
- Clearly documenting and implementing QA/QC processes (see Section 16.3.5)
- Using appropriate technology to assist QA/QC, data collection, and data processing, as well as participant and community engagement
- For shark and ray photograph identification projects, creating the mechanisms to solicit, store, and disseminate contributed images
- Ensuring participant health and safety by providing training and support, especially if the project involves shark or ray species that could pose a risk of injury to participants
- Coordinating efforts with existing aligned groups to share resources and information, coordinate activities, and avoid community consultation fatigue
- Considering opportunities to make data identifiable and retrievable for the long term by archiving data within online data repositories (e.g., iNaturalist, Atlas of Living Australia).

16.4.3 Working with Communities

Implementing successful citizen science projects means working with individuals and their communities toward achieving shared goals on the basis of mutual understanding and trust. Researchers should remember to

- Ensure that the most appropriate people are involved. Successful citizen science projects may have community champions who are fully engaged in the project and actively support and promote it within their community.
- Consider involving key community leaders—specifically, individuals who have legitimacy and wider community support. These individuals may be identified by consulting with government agencies, industry groups, and non-governmental organizations to identify individuals who represent communities, and perhaps becoming familiar with community or industry publications to find out about existing community-based activities.
- Work with honesty and integrity and implement previously described processes for communication and engagement, governance, and ethics. Keeping to these values will help build trust and good relationships between scientists and the community, essential components of successful citizen science projects (Theobald et al., 2015).

- Recognize that participants and communities may have different values, beliefs, and world views and that researchers need to consider and accommodate these views. For example, fishers may have a general distrust of science and conservation, so engaging fishers to collect data on sharks and rays may require respectful and sustained dialog to build trust and shared understanding.

16.5 CONCLUSION: THE FUTURE OF CITIZEN SCIENCE

It seems likely that citizen science will be an increasingly significant part of science and research into the future, with government policy and strategy already espousing the role of community-based scientists (Mackechnie et al., 2011; Pecl et al., 2015). Although the growth of shark and ray citizen science may be slowed due to logistical constraints, advances in technology are fast increasing the capability of citizen scientists to collect high-quality, robust data at massive scales (Newman et al., 2012). For example, the eBird project now receives over 5 million data reports per month and has generated close to 100 scientific publications (Bonney et al., 2014). Lower cost components and electronics are enabling citizen scientists to collect richer and more technical datasets that can be processed further (Raoult et al., 2016) and to rapidly collect data that was once only possible to obtain with highly specialized equipment. This enhanced capability can have immense community benefit in emergency response and management, as became evident with the Safecast project that allowed community members to monitor radiation following the Fukushima disaster (Azby et al., 2016). The availability of drones has created a new means of data collection (see Chapter 4 in this volume), and projects such as Open Reef in Florida and Belize are exploring the potential of volunteer drone operators in community geography and geographic information system (GIS) applications. The Nature Conservancy is exploring the potential of community-based drone imagery to map flooding during El Niño events. Furthermore, the Internet allows community members to access increasingly sophisticated datasets, and citizen science is now moving beyond research and into active compliance and management. The Global Fishing Watch project engages a global network of volunteers, who, together with machine-learning algorithms, monitor fishing vessel movement data sourced from vessel monitoring systems to identify and flag suspicious vessel activity, such as potential fishing in no-take marine protected areas. In addition to providing valuable data and information, these programs enable ordinary people to participate in science and to be empowered with new knowledge, experiences, social networks, and awareness to drive positive changes in their respective communities. Given the power and potential of citizen science to collect large amounts of useful data and

the opportunities it provides to connect researchers and communities, it seems highly likely that citizen science will become an accepted, established, and mainstream methodology used by contemporary scientists well into the future.

REFERENCES

Araujo G, Snow S, So CL, Labaja J, Murray R, Colucci A, Ponzo A (2017) Population structure, residency patterns and movements of whale sharks in Southern Leyte, Philippines: results from dedicated photo-ID and citizen science. *Aquat Conserv* 27(1):237–252.

Arzoumanian Z, Holmberg J, Norman B (2005) An astronomical pattern-matching algorithm for computer-aided identification of whale sharks *Rhincodon typus*. *J Appl Ecol* 42(6):999–1011.

Azby B, Pieter F, Sean B, Nick D, Joe M (2016) Safecast: successful citizen-science for radiation measurement and communication after Fukushima. *J Radiol Prot* 36(2):S82.

Baker B (2016) Frontiers of citizen science: explosive growth in low-cost technologies engage the public in research. *Bioscience* 66(11):921–927.

Bird TJ, Bates AE, Lefcheck JS, Hill NA, Thomson RJ, Edgar GJ, Stuart-Smith RD, et al. (2014) Statistical solutions for error and bias in global citizen science datasets. *Biol Conserv* 173:144–154.

Bonney R, Ballard H, Jordan R, McCallie E, Phillips T, Shirk J, Wilderman CC (2009) *Public Participation in Scientific Research: Defining the Field and Assessing Its Potential for Informal Science Education*, a CAISE Inquiry Group Report. Center for Advancement of Informal Science Education (CAISE), Washington, DC.

Bonney R, Shirk JL, Phillips TB, Wiggins A, Ballard HL, Miller-Rushing AJ, Parrish JK (2014) Next steps for citizen science. *Science* 343(6178):1436–1437.

Bonter DN, Cooper CB (2012) Data validation in citizen science: a case study from Project FeederWatch. *Front Ecol Environ* 10(6):305–307.

Brunnschweiler JM, Barnett A (2013) Opportunistic visitors: long-term behavioural response of bull sharks to food provisioning in Fiji. *PLoS ONE* 8(3):e58522.

Buesching CD, Newman C, Macdonald DW (2014) How dear are deer volunteers: the efficiency of monitoring deer using teams of volunteers to conduct pellet group counts. *Oryx* 48(4):593–601.

Buray N, Mourier J, Planes S, Clua E (2009) Underwater photo-identification of sicklefin lemon sharks, *Negaprion acutidens*, at Moorea (French Polynesia). *Cybium* 33(1):21–27.

Burrows MT, Schoeman DS, Richardson AJ, Molinos JG, Hoffmann A, Buckley LB, Moore PJ, et al. (2014) Geographical limits to species-range shifts are suggested by climate velocity. *Nature* 507(7493):492–495.

Chin A (2013) *Citizen Science in the Great Barrier Reef: A Scoping Study*. Great Barrier Reef Foundation, Brisbane, Australia.

Chin A (2014) Hunting porcupines: citizen scientists contribute new knowledge about rare coral reef species. *Pac Conserv Biol* 20(1):48–53.

Chin A, Kyne PM, Walker TI, McAuley RB (2010) An integrated risk assessment for climate change: analysing the vulnerability of sharks and rays on Australia's Great Barrier Reef. *Glob Change Biol* 16(7):1936–1953.

Cigliano JA, Meyer R, Ballard HL, Freitag A, Phillips TB, Wasser A (2015) Making marine and coastal citizen science matter. *Ocean Coast Manage* 115:77–87.

Cooper CB, Shirk J, Zuckerberg B (2014) The invisible prevalence of citizen science in global research: migratory birds and climate change. *PLoS ONE* 9(9):e106508.

Couturier LIE, Jaine FRA, Kashiwagi T (2015) First photographic records of the giant manta ray Manta birostris off eastern Australia. *PeerJ* 3:e742.

Crall AW, Newman GJ, Jarnevich CS, Stohlgren TJ, Waller DM, Graham J (2010) Improving and integrating data on invasive species collected by citizen scientists. *Biol Invasions* 12(10):3419–3428.

Crall AW, Newman GJ, Stohlgren TJ, Holfelder KA, Graham J, Waller DM (2011) Assessing citizen science data quality: an invasive species case study. *Conserv Lett* 4(6):433–442.

Darwall WRT, Dulvy NK (1996) An evaluation of the suitability of non-specialist volunteer researchers for coral reef fish surveys. Mafia Island, Tanzania—a case study. *Biol Conserv* 78(3):223–231.

Davies TK, Stevens G, Meekan MG, Struve J, Rowcliffe JM (2012) Can citizen science monitor whale-shark aggregations? Investigating bias in mark–recapture modelling using identification photographs sourced from the public. *Wildl Res* 39(8):696–704.

Delaney DG, Sperling CD, Adams CS, Leung B (2008) Marine invasive species: validation of citizen science and implications for national monitoring networks. *Biol Invasions* 10(1):117–128.

Dickinson JL, Zuckerberg B, Bonter DN (2010) Citizen science as an ecological research tool: challenges and benefits. *Annu Rev Ecol Evol Syst* 41:149–172.

Dickinson JL, Shirk J, Bonter D, Bonney R, Crain RL, Martin J, Phillips T, Purcell K (2012) The current state of citizen science as a tool for ecological research and public engagement. *Front Ecol Environ* 10(6):291–297.

Dulvy NK, Fowler SL, Musick JA, Cavanagh RD, Kyrne PM, Harrison LR, Carlson JK, et al. (2014) Extinction risk and conservation of the world's sharks and rays. *eLife* 3:e00590.

Edgar GJ, Stuart-Smith RD, Cooper A, Jacques M, Valentine J (2017) New opportunities for conservation of handfishes (family Brachionichthyidae) and other inconspicuous and threatened marine species through citizen science. *Biol Conserv* 208:174–182.

Fitzpatrick R, Abrantes KG, Seymour J, Barnett A (2011) Variation in depth of whitetip reef sharks: does provisioning ecotourism change their behaviour? *Coral Reefs* 30(3):569–577.

Forrester TD, Baker M, Costello R, Kays R, Parsons AW, McShea WJ (2017) Creating advocates for mammal conservation through citizen science. *Biol Conserv* 208:98–105.

Frusher SD, Hobday AJ, Jennings SM, Creighton C, D'Silva D, Haward M, Holbrook NJ, et al. (2014) The short history of research in a marine climate change hotspot: from anecdote to adaptation in south-east Australia. *Rev Fish Biol* 24:593–611.

Gallo T, Waitt D (2011) Creating a successful citizen science model to detect and report invasive species. *BioScience* 61(6):459–465.

Ganzevoort W, van den Born RJG, Halffman W, Turnhout S (2017) Sharing biodiversity data: citizen scientists' concerns and motivations. *Biodivers Conserv* 26(12):2821–2837.

Gordon CA, Hood AR, Ellis JR (2016) Descriptions and revised key to the eggcases of the skates (Rajiformes: Rajidae) and catsharks (Carcharhiniformes: Scyliorhinidae) of the British Isles. *Zootaxa* 4150(3):255–280.

Greenwood JJD (2007) Citizens, science and bird conservation. *J Ornithol* 148(Suppl 1):S77–S124.

Gubili C, Johnson R, Gennari E, Oosthuizen, WH, Kotze D, Meyer M, Sims DW, et al. (2009) Concordance of genetic and fin photo identification in the great white shark, *Carcharodon carcharias*, off Mossel Bay, South Africa. *Mar Biol* 156(10):2199–2207.

Hobday AJ, Pecl GT (2014) Identification of global marine hotspots: sentinels for change and vanguards for adaptation action. *Rev Fish Biol Fish* 24(2):415–425.

Hochachka WM, Fink D, Hutchinson RA, Sheldon D, Wong W-K, Kelling S (2012) Data-intensive science applied to broad-scale citizen science. *Trends Ecol Evol* 27(2):130–137.

Holland KN, Lowe CG, Peterson JD, Gill A (1992) Tracking coastal sharks with small boats: hammerhead shark pups as a case study. *Aust J Mar Freshwater Res* 43:61–66.

Holmberg J, Norman B, Arzoumanian Z (2008) Robust, comparable population metrics through collaborative photomonitoring of whale sharks Rhincodon typus. *Ecol Appl* 18(1):222–233.

Hylton S, White WT, Chin A (2017) The sharks and rays of the Solomon Islands: a synthesis of their biological diversity, values and conservation status. *Pac Conserv Biol* 23(4):324–334.

Jaine FRA, Couturier LI, Weeks SJ, Townsend KA, Bennett MB, Fiora K, Richardson AJ (2012) When giants turn up: sighting trends, environmental influences and habitat use of the manta ray Manta alfredi at a Coral Reef. *PLoS ONE* 7(10):e46170.

Johnson KG, Brooks SJ, Fenberg PB, Glover AG, James KE, Lister AM, Michel E, et al. (2011) Climate change and biosphere response: unlocking the collections vault. *BioScience*, 61(2):147–153.

Last PR, White WT, Gledhill DC, Hobday AJ, Brown R, Edgar GJ, Pecl G (2011) Long-term shifts in abundance and distribution of a temperate fish fauna: a response to climate change and fishing practices. *Glob Ecol Biogeogr* 20(1):58–72.

Lewandowski EJ, Oberhauser KS (2017) Butterfly citizen scientists in the United States increase their engagement in conservation. *Biol Conserv* 208:106–112.

Li O (2016) Employing Informal Learning Theory and Network Analysis to Improve the Way We Communicate Scientific Information to Fisheries Stakeholders, PhD thesis, James Cook University, Townsville, Queensland, Australia.

Lindenmayer D, Scheele B (2017) Do not publish. *Science* 356(6340):800–801.

Lindfield SJ, Harvey ES, McIlwain JL, Halford AR (2014) Silent fish surveys: bubble-free diving highlights inaccuracies associated with SCUBA-based surveys in heavily fished areas. *Meth Ecol Evol* 5(10):1061–1069.

Mackechnie C, Maskell L, Norton L, Roy D (2011) The role of 'big society' in monitoring the state of the natural environment. *J Environ Monit* 13(10):2687–2691.

Marshall AD, Pierce SJ (2012) The use and abuse of photographic identification in sharks and rays. *J Fish Biol* 80(5):1361–1379.

Martin VY, Christidis L, Lloyd DJ, Pecl GT (2016a) Understanding drivers, barriers and information sources for public participation in marine citizen science. *J Sci Commun* 15(2).

Martin VY, Smith L, Bowling A, Christidis L, Lloyd DJ, Pecl GT (2016b) Citizens as scientists: what influences public contributions to marine research? *Sci Commun* 38(4):495–522.

McKinley DC, Miller-Rushing AJ, Ballard HL, Bonney R, Brown H, Cook-Patton SC, Evans DM, et al. (2017) Citizen science can improve conservation science, natural resource management, and environmental protection. *Biol Conserv* 208:15–28.

Meekan MG, Trevitt L, Simpfendorfer CA, White W (2016) The piggybacking stingray. *Coral Reefs* 35(3):1011.

Miller-Rushing A, Primack R, Bonney R (2012) The history of public participation in ecological research. *Front Ecol Environ* 10(6):285–290.

Mourier J, Vercelloni J, Planes S (2012) Evidence of social communities in a spatially structured network of a free-ranging shark species. *Anim Behav* 83(2):389–401.

Nelson DR, Johnson RH (1970) Diel activity rhythms in the nocturnal, bottom-dwelling sharks, *Heterodontus francisci* and *Cephaloscyllium ventriosum. Copeia* 1970(4):732–739.

Nerbonne JF, Nelson KC (2008) Volunteer macroinvertebrate monitoring: tensions among group goals, data quality, and outcomes. *Environ Manage* 42(3):470–479.

Newman G, Wiggins A, Crall A, Graham E, Newman S, Crowston K (2012) The future of citizen science: emerging technologies and shifting paradigms. *Front Ecol Environ* 10(6):298–304.

Nisbet MC (2009) Communicating climate change: why frames matter for public engagement. *Environment* 51(2):12–23.

Norton SL, Wiley TR, Carlson JK, Frick AL, Poulakis GR, Simpfendorfer CA (2012) Designating critical habitat for juvenile endangered smalltooth sawfish in the United States. *Mar Coastal Fish* 4(1):473–480.

Nursey-Bray M, Palmer R, Pecl G (2018) Spot, log, map: assessing a marine virtual citizen science program against Reed's best practice for stakeholder participation in environmental management. *Ocean Coast Manage* 151:1–9.

Peckham SH, Diaz DM, Walli A, Ruiz G, Crowder LB, Nichols WJ (2007) Small-scale fisheries bycatch jeopardizes endangered Pacific loggerhead turtles. *PLoS ONE* 2(10):e1041.

Pecl G, Gillies C, Sbrocchi C, Roetman P (2015) *Building Australia Through Citizen Science*, Occasional Paper Series. Office of the Chief Scientist, Canberra, Australia.

Phillips SJ, Elith J (2010) POC plots: calibrating species distribution models with presence-only data. *Ecology* 91(8):2476–2484.

Porcher IF (2005) On the gestation period of the blackfin reef shark, *Carcharhinus melanopterus*, in waters off Moorea, French Polynesia. *Mar Biol* 146(6):1207–1211.

Pulsifer PL, Huntington HP, Pecl GT (2014) Introduction: local and traditional knowledge and data management in the Arctic. *Polar Geogr* 37(1):1–4.

Raoult V, David PA, Dupont SF, Mathewson CP, O'Neill SJ, Powell NN, Williamson JE (2016) GoPros™ as an underwater photogrammetry tool for citizen science. *PeerJ* 4:e1960.

Resnik DB, Elliott KC, Miller AK (2015) A framework for addressing ethical issues in citizen science. *Environ Sci Policy* 54:475–481.

Riesch H, Potter C (2014) Citizen science as seen by scientists: methodological, epistemological and ethical dimensions. *Public Underst Sci* 23(1):107–120.

Robinson LM, Gledhill DC, Moltschaniwskyj NA, Hobday AJ, Frusher SD, et al. (2015) Rapid assessment of an ocean warming hotspot reveals "high" confidence in potential species' range extensions. *Glob Environ Change* 31:28–37.

Seitz JC, Poulakis GR (2006) Anthropogenic effects on the smalltooth sawfish (*Pristis pectinata*) in the United States. *Mar Pollut Bull* 52(11):1533–1540.

Sequeira AMM, Mellin C, Meekan MG, Sims DW, Bradshaw CJA (2013) Inferred global connectivity of whale shark *Rhincodon typus* populations. *J Fish Biol* 82(2):367–389.

Silvertown J (2009) A new dawn for citizen science. *Trends Ecol Evol* 24(9):467–471.

Stuart-Smith RD, Edgar GJ, Barrett NS, Bates AE, Baker SC, Bax NJ, Becerro MA, et al. (2017) Assessing national biodiversity trends for rocky and coral reefs through the integration of citizen science and scientific monitoring programs. *BioScience* 67:134–146.

Sullivan BL, Wood CL, Iliff MJ, Bonney RE, Fink D, Kelling S (2009) eBird: a citizen-based bird observation network in the biological sciences. *Biol Conserv* 142(10):2282–2292.

Sullivan BL, Aycrigg JL, Barry JH, Bonney RE, Bruns N, Cooper CB, Damoulas T, et al. (2014) The eBird enterprise: an integrated approach to development and application of citizen science. *Biol Conserv* 169:31–40.

Theiss SM, Kyne PM, Chisholm LA (2010) Distribution of the porcupine ray *Urogymnus asperrimus* (Bloch & Schneider, 1801) in Australian waters, with new records from Queensland. *Mem Queensl Mus* 55(1):101–105.

Theobald EJ, Ettinger AK, Burgess HK, DeBey LB, Schmidt NR, Froehlich HE, Wagner C, et al. (2015) Global change and local solutions: tapping the unrealized potential of citizen science for biodiversity research. *Biol Conserv* 181:236–244.

Thiel M, Penna-Díaz MA, Luna-Jorquera G (2014) Citizen scientists and marine research: volunteer participants, their contributions, and projection for the future. *Oceanogr Mar Biol Annu Rev* 52:257–314.

Tian H, Stige LC, Cazelles B, Kausrud KL, Svarverud R, Stenseth NC, Zhang Z (2011) Reconstruction of a 1,910-y-long locust series reveals consistent associations with climate fluctuations in China. *Proc Natl Acad Sci USA* 108(35):14521–14526.

Town C, Marshall A, Sethasathien N (2013) Manta Matcher: automated photographic identification of manta rays using keypoint features. *Ecol Evol* 3(7):1902–1914.

Tulloch AIT, Possingham HP, Joseph LN, Szabo J, Martin TG (2013) Realising the full potential of citizen science monitoring programs. *Biol Conserv* 165:128–138.

Vianna GMS, Meekan MG, Bornovski TH, Meeuwig JJ (2014) Acoustic telemetry validates a citizen science approach for monitoring sharks on coral reefs. *PLoS ONE* 9(4):e95565.

Ward-Paige C, Flemming JM, Lotze HK (2010a) Overestimating fish counts by non-instantaneous visual censuses: consequences for population and community descriptions. *PLoS ONE* 5(7):e11722.

Ward-Paige CA, Mora C, Lotze HK, Pattengill-Semmens C, McClenachan L, Arias-Castro E, Myers RA (2010b) Large-scale absence of sharks on reefs in the greater-Caribbean: a footprint of human pressures. *PLoS ONE* 5(8):e11968.

Warton DI, Renner IW, Ramp D (2013) Model-based control of observer bias for the analysis of presence-only data in ecology. *PLoS ONE* 8(11):e79168.

Waters JD, Coelho R, Fernandez-Carvalho J, Timmers AA, Wiley T, Seitz JC, Mcdavitt MT, et al. (2014) Use of encounter data to model spatio-temporal distribution patterns of endangered smalltooth sawfish, *Pristis pectinata*, in the western Atlantic. *Aquat Conserv* 24(6):760–776.

Wernberg T, Bennett S, Babcock RC, de Bettignies T, Cure K, Depczynski M, Dufois F, et al. (2016) Climate-driven regime shift of a temperate marine ecosystem. *Science* 353(6295):169–172.

White WT, Appleyard SA, Sabub B, Kyne PM, Harris M, Lis R, Baje L, et al. (2015) Rediscovery of the threatened river sharks, *Glyphis garricki* and *G. glyphis*, in Papua New Guinea. *PLoS ONE* 10(10):e0140075.

Whitelaw G, Vaughan H, Craig B, Atkinson D (2003) Establishing the Canadian Community Monitoring Network. *Environ Monit Assess* 88(1):409–418.

Whitney N, Pyle R, Holland K, Barcz J (2012) Movements, reproductive seasonality, and fisheries interactions in the whitetip reef shark (*Triaenodon obesus*) from community-contributed photographs. *Environ Biol Fish* 93(1):121–136.

Wiley TR, Simpfendorfer CA (2010) Using public encounter data to direct recovery efforts for the endangered smalltooth sawfish *Pristis pectinata*. *Endang Species Res* 12:179–191.

Willis TJ, Millar RB, Babcock RC (2000) Detection of spatial variability in relative density of fishes: comparison of visual census, angling, and baited underwater video. *Mar Ecol Prog Ser* 198:249–260.

CHAPTER 17

Social Science and Its Application to the Studies of Shark Biology

Karin Gerhardt
College of Science and Engineering and Centre for Sustainable Tropical Fisheries and Aquaculture,
James Cook University, Townsville, Queensland, Australia

Amy Diedrich
College of Science and Engineering and Centre for Sustainable Tropical Fisheries and Aquaculture,
James Cook University, Townsville, Queensland, Australia

Vanessa Jaiteh
Centre for Fish and Fisheries Research and Asia Research Centre, Murdoch University, Perth, Western Australia,
Australia; Coral Reef Research Foundation, Koror, Palau

CONTENTS

17.1 INTRODUCTION: INCORPORATING THE SOCIAL DIMENSION INTO SHARK MANAGEMENT AND CONSERVATION

Meeting the needs of people while sustaining ecosystems and the benefits they deliver is a global challenge. Coastal marine systems present a particularly important case, given that over half of the world's population lives within 100 km of the ocean, and fisheries provide the primary source of protein for over a billion people worldwide (Leslie et al., 2015). Sharks and rays play important roles in many of these fisheries and coastal communities, as not only do they provide food and income but they can also have strong social and cultural value (Dulvy et al., 2017). However, pressures on sharks and rays from fishing, habitat loss, and other factors are increasing, resulting in global declines in some species and raising concerns that up to a quarter of the world's sharks and rays are threatened with extinction (Dulvy et al., 2008).

At a time when many shark and ray species are experiencing ongoing population declines from these growing pressures, the need to engage in research that bridges the human–shark interface to better inform conservation and fisheries management and policy is becoming increasingly recognized (Jacques, 2010; Simpfendorfer et al., 2011). With the possible exception of work related to "shark attacks" on humans, social science research that focuses on sharks has not kept pace with biophysical science, despite knowledge that understanding people is pivotal to effective natural resource management (Gutiérrez et al., 2011; Reid, 2016; Twyman, 2017).

It is acknowledged that the majority of shark fishing and mortality occur in developing countries, where there may not be adequate fisheries data, legislation, or even capacity for management plans to be considered (Dulvy et al., 2014). Regardless of location, the management of shark resources must incorporate different management strategies that are tailored to local social, cultural, and economic circumstances. These management strategies could include methods to improve the global regulation of fisheries and ways to improve global conservation ethics and encourage active community participation in management, in addition to identifying ways to maximize adoption of specific management strategies (Barker and Schluessel, 2005). Whatever the desired outcome, the social science required to inform these management decisions plays a critical part in ensuring that they are relevant and appropriate to the country and context in which they are implemented.

In this chapter, we showcase key social science methods that, if used in combination with biophysical science research, could support a holistic approach to improving outcomes of shark conservation and management. This approach centers around the concept of livelihood, which is central to the shark–human interface and reflects people's capabilities and means of living, including food, income, and assets (Chambers and Conway, 1992). The livelihood concept encompasses all aspects of shark use by humans, including fishing (including finning), recreation, and culture. The concept of sustainable livelihoods is particularly relevant to poverty alleviation in fishery-dependent communities (Blythe, 2015), which may contribute to shark declines from fishing.

The need to balance sustainable livelihoods and shark conservation should be central to the emerging efforts to integrate social science research into the shark research agenda. Social–ecological approaches that address conservation and management (e.g., those that recognize the inextricable links and feedbacks between social and ecological systems) are increasingly prevalent in the broader fisheries literature (Cinner et al., 2012; McClanahan et al., 2009; Ostrom, 2009; Pollnac et al., 2010), as is research related to the concept of sustainable livelihoods (Blythe, 2015; Pomeroy et al., 2017). Surprisingly, for the most part, livelihood-based research approaches have yet to find their way into broader shark and ray research and conservation. Instead, when social dimensions are addressed, they are often presented as conflicting with biodiversity goals, with relatively few case studies being provided to present the potential for synergies between the two areas (Persha et al., 2011). Potential exceptions to this conflict exist within the shark tourism literature, where several studies suggest that tourism can be a vehicle for shark conservation (Gallagher and Hammerschlag, 2011; Topelko and Dearden, 2005; Vianna et al., 2012). In this chapter, we present methodological approaches that will help to identify ways to balance livelihoods and shark conservation outcomes. We also examine how conservation and livelihood issues play out in real-world examples with a detailed examination of social science methods used to investigate shark fishing and finning in Indonesia (see Case Study 1 in Section 17.6).

The cultural significance of sharks and rays and its role in achieving conservation outcomes is an important consideration not only for social scientists but also for biophysical scientists, managers, and policymakers. This is partially because of the important cultural values expressed by people for sharks and rays, but also because working with indigenous peoples regarding cultural values and indigenous knowledge evokes complex ethical considerations regarding intellectual property rights of knowledge (and its use) (see Case Study 2 in Section 17.6).

Whether we are speaking about livelihood attributes or cultural considerations, there is no contention that a single feature operates in isolation. Social, ecological, and cultural systems all intertwine, with dependencies, linkages, and influences found in some of the most unusual places; for example, see Mills et al. (2011), who found that successful poverty interventions in fishing communities may not be linked primarily to fish resources but to other more fundamental sources (e.g., access to farmland, treatment of waterborne disease). It is through this multifaceted lens that we acknowledge that systems of people and nature are linked, emphasizing that humans must be seen as a part of, not apart from, the shark management agenda. We recommend a much greater consideration for and integration of social science in shark and ray research and conservation. This poses a challenge for project funders and research team leaders, as a broad scope of social science disciplines and methods may be applicable.

17.2 SHARK SOCIAL SCIENCE OVERVIEW

17.2.1 Types of Social Science Research

This section provides a basic overview of what social science is—the theories that underpin it, the methods that are often used—and why it is critical that teams seeking to conduct research that delivers tangible benefits for shark and ray conservation or management must consider including a social scientist in projects. Social scientists study all aspects of society, from historical events and accomplishments to human behaviors and relationships among groups. Through their studies and analyses they offer insight into the physical, social, and cultural development of humans, as well as the links between human activity and the environment

(Aswani and Hamilton, 2004; Ban et al., 2013; Kittinger et al., 2013). The greater the linkage between humans and natural resources (such as sharks), the more incentive users and management have to maintain not only ecosystem integrity but also productivity (Salafsky and Wollenberg, 2000).

Social science draws on multiple disciplines, theoretical perspectives, and methodological approaches. Just as in the biophysical sciences, the social sciences are extremely diverse, and it is useful to understand and capitalize on this diversity when designing and implementing social science research. Both the classic and applied social sciences are used to study a range of occurrences produced by social phenomena, processes, or individual attributes (Bennett et al., 2017). To illustrate a small section of the social science field, some of the classic and applied social science disciplines are illustrated in Figure 17.1, which shows the range of social attributes that can be explored through qualitative and quantitative research studies. Whether a classic or applied social science perspective is chosen, they both tend to concentrate on social phenomena, social processes, or individual attributes. The complexity and breadth of the social science field mean that there is often overlap between these areas and it is rare that social scientists focus on only one area.

17.2.2 Understanding the Theoretical Foundations of Social Science

The development of good social research depends on paying attention to certain theoretical foundations. If such factors are ignored or overlooked, the research will be open to criticism, and serious questions may be raised about the quality of the findings. It is important to recognize that there is no single, accepted approach to social science research. Indeed, how researchers approach research problems depends on a range of factors, including the basic set of beliefs that guide actions (i.e., philosophical perspectives) (Guba and Lincoln, 1994); beliefs about the nature of the world and what can be known about it (ontology); the nature of knowledge and how it can be acquired (epistemology); and the purposes and goals of the research (Ormston et al., 2014). In addition to this, social scientists must also consider the characteristics of the research participants, the audience for the research, the funders of the research, and the position and environment of the researchers themselves as they develop their methods for answering the project questions (Ormston et al., 2014; Ritchie et al., 2013).

All scientists (social and biophysical) base their research on a philosophical approach that provides the context and boundaries within which data collection and analysis techniques should be selected. A philosophical perspective—also called *paradigm* (Kuhn, 1970) or *world view* (Creswell and Poth, 2017)—is something personal that drives the way research is conducted and influences how a researcher creates knowledge and derives meaning from the data.

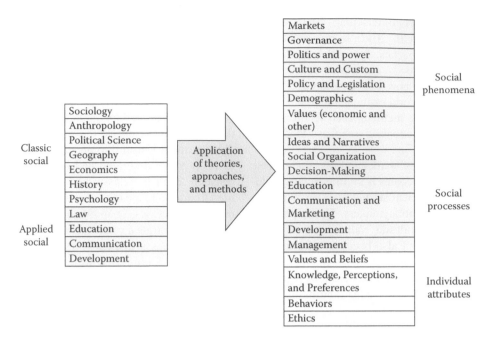

Figure 17.1   The social sciences and related topics of study. (Adapted from Bennett, N.J. et al., *Biol. Conserv.*, 205, 93–108, 2017.)

Perhaps one of the most significant differences between the social sciences and biophysical sciences is the diversity of philosophical assumptions that can be selected to underpin social research. The biophysical sciences tend to be dominated by a single philosophy known as *positivism*. Positivism is based on the belief that only knowledge gained through the scientific method via unprejudiced use of the senses can be accurate and true (Comte, 1975), but Huff (1984) explained that a cause-and-effect type of question in which certain variables are predicted to explain an outcome (positivist approach) is different from an exploration of a single phenomenon, as can be found in social science research. Take, for example, the philosophical approach of postmodernism, which basically exists at the opposite end of the spectrum from positivism. The underlying premise of postmodernism is that no definite terms, boundaries, or absolute truths exist, and it might not be possible to arrive at any conclusive definition of reality (Smircich and Calas, 1987).

Although some of us may view the descriptions and records of scientists undertaking positivist-based research as objectively true or false (in principle), the postmodern viewpoint, which follows from the rejection of an objective natural reality, can be expressed as "there is no such thing as truth." With such philosophical extremes present in the field of social science, scientists tend to find a balance that works for the area of research that they are in or embraces a number of philosophical perspectives. However, to clearly understand the outcomes of the social research being presented, some explanation of the theoretical foundations being applied must be provided.

Saunders et al. (2007) developed a model to illustrate the different stages that people move through when developing a research strategy that helps consolidate their thinking regarding a methodological approach. The process is diagrammatically represented as a research "onion" (Figure 17.2). Researchers begin at the outside layer (containing the researcher's philosophical perspectives) and then move through each layer, deciding on each attribute that will ultimately make up their research methodology. There are no fixed or linear relationships among philosophical approaches and research strategies, approaches, timelines, etc., so the different methodologies that emerge can be unique to each researcher or even project. It is through these different disciplinary perspectives that particular social science areas find their strengths and establish their thinking.

So, what are the philosophical orientations and assumptions made by researchers when they undertake a social science study? To help explain how research philosophy paradigms can drive research questions, Table 17.1 illustrates how these paradigms or world views are matched with a type of research question that might be asked in the realm of shark science. Although a large number of examples are provided, they are by no means the only paradigms on which social science research is based. Also, it is quite common for large research project to span a number of paradigms and have questions and analysis grouped into areas. For example, a positivist approach may support the first part of a research project where researchers are trying to ascertain the demographics of a shark fishing community, but then a constructivist approach might be applied to provide a basis for understanding the motivation of fishers to target sharks.

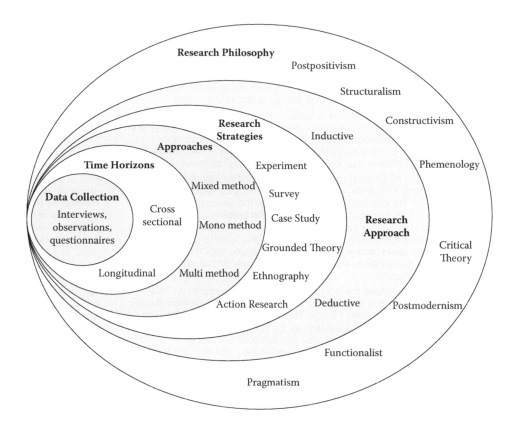

Figure 17.2 Different layers that compose a research methodology (Adapted from Saunders, M. et al., *Research Methods for Business Students*, 6th ed., Pearson Education, London, 2007.)

17.2.3 How Social Science Research Can Contribute to the Shark Research Knowledge Base

The field of social science is so incredibly diverse that it is not feasible to list every potential application and benefit to shark and ray research and conservation. However, by applying a social–ecological viewpoint that focuses on livelihood and cultural attributes, we highlight six potential applications of social science aimed at improving the shark research knowledge base and its potential to contribute to policy, legislation, and management. These six applications show how social science provides the means to unpack environmental and sustainability issues to see what they are about or what lies inside and to explore how they are understood by those connected with them.

17.2.3.1 Social Science Can Identify and Examine Relationships and Trends Among Social and Ecological Factors

Publicly available data such as population statistics and information derived from community surveys can be used to create social, economic, or demographic variables at a variety of spatial scales (e.g., country, state, community) and data formats (e.g., scale, ordinal, categorical, binomial,

spatial). These social data can be included in conceptual or quantitative models with biophysical data to explore research questions such as the following:

- Which socioeconomic and cultural factors are most likely to influence shark conservation or fisheries management outcomes?
- What is the value of shark meat to food security and what is the resulting impact on shark populations?
- Is there any link between the use of indigenous knowledge in conservation planning and positive species management outcomes?

17.2.3.2 Social Science Provides the Means to Analyze Values, Attitudes, and Perceptions Related to Sharks and Their Management

This area is of particular interest to conservation and management practitioners, as values, attitudes, beliefs, and perceptions guide and influence human behavior. Attitude-based research may focus either on general values or on very specific management issues. How people perceive issues about sharks or fishing or conservation measures can be critically important to the effective implementation of any management plans (Agardy, 2000; Hilborn, 2007). Some examples of research questions for this area include the following:

Table 17.1   Examples of Philosophical Perspectives Encountered in the Field of Social Science

Philosophical Orientation	Description and Assumptions That Researchers May Make	Example Research Questions
1 Positivist	Natural science method (posit, observe, derive logical truths) *Assumption:* Only knowledge gained through the scientific method through unprejudiced use of the senses is accurate and true.	What are the demographic features of a community where shark fishing occurs?
2 Post-positivism	Researchers view inquiry as a series of logically related steps, believe in multiple perspectives from participants rather than a single reality, and use rigorous methods of qualitative data collection and analysis. *Assumption:* There is no strict cause-and-effect relationship but every cause and effect is a probability that may or may not occur.	What is the economic benefit that shark finning provides to communities in Melanesia?
3 Structuralism	Elements of human culture must be understood by way of their relationship to a larger, overarching system or structure. Research based on this philosophy works to uncover the structures that underlie all the things that people do, think, feel, and perceive. *Assumption:* When a systematic structure of social classes and relationships has been ascertained (through understanding objects, concepts, ideas, and words as they relate to one another), then it is possible to generalize the knowledge and apply it to all aspects of human culture.	What is the purpose of the (social) structural relationships within this community (e.g., social classes, governments), and how do they influence shark fishing practices here and elsewhere?
4 Constructivism	Seeks to understand the world in which people live and work: They develop subjective meanings of their experiences—meanings directed toward certain objects of things. There can be many meanings, and researchers in this area will look for the complexity of views rather than try to narrow the meanings into a few categories of ideas. *Assumption:* Individuals frame problems in their own way and these differences must be understood to evaluate the system.	What currently motivates individuals in this community to fish for sharks?
5 Interpretivism	Interpretive theories are orientations to social reality based on the goal of understanding. It is more accepting of free will and sees human behavior as the outcome of the subjective interpretation of the environment. Interpretations of reality are generally derived from a cultural and historical base. Two focused examples of interpretivism are phenomenology and symbolic interactions.	—
5a Phenomenology	Phenomenology is the study of structures of experience, or consciousness. It literally means the study of phenomena: the appearance of things (or things as they appear in our experience) or the way we experience things and what that means. *Assumption:* Researchers can put aside their own systems of meaning (or reality) and interpret what is currently happening and thus give rise to a new revitalized or enhanced meaning of the phenomenon.	Why do people fish for sharks?

- What sociocultural values do locals associate with sharks and how do these influence their behavior toward sharks?
- What value do tourists place on sharks and how can this be translated into local support for conservation?
- What are the drivers of noncompliance with shark-related regulations (e.g., ban on shark finning, gear regulations)?

17.2.3.3   *Social Scientists Can Determine Which Actions Lead to Effective Programs, Policies, or Services by Describing and Evaluating Decision-Making Processes*

Having a better understanding of decision-making processes helps managers and planners avoid negative outcomes and focus on the developments that are likely to succeed. This could be applicable in the realm of marine park planning for shark conservation. Marine protected areas for sharks are a type of management tool that is beginning to see uptake around the world; however, the planning and development of marine parks designed for shark conservation are heavily influenced by the available biological and movement data

of shark species (Knip et al., 2012; White et al., 2017). A large gap in the research and management agenda appears to be the human dimensions associated with developing and implementing effective shark conservation approaches. This kind of research can examine the many different social perspectives and decision-making processes held by different stakeholders, allowing concepts of equity, legitimacy, effectiveness, and satisfaction to be analyzed from the perspective of different communities and industries, stakeholders, legislators, and the wider public (Bennett, 2016; Bennett et al., 2017). This knowledge can ensure that approaches used to develop, plan, and implement marine parks or other regulations are appropriate for the surrounding communities and hence effective in achieving their intended outcomes. When looking at this from a social perspective, the types of issues that could be researched include the following:

- What are the most effective ways to involve local communities in the planning and establishment of protected areas? Does this lead to an increase in local support and compliance?

Table 17.1 (continued) Examples of Philosophical Perspectives Encountered in the Field of Social Science

Philosophical Orientation		Description and Assumptions That Researchers May Make	Example Research Questions
5b	Symbolic interactions	The symbolic interactions approach investigates society by addressing the subjective meanings that people impose on objects, events, and behaviors. *Assumption:* The meaning of objects arises out of social interaction (language) between people, and people interact with and interpret objects based on that. People are conscious of how they interact and can change their behaviors.	How do different individuals' descriptions, definitions, and metaphors of the sharks affect fishing outcomes in this community (e.g., are the sharks considered part of the ocean ecosystem or are they considered a resource)?
6	Critical theory	Critical theory perspectives are concerned with empowering human beings to surpass the constraints placed on them by race, class, and gender. Critical theory can have areas of specialty, as well. The three main specialties are emancipatory, advocacy/participatory, and feminism.	—
6a	Emancipatory	The subjects of social inquiry should be empowered. *Assumption:* The researcher wants to establish a mutual interdependence among the research participants and to transform structures that exploit people.	How can we ensure that the community shares in the benefits of shark fishing or alternatives to fishing?
6b	Advocacy/ participatory	This approach to research in communities emphasizes participation and action. It seeks to understand the world by trying to change it collaboratively and following reflection. *Assumption:* Collaborating with people in the community rather than conducting research may create an agenda for active change or political reform.	How can we garner support and develop effective governance structures to enable sustainable livelihoods in this community?
6c	Feminism	Feminism aims to understand the nature of gender inequality. It examines women's and men's social roles, experience, interests, chores, and feminist politics in a variety of fields. *Assumption:* There is a belief that shark fishing is a male-dominated activity and reflects a patriarchal world view and culture.	Does examining shark fishing from a feminist perspective offer alternative understandings of the dynamics and power relations among and between stakeholders?
7	Postmodernism	Postmodernistic approaches dispute the underlying assumptions of mainstream social science and tend to be orientated more toward cultural critique (Rosenau, 1991). *Assumption:* Approaches to generating knowledge should be done with skepticism, and research should scrutinize, contest, deconstruct, and make visible the (invisible) origins, assumptions, and effects of meaning.	Why is it assumed that fishing is a problem?

Note: Having example questions matched against individual philosophies is quite arbitrary, and, as social scientists will quickly point out, many researchers do not necessarily operate under a single paradigm. This table has been included to show (in a simple form) how the philosophical orientation may drive the way a research question is presented.

- What makes some public involvement processes succeed?
- Are there cultural considerations that need to be incorporated into decision-making processes that would provide better social and environmental outcomes?
- What level of satisfaction is there with the marine park consultation process?

17.2.3.4 *Social Science Can Investigate Inequality and Facilitate the Application of More Socially Equitable Processes to Conservation and Fishery Management*

Questions regarding social equity can be approached from high-level spatial scales (e.g., country or region) but will more often be addressed at a more focused scale (e.g., community or group level), especially when exploring gender issues or indigenous studies. Questions could include the following:

- Are fishing resources shared or accessed equally by community members?

- Within a community, are there gender differences in fishing practices?
- Why are specific social groups marginalized in a community?
- How do gender issues affect community decision-making and consultation processes?
- How might inequality within a community be addressed?
- How does inequality affect the intended outcomes of shark conservation or management?

17.2.3.5 *Social Science Contributes to Understanding Different Knowledge Systems, Cultures, and Alternative Value Systems*

Social scientists can work with indigenous fishers and coastal communities to understand a wide range of issues that are related to sharks and rays. Knowing which resource management arrangements function well within cultural and belief systems can be important for resource sustainability or community cohesion. For example, religious beliefs may provide proxy conservation protection for sharks in some parts of the Solomon Islands (Hylton et al., 2017), while indigenous knowledge systems may provide crucial insights about designing effective conservation or management plans based on cultural values attributed to sharks (Nirmale et al., 2007). Questions that could be explored in this area include the following:

- What is the existence value of a species (e.g., culturally significant, perhaps endangered)?
- What are the customary management systems in place for subsistence fishing in coastal communities?
- What contributions can indigenous knowledge have for the conservation of a shark species?

17.2.4   Social and Economic Impacts

Fishing and other livelihood means are often considered in terms of social and economic impact, and this is perhaps one area that people may look to first when considering shark management and conservation. Studies in this area could examine the impacts and consequences of management actions or economic investment activities on the social or economic fabric of a community or region (see Case Study 1 in Section 17.6). This research area may include the compilation of community profiles that document social indicators such as gender, age, income level, livelihood diversity, education, mobility, and debt. Social impact questions could include the following:

- What are the effects of implementing a marine protected area for sharks on the socioeconomic situation of a community or region that participates in shark fishing?
- If alternative fishing practices were introduced to a community, what are the effects to livelihood factors?

Economic impact questions may include the following:

- What are the economic impacts on small coastal communities from collapsed or closed shark fisheries?
- What are the predicted economic impacts of a legislative change that bans all exports of shark fin?
- How does access to markets influence people's engagement in illegal shark fishing?

17.3   SOCIAL SCIENCE PLANNING AND RESEARCH DESIGN

17.3.1   Special Considerations for Social Science Methodologies

In addition to standard scientific procedures related to permits, sampling, experimental design, and data management (including storage), social scientists have other idiosyncratic considerations to take into account. These include ethical implications specific to working with people (see Section

17.5), gaining support and access to communities, language and literacy considerations, adhering to cultural protocols, and other more subtle but important social and cultural factors (Creswell and Poth, 2017). Unlike many other types of science, social scientists are direct participants in their research and also act as the primary data collection and processing tool. In many types of projects, social scientists live the research experience, and this can make handling subjectivity particularly challenging. To address the challenges of social science, various research design criteria are available to help navigate these complexities.

17.3.2   Defining Research Questions

When projects are generated by the researchers themselves, defining the research questions may involve an initial idea or topic. But, as researchers on multidisciplinary teams, social scientists might have existing problems or situations presented to them (e.g., overfishing of sharks in the region) and the specific questions for the study may emerge through an iterative process. In this situation, it helps to have a good sense of the issues that the research topic involves and to be clear how new questions build on, and might add to, previous research. Although research might start out with specific questions and data collection plans, the relationship between social science design, data, and theory should be considered to be multidirectional (Ritchie et al., 2013).

It is important to develop clear and unambiguous questions, but it is also essential that the questions can feasibly be answered through data collection, can be addressed given the resources that are available, and are informed by existing research or theory (with the potential to fill any knowledge gaps). Even though it might particularly relate to participatory action research (PAR), which is a form of self-reflective inquiry that researchers and participants undertake together in order to understand not only the situations they are in but also the processes that they participate in (Baum et al., 2006), much benefit can be gained from working with a community or stakeholder group to have them included in a research project from the earliest stages, including in the development of the questions. It is through this collaborative approach that scientists are able to gauge what is important to people (or not) and establish what is important to ask. It also minimizes the risk that a researcher will impose a world view that is irrelevant or inappropriate to the cultural context in which they are working.

17.3.3   Understanding Research Settings and Fieldwork Requirements

Fieldwork, as a part of social science research, usually involves interacting with or observing people (as the subject of that research). It is a dynamic process where there is an exchange among the researchers, participants, stakeholders, gatekeepers, community, and larger sociopolitical context in which the research problem is located. This may not always be the case, though. In studies that are looking at people's perspectives or opinions, especially at a large scale, the data collection methods may involve no physical contact with people but may instead require online or phone surveys to be undertaken to collect both demographic and perspective information.

Of course, fieldwork for all researchers will be very different, but the thinking and planning invested in this process are reasonably consistent. There are lists to make, equipment to gather, travel to book, and people to liaise with. Time, budgetary, and other resource constraints all must be considered as part of the research design, as do the data collection methods and the staff or volunteers available to the project.

A good place to begin when establishing the context of a research setting is to consider what permissions are required to work there. Those permissions might have to run across multiple levels and can take a long time to get through, so appropriate time and scope must be allowed for this step. In Australia, for example, if research involves conducting surveys with fishers along the east coast, university human ethics approval must be in place before data collection can commence. This single permission contrasts with the multiple permissions that one may need in order to conduct very similar surveys in Papua New Guinea (PNG). Along with human ethics approval, researchers may also need PNG government research permits and local community permission from the village chief or elder (that is, before permission is sought from potential interview or survey participants).

Having to gain permission from multiple levels of governance (e.g., university, country, community, indigenous group) is becoming common practice in today's research environment and free prior-informed consent is taken very seriously by ethics committees, legal practitioners, and publishers. If researchers are not familiar with their fieldwork location, it pays to discuss the practicalities of working in locations with people who have done similar work in that space, or—if that is not possible—planning for a scoping trip to introduce yourself and look at the conditions that will be encountered is definitely beneficial.

Many location-based questions must be asked, and here are some of the key factors that have come up based on our collective experiences:

- Is it safe? Can individual researchers go into places by themselves and undertake ethnographic-styled observations, or would it be more appropriate to take a team and plan for different data collection tools? Look around at the political and security climate of the area—should researchers be going there at all?
- What amenities are there? Are the only accommodation options provided by community members, or are hotels available? Where would food be purchased? Is money accepted as a payment method?
- Is it necessary to have a gender-balanced team to speak to men and women in the community? This must be considered from both a custom point of view (women may

not be allowed to talk to non-family males) or from a topic point of view (hunting may be viewed as a male-only activity and cannot be discussed in front of women).
- Are there language barriers? Are interpreters necessary?
- What are the cultural customs that researchers should be familiar with before entering a community? They may be obvious, such as clothing protocols in Muslim countries, or less obvious, such as not asking a man where the toilet is in the Solomon Islands or blowing one's nose in Indonesia.

17.3.4 Fieldwork Strategies

How researchers undertake field work is very much influenced by the information gathered from the research setting (see above), as well as all the other aspects of research that everyone has to deal with (money, time, scope, etc.). Scientists across any domain will encounter the practical considerations of actually getting out and collecting data; however, the importance of the human relationships that are established as part of the research process heavily influences the success (or failure) of social research and cannot be underestimated. Here are some of the on-ground considerations that social scientists must work through to conduct their research:

- With regard to access and field trip length, a lot of applied social science research is dependent on the relationships that can be formed with the local community, so it is important to factor in multiple trips (if possible). Initial travel may consist of a scoping trip to introduce the research team and request community permission to conduct research, but some trips may require longer field stays to gain people's trust, to encourage better participation. It is good to allocate extra time to train local research assistants and to plan return trips to give back the results to the community.
- Time allowed must reflect the sample generation approach (see Section 17.4.5 for sampling), such as the likely duration of interviews and group discussions.
- Some of the data collection methods for social science require working in pairs, so it is important to consider, for example, management of research assistants and volunteers, as well as the training required for everyone to be consistent in their data collection approaches.
- Debriefing of research team and liaising with any local team members that need extra time for training or language interpretation should be planned for.
- The scope for integrating early analysis with later fieldwork should be determined.
- Pilot testing is an often overlooked but extremely important part of the research process. It helps detect potential problems in research design or instrumentation (e.g., whether the questions asked are intelligible to the targeted sample) and ensures that the data collection tools used in the study are reliable and valid measures of the constructs of interest (Bhattacherjee, 2012). How people interpret the question both linguistically and culturally (how they react to it) is really critical, and such testing

allows researchers to determine if any questions generate unwanted reactions or suspicions.

- With regard to transcribing, if data (notes) are collected via video or audio, then those words need to be typed up. Adequate time must be allocated for this task, and the complexity of the job increases when the responses must be translated and transcribed.
- It is necessary to find and work with gatekeepers (i.e., the people you need to keep onside to maintain communication between you and your subjects). The challenge in working with gatekeepers is maintaining neutrality and ensuring that you do not accidentally align with people or organizations that would alienate you from certain other groups in the community. It is essential to be objective as possible and not push any agenda or perspective on people. This is difficult when you are trying to be open and ask people to trust you.
- Continuing on from gatekeepers, do you need a cultural broker or project champion—someone from the location where you are doing research who can introduce you, speak for your credibility, and encourage participation? Relationships enable fieldwork. Multiple relationships can be developed with a wide range of stakeholders: non-governmental organizations, academics, community members, politicians, translators, friends—the list goes on. This can be a critical feature of your research setup if things begin to go wrong (e.g., rumors are generated about your intentions, conflicts arise related to the project).
- How the research will be shared with the people or community involved should be considered.

17.4 DATA TYPES AND DATA COLLECTION METHODS

The domain of social science is commonly linked with the production of qualitative data; however, quantitative data have an equally valid role in this space and can be especially useful when applying a mixed-methods approach (use of multiple methods to data collection) to the research questions. A researcher needs to decide what data will best answer the question and look for the contextually appropriate data collection methods during the planning phase of the research. Table 17.2 provides a brief overview of the features of quantitative and qualitative data types, including a list of commonly used tools that are associated with data collection.

17.4.1 Quantitative Data in Social Science

Quantitative social science data are useful because they can be analyzed statistically. This allows for detection and monitoring of trends across large samples at multiple scales (e.g., individuals, households, communities, regions). Similarly, quantitative social science data can be analyzed or modeled in conjunction with ecological data, thus gaining insight into the feedbacks between social and ecological systems. Quantitative social science data can range from ratio to ordinal scales. Ratio or continuous data can represent demographic trends (e.g., age, population), whereas ordinal scales are generally used to capture people's beliefs and perceptions related to different aspects of their lives. Ordinal scales are numeric, but because the distance between each unit is subjective and may vary among respondents—for example, on a scale of 1 (strongly disagree) to 10 (strongly agree), how much do you agree with the statement that sharks are declining in the waters surrounding your community)—they must be analyzed using nonparametric statistical approaches. Such approaches can also accommodate categorical variables (e.g., five discrete categories of fishing gears affecting sharks) or dichotomous variables (e.g., yes/no, present/absent), which means that multiple scales of social science data can be analyzed quantitatively.

As mentioned previously, an enduring feature of quantitative methods comes from its replicability and its ability to illustrate broader trends (Spoon, 2014). For example, the use of quantitative surveys to determine the support of Costa Rican fishermen for conservation measures toward sharks found that almost all surveyed fishermen (97%) would be willing to support shark conservation. However, their support decreased to 67% if their current fishing practices were impacted by any conservation measures. Interestingly, a large number (86%) of fishermen indicated they would be willing to potentially support shark conservation if fishing communities were included in the overall decision-making process (O'Bryhim et al., 2016). Conversely, quantitative methods can have limited depth because of their quantified, positivist approach. Variables must be reduced to measurable units, which can simplify or misinterpret the dynamics of cultural phenomena without proper context; thus, qualitative methods may offer a more flexible option to inductively explore a knowledge domain or practice (Spoon, 2014).

Table 17.2 Quantitative and Qualitative Data Types

Quantitative Data	Qualitative Data
Numerical data or data that can be transformed into usable statistics; the data are used to quantify attitudes, opinions, behaviors, demographics, economic trends, etc. Measurable data are used to test or develop hypotheses, identify trends and patterns, make comparisons or correlations among social groups, test dependent variables against a set of independent variables.	Data are used to gain an understanding of underlying drivers and motivations of behavior, to uncover trends in thought and opinion, and to delve deeper into the problem.
Tools—Quantitative data collection methods are usually much more structured than those for qualitative data (e.g., population and sample surveys, online polls, systematic observations, experiments, longitudinal studies, website interceptors)	Tools—A variety of unstructured or semi-structured techniques are used (e.g., semi-structured and open-ended interviews, life histories, focus groups, participant observation, content analysis)

17.4.2 Understanding Qualitative Data

Social science research is often referred to as "qualitative research" due to the predilection of scientists to use qualitative methods. Approaches to collecting qualitative data can be divided into two very broad groups: those that focus on data that already exists (naturally occurring) and those that generate data through the research process (generated data).

17.4.2.1 Naturally Occurring

Naturally occurring data are what happens in the world without any kind of researcher intervention (e.g., they are not dependent on a researcher asking questions or organizing a focus group). The methods can be applied when behaviors and interactions of people must be understood in real-world contexts, which can be particularly useful in research hoping to understand a particular culture or community and the governance, customs, or rules that exist there. The main methods involved in working with naturally occurring data are observation, participant observation (where the researcher takes part in the group or community to record actions, interactions, and events that occur), document analysis, conversation analysis, and discourse analysis.

17.4.2.2 Generated Data

Generated data give insight into people's own perspectives on and interpretation of their beliefs and behaviors—and, most importantly, an understanding of the meaning that they attach to them. Generated data are required in an assortment of research settings, partly because they provide the only means of understanding things such as motivations, beliefs, and decision processes, but also because they allow participants' reflections on, and understanding of, social phenomena to be documented. Numerous qualitative methods generate data. They include, but are not limited to, biographical methods, which can cover a large range of material, both written and spoken, including life and oral histories, biographical accounts, and documents from peoples' lives; interviews (semi-structured, open ended); and focus or discussion groups.

17.4.3 Combining Qualitative and Quantitative Methods

Mixed-methods research is a methodology for conducting research that involves collecting, analyzing, and integrating both quantitative and qualitative data. Qualitative- and quantitative-based research should not be seen as competing and contradictory, but rather as complementary strategies appropriate to different types of research questions or issues (Ritchie et al., 2013). It can be a useful approach to research when a scientist requires a deeper understanding of the phenomenon, while accounting for the weakness of any single method (Ritchie et al., 2013). Mixed methods are particularly useful in a number of instances. For example, if very little is known about a community and its association with sharks, a qualitative study could be done initially to learn about what variables exist. A researcher could then study those identified variables with a large sample of individuals from a community or a coastal region using quantitative research methods.

Another advantage of linked quantitative and qualitative methods is that they can assist in the translation of social science information to predominantly biophysical science contexts, which can subsequently be applied to policy and management (Spoon, 2014). It should be mentioned that mixed methods can bring some limitations or disadvantage to the research. For some situations, the research design can be very complex, requiring more resources or time than what might be available. In addition, the interpretation of the results can also be tricky if discrepancies arise in the results. It should be noted that researchers who are experienced in applying mixed methods to research questions are able to work through or account for these issues.

17.4.4 Data Collection Methods

The methods used by a social scientist will be heavily influenced by the aims of the research and the specific questions that must be answered. Although all methods can provide important insights into the social realm, the limitations of each method will need to be acknowledged. As explained in the section above, the use of multiple or mixed methods can be applied, and the option of using both qualitative and quantitative analytical techniques for the same data is possible. The purpose of the research, as well as the geographical scale at which the study is to be conducted will also influence which methods are suitable for use in a study. In Table 17.3, we have arranged an array of social science methods by their data type to highlight the large range of choice that can exist for an area. Some of the more commonly used data collection tools in social science research are surveys, observations, and interviews. Because they are so widely used, these three tools are discussed in more detail below.

Table 17.3 Methods Toolbox for Social Scientists

Method Category	Examples of Social Science Methods
Qualitative data	Interviews, focus groups, participant observation, discourse and textual analysis, document analysis, free list and pile sorts, ethnographies, photograph elicitation, case studies, image and video analysis, photovoice, participatory videography, scenarios, policy analysis, case analysis
Quantitative data	Surveys, economic valuations, cost–benefit analysis, modeling, scanner data gathering

17.4.4.1 Surveys

Surveys are a popular data collection tool and are often used in conjunction with interviews or other methods. They are concerned with current phenomena (as opposed to historical research, which is concerned with past events) and can be used to gather data from people using a (generally) structured instrument in the form of a questionnaire. Surveys can be used to establish people's views of what they think, believe, value, or feel, and it is possible to sample a population of potential respondents in order to generalize conclusions more widely (Jankowicz, 2000). Surveys can also explore the frequency with which attitudes, events, etc. occur or issues at a particular point in time (cross-sectional) and over a particular time period (longitudinal). There are a great many situations that lend themselves to surveys—for example, when establishing catch composition, fishing equipment, and use of caught sharks in artisanal fisheries. Glaus et al. (2015) used a questionnaire that included dichotomous and pre-categorized questions to establish what fishing gear was prevalent in a shark fishery and to establish motivations to retain or release sharks. In another study that investigated insights into shark tourism, Richards et al. (2015) used surveys to ascertain shark species encountered on tourist dives, type of habitat sharks were seen in, whether sharks were fed on dives, and whether dive operators had a code of conduct for shark encounters. Another aspect of the shark diving industry was investigated by Vianna et al. (2012), who conducted socioeconomic surveys based on four different questionnaires to collect information from people who were directly or indirectly affected by the shark diving industry in Palau.

17.4.4.2 Observation

Observation and participant observation are types of data collection methods typically used in qualitative research. Observation is widely used in many disciplines, particularly cultural anthropology, but also in sociology, ethnobiology, sociology, communication studies, human geography, and social psychology. Observation can be defined as the systematic description of events, behaviors, and artifacts in the social setting chosen for study (Marshall and Rossman, 2014). Observations enable the researcher to describe existing situations using the five senses and provide a written account of everything that is occurring. Participant observation, on the other hand, is the process of enabling researchers to learn about the activities of the people under study in the natural setting through observing and participating in those activities (DeWalt and DeWalt, 1998). So, it is the process of learning through exposure to or involvement in the day-to-day or routine activities of participants in the research setting. Anthropological insights into the cultural practices of shark fishing by the Sama Dilaut people of the

Tawi Tawi Island in the Southern Philippines were achieved through data collected via observation (Nimmo, 2001). This research spanned 40 years and synthesized years of notes collected while the researchers lived with and observed these maritime people.

17.4.4.3 Interviews

The two most common qualitative data collection tools are surveys and interviews. The latter are generally considered more useful in eliciting narrative data that allow researchers to investigate people's views in great depth (Kvale and Brinkmann, 2009). That is, the value of the interview lies in its ability not only to build a complete picture, allow for the analysis of words, and report detailed views of participants but also to enable people to tell their own story and express their own thoughts and feelings (Berg, 2004).

There are four types of interviews that are frequently employed in social sciences. The first is the *structured interview*, which as the name suggests is organized around a list of predetermined questions that require immediate (generally yes or no) answers. There is very little freedom in this interview type (Berg, 2004), and it is sometimes likened to questionnaires in both its form and underlying assumptions. The second type of interview is the *open-ended (unstructured) interview*. This kind of interview is an open situation that offers much greater flexibility and freedom to both the interviewee and interviewer, which allows for the elaboration of various issues and closer follow-up and investigation of interesting topics (Gubrium and Holstein, 2002). Third is the *semi-structured interview*, a more flexible version of the structured interview that provides the opportunity for the interviewer to probe and further expand the interviewee's responses (Rubin and Rubin, 2011). A number of examples of social science research in the shark field have used semi-structured interviews. Stacey et al. (2012) conducted field research in eastern Indonesia using observations and semi-structured interview collection methods to understand customary practices and beliefs concerning whale sharks, as well as their geographic locations, migrations, aggregations, seasonal patterns, and threats they face. This information was collected to assess whether suitable biological conditions existed to support ecotourism based on whale sharks in the region. Semi-structured interviews were also used by Barbosa-Filho et al. (2014) to interview shark fishing specialists in northeast Brazil. This research was able to document demographic data about fishermen, their shark fishing techniques, and their knowledge about shark behavior, and it was able to demonstrate the complexity and robustness of artisanal fishermen's knowledge of sharks. Finally, interviewing using a *focus group* is a useful data collection method suitable for studying processes that have a strong social element, such as those that depend on group interaction and where it is important to take account of assorted views and perspectives.

Table 17.4 Key Terms in Sampling

Term	Description
Sampling frame	List of accessible population from which you would draw your sample (e.g., everyone listed on the electoral roll, all permanent residents of a village, all shark dive tour operators based in a specific port)
Sample	Group of people from your sampling frame whom you select to be in your study (selection is via a sampling strategy described in Sections 17.4.5.1 and 17.4.5.2)
Subsample	Group that actually completes your study (not everyone in the sample population may consent to be in the study or they may not finish)

17.4.5 Sampling

When conducting any research, it is not often possible to study the entire population of interest. This is the case whether the research is qualitative or quantitative. Even if research involves very small numbers of people or a single case study, decisions have to be made about the participant, the environment (context) in which the research is occurring, and the research steps that must occur. In some cases, a sample will represent a larger population and can be used to draw inferences about that population; however, in other research, a sample may not represent a population but instead a phenomenon (or feature) of a person or community. The sample size that a researcher determines as suitable for the research will vary with methodological approach and the time available (Ritchie et al., 2013). Specific terms are used when discussing sampling in social inquiry, and it is important to use the correct terms when detailing the research methods so results can be interpreted correctly (see Table 17.4). Within social science, the two main types of sampling techniques are those based on probability and those that are not. This sounds straightforward, but sampling is a difficult multistep process and requires thought and planning to avoid issues of introducing systematic error or bias. With that in mind, we will briefly describe the different kinds of samples that can be created using both techniques.

17.4.5.1 Non-Probability Sampling Techniques

Samples collected through this technique are gathered in a process that does not give all individuals in the population equal chances of being selected. Three main types of sample can be created in this way:

- *Convenience sampling*—Available subjects must be accessible, such as people walking through a shopping center. This does not allow the researcher to have any control over the representativeness of the sample; however, it is useful to study the characteristics of people going by. This approach is not often used if the results are to be generalized to a wider population.
- *Purposive sampling*—Participants are chosen because they have particular features or characteristics that will allow detailed exploration and understanding of the main ideas that the researcher wishes to study. These

may be sociodemographic characteristics, or they may relate to specific experiences or behaviors. For example, to understand the motivations behind people who undertake shark finning, the sample would exclusively include people who fish for sharks. Many different types of purposive sampling are available (e.g., heterogeneous samples, extreme case or deviant sampling, intensity sampling, stratified purposive sampling) (Patton, 2002).
- *Snowball sampling*—This technique is appropriate to use in research when the members of population are difficult to locate or might have specific attributes that are known to the person being interviewed. For example, a researcher who wants to interview all scientists who have studied sharks along the east coast of Australia in the past 10 years might ask the person they are currently interviewing if he or she knows of anyone else who is doing or has done such research on sharks in Australia and if that person could help contact them.

17.4.5.2 Probability Sampling Techniques

In this technique, all the samples are gathered in a process that gives all of the individuals in the population an equal chance of being selected. Some people consider this a more methodologically rigorous approach to sampling because it eliminates social biases that could influence the research sample. Two types of probability sampling techniques are

- *Simple random sample*—This basic sampling method is used in statistical methods. The simplest example of this type of sampling is when every person of the target population is assigned a number. A set of random numbers is then generated and the people having those numbers are included in the sample. This may not always work if the population differs much in age, race, education, or social class because this technique cannot take demographic differences into account.
- *Stratified sampling*—With this method, the researcher divides the entire target population into different subgroups (or strata) and then randomly selects the final subjects from each of the different areas. For example, a researcher who wants to know how people view shark meat as a resource (food source) could organize people into groups based on their age and gender and then select appropriate numbers from each of the groups. This would ensure that the researcher has adequate amounts of subjects from each class in the final sample.

17.5 ETHICS

Within the realm of human interactions, human ethics (especially those related to indigenous peoples) move into the forefront of thinking. The ethical issues that might surface during planning, designing, and conducting a qualitative study must be considered and addressed at each level. It is a common misconception that ethics only pertain to the data collection phase of research; however, they occur through several phases of the research process, and a certain level of sensitivity is required to work with participants, stakeholders, publishers of research, and holders of indigenous knowledge. This ethical behavior can even extend beyond the conduct of the research to the use of the data (e.g., how the data are identified/de-identified, reported, or analyzed) and may involve the ethical fairness of the benefits of knowledge generation. In addition, working in complex situations or at culturally significant sites requires an informed and respectful approach.

Some excellent scholarly literature is available on how ethical consideration should be applied to different phases of the social science research process. Creswell and Poth (2017), Lincoln (2009), and Mertens and Ginsberg (2009) have provided overviews on how ethical issues in social science can be described as occurring prior to and at the beginning of the research, during data collection, in data analysis, and in the reporting and publishing of the research outcomes. Following are two aspects of ethics warranting special mention in this chapter:

- All human interactions, including the interactions involved in human research, have ethical dimensions; however, ethical conduct is more than simply doing the right thing.
- All researchers working with people should consider the ethical fairness of the benefits of knowledge generation. This is an important issue and is raised often in research that involves indigenous people. Much work has taken place in explaining the intellectual property rights pertaining to indigenous knowledge, at both the community and individual level (Janke, 2005). Knowledge generation pertains to both social and biophysical scientists, as it can relate to the use of knowledge for social research as well as for research on sites or species. An example of this would be scientists who work with indigenous communities to understand the presence or absence of animals and to determine temporal or spatial distributions to focus their tracking or biodiversity surveys. Information may be shared about a species' location, distribution, or habitat preference, and critical local knowledge about the area might be used to conduct the research safely. But, although the people involved in the work may be acknowledged, there is not often more consideration given to the knowledge input (i.e., they very rarely see co-authorship on papers despite significant intellectual input).

17.6 CASE STUDIES

Case Study 1. Mixed Methods Research

SHARK FISHING LIVELIHOODS IN THE CROSSFIRE: A CASE STUDY FROM EASTERN INDONESIA

A case study conducted over the course of 2 years examined how livelihoods in three eastern Indonesian fishing communities had changed over a 20-year period, stretching from the beginning of commercial shark fishing and finning through to the Asian financial crisis and more recent impacts. These impacts included marine conservation initiatives, declines in shark numbers, forced and voluntary changes in fishing grounds, fisher–patron relationships involving increasing debt, and the effects of changes in consumer awareness and trade on shark fishers in Indonesia (Jaiteh et al., 2017a,b). Shark fishers in each community—two in Maluku province and one in Nusa Tenggara Timur province—were involved in data collection through various qualitative and quantitative research methods. The principal researcher lived in each community for at least 3 months to allow for familiarity and trust to be established between the researcher and community members. At the beginning of each stay, the researcher sought permission from the *Kepala desa* (head of the village) and, if applicable, other persons or groups, such as the *Adat* (village council), to approach and work with fishers. Active shark fishers were approached first and asked if they would like to contribute to data collection for the research project. The project was described in detail to each fisher, and it was stressed that participation was entirely voluntary.

Fishers who agreed to take part were asked to collect data on their shark catch, including the fork and/or total lengths, local species names, latitude and longitude of capture, sex, and a number of sex-specific parameters. Active fishers from the three communities recorded information for almost 2000 individuals within a relatively short time, which provided the basis for the most comprehensive dataset recorded to date in eastern Indonesia, part of the world's biggest shark producing and exporting country. This dataset led to a number of interesting findings, including the likely status of the shark fishery and the diversity of species being caught (Jaiteh et al., 2017a). However, the most valuable information about the fishery and the livelihoods it supports was arguably obtained through extensive participant observation and detailed semi-structured interviews.

Participant observation began as soon as researchers entered a community and continued throughout their stay and included land-based and sea-based activities that were not necessarily fishing related. Interviews were always conducted toward the end of a stay in a community, after relationships had been established and most respondents had met the researcher. Respondents were chosen opportunistically, taking into consideration that no two respondents could live in the same household. Also, respondents included at least 30 active fishers, 30 retired fishers (those who had not fished for at least a year and declared to have no intention of returning to shark fishing), and 20 non-fishing community members. Interview questions spanned the respondent's household condition, financial status, the history of shark fishing in each community, current fishing practices and beliefs, changes in catch and shark fin prices over the previous 20 years, alternative livelihood options and aspirations, and thoughts on the future of shark fishing.

The results from this research brought to light some important perspectives from the shark fishers that may be useful for community development, fisheries management, and shark conservation. For example, fishers in each of the three case study sites had noticed declines in shark populations over the last two decades and particularly the most recent ten years (Jaiteh et al., 2017a; Whitcraft et al., 2014). In the two Moluccan sites, fishers responded to these declines by changing their fishing grounds (Jaiteh et al., 2017b). One of those fishing grounds encompassed the rich waters of the Raja Ampat regency which were made a shark sanctuary in 2013 after a decree was passed by parliament that banned all commercial and artisanal fishing for sharks and rays in the regency. Shortly before the sanctuary was officially established, one of the shark fishing boats from this study site (along with several others from elsewhere in the region) was apprehended (Jaiteh et al., 2017b). Their catch, 6 weeks' worth of fishing, was confiscated along with their gear; the crew was given a warning and told not to return. This single incident caused the majority of fishers from the community to stop shark fishing altogether; in interviews, they stated that it was not worth the risk of losing all their catch and gear. In the other two study sites, most fishers worked on boats owned and financed by a "boss" (or *bos* in Indonesian), which in many cases led to substantial debt. As such, the majority of fishers were not free to change their livelihood until they had paid off their debt. Many of these fishers made it clear that if there were viable livelihood alternatives for them they would not hesitate to take advantage of them. However, because these fishers lived on remote islands with limited infrastructure, and often a relatively high proportion of them had received only minimal education, livelihood enhancements and alternatives were not readily available. It remains to be seen how fishers in each of the case study sites, as well as in Indonesia and in source countries elsewhere, adapt and respond to the ongoing challenges of their precarious, rapidly changing livelihoods.

Case Study 2. Ethics and Approach

INDIGENOUS KNOWLEDGE IN SHARK RESEARCH AND MANAGEMENT

This example of mixed-methods research is an Australia-based project that provides an example of ethical considerations and engagement in community based research. The project used a case study approach to observe and document indigenous knowledge about sharks and rays. In collaboration with a traditional owner group (Yuku Baja Muliku) in far north Queensland, the researchers designed the study in a culturally respectful way that encouraged involvement and feedback from the community from the very inception of the study (K. Gerhardt, unpublished data). Some of the objectives of the project were as follows:

1. Establish a culturally appropriate protocol for the management of indigenous knowledge that was documented during the project (a protocol that was specific to the group and met all of their concerns and needs).
2. Document the traditional and contemporary practices of shark and ray fishing and use in traditional owner sea country.
3. Document and understand the cultural and social roles that sharks and rays play within the traditional and contemporary society.
4. Agree upon two-way sharing benefits and deliver them as part of the project.

In accordance with national and international best practice protocols for research in indigenous communities, this project used a collaborative research methodology based on the Australian Institute of Aboriginal and Torres Strait Islander Studies (AIATSIS) Guidelines for Ethical Research and the International Society for Ethnobiologists (ISE) Code of Ethics, thus ensuring that data sharing agreements (including benefit sharing agreements) were in place, that research methods were agreed upon by the community, and that community feedback loops were built into the project to allow for changes suggested by the traditional owners.

Ethics for the project were provided through the James Cook University human research ethics committee (Aboriginal and Torres Strait Islander ethics approval), and permission and endorsement were provided by the Yuku Baja Muliku traditional owner negotiating committee. Two project managers were identified within the traditional owner group to help champion the project and provide on-ground communication pathways for the researcher and community members. It was also an opportunity for traditional owners to gain valuable project management experience and skill in the area of research projects and collaborating with scientists.

Reciprocity (two-way sharing) was an important feature during the initial design phase of the project. The proposition that working together on the research meant that there would be benefits for both the researcher and the traditional owner group was important to document before the research was underway. Project deliverables that were included in the benefit sharing agreement (a memorandum of understanding between the researcher and the traditional owner group) were as follows:

- Traditional owner fact sheets on sharks and rays describing species, cultural values, traditional ecological knowledge, and relevant Western scientific knowledge
- Cultural knowledge collected and returned to the traditional owner group (in raw form and in collated form that could be uploaded into their cultural database)
- Shark and ray posters with pictures of each species found in traditional owner sea country, with traditional and scientific names (e.g., for use in Junior Ranger programs or to share with schools)
- Final report back to each group on outcomes of the project that can be used to support further work or research that the groups may wish to undertake
- Research data and reports (for the researcher)
- Scientific publications arising from the project with co-authorship offered to traditional owners who participated in the research

17.7   SUMMARY AND CONCLUSION

Human-driven pressures on sharks and rays are one of the most important factors in the collapse of top order predator populations (Davidson et al., 2016; Dulvy et al., 2014); yet, social science related to shark fishing, conservation, and management is still in its infancy and has not caught up with the ground swell of biophysical shark science. As many shark and ray species around the world continue to decline, the research needs of species and fisheries managers, conservationists, policymakers, and communities will require a more comprehensive understanding of socioecological systems that sharks exist within. Better incorporation of social science research in the form of economic, governance, cultural, or social findings has the potential to produce initiatives that are more suited to local contexts and may find improved outcomes for sharks, simply because the management fit is more suited to the people involved.

Support for the use of interdisciplinary approaches to shark research and management is critical to understanding and dealing effectively with the complexity of livelihood issues associated with sharks (whether through fishing, trade, conservation, or custom). It is hoped that this acceptance of interdisciplinary teams as a standard approach to shark research will lead to increased collaboration between biophysical scientists and social scientists—as well as

greater interaction between managers and policymakers—through all stages of research. As we expand our research teams, it is also important to recognize that different types of social research will provide different insights into the same issues, simply because the social reality is complex and one social science methodology may not have the scope to provide all the insights necessary to explain what is occurring.

REFERENCES

Agardy T (2000) Information needs for marine protected areas: scientific and societal. *Bull Mar Sci* 66(3):875–888.

Aswani S, Hamilton RJ (2004) Integrating indigenous ecological knowledge and customary sea tenure with marine and social science for conservation of bumphead parrotfish (*Bolbometopon muricatum*) in the Roviana Lagoon, Solomon Islands. *Environ Conserv* 31(01):69–83.

Ban NC, Mills M, Tam J, Hicks CC, Klain S, Stoeckl N, Bottrill MC, et al. (2013) A social–ecological approach to conservation planning: embedding social considerations. *Front Ecol Environ* 11(4):194–202.

Barbosa-Filho MLV, Schiavetti A, Alarcon DT, Costa-Neto EM (2014) "Shark is the man!": ethnoknowledge of Brazil's South Bahia fishermen regarding shark behaviors. *J Ethnobiol Ethnomed* 10(1):54.

Barker MJ, Schluessel V (2005) Managing global shark fisheries: suggestions for prioritizing management strategies. *Aquat Conserv Mar Freshw Ecosyst* 15(4):325–347.

Baum F, MacDougall C, Smith D (2006) Participatory action research. *J Epidemiol Commun Health* 60(10):854–857.

Bennett NJ (2016) Using perceptions as evidence to improve conservation and environmental management. *Conserv Biol* 30(3):582–592.

Bennett NJ, Roth R, Klain SC, Chan K, Christie P, Clark DA, Cullman, G, et al. (2017) Conservation social science: Understanding and integrating human dimensions to improve conservation. *Biol Conserv* 205:93–108.

Berg BL (2004) *Methods for the Social Sciences.* Pearson Education, London.

Bhattacherjee A (2012) *Social Science Research: Principles, Methods, and Practices*, Textbooks Collection 3. University of South Florida Scholar Commons, Tampa (http://scholarcommons.usf.edu/oa_textbooks/3).

Blythe JL (2015) Resilience and social thresholds in small-scale fishing communities. *Sustain Sci* 10(1):157–165.

Chambers R, Conway G (1992) *Sustainable Rural Livelihoods: Practical Concepts for the 21st Century.* Institute of Development Studies, East Sussex, UK.

Cinner JE, McClanahan TR, MacNeil MA, Graham NAJ, Daw TM, Mukminin A, Feary DA, et al. (2012) Comanagement of coral reef social-ecological systems. *Proc Natl Acad Sci USA* 109(14):5219–5222.

Comte A (1975) *Auguste Comte and Positivism: The Essential Writings.* Transaction Publishers, Piscataway, NJ.

Creswell JW, Poth CN (2017) *Qualitative Inquiry and Research Design. Choosing Among Five Approaches.* Sage Publications, Thousand Oaks, CA.

Davidson LN, Krawchuk MA, Dulvy NK (2016) Why have global shark and ray landings declined: improved management or overfishing? *Fish Fish* 17(2):438–458.

DeWalt K, DeWalt B (1998) Participant observation. In: Bernard HR (ed) *Handbook of Methods in Cultural Anthropology*. AltaMira Press, Lanham, MD, pp 259–300.

Dulvy NK, Baum JK, Clark S, Compagno LJV, Cortés E, Domingo A, Fordham S, et al. (2008) You can swim but you can't hide: the global status and conservation of oceanic pelagic sharks and rays. *Aquat Conserv Mar Freshw Ecosyst* 18(5):459.

Dulvy NK, Fowler SL, Musick JA, Cavanagh RD, Kyne PM, Harrison LR, Carlson JK, et al. (2014) Extinction risk and conservation of the world's sharks and rays. *eLife* 3:e00590.

Dulvy NK, Simpfendorfer CA, Davidson LNK, Fordham SV, Bräutigam A, Sant G, Welch DJ (2017) Challenges and priorities in shark and ray conservation. *Curr Biol* 27(11):R565–R572.

Gallagher AJ, Hammerschlag N (2011) Global shark currency: the distribution, frequency, and economic value of shark ecotourism. *Curr Issues Tour* 14(8):797–812.

Glaus KB, Adrian-Kalchhauser I, Burkhardt-Holm P, White WT, Brunnschweiler JM (2015) Characteristics of the shark fisheries of Fiji. *Sci Rep* 5:17556.

Guba EG, Lincoln YS (1994) Competing paradigms in qualitative research. In: Denzin NK, Lincoln YS (eds) *Handbook of Qualitative Research*. Sage Publications, Thousand Oaks, CA, pp 105–117.

Gubrium JF, Holstein JA (2002) *Handbook of Interview Research: Context and Method*. Sage Publications, Thousand Oaks, CA.

Gutiérrez NL, Hilborn R, Defeo O (2011) Leadership, social capital and incentives promote successful fisheries. *Nature* 470(7334):386–389.

Hilborn R (2007) Managing fisheries is managing people: what has been learned? *Fish Fish* 8(4):285–296.

Huff TE (1984) *Max Weber and the Methodology of the Social Sciences*. Transaction Publishers, Piscataway, NJ.

Hylton S, White W, Chin A (2017) The sharks and rays of the Solomon Islands: a synthesis of their biological diversity, values and conservation status. *Pac Conserv Biol* 23:324–334.

Jacques PJ (2010) The social oceanography of top oceanic predators and the decline of sharks: a call for a new field. *Prog Oceanogr* 86(1):192–203.

Jaiteh VF, Hordyk AR, Braccini M, Warren C, Loneragan NR (2017a) Shark finning in eastern Indonesia: assessing the sustainability of a data-poor fishery. *ICES J Mar Sci* 74(1):242–253.

Jaiteh VF, Lindfield SJ, Mangubhai S, Warren C, Fitzpatrick B, Loneragan NR (2017b) Higher abundance of marine predators and changes in fishers' behavior following spatial protection within the world's biggest shark fishery. *Front Mar Sci* 3:43.

Janke T (2005) Managing indigenous knowledge and indigenous cultural and intellectual property. *AARL* 36(2):95–107.

Jankowicz D (2000) From 'learning organization' to 'adaptive organization.' *Manage Learn* 31(4):471–490.

Kittinger JN, Finkbeiner EM, Ban NC, Broad K, Carr MH, Cinner JE, Gelcich S, et al. (2013) Emerging frontiers in social-ecological systems research for sustainability of small-scale fisheries. *Curr Opin Environ Sustain* 5(3):352–357.

Knip DM, Heupel MR, Simpfendorfer CA (2012) Evaluating marine protected areas for the conservation of tropical coastal sharks. *Biol Conserv* 148(1):200–209.

Kuhn TS (1970) *The Structure of Scientific Revolution*, 2nd ed. University of Chicago Press, Chicago, IL.

Kvale S, Brinkmann S (2009) *Learning the Craft of Qualitative Research Interviewing*. Sage Publications, Thousand Oaks, CA.

Leslie HM, Basurto X, Nenadovic M, Sievanen L, Cavanaugh KC, Cota-Nieto JJ, Erisman BE, et al. (2015) Operationalizing the social-ecological systems framework to assess sustainability. *Proc Natl Acad Sci USA* 112(19):5979–5984.

Lincoln YS (2009) Ethical practices in qualitative research. In: Ginsberg P, Mertens DM (eds) *The Handbook of Social Research Ethics*. Sage Publications, Thousand Oaks, CA, pp 150–169.

Marshall C, Rossman GB (2014) *Designing Qualitative Research*. Sage Publications, Thousand Oaks, CA.

McClanahan TR, Castilla JC, White AT, Defeo O (2009) Healing small-scale fisheries by facilitating complex socio-ecological systems. *Rev Fish Biol Fish* 19(1):33–47.

Mertens DM, Ginsberg PE (eds) (2009) *The Handbook of Social Research Ethics*. Sage Publications, Thousand Oaks, CA.

Mills D, Béné C, Ovie S, Tafida A, Sinaba F, Kodio A, Russell A, et al. (2011) Vulnerability in African small-scale fishing communities. *J Int Dev* 23(2):308–313.

Nimmo H (2001) *Magosaha: An Ethnology of the Tawi-Tawi Sma Dilaut*. Ateneo de Manila University Press, Manilla, Phillipines.

Nirmale V, Sontakki B, Biradar R, Metar S, Charatkar S (2007) Use of indigenous knowledge by coastal fisher folk of Mumbai district in Maharashtra. *Indian J Tradit Knowl* 6(2):378–382.

O'Bryhim JR, Parsons E, Gilmore MP, Lance SL (2016) Evaluating support for shark conservation among artisanal fishing communities in Costa Rica. *Mar Policy* 71:1–9.

Ormston R, Spencer L, Barnard M, Snape D (2014) The foundations of qualitative research. In: Ritchie J, Lewis J, McNaughton NC, Ormston R (eds) *Qualitative Research Practice: A Guide for Social Science Students and Researchers*. Sage Publications, Thousand Oaks, CA, pp 1–25.

Ostrom E (2009) A general framework for analyzing sustainability of social-ecological systems. *Science* 325(5939):419–422.

Patton MQ (2002) *Qualitative Research and Evaluation Methods*, 3 ed. Sage Publications, Thousand Oaks, CA.

Persha L, Agrawal A, Chhatre A (2011) Social and ecological synergy: local rulemaking, forest livelihoods, and biodiversity conservation. *Science* 331(6024):1606–1608.

Pollnac R, Christie P, Cinner JE, Dalton T, Daw TM, Forrester GE, Graham NA, et al. (2010) Marine reserves as linked social–ecological systems. *Proc Natl Acad Sci USA* 107(43):18262–18265.

Pomeroy R, Ferrer AJ, Pedrajas J (2017) An analysis of livelihood projects and programs for fishing communities in the Philippines. *Mar Policy* 81:250–255.

Reid H (2016) Ecosystem-and community-based adaptation: learning from community-based natural resource management. *Clim Dev* 8(1):4–9.

Richards K, O'Leary BC, Roberts CM, Ormond R, Gore M, Hawkins JP (2015) Sharks and people: insight into the global practices of tourism operators and their attitudes to shark behaviour. *Mar Pollut Bull* 91(1):200–210.

Ritchie J, Lewis J, Nicholls CM, Ormston R (2013) *Qualitative Research Practice: A Guide for Social Science Students and Researchers.* Sage Publications, Thousand Oaks, CA.

Rosenau PM (1991) *Post-modernism and the Social Sciences: Insights, Inroads and Intrusions.* Princeton University Press, Princeton, NJ, 248 pp.

Rubin HJ, Rubin IS (2011) *Qualitative Interviewing: The Art of Hearing Data.* Sage Publications, Thousand Oaks, CA.

Salafsky N, Wollenberg E (2000) Linking livelihoods and conservation: a conceptual framework and scale for assessing the integration of human needs and biodiversity. *World Dev* 28(8):1421–1438.

Saunders M, Lewis P, Thornhill A (2007) *Research Methods for Business Students,* 6th ed. Pearson Education, London.

Simpfendorfer C, Heupel M, White W, Dulvy N (2011) The importance of research and public opinion to conservation management of sharks and rays: a synthesis. *Mar Freshwater Res* 62(6):518–527.

Smircich L, Calas MB (1987) Organizational culture: a critical assessment. In: Jablin F, Putnam L, Roberts K, Porter L (eds) *Handbook of Organisational Communication.* Sage Publications, Thousand Oaks, CA, pp 228–263.

Spoon J (2014) Quantitative, qualitative, and collaborative methods: approaching indigenous ecological knowledge heterogeneity. *Ecol Soc* 19(3):33.

Stacey NE, Karam J, Meekan MG, Pickering S, Ninef J (2012) Prospects for whale shark conservation in Eastern Indonesia through Bajo traditional ecological knowledge and community-based monitoring. *Conserv Soc* 10(1):63.

Topelko KN, Dearden P (2005) The shark watching industry and its potential contribution to shark conservation. *J Ecotour* 4(2):108–128.

Twyman C (2017) Community-based natural resource management. In: Richardson D, Castree N, Goodchild MF, Kobayashi A, Liu W, Marson RA (eds) *International Encyclopedia of Geography: People, the Earth, Environment and Technology.* John Wiley & Sons, New York, pp 873–883.

Vianna G, Meekan M, Pannell D, Marsh S, Meeuwig J (2012) Socio-economic value and community benefits from shark-diving tourism in Palau: a sustainable use of reef shark populations. *Biol Conserv* 145(1):267–277.

Whitcraft S, Hofford A, Hilton P, O'Malley M, Jaiteh V, Knights P (2014) *Evidence of Declines in Shark Fin Demand: China.* WildAid, San Francisco, CA.

White TD, Carlisle AB, Kroodsma DA, Block BA, Casagrandi R, De Leo GA, Gatto M, et al. (2017) Assessing the effectiveness of a large marine protected area for reef shark conservation. *Biol Conserv* 207:64–71.

CHAPTER 18

Network Analysis and Theory in Shark Ecology—Methods and Applications

Johann Mourier
PSL Research University, EPHE–UPVD–CNRS, CRIOBE USR 3278, Perpignan, France

Elodie Lédée
Fish Ecology and Conservation Physiology Lab, Carleton University, Ottawa, Ontario, Canada

Tristan Guttridge
Bimini Biological Field Station Foundation, South Bimini, Bahamas

David M.P. Jacoby
Institute of Zoology, Zoological Society of London, London, United Kingdom

CONTENTS

18.1 INTRODUCTION

In recent decades, network analyses have become ubiquitous in ecology, facilitating our understanding of linkages between paired entities, whether it be genes, proteins, individuals, species, or habitats (Blüthgen et al., 2008; Croft et al., 2008; Krause et al., 2007; Proulx et al., 2005; Wey et al., 2008). Network theory (also known as graph theory) originates from the mathematical and social sciences but has developed concurrently across many disciplines, including computational science, physics, management, genetics, and epidemiology (Newman, 2010), to name but a few.

Widespread uptake of these developments in behavioral ecology has ensured that network analyses, and in particular social network analysis (SNA), are now among the go-to tool kits for researchers wishing to measure animal association and aggregation, species interactions, or animal-mediated habitat connectivity (Farine and Whitehead, 2015; Fletcher et al., 2011; Jacoby et al., 2012a; James et al., 2009; Krause et al., 2009a). Given the challenges that face researchers studying sharks (e.g., their slow growth and low fecundity, their tendency to be wide ranging and cryptic) and indeed behavior in any subsurface marine organism, it is perhaps not surprising that this exciting branch of ecology has only really found a home in shark biology in the last decade (Krause et al., 2015). Before this, records of shark social behavior were rare and often anecdotal (Jacoby et al., 2012b).

Guided by the pioneering work on teleost fishes, such as the three-spined stickleback (*Gasterosteus aculeatus*) and the guppy (*Poecilia reticulata*), and by work on marine mammal societies (predominantly cetaceans), network methods now offer a robust framework to quantify and analyze components of shark behavior that until recently have proven extremely difficult (Croft et al., 2008; Krause and Ruxton, 2002; Ward et al., 2002; Whitehead, 2008). Do individual sharks have preferential social partners? Are shark groups assorted by phenotypic traits and, if so, over what spatial and temporal scales? Are sharks capable of learning from social partners? Observational experiments on small groups of captive or semi-wild individuals have driven our initial understanding of social networks in sharks (Guttridge et al., 2009; Jacoby et al., 2010); for some of the first wild examples, see Krause et al. (2009b) and Mourier et al. (2012). Technological advances are now driving the progress of new analytical techniques that can handle very large datasets, such as those obtained from biotelemetry. Thus, technology in combination with network approaches has recently helped to facilitate the scaling of some of these questions to wild sharks at the population level (Jacoby and Freeman, 2016; Jacoby et al., 2016; Krause et al., 2013).

Although widely used and often easily implemented in bespoke programs or R packages, network analyses are highly nuanced and should be tailored specifically to a species or study system. In this chapter, we explore the two core principles of network application in shark ecology: (1) shark social networks (i.e., the how, why, and with whom sharks associate), and (2) shark spatial networks (i.e., understanding how the movements of individuals can link discrete locations as a movement network). These components are far from mutually exclusive, reflecting how social processes are inextricably linked to the distribution of sharks in space. Network analyses offer a unique set of statistical tools that help researchers to understand how individual behavioral patterns can influence group and population-level processes, how overall network structure can select for behavior at the individual level, and how both direct and indirect connections within a population can matter greatly (Croft et al., 2016; Krause et al., 2009a).

18.1.1 What Is a Network? Basics of Network Theory

A network (or graph) consists of a set of nodes and edges. As a visual illustration, the simple network depicted in Figure 18.1 has six nodes (i.e., six individual sharks for social networks and six locations for spatial networks) and ten edges (i.e., ten associating pairs for social networks and ten movement paths between locations for spatial networks), and the interactions between all possible pairs are represented by an accompanying adjacency matrix, A (a quantitative representation of the network). These interactions can be represented in various ways depending on which components of the shark's behavior are being measured. For example, in Figure 18.1A, an unweighted (i.e., binary) directed network is presented in which an edge represents directed interactions such as "1 is dominant over 2" (social) or movement from location 1 to location 2 (spatial). Figure 18.1B shows a binary undirected network with a 1 or a 0 to indicate presence or absence of an interaction within the adjacency matrix. Figure 18.1C shows the same network but with weighted edges proportional to the frequency or strength of association generally between 0 (no association) and 1 (constant association). Note that each edge appears twice in the adjacency matrix of an undirected network (i.e., symmetric), whereas upper and lower triangles of the matrix are different in directed networks.

Network analyses offer a set of quantitative metrics or test statistics that allow researchers to characterize and analyze its structure (Table 18.1). These are used to measure structural properties at the node, group, or network level; for example, one can assess the centrality of individual sharks to differentiate the social importance or influence of members of the population. Node degree (unweighted) or the strength, also known as weighted degree, can help identify the most gregarious individuals in the population based on the number of associations an individual has with conspecifics (Croft et al., 2008). For binary directed networks (see Figure 18.1A), out-degree and in-degree (the number of edges leaving and arriving at a node, respectively) can be used to better determine the centrality of nodes (e.g., identify hubs in the network) and identify the directionality of interactions or movements. Such node-based metrics are useful for understanding the position and relative importance of sharks in their network (Jacoby et al., 2010; Mourier et al., 2017b). Such centrality metrics are also useful for movement networks to better identify the central locations most pertinent to conservation and management (Jacoby and Freeman, 2016; Jacoby et al., 2012a). Beyond the individual-level metrics, many networks contain groups of nodes that are better

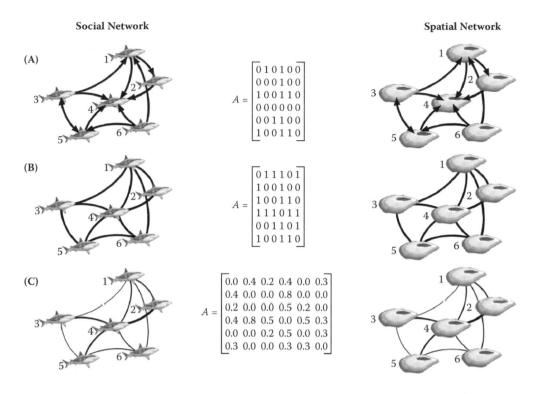

Figure 18.1 Examples of social (left) and spatial (right) networks. Different representations are displayed: (A) directed binary networks with an associated nonsymmetrical adjacency matrix, (B) undirected binary networks with symmetrical adjacency matrix, and (C) weighted undirected networks with symmetrical adjacency matrix.

connected among themselves than they are to the rest of the network, and these clusters of well-connected nodes are usually referred as communities or cliques. Community detection analyses, for example, were used to demonstrate that blacktip reef sharks (*Carcharhinus melanopterus*) in French Polynesia could form well-defined, mixed-sex communities within a small portion of the reef (Mourier et al., 2012).

It is important to note that social and spatial networks are often correlated, as social networks are usually derived from spatial proximity (co-occurrences) and therefore are dependent on the movement of sharks between locations (Figure 18.2) (Jacoby and Freeman, 2016; Jacoby et al., 2016). Although social networks are to an extent somewhat dependent on the spatial proximity of their members, non-random, preferential associations can emerge beyond those predicted through overlap of spatial ranges (Mourier et al., 2012). This interplay can be captured by bimodal networks that consist of links between two sets of nodes belonging to different classes such as individuals linked to locations (Figure 18.2). As before, these bimodal networks can also be (1) unweighted or binary, only showing the presence or absence of the interactions; or (2) weighted if cells in a matrix represent, for example, the number of visits by animal species to a monitoring receiver. Bimodal networks are often analyzed after projecting them into unimodal ones in which individuals are linked if they share locations (Figure 18.2).

Networks can have very different properties but can sometimes be defined by characteristic structural properties (Figure 18.3). In regular networks, all nodes have the same degree. For example, a circular network is a type of regular network where all nodes in the network have a degree of 2. It also displays additional characteristics such as no clustering coefficient and long average path length (Table 18.1), indicating that most nodes must pass through many other nodes to reach anything other than their immediate network neighbors (Csárdi and Nepusz, 2006). In such a network, the direct influence of any single node on the network overall is low. Random networks are characterized by a normal node degree distribution (Erdös and Rényi, 1959), whereas small-world networks are characterized by a small diameter (longest path between any pair of nodes; see Table 18.1) relative to the number of nodes, as well as a higher clustering coefficient and a smaller average path length compared with random network (Watts and Strogatz, 1998). Other networks can have scale-free properties characterized by a power law node degree distribution (or right skewed distribution) where just a few nodes have a disproportionately high degree (many connections) but the majority have a low degree (few connections) (Barabási and Albert, 1999). In scale-free networks, the rapid diffusion of information or disease across the network is facilitated by this small number of very well-connected nodes (hubs).

Table 18.1 Definition of Network Metrics and Their Application in Shark Social and Spatial Networks

Network Metric		Definition	Social Network	Spatial Network
Network level	Path	A route between any two nodes in a network	—	—
	Average path length	Mean shortest path between all nodes in a network	Low average path length can indicate potentially fast information flow across a network.	Low average path length means that an individual travels more rapidly and directly across their activity space because of the greater presence of shortcuts in the network.
	Diameter	Longest path between any pair of nodes in the network (maximum path length)	—	—
	Density	Number of edges present in the network as a proportion of the total possible number of edges (i.e., a network of density 1 has all nodes connected to all others)	—	—
Substructure	Component	Group of nodes that are interconnected but with no connection/edge to the rest of the network	A population can be divided into two independent social groups (two components) with no connection between them.	A seascape can be divided into two independent areas that are not connected by the movement of any individual sharks.
	Cluster or community	Subnetwork of interconnected nodes that are closer to each other than to other nodes in network space	A group of sharks is more connected to one another than to others in the network.	Locations are more frequently connected by shark movements than they are with others from the seascape (perhaps because they are closer together).
Local level	Node degree	Number of other nodes connected to a node	Number of a node's social partners	Number of locations that a node is connected to via movements
	Node strength or weighted degree	Sum of all edge weights connected to the node	Sum of association indices	Total number of incoming/outgoing movements from a node
	Betweenness	The proportion of all shortest paths between pairs of nodes on the network that pass through a given node	An individual has particularly high betweenness centrality if it is the only bridge between two subgroups. If removed the network can split into two separate components.	A location with high betweenness indicates that it is a crucial site for population connectivity.
	Eigenvector centrality	—	Sum of the centralities of an individual's neighbors	Sum of incoming/outgoing movements from a node weighted by the node strength of the node it is connected to; nodes with a high eigenvector centrality value have high node strength values and are connected to nodes with similarly high node strength values.
	Clustering coefficient	—	Tendency for an individual's close social partners to be associates with one another as well	Tendency of locations to be connected with other well connected locations

18.1.2 Sampling a Network: Data Collection Methods

Building networks is not an easy process and requires collecting a large amount of relational data (interactions or movements) under a robust sampling design. Many methods and approaches are available to collect data to construct a network, all of which hinge on the ability to individually identify sharks. Sharks can be identified using body coloration, patterns, or fin notches, which are specific to each individual (Figure 18.4A). This technique of photo-identification is non-invasive and has been used for many elasmobranch species (Marshall and Pierce, 2012; see also Chapter 12 in this volume) and to track shark and ray associations and movements, including blacktip reef sharks (Mourier et al., 2012), spotted eagle rays (*Aetobatus narinari*) (Krause et al., 2009b), and even the sicklefin lemon shark (*Negaprion acutidens*), which has a rather homogeneous body coloration (Buray et al., 2009). However, other studies on species in which individuals are difficult to identify used externally attached visual color-coded tags (Figure 18.4C) (Guttridge et al., 2011; Jacoby et al., 2010) or fluorescent visible implant

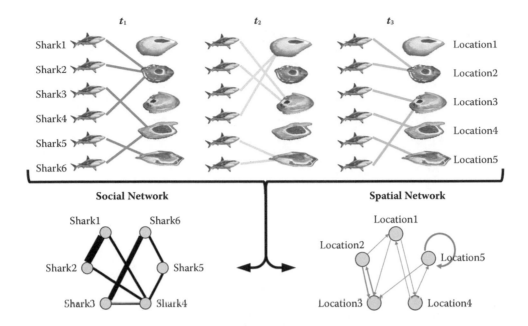

Figure 18.2 The emergence of social and spatial networks from shark movements. Sharks moving between reefs from time t_1 to time t_3 create a directed and aggregated movement network made up of the frequency of movements of all individuals accumulated between reefs and through time. As sharks move in the seascape, their co-occurrences can be used to create a social network defined by the frequencies of associations between individuals in space and time.

elastomer tags inserted subcutaneously on the dorsal surface (Figure 18.4B) (Jacoby et al., 2012c, 2014). Each species comes with its own challenges. Nevertheless, the difficulty in tracking sharks in the wild and the necessity to record repeated interactions has pushed the development of autonomous tracking devices for inferring contact rates between individuals. The use of proximity loggers attached to individual sharks has been employed in several studies (Guttridge

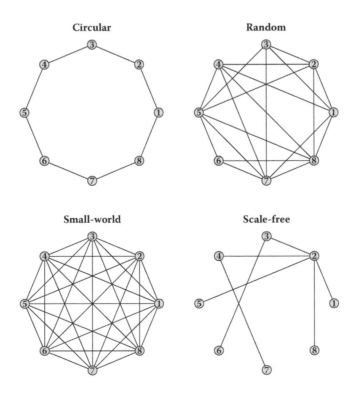

Figure 18.3 Examples of theoretical networks using a circle layout.

Figure 18.4 Example methods for identifying individual sharks and tracking them in space and time, thus enabling networks to be constructed: (A) coloration patterns or notches on a blacktip reef shark's body used for photo-identification (see Chapter 12 in this volume); (B) fluorescent elastomer tag inserted under the skin of a catshark to follow individuals in a captive environment; (C) a small spaghetti tag inserted externally on the dorsal fin of a juvenile lemon shark with a unique code; and (D) a proximity logger attached to the dorsal fin of a juvenile lemon shark to measure fine-scale social interaction with tagged conspecifics.

et al., 2010; Haulsee et al., 2016; Holland et al., 2009; Mourier et al., 2017a) to record every tagged shark encountered within a certain proximity (Figure 18.4D). However, it is challenging to recapture the shark and retrieve the tag and data, so this method is often confined to semi-captive or more resident shark species. Deploying and retrieving proximity loggers on enough individuals to explore population-level social structure still remains a considerable challenge in this field. Acoustic telemetry has been proposed to track movements of tagged sharks and build social networks from their spatiotemporal co-occurrences within arrays of fixed acoustic receivers (Jacoby and Freeman 2016; Jacoby et al., 2016). This method, however, requires careful interpretation of the data and robust analyses to tease apart preferred associations from random encounters under quite wide and often variable detection ranges (Mourier et al., 2017a). To date, acoustic telemetry (see Chapter 8 in this volume) has mainly been employed to build movement networks in sharks and analyze the properties of the movement patterns in a population (Espinoza et al., 2015; Jacoby and Freeman, 2016; Jacoby et al., 2012a; Lea et al., 2016; Lédée et al., 2015; Papastamatiou et al., 2015; Stehfest et al., 2015).

Methods of sampling and constructing a social network can be either direct or indirect. Observers can identify and follow a focal individual within the population and record its interactions with others forming an egocentric social network. Separate egocentric networks can also be combined into a global network to explore population structure. This approach was taken by Wilson et al. (2015) with juvenile lemon sharks (*Negaprion brevirostris*) in a mesocosm in the Bahamas. For each network session, an individual shark was randomly chosen as a focal individual and tracked continuously for a predefined amount of time (i.e., 100 seconds), and its associations with the nearest group member (if present) were recorded every 10 seconds. Sharks were considered to be associating if they were within one body length of each other during the sampling interval. After an observation period was finished, another shark was chosen as the next focal individual until all individuals had been recorded so that they were all recorded for every session.

Alternatively, a social network can be built on repeated samples of associations (co-occurrences) or interactions of dyads and can follow the "gambit of the group" approach, where groups are defined as co-occurrences of individuals within a defined distance and sampling period. Such approaches have been used to construct social networks in blacktip reef sharks, where individual sharks were considered as part of the same group if observed together during a dive (Mourier et al., 2012), or for constructing a network of white sharks (*Carcharodon carcharias*) aggregating around a chumming boat (Findlay et al., 2016). How associations are defined; the frequency, duration, and interval between sampling periods; and the spatial area over which sampling takes place will be highly dependent on the characteristics of the species and system and require considerable thought, because all of these factors will directly influence the eventual structure of the network. For visual observations, various association indices are designed to account for different sampling methodologies (e.g., the probability of encountering one individual might differ from another at specific locations). As an example, the simple ratio index is used when observations are rarely missed and the half-weight index when individuals are frequently missed in samples. Using an inappropriate association index for the sampling method used might result in the artificial over- or underrepresentation of individuals in the network. Considerable developments have been made in recent years to control for biases associated with different sampling regimes and association indices. Haddadi et al. (2011) discussed how best to define an appropriate spatiotemporal sampling rate, and Mann et al. (2012) used social affinity indices to control for demographic population changes and a 10-m chain rule to determine associations in a very long-term (22-year) dolphin social network. A thorough discussion of all appropriate indices goes beyond the scope of this chapter; however, more detail on how to choose an appropriate index can be found in Farine and Whitehead (2015), Whitehead and James (2015), and Hoppitt and Farine (2018).

An intermediate approach can also be adapted using automated recording of individual movements, which offers a means to reconstruct social structure in intractable species from the frequency of paired spatial associations between tracked individuals. Using Bayesian inference, specifically Gaussian mixture modeling (GMM) approaches, an automated approach explores the inherent structure present in the visitation profile of tagged animals. This machine learning approach is used to detect the most likely clustering events in the time-series of tag detections (at an acoustic receiver, for example), and these clusters are then used as a basis for constructing a social network. Crucially, these clusters can vary in size temporally, reflecting the variation expected in dynamic animal societies, and code offers built-in permutation tests to remove random co-occurrences. Jacoby et al. (2016) explored the utility of GMMs for retrieving inference on social network structure from telemetry data of spatiotemporal co-occurrences of tagged gray reef sharks (*Carcharhinus amblyrhynchos*), modifying the available methodologies to extract additional behavioral information on the timing and directionality of dyadic interactions.

Another important factor to consider alongside measures of association is information on the attributes of individuals, such as phenotypic traits (e.g., sex, size, age) or details about individual state (e.g., personality, dominance rank, maturity). These attributes can be used to determine how

sociality and network structure are mediated by individual traits. Mourier et al. (2012) showed that blacktip reef sharks tended to group with individuals of a similar size and sex (homophily), similar to lemon sharks, which preferentially associate with size-match conspecifics (Guttridge et al., 2009, 2011).

18.1.3 Assumptions and Randomization Analysis

When relational data have been gathered using the most appropriate sampling method for the question and species of interest, it is extremely useful to visualize the data to help guide further, more quantitative analyses. Network visualization is particularly compelling and intuitive and in some instances can generate informative patterns. It is important to remember that these observed networks typically represent a subset of the true underlying relationships between individuals because network sampling rarely captures all of the nuances that dictate how individuals are likely to interact (Farine and Whitehead, 2015). In highly connected systems, there might be the temptation to threshold the data above a specific edge weighting in order to help reveal underlying structural characteristics that might be masked by a bird's nest of connected nodes. It is crucial, however, to only use thresholding for visual exploration, in order to avoid misleading and inaccurate metrics or measures of community structuring (Farine and Whitehead, 2015).

Because network data are inherently non-independent (i.e., the link between any two nodes is directly influenced by other links in the network), it is important to consider how to approach null hypothesis significance testing (Croft et al., 2011). To determine whether the patterns observed in visualizations are significant, it is important to compare the observed result (e.g., one or several metrics calculated from the observed data) to a distribution of that same metric from a large number of network permutations (Farine, 2017). For example, the movements of pigeye shark (*Carcharhinus amboinensis*) and spot-tail sharks (*Carcharhinus sorrah*) between acoustic receivers in Cleveland Bay, Eastern Australia, were randomized to generate null spatial networks against which the coefficient of variation in the observed movement network was compared (Lédée et al., 2015). In spatial networks, these methods are designed to control for the bias introduced by the physical layout of the receivers from which data are generated. The subtleties of deciphering the most appropriate null model to use and deciding at which stage to permute the data (matrix permutation vs. data stream permutation), in addition to how to statistically deal with overlapping spatial and social processes, are explored in detail in a number of excellent papers that discuss the pros and cons of different statistical approaches (Croft et al., 2011; Farine, 2015, 2017; Farine and Whitehead, 2015; Furmston et al., 2015; James et al., 2009; Spiegel et al., 2016). Given that network structure can inform the emergence of crucial ecological and evolutionary processes, such as sexual segregation or cooperative hunting, developing rigorous null models and factoring in the violation of statistical assumptions in network data analysis is of upmost importance, as is the interpretation of the network.

18.1.4 Interpretation and Output of Network Analysis

The interpretation of any network should be approached with caution. First, is the sample size of the network sufficient to make population-level inferences? This is particularly important for SNA because networks with low numbers of individuals can be unreliable (Silk et al., 2015). Ideally, social analyses should incorporate as many members of a population as possible to capture the variability within social dynamics; indeed, missing individuals can have important consequences for the description of the structure of the population. Given that spatial network analyses are often driven by very different questions, robust results can sometimes be obtained from a smaller representative sample of the population. It is increasingly being found, however, that individuals within populations of sharks can exhibit substantial differences in movements and feeding behavior (Matich and Heithaus, 2015) as well as personality traits (Finger et al., 2016, 2017; Jacoby et al., 2014), which may necessitate larger sample sizes.

Second, before interpreting the outputs of a network analysis, it is important to return to the original research question of the study and interpret the results in the context of this. For example, a network is simply a graphical representation of links between individuals or locations, and it can be built from any kind of random data. Therefore, before drawing conclusions about the social structure of a shark population it is crucial to verify whether or not the network is based on non-random associations between some individuals (i.e., preference and avoidance behavior). Random associations can result in considerable variation in association strengths among individuals and apparent social structure. For this reason, it is critical to compare observed patterns to the random model to test for true structuring within the population. Every interpretation should therefore be made in light of the initial ecological question or hypothesis being tested.

For spatial networks, the resulting structure of aggregated networks, or the average of all individual movement networks, can be highly dependent on the number of individuals included in the study. Indeed, the density of the network (i.e., the number of edges in the network divided by the total possible edges) or the centrality of a location is dependent on the diversity of individual movement strategies. Increasing the number of individuals in the analysis will consequently increase the number of different edges constituting the network. We envisage that developments in multilayer network approaches, which help to disentangle the role of individual variability on connectivity, are likely to be an area of future interest in the field of networks in movement ecology.

18.2 NETWORK APPLICATIONS IN SHARK BIOLOGY

18.2.1 Shark Social Networks

Over the past decade, our understanding of the mechanisms and functions that shape shark social lives has progressed rapidly (Jacoby et al., 2012b; Wilson et al., 2014). This is in part due to advances in remote monitoring devices, such as biotelemetry (e.g., acoustic telemetry) (see Chapter 8 in this volume) and biologging (e.g., archival loggers) (Hussey et al., 2015; see also Chapter 3 in this volume), but also to our ability to collect, handle, and analyze ever larger datasets. The use of SNA has made important contributions to this improved understanding by providing a framework to quantify associations. To date, 18 studies have used SNA, with questions typically focused on whether sharks have non-random associations (see Table 18.2), and, if so, what attributes (e.g., sex, size) influence group joining decisions. Do these associations persist temporally or spatially? Just ten species have been used in these studies, but the focal species have varied considerably in body size, ranging from 12-cm small-spotted catsharks (*Scyliorhinus canicula*) to 400-cm great white sharks; life stage (neonates to adults); reproductive mode (oviparous to viviparous); and habitat types (temperate benthic to tropical reef-associated or pelagic). Importantly, the versatility of the SNA approach has also prompted the use of diverse data collection methods, ranging from direct observations of social behaviors, such as nose to tail following (Guttridge et al., 2011) or tactile resting (Jacoby et al., 2010), to co-occurrences at provisioning or dive sites (Findlay et al., 2016; Mourier et al., 2012).

More recently, triangulation of acoustic detections (Armansin et al., 2016), use of sharkborne proximity receivers (Mourier et al., 2017a), and machine learning algorithms have been used to infer social associations from time-series data (Jacoby et al., 2016). Further, the temporal and spatial scales used to examine shark social networks have varied considerably. For example, the focal follows conducted by Wilson et al. (2015) on ten juvenile lemon sharks in a mesocosm recorded their nearest neighbor every 10 seconds for a 100-second sampling period; the study was completed in 8 days. By contrast, Mourier et al. (2012) completed 190 dives over 2 years, recording co-occurrences of 133 blacktip reef sharks across seven locations spanning 10 km. To further explore the outcomes and SNA tool kits used by researchers working with sharks, the following section considers lab-based and semi-captive studies separately from those conducted on free-ranging species.

Table 18.2   Overview of Species of Elasmobranchs and Topics That Have Been Investigated Using Network Analysis

Species	Topic	Method	Conditions	Refs.
Social Network				
Blacktip reef shark (*Carcharhinus melanopterus*)	Preferred associations, communities, assortment, robustness	Observations, photo-ID	Wild	Mourier et al. (2012, 2017b)
Bull shark (*Carcharhinus leucas*)	Preferred associations	Observations, photo-ID, video recording	Wild	Loiseau et al. (2016)
Gray reef shark (*Carcharhinus amblyrhynchos*)	Preferred associations, leadership, assortment	Acoustic telemetry	Wild	Jacoby et al. (2016)
Lemon sharks (*Negaprion brevirostris*)	Preferred associations, assortment, familiarity, leadership	Observations, visual tags, accelerometers, video recording, proximity logger	Semi-captive and wild	Guttridge et al. (2010, 2011), Keller et al. (2017), Wilson et al. (2015)
Port-Jackson shark (*Heterodontus portusjacksoni*)	Preferred associations	Acoustic telemetry, proximity logger	Wild	Mourier et al. (2017a)
Sand tiger shark (*Carcharias taurus*)	Interactions	Acoustic telemetry, proximity logger	Wild	Haulsee et al. (2016)
Sicklefin lemon shark (*Negaprion acutidens*)	Preferred associations, communities, dominance hierarchy	Observations, photo-ID, video recording	Wild	Brena et al. (2018), Clua et al. (2010)
Small-spotted catshark (*Scyliorhinus canicula*)	Preferred associations, personality, familiarity	Observations	Captive	Jacoby et al. (2010, 2012b, 2014)
Spotted-eagle ray (*Aetobatus narinari*)	Preferred associations	Observations, photo-ID	Wild	Krause et al. (2009a)
Spotted wobbegong shark (*Orectolobus maculatus*)	Preferred associations	Acoustic telemetry	Wild	Armansin et al. (2016)
White shark (*Carcharodon carcharias*)	Preferred associations	Observations, photo-ID	Wild	Findlay et al. (2016)
Spatial Network				
Australian weasel shark (*Hemigaleus australiensis*)	Network properties Spatial partitioning	Acoustic telemetry	—	Heupel et al. (2017)
Blacktip reef sharks (*Carcharhinus melanopterus*)	Node centrality Network properties Bimodal network Spatial partitioning	Acoustic telemetry	—	Heupel et al. (2017), Lea et al. (2016)
Bull shark (*Carcharhinus leucas*)	Network properties	Acoustic telemetry	—	Espinoza et al. (2015)
Caribbean reef shark (*Carcharhinus perezi*)	Node centrality, ontogenetic shifts in network structure	Acoustic telemetry	—	Jacoby et al. (2012a)
Galapagos sharks (*Carcharhinus galapagensis*)	Node centrality	Acoustic telemetry	—	Papastamatiou et al. (2015)
Gray reef shark (*Carcharhinus amblyrhynchos*)	Sex differences in network properties, nutrient dynamics Spatial partitioning	Acoustic telemetry	—	Espinoza et al. (2015), Heupel et al. (2017), Lea et al. (2016), Williams et al. (2018)
Pigeye shark (*Carcharhinus amboinensis*)	Node centrality Bimodal network	Acoustic telemetry	—	Lédée et al. (2015)
Silvertip shark (*Carcharhinus albimarginatus*)	Network properties Spatial partitioning	Acoustic telemetry	—	Espinoza et al. (2015)
Sicklefin lemon shark (*Negaprion acutidens*)	Network properties	Acoustic telemetry	—	Lea et al. (2016)
Small spotted catshark (*Scyliorhinus canicula*)	Node centrality	Acoustic telemetry	—	Jacoby et al. (2012a)
Spottail shark (*Carcharhinus sorrah*)	Node centrality Bimodal network	Acoustic telemetry	—	Lédée et al. (2015)
Tawny nurse shark (*Nebrius ferrugineus*)	Network properties	Acoustic telemetry	—	Lea et al. (2016)
Tiger shark (*Galeocerdo cuvier*)	Network properties Spatial partitioning	Acoustic telemetry	—	Heupel et al. (2017)
Whitetip reef shark (*Triaenodon obesus*)	Network properties Spatial partitioning	Acoustic telemetry	—	Heupel et al. (2017)

18.2.1.1 Lab and Semi-Captive Studies

To date, two shark species, the lemon shark and small-spotted catshark, have been used in lab or semi-captive (i.e., exposed to ambient conditions with water exchange) experiments. Both species form groups, are abundant, and can be easily maintained in captivity (Guttridge et al., 2009; Sims, 2003). Jacoby et al. (2010, 2012c, 2014) conducted a series of experiments using juvenile and adult small-spotted catsharks. In the first study, social network structure, temporal stability, and activity profiles were analyzed to examine the impact of introduced males on the social structure of four captive groups of mature female catsharks. Social networks were constructed from symmetric tactile association behaviors (i.e., sharks resting in contact with each other) and by examining network measures (eigenvector centrality, weighted degree, and average path length; see Table 18.1) before and during male introduction; results showed that shark groups differed in their tendency to aggregate in a unisex environment and in social responses to male presence. In their next experiments, juvenile catsharks hatched in captivity provided greater numbers (N = 300) and the rare opportunity to manipulate the social environment and habitat of treatment tanks to examine preferred associations, repeatability in social behavior, and the role of familiarity and habitat type in aggregation formation. Randomizations revealed nonrandom associations, with familiar sharks forming more groups of greater size. Finally, network measures (e.g., clustering coefficient, reach) were employed to characterize individual repeatability of social traits across habitats to explore social personality types. Keller et al. (2017) also investigated the potential role of familiarity in group formation and social behavior of juvenile lemon sharks using a remote camera and an automated tracking system, which allowed the inference of interactions between individual juvenile sharks moving around a holding pen. These experiments showed that juvenile lemon sharks preferred social interactions with familiar individuals.

Building on this idea of behavioral phenotypes in shark social behavior, Wilson and colleagues (2015) used a novel fission–fusion model based on Markov chains to explain juvenile lemon shark social dynamics. Individual-level differences in sociality (leadership and network measures, such as node strength, weighted node betweenness, and weighted clustering coefficient) were determined for ten sharks across 8 days of observations (Wilson et al., 2015). In addition, sharks were fitted with triaxial accelerometers to provide locomotor profiles (e.g., time spent fast/steady swimming) in a rare example of multi-tool approaches to quantifying social behavior. Interestingly, lemon sharks did not show consistency in their social network positions but preferred to associate with other individuals of similar locomotor profiles. Although a small sample size and short study duration likely limited the conclusions that could be drawn from

these results, the integration of SNA tools with accelerometers holds considerable promise for exploring the energetic benefits or costs to grouping.

18.2.1.2 Free-Ranging Studies

Patterns of association for wild sharks are difficult to quantify due to the concealing nature of their environment (Jacoby et al., 2012b). Taking advantage of shallow water and sheltered mangrove inlets, Guttridge et al. (2011) explored the social structure of a population of juvenile lemon sharks in Bimini, Bahamas. Across 2 years, the social behavior (e.g., nose-to-tail following, circling) of 38 sharks was observed from wooden platforms and recorded at 2-minute intervals. Networks constructed at 10- and 60-minute sampling periods (to avoid issues with independence) revealed that juvenile lemon sharks showed repeated social interactions, with group structure primarily explained by body length and possibly by preference for relatives but not sex. In addition, the researchers also documented differences in leadership tendencies of sharks, with lead individuals usually being significantly larger than other group members (Guttridge et al., 2011).

Mourier and colleagues (2012, 2017b) conducted an extensive study examining the social structure of a population of free-ranging blacktip reef sharks. Unique fin markings were used to identify individuals, and dive surveys were completed at provisioned and non-provisioned sites. They incorporated community analyses (e.g., modularity matrix clustering technique) (Whitehead, 2008) and further use of lagged association rates and egocentric network measures (e.g., strength, eigenvector centrality, reach, clustering coefficient, affinity). Findings revealed the first evidence for communities in sharks which were characterized by non-random associations, with size and sex driving preferences in some locations. Interestingly, when spatial overlap was included in the analysis this explained much of the community separation; however, this was not exclusive, suggesting active social preferences probably influenced associative patterns within communities (Mourier et al., 2012). More recent exploration of the data quantified impacts of node removal on the network properties and robustness to catch and release fishing. These simulations revealed that the global network was resilient and did not fragment, even when 25% of the individuals were removed. Catch-and-release fishing conducted for 30 minutes after dives provided an interesting experimental component to this study, showing that the probability of capture of individual sharks decreased with an increasing number of sightings as well as with the individual's experience of capture (Mourier et al., 2017b).

Direct observations were also used by Findlay et al. (2016) to monitor co-occurrences of 323 great white sharks in six locations across 6 years (2008 to 2013). This was the first attempt to examine the social preferences of a highly migratory pelagic species. Despite finding random associations in white shark social networks, the study further highlighted the applicability of SNA to co-occurrence data and the opportunities for generating this type of data during ecotourism operations (Gallagher et al., 2015). A similar approach was also used to demonstrate non-random communities based on affinities and residency patterns of sicklefin lemon sharks visiting a provisioning site in French Polynesia (Clua et al., 2010).

Similar to the semi-captive experiments by Keller et al. (2017), remote cameras have recently been employed to track interactions between free-ranging individual adult sicklefin lemon sharks under an artificial food stimulus (Brena et al., 2018). This technique allowed not only recording associations between individuals around a food source but also the construction of a social hierarchy of the members of the network without human interference. In this case, individual sharks were free to come and interact with the food and other individuals.

More recently, a handful of studies have explored the use of SNA to provide insights into shark social structure by examining data collected from acoustic tracking. Armansin et al. (2016) used spatial data obtained from fine-scale passive acoustic telemetry (VEMCO Positioning System, or VPS) to infer association preferences of 15 tagged wobbegong sharks (*Orectolobus maculatus*) over a 15-month period. Despite being presumed to be solitary, this species showed non-random casual and long-term associations. Home-range overlap did not correlate with associations, but changes in social cohesion were documented before and during the breeding season. Similarly, Haulsee et al. (2016) found that male sand tiger shark (*Carcharias taurus*) interactions varied seasonally. Using implanted acoustic transceivers, they generated egocentric networks of con- and heterospecific interactions across a year. Networks were only visually depicted, however, and neither randomization testing nor exploration of the network properties was conducted.

Another exciting approach used Gaussian mixture models (GMMs) (see Section 18.1.2) (Jacoby et al., 2016). Using the number of times individuals co-occurred and the duration of these co-occurrences at different locations, it was possible to make inferences about the leadership patterns within populations of wild gray reef sharks. SNA was used to analyze co-occurrence count and duration data, and leadership scores were based on the proportion of each individual's degree, which was represented by an *in degree* value (for details, see Jacoby et al., 2016). This novel method for extracting social structure from acoustic tracking data would benefit from validation with direct observations. Given the vast passive acoustic receiver arrays that are maintained globally and the great diversity of species that can be equipped with tracking devices (Hussey et al., 2015), this method holds tremendous promise.

Finally, Mourier et al. (2017a) explored the efficacy of three types of receivers (VEMCO VR2W; Sonotronics miniSUR and proximity receivers) differing in detection range to generate co-occurrence networks for a benthic shark species. By using SNA, it was possible to compare networks across receivers effectively by examining the correlation between association indices and whether centrality rank was consistent across methods. Results revealed that VR2W receivers were not able to capture co-occurrences at an appropriate spatial scale to infer social associations for a relatively immobile species, the Port Jackson shark (*Heterodontus portusjacksoni*). Further, the consistency of individual social rank was not significant when comparing the network produced by the receivers with small ranges (10 to 60 m) with those constructed with a larger range (400 m). This study highlights the importance of considering the ecology of the study species and defining the scale of biologically meaningful interactions between individuals.

18.2.2 Shark Movement Networks

Recently, the use of network analysis to study animal spatial ecology has gained momentum (Jacoby and Freeman, 2016). This section reviews researchers who have examined shark space use, movement, habitat use, and drivers of shark movement using various network analysis techniques. It is worth noting that, to date, most studies have used network analysis in combination with passive acoustic telemetry to understand shark movement networks. Network analysis enables the exploration of shark movement and space use particularly, as the visualization tools associated with network packages are so intuitive and versatile. For example, Jacoby et al. (2012a) and Lédée et al. (2015) used spring embedding algorithms, which sort randomly placed nodes into a desirable layout that satisfies the aesthetics for visual presentation (e.g., symmetry, non-overlapping nodes) to visually compare changes in Caribbean reef shark (*Carcharhinus perezi*) and pigeye shark (*Carcharhinus amboinensis*) space use, respectively. Ontogeny was identified as a possible explanation for the observed changes in Caribbean reef shark space use, whereas responses to acute changes such as the influence of freshwater was a prominent feature of pigeye shark space use. Using centrality metrics, researchers can further capture distinct aspects of a location's importance in network space and distinct patch use (Jacoby and Freeman, 2016; Nicol et al., 2016). Single or multiple metrics might be used to determine the most important patches and the differential use of patches in the network depending on the research question.

Three studies have used descriptive network statistics to examine the space use of sharks. Degree (Jacoby et al., 2012a), eigenvector (Stehfest et al., 2015), and a combination of centrality metrics (i.e., node strength, closeness, and eigenvector; see Table 18.1) (Lédée et al., 2015) are among the tools that have been used to determine the most important patches and examine their use within networks. Using degree, Jacoby et al. (2012a) demonstrated segregation in core patches and movements between male and female small-spotted catsharks. Stehfest et al. (2015) used the eigenvector to examine the movement network of broadnose sevengill sharks (*Notorynchus cepedianus*) and found male and female spatial segregation, with each sex using a different core area. Finally, by combining node strength, closeness, and eigenvector, Lédée et al. (2015) defined the core and general use areas of two nearshore shark species and identified the importance of movement corridors within core areas for both species.

Observed individual spatial networks can also be compared to simulated networks that have known structural properties (i.e., circular, small-world, and scale-free networks; see Figure 18.3) to examine individual movement patterns within the landscape. To date, only one study has used this technique (Heupel et al., 2017). This study found that the movement of gray reef (*Carcharhinus amblyrhynchos*), blacktip reef (*Carcharhinus melanopterus*), whitetip reef (*Triaenodon obesus*), tiger (*Galeocerdo cuvier*), and Australian weasel (*Hemigaleus australiensis*) sharks within the Great Barrier Reef exhibited small-world and scale-free properties. These characteristics facilitate dispersal through alternative pathways (small-world) and enhance resilience to random disturbances (scale-free) (Fortuna et al., 2006; Minor, and Urban, 2008).

In the context of habitat use, two types of habitat network can be created—unimodal or bimodal—which have been touched on briefly in previous sections. Unimodal habitat networks represent the movement of individuals, a population, or species between habitat types and may be used to examine habitat use. Bimodal habitat networks represent how frequently habitat types (i.e., first set of nodes) are used during a specified period (i.e., second set of nodes, such as monthly, seasonally; see Figure 18.2) (Borgatti, 2012; Opsahl, 2013), thus allowing the examination of habitat preferences. To date, three studies have used network analysis to examine how habitats can drive the movement network structure of shark species and wider ecological processes. Williams et al. (2018) constructed the movement networks of gray reef sharks at Palmyra Atoll in order to quantify the distribution of pelagic-derived nitrogen onto the reef ecosystem and track where precisely these nutrients are deposited thereafter. Using tagged sharks to extrapolate to the population, gray reefs appear to contribute substantially to reef primary productivity, particularly on the forereef. Papastamatiou et al. (2015) quantified the habitat use of Galapagos sharks (*Carcharhinus galapagensis*) at a Hawaiian atoll by measuring the degree and betweenness centrality metrics in unimodal habitat networks. Deep habitats within the atoll were found to be more important for Galapagos sharks than the shallow habitat surrounding the atoll. Finally, Lea et al. (2016) measured node strength and betweenness and edge density from unimodal habitat networks to examine the habitat use of silvertip (*Carcharhinus albimarginatus*), gray reef (*Carcharhinus amblyrhynchos*), blacktip reef (*Carcharhinus melanopterus*), tawny nurse (*Nebrius ferrugineus*), and sicklefin lemon sharks

(*Negaprion acutidens*) in the Seychelles. Habitat use varied among species, with blacktip reef and lemon sharks using mostly lagoon areas and gray reef and silvertip sharks using mainly coastal reefs and drop-offs. Tawny nurse sharks showed habitat segregation by size within the atoll, as small individuals were found inside the lagoon and large individuals outside (Lea et al., 2016).

Approaches such as the multiple regression quadratic assignment procedure (MRQAP), a variant of the Mantel test with multiple factors, and mixed effect models can help to evaluate the influence of biological and environmental factors on movement and habitat network structures by incorporating node attributes (Dekker et al., 2007; Pinter-Wollman et al., 2014) and network metrics into the models, respectively. Only one study has used MRQAP (with the double-Dekker semi-partialing method) to examine factors influencing shark movement. Jacoby et al. (2012a) used MRQAP to study the influence of inshore vs. offshore locations, mean depth, and habitat complexity on the movement of female and male small-spotted catsharks.

18.2.3 Limitations of Network Analyses

Despite the successful contribution of network analysis to understanding shark ecology, it is important to acknowledge the current limitations. For example, as automated and indirect methods are being used increasingly to infer sociality rather than measuring it with direct observations, it is critical to understand and test the assumptions underlying the different approaches. Importantly, network-based tools are being used in studies of movement ecology (see Section 18.2.2), and statistical methods are already emerging (Jacoby et al., 2016; Spiegel et al., 2016). Most spatial network studies reviewed here used passive acoustic monitoring to examine shark movement, which is well suited to network analysis due to the use of discrete moored acoustic receivers as nodes, but that limits interpretations to specific or local areas. For large migratory species of sharks, acoustic monitoring studies only provide a local snapshot of their movements, with some critical behavior taking place beyond the limits of the acoustic array in place. For more localized species or those that show considerable site fidelity, careful consideration should be taken when choosing what a node represents to allow comparison between individuals or between species. Therefore, it is crucial to ensure that the design of any array of receivers is tailored to the species studied and the ecological question investigated (e.g., small-scale social behavior vs. large-scale migratory routes that require acoustic gates), in addition to accounting for landmasses that will influence the timing and directionality of transitions between some of the receivers.

Network analysis provides a simple way to display complex processes that instantly reveal information on spatial and temporal changes in animal space use (Jacoby et al., 2012, Lédée et al., 2015). However, compared to traditional analyses, network analysis does not estimate activity space or provide an exact match of individual core use areas measured using kernel utilization distribution or Brownian bridge, as shown in Lédée et al. (2015). Therefore, although network analysis alone is useful in providing information on animal patterns, combining traditional and network analyses will provide a more comprehensive picture of animal movement (Bascompte, 2007). Furthermore, caution should be used when selecting metrics to answer specific questions and interpreting results from networks with low numbers of nodes and connections; for example, to reiterate, the precision of betweenness and clustering coefficient declines as the number of nodes decrease (Silk et al., 2015). Also, missing data (e.g., low acoustic receiver coverage in acoustic monitoring studies, unknown habitat use in mark–recapture studies) may influence measures of movement between locations or habitat types (Silk et al., 2015). Thus, with limited or missing data the use of network analysis may not adequately represent animal movement, and traditional analyses may be more suited (Whitehead, 2008). It is also worth remembering that interactions are likely to occur between tagged and untagged individuals, and, although these cannot be measured, some discussion about how representative a sample of tagged sharks can be for a given species or location is useful.

Networks are a static representation of movement or habitat use ignoring the temporal dynamics of movement (Cumming et al., 2010; Stehfest et al., 2015). Where possible, temporal dynamics must be taken into consideration when examining the movement of animals, and comparison with other methods may be crucial to validating each approach. Within a movement network, pathways (edges) between acoustic receivers (nodes) are created regardless of the time taken for a shark to travel from one receiver to the next, which is misleading if data are missing for long periods (e.g., outside of receiver range). Furthermore, approaches such as MRQAP require detailed information at the node level. Telemetry data often lack information on environmental factors where and when the individual was detected given the often large detection range. Providing more accurate information about movement and environmental factors at the time the individual was present in the area could be used to refine conservation and management measures (Hastings et al., 2011).

Movements in the marine environment are three dimensional (i.e., include depth); they are constrained by spatial features in both the horizontal and vertical plane and therefore rarely follow a straight path. Movement between two locations or habitat types within a network are shown as a straight path (Stehfest et al., 2015; Tremblay et al., 2006) and so are unrealistic in most situations. Improved visualizations that demonstrate or account for the variability in path orientation or duration will no doubt help resolve this issue to some extent, although quantitative analyses will also have to factor this in.

18.3 FUTURE DEVELOPMENTS AND RESEARCH DIRECTIONS

18.3.1 Technological Challenges and Developments

The field of network analysis is progressing rapidly with improvements in technology and analytical methods. For example, much more detailed inferences about the social networks of sharks and rays would be obtained using a combination of mobile receivers (e.g., VEMCO Mobile Transceiver, or VMT) recording the tagged individuals encountered within a reduced range (Mourier et al., 2017a) and efficient data transfer to fixed listening stations (Holland et al., 2009). Alternatively, proximity receivers could communicate with other animal-borne devices to provide the location of the animal in addition to remotely transferring the data from the proximity logger. This idea is currently under development such that a VMT communicates with Service Argos via Bluetooth to remotely transmit data (Lidgard et al., 2014), offering an exciting opportunity for future research on shark sociality. Proximity loggers are likely the most accurate method to infer small-scale interactions in marine animals, but technological drawbacks (e.g., battery life, compatibility between manufacturers, retrieval, download duration) still weaken their effectiveness.

The use of multisensor tagging will provide complementary information that can be integrated into more detailed network analyses. Promising technological developments such as Encounternet's adaptation to the marine environment (Tentelier et al., 2016) will benefit social interaction studies of sharks. This will require the development of (1) a set of small acoustic tags that emit individually coded signals and record signal and distance from other tags; (2) a set of fixed base stations that record encounters with tags, upload the logs stored in the tags' memory, and transmit information between tags and the third component of the system; and (3) an interface between the user and the system (e.g., a WiFi system transmitting the data to a land-based computer) that collects the data from the base stations (Figure 18.5A).

New technology can also help in developing remote monitoring of interactions between sharks. Automated underwater vehicles (AUVs) (see Chapter 6 in this volume), such as gliders or drifting robots (Figure 18.5B,C), can record interactions at sea if they are fitted with acoustic receivers (Blonder et al., 2012; Haulsee et al., 2015, 2016). The recent development of unmanned aerial vehicles (UAVs) (Kiszka et al., 2016; see also Chapter 4 in this volume) can provide a means to track shark movements and interactions (Figure 18.5D), although battery life issues limit the time scales for sampling as well as the ability to clearly identify individuals. The use of underwater animal-borne cameras (see Chapter 5 in this volume) can provide data on interactions with other individuals as well as the environmental context of interactions using on-board sensors. This technology will be especially useful

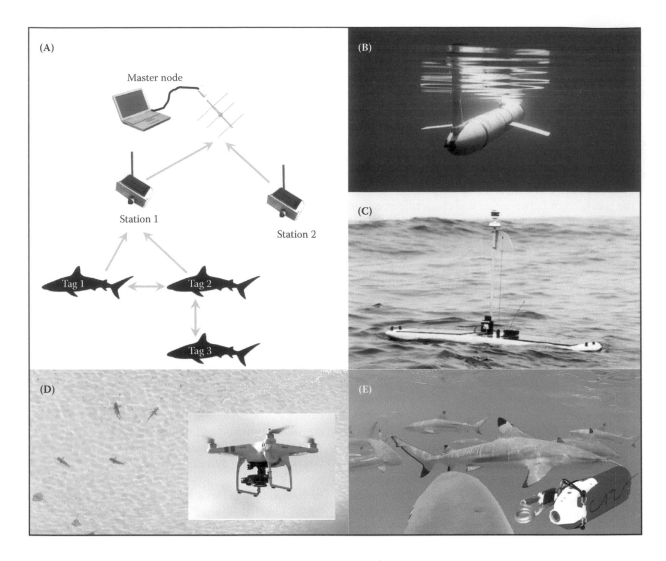

Figure 18.5 Emerging technologies for shark interaction monitoring: (A) Encounternet-like system would track shark encounters automatically; (B) glider automated underwater vehicle able to record tagged sharks at sea; (C) a surfing robot with similar abilities; (D) unmanned automated vehicle, a drone to track shark interactions from the air; and (E) animal-borne cameras attached to a dorsal fin potentially able to capture interactions with other known sharks using photo-identification.

in species where individuals can be distinguished by photo-identification, such as blacktip reef sharks (Figure 18.5E) (see Chapter 12 in this volume), providing a novel method for obtaining an egocentric network. Finally, citizen science programs at dive tourism sites can help in collecting data on interactions within shark aggregations, especially for species amenable to photo-identification (Andrzejaczek et al., 2016; see also Chapter 16 in this volume).

18.3.2 Analytical Challenges and Developments

Network approaches are still relatively new in shark ecology, and the field remains predominantly focused on determining appropriate sampling methodologies, but many other challenges remain. Analytical developments will be critical for allowing researchers to make more of the data (and technologies) that are currently available. This is in part because some

of the logistic limitations, such as the speed of remote download from the animal to the receiver of logged environmental or social data, will likely remain for some time to come.

There are two key analytical considerations regarding the aggregation of data prior to constructing a network. First is the temporal aggregation of data, and there are several approaches that can be used to compensate for the aggregation of samples through time (i.e., data gathered over months or years represented as a single network), including using intervals more relevant to the biology and ecology of the species studied or using time-ordered networks that consider repeatable patterns in network structure through time (for details, see Blonder et al., 2012; Snijders et al., 2010). For example, networks can be created at different temporal scales to incorporate the relevant temporal dynamics of the species' movements. A recent instructive paper by Farine (2018) clearly details when and why a researcher might wish to use a dynamic over a static network approach, and we hope that such guidelines like this will encourage greater use of dynamic networks in shark ecology.

The second consideration is the aggregation of subsets of individuals, which is predominantly an issue for movement networks. For example, shark movements can be highly variable both within individuals during ontogeny (Matich and Heithaus, 2015) and across individuals within a population (e.g., Espinoza et al., 2016), but movements are often aggregated across the full or subsets of the population (e.g., sexes). Multilayer networks will likely play a role in providing new network metrics for networks that operate at multiple levels, such as interspecific networks of movement and social behavior. By considering each shark as a separate layer in a multilayer network we can explore the role of individual movements on habitat connectivity and flow using a more suitable framework. Indeed, these approaches should have considerable impact on the burgeoning research on individual specialization, personality traits, and cognitive variation in sharks (Finger et al., 2017; Guttridge et al., 2013; Jacoby et al., 2014; Matich et al., 2011) in addition to defining new research directions. One promising avenue might consider the role of social behavior in the development of cooperation and social hunting strategies among pelagic sharks relying on diffuse prey fields (Lang and Farine, 2017).

18.3.3 New Ideas and Future Research Questions

18.3.3.1 Social Networks

The use of SNA has revealed complexities in the social lives of sharks; however, we have only just scratched the surface of what is possible given integration of SNA with other tools. For example, in addition to the biotelemetry and biologging techniques discussed above, stable isotope analysis (SIA) (see Chapter 1 in this volume) could be incorporated to add isotopic niche as a node attribute within a social network, helping unlock the role of foraging in the social behavior of sharks. Tissues (e.g., whole blood, plasma, muscle, skin) have different turnover rates and can inform variation in resource or habitat use across short and long time scales (Hussey et al., 2012). Where and what an animal eats can have important implications for social interactions, especially when considered in parallel with body condition or nutrition (Senior et al., 2016). In addition, Wilson et al. (2015) used accelerometers (see Chapter 3 in this volume) to generate locomotor profiles for juvenile lemon sharks simultaneously while collecting social data. This is a powerful approach as sharks could assort by energetic profiles or energy budgets, and it is now possible to use acoustic tracking devices that have inbuilt accelerometers and pressure sensors (Shipley et al., 2017). Thus, in theory it would be possible to add some context to social interactions, such as during resting, fast swimming, or steady swimming.

Further, with improved husbandry and careful selection of the study species it is possible to have enough subjects to manipulate and replicate networks. Only through controlled experiments will social networks be able to provide definitive causative evidence for socially mediated mechanisms underpinning evolutionary processes (Farine and Whitehead, 2015). Numerous small-bodied, abundant species could provide tractable models of behavior, such as Port Jackson sharks (Mourier et al., 2017a) and gummy sharks (Frick et al., 2010). Furthermore, experimental studies that validate the indirect methods for assessing social interactions and elucidate the mechanisms underpinning associations are particularly important. Species such as juvenile lemon, Port Jackson, and blacktip reef sharks are accessible for direct observation as well as acoustic tracking methods, allowing dedicated validation studies. As discussed in Mourier et al. (2017a) sharks can socialize in different ways, so testing methods on species that exhibit variation in sociality (e.g., resting, schooling) will ensure that inference methods are applicable to a broader number of species.

18.3.3.2 Spatial Networks

Recent advances in telemetry have allowed researchers to monitor long-term social behavior and movement patterns of multiple species over vast areas (Espinoza et al., 2015). Animal movement and space and habitat use are often explained only using biological and environmental factors, rarely including individual variation (Nathan et al., 2008). Behavior, fitness, and social position within the population can influence individual movement and generate a more comprehensive picture of how populations may respond to changes to their environment and what this means for their conservation (Snijders et al., 2017; Spiegel et al., 2017). Therefore, combining social and spatial network analysis using movement multilayer networks, for example, can provide a better understanding of spatial patterns in shark ecology, allowing differentiation between spatially driven social processes and social-mediated movement patterns.

Recent advances in biological and environmental sensors provide new opportunities to combine these sensors with tags or receivers. Receivers could record not only individual identification and time and date of detection but also the environmental conditions at the time of detection. Several (acoustic) tags that enable this are already commercially available (e.g., VEMCO V16TP). Furthermore, a habitat or video survey or remotely sensed environmental data at node locations could be gathered to obtain more information about an individual's habitat which could then be included in the analyses.

Networks often ignore the temporal dynamics or three-dimensionality of movement. To allow for this, information on maximum speed of a species could be used to create the network. Observed speed can be calculated for each edge and added to attributes. Then, using an edge threshold analysis any edge with a value greater than the maximum speed of a species could be removed to obtain a more realistic network. Furthermore, standardizing edge length with actual distance between nodes constrained by spatial features (e.g., land, coral reefs) or the use of multilayer networks to incorporate depth information from tags might provide a better representation of paths used by individuals.

Finally, most studies often focus on a single species of shark or areas without considering interactions among species (for a rare example, see Espinoza et al., 2015) or the threat status of the species. The affordability of the tracking technologies and easier online access to the data greatly facilitate the establishment of collaborative efforts over larger areas (Hussey et al., 2015). The challenge here is how best to standardize the design of the movement networks across the different study areas and array designs for useful comparisons. One possible solution is to create a grid that covers all of the areas of interest (Dilts et al., 2016), with each grid representing a node within the movement network. Another possible solution would be to create networks for each area of interest separately and compare standardized network properties, looking at degree correlations for example.

18.3.4 Applications for Management and Conservation

Network analysis provides a toolbox of methods that can be used to assess and model risks such as habitat loss and fragmentation, climate change, and fisheries exposure and can help design and evaluate the effectiveness of management, thus guiding improved conservation practices (Borrett et al., 2014; Cummings et al., 2010; Galpern et al., 2011). Using centrality metrics, the importance of each patch (node) or corridor (edge) in maintaining or contributing to landscape connectivity can be determined to help prioritize areas for management and conservation (Nicol et al., 2016; Rayfield et al., 2016). For example, species habitat fragmentation can be identified by looking at communities using metrics such as component and cluster (Table 18.1). Knowing how habitats are connected or fragmented can help inform management plans to protect clusters of habitat, stepping stones, and corridors (Bodin et al., 2006; Thomas et al., 2014). Degree and node removal analyses can be used to examine population source and sink dynamics to identify potential corridors that may aid in species restoration (Treml et al., 2008). Finally, seascape connectivity can be measured by determining habitat availability and dispersal probabilities between habitat patches to help design or evaluate effectiveness of marine protected area networks (Espinoza et al., 2015; Engelhard et al., 2017). Using patch and edge removal or edge thresholding analyses, the role these patches and corridors have in maintaining connectivity in the landscape can be examined under different patch- and edge-loss scenarios. The advantage of these methods is that researchers can simulate the destruction of patches or corridors and rank them by their contributions to landscape/seascape connectivity, thereby allowing managers to make decisions based on which patches and corridors are most critical to connectivity (Kurvers et al., 2014), not just for one species but multiple species within the same landscape.

Alternatively, network analysis can be used to assess management and conservation plans. For example, network analysis can inform managers about fishing activity pattern. Martin et al. (2017) used reservoir (i.e., node) removal analysis to examine the differences in participation among anglers within a regional fishery and assess resilience of the regional fishery to disturbance (e.g., disease, invasive species etc.). Network analysis can also evaluate the efficacy of species protection efforts across borders. Treml et al. (2015) compared species dispersal networks with institutional networks across multiple countries within the Coral Triangle to determine if species dispersal was adequately protected in the institutional network. In doing so, they identified discrepancies between ecological processes and their governance. Finally, network theory can be applied in a social science capacity to better understand the spread of information among fishermen and better manage resources (see Chapter 17 in this volume). Using these approaches, Barnes et al. (2016) found that enhanced communication channels across segregated fisher groups could have prevented the incidental catch of over 46,000 sharks between 2008 and 2012 in a single commercial fishery.

The use of social network analysis can be profitably applied by wildlife managers and conservationists to, for example, identify changes in animal interactions or connectivity within a population and identify warning signs of detrimental changes (Snijders et al., 2017). Spatial tracking in combination with network analysis can help identify both locations and individuals that form crucial social bridges between subpopulations, as social connectivity likely contributes to effective spatial connectivity. The fragmentation of habitats after an anthropogenic perturbation can reduce encounter rates, which can likely induce changes in social interactions or mate choice options. As such, monitoring the interaction network within a shark population before and after a perturbation can help managers to predict potential changes in population viability. Additionally, managers can test whether a shark population is resilient to selective harvesting. Mourier et al. (2017b), for example, tested whether the social connectivity within a population of blacktip reef sharks would be robust to both selective harvesting and random individual removal. They found that the structure of the networks makes the population relatively robust to such perturbation.

Network analysis is advantageous for developing, guiding, and assessing management measures. It allows for the assessment of species movement and behavior and for predicting the consequences of anthropogenic and natural disturbances by testing and experimenting on a variety of species at different scales and under multiple scenarios. Finally, it allows for the assessment of management and conservation plans across borders.

18.4 SUMMARY AND CONCLUSION

Network analyses are increasingly being used in shark behavioral ecology. Here, we present a broad overview of the utility and challenges of applying network analyses and describe the early progress this field has had in informing shark behavior. Developments in SNA across other taxonomic groups (particularly in terrestrial systems) continue to help guide and inform the tools available for studying underwater social networks in conjunction with advances in tracking technologies. At present, the number of studies adopting these approaches remains relatively small; however, we hope that this chapter helps stimulate ideas and research directions that continue to push for developments in shark network ecology. Importantly, new social or spatial network studies on different species and systems will certainly contribute to improving the understanding of the main drivers affecting the evolution of social behavior in sharks and rays as well as their spatial ecology. This in turn will provide researchers with much-needed information to take a more informed and location-specific approach to their conservation. We are excited by the burgeoning developments in this area, by how these developments might guide the design of new technologies, and ultimately by the impact this holistic approach might eventually have on shark conservation.

REFERENCES

Andrzejaczek S, Meeuwig J, Rowat D, Pierce S, Davies T, Fisher R, Meekan M (2016) The ecological connectivity of whale shark aggregations in the Indian Ocean: a photo-identification approach. *R Soc Open Sci* 3:160455.

Armansin NC, Lee KA, Huveneers C, Harcourt RG (2016) Integrating social network analysis and fine-scale positioning to characterize the associations of a benthic shark. *Anim Behav* 115:245–258.

Barabási A-L, Albert R (1999) Emergence of scaling in random networks. *Science* 286:509–512.

Barnes ML, Lynham J, Kalberg K, Leung P (2016) Social networks and environmental outcomes. *Proc Natl Acad Sci USA* 113:6466–6471.

Bascompte J (2007) Networks in ecology. *Basic Appl Ecol* 8:485–490.

Blonder B, Wey TW, Dornhaus A, James R, Sih A (2012) Temporal dynamics and network analysis. *Meth Ecol Evol* 3:958–972.

Blüthgen N, Fründ J, Vazquez DP, Menzel F (2008) What do interaction network metrics tell us about specialization and biological traits? *Ecology* 89:3387–3399.

Bodin Ö, Tengö M, Norman A, Lundberg J, Elmqvist T (2006) The value of small size: loss of forest patches and ecological thresholds in Southern Madagascar. *Ecol Appl* 16:440–451.

Borgatti SP (2012) Social network analysis, two-mode concepts. In: Meyers RA (ed) *Computational Complexity*. Springer, New York, pp 2912–2924.

Borrett SR, Moody J, Edelmann A (2014) The rise of network ecology: maps of the topic diversity and scientific collaboration. *Ecol Model* 293:111–127.

Brena P, Mourier J, Planes S, Clua E (2018) Concede or clash? Solitary sharks competing for food assess rivals to decide. *Proc R Soc Lond B Biol Sci* 285(1875):20180006.

Buray N, Mourier J, Planes S, Clua E (2009) Underwater photoidentification of sicklefin lemon sharks, *Negaprion acutidens*, at Moorea (French Polynesia). *Cybium* 33:21–27.

Clua E, Buray N, Legendre P, Mourier J, Planes S (2010) Behavioural response of sicklefin lemon sharks *Negaprion acutidens* to underwater feeding for ecotourism purposes. *Mar Ecol Prog Ser* 414:257–266.

Croft DP, James R, Krause J (2008) *Exploring Animal Social Networks*. Princeton University Press, Princeton, NJ.

Croft DP, Madden JR, Franks DW, James R (2011) Hypothesis testing in animal social networks. *Trends Ecol Evol* 26:502–507.

Croft DP, Darden SK, Wey TW (2016) Current directions in animal social networks. *Curr Opin Behav Sci* 12:52–58.

Csárdi G, Nepusz T (2006) The igraph software package for complex network research. *Int J Complex Syst* 1695:1–9.

Cumming GS, Bodin Ö, Ernstson H, Elmqvist T (2010) Network analysis in conservation biogeography: challenges and opportunities. *Divers Distrib* 16:414–425.

Dekker D, Krackhardt D, Snijders TAB (2007) Sensitivity of MRQAP tests to collinearity and autocorrelation conditions. *Psychometrika* 72:563–581.

Dilts TE, Weisberg PJ, Leitner P, Matocq MD, Inman RD, Nussear KE, Esque TC (2016) Multiscale connectivity and graph theory highlight critical areas for conservation under climate change. *Ecol Appl* 26:1223–1237.

Engelhard SL, Huijbers CM, Stewart-Koster B, Olds AD, Schlacher TA, Connolly RM (2017) Prioritising seascape connectivity in conservation using network analysis. *J Appl Ecol* 54:1130–1141.

Erdös P, Rényi A (1959) On random graphs. *Publ Math* 6:290–297.

Espinoza M, Lédée EJI, Simpfendorfer CA, Tobin AJ, Heupel MR (2015) Contrasting movements and connectivity of reefassociated sharks using acoustic telemetry: implications for management. *Ecol Appl* 25:2101–2118.

Espinoza M, Heupel MR, Tobin AJ, Simpfendorfer CA (2016) Evidence of partial migration in a large coastal predator: opportunistic foraging and reproduction as key drivers? *PLoS ONE* 11:e0147608.

Farine DR (2015) Proximity as a proxy for interactions: Issues of scale in social network analysis. Anim Behav 104:e1–e5. doi: 10.1016/j.anbehav.2014.11.019.

Farine DR (2017) A guide to null models for animal social network analysis. *Meth Ecol Evol* 8:1309–1320.

Farine DR (2018) When to choose dynamic vs. static social network analysis. *J Anim Ecol* 87:128–138.

Farine DR, Whitehead H (2015) Constructing, conducting, and interpreting animal social network analysis. *J Anim Ecol* 84:1144–1163.

Findlay R, Gennari E, Cantor M, Tittensor DP (2016) How solitary are white sharks: social interactions or just spatial proximity? *Behav Ecol Sociobiol* 70:1735–1744.

Finger JS, Dhellemmes F, Guttridge TL, Kurvers RHJM, Gruber SH, Krause J (2016) Rate of movement of juvenile lemon sharks in a novel open field, are we measuring activity or reaction to novelty? *Anim Behav* 116:75–82.

Finger JS, Dhellemmes F, Guttridge TL (2017) Personality in elasmobranchs with a focus on sharks: early evidence, challenges, and future directions. In: Vonk J, Weiss A, Kuczaj SA (eds) *Personality in Nonhuman Animals*. Springer, Cham, Switzerland, pp 129–152.

Fletcher RJ, Acevedo MA, Reichert BE, Pias KE, Kitchens WM (2011) Social network models predict movement and connectivity in ecological landscapes. *Proc Natl Acad Sci USA* 108:19282–19287.

Fortuna MA, Gómez-Rodríguez C, Bascompte J (2006) Spatial network structure and amphibian persistence in stochastic environments. *Proc R Soc Lond B Biol Sci* 273:1429–1434.

Frick LH, Walker TI, Reina RD (2010) Trawl capture of Port Jackson sharks, *Heterodontus portusjacksoni*, and gummy sharks, *Mustelus antarcticus*, in a controlled setting: effects of tow duration, air exposure and crowding. *Fish Res* 106:344–350.

Furmston T, Morton AJ, Hailes S (2015) A significance test for inferring affiliation networks from spatio-temporal data. *PLoS ONE* 10:e0132417.

Gallagher AJ, Vianna GMS, Papastamatiou YP, Macdonald C, Guttridge TL, Hammerschlag N (2015) Biological effects, conservation potential, and research priorities of shark diving tourism. *Biol Conserv* 184:365–379.

Galpern P, Manseau M, Fall A (2011) Patch-based graphs of landscape connectivity: a guide to construction, analysis and application for conservation. *Biol Conserv* 144:44–55.

Guttridge TL, Gruber SH, Gledhill KS, Croft DP, Sims DW, Krause J (2009) Social preferences of juvenile lemon sharks, *Negaprion brevirostris*. *Anim Behav* 78:543–548.

Guttridge TL, Gruber SH, Krause J, Sims DW (2010) Novel acoustic technology for studying free-ranging shark social behaviour by recording individuals' interactions. *PLoS ONE* 5:e9324.

Guttridge TL, Gruber SH, DiBattista JD, Feldheim KA, Croft DP, Krause S, Krause J (2011) Assortative interactions and leadership in a free-ranging population of juvenile lemon shark *Negaprion brevirostris*. *Mar Ecol Prog Ser* 423:235–245.

Guttridge TL, Dijk S van, Stamhuis EJ, Krause J, Gruber KJ, Brown C (2013) Social learning in juvenile lemon sharks, *Negaprion brevirostris*. *Anim Cogn* 16:55–64.

Haddadi H, King AJ, Wills AP, Fay D, Lowe J, Morton AJ, Hailes S, Wilson AM (2011) Determining association networks in social animals: choosing spatial–temporal criteria and sampling rates. *Behav Ecol Sociobiol* 65:1659–1668.

Hastings A, Petrovskii S, Morozov A (2011) Spatial ecology across scales. *Biol Lett* 7:163–165.

Haulsee D, Breece M, Miller D, Wetherbee BM, Fox DA, Oliver MJ (2015) Habitat selection of a coastal shark species estimated from an autonomous underwater vehicle. *Mar Ecol Prog Ser* 528:277–288.

Haulsee DE, Fox DA, Breece MW, Brown LM, Kneebone J, Skomal GB, Oliver MJ (2016) Social network analysis reveals potential fission–fusion behavior in a shark. *Sci Rep* 6:34087.

Heupel MR, Lédée EJI, Simpfendorfer CA (2017) Telemetry reveals spatial separation of co-occurring reef sharks. *Mar Ecol Prog Ser* 589:179–192.

Holland K, Meyer C, Dagorn L (2009) Inter-animal telemetry: results from first deployment of acoustic 'business card' tags. *Endanger Species Res* 10:287–293.

Hoppitt WJE, Farine DR (2018) Association indices for quantifying social relationships: how to deal with missing observations of individuals or groups. *Anim Behav* 136:227–238.

Hussey NE, Kessel ST, Aarestrup K, Cooke SJ, Cowley PD, Fisk AT, Harcourt RG, et al. (2015) Aquatic animal telemetry: a panoramic window into the underwater world. *Science* 348:1255642.

Hussey NE, MacNeil MA, Olin JA, McMeans BC, Kinney MJ, Chapman, DD, Fisk AT (2012) Stable isotopes and elasmobranchs: tissue types, methods, applications and assumptions. *J Fish Biol* 80:1449–1484.

Jacoby DMP, Freeman R (2016) Emerging network-based tools in movement ecology. *Trends Ecol Evol* 31:301–314.

Jacoby DMP, Busawon DS, Sims DW (2010) Sex and social networking: the influence of male presence on social structure of female shark groups. *Behav Ecol* 21:808–818.

Jacoby DMP, Brooks EJ, Croft DP, Sims DW (2012a) Developing a deeper understanding of animal movements and spatial dynamics through novel application of network analyses. *Methods Ecol Evol* 3:574–583.

Jacoby DMP, Croft DP, Sims DW (2012b) Social behaviour in sharks and rays: analysis, patterns and implications for conservation. *Fish Fish* 13:399–417.

Jacoby DMP, Sims DDW, Croft DP (2012c) The effect of familiarity on aggregation and social behaviour in juvenile small spotted catsharks *Scyliorhinus canicula*. *J Fish Biol* 81:1596–1610.

Jacoby DMP, Fear LN, Sims DW, Croft DP (2014) Shark personalities? Repeatability of social network traits in a widely distributed predatory fish. *Behav Ecol Sociobiol* 68:1995–2003.

Jacoby DMP, Papastamatiou YP, Freeman R (2016) Inferring animal social networks and leadership: applications for passive monitoring arrays. *J R Soc Interface* 13:20160676.

James R, Croft DP, Krause J (2009) Potential banana skins in animal social network analysis. *Behav Ecol Sociobiol* 63:989–997.

Keller BA, Finger J-S, Gruber SH, Abel DC, Guttridge TL (2017) The effects of familiarity on the social interactions of juvenile lemon sharks, *Negaprion brevirostris*. *J Exp Mar Biol Ecol* 489:24–31.

Kiszka JJ, Mourier J, Gastrich K, Heithaus MR (2016) Using unmanned aerial vehicles (UAVs) to investigate shark and ray densities in a shallow coral lagoon. *Mar Ecol Prog Ser* 560:237–242.

Krause J, Ruxton GD (2002) *Living in Groups*. Oxford University Press, Oxford.

Krause J, Croft DP, James R (2007) Social network theory in the behavioural sciences: potential applications. *Behav Ecol Sociobiol* 62:15–27.

Krause J, Lusseau D, James R (2009a) Animal social networks: an introduction. *Behav Ecol Sociobiol* 63:967–973.

Krause S, Mattner L, James R, Guttridge T, Corcoran MJ, Gruber SH, Krause J (2009b) Social network analysis and valid Markov chain Monte Carlo tests of null models. *Behav Ecol Sociobiol* 63:1089–1096.

Krause J, Krause S, Arlinghaus R, Psorakis I, Roberts S, Rutz C (2013) Reality mining of animal social systems. *Trends Ecol Evol* 28:541–551.

Krause J, Croft DP, Wilson ADM (2015) The network approach in teleost fishes and elasmobranchs. In: Krause J, James R, Franks D, Croft D (eds) *Animal Social Networks*. Oxford University Press, Oxford, pp 150–159.

Kurvers RHJM, Krause J, Croft DP, Wilson ADM, Wolf M (2014) The evolutionary and ecological consequences of animal social networks: emerging issues. *Trends Ecol Evol* 29:326–335.

Lang SDJ, Farine DR (2017) A multidimensional framework for studying social predation strategies. *Nat Ecol Evol* 1:1230–1239.

Lea JSE, Humphries NE, Brandis RG von, Clarke CR, Sims DW (2016) Acoustic telemetry and network analysis reveal the space use of multiple reef predators and enhance marine protected area design. *Proc R Soc Lond B Biol Sci* 283:20160717.

Lédée EJI, Heupel MR, Tobin AJ, Knip DM, Simpfendorfer CA (2015) A comparison between traditional kernel-based methods and network analysis: an example from two nearshore shark species. *Anim Behav* 103:17–28.

Lidgard DC, Bowen WD, Jonsen ID, McConnell BJ, Lovell P, Webber DM, Stone T, Iverson SJ (2014) Transmitting species-interaction data from animal-borne transceivers through Service Argos using Bluetooth communication. *Methods Ecol Evol* 5:864–871.

Loiseau N, Kiszka JJ, Bouveroux T, Heithaus MR, Soria M, Chabanet P (2016). Using an unbaited stationary video system to investigate the behaviour and interactions of bull sharks Carcharhinus leucas under an aquaculture farm. *Afr J Mar Sci* 38:73–79.

Mann J, Stanton MA, Patterson EM, Bienenstock EJ, Singh LO (2012) Social networks reveal cultural behaviour in tool-using dolphins. *Nat Commun* 3:980.

Marshall AD, Pierce SJ (2012) The use and abuse of photographic identification in sharks and rays. *J Fish Biol* 80:1361–1379.

Martin DR, Shizuka D, Chizinski CJ, Pope KL (2017) Network analysis of a regional fishery: implications for management of natural resources, and recruitment and retention of anglers. *Fish Res* 194:31–41.

Matich P, Heithaus MR (2015) Individual variation in ontogenetic niche shifts in habitat use and movement patterns of a large estuarine predator (*Carcharhinus leucas*). *Oecologia* 178:347–359.

Matich P, Heithaus MR, Layman CA (2011) Contrasting patterns of individual specialization and trophic coupling in two marine apex predators. *J Anim Ecol* 80:294–305.

Minor ES, Urban DL (2008) A graph-theory framework for evaluating landscape connectivity and conservation planning. *Conserv Biol* 22:297–307.

Mourier J, Vercelloni J, Planes S (2012) Evidence of social communities in a spatially structured network of a free-ranging shark species. *Anim Behav* 83:389–401.

Mourier J, Bass NC, Guttridge TL, Day J, Brown C (2017a) Does detection range matter for inferring social networks in a benthic shark using acoustic telemetry? *R Soc Open Sci* 4:170485.

Mourier J, Brown C, Planes S (2017b) Learning and robustness to catch-and-release fishing in a shark social network. *Biol Lett* 13:20160824.

Nathan R, Getz WM, Revilla E, Holyoak M, Kadmon R, Saltz, D, Smouse PE (2008) A movement ecology paradigm for unifying organismal movement research. *Proc Natl Acad Sci* 105:19052–19059.

Newman M (2010) *Networks: An Introduction*. Oxford University Press, Oxford.

Nicol S, Wiederholt R, Diffendorfer JE et al. (2016) A management-oriented framework for selecting metrics used to assess habitat- and path-specific quality in spatially structured populations. *Ecol Indic* 69:792–802.

Opsahl T (2013) Triadic closure in two-mode networks: redefining the global and local clustering coefficients. *Soc Networks* 35:159–167.

Papastamatiou Y, Meyer C, Kosaki R, Wallsgrove NJ, Popp BN (2015) Movements and foraging of predators associated with mesophotic coral reefs and their potential for linking ecological habitats. *Mar Ecol Prog Ser* 521:155–170.

Pinter-Wollman N, Hobson EA, Smith JE, Edelman AJ, Shizuka D, de Silva S, Waters JS, et al. (2014) The dynamics of animal social networks: analytical, conceptual, and theoretical advances. *Behav Ecol* 25:242–255.

Proulx SR, Promislow DEL, Phillips PC (2005) Network thinking in ecology and evolution. *Trends Ecol Evol* 20:345–353.

Rayfield B, Pelletier D, Dumitru M, Cardille JA, Gonzalez A (2016) Multipurpose habitat networks for short-range and long-range connectivity: a new method combining graph and circuit connectivity. *Methods Ecol Evol* 7:222–231.

Senior AM, Lihoreau M, Charleston MA, Buhl J, Raubenheimer D, Simpson SJ (2016) Adaptive collective foraging in groups with conflicting nutritional needs. *R Soc Open Sci* 3:150638.

Shipley ON, Brownscombe JW, Danylchuk AJ, Cooke SJ, O'Shea OR, Brooks EJ (2017) Fine-scale movement and activity patterns of Caribbean reef sharks (*Carcharhinus perezi*) in the Bahamas. *Environ Biol Fish* 1–8.

Silk MJ, Jackson AL, Croft DP, Colhoun K, Bearhop S (2015) The consequences of unidentifiable individuals for the analysis of an animal social network. *Anim Behav* 104:1–11.

Sims DW (2003) Tractable models for testing theories about natural strategies: foraging behaviour and habitat selection of free-ranging sharks. *J Fish Biol* 63:53–73.

Snijders L, Blumstein DT, Stanley CR, Franks DW (2017) Animal social network theory can help wildlife conservation. *Trends Ecol Evol* 32:567–577.

Snijders TAB, van de Bunt GG, Steglich CEG (2010) Introduction to stochastic actor-based models for network dynamics. *Soc Netw* 32:44–60.

Spiegel O, Leu ST, Sih A, Bull CM (2016) Socially interacting or indifferent neighbours? Randomization of movement paths to tease apart social preference and spatial constraints. *Methods Ecol Evol* 7:971–979.

Spiegel O, Leu ST, Bull CM, Sih A (2017) What's your move? Movement as a link between personality and spatial dynamics in animal populations. *Ecol Lett* 20:3–18.

Stehfest KM, Patterson TA, Barnett A, Semmens JM (2015) Markov models and network analysis reveal sex-specific differences in the space-use of a coastal apex predator. *Oikos* 124:307–318.

Tentelier C, Aymes J-C, Spitz B, Rives J (2016) Using proximity loggers to describe the sexual network of a freshwater fish. *Environ Biol Fish* 99:621–631.

Thomas CJ, Lambrechts J, Wolanski E, Traag VA, Blondel VD, Deleersnijder E, Hanert E (2014) Numerical modelling and graph theory tools to study ecological connectivity in the Great Barrier Reef. *Ecol Model* 272:160–174.

Tremblay Y, Shaffer SA, Fowler SL, Kuhn CE, McDonald BI, Weise MJ, Bost C-A, et al. (2006) Interpolation of animal tracking data in a fluid environment. *J Exp Biol* 209:128–140.

Treml EA, Halpin PN, Urban DL, Pratson LF (2008) Modeling population connectivity by ocean currents, a graph-theoretic approach for marine conservation. *Landsc Ecol* 23:19–36.

Treml EA, Fidelman PIJ, Kininmonth S, Ekstrom JA, Bodin O (2015) Analyzing the (mis)fit between the institutional and ecological networks of the Indo-West Pacific. *Glob Environ Change* 31:263–271.

Ward AJW, Botham MS, Hoare DJ, James R, Broom M, Godin J-GJ, Krause J (2002) Association patterns and shoal fidelity in the three-spined stickleback. *Proc R Soc Lond B Biol Sci* 269:2451–2455.

Watts DJ, Strogatz SH (1998) Collective dynamics of "small-world" networks. *Nature* 393:440–442.

Wey T, Blumstein DT, Shen W, Jordán F (2008) Social network analysis of animal behaviour: a promising tool for the study of sociality. *Anim Behav* 75:333–344.

Whitehead H (2008) *Analyzing Animal Societies: Quantitative Methods for Vertebrate Social Analysis*. University of Chicago Press, Chicago, IL.

Whitehead H, James R (2015) Generalized affiliation indices extract affiliations from social network data. *Methods Ecol Evol* 6:836–844.

Williams JJ, Papastamatiou YP, Caselle JE, Bradley D, Jacoby DM (2018) Mobile marine predators: an understudied source of nutrients to coral reefs in an unfished atoll. *Proc R Soc Lond B Biol Sci* 285(1875):20172456.

Wilson ADM, Croft DP, Krause J (2014) Social networks in elasmobranchs and teleost fishes. *Fish Fish* 15:676–689.

Wilson ADM, Brownscombe JW, Krause J, Krause S, Gutowsky LFG, Brooks EJ, Cooke SJ (2015) Integrating network analysis, sensor tags, and observation to understand shark ecology and behavior. *Behav Ecol* 26:1577–1586.

CHAPTER 19

Satellite Tracking Technologies and Their Application to Shark Movement Ecology

Luciana C. Ferreira
Australian Institute of Marine Science, Indian Ocean Marine Research Centre,
University of Western Australia, Crawley, Western Australia, Australia

Kate L. Mansfield
Marine Turtle Research Group, University of Central Florida, Orlando, Florida

Michele Thums
Australian Institute of Marine Science, Indian Ocean Marine Research Centre,
University of Western Australia, Crawley, Western Australia, Australia

Mark G. Meekan
Australian Institute of Marine Science, Indian Ocean Marine Research Centre,
University of Western Australia, Crawley, Western Australia, Australia

CONTENTS

19.1 INTRODUCTION

Telemetry and biologging approaches are commonly used to understand the movement ecology, spatial use, and behavior of marine vertebrates (Hays et al., 2016). For species with wide-ranging and unpredictable movement patterns (e.g., many pelagic sharks), satellite-linked telemetry devices are usually required to remotely monitor their movements through time and in the three-dimensional space of the open ocean or coastal areas they inhabit. By instrumenting an animal, researchers make an assumption that the data collected by the tag are representative of the "true" or "normal" behavior of the targeted species in the wild. However, issues associated with the technology (e.g., spatial error, uplink rate), sample design (e.g., whether or not instrumented animals are representative of the population as a whole), behavior of animals in response to the presence of the tag, and analytical methods applied to the data can affect the accuracy and robustness of results and ultimately our interpretation of biological patterns. In this chapter, we review the range of satellite tags available for sharks and rays, methods of attachment, study design, types of data these tags provide, and the methodologies for analysis and application of data collected by satellite telemetry. This information will assist researchers in tag selection, data processing, and analysis and provide insight into emerging technologies and new directions for the field of satellite telemetry applied to sharks and rays.

19.2 TYPES OF SATELLITE TAGS

19.2.1 Satellite-Linked Transmitters

Satellite-linked transmitters are widely used to track the large-scale (hundreds to thousands of kilometers) horizontal movements of a range of animals in marine systems. Because satellite tags rely on radio signals to communicate with satellites, and these signals are rapidly attenuated in water, this approach is often focused on species that spend time at the sea surface. However, tag models that archive data and delay transmission to satellite networks until they detach from animals and float to the surface (such as PSAT tags, discussed below) are also often used for animals such as sharks and tunas that may only rarely or never break the surface during day-to-day movements. Satellite tags deployed on animals frequenting the surface of the ocean use polar-orbiting satellites from the Argos system (http://www.clsamerica.com/) to relay locations at near real-time. The satellites in the Argos system orbit at an altitude of 850 km, have a footprint of 5000 km, and complete their orbital revolution in approximately 100 minutes, passing over the poles approximately 14 times per day (http://www.argos-system.org). Wet–dry sensors incorporated into the tag indicate when the tag is out of the water (dry), triggering periodic message transmission of

short radio signals (typically for 360 to 920 msec at an interval of around 60 sec and frequency of 401.650 mHz) that are received by passing satellites (Argos, 2016; Hays et al., 2001). Locations are calculated from all messages received during a satellite pass using the Doppler shift, which is the change in frequency of transmissions received by a satellite during its approach and when the satellite moves away from the transmitter (Argos, 2016).

Prior to 2011, Argos used a position algorithm based on least-squares analysis that provided locations and an associated estimated error radius for positions received from four or more messages. When fewer than four messages were received, the Argos system was unable to estimate accuracy for the location provided. In 2011, Argos implemented the Kalman filter (Harvey, 1990) algorithm for location processing that processes locations based on one message per satellite pass, providing an error ellipse for all locations. This accounts for the anisotropy of errors as a result of the satellite polar orbit and provides greater location accuracy (McClintock et al., 2015). Users may now choose between the two algorithms; however, Argos recommends using the Kalman filter for its improved accuracy and increased number of usable locations, although noting that the least-square methods is still useful for very long time series spanning many years (Argos, 2016).

Argos locations are assigned a location class (LC) of 3, 2, 1, 0, A, B, or Z, indicating the level of spatial accuracy in the estimates of latitude and longitude. Location classes 3, 2, 1, and 0 are only provided when four or more messages are received from the tag. Location class A is provided when three messages are received and location class B when only one message is received, and Z represents a failed attempt to obtain a location, although a location estimate is provided. The Argos system indicates that the estimated errors in latitude and longitude for each LC are <250 m for LC 3, between 250 and 500 m for LC 2, between 500 m and 1500 m for LC 1 and >1500 m for LC 0, with a non-guaranteed accuracy for LC A and LC B (Argos, 2016). However, field tests of Argos location error in double-tagging (GPS and Argos) experiments show that estimated error values may vary greatly (Costa et al., 2010; Hoenner, 2012; Vincent et al., 2002). The short and intermittent durations of surface behavior mean that satellite-linked transmitters deployed on sharks report location estimates with a high proportion of the least accurate location classes (LC 0, A, B), corresponding to mean errors of 3.03 to 11.48 km (Costa et al., 2010; Hays et al., 2001; Hazel, 2009; Hoenner, 2012; Vincent et al., 2002). Additionally, location estimates are often sparse in time.

Other common issues in deployments of tags on sharks include low coverage of satellites in some areas, such as the tropics, where satellite passes are much less frequent than in higher latitudes (Hoenner et al., 2012; Vincent et al., 2002); sensor and antenna damage and biofouling (Hays et al., 2007); and premature tag shedding due to failure of the tag attachment (Hays et al., 2007), animals rubbing against structures or the bottom, other animals attacking tags, or deep-diving and activating the release mechanism (see Section 19.3). Together, these issues tend to result in datasets with large spatial error (low-accuracy LC) and large temporal data gaps. For example, satellite tag deployments on tiger sharks on average provided just one or fewer location estimates per day across all deployments (Ferreira et al., 2015; Fitzpatrick et al., 2012). Fastloc® Global Positioning System (GPS) transmitters (see description in Section 19.2.3) may resolve some of these issues with faster transmission rates (<100 msec) and more accurate location estimates (<50 m); however, the high cost of this technology (approximately double the cost of traditional Argos tags) remains prohibitive for many researchers. The typically low spatial and temporal resolution of shark tracking data must be addressed during processing and analysis (see Sections 19.4 and 19.5). The problem of biofouling can be minimized with nontoxic, nonmetallic antifouling agents (Hammerschlag et al., 2012; Hays et al., 2007; Meyer et al., 2010), which is particularly essential in the tropics (Figure 19.1E).

In addition to providing location estimates, satellite-linked tags can also collect data on diving depth and temperature that are sampled and archived during deployment and summarized into bins for transmission. The sampling rate and bins are programmed by the user prior to tag deployment. The binned data are then used to construct histograms of dive duration, maximum dive depth, and time-at-depth/temperature. Coarse profiles of depth and temperature (e.g., 15-minute to 1-hour sample intervals) can also be transmitted by some tags. This compression of raw data is necessary due to the limited bandwidth available in the Argos system; however, complete archives of data can be downloaded if the tag is recovered after deployment.

19.2.2 Pop-Up Satellite Archival Transmitters

Pop-up satellite archival transmitters (PSATs) are designed to track animals that do not spend enough time at the surface to ensure adequate time for communication between the transmitter and the Argos systems while the tag is attached to an animal. PSATs include sensors for depth, temperature, and ambient light level (used to provide coarse location estimates by geolocation with errors of hundreds to thousands of kilometers) (Hill and Braun, 2001), which are recorded at preprogrammed sampling rates that can vary from a few seconds to a few minutes. These tags archive all sensor data during deployment in the memory for a programmable period of deployment time (ranging from a few months to a year). When this time period elapses, the tags release from the animal and float to the surface, and the archived data are summarized and transmitted through the Argos satellite network.

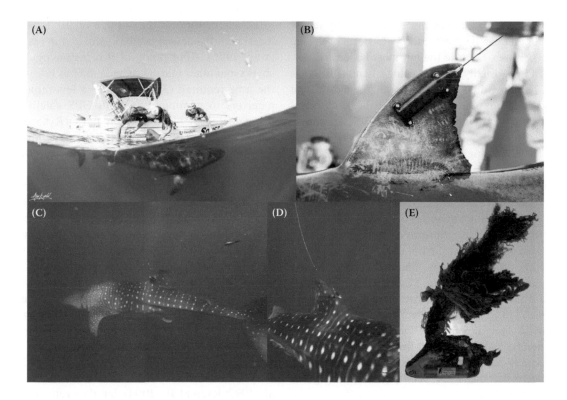

Figure 19.1 Tag type and attachment methods. (A) Tagging of a tiger shark (photograph © Alex Kydd; used with permission). (B) Fin-mounted satellite-linked transmitter (photograph © OCEARCH; used with permission). (C) Fastloc® GPS tag attached to a CATs tag being towed by a whale shark (photograph © AIMS; used with permission). (D) Detail of a fin clamp attachment to a whale shark's first dorsal fin (photograph © AIMS; used with permission). (E) Biofouling of a satellite-linked transmitter deployed on a tiger shark at Raine Island, Australia (photograph © Richard Fitzpatrick/Biopixel; used with permission).

Similarly, the full data archive can also be downloaded from the tag if it is recovered. The period of data transfer while the tag is floating typically spans 1 to 2 weeks, depending on deployment length and power consumption or battery depletion. The mechanism for tag detachment from the shark occurs via a corrodible link that releases the transmitter from the shark or when certain conditions are met indicating that the tag is no longer attached to an animal, that the animal may have died (e.g., the transmitter remains at a constant depth, sea bottom, or surface), or that the animal has gone below the maximum set depth for operation of the tag (usually not more than 1000 m but up to 2000 m). Other parameters collected by the tag include time in the mixed layer and maximum depth of the mixed layer, temperature recordings (maximum, minimum, and average) of the mixed layer and surface, overall depth and temperature, and daily messages of light level curves (dawn and dusk transitions) (Musyl and McNaughton, 2007; Sims, 2010).

Geolocations are estimated from light-level measurements collected by the tag. Latitude is determined by the time between sunrise and sunset, and longitude by the time of the midpoint between sunrise and sunset (Hill and Braun, 2001). Many environmental and biological factors can influence the quality of the light-level measures. As a result, large spatial errors (hundreds to thousands of kilometers) are commonly associated with geolocation estimates (see Section 19.4.1). As mentioned above, the full record of data collected is archived in the memory so that if the transmitter is physically recovered the full dataset can be downloaded. Retrieval of the tag at sea can be complex because the tag must be located during a brief window (usually 1 to 2 weeks) when the PSAT is transmitting. For a tag floating in the ocean, the most recent Argos uplink is taken as a start position (with the best LC usually taken); where very high frequency (VHF) transmitters have been incorporated into the tag, handheld VHF receivers or radio receivers tuned to the appropriate frequency can then be used to pinpoint the tag position for recovery. Strong winds and poor weather conditions (currents, waves, white caps) reduce the chances of recovery by attenuating the VHF signal, thus inducing variability in signal strength and making spotting the floating transmitter more difficult. For tags deployed on animals that inhabit coastal and nearshore habitats or where prevailing local wind and current circulation are directed toward the shore, PSATs can be recovered after being washed ashore and discovered by beachcombers (Økland et al., 2013). Programs to inform fishers about tag recovery may also be implemented for species that are targets of commercial and recreational fisheries. Reward offers and contact details printed on the tags at their manufacture assist with the likelihood of recovery (e.g., http://tagagiant.org/science/tag-rewards).

When first developed in the late 1980s, PSAT tags revolutionized the way we acquire data from highly mobile and migratory species that remain below the sea surface. The use of these transmitters revealed novel three-dimensional behaviors and large-scale two-dimensional movements for many species (Sims, 2010). But, unless these tags are physically recovered, the datasets remain limited to the amount of information that is transferred during the pop-off transmission period. The large error associated with geolocations (hundreds to thousands of kilometers) (Teo et al., 2004) and the fact that data are summarized for transmission prevent fine-scale understanding of both horizontal and vertical movement behavior by target animals. For cryptic species where knowledge of vertical movements is limited, and for pelagic sharks that do not spend time at the water surface, PSATs might be the best option available. Additionally, most studies also report some proportion of premature PSAT releases and transmitters that fail to ever transmit data, perhaps due to predation, tag shedding (drag from the tag, inadequate healing, or inadequate attachment of the tag), damage to the tag housing or flotation mechanism, or tag aerial and/or system failure (Lutcavage et al., 2015; Musyl et al., 2011).

19.2.3 Fastloc® GPS

The Global Positioning System (GPS) uses its own constellation of orbiting satellites flying in medium Earth orbit (altitude of approximately 20,200 km) that provide location and time information (ephemeris and almanac data) to GPS receivers on the surface of the Earth so that the location of the device can be calculated in real-time (Moen et al., 1996). Application of this technology to studies of sharks and other marine megafauna has been hampered by the time required (30 seconds to a few minutes) for GPS receivers to communicate with GPS satellites to obtain the data to calculate location fixes (Ryan et al., 2004). However, the development of Fastloc® technology and its integration into Argos transmitters has removed this barrier to the use of GPS to track marine species that are only present at the surface very briefly (Hazel, 2009; Witt et al., 2010). The GPS receiver in Fastloc® tags is able to record a quick (fraction of a second) snapshot (i.e., satellite ID numbers, pseudo ranges, and a time stamp) of the signals from GPS satellites (Bryant, 2007). The data from GPS signals are processed, compressed, and stored in the onboard memory and relayed by the Argos system (Hazel, 2009; Sims et al., 2009a). These data are then processed into location estimates via the tag manufacturer's portal or software. Fastloc® GPS transmitters also offer higher location accuracy than regular Argos transmitters, with spatial errors generally less than 50 m with seven satellites or more (Bryant, 2007; Dujon et al., 2014; Hazel, 2009; Moen et al., 1997; Ryan et al., 2004), but they remain dependent on the number of GPS satellites detected by the receiver (i.e., the greater the number of GPS satellites, the lower error associated with the location estimate) (Dujon et al., 2014;

Hazel, 2009). Although the higher resolution data provided by Fastloc® GPS tracking allows for finer scale studies of behavior, movement patterns, and space use (Dujon et al., 2014; Hoenner et al., 2012; Sims et al., 2009a), there are still trade-offs associated with the technology, most notably purchase costs and lower tracking durations due to greater power consumption.

19.3 TAGGING PROCEDURES AND TAG ATTACHMENTS

There are many approaches to deploying satellite transmitters on sharks. Most commonly sharks are captured, restrained, and tagged from a boat (Figure 19.1). Shark capture can be accomplished using handlines, rod and reel, drumlines, gillnets, and longlines. Once captured, sharks are brought alongside the boat and restrained in the water with ropes or harnesses (Figure 19.1A), or they may be brought onboard the boat or placed in stretchers for measurements and tagging procedures. For species where the physical handling of animals is not possible due to their large size or poor recovery from capture, tagging is sometimes conducted on free-swimming animals. White sharks, for example, due to their large size, are often lured to boats with chum and are tagged with a handheld pole (Jorgensen et al., 2009). Filter-feeding species, such as whale sharks, mantas (*Mobula* spp.), and basking sharks (*Cetorhinus maximus*), can be tagged with spearguns or tagging poles from a vessel approaching the animal as it surface-feeds (Sims et al., 2003) or by free-divers approaching sharks in water (Hearn et al., 2013; Thums et al., 2012; Wilson et al., 2006). Free-diving is also used to deploy tags on species such as hammerhead sharks (*Sphyrna* sp.) (Bessudo et al., 2011) that show high vulnerability to fishing stress (Gallagher et al., 2014) and high post-release mortality (Butcher et al., 2015; Coelho et al., 2012). Commercial and recreational fishing operations can be used for deploying satellite transmitters when species are the targets of fisheries, as is the case for blue sharks (*Prionace glauca*) (Queiroz et al., 2010), makos (*Isurus* sp.) (Rogers et al., 2015), and other pelagic sharks (Carlson and Gulak, 2012).

Satellite transmitters can be mounted onto the dorsal fin of a shark (Argos and Fastloc® GPS) or attached using an anchor with a tether (typically around 0.5 to 1 m in length) that tows the tag behind the animal (Figure 19.1D). These modes of attachment allow the wet/dry sensors to emerge from the water if the animal is near or at the surface. PSAT tethers are usually much shorter (a few centimeters) than those used with GPS or satellite-linked tags because they are not required to transmit location data until after detachment. The fin-mount is the most common attachment for satellite-linked transmitters (Figure 19.1B,C). The tag is affixed to the first dorsal fin by placing corrodible titanium or stainless steel bolts through holes pierced near the tip, and it is secured with neoprene and steel washers and steel nuts (Figure 19.1B). The use of neoprene protects the shark's skin from contact with metallic corrosion of steel nuts that results in the eventual detachment of the transmitter. For species that may not break the water surface or in situations where the capture and restraint of sharks is not possible, satellite tags (usually purpose-made, hydrodynamic models such as Wildlife Computers SPLASH10-F-321) can be deployed with tethers and towed. Towed satellite-linked tags are usually deployed by free-divers. The tether (braided stainless steel or monofilament) is attached to a stainless steel or titanium dart that is embedded into the shark's dorsal musculature near the base of dorsal fin by spearguns or hand spears, or it is affixed to the first dorsal fin by a fins clasp such as those manufactured by CATS (Customized Animal Tracking Solutions, http://www.cats.is/) (see Figure 19.1) (Chapple et al., 2015; Gleiss et al., 2009). Towed tags also have buoyancy, so that when the animal is near the surface it floats up and can breach the surface to make connection with satellites.

Pop-up satellite transmitters are typically deployed as towed tags with short tethers (depending on animal size) to allow the detachment mechanism to work after the pre-programmed time. Towed PSATs are affixed with a dart inserted into the shark's dorsal musculature or looped through a hole on the first dorsal fin (Hazin et al., 2013). The towed tag attachment (for PSAT and satellite-linked transmitters) in the musculature of an animal consists of a medical-grade nylon umbrella tip dart, stainless steel T-bar arrowhead anchor, or flat titanium or stainless-steel dart that is inserted close to the posterior end and near the base of the first dorsal fin. The anchor is inserted at a 45° angle to engage the cartilaginous radials that are located underneath the dorsal fin, reducing the likelihood of a premature release due to drag caused by the transmitter. The anchor is connected to the tag by a monofilament leader with high strength but low stretch (Campana et al., 2009a; Carlson et al., 2010; Queiroz et al., 2010), or braided stainless wire. The latter should be sheathed with heat-shrink tubing to minimize abrasion to the shark's skin.

Deployments of tags can last up to 12 months for PSATs and over 3 years for satellite-linked transmitters before tag detachment or battery exhaustion, but much shorter durations are common (Domeier and Nasby-Lucas, 2013; Hammerschlag et al., 2011; Weng et al., 2008). Although little information is available on the effects of long-term tag attachment, damage to dorsal fins from fin-mounted tags and parasite infection at the tag attachment site may affect the behavior, swimming efficacy, healing, and possibly survival of individuals (Hammerschlag et al., 2011; Jewell et al., 2011). Hence, ethical issues and reduction of negative impacts, although challenging to quantify, must be considered by researchers when conducting animal tracking experiments (Hays et al., 2016). For a review of these issues, see Hammerschlag et al. (2011).

19.4 DATA PROCESSING

Appropriate methods for processing of data should be selected in order to ensure correct interpretation of results from satellite telemetry. The accuracy of processing techniques such as filtering and interpolation will depend greatly on the spatial error associated with location estimates and the uplink rate from the satellite transmitter (Tremblay et al., 2006).

19.4.1 Geolocation

Initial position at tagging (GPS point recorded by user) and point of release from the animal (provided by PSAT along with summarized, archived data transmitted data after release) are the start and end points used for track reconstruction. Light-level data are analyzed to provide raw geolocations with software provided by tag manufacturers such as the WC-GPE: Global Position Estimator Program Suite (Wildlife Computers, Inc.), Sea Track (Desert Star Systems LLC), or LAT Viewer Studio (Lotek Wireless, Inc.), or they are processed onboard with the manufacturer's proprietary algorithm (Microwave Telemetry, Inc.). However, many factors can influence the accuracy of estimates of latitude and longitude. These may be related to variability in the sensitivity of the light-level sensor in the PSAT, physical conditions (water turbidity, weather, cloud cover), the animal's location (low latitudes or around an equinox, when day lengths are similar in all latitudes), behavioral and biological factors (diving behavior, demersal lifestyle), and biofouling (Hill and Braun, 2001; Lam et al., 2008).

The most probable track can be reconstructed from raw geolocations by applying a state–space unscented Kalman filter (Lam et al., 2008) that considers the movement of tagged animals as biased random walks, with raw geolocations being a representation of true location with some measurement-associated error (Lam et al., 2008; Nielsen and Sibert, 2007; Nielsen et al., 2009; Sibert et al., 2003). Post-processing models also combine satellite-derived sea-surface temperature measurements to correct raw geolocations and obtain the most probable track with improved accuracy (Lam et al., 2008, 2010). These models can be applied using packages such as *TrackIt* (Lam et al., 2010; Nielsen and Sibert, 2011) or *ukfsst* (Nielsen et al., 2009), available for the software R (R Core Team, 2017).

19.4.2 Filtering and Interpolation

19.4.2.1 Location Accuracy Filtering

All data, particularly Argos data, should be filtered prior to movement analyses. The large error typically associated with satellite tag deployments on sharks prevents the identification or may result in the erroneous interpretation of specific behaviors associated with search and foraging (high turn rates and slower speed) compared to migratory movements (directed, faster movement) (Bradshaw et al., 2007). Filtering techniques address issues about the quality of locations (e.g., Argos LC error) and are widely applied to data collected by satellite telemetry. The most common filtering technique for locations relayed by Argos satellites is the removal of low-accuracy location classes. Locations with LC Z are usually removed from a dataset because they indicate the inability of a satellite to obtain an uplink (Hays et al., 2001), but LC A and B may also be removed due to their low accuracy (no error estimation by Argos). However, some argue that keeping the low-accuracy LC A, B, and Z and incorporating the information of the location error in the data analysis (many movement-based models can incorporate information of error; see Section 19.5) can minimize data loss, especially important for species or areas where uplink rate is low (Sumner et al., 2009).

Erroneous locations can be removed by applying point-to-point speed filters that consider the average swimming speed for the focal species (usually obtained from the literature) to identify successive position estimates that require unrealistic travel speeds and thus should be excluded from the dataset. Hays et al. (2001) recommended that a maximum distance between successive locations should be considered when calculating speeds for filtering because over short distances large errors in position estimates might confound calculated speeds. For this reason, speed should be calculated between successive points over a distance sufficient to account for the lowest location accuracy from the track. It is important to note that Argos reports secondary, alternative estimates of location, and in some cases these can provide a more plausible estimate of position than primary estimates (Ferreira et al., 2015). This is particularly useful for tracking datasets where locations are sparse and removal of data could strongly influence subsequent analyses.

Processing of tracking data from Fastloc® GPS transmitters is still in its infancy, and standard methods have yet to be identified. The confidence in the accuracy of Fastloc® GPS locations is influenced by the number of GPS satellites used to obtain the snapshot. Each GPS location is accompanied by information on how many satellites were used and a "residual" that defines relative spatial accuracy (Witt et al., 2010). Hazel (2009) suggested filtering out locations derived from fewer than six satellites as the best option to reduce the proportion of locations with potentially large errors, assuming that this results in minimal data loss.

19.4.2.2 Movement Models

A combination of track filtering and interpolation methods is used to address the patchy and sporadic nature of locations within tracks that result from deployment of satellite tags on sharks. Interpolation is used to correct irregular sampling and ensure that locations are spaced equally in time, a requirement for some types of data analysis (Tremblay et al., 2006). Although linear interpolation is the simplest method, continuous-time models can account for more realistic representation of marine animal movement (Johnson et al., 2008). Continuous-time correlated random walk (CTCRW) models assume that movement is a stochastic process in continuous time, where the inertia in the movement of an animal will keep it moving in a similar way at successive times (Johnson et al., 2008; Jonsen et al., 2005). Given this assumption, statistical inference can be used to estimate movement path, rate of movement and speed, and the associated standard error at each location along the track (Johnson et al., 2008). CTCRW models can be implemented in R to interpolate tracks with the package *crawl* (Johnson, 2014).

State–space models (SSMs) encompass a range of rigorous time-series models that have become widely applied filtering tools for animal telemetry data (Jonsen et al., 2013). SSMs fitted to location data obtained from satellite tracking are able to predict unobserved true states (spatial location and possibly behavioral mode) from the observed track via a process model, accounting for uncertainty in the data (location error = Argos LC from SAT or geolocation from PSAT) and stochasticity of animal movement (Jonsen et al., 2003, 2005; Patterson et al., 2008). The mathematical assumption that future state can be inferred from the present state is known as a Markov condition. SSMs will produce a most probable track that is obtained from the means, medians, or modes of the posterior distributions of the observed data and the associated behavior (Jonsen et al., 2003, 2005) (Figure 19.2). The SSM can also fit a mode of behavior to the data that is derived from simple movement metrics such as the autocorrelation between successive speeds and turning angles (Jonsen et al., 2005; Patterson et al., 2008, 2009) to identify modes of behavior such as resident and transient

movement. The mode of residency is often associated with foraging or searching, as an animal foraging is assumed to make more and sharper turns (changes in direction) and shorter movements than an animal transiting, which will move in a more directed and rapid way. This notion can be better examined with the use of accessory data that might be obtained from devices such as accelerometers (see Chapter 4 in this volume). Although applied by many in animal movement studies (Jonsen et al., 2013; Patterson et al., 2008) with analysis packages freely accessible in the software R, such as *bsam* (Jonsen et al., 2017), SSM fitting is often complex, computationally demanding, and time consuming, and diagnostic assessment may not be straightforward (Jonsen et al., 2013; Patterson et al., 2008). Model convergence can be difficult for coarse, highly imprecise, and irregular data (Jonsen et al., 2013), as is often the case for deployments on sharks.

19.5 ANALYSIS AND APPLICATIONS

Satellite telemetry techniques have been widely applied in many studies of both horizontal and vertical movement patterns of sharks that can be used to describe important ecological aspects of a shark's spatial behavior, such as foraging, mating, and pupping areas; migratory pathways; and habitat and environmental preferences. Several methods can be applied to analyze animal movement data generated by satellite tracking (Hooten et al., 2017). Here, we examine some of the most common approaches used to describe the vertical and horizontal movements of sharks.

19.5.1 Movement Patterns

Although mark–recapture studies have provided valuable information on the scale of movements of several species of shark (Kohler and Turner, 2001; Kohler et al., 1998) by calculating the distance between tagging and recapture locations, the advance of satellite telemetry has allowed better description of movement patterns (areas where residency is high vs. low) and migratory routes, as well as identification of environmental drivers and habitat preferences—all traits that could not be described with mark–recapture studies. Simple metrics can be extracted from tracking datasets to describe and characterize the movement of tracked animals before more complex data analyses are employed. Understanding the scale of movements can be accomplished by simply calculating the maximum or mean (per day, for example) distance traveled, although such measurements of maximum distances traveled will be greatly affected by the duration of tag deployment, particularly for species that display partial migration such as tiger sharks (Ferreira et al., 2015; Lea et al., 2018; Papastamatiou et al., 2013). The swimming speed of sharks is also commonly reported in tracking studies and is calculated by dividing the average distance traveled between locations over time; however, calculation

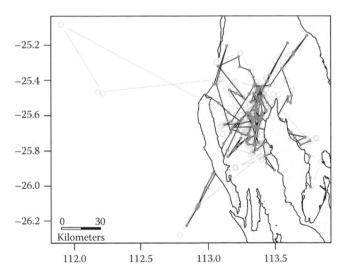

Figure 19.2 Plot of tiger shark track data from Shark Bay, Australia, with observed Argos locations as open circles and state estimates from the state–space model as blue-filled circles. The gray line is the straight-line path between Argos locations, and the black line is the straight-line path between state estimates.

of speed can be problematic if the distance between locations is smaller than the largest spatial error associated with these locations (Hays et al., 2001). Data quality, which may affect the robustness of data analyses, can be investigated by calculating deployment length and daily uplink rate. The former is often influenced by tag/sensor failure and tag shedding and the latter by the surface behavior of the animal, cloud cover, and satellite orbits, with both metrics influencing the choice of data analysis and the interpretation of results. Although analyses of movement patterns are often descriptive, they provide important insight into the spatial behavior of sharks. For example, whereas early studies described white sharks as coastal residents (Compagno, 1984), satellite tracking of these animals revealed that they occupy both coastal and oceanic habitats with distinct behaviors of seasonal residency and active swimming, and that they may even cross an entire ocean basin (Bonfil et al., 2005; Boustany et al., 2002; Bruce et al., 2006; Jorgensen et al., 2009). Long-term satellite tracking of white sharks enabled the description of migratory routes related to the reproductive cycle of females, with sharks moving between offshore gestation areas and pupping grounds near the coast (Domeier and Nasby-Lucas, 2013).

19.5.2 Habitat Use

Habitat use describes how an animal uses specific areas within its home range, which may vary temporally and spatially according to resources available and habitat quality (Benhamou, 2011; Benhamou and Riotte-Lambert, 2012). Habitat use is thus a consequence of multiple factors such as accessibility and quality of food, environmental conditions, predator avoidance, and physiological state (mating, gestation, and reproductive cycle), among others. Quantifying the area occupied by the animal (its home range) and the areas of high use within its home range is the first step in developing an understanding of habitat use. These metrics also provide the basis for an understanding of the physical, biological, and behavioral factors that might explain areas of high use when analyzed using statistical models such as generalized linear mixed models and generalized additive mixed models (Zuur et al., 2009), for example.

Kernel density analysis techniques have been used extensively to determine habitat use from tracking data of sharks for the characterization of important areas at individual, population, and species levels (Heupel et al., 2004). As a statistical concept, home range can be defined as a fixed percentage (commonly 95%, 50%, or 25%) of the confidence region obtained from the relative frequency distribution of an animal's location over time (Worton, 1987). This mathematical definition allows researchers to calculate utilization distributions (UDs) from locations obtained by satellite tracking an animal through time by predicting the probability of an animal being in an area during the time it spent

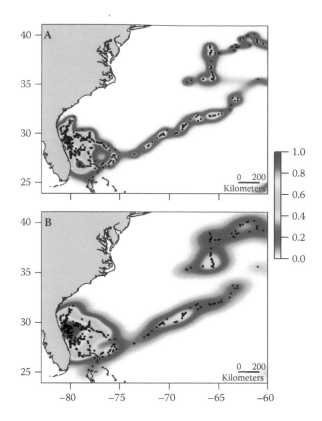

Figure 19.3 Representative utilization distribution (UD) of a tiger shark track (black points) analyzed with (A) kernel density estimation and (B) biased random bridges, showing both residency cores (>0.75 UD) and migration (<0.4 UD). Color scale represents the utilization distribution. (Track provided by N. Hammerschlag, Shark Research & Conservation Program, University of Miami, Miami, FL.)

there (Figure 19.3). Kernel density estimation (KDE) determines the underlying probability function from the data (Silverman, 1986) by placing a probability density function over each data point. Values are estimated by adding the contribution from all components (Horne and Garton, 2006; Worton, 1989). A smoothing parameter (h) or bandwidth controls the amount of variation incorporated in each component and has a large effect on resulting kernel estimates by controlling the smoothness of the kernel density estimation around the data points (Horne and Garton, 2006; Koláček and Horová, 2017; Duong, 2013; Worton, 1987, 1989). Although bandwidth can be selected visually based on successive trials (Silverman, 1986; Wand and Jones, 1995), automated methods, such as plug-in bandwidth selectors utilizing least squares cross-validation and maximum likelihood methods (Calenge, 2006, 2015, 2016; Calenge and Royer, 2018; Duong, 2007, 2013), in combination with the packages *ks*, *adehabitatLT*, and *adehabitatHR* in the software R (R Core Team, 2017), offer a more objective way to determine bandwidth and calculate KDE. Geographic information system (GIS) tools are also commonly used to calculate home ranges from telemetry data (Simpfendorfer and Heupel, 2004), using the Animal Movement Extension for ArcView (Hooge and Eichenlaub, 1997).

The concentration or rarity of data points will determine the value of the kernel density estimation, ignoring the temporal element of tracking data, a key component of animal movement and space use. Brownian bridges are continuous-time, stochastic movement-based models that take into account the path traveled and time spent between each successive location to model utilization distributions (Bullard, 1991; Horne et al., 2007). The method also incorporates smoothing parameters in the computation of utilization density; one, the Brownian bridge motion variance parameter, controls the width of the Brownian bridge (Horne et al., 2007), and another is related to the uncertainty around locations, which can be determined with the location error from the type of telemetry used (Argos LC error, spatial errors of geolocations or GPS location).

Similarly, biased random bridges are also movement-based kernels but include a component of advection that affects the orientation and shape of the bridge, where stronger advection (e.g., during migration) generates longer and narrower bridges (Benhamou, 2011), allowing for a better definition of habitat use during migration (Figure 19.3). This method also sets a maximum duration for each step, so that steps longer than this threshold will not be accounted for in the computation of utilization distribution (Benhamou, 2011). This can facilitate the application of this method to shark tracking data where many temporal gaps are present that should not be accounted for in the computation of habitat use to avoid overestimating areas between locations that are temporally sparse (Ferreira, 2017). Brownian bridges and biased random bridges can be applied to telemetry data using packages in R such as *BBMM* (Nielson et al., 2013) and *adehabitatHR* (Calenge, 2006, 2015). Brownian bridge kernel density applied to tracking data from tiger sharks tagged at Ningaloo Reef, Australia, showed large areas (634,944 km²) of seasonal high use that encompassed both tropical and temperate habitats (Ferreira et al., 2015). In Hawaii, estimates of utilization density combined with generalized additive mixed models indicated movement between islands by female tiger sharks that was influenced by environmental factors such as water temperature and was associated with partial migration (Papastamatiou et al., 2013).

Habitat use can also be described as spatial occupancy, measured as the time spent in a grid cell when the tracking area is divided into a grid. Higher occupancy can indicate areas of importance (Figure 19.4), such as foraging areas. Unlike utilization distributions, occupancy (time spent) will only be calculated for cells that contain one or more locations. The measure of time spent can be calculated on a per-individual basis or for all individuals combined. When pooling data across multiple individuals, calculations of habitat use can be biased either toward short tracks with movement concentrated in a small area or by tracks that are much longer than the average deployment length that may encompass much larger areas. This problem is shared by analyses of utilization density and can be remedied by removing very short tracks (in relation to the average track length of all individuals) or weighting the occupancy of individual tracks in relation to their length or number of cells occupied prior to calculating overall occupancy for multiple individuals. These biases can be minimized by overlaying the analyses of individuals and calculating the number (or proportion) of animals using a grid cell or the average time spent (or utilization distribution) across all individuals. Time spent per unit area can be calculated using the R package *trip* (Sumner and Luque, 2016) which will create a spatial grid containing the information of time spent in each cell (Figure 19.4).

Analytical tools from the field of spatial statistics can also be applied to calculations of habitat use. Getis-Ord Gi* hotspot analysis can define spatial patterns of significance based on clusters of locations, allowing identification of areas with higher or lower utilization values than predicted by a random pattern. The analysis determines the correlation of a point value (time spent in a grid, relative abundance) in the context of neighboring areas so that a statistically significant hotspot or coldspot can be identified (Figure 19.4) (Getis and Ord, 1992).

One of the main goals of satellite tracking studies is to define population-level estimates of habitat use that are ecologically significant and can have management and conservation applications. A sample size large enough to be representative of the population is essential for this analysis, and this can be determined by calculating the cumulative area used by each additional individual. If an asymptote is achieved, the dataset contains a sufficient number of samples (e.g., tracks) to adequately characterize the area used by that population (Soanes et al., 2013). For species for which most individuals utilize similar areas, a smaller sample size (number of tracks) will be required to predict population-level habitat use. However, for populations or species characterized by individuals that display large variability or roam over very distinct areas, population estimation of habitat use may only be achieved with very large sample sizes to incorporate such variability (Soanes et al., 2013; Thums et al., in review). Although the approach above can help determine sample sizes needed, it is important to consider whether the subsample of a population included in analyses is likely representative of broader populations. For example, for a study focusing on a population with a large and continuous range, if all tags are deployed in a similar area it is possible that analyses may indicate that sample sizes are adequate for characterizing movements. This analysis, however, only applies to individuals in the area where tracks began and may or may not be applicable to larger populations. Because of the tendency

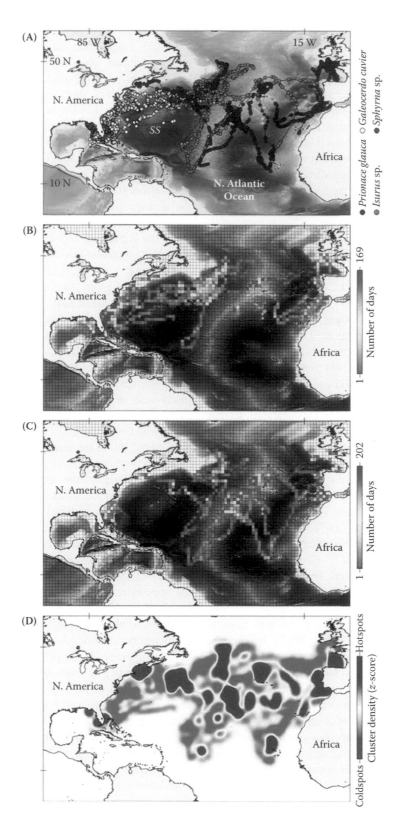

Figure 19.4 Spatial distributions of satellite-tracked pelagic sharks. Geolocations of sharks in the North Atlantic Ocean are shown in (A). Space use was calculated as the effort-corrected index of occurrence per unit area (number of mean days per grid cell) between (B) March and August and (C) September and February. Map of the calculated high (hotspot; red) and low (coldspot; blue) use habitats of tracked sharks are shown in (D). (From Queiroz, N. et al., *Proc. Natl. Acad. Sci. USA*, 113, 1582–1587, 2016. With permission.)

for satellite-based studies to focus on wide-ranging species with large distributions, a failure to consider spatiotemporal variation in behavior across a contiguous range could lead to improper conclusions about habitat use or interspecific interactions.

19.5.3 Environmental Drivers

Defining environmental and habitat preferences during residency and along migration routes can identify critical habitats (Block et al., 2011; Cooke, 2008). Multiple methods are available to predict or describe the relationship between space use and environmental parameters. Resource selection functions (RSFs) and resource selection probability functions (RSPFs) can be used to relate spatial distributions of species to environmental and habitat variables (Aarts et al., 2008; Manly et al., 2002). These methods assume that animals will select high-quality resources more often than low-quality resources; therefore, patterns of resource use can be compared to their availability in the environment to determine selectively the probability of use of a resource (or variable) (Lele and Keim, 2006; Manly et al., 2002). For tracking data, defining what resources are available and determining the areas that are not used or where there is an absence of use of a resource can be challenging because of the multitude of options of unused resources that are available around locations in a track (Boyce et al., 2002). Consequently, presence must be compared to a random sample of available habitats generated around the observed points (Boyce et al., 2002).

This issue of estimating unused habitats for tracking data has led to the development of step-selection functions (SSFs) (Fortin et al., 2005). In this model, consecutive locations, considered as steps, are paired with pseudo-absences generated by random steps with the same starting point as the "real" location (Thurfjell et al., 2014). Real locations are given a value of 1 (presence) and random locations are given a value of 0 (absence or unused), which are then used as a dependent variable in models such as conditional logistic regressions to determine which variables influence the choice of habitat (Thurfjell et al., 2014). Temporal autocorrelation between successive locations will cause issues in the model estimation, but data may be thinned to assist here. These spatial analyses can be conducted with ArcGis tools and with R packages such as *ResourceSelection* (Lele et al., 2017) or *hab* (Basille, 2015).

A wide range of statistical models can be applied to investigate species distributions and modeling of relationships between animal distribution and internal and external drivers, each with their advantages and pitfalls (see review by Austin, 2002). For example, the combination of utilization density calculated with biased random bridges and generalized additive mixed models (Zuur et al., 2009) was used to identify the likely environmental factors driving the

utilization density of tiger sharks in Hawaii and identified the importance of water temperature and chlorophyll in the probability of a shark using the waters around any island of the island chain of Hawaii (Papastamatiou et al., 2013). Globally, the same approach identified water temperature and bathymetry as common environmental predictors of habitat use for the species (Ferreira, 2017). As with the other analyses discussed so far, low spatial and temporal resolution of the tracking data may greatly affect the results of statistical modeling, possibly leading to a lack of convergence in models, misinterpretation of results, and incorrect study conclusions (Bradshaw et al., 2007; Patterson et al., 2008). Development of new satellite transmitter technology (see Section 19.6) and the application of appropriate modeling techniques and incorporation of location error estimates in modeling of habitat use can assist with extracting meaningful biological and ecological signals from poor-quality data (Bradshaw et al., 2007; Ferreira, 2017). Temporal autocorrelation of tracking datasets must also be dealt with by rarefying the data and using a subset of locations randomly sampled so that locations are not autocorrelated, by including correlation structures in the models (Zuur et al., 2009), or by using matched-block bootstrap sampling (Carlstein et al., 1998; Politis and White, 2004).

19.5.4 Foraging

The theory of optimal search strategy states that when encountering patchily distributed resources, which is usually the case in the ocean, a predator should slow down and increase the frequency of turning angles to maximize the probability of prey encounter (Fauchald, 1999; Kareiva and Odell, 1987). This is known as area-restricted search. Because direct observations of feeding are exceedingly rare for most species, area-restricted search patterns have been used as a proxy of foraging. The benefit of this framework is the ability to apply objective methods to classify area-restricted movement behavior in tracking data by identifying behavioral switches using state–space models (Bestley et al., 2013; Jonsen et al., 2005) or first-passage time (Fauchald and Tveraa, 2003). In addition, areas of high use (identified by time spent and kernel analyses described above) can also be useful proxies to identify areas that may be related to foraging.

The first-passage time (FPT) is defined as the time an animal takes to cross a circle of specific radius (Johnson et al., 1992). Assuming that all locations in the path are associated with a circle of specific radius, this method considers the time between the first passage of the circle backwards and forwards along the path as a measure of search effort. The circle is then moved along the path with increasing radii so that more of the tortuous path is covered until the circle is large enough to capture all the tortuosity of the movement (Fauchald and Tveraa, 2003; Pinaud, 2008). As a result, FPT will increase as the radius of the circle increases, and by plotting the variance of FPT against different radii values the spatial scale where search effort is concentrated can be identified (Fauchald and Tveraa, 2003; Pinaud, 2008). If an animal's movement path shows high tortuosity and the animal is spending considerable time in one area, that would suggest higher search effort corresponding to a longer FPT in relation to an animal transiting. When an animal is transiting, the time spent in an area will be reduced and the path will become more linear, resulting in a shorter FPT. Area-restricted search behavior identified by FPT can then be used to define likely foraging areas and can be modeled against environmental variables at different spatial scales to identify correlates of likely foraging that can be scale dependent (Fritz et al., 2003; Pinaud and Weimerskirch, 2005).

Although FPT has been used to determine area-restricted search behavior of salmon sharks in productive ecoregions in the eastern North Pacific (Figure 19.5) (Weng et al., 2008), the method has been most widely applied to data from GPS tracking of seabirds (Fritz et al., 2003; Pinaud and Weimerskirch, 2005; Weimerskirch et al., 2000) and tracking of seals (Thums et al., 2011), which are characterized by high-resolution, regular sampling rates, and large volumes of data. Limitations of the approach have also been identified, such as issues with fitting the model for sparse data and the need for interpolation techniques (Johnson et al., 2006; Patterson et al., 2008; Pinaud, 2008), and state–space switching models as discussed earlier (Section 19.4.2.2) are considered to be more appropriate for identifying area-restricted searches (Patterson et al., 2008).

19.5.5 Post-Release Behaviors

Although primarily used for ecological studies, satellite transmitters can provide important data on post-release behavior and mortality rates that can reveal the risk of capture and release by commercial and recreational fishers. Information on survivorship of discarded elasmobranchs is helpful in assessing the sustainability of fisheries and ensuring the efficacy of management actions (Ellis et al., 2016). For example, there was no observed post-release mortality of healthy blue sharks caught by both Atlantic and Pacific longline fisheries (Campana et al., 2009b; Moyes et al., 2006), but tagging revealed 33% mortality rates among injured sharks in an Atlantic fishery (Campana et al., 2009a). In contrast, a Canadian longline fishery for porbeagle sharks (*Lamna nasus*) resulted in 10% and 75% post-release mortality for healthy and injured sharks, respectively, whereas shortfin mako sharks exhibited mortality rates of 30% and 33%, respectively. Thus, a large proportion of mortality caused by this fishery was not accounted for by measures of landed catch (Campana et al., 2016). However, post-release behavioral responses and modification, however, will be a result of not only the stress and trauma of capture and handling but also the tagging procedure itself and from carrying tags (Hoolihan et al., 2011).

Pop-up transmitters have been used to monitor the vulnerability of common thresher sharks (*Alopias vulpinus*) to fishing gear of recreational fishers in southern California. The study assessed the depth profiles and horizontal net displacement of tagged individuals, revealing that two-thirds of individuals released with trailing gear (e.g., when fishing line is parted during the fight) died within 5 days of release (Sepulveda et al., 2015). Afonso and Hazin (2014) investigated the effects of a longline survey on survivorship of juvenile tiger sharks tagged with PSATs in the South Atlantic and found that, although some sharks demonstrated altered behavior during a period of 12 days after capture, there was no evidence for post-release mortality. The authors suggested that mandatory release of live animals caught by fishing gear would likely reduce the fishing mortality of tiger sharks. Results from post-release studies reinforce the need for increased ethical considerations of the effects of tagging when designing tracking experiments (Hays et al., 2016), particularly for those species more vulnerable to the stress of capture, handling, and tagging procedures.

19.5.6 Management and Conservation of Sharks

Understanding habitat preferences and critical habitats is essential for species conservation and implementation of effective management strategies (Maxwell et al., 2011). The high mobility of large sharks, like other marine and terrestrial megafauna, presents one of the greatest challenges for conservation (Campana, 2016; Heupel et al., 2015; Runge et al., 2014; Shuter et al., 2011). Despite the potential for conservation and management planning, telemetry data are still largely underutilized due to misalignment of individual research objectives and management needs (McGowan et al., 2016). For example, satellite tracking of sharks in the North Atlantic has been used to define areas around oceanic frontal systems that present high risk to species due to the overlap between sharks and fishing vessels that results in high catch rates of oceanic sharks (Queiroz et al., 2016). However, despite the recommendation to reduce catch quotas for mako sharks based on scientific data (tracking and stock assessment), regulatory bodies of fisheries have yet to incorporate this information into their recommendations and management actions to restrict quotas (Sims et al., 2018). Although marine protected areas (MPAs) are a key strategy for ecosystem-based management (Abecasis et al., 2014; Hooker et al., 2011; O'Leary et al., 2016) and have proven to aid shark population recovery when well-managed and properly enforced (Speed et al., 2018), most MPAs are still too small when compared to the scale of movements and home ranges of large, highly mobile species (Heupel et al., 2015; McCauley et al., 2015). However, characterization of the relationships among habitat use, environmental conditions, and oceanographic features

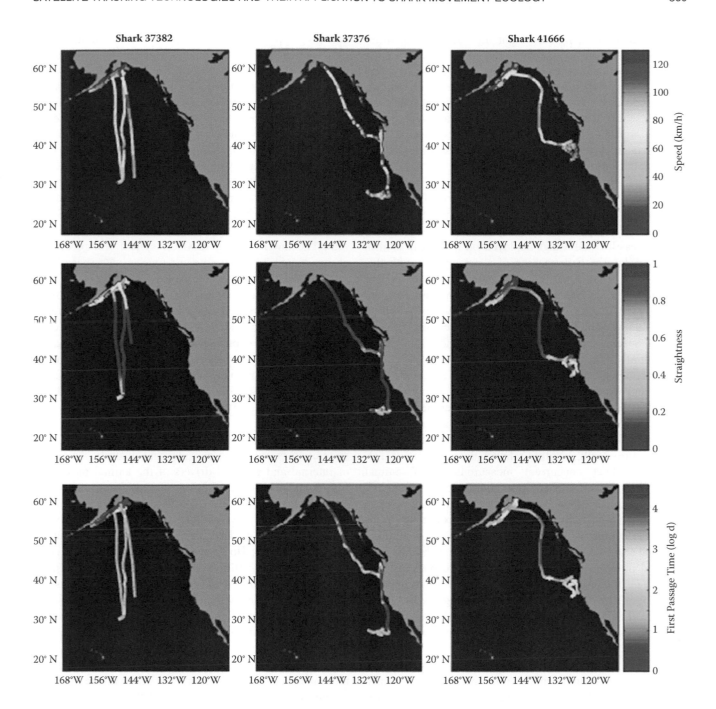

Figure 19.5 Representation of (A) speed, (B) straightness, and (C) first passage time of three salmon sharks (*Lamna ditropis*) in the eastern North Pacific Ocean indicating an area-restricted search. Speed is km/day, straightness is a dimensionless index (1 = straight line), and first passage time is in log d. (From Weng, K.C. et al., *Mar. Ecol. Prog. Ser.*, 372, 253–264, 2008. With permission.)

(e.g., mesoscale features such as upwelling areas, fronts, and eddies), when incorporated, for example, into ecosystem-based fisheries management (Pikitch et al., 2004) and dynamic ocean management (Hobday et al., 2014; Maxwell et al., 2015), could allow for the development of management actions to adapt to the spatial and temporal dynamics of species distribution.

19.6 NEW TECHNOLOGIES AND FUTURE DIRECTIONS

Recent advances in telemetry systems herald what some researchers are referring to as a "Golden Age" for tracking animals (Hays et al., 2016; Kays et al., 2015). Technological boundaries are expanding with the development of less expensive, smaller, lighter-weight tags designed to minimize handling and attachment time and to reduce the energetic costs of drag when the tag has been deployed on an animal. Less expensive tags will promote larger sample sizes and more robust analyses of movement, habitat use, and behavior. However, despite the increasing ability to obtain large sample sizes, there remains a critical need to appropriately develop *a priori* sampling designs to address relevant questions.

Despite their potential to resolve fine-scale movement and space use and their increasing popularity for tracking other marine taxa such as marine turtles (Hays et al., 2013), Fastloc® GPS tags are not widely used in shark tracking studies. Fastloc® GPS tags deployed on the ocean sunfish (*Mola mola*), the first tracking of a pelagic fish with this technology, showed the tracking potential that these tags offer (Sims et al., 2009a). This attachment method or adaptations of it (Figure 19.1) could be easily transferrable to shark studies using Fastloc® GPS tags, particularly for large pelagic species that swim regularly at or near the surface (Sims et al., 2009a,b).

Although the main goal of tracking is to describe movements and space use, it also offers a unique opportunity to use animals as sampling units to obtain detailed environmental and oceanographic data in remote and data-sparse or unsurveyed regions (Boehme et al., 2009). Conductivity–temperature–depth, satellite-relayed data loggers (CTD-SRDLs) collect concurrent measurements of animal movement and behavior and oceanographic data including vertical profiles of conductivity, temperature, and pressure as the animals dive that are relayed through the Argos system (Boehme et al., 2009). Tags deployed on southern elephant seals (*Mirounga leonina*) for the last 15 years have resulted in a better understanding of the response of animals to *in situ* environmental conditions as well as detailed mapping of the oceanography and hydrological profiles of areas in the Southern Ocean that would otherwise be difficult to observe due to the seasonal cover of sea ice (Biuw et al., 2007; Roquet et al., 2014).

These tag designs are currently not appropriate for sharks, although future models may be. Just as for seals, deployments on large pelagic sharks could offer a quasi-synoptic, regional view of boundary currents and mesoscale fronts and features that have been shown to be important for many pelagic taxa (Croll et al., 2005; Doniol-Valcroze et al., 2007; Klimley and Butler, 1988; Queiroz et al., 2016; Royer et al., 2004; Scales et al., 2014). A better understanding of how sharks utilize frontal habitats and respond to *in situ* oceanographic conditions would provide information urgently needed for conservation and management planning due to the high overlap with intense fishing pressure in these areas (Queiroz et al., 2016). Tracking animals using airborne drones or autonomous underwater vehicles equipped with abiotic sensors and cameras is another emerging approach to assessing microscale habitat use and localized environmental conditions (Figure 19.6) (Clark et al., 2013; Haulsee et al., 2015; Kiszka et al., 2016).

Across taxa, smaller more hydro- or aerodynamic tag designs are in development or newly available such as miniaturized GPS-Argos satellite transmitters (Scarpignato et al., 2016). Historically, one of the greatest limitations to marine telemetry has been a lack of appropriate technology to track the smallest taxa and youngest life stages (Hays et al., 2016; Mansfield et al., 2012). In October 2017 and February 2018 the components for a new, lower-orbiting telemetry receiver were launched to the International Space Station (ISS) in association with the ICARUS Initiative (icarusinitiative.org). The lower-orbiting ICARUS system will reduce the energy required by tags to communicate with space-based satellite receivers such as the Argos system (Pennisi, 2011), thus enabling a reduction in tag size. Small (<4 to 5 g) solar-powered GPS tags at a cost of under €500 will be tested as part of this new system and will include accelerometers, temperature and depth sensors, and land-based tracking capabilities. This new alternative to the Argos system has the potential to revolutionize animal movement studies, opening up the possibility of tracking smaller or younger animals, particularly those that spend time at or very near the sea surface.

When inferring behavior from tracking data in marine systems, it is critical to consider and quantify the impact of internal and external drivers of the animal movement. Surprisingly, ecological studies of shark movement using satellite tags are still largely focused on where animals go, rather than quantitative analyses of what drives their movement patterns (e.g., Bruce and Bradford, 2008; Carlson et al., 2010, 2014; Hoffmayer et al., 2014; Ketchum et al., 2014; Lea et al., 2015; Sepulveda et al., 2004; Weng et al., 2007). Evolution toward more quantitative analyses is hampered by issues with autocorrelation, data quality, and the low spatial and temporal resolution of remotely sensed environmental variables (Bradshaw et al., 2007; Ferreira 2017; Patterson et al., 2016). In addition, the development of modeling techniques and statistical expertise for animal movement is only now beginning to deal with the increased volume of data being generated (Patterson et al., 2016). However, progress is required for the statistical inference of animal–environment interactions for shark movement studies. For example, kernel density estimates, utilization densities, time spent in a grid cell, and first-passage time can all be used as dependent variables in models such as generalized linear mixed models or generalized additive mixed models (Zuur et al., 2009), which can be applied with R packages such as *gamm4* (Wood and Scheipl, 2017), *mgcv* (Wood, 2018), *MuMIn* (Barton, 2013), and *FSSgam* (Fisher et al., 2018). These variables can be modeled against a set of ecologically relevant predictors, including environmental (e.g., sea surface temperature, salinity, depth, bathymetry,

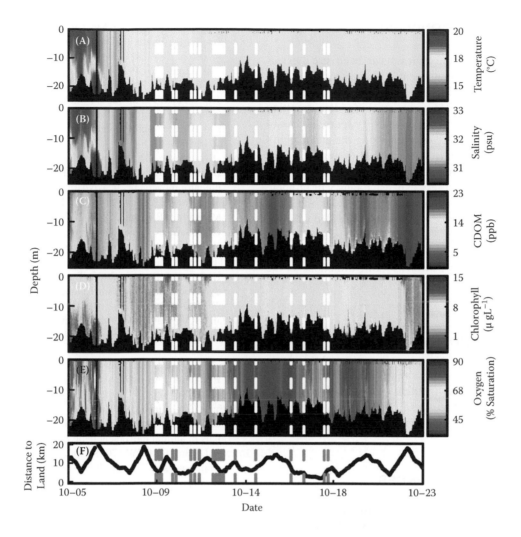

Figure 19.6 Environmental variables measured by autonomous underwater vehicle (AUV) sensors: (A) temperature (°C), (B) salinity (psu), (C) colored dissolved organic matter (CDOM) (ppb), (D) chlorophyll-*a* (μg/L), and (E) dissolved oxygen (% saturation) relative to water depth (m). (F) Distance to land (km). White and gray dashed vertical lines indicate the median time point of detection for each of the 23 sand tigers (*Carcharias taurus*). Black areas indicate where no data were collected. Dates are given as mm/dd/yy. (From Haulsee, D.E. et al., *Mar. Ecol. Prog. Ser.*, 528, 277–288, 2015.With permission.)

chlorophyll *a*), biological (e.g., size, sex, maturity stage), and temporal (e.g., season, month, Julian day) variables, to determine the importance of such predictors to animal behavior (Block et al., 2011; Ferreira et al., 2015; Lea et al., 2018; Papastamatiou et al., 2013). In these models, animal identification is often coded as a random effect to account for repeated measures on an individual and allow for population-level inference (Zuur et al., 2009).

The rapid uptake of animal telemetry over the last few decades means that big data approaches to understand movement are now being realized (Block et al., 2011; Meekan et al., 2017; Queiroz et al., 2016; Raymond et al., 2015; Rodríguez et al., 2017; Sequeira et al., 2018). These can provide more powerful insights into changes in movement and behavior due to environmental factors (Block et al., 2011; Raymond et al., 2015), inter- and intraspecific interactions,

anthropogenic threats, and, by searching for and describing universal patterns, collective behaviors and emergent properties (Rodríguez et al., 2017; Sequeira et al., 2018). Many online repositories collectively document the movements of tens to hundreds of thousands of wild animals across diverse taxa spanning all continents and biomes (for a full list of repositories, see Campbell et al., 2016). For the most part, though, these repositories are not open access, thus hindering any attempt to search for general patterns in animal movement and document large-scale movement (Thums et al., 2018). Collaborative initiatives that have facilitated data synthesis (e.g., TOPP, CLIOTOP, MMMAP, MEOP) are enabling researchers to overcome some of these issues, thus showcasing the great potential of "big data" and collaborations to drive the field of animal movement forward (Thums et al., 2018).

19.7   CONCLUSION

Satellite transmitter technologies have been used extensively to understand the movement ecology, spatial use, and behavior of sharks, particularly those species that display large-scale movements. However, tracking data resulting from deployments on sharks are often very sparse and associated with large spatial errors, hampering the applicability of robust models and statistical methods to analyze shark movement data. Emerging technologies and the development of new, smaller, and less expensive tags will promote larger sample sizes across broader spatial scales and more robust analyses of tracking data. We have provided a framework for the use of satellite tracking approaches and analytical methods to describe important ecological aspects of the spatial behavior of sharks, as well as future directions for the field of shark movement ecology.

ACKNOWLEDGMENTS

We thank Samantha Andrzejaczek, OCEARCH, Richard Fitzpatrick, the Australian Institute of Marine Science, and Kim Brooks for field photos and Neil Hammerschlag and Mike Heithaus for data used in Figures 19.2 and 19.3.

REFERENCES

Aarts G, MacKenzie M, McConnell B, Fedak M, Matthiopoulos J (2008) Estimating space-use and habitat preference from wildlife telemetry data. *Ecography* 31(1):140–160.

Abecasis D, Afonso P, Erzini K (2014) Combining multispecies home range and distribution models aids assessment of MPA effectiveness. *Mar Ecol Prog Ser* 513:155–169.

Afonso AS, Hazin FHV (2014) Post-release survival and behavior and exposure to fisheries in juvenile tiger sharks, *Galeocerdo cuvier*, from the South Atlantic. *J Exp Mar Biol Ecol* 454:55–62.

Argos (2016) *Argos User's Manual*. CLS/Service Argos, Toulouse, France.

Austin MP (2002) Spatial prediction of species distribution: an interface between ecological theory and statistical modelling. *Ecol Model* 157:101–118.

Barton K (2013) *Package 'MuMIn'*. R package version 1.9.0.

Basille M (2015) *Package hab*. R package version 1.10.4.

Benhamou S (2011) Dynamic approach to space and habitat use based on biased random bridges. *PLoS ONE* 6:e14592.

Benhamou S, Riotte-Lambert L (2012) Beyond the utilization distribution: identifying home range areas that are intensively exploited or repeatedly visited. *Ecol Model* 227:112–116.

Bessudo S, Soler GA, Klimley PA, Ketchum J, Arauz R (2011) Vertical and horizontal movements of the scalloped hammerhead shark (*Sphyrna lewini*) around Malpelo and Cocos islands (tropical eastern Pacific) using satellite telemetry. *Bol Invest Mar Cost* 40:91–106.

Bestley S, Jonsen ID, Hindell Ma, Guinet C, Charrassin J-B (2013) Integrative modelling of animal movement: incorporating *in situ* habitat and behavioural information for a migratory marine predator. *Proc R Soc Lond B Biol Sci* 280:2012–2262.

Biuw M, Boehme L, Guinet C, Hindell M, Costa D, Charrassin JB, Roquet F, et al. (2007) Variations in behavior and condition of a Southern Ocean top predator in relation to in situ oceanographic conditions. *Proc Natl Acad Sci USA* 104:13705–13710.

Block BA, Jonsen ID, Jorgensen SJ, Winship AJ, Shaffer SA, Bograd SJ, Hazen EL, et al. (2011) Tracking apex marine predator movements in a dynamic ocean. *Nature* 475:86–90.

Boehme L, Lovell P, Biuw M, Roquet R, Nicholson J, Thorpe SE, Meredith MP, Fedak M (2009) Technical note: animal-borne CTD-Satellite Relay Data Loggers for real-time oceanographic data collection. *Ocean Sci* 5:685–695.

Bonfil R, Meÿer M, Scholl MC, Johnson R, O'Brien S, Oosthuizen H, Swanson S, et al. (2005) Transoceanic migration, spatial dynamics, and population linkages of white sharks. *Science* 310:100–103.

Boustany AM, Davis SF, Pyle P, Anderson SD, Le Boeuf BJ, Block BA (2002) Expanded niche for white sharks. *Nature* 415:35–36.

Boyce MS, Vernier PR, Nielsen SE, Schmiegelow FKA (2002) Evaluating resource selection functions. *Ecol Model* 157:281–300.

Bradshaw CJA, Sims DW, Hays GC (2007) Measurement error causes scale-dependent threshold erosion of biological signals in animal movement data. *Ecol Appl* 17:628–638.

Bruce BD, Bradford RW (2008) *Spatial Dynamics and Habitat Preferences of Juvenile White Sharks—Identifying Critical Habitat and Options for Monitoring Recruitment*. CSIRO Marine and Atmospheric Research, Hobart, Tasmania, Australia.

Bruce BD, Stevens JD, Malcolm H (2006) Movements and swimming behaviour of white sharks (*Carcharodon carcharias*) in Australian waters. *Mar Biol* 150:161–172.

Bryant E (2007) *2D Location Accuracy Statistics for Fastloc® Cores Running Firmware Versions 2.2 and 2.3*, Technical Report TR01. Wildtrack Telemetry Systems Ltd, Leeds, UK.

Bullard F (1991) Estimating the Home Range of an Animal, a Brownian Bridge Approach, masters thesis, University of North Carolina, Chapel Hill, NC.

Butcher PA, Peddemors VM, Mandelman JW, McGrath SP, Cullis BR (2015) At-vessel mortality and blood biochemical status of elasmobranchs caught in an Australian commercial longline fishery. *Glob Ecol Conserv* 3:878–889.

Calenge C (2006) The package "adehabitat" for the R software: a tool for the analysis of space and habitat use by animals. *Ecol Model* 197(3–4):516–519.

Calenge C (2015) *Home Range Estimation in R: The adehabitatHR Package*. R package version 0.4.15.

Calenge C (2016) *Analysis of Animal Movements in R: The adehabitatLT*. R package version 0.3.23.

Calenge C, Royer M (2018) *Package 'adehabitatLT'*. R packege version 0.3.23.

Campana SE (2016) Transboundary movements, unmonitored fishing mortality, and ineffective international fisheries management pose risks for pelagic sharks in the Northwest Atlantic. *Can J Fish Aqua Sci* 73:1599–1607.

Campana SE, Joyce W, Manning M (2009a) Bycatch and discard mortality in commercially caught blue sharks *Prionace glauca* assessed using archival satellite pop-up tags. *Mar Ecol Prog Ser* 387:241–253.

Campana SE, Joyce W, Francis MP, Manning MJ (2009b) Comparability of blue shark mortality estimates for the Atlantic and Pacific longline fisheries. *Mar Ecol Prog Ser* 396:161–164.

Campana SE, Joyce W, Fowler M, Showell M (2016) Discards, hooking, and post-release mortality of porbeagle (*Lamna nasu*), shortfin mako (*Isurus oxyrinchus*), and blue shark (*Prionace glauca*) in the Canadian pelagic longline fishery. *ICES J Mar Sci* 73(2):520–528.

Campbell HA, Urbano F, Davidson S, Dettki H, Cagnacci F (2016) A plea for standards in reporting data collected by animal-borne electronic devices. *Anim Biotelem* 4(1):1.

Carlson JK, Gulak SJB (2012) Habitat use and movements patterns of oceanic whitetip, bigeye thresher and dusky sharks based on archival satellite tags. *Collect Vol Sci Pap ICCAT* 68(5):1922–1932.

Carlson JK, Ribera MM, Conrath CL, Heupel MR, Burgess GH (2010) Habitat use and movement patterns of bull sharks *Carcharhinus leucas* determined using pop-up satellite archival tags. *J Fish Biol* 77(3):661–675.

Carlson JK, Gulak SJB, Simpfendorfer CA, Grubbs RD, Romine JG, Burgess GH (2014) Movement patterns and habitat use of smalltooth sawfish, *Pristis pectinata*, determined using pop-up satellite archival tags. *Aquat Conserv Mar Freshw Ecosyst* 24:104–117.

Carlstein E, Do K-A, Hall P, Hesterberg T, Künsch HR (1998) Matched-block bootstrap for dependent data. *Bernoulli* 4:305–328.

Chapple TK, Gleiss AC, Jewell OJ, Wikelski M, Block BA (2015) Tracking sharks without teeth: a non-invasive rigid tag attachment for large predatory sharks. *Anim Biotelem* 3:14.

Clark CM, Forney C, Manii E, Shinzaki D, Gage C, Farris M, Lowe CG, Moline M (2013) Tracking and following a tagged leopard shark with an autonomous underwater vehicle. *J Field Robot* 30(3):309–322.

Coelho R, Fernandez-Carvalho J, Lino PG, Santos MN (2012) An overview of the hooking mortality of elasmobranchs caught in a swordfish pelagic longline fishery in the Atlantic. *Ocean Aquat Living Resour* 25(4):311–319.

Compagno LJV (1984) *Sharks of the World: An Annotated and Illustrated Catalogue of Shark Species Known to Date*. Food and Agriculture Organization, Rome.

Cooke SJ (2008) Biotelemetry and biologging in endangered species research and animal conservation: relevance to regional, national, and IUCN Red List threat assessments. *Endanger Species Res* 4:165–185.

Costa DP, Robinson PW, Arnould JP, Harrison AL, Simmons SE, Hassrick JL, Hoskins AJ, et al. (2010) Accuracy of ARGOS locations of pinnipeds at-sea estimated using Fastloc GPS. *PLoS ONE* 5:e8677.

Croll DA, Marinovic B, Benson S, Chavez FP, Black N, Ternullo R, Tershy BR (2005) From wind to whales: trophic links in a coastal upwelling system. *Mar Ecol Prog Ser* 289:117–130.

Domeier ML, Nasby-Lucas N (2013) Two-year migration of adult female white sharks reveals widely separated nursery areas and conservation concerns. *Anim Biotelem* 1:2.

Doniol-Valcroze T, Berteaux D, Larouche P, Sears R (2007) Influence of thermal fronts on habitat selection by four rorqual whale species in the Gulf of St. Lawrence. *Mar Ecol Prog Ser* 335:207–216.

Dujon AM, Lindstrom RT, Hays GC, Backwell P (2014) The accuracy of Fastloc-GPS locations and implications for animal tracking. *Methods Ecol Evol* 5(11):1162–1169.

Duong T (2007) ks: kernel density estimation and kernel discriminant analysis for multivariate data in R. *J Stat Software* 21(7):1–16.

Duong T (2013) *Package ks—Kernel Smoothing*. R software version 1.11.0.

Ellis JR, McCully Phillips SR, Poisson F (2016) A review of capture and post-release mortality of elasmobranchs. *J Fish Biol* 90(3):653–722.

Fauchald P (1999) Foraging in a hierarchical patch system. *Am Nat* 153:603–613.

Fauchald P, Tveraa T (2003) Using first-passage time in the analysis of area-restricted search and habitat selection. *Ecology* 84:282–288.

Ferreira LC (2017) Spatial Ecology of a Top-Order Marine Predator, the Tiger Shark (*Galeocerdo cuvier*), doctoral thesis, University of Western Australia, Perth, Western Australia, Australia.

Ferreira LC, Thums M, Meeuwig JJ, Vianna GMS, Stevens J, McAuley R, Meekan MG (2015) Crossing latitudes—long-distance tracking of an apex predator. *PLoS ONE* 10:e0116916.

Fisher R, Wilson SK, Sin TM, Lee AC, Langlois TJ (2018) A simple function for full-subsets multiple regression in ecology with R. *Ecol Evol* doi:10.1002/ece3.4134.

Fitzpatrick R, Thums M, Bell I, Meekan MG, Stevens JD, Barnett A (2012) A comparison of the seasonal movements of tiger sharks and green turtles provides insight into their predator-prey relationship. *PLoS ONE* 7:e51927.

Fortin D, Beyer HL, Boyce MS, Smith DW, Duchesne T, Mao JS (2005) Wolves influence elk movements: behavior shapes a trophic cascade in Yellowstone National Park. *Ecology* 86:1320–1330.

Fritz H, Said S, Weimerskirch H (2003) Scale-dependent hierarchical adjustments of movement patterns in a long-range foraging seabird. *Proc Biol Sci* 270:1143–1148.

Gallagher AJ, Serafy JE, Cooke SJ, Hammerschlag N (2014) Physiological stress response, reflex impairment, and survival of five sympatric shark species following experimental capture and release. *Mar Ecol Prog Ser* 496:207–218.

Getis A, Ord JK (1992) The analysis of spatial association by use of distance statistics. *Geogr Anal* 24(3):189–206.

Gleiss AC, Norman B, Liebsch N, Francis C, Wilson RP (2009) A new prospect for tagging large free-swimming sharks with motion-sensitive data-loggers. *Fish Res* 97(1–2):11–16.

Hammerschlag N, Gallagher AJ, Lazarre DM (2011) A review of shark satellite tagging studies. *J Exp Mar Biol Ecol* 398:1–8.

Hammerschlag N, Gallagher AJ, Wester J, Luo J, Ault JS (2012) Don't bite the hand that feeds: assessing ecological impacts of provisioning ecotourism on an apex marine predator. *Funct Ecol* 26:567–576.

Harvey AC (1990) *Forecasting, Structural Time Series Models and the Kalman Filter*. Cambridge University Press, Cambridge, UK.

Haulsee DE, Breece MW, Miller DC, Wetherbee BM, Fox DA, Oliver MJ (2015) Habitat selection of a coastal shark species estimated from an autonomous underwater vehicle. *Mar Ecol Prog Ser* 528:277–288.

Hays GC, Åkesson S, Godley BJ, Luschi P, Santidrian P (2001) The implications of location accuracy for the interpretation of satellite-tracking data. *Anim Behav* 61:1035–1040.

Hays GC, Bradshaw CJA, James MC, Lovell P, Sims DW (2007) Why do Argos satellite tags deployed on marine animals stop transmitting? *J Exp Mar Biol Ecol* 349:52–60.

Hays GC, Scott R, Higham T (2013) Global patterns for upper ceilings on migration distance in sea turtles and comparisons with fish, birds and mammals. *Funct Ecol* 27(3):748–756.

Hays GC, Ferreira LC, Sequeira AMM, Meekan MG, Duarte CM, Bailey H, Bailleul F, et al. (2016) Key questions in marine megafauna movement ecology. *Trends Ecol Evol* 31:463–475.

Hazel J (2009) Evaluation of fast-acquisition GPS in stationary tests and fine-scale tracking of green turtles. *J Exp Mar Biol Ecol* 374:58–68.

Hazin FH, Afonso AS, De Castilho PC, Ferreira LC, Rocha BC (2013) Regional movements of the tiger shark, *Galeocerdo cuvier*, off Northeastern Brazil: inferences regarding shark attack hazard. *An Acad Bras Ciênc* 85(3):1053–1062.

Hearn AR, Green JA, Espinoza E, Penaherrera C, Acuna D, Klimley AP (2013) Simple criteria to determine detachment point of towed satellite tags provide first evidence of return migrations of whale sharks (*Rhincodon typus*) at the Galapagos Islands, Ecuador. *Anim Biotelem* 1:11.

Heupel MR, Simpfendorfer CA, Hueter RE (2004) Estimation of shark home ranges using passive monitoring techniques. *Environ Biol Fish* 71:135–142.

Heupel MR, Simpfendorfer CA, Espinoza M, Smoothey AF, Tobin A, Peddemors V (2015) Conservation challenges of sharks with continental scale migrations. *Front Mar Sci* 2:1–7.

Hill RD, Braun MJ (2001) Geolocation by light level. The next step: latitude. In: Sibert J, Nielsen J (eds) *Electronic Tagging and Tracking in Marine Fisheries*. Kluwer Academic Press, Dordrecht, pp 315–330.

Hobday AJ, Maxwell SM, Forgie J, McDonald J, Darby M, Seto K, Bailey H, et al. (2014) Dynamic Ocean Management: Integrating scientific and technological capacity with law, policy and management. *Stanford Environ Law J* 33:125–165.

Hoenner X (2012) Spatial and Behavioural Ecology of Hawksbill Turtles Nesting on Groote Eylandt, Northern Australia, doctoral thesis, Charles Darwin University, Alice Springs, Northern Territory, Australia.

Hoenner X, Whiting SD, Hindell MA, McMahon CR (2012) Enhancing the use of Argos satellite data for home range and long distance migration studies of marine animals. *PLoS ONE* 7(7):e40713.

Hoffmayer ER, Franks JS, Driggers III WB, McKinney JA, Hendon JM, Quattro JM (2014) Habitat, movements and environmental preferences of dusky sharks, *Carcharhinus obscurus*, in the northern Gulf of Mexico. *Mar Biol* 161:911–924.

Hooge PN, Eichenlaub B (1997) *Animal movement extension to ArcView, Version 1.1*. Alaska Science Center, Biological Science Office, U.S. Geological Survey, Anchorage, AK.

Hooker SK, Cansadas A, Hyrenbach KD, Corrigan C, Polovina JJ, Reeves RR (2011) Making protected area networks effective for marine top predators. *Endang Species Res* 13:203–218.

Hoolihan JP, Luo J, Abascal FJ, Campana SE, De Metrio G, Dewar H, Domeier ML, et al. (2011) Evaluating post-release behaviour modification in large pelagic fish deployed with pop-up satellite archival tags. *ICES J Mar Sci* 68(5):880–889.

Hooten MB, Johnson DS, McClintock BT, Morales JM (2017) *Animal Movement: Statistical Models for Telemetry Data*. CRC Press, Boca Raton, FL.

Horne JS, Garton EO (2006) Likelihood cross-validation versus least square cross-validation for choosing the smoothing parameter in kernel home-range analysis. *J Wildl Manage* 70:641–648.

Horne JS, Garton EO, Krone SM, Lewis JS (2007) Analyzing animal movements using Brownian bridges. *Ecology* 88:2354–2363.

Jewell OJD, Wcisel MA, Gennari E, Towner AV, Bester MN, Johnson RL, Singh S (2011) Effects of Smart Position Only (SPOT) tag deployment on white sharks *Carcharodon carcharias* in South Africa. *PLoS ONE* 6:e27242.

Johnson AR, Wiens JA, Milne BTI, Crist TO (1992) Animal movements and population dynamics in heterogeneous landscapes. *Landscape Ecol* 7:63–75.

Johnson CJ, Parker KL, Heard DC, Gillingham MP (2006) Unrealistic animal movement rates as behavioural bouts: a reply. *J Anim Ecol* 75:303–308.

Johnson DS (2014) *Fit Continuous-Time Correlated Random Walk Models to Animal Movement Data*. R package version 1.4.1.

Johnson DS, London JM, Lea MA, Durban JT (2008) Continuous-time correlated random walk model for animal telemetry data. *Ecology* 89(5):1208–1215.

Jonsen ID, Myers RA, Flemming JM (2003) Meta-analysis of animal movement using state-space models. *Ecology* 84(11):3055–3063.

Jonsen ID, Flemming JM, Myers RA (2005) Robust state—space modeling of animal movement data. *Ecology* 86(11):2874–2880.

Jonsen ID, Basson M, Bestley S, Bravington MV, Patterson TA, Pedersen MW, Thomson R, et al. (2013) State-space models for bio-loggers: a methodological road map. *Deep Sea Res II* 88–89:34–46.

Jonsen I, Bestley B, Wotherspoon S, Sumner M, Flemming JM (2016) *Bayesian State-Space Models for Animal Movement*. R package version 1.1.2.

Jorgensen SJ, Reeb CA, Chapple TK, Anderson S, Perle C, Van Sommeran SR, Fritz-Cope C, et al. (2009) Philopatry and migration of Pacific white sharks. *Proc R Soc B Biol Sci* 277:679–688.

Kareiva P, Odell G (1987) Swarms of predators exhibit "preytaxis" if individual predators use area-restricted search. *Am Nat* 130:233–270.

Kays R, Crofoot MC, Jetz W, Wikelski M (2015) Terrestrial animal tracking as an eye on life and planet. *Science* 348(6240):aaa2478.

Ketchum JT, Hearn A, Klimley AP, Espinoza E, Peñaherrera C, Largier JL (2014) Seasonal changes in movements and habitat preferences of the scalloped hammerhead shark (*Sphyrna lewini*) while refuging near an oceanic island. *Mar Biol* 161:755–767.

Kiszka JJ, Mourier J, Gastrich K, Heithaus MR (2016) Using unmanned aerial vehicles (UAVs) to investigate shark and ray densities in a shallow coral lagoon. *Mar Ecol Prog Ser* 560:237–242.

Klimley AP, Butler SB (1988) Immigration and emigration of a pelagic fish assemblage to seamounts in the Gulf of California related to water mass movements using satellite imagery. *Mar Ecol Prog Ser* 49:11–20.

Kohler NE, Turner PA (2001) Shark tagging: a review of conventional methods and studies. *Environ Biol Fish* 60:191–223.

Kohler NE, Casey JG, Turner PA (1998) NMFS Cooperative Shark Tagging Program, 1962–93: an atlas of shark tag and recapture data. *Mar Fish Rev* 60:1–87.

Koláček J, Horová I (2017) Bandwidth matrix selectors for kernel regression. *Comput Stat* 32(3):1027–1046.

Lam CH, Nielsen A, Sibert JR (2008) Improving light and temperature based geolocation by unscented Kalman filtering. *Fish Res* 91:15–25.

Lam CH, Nielsen A, Sibert JR (2010) Incorporating sea-surface temperature to the light-based geolocation model TrackIt. *Mar Ecol Prog Ser* 419:71–84.

Lea JS, Wetherbee BM, Queiroz N, Burnie N, Aming C, Sousa LL, Mucientes GR, et al. (2015) Repeated, long-distance migrations by a philopatric predator targeting highly contrasting ecosystems. *Sci Rep* 5:11202.

Lea JSE, Wetherbee BM, Sousa LL, Aming C, Burnie N, Humphries NE, Queiroz N, et al. (2018) Ontogenetic partial migration is associated with environmental drivers and influences fisheries interactions in a marine predator. *ICES J Mar Sci* doi:10.1093/icesjms/fsx238.

Lele SR, Keim JL (2006) Weighted distributions and estimation of resource selection probability functions. *Ecology* 87(12):3012–3028.

Lele ASR, Keim JL, Solymos P (2017) *Package 'ResourceSelection'*. R package version 0.3-2.

Lutcavage ME, Lam CH, Galuardi B (2015) *Seventeen Years and $3 Million Dollars Later: Performance of PSAT Tags Deployed on Atlantic Bluefin and Bigeye Tuna*, SCRS/2014/178. International Commission for the Conservation of Atlantic Tunas, Madrid, Spain.

Manly BFJ, McDonald LL, Thomas DL, McDonald TL, Erickson WP (2002) *Resource Selection by Animals: Statistical Design and Analysis for Field Studies*, 2nd ed. Springer Science+Business, Dordrecht.

Mansfield KL, Wyneken J, Rittschof D, Walsh M, Lim CW, Richards PM (2012) Satellite tag attachment methods for tracking neonate sea turtles. *Mar Ecol Prog Ser* 457:181–192.

Maxwell SM, Breed GA, Nickel BA, Makanga-Bahouna J, Pemo-Makaya E, et al. (2011) Using satellite tracking to optimize protection of long-lived marine species: olive ridley sea turtle conservation in Central Africa. *PLoS ONE* 6(5):e19905.

Maxwell SM, Hazen EL, Lewison RL, Dunn DC, Bailey H, Bograd SJ, Briscoe DK, et al. (2015) Dynamic ocean management: defining and conceptualizing real-time management of the ocean. *Mar Policy* 58:42–50.

McCauley DJ, Pinsky M, Palumbi S, Estes JA, Joyce F, Warner RR (2015) Marine defaunation: animal loss in the global ocean. *Science* 347(6219):1255641.

McClintock BT, London JM, Cameron MF, Boveng PL, Gimenez O (2015) Modelling animal movement using the Argos satellite telemetry location error ellipse. *Methods Ecol Evol* 6(3):266–277.

McGowan J, Beger M, Lewison RL, Harcourt R, Campbell H, Priest M, Dwyer RG, et al. (2016) Integrating research using animal-borne telemetry with the needs of conservation management. *J Appl Ecol* 54(2):423–429.

Meekan MG, Duarte CM, Fernández-Gracia J, Thums M, Sequeira AMM, Harcourt R, Eguíluz VM (2017) The ecology of human mobility. *Trends Ecol Evol* 32(3):198–210.

Meyer CG, Papastamatiou YP, Holland KN (2010) A multiple instrument approach to quantifying the movement patterns and habitat use of tiger (*Galeocerdo cuvier*) and Galapagos sharks (*Carcharhinus galapagensis*) at French Frigate Shoals, Hawaii. *Mar Biol* 157:1857–1868.

Moen R, Pastor J, Cohen Y, Schwartz CC (1996) Effects of moose movement and habitat use on GPS collar performance. *J Wildl Manage* 60(3):659–668.

Moen R, Pastor J, Cohen Y (1997) Accuracy of GPS telemetry collar locations with differential correction. *J Wildl Manage* 61(2):530–539.

Moyes CD, Fragoso N, Musyl MK, Brill RW (2006) Predicting postrelease survival in large pelagic fish. *Trans Am Fish Soc* 135(5):1389–1397.

Musyl MK, McNaughton LM (2007) *Report on Pop-Up Satellite Archival Tag (PSAT) Operations Conducted on Sailfish, Istiophorus platypterus, by Research Scientists of the Fisheries Research Institute, Eastern Marine Biology Research Center, and Institute of Oceanography, College of Science, National Taiwan University*. Chengkong, Taiwan.

Musyl MK, Domeier ML, Nasby-Lucas N, Brill RW, McNaughton LM, Swimmer JY, Lutcavage MS, et al. (2011) Performance of pop-up satellite archival tags. *Mar Ecol Prog Ser* 433:1–28.

Nielsen A, Sibert JR (2007) State–space model for light-based tracking of marine animals. *Can J Fish Aquat Sci* 64:1055–1068.

Nielsen A, Sibert JR (2011) *Track Tagged Individuals from Light Measurements*. R package version 0.2-6.

Nielsen A, Sibert JR, Kohin S, Musyl MK (2009) State space model for light based tracking of marine animals: validation on swimming and diving creatures. In: Nielsen JL, Arrizabalaga H, Fragoso N, Hobday A, Lutcavage M, Sibert J (eds) *Tagging and Tracking of Marine Animals with Electronic Devices*. Springer Science+Business, Dordrecht, pp 295–309.

Nielson RM, Sawyer H, McDonald TL (2013) *Package 'BBMM'*. R package version 3.0.

O'Leary BC, Winther-Janson M, Bainbridge JM, Aitken J, Hawkins JP, Roberts CM (2016) Effective coverage targets for ocean protection. *Conserv Lett* 9(6):398–404.

Økland F, Thorstad EB, Westerberg H, Aarestrup K, Metcalfe JD (2013) Development and testing of attachment methods for pop-up satellite archival transmitters in European eel. *Anim Biotelem* 1(1):3.

Papastamatiou YP, Meyer CG, Carvalho F, Dale JJ, Hutchinson MR, Holland KN (2013) Telemetry and random-walk models reveal complex patterns of partial migration in a large marine predator. *Ecology* 94:2595–2606.

Patterson TA, Thomas L, Wilcox C, Ovaskainen O, Matthiopoulos J (2008) State-space models of individual animal movement. *Trends Ecol Evol* 23:87–94.

Patterson TA, Basson M, Bravington MV, Gunn JS (2009) Classifying movement behaviour in relation to environmental conditions using hidden Markov models. *J Anim Ecol* 78:1113–1123.

Patterson TA, Parton A, Langrock R, Blackwell PG, Thomas L, King R (2016) Statistical modelling of animal movement: a myopic review and a discussion of good practice. *arXiv* 1603:07511.

Pennisi E (2011) Global tracking of small animals gains momentum. Science 334(6059):1042.

Pikitch EK, Santora C, Babcock EA, Bakun A, Bonfil R, Conover DO, Dayton P, et al. (2004) Ecosystem-based fishery management. *Science* 305(5682):346–347.

Pinaud D (2008) Quantifying search effort of moving animals at several spatial scales using first-passage time analysis: effect of the structure of environment and tracking systems. *J Appl Ecol* 45:91–99.

Pinaud D, Weimerskirch H (2005) Scale-dependent habitat use in a long-ranging central place predator. *J Anim Ecol* 74:852–863.

Politis DN, White H (2004) Automatic block-length selection for the dependent bootstrap. *Econ Rev* 23:53–70.

Queiroz N, Humphries NE, Noble LR, Santos AM, Sims DW (2010) Short-term movements and diving behaviour of satellite-tracked blue sharks *Prionace glauca* in the northeastern Atlantic Ocean. *Mar Ecol Prog Ser* 406:265–279.

Queiroz N, Humphries NE, Mucientes G, Hammerschlag N, Lima FP, Scales KL, Miller PI, et al. (2016) Ocean-wide tracking of pelagic sharks reveals extent of overlap with longline fishing hotspots. *Proc Natl Acad Sci USA* 113:1582–1587.

R Core Team (2017) *The R Project for Statistical Computing*. The R Foundation for Statistical Computing, Vienna, Austria (http://www.R-project.org/).

Raymond B, Lea M-A, Patterson T, Andrews-Goff V, Sharples R, Charrassin J-B, Cottin M, et al. (2015) Important marine habitat off east Antarctica revealed by two decades of multi-species predator tracking. *Ecography* 38:121–129.

Rodríguez JP, Fernández-Gracia J, Thums, M, Hindell MA, Sequeira AMM, Meekan MG, Costa DP, et al. (2017) Big data analyses reveal patterns and drivers of the movements of southern elephant seals. *Sci Rep* 7(1):112.

Rogers PJ, Huveneers C, Page B, Goldsworthy SD, Coyne M, Lowther AD, Mitchell JG, Seuront L (2015) Living on the continental shelf edge: habitat use of juvenile shortfin makos *Isurus oxyrinchus* in the Great Australian Bight, southern Australia. *Fish Oceanogr* 24:205–218.

Roquet F, Williams G, Hindell MA, Harcourt R, McMahon C, Guinet C, Charrassin J-B, et al. (2014) A Southern Indian Ocean database of hydrographic profiles obtained with instrumented elephant seals. *Sci Data* 1:140028.

Royer F, Fromentin JM, Gaspar P (2004) Association between bluefin tuna schools and oceanic features in the western Mediterranean. *Mar Ecol Prog Ser* 269:249–263.

Runge CA, Martin TG, Possingham HP, Willis SG, Fuller RA (2014) Conserving mobile species. *Front Ecol Environ* 12:395–402.

Ryan PG, Petersen SL, Peters G, Grémillet D (2004) GPS tracking a marine predator: the effects of precision, resolution and sampling rate on foraging tracks of African penguins. *Mar Biol* 145(2):215–223.

Scales KL, Miller PI, Embling CB, Ingram SN, Pirotta E, Votier SC (2014) Mesoscale fronts as foraging habitats: composite front mapping reveals oceanographic drivers of habitat use for a pelagic seabird. *J R Soc Interface* 11:20140679.

Scarpignato AL, Harrison A-L, Newstead DJ, Niles LJ, Porter RR, van den Tillaart M, Marra PP (2016) Field-testing a new miniaturized GPS-Argos satellite transmitter (3.5 g) on migratory shorebirds. *Wader Study* 123(3):240–246.

Sepulveda CA, Kohin S, Chan C, Vetter R, Graham JB (2004) Movement patterns, depth preferences, and stomach temperatures of free-swimming juvenile mako sharks, *Isurus oxyrinchus*, in the Southern California Bight. *Mar Biol* 145:191–199.

Sepulveda CA, Heberer C, Aalbers SA, Spear N, Kinney M, Bernal D, Kohin S (2015) Post-release survivorship studies on common thresher sharks (*Alopias vulpinus*) captured in the southern California recreational fishery. *Fish Res* 161:102–108.

Sequeira AMM, Rodríguez JP, Eguíluz VM, Harcourt R, Hindell M, Sims DW, Duarte CM, et al. (2018) Convergence of marine megafauna movement patterns in coastal and open oceans. *Proc Natl Acad Sci USA* 115(12):3072–3077.

Shuter JL, Broderick AC, Agnew DJ, Jonzen N, Godley BJ, Milner-Gulland EJ, Thirgood S (2011) Conservation and management of migratory species. In: Milner-Gulland EJ, Fryxell JM, Sinclair ARE (eds) *Animal Migration: A Synthesis*. Oxford University Press, Oxford, pp 172–206.

Sibert JR, Musyl MK, Brill RW (2003) Horizontal movements of bigeye tuna (*Thunnus obesus*) near Hawaii determined by Kalman filter analysis of archival tagging data. *Fish Oceanogr* 12:141–151.

Silverman BW (1986) *Density Estimation for Statistics and Data Analysis*. Chapman & Hall, London.

Simpfendorfer CA, Heupel MR (2004) Assessing habitat use and movement. In: Carrier JC, Musick JA, Heithaus MR (eds) *Biology of Sharks and Their Relatives*. CRC Press, Boca Raton, FL, pp 553–572.

Sims DW (2010) Tracking and analysis techniques for understanding free-ranging shark movements and behavior. In: Carrier JC, Musick JA, Heithaus MR (eds) *Sharks and Their Relatives II: Biodiversity, Adaptive Physiology, and Conservation*. CRC Press, Boca Raton, FL, pp 351–386.

Sims DW, Southall EJ, Richardson AJ, Reid PC, Metcalfe JD (2003) Seasonal movements and behaviour of basking sharks from archival tagging: no evidence of winter hibernation. *Mar Ecol Prog Ser* 248:187–196.

Sims DW, Queiroz N, Doyle TK, Houghton JDR, Hays GC (2009a) Satellite tracking of the world's largest bony fish, the ocean sunfish (*Mola mola* L.) in the North East Atlantic. *J Exp Mar Biol Ecol* 370:127–133.

Sims DW, Queiroz N, Humphries NE, Lima FP, Hays GC (2009b) Long-term GPS tracking of ocean sunfish *Mola mola* offers a new direction in fish monitoring. *PLoS ONE* 4(1):e7351.

Sims DW, Mucientes G, Queiroz N (2018) Shortfin mako sharks threatened by inaction. *Science* 359(6382):1342.

Soanes LM, Arnould JPY, Dodd SG, Sumner MD, Green JA, Frederiksen M (2013) How many seabirds do we need to track to define home-range area? *J Appl Ecol* 50(3):671–679.

Speed CW, Cappo M, Meekan MG (2018) Evidence for rapid recovery of shark populations within a coral reef marine protected area. *Biol Conserv* 220:308–319.

Sumner MD, Luque S (2016) *Package 'trip'*. R package version 1.5.0.

Sumner MD, Wotherspoon SJ, Hindell MA (2009) Bayesian estimation of animal movement from archival and satellite tags. *PLoS ONE* 4(10):e7324.

Teo S, Boustany A, Blackwell S, Walli A, Weng K, Block B (2004) Validation of geolocation estimates based on light level and sea surface temperature from electronic tags. *Mar Ecol Prog Ser* 283:81–98.

Thums M, Bradshaw CJA, Hindell MA (2011) *In situ* measures of foraging success and prey encounter reveal marine habitat-dependent search strategies. *Ecology* 92:1258–1270.

Thums M, Meekan M, Stevens J, Wilson S, Polovina J (2012) Evidence for behavioural thermoregulation by the world's largest fish. *J R Soc Interface* 10(78):20120477.

Thums M, Fernández-Gracia J, Sequeira AMM, Eguíluz VM, Duarte CM, Meekan MG (2018) How big data fast tracked human mobility research and the lessons for animal movement ecology. *Front Mar Sci* 5(21):1–12.

Thums M, Rossendell J, Guinea M, Ferreira LC (in review) Horizontal and vertical movement behaviour of adult flatback turtles and spatial overlap with industrial development. *Mar Ecol Prog Ser.*

Thurfjell H, Ciuti S, Boyce MS (2014) Applications of step-selection functions in ecology and conservation. *Mov Ecol* 2:4.

Tremblay Y, Shaffer SA, Fowler SL, Kuhn CE, McDonald BI, Weise MJ, Bost CA, et al. (2006) Interpolation of animal tracking data in a fluid environment. *J Exp Biol* 209(Pt 1):128–140.

Vincent C, McConnell BJ, Ridoux V, Fedak MA (2002) Assessment of Argos location accuracy from satellite tags deployed on captive grey seals. *Mar Mam Sci* 18(1):156–166.

Wand MP, Jones MC (1995) *Kernel Smoothing*. Chapman & Hall, London.

Weimerskirch H, Guionnet T, Martin J, Shaffer SA, Costa DP (2000) Fast and fuel efficient? Optimal use of wind by flying albatrosses. *Proc Biol Sci* 267:1869–1874.

Weng KC, Sullivan JBO, Lowe CG, Winkler CE, Dewar H, Block BA (2007) Movements, behavior and habitat preferences of juvenile white sharks *Carcharodon carcharias* in the eastern Pacific. *Mar Ecol Prog Ser* 338:211–224.

Weng KC, Foley DG, Ganong JE, Perle C, Shillinger GL, Block BA (2008) Migration of an upper trophic level predator, the salmon shark *Lamna ditropis*, between distant ecoregions. *Mar Ecol Prog Ser* 372:253–264.

Wilson SG, Polovina JJ, Stewart BS, Meekan MG (2006) Movements of whale sharks (*Rhincodon typus*) tagged at Ningaloo Reef, Western Australia. *Mar Biol* 148(5):1157–1166.

Witt MJ, Åkesson S, Broderick AC, Coyne MS, Ellick J, Formia A, Hays GC, et al. (2010) Assessing accuracy and utility of satellite-tracking data using Argos-linked Fastloc-GPS. *Anim Behav* 80:571–581.

Wood S (2018) *Mixed GAM Computation Vehicle with GCV/AIC/REML Smoothness Estimation*. R package version 1.8-23.

Wood S, Scheipl F (2017) *gamm4: Generalized Additive Mixed Models Using 'mgcv' and 'lme4'*. R package version 0.2-5.

Worton BJ (1987) A review of models of home range for animal movement. *Ecol Model* 38:277–298.

Worton BJ (1989) Kernel methods for estimating the utilization in home-range studies. *Ecology* 70:164–168.

Zuur A, Ieno EN, Walker N, Saveliev AA, Smith GM (2009) *Mixed Effects Models and Extensions in Ecology with R*. Springer Science+Business, New York.

Index